Springer Monographs in Mathematics

Springer
Berlin
Heidelberg
New York
Barcelona
Hong Kong
London
Milan
Paris
Singapore
Tokyo

Sibe Mardešić

Strong Shape and Homology

Springer

Sibe Mardešić
Department of Mathematics
University of Zagreb
Bijenička cesta 30
10000 Zagreb, Croatia
e-mail: smardes@math.hr

Library of Congress Cataloging-in-Publication Data

Mardešić, S. (Sibe), 1927-
Strong shape and homology / Sibe Mardešić.
p.cm. -- (Springer monographs in mathematics)
Includes bibliographical references and index.
ISBN 3540661980 (alk. paper)
1. Shape theory (Topology) 2. Homology theory. I. Title. II Series

QA612.7 .M353 1999
514'.24--dc21 99-047673

Mathematics Subject Classification (1991): 55NXX, 55PXX, 18GXX

ISBN 3-540-66198-0 Springer-Verlag Berlin Heidelberg New York

Cover design: *Erich Kirchner, Heidelberg*
Typesetting by the author using a Springer TeX macro package
Printed on acid-free paper SPIN 10732798 41/3143AT-5 4 3 2 1 0

Preface

It is well known that standard notions of homotopy theory are not adequate to study global properties of spaces with bad local behavior. For instance, all homotopy groups of the dyadic solenoid vanish in spite of the fact that the solenoid is globally a non-trivial object. Shape theory is designed to correct these shortcomings of homotopy theory. When restricted to spaces with good local behavior, like the ANR's, polyhedra or CW-complexes, shape theory coincides with homotopy theory, therefore, it can be viewed as the appropriate extension of homotopy theory to general spaces.

Many constructions in topology lead naturally to spaces with bad local behavior even if one initially considers locally good spaces, e.g., manifolds. Standard examples include fibers of mappings, sets of fixed points, attractors of dynamical systems, spectra of operators, boundaries of certain groups. In all these areas shape theory has proved useful.

It took some time to realize that beside ordinary shape, introduced in 1968 by K. Borsuk, there exists a finer theory, presently called strong shape theory, which has various advantages over ordinary shape. Its position is intermediate, between homotopy and ordinary shape. One encounters a similar situation in homology theory. Beside singular homology, which is a homotopy invariant, and Čech homology, which is a shape invariant, there exists strong homology, which is a strong shape invariant. In the special case of metric compacta, this homology was introduced by N.E. Steenrod in 1940 and is often referred to as the Steenrod homology.

The main purpose of the present book is to develop in detail strong shape theory and (ordinary) strong homology groups for arbitrary spaces. To date there exist five books considering shape theory: (Borsuk 1975), (Edwards, Hastings 1976), (Dydak, Segal 1978), (Mardešić, Segal 1982) and (Cordier, Porter 1989). However, only the second one includes considerations on strong shape (of metric compacta). There exist numerous books on homology theory, but only a few ((Bredon 1967), (Edwards, Hastings 1976), (Massey 1978) and (Sklyarenko 1989a, 1989b)) consider strong homology (mostly restricted to compact or locally compact spaces).

In the present book the approach to strong shape uses the technique of inverse systems, already successfully applied to ordinary shape, e.g., in (Mardešić, Segal 1982). One first generalizes the homotopy theory of spaces to a homotopy theory of inverse systems of spaces. The second step consists of approximating spaces by polyhedra (or ANR's), i.e., of replacing spaces by suitable inverse systems of polyhedra (or ANR's), associated with these spaces. Finally, one applies the developed homotopy theory to the associated systems and one proves that results depend only on the spaces and not on the choice of the approximating systems. In contrast to the situation in ordinary shape, in strong shape there are considerable difficulties in carrying out this program. The right modification of homotopy to systems of spaces is the rather involved coherent homotopy. Moreover, the right approximation process consists of delicate constructions like resolutions and strong expansions.

Similarly, strong homology groups are first defined for inverse systems of spaces. An essential feature of the definition is that one defines a complex of strong chains. Strong homology groups of a system are just homology groups of this chain complex. This insures exactness of strong homology, a property lacked by Čech homology. For compact metric spaces, our strong homology coincides with the Steenrod homology. In general, strong homology does not have compact supports.

The content of the present book is almost disjoint from the content of other books on shape. However, acquaintance with the shape theory book (Mardešić, Segal 1982), to which we often refer, will facilitate the reading. Results forming the core of the book are fully proved, even when this requires performing lengthy computations. The reader is advised to skip at first reading these computations and concentrate on the nature and structure of the formulae in question, which are relatively simple and pretty. Additional information and bibliographic data can be found in Bibliographic notes following each section, except for Sections 10 and 22, which are surveys of results without proofs. In an effort to make the book as selfcontained as possible, I have included some material from homological algebra. In particular, I discuss at length derived functors of the functor lim, which play an essential role in strong homology of non-compact spaces. The only book on this subject is (Jensen 1972). I also included an introduction to spectral sequences and some material on abelian groups.

The book consists of four chapters, the first two being devoted to strong shape and the last two to strong homology. Chapters are divided in sections, whose numbering runs throughout the whole book. Subsections are numbered within sections. The same applies to Theorems, Corollaries, Lemmas, Remarks and Examples, which use the same counter, e.g., Theorem 3.4 refers to the fourth of these items in Section 3. Formulae are numbered by Subsections. When reference is made to a formula from a different Subsection, its number is added. E.g., (2.3.4) refers to the fourth formula in Subsection 2.3. External referencing is by author and year of publication, e.g., (Borsuk 1968).

In my work on strong shape and homology I benefited very much from contacts with a number of colleagues in various parts of the world. This is especially true of Ju.T. Lisica (Moscow), Z. Miminoshvili (Tbilisi), A.V. Prasolov (Minsk and Tromsø) and T. Watanabe (Yamaguchi) with whom I wrote joint papers on the subject. I am grateful to my colleagues at the Universities of Zagreb, Split and Ljubljana who attended many of my seminar talks based on various parts of the manuscript. I would also like to thank my colleague Šime Ungar (Zagreb) for having patiently guided me through delicate points of the LaTeX typesetting.

Zagreb, June 1999

Sibe Mardešić

Table of Contents

II. STRONG SHAPE

III. HIGHER DERIVED LIMITS

IV. HOMOLOGY GROUPS

Introduction

It is generally considered that K. Borsuk founded shape theory in 1968, when he defined the shape category of compact metric spaces Sh(CM) and the shape functor S, which maps the homotopy category H(CM) to Sh(CM) (Borsuk 1968). After Borsuk's work the general notion of shape gradually evolved in a series of papers of several authors ((Mardešić, Segal 1971), (Fox 1972), (Mardešić 1973), (Morita 1975a, 1975b)). In particular, in the first of these papers it was realized that there is a rather categorical description of shape for compact Hausdorff spaces, which defines shape morphisms using inverse systems of ANR's. From this point of view, the beginnings of shape theory can be traced back to the early papers on inverse limits and Čech homology ((Alexandroff 1926, 1927, 1929), (Vietoris 1927), (Lefschetz 1931), (Čech 1932), (Freudenthal 1937)). Moreover, one encounters early ideas of shape categories already in (Christie 1944) and (Lima 1959).

The strong shape category for metric compacta SSh(CM) was first described in a rather elementary way in (Quigley 1973). A more sophisticated way used localization of the category of inverse systems pro-Top at weak homotopy equivalences (Edwards, Hastings 1976a). This procedure yields a homotopy category in the sense of (Quillen 1967), denoted in the literature by Ho(pro-Top). Restriction of this category to sequences of compact polyhedra enabled D.A. Edwards and H.M. Hastings to give a new description of SSh(CM). Moreover, these authors proved that the restriction of the latter category to Z-sets X of the Hilbert cube Q is isomorphic to the proper homotopy category of their complements $Q \backslash X$. This result demonstrated the advantage of strong shape over ordinary shape, because the analogous result for ordinary shape involves the less natural weak proper homotopy category (Chapman 1972). For a while the difference between shape and strong shape was not generally understood, because Chapman's complement theorem implies that two metric compacta have the same strong shape if and only if they have the same shape. To see that the two theories are distinct, one must look at morphisms. Further progress in strong shape of metric compacta was obtained by the systematic use of contractible telescopes and strong shape deformation retracts (Dydak, Segal 1978a, 1981), (Cathey 1981).

F.W. Bauer was the first to define and study strong shape for arbitrary spaces (Bauer 1976). He used a universal construction based on the notion of a

2 - category (later generalized to an ∞ - category). His ideas, restated in terms of simplicial categories, assumed a final form much later (Günther 1992b). Construction of the strong shape category for arbitrary spaces SSh(Top), using the method of inverse systems, first appeared in (Cathey, Segal 1983) and (Lisica, Mardešić 1983, 1984a, 1984b). The philosophy behind this approach is the same as behind the inverse system approach to ordinary shape (Morita 1975a). Therefore, we first briefly outline the construction of the ordinary shape category Sh(Top).

One begins by considering the category pro - H(Top). Its objects are inverse systems $\boldsymbol{X} = (X_{\lambda_0}, [p_{\lambda_0\lambda_1}], \Lambda)$ in the homotopy category H(Top). Morphisms $\boldsymbol{X} \to \boldsymbol{Y} = (Y_{\mu_0}, [q_{\mu_0\mu_1}], M)$ are given by increasing functions $f: M \to \Lambda$ and by homotopy classes $[f_{\mu_0}]$ of mappings $f_{\mu_0}: X_{f(\mu_0)} \to Y_{\mu_0}$ which, for $\mu_0 \le \mu_1$, make the following diagram commutative up to homotopy,

$$\begin{array}{ccc} X_{f(\mu_0)} & \xleftarrow{\;p\;} & X_{f(\mu_1)} \\ {\scriptstyle f_{\mu_0}}\downarrow & & \downarrow{\scriptstyle f_{\mu_1}} \\ Y_{\mu_0} & \xleftarrow[\;q\;]{} & Y_{\mu_1}. \end{array} \tag{1}$$

The next step consists in associating with every space X a polyhedral (or ANR) inverse system $\boldsymbol{X}$ from H(Top) together with a collection $\boldsymbol{p} = (p_\lambda): X \to \boldsymbol{X}$, where $p_\lambda: X \to X_\lambda$ are mappings such that $p_{\lambda_0\lambda_1}p_{\lambda_1} \simeq p_{\lambda_0}$, for $\lambda_0 \le \lambda_1$. One also imposes certain additional conditions (Morita's conditions (M1), (M2)). In the third step one proves that these conditions imply the following universal property: If $\boldsymbol{Y}$ is a polyhedral system, then every morphism $[\boldsymbol{f}]: X \to \boldsymbol{Y}$ of pro - H(Top) admits a unique factorization

$$[\boldsymbol{f}] = [\boldsymbol{g}][\boldsymbol{p}]. \tag{2}$$

Shape morphisms $F: X \to Y$ are now determined by triples $([\boldsymbol{p}], [\boldsymbol{q}], [\boldsymbol{f}])$, where $\boldsymbol{X}$ and $\boldsymbol{Y}$ are polyhedral systems associated with X and Y, respectively.

$$\begin{array}{ccc} \boldsymbol{X} & \xleftarrow{\;[\boldsymbol{p}]\;} & X \\ {\scriptstyle [\boldsymbol{f}]}\downarrow & & \\ \boldsymbol{Y} & \xleftarrow[\;[\boldsymbol{q}]\;]{} & Y \end{array} \tag{3}$$

Strong shape is defined in three analogous (but more sophisticated) steps. To give a detailed account of the construction of the category SSh(Top) and to establish its main properties is the aim of Chapters I and II of this book.

(i). Instead of pro - H(Top), one considers the coherent homotopy category CH(pro - Top). Its objects are cofinite inverse systems of spaces, hence, objects of pro - Top. Morphisms are homotopy classes of coherent mappings

$\boldsymbol{f}: \boldsymbol{X} \to \boldsymbol{Y}$. The latter consist of increasing functions $f: M \to \Lambda$ and of mappings $f_{\mu_0}: X_{f(\mu_0)} \to Y_{\mu_0}$, which also make diagram (1) commutative up to homotopy. However, in this case one must choose a homotopy $f_{\mu_0\mu_1}: X_{f(\mu_1)} \times I \to Y_{\mu_0}$, which realizes the commutativity of (1). Furthermore, for three indices $\mu_0 \leq \mu_1 \leq \mu_2$, one has a homotopy of order 2, which connects the mappings $f_{\mu_1\mu_2}$, $f_{\mu_0\mu_2}$ and $f_{\mu_0\mu_1}$, i.e., one has a mapping $f_{\mu_0\mu_1\mu_2}: X_{f(\mu_2)} \times \Delta^2 \to Y_{\mu_0}$, which on the three faces of the standard 2-simplex Δ^2 is given by those three mappings. One requires that analogous conditions hold for arbitrarily long sequences of indices $\mu_0 \leq \ldots \leq \mu_n$.

(ii). With every space X one associates inverse systems of polyhedra (or ANR's) $\boldsymbol{X} \in$ pro-Top and collections of mappings $\boldsymbol{p} = (p_\lambda: X \to X_\lambda)$, where $p_{\lambda_0\lambda_1} p_{\lambda_1} = p_{\lambda_0}$, for $\lambda_0 \leq \lambda_1$. The additional requirement needed to obtain the right type of approximation consists in the appropriate strengthening of Morita's conditions, which leads to the notion of a strong expansion.

(iii). The final step consists in proving the universal property (2), where $[\boldsymbol{f}]$ and $[\boldsymbol{g}]$ are now morphisms of CH(pro-Top).

In defining the strong shape category SSh(Top), Cathey and Segal used the Edwards – Hastings category Ho(pro-Top) instead of CH(pro-Top). However, it has been proved that these two categories are isomorphic (see (Porter 1988) and (Cordier 1989), (Günther 1991a) or (Mardešić 1999a)). Consequently, the Cathey – Segal strong shape theory coincides with the Lisica – Mardešić theory.

In 1940 N.E. Steenrod defined homology groups for compact metric spaces X using infinite cycles of the contractible telescope, associated with X (Steenrod 1940). He proved that these groups, in contradistinction to singular or Čech homology groups, satisfy the Alexander duality law for compacta $X \subseteq S^n$. Extended to pairs of compacta, Steenrod homology groups also had the important property of being exact (which was not the case with Čech groups). Steenrod homology for metric compacta was characterized by the Eilenberg – Steenrod axioms (using a stronger form of the excision axiom) and the cluster axiom (Milnor 1960). Subsequently, Steenrod type homology groups were defined also for compact Hausdorff spaces and then extended to locally compact spaces using the direct system of compact subsets (Sitnikov 1954), (Borel, Moore 1960), (Sklyarenko 1969, 1971), (Massey 1978). For compact Hausdorff spaces, they are characterized by the Eilenberg – Steenrod axioms and the universal coefficient formula (Berikashvili 1984). For more general spaces, exact homology groups were constructed in (Deheuvels 1962), (Bauer 1976), (Miminoshvili 1981), (Lisitsa 1983a) and (Lisica, Mardešić 1983, 1985d, 1985e, 1985f, 1986). In the present book (Chapter IV) we describe in detail the Lisica – Mardešić construction and call the obtained groups, strong homology groups.

In accordance with the philosophy of shape theory, one first defines strong homology groups $\overline{H}_m(\boldsymbol{X}; G)$, for inverse systems of spaces $\boldsymbol{X}$. Strong groups for spaces X are then defined by putting $\overline{H}_m(X; G) = \overline{H}_m(\boldsymbol{X}; G)$, where

$\boldsymbol{X}$ is a polyhedral (ANR) strong expansion of X. To define $\overline{H}_m(\boldsymbol{X};G)$, one associates with $\boldsymbol{X}$ and G the pro-chain complex $\boldsymbol{C} = S(\boldsymbol{X}) \otimes G$, where $S(\boldsymbol{X}) = (S(X_\lambda), S(p_{\lambda\lambda'}), \Lambda)$ is the pro-chain complex of singular complexes of the terms of $\boldsymbol{X}$. The fact that $\boldsymbol{X}$ is a genuine inverse system (and not some kind of coherent system) insures that $\boldsymbol{C}$ is well defined. One defines strong homology groups of $\boldsymbol{X}$ with coefficients in G by putting $\overline{H}_m(\boldsymbol{X};G) = \overline{H}_m(\boldsymbol{C})$.

The last step consists in defining $\overline{H}_m(\boldsymbol{C})$, for pro-chain complexes $\boldsymbol{C}$. One first associates with $\boldsymbol{C}$ a bicomplex $K(\boldsymbol{C}) = (K^{ns}, \partial^{ns}, \delta^{ns})$. Here K^{ns} is given by

$$K^{ns} = \prod_{\lambda_0 \leq \ldots \leq \lambda_s} C^n_{\lambda_0}. \tag{4}$$

The operator $\partial^{ns}\colon K^{ns} \to K^{n-1,s}$ is induced by the boundary operators $\partial^n_{\lambda_0}$, while $\delta^{ns}\colon K^{n,s-1} \to K^{ns}$ is the Nöbeling – Roos operator, used in the explicit formulae for $\lim^s$ (the derived functors of the functor lim),

$$(\delta^{ns} c)_{\lambda_0 \ldots \lambda_s} = p^n_{\lambda_0\lambda_1}(c_{\lambda_1 \ldots \lambda_s}) + \sum_{j=1}^{s} (-1)^j c_{\lambda_0 \ldots \lambda_{j-1}\lambda_{j+1} \ldots \lambda_s}. \tag{5}$$

By definition, $\overline{H}_m(\boldsymbol{C})$ is the m - th homology group of the total chain complex of the bicomplex $K(\boldsymbol{C})$, called the complex of strong chains of $\boldsymbol{C}$. In general, with a double complex K one can associate two different total complexes. The one commonly used has as its terms direct sums along the diagonals of K. In the case of strong homology, one uses the second type of total complex, whose terms are direct products of terms along the diagonals of K. This fact is responsible for many desirable properties of strong homology groups.

Strong homology groups of spaces satisfy all the Eilenberg – Steenrod axioms. Only the exactness axiom requires the rather mild restriction to pairs (X, A), where A is normally embedded in X. Note that this is the case if X is paracompact and A is closed. For polyhedra or ANR's, strong groups coincide with singular groups. The homotopy axiom assumes a much stronger form, because strong homology groups are invariants of strong shape.

The Miminoshvili exact sequences relate strong homology groups to homology pro-groups $H_m(\boldsymbol{X};G) = (H_m(X_\lambda;G), p_{\lambda_0\lambda_1}, \Lambda)$. They involve the derived limits $\lim^s$ of the pro-groups $H_m(\boldsymbol{X};G)$. This is the reason why all of Chapter III is devoted to homological algebra, in particular to the functors $\lim^s$. Not only do we consider the usual vanishing theorems for $\lim^s$, but we also establish relevant non-vanishing theorems. Using these results (in Chapter IV) we construct examples of spaces X with $\lim^s H_m(\boldsymbol{X};G) \neq 0$, for $s \geq 2$ (Mardešić 1996a).

A large section of Chapter IV is devoted to strong homology of compact Hausdorff spaces (based on (Mardešić, Prasolov 1998)). We establish the Milnor sequence (which relates strong and Čech homology) as well as the universal coefficient theorem. In view of Berikashvili's characterization, for compact Hausdorff spaces, our strong homology coincides with most of the previously developed theories. Key to the results on strong homology of compact Hausdorff spaces is the fact that for such spaces $\lim^s H_m(\boldsymbol{X}; G) = 0$, for $s \geq 2$ (Kuz'minov 1971). The proof of this theorem uses some spectral sequences due to J.-E. Roos (Roos 1961). To make the book as selfcontained as possible, we develop the necessary machinery of spectral sequences and give a complete proof of the results mentioned. We also consider strong homology groups with compact supports and discuss their properties. However, these groups are not strong shape invariant. The book closes with a short survey of generalized (extraordinary) strong homology theories.

For the historical development of shape theory, the reader is referred to (Mardešić, Segal 1982) and to the survey articles (Mardešić 1999b) and (Mardešić, Segal to appear). In recent years shape theory found many interesting applications both in topology and in other areas of mathematics, especially in dynamical systems and C^* - algebras. Some of these applications are described also in (Mardešić, Segal to appear).

I. COHERENT HOMOTOPY

1. Coherent mappings

A mapping between two inverse systems $\boldsymbol{f} : \boldsymbol{X} \to \boldsymbol{Y}$, consisting of spaces X_λ and Y_μ, respectively, is a commutative diagram, which contains the two systems $\boldsymbol{X}$, $\boldsymbol{Y}$ as subdiagrams, and also contains a collection of mappings $f_\mu: X_\lambda \to Y_\mu$. Coherent mappings modify mappings of inverse systems. They include all the data of a mapping of systems. However, instead of requiring commutativity of the diagram, one has, as additional data, homotopies which relate the mappings f_μ and the bonding mappings of the systems $\boldsymbol{X}$, $\boldsymbol{Y}$, making the diagram commutative up to homotopy. Moreover, these homotopies are related by homotopies of higher order, which also make part of the data of a coherent mapping. Composition of coherent mappings and the identity mappings are defined, but they do not form a category.

1.1 Mappings of inverse systems

A *preordering* on a set Λ is a binary relation $\leq$ on Λ, which is *reflexive*, i.e., $\lambda \leq \lambda$ and *transitive*, i.e., $\lambda \leq \lambda'$ and $\lambda' \leq \lambda''$ imply $\lambda \leq \lambda''$. If $\leq$ is also *antisymmetric*, i.e., $\lambda \leq \lambda'$ and $\lambda' \leq \lambda$ imply $\lambda = \lambda'$, then $\leq$ is an *ordering*. If for an ordering $\lambda \leq \lambda'$, but $\lambda \neq \lambda'$, we write $\lambda < \lambda'$.

An *inverse system* $\boldsymbol{X} = (X_\lambda, p_{\lambda\lambda'}, \Lambda)$ in a category $\mathcal{C}$ consists of a preordered set $(\Lambda, \leq)$, of objects X_λ, for $\lambda \in \Lambda$, and of morphisms $p_{\lambda\lambda'}: X_{\lambda'} \to X_\lambda$, for $\lambda \leq \lambda'$. One requires that

$$p_{\lambda\lambda} = \mathrm{id}, \ \ \lambda \in \Lambda, \tag{1}$$

$$p_{\lambda\lambda'} p_{\lambda'\lambda''} = p_{\lambda\lambda''}, \ \ \lambda \leq \lambda' \leq \lambda''. \tag{2}$$

X_λ are the *terms* of $\boldsymbol{X}$ and $p_{\lambda\lambda'}$ are the *bonding morphisms* of $\boldsymbol{X}$.

If $\boldsymbol{X} = (X_\lambda, p_{\lambda\lambda'}, \Lambda)$ and $\boldsymbol{Y} = (Y_\lambda, q_{\lambda\lambda'}, \Lambda)$ are inverse systems over the same index set $(\Lambda, \leq)$, then a *level-preserving morphism*, shorter a *level morphism*, $\boldsymbol{f} = (f_\lambda): \boldsymbol{X} \to \boldsymbol{Y}$ consists of morphisms $f_\lambda: X_\lambda \to Y_\lambda$ such that

$$f_\lambda p_{\lambda\lambda'} = q_{\lambda\lambda'} f_{\lambda'}, \ \ \lambda \leq \lambda', \tag{3}$$

i.e., such that the following diagram commutes.

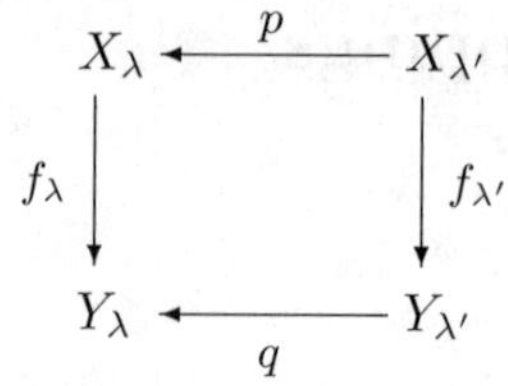

(4)

In the diagrams and formulae we will sometimes simplify the notation of various mappings by suppressing some of the indices. E.g., in the above diagram we used p and q in place of the full notation $p_{\lambda\lambda'}$ and $q_{\lambda\lambda'}$. The suppressed indices can always be recovered by considering the domain and codomain of the mapping in question.

Composition $\boldsymbol{h} = \boldsymbol{gf}$ of level morphisms $\boldsymbol{f} = (f_\lambda)\colon \boldsymbol{X} \to \boldsymbol{Y}$, $\boldsymbol{g} = (g_\lambda)\colon \boldsymbol{Y} \to \boldsymbol{Z}$ is the level morphism $\boldsymbol{h} = (h_\lambda)\colon \boldsymbol{X} \to \boldsymbol{Z}$, given by the morphisms

$$h_\lambda = g_\lambda f_\lambda, \ \lambda \in \Lambda. \tag{5}$$

The *identity morphism* $\mathbf{1}_{\boldsymbol{X}}\colon \boldsymbol{X} \to \boldsymbol{X}$ is given by the identity morphisms $1_\lambda\colon X_\lambda \to X_\lambda$, $\lambda \in \Lambda$. Inverse systems in $\mathcal{C}$ indexed by Λ and level morphisms between such systems form a category, denoted by $\mathcal{C}^\Lambda$.

One often encounters a more general situation, where inverse systems are indexed by different index sets. If $\boldsymbol{X} = (X_\lambda, p_{\lambda\lambda'}, \Lambda)$ and $\boldsymbol{Y} = (Y_\mu, q_{\mu\mu'}, M)$, then a *morphism* between systems $\boldsymbol{f} = (f, f_\mu)\colon \boldsymbol{X} \to \boldsymbol{Y}$ consists of a function $f\colon M \to \Lambda$ and of morphisms $f_\mu\colon X_{f(\mu)} \to Y_\mu$ from $\mathcal{C}$, for $\mu \in M$. One requires that the *index function* f is *increasing*, i.e., $\mu \le \mu'$ implies $f(\mu) \le f(\mu')$. Moreover, one requires that the mappings f_μ satisfy

$$f_\mu p_{f(\mu)f(\mu')} = q_{\mu\mu'} f_{\mu'}, \ \mu \le \mu', \tag{6}$$

i.e., the following diagram commutes.

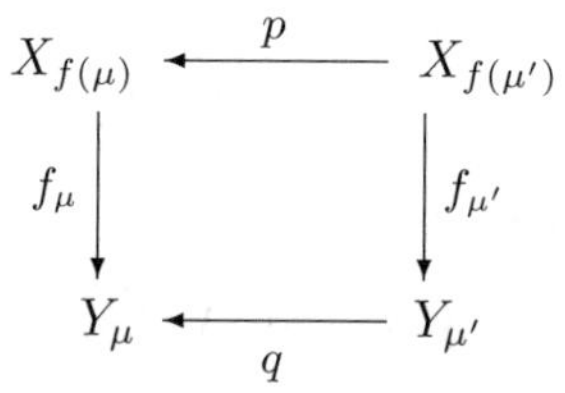

(7)

Composition $\boldsymbol{gf}\colon \boldsymbol{X} \to \boldsymbol{Z}$ of morphisms $\boldsymbol{f} = (f, f_\mu)\colon \boldsymbol{X} \to \boldsymbol{Y}$ and $\boldsymbol{g} = (g, g_\nu)\colon \boldsymbol{Y} \to \boldsymbol{Z} = (Z_\nu, r_{\nu\nu'}, N)$ is the morphism $\boldsymbol{h} = (h, h_\nu)$, given by the function $h = fg$ and by the mappings

$$h_\nu = g_\nu f_{g(\nu)}, \ \nu \in N. \tag{8}$$

Note that level morphisms are just morphisms with the index function $f = \mathrm{id}$. Such is the identity morphism $\mathbf{1}_{\boldsymbol{X}}\colon \boldsymbol{X} \to \boldsymbol{X}$ defined before. Inverse systems in $\mathcal{C}$ and morphisms between them form a category, denoted by inv-$\mathcal{C}$.

REMARK 1.1. A preordered set $(\Lambda, \leq)$ can be viewed as a small category Λ. Its objects are the elements of Λ. The set of morphisms $\Lambda(\lambda, \lambda')$ in Λ is non-empty if and only if $\lambda \leq \lambda'$, and in that case, $\Lambda(\lambda, \lambda')$ consists of a single morphism $\lambda \to \lambda'$. An inverse system $\boldsymbol{X}$, indexed by Λ, is just a contravariant functor $\boldsymbol{X}: \Lambda \to \mathcal{C}$. An increasing function $f: M \to \Lambda$ can be viewed as a functor. If $\boldsymbol{f} = (f, f_\mu): \boldsymbol{X} \to \boldsymbol{Y}$ is a morphism of inv-$\mathcal{C}$, then the collection of morphisms $f_\mu: X_{f(\mu)} \to Y_\mu$, $\mu \in M$, is a natural transformation from the composed functor $\boldsymbol{X}f: M \to \mathcal{C}$ to the functor $\boldsymbol{Y}: M \to \mathcal{C}$.

Already at the next level of generality, i.e., in defining the category pro-$\mathcal{C}$, we need to impose some restrictions on the indexing sets Λ. A preordering $\leq$ on Λ is *directed* provided, for any $\lambda', \lambda'' \in \Lambda$, there is a $\lambda \in \Lambda$ such that $\lambda \geq \lambda', \lambda''$. A preordering $\leq$ is *cofinite* provided it is an ordering and each element $\lambda \in \Lambda$ admits only finitely many predecessors $\lambda_1, \ldots, \lambda_n \leq \lambda$. Cofiniteness often makes possible proofs by induction on the number of predecessors. The following simple, but useful lemma is an example.

LEMMA 1.2. *Let $f: M \to \Lambda$ be an arbitrary function between preordered sets. If M is cofinite and Λ is directed, then there exists an increasing function $f': M \to \Lambda$ such that $f' \geq f$.*

Proof. The construction of f' is by induction on the number $k(\mu), \mu \in M$, of predecessors $\nu < \mu$ of $\mu \in M$. If $k(\mu) = 0$, we choose for $f'(\mu)$ any value in Λ such that $f'(\mu) \geq f(\mu)$. Now assume that we have already defined $f'(\mu)$, for $k(\mu) < n, n > 0$. If for a given μ the number $k(\mu) = n$, we choose for $f'(\mu)$ a value in Λ such that $f'(\mu) \geq f(\mu)$ and $f'(\mu) \geq f(\nu)$, for all $\nu < \mu$. This is possible, because the number of such predecessors is finite and Λ is directed. □

REMARK 1.3. If Λ is a directed set which has no terminal element, then every finite set of its elements $\{\mu_1, \ldots, \mu_k\}$ admits an element $\lambda' \in \Lambda$ such that $\lambda' > \mu_1, \ldots, \mu_k$. Indeed, by directedness, there exists an element $\mu \geq \mu_1, \ldots, \mu_k$. Since Λ has no terminal element, there exists an element λ such that $\lambda \not\leq \mu$. Let $\lambda' \in \Lambda$ be such that $\lambda' \geq \lambda, \mu$. Clearly, $\mu_i \leq \mu \leq \lambda'$, for every $i \in \{1, \ldots, k\}$. One cannot have $\lambda' = \mu_i$, for some i, because this would imply $\lambda \leq \lambda' = \mu_i \leq \mu$, which is in contradiction with $\lambda \not\leq \mu$. Consequently, $\mu_i < \lambda'$, for all $i \in \{1, \ldots, k\}$.

Let $\boldsymbol{f} = (f, f_\mu): \boldsymbol{X} \to \boldsymbol{Y}$ be a morphism of inv-$\mathcal{C}$. For any increasing function $g: M \to \Lambda$, such that $g \geq f$, one can consider the morphisms $g_\mu: X_{g(\mu)} \to Y_\mu$, given by

$$g_\mu = f_\mu p_{f(\mu)g(\mu)}. \tag{9}$$

It is readily seen that $\boldsymbol{g}: \boldsymbol{X} \to \boldsymbol{Y}$ is also a morphism of inv-$\mathcal{C}$. We say that $\boldsymbol{g}$ is the *shift* of $\boldsymbol{f}$ by g. Clearly, if $f \leq g \leq h$ and $\boldsymbol{g}$ is the shift of $\boldsymbol{f}$ by g, then the shift of $\boldsymbol{g}$ by h coincides with the shift of $\boldsymbol{f}$ by h. Two morphisms $\boldsymbol{f}, \boldsymbol{f}': \boldsymbol{X} \to \boldsymbol{Y}$ of inv-$\mathcal{C}$ are said to be *congruent*, $\boldsymbol{f} \equiv \boldsymbol{f}'$, provided they have a common shift $\boldsymbol{g}$.

LEMMA 1.4. *If M is cofinite and Λ is directed, then the congruence $\equiv$ is an equivalence relation on the set of morphisms $\boldsymbol{f}: \boldsymbol{X} \to \boldsymbol{Y}$ of* inv-$\mathcal{C}$.

Proof. Assume that $\boldsymbol{f} \equiv \boldsymbol{f}'$, $\boldsymbol{f}' \equiv \boldsymbol{f}''$. Then $\boldsymbol{f}$ and $\boldsymbol{f}'$ have a common shift $\boldsymbol{g} = (g, g_\mu)$ and $\boldsymbol{f}'$ and $\boldsymbol{f}''$ have a common shift $\boldsymbol{g}' = (g', g'_\mu)$. By directedness of Λ, there exists a function $h: M \to \Lambda$ such that $h \geq g, g'$. By Lemma 1.2, one can assume that h is an increasing function. Let $\boldsymbol{h}$ denote the shift of $\boldsymbol{f}'$ by h. Since $f' \leq g \leq h$, $\boldsymbol{h}$ is also the shift of $\boldsymbol{g}$ by h, and thus, the shift of $\boldsymbol{f}$ by h. The same argument shows that $\boldsymbol{h}$ is also the shift of $\boldsymbol{f}''$ by h. Consequently, $\boldsymbol{h}$ is a common shift of $\boldsymbol{f}$ and $\boldsymbol{f}''$, i.e., $\boldsymbol{f} \equiv \boldsymbol{f}''$. □

The objects of pro-$\mathcal{C}$ are inverse systems in $\mathcal{C}$ (Λ directed and cofinite). Morphisms are classes $[\boldsymbol{f}]$ of morphisms $\boldsymbol{f}: \boldsymbol{X} \to \boldsymbol{Y}$ with respect to $\equiv$. It is an easy consequence of the transitivity of $\equiv$ that $\boldsymbol{f} \equiv \boldsymbol{f}'$ and $\boldsymbol{g} \equiv \boldsymbol{g}'$ imply $\boldsymbol{g}\boldsymbol{f} \equiv \boldsymbol{g}'\boldsymbol{f}'$. Indeed, $\boldsymbol{g}'\boldsymbol{f}$ is a shift of $\boldsymbol{g}\boldsymbol{f}$ by $fg' \geq fg$ and thus, $\boldsymbol{g}\boldsymbol{f} \equiv \boldsymbol{g}'\boldsymbol{f}$, while $\boldsymbol{g}'\boldsymbol{f}'$ is a shift of $\boldsymbol{g}'\boldsymbol{f}$ by $f'g' \geq fg'$ and thus, $\boldsymbol{g}'\boldsymbol{f} \equiv \boldsymbol{g}'\boldsymbol{f}'$. Therefore, composition in pro-$\mathcal{C}$ is well defined by putting $[\boldsymbol{g}][\boldsymbol{f}] = [\boldsymbol{g}\boldsymbol{f}]$. The identity on $\boldsymbol{X}$ is the class $[\mathbf{1}_{\boldsymbol{X}}]$.

REMARK 1.5. A more general approach allows one to define pro-$\mathcal{C}$ also for systems indexed by directed sets which are not cofinite. In this case the index functions f of morphisms $\boldsymbol{f} = (f, f_\mu)$ need not be increasing. However, the category thus obtained is equivalent to the one of above (see (Mardešić, Segal 1982), I.1.2 and 1.3).

A single object X of $\mathcal{C}$ can be viewed as a *rudimentary system*, i.e., a system indexed by a singleton $\Lambda = \{*\}$. In this case, a morphism $\boldsymbol{f}: X \to \boldsymbol{Y} = (Y_\mu, q_{\mu\mu'}, M)$ of inv-$\mathcal{C}$ is at the same time a morphism of pro-$\mathcal{C}$ and consists of a collection of mappings $f_\mu: X \to Y_\mu$ such that $q_{\mu\mu'} f_{\mu'} = f_\mu$, for $\mu \leq \mu'$. Omitting the constant index function $f: M \to \{*\}$, we use the notation $\boldsymbol{f} = (f_\mu): X \to \boldsymbol{Y}$.

An *inverse limit* of an inverse system $\boldsymbol{X}$ is a morphism $\boldsymbol{p}: X \to \boldsymbol{X}$ of inv-$\mathcal{C}$, where X is an object from $\mathcal{C}$. Among all such morphisms, the limit is characterized by the following universal property. If $\boldsymbol{p}': X' \to \boldsymbol{X}$ is an arbitrary morphism of inv-$\mathcal{C}$, there exists a unique morphism $f: X' \to X$ of $\mathcal{C}$ such that $\boldsymbol{p}f = \boldsymbol{p}'$. Consequently, if X exists, it is unique up to natural isomorphism. We write $X = \lim \boldsymbol{X}$ and we refer to the morphisms $p_\lambda: X \to X_\lambda$ as to *natural projections*. A morphism $\boldsymbol{f}: \boldsymbol{X} \to \boldsymbol{Y}$ induces a morphism $\lim \boldsymbol{f}: \lim \boldsymbol{X} \to \lim \boldsymbol{Y}$, making lim a functor lim: inv-$\mathcal{C} \to \mathcal{C}$. Congruent morphisms induce the same limit morphism making lim also a functor lim: pro-$\mathcal{C} \to \mathcal{C}$.

Of particular interest to us is the case when $\mathcal{C}$ is the *category of topological spaces* Top. Its objects are topological spaces and morphisms are continuous mappings. The corresponding categories of inverse systems are Top^Λ, inv-Top and pro-Top. We will usually refer to their morphisms as *level mappings, mappings* and *congruence classes of mappings*, respectively. In the category

Top limits of inverse systems always exist. It suffices to take for X the subset of the direct product $\prod X_\lambda$, consisting of all points $x = (x_\lambda)$, whose coordinates x_λ satisfy the condition $x_\lambda = p_{\lambda\lambda'}(x_{\lambda'})$, for $\lambda \leq \lambda'$. As natural projections p_λ one takes the restrictions to X of the natural projections $\prod X_\lambda \to X_\lambda$.

REMARK 1.6. The above construction yields limits in many categories in particular in the category of sets Set, of groups Grp, of abelian groups Ab, etc. The *homotopy category* H(Top), whose objects are topological spaces and whose morphisms are homotopy classes $[f]$ of mappings f, is an example of a category in which there exist inverse systems with no limit (see e.g., (Mardešić, Segal 1982), I.5.1).

1.2 Coherent mappings of inverse systems

Fundamental to the development of ordinary shape theory is the category pro - H(Top) (see I.4 of (Mardešić, Segal 1982)). Note that its objects, as well as the objects of inv - H(Top), are inverse systems $(X_\lambda, [p_{\lambda\lambda'}], \Lambda)$ in the homotopy category H(Top). A morphism of inv - H(Top) between systems $(X_\lambda, [p_{\lambda\lambda'}], \Lambda)$ and $(Y_\mu, [q_{\mu\mu'}], M)$ consists of an increasing function $f: M \to \Lambda$ and of homotopy classes of mappings $[f_\mu]: X_{f(\mu)} \to Y_\mu$, $\mu \in M$, such that

$$[f_{\mu_0}][p_{f(\mu_0)f(\mu_1)}] = [q_{\mu_0\mu_1}][f_{\mu_1}],\ \mu_0 \leq \mu_1. \tag{1}$$

Replacing morphisms of inv - H(Top) by their congruence classes, one obtains morphisms of pro - H(Top).

We will denote by H: Top $\to$H(Top) the *homotopy functor.* It keeps spaces fixed and sends every mapping f to its homotopy class $H(f) = [f]$. Application of H to an inverse system $\boldsymbol{X} = (X_\lambda, p_{\lambda\lambda'}, \Lambda)$ in Top yields the system $[\boldsymbol{X}] = (X_\lambda, [p_{\lambda\lambda'}], \Lambda)$ in H(Top).

Note that a morphism $[\boldsymbol{X}] \to [\boldsymbol{Y}]$ from inv - H(Top) is determined by an increasing function $f: M \to \Lambda$ and by mappings $f_{\mu_0}: X_{f(\mu_0)} \to Y_{\mu_0}$, $\mu_0 \in M$, which satisfy the following condition.

$$f_{\mu_0} p_{f(\mu_0)f(\mu_1)} \simeq q_{\mu_0\mu_1} f_{\mu_1},\ \mu_0 \leq \mu_1. \tag{2}$$

We refer to $\boldsymbol{f} = (f, f_{\mu_0}): \boldsymbol{X} \to \boldsymbol{Y}$ as to a *homotopy mapping.*

The notion of a coherent homotopy mapping, shorter coherent mapping, $\boldsymbol{f}: \boldsymbol{X} \to \boldsymbol{Y}$ is an enrichment of the notion of a homotopy mapping. Instead of just requiring existence of homotopies which realize (2), such homotopies are given as part of the data, and are denoted by $f_{\mu_0\mu_1}: X_{f(\mu_1)} \times I \to Y_{\mu_0}$. Since a sequence of three indices $\mu_0 \leq \mu_1 \leq \mu_2$ determines the three pairs $\mu_1 \leq \mu_2$, $\mu_0 \leq \mu_2$ and $\mu_0 \leq \mu_1$, one also has the corresponding homotopies $f_{\mu_1\mu_2}, f_{\mu_0\mu_2}$ and $f_{\mu_0\mu_1}$. A further requirement is that these three first-order homotopies be related by a second-order homotopy $f_{\mu_0\mu_1\mu_2}$. It is convenient to view $f_{\mu_0\mu_1\mu_2}$, not as a mapping $X_{f(\mu_2)} \times I \times I \to Y_{\mu_0}$, but as a mapping $X_{f(\mu_2)} \times \Delta^2 \to Y_{\mu_0}$, where Δ^2 is the standard 2-simplex. The reason is that (in contrast to the square $I \times I$), the 2-simplex Δ^2 has three faces, which can be put in correspondence with the three above mentioned first-order homotopies. Proceeding in this way, one considers homotopies $f_{\mu_0\dots\mu_n}\colon X_{f(\mu_n)} \times \Delta^n \to Y_{\mu_0}$ of all orders $n \geq 0$ as further data forming the coherent morphism $\boldsymbol{f}$.

In order to make these notions precise, for a preordered set M, we consider the sets M_n of all increasing $(n+1)$-tuples $\boldsymbol{\mu} = (\mu_0, \dots, \mu_n)$, $n \geq 0$, where $\mu_i \in M$ and $\mu_0 \leq \dots \leq \mu_n$. We call $\boldsymbol{\mu}$ a *multiindex* of *length* n. We also define *face operators* $d_n^j\colon M_n \to M_{n-1}$, $j = 0, \dots, n$, $n \geq 1$, and *degeneracy operators* $s_n^j\colon M_n \to M_{n+1}$, $j = 0, \dots, n$, $n \geq 0$, by putting

$$d_n^j(\mu_0, \dots, \mu_n) = (\mu_0, \dots, \mu_{j-1}, \mu_{j+1}, \dots, \mu_n). \tag{3}$$

$$s_n^j(\mu_0, \dots, \mu_n) = (\mu_0, \dots, \mu_j, \mu_j, \dots, \mu_n). \tag{4}$$

Furthermore, we consider the *standard* n*-simplex* Δ^n, $n \geq 0$, defined as the set of all points $t = (t_0, \dots, t_n) \in \mathbb{R}^{n+1}$, where the *barycentric coordinates* t_i satisfy $t_0 \geq 0, \dots, t_n \geq 0$ and $t_0 + \dots + t_n = 1$. We then define *face operators* $d_j^n\colon \Delta^{n-1} \to \Delta^n$, $0 \leq j \leq n$, $n \geq 1$, and *degeneracy operators* $s_j^n\colon \Delta^{n+1} \to \Delta^n$, $j = 0, \dots, n$, $n \geq 0$, by putting

$$d_j^n(t_0, \dots, t_{n-1}) = (t_0, \dots, t_{j-1}, 0, t_j, \dots, t_{n-1}). \tag{5}$$

$$s_j^n(t_0, \dots, t_{n+1}) = (t_0, \dots, t_{j-1}, t_j + t_{j+1}, t_{j+2}, \dots, t_{n+1}). \tag{6}$$

If the vertices of Δ^n are denoted by $e_0 = (1, 0, \dots 0), \dots, e_n = (0, \dots, 0, 1)$, then d_j^n is the affine mapping, which sends the vertex e_i, $0 \leq i \leq n-1$, to e_i, for $i < j$, and to e_{i+1}, for $i \geq j$. Similarly, s_j^n is the affine mapping, which sends the vertex $e_i \in \Delta^{n+1}$ to $e_i \in \Delta^n$, for $i \leq j$ and to $e_{i-1} \in \Delta^n$, for $i > j$. We will often omit the dimensional index n of d_n^j, s_n^j and d_j^n, s_j^n and just write d^j, s^j and d_j, s_j, respectively.

A *coherent mapping* $\boldsymbol{f} = (f, f_{\boldsymbol{\mu}})\colon \boldsymbol{X} \to \boldsymbol{Y}$ is now formally defined as a collection consisting of an increasing function $f\colon M \to \Lambda$, called the *index function* and of mappings $f_{\boldsymbol{\mu}}\colon X_{f(\mu_n)} \times \Delta^n \to Y_{\mu_0}$, for $\boldsymbol{\mu} = (\mu_0, \dots, \mu_n) \in M_n$, such that

$$f_{\boldsymbol{\mu}}(x, d_j t) = \begin{cases} q_{\mu_0\mu_1} f_{d^0\boldsymbol{\mu}}(x, t), & j = 0, \\ f_{d^j\boldsymbol{\mu}}(x, t), & 0 < j < n, \\ f_{d^n\boldsymbol{\mu}}(p_{f(\mu_{n-1})f(\mu_n)}(x), t), & j = n. \end{cases} \tag{7}$$

$$f_{\boldsymbol{\mu}}(x, s_j t) = f_{s^j\boldsymbol{\mu}}(x, t), \ 0 \leq j \leq n. \tag{8}$$

Condition (7) applies only when $n > 0$. Omitting the indices of p and q, it assumes the simple form

$$f_{\boldsymbol{\mu}}(x, d_j t) = q f_{d^j\boldsymbol{\mu}}(p(x), t), \ 0 \leq j \leq n. \tag{9}$$

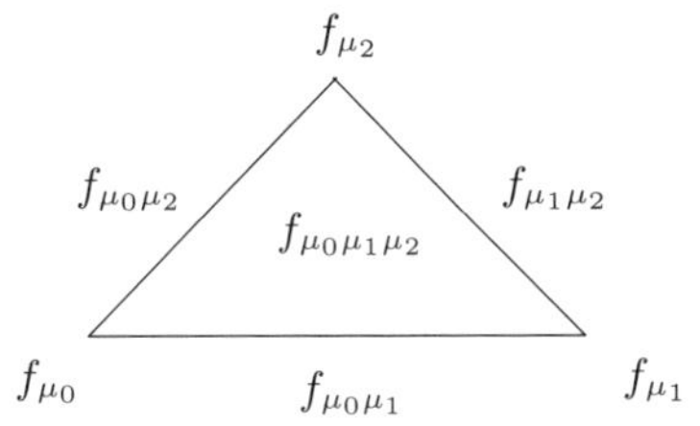

Fig. 1.1. Boundary conditions for $n = 2$

Fig. 1.1. illustrates condition (7), for $n = 2$.

If $\boldsymbol{X}$ and $\boldsymbol{Y}$ are inverse systems over the same index set Λ and $\boldsymbol{f} = (f, f_{\boldsymbol{\mu}}) \colon \boldsymbol{X} \to \boldsymbol{Y}$ is a coherent mapping with $f = \mathrm{id}$, we speak of a *level coherent mapping*. In this case we simplify the notation to $\boldsymbol{f} = (f_{\boldsymbol{\lambda}})$, where $f_{\boldsymbol{\lambda}} \colon X_{\lambda_n} \times \Delta^n \to Y_{\lambda_0}$.

REMARK 1.7. The codomain of the mapping $f_{\boldsymbol{\mu}}$ is Y_{μ_0}, while the codomain of the mapping $f_{d^0\boldsymbol{\mu}}$ is Y_{μ_1}. Therefore, in (7), $f_{d^0\boldsymbol{\mu}}$ must be composed with $q_{\mu_0\mu_1}$ in order to become comparable with $f_{\boldsymbol{\mu}}$. Similarly, comparison of the domains explains composition with $p_{f(\mu_{n-1})f(\mu_n)}$, for $j = n$. Inverse systems used in shape theory are often *inclusion systems*, i.e., have inclusions for bonding mappings. In such cases formula (7) assumes the simple form $f_{\boldsymbol{\mu}}(x, d_j t) = f_{d^j\boldsymbol{\mu}}(x, t)$, for $0 \le j \le n$.

REMARK 1.8. We identify $(x, 1) \in X_{f(\mu_0)} \times \Delta^0$ with $x \in X_{f(\mu_0)}$ and thus, identify $f_{\mu_0} \colon X_{f(\mu_0)} \times \Delta^0 \to Y_{\mu_0}$ with a mapping $f_{\mu_0} \colon X_{f(\mu_0)} \to Y_{\mu_0}$. Identifying $(t_0, t_1) \in \Delta^1$ with $t_1 \in I$, $f_{\mu_0\mu_1} \colon X_{f(\mu_1)} \times \Delta^1 \to Y_{\mu_0}$ becomes a homotopy $f_{\mu_0\mu_1} \colon X_{f(\mu_1)} \times I \to Y_{\mu_0}$, which connects $f_{\mu_0} p_{f(\mu_0)f(\mu_1)}$ to $q_{\mu_0\mu_1} f_{\mu_1}$, because $t_1 = 0$ implies $(t_0, t_1) = (1, 0) = d_1(1)$ and $t_1 = 1$ implies $(t_0, t_1) = (0, 1) = d_0(1)$.

REMARK 1.9. In the above definitions we allowed repetition of elements μ_j in the multiindices $\boldsymbol{\mu}$, i.e., we allowed *degenerate* multiindices. Clearly, $\boldsymbol{\mu} \in M_n$ is degenerate if and only if it is of the form $\boldsymbol{\mu} = s^j \boldsymbol{\nu}$, where $\boldsymbol{\nu} \in M_{n-1}$. Our requirement (8) insures that mappings $f_{\boldsymbol{\mu}}$, for degenerate $\boldsymbol{\mu}$, do not carry new geometrically relevant information (cf. Lemma 1.13). E.g., the mapping $f_{\mu\mu\mu'} \colon X_{f(\mu')} \times \Delta^2 \to Y_\mu$ is the composition of the projection $1 \times s_0 \colon X_{f(\mu')} \times \Delta^2 \to X_{f(\mu')} \times \Delta^1$ and the mapping $f_{\mu\mu'} \colon X_{f(\mu')} \times \Delta^1 \to Y_\mu$.

One can generalize the face and degeneracy operators as follows. For $n \in \{0, 1, \ldots\}$, let $[n]$ denote the set $\{0, 1, \ldots, n\}$, endowed with its natural ordering $\le$. With every increasing function $u \colon [k] \to [n]$ one associates an operator $u^* \colon M_n \to M_k$, defined as the composition

$$u^*(\boldsymbol{\mu}) = \boldsymbol{\mu} u, \;\; \boldsymbol{\mu} \in M_n, \tag{10}$$

where $\boldsymbol{\mu} = (\mu_0, \ldots, \mu_n)$ is interpreted as the increasing function $\boldsymbol{\mu} \colon [n] \to M$, given by $\boldsymbol{\mu}(j) = \mu_j$.

In particular, let $\delta_j^n\colon [n-1] \to [n]$, $n \geq 1$, and $\sigma_j^n\colon [n+1] \to [n], n \geq 0$, $0 \leq j \leq n$, be the increasing functions, defined by

$$\delta_j^n(i) = \begin{cases} i, & i < j, \\ i+1, & i \geq j, \end{cases} \tag{11}$$

$$\sigma_j^n(i) = \begin{cases} i, & i \leq j, \\ i-1, & i > j, \end{cases} \tag{12}$$

or in the multiindex notation, defined by $\delta_j^n = (0, \ldots, j-1, j+1, \ldots, n)$, $\sigma_j^n = (0, \ldots, j, j, \ldots, n)$. Then

$$(\delta_j^n)^* = d_n^j,\ (\sigma_j^n)^* = s_n^j, \tag{13}$$

i.e., one obtains the face and degeneracy operators $d_n^j\colon M_n \to M_{n-1}$, $s_n^j\colon M_n \to M_{n+1}$.

Every increasing function $u\colon [k] \to [n]$ admits a unique factorization

$$\begin{gathered} u = \delta_{i_1} \ldots \delta_{i_p} \sigma_{j_1} \ldots \sigma_{j_q}, \\ 0 \leq i_p < \ldots < i_1 \leq n,\ 0 \leq j_1 < \ldots < j_q < k,\ k+p = n+q \end{gathered} \tag{14}$$

(the dimension indices omitted). The set $\{i_1, \ldots, i_p\}$ is the complement of the image $u[k]$ in $[n]$, while $\{j_1, \ldots, j_q\}$ is the set of all $j \in [k]$, for which $u(j) = u(j+1)$.

For increasing functions $u\colon [k] \to [n]$, $v\colon [n] \to [m]$ and for the identity function $\mathrm{id}\colon [n] \to [n]$, (10) implies

$$(vu)^* = u^* v^*,\ \mathrm{id}^* = \mathrm{id}. \tag{15}$$

Hence, by (14) and (13), all the operators u^* are completely determined by the face and degeneracy operators d_n^j, s_n^j, which explains their importance.

A straightforward verification shows that the functions δ_j^n and σ_j^n satisfy the following conditions.

$$\delta_i \delta_j = \delta_j \delta_{i-1},\ j < i, \tag{16}$$

$$\sigma_i \sigma_j = \sigma_{j-1} \sigma_i,\ i < j, \tag{17}$$

$$\sigma_i \delta_j = \begin{cases} \delta_j \sigma_{i-1}, & j < i, \\ \mathrm{id}, & j = i, i+1, \\ \delta_{j-1} \sigma_i, & i+1 < j. \end{cases} \tag{18}$$

An immediate consequence of (13) and (15) - (18) is the fact that the face and degeneracy operators $d_n^j\colon M_n \to M_{n-1}$ and $s_n^j\colon M_n \to M_{n+1}$, $j = 0, \ldots, n$, satisfy the following conditions, dual to conditions (16) - (18).

$$d^j d^i = d^{i-1} d^j,\ j < i, \tag{19}$$

$$s^j s^i = s^i s^{j-1},\ i < j, \tag{20}$$

$$d^j s^i = \begin{cases} s^{i-1} d^j, & j < i, \\ \mathrm{id}, & j = i, i+1 \\ s^i d^{j-1}, & i+1 < j. \end{cases} \qquad (21)$$

Similar arguments apply to affine mappings between standard simplices. With every increasing function $u: [k] \to [n]$ one associates the affine mapping $u_*: \Delta^k \to \Delta^n$, which maps the vertices e_j of Δ^k to the vertices $e_{u(j)}$ of Δ^n, $j = 0, \ldots, k$. In particular, with the functions $\delta_j^n: [n-1] \to [n]$ and $\sigma_j^n: [n+1] \to [n]$, defined by (11) and (12), are associated the face and degeneracy operators $d_j^n: \Delta^{n-1} \to \Delta^n$ and $s_j^n: \Delta^{n+1} \to \Delta^n$, respectivaly, i.e.,

$$(\delta_j^n)_* = d_j^n, \quad (\sigma_j^n)_* = s_j^n. \qquad (22)$$

Also note that

$$(vu)_* = v_* u_*, \;\; \mathrm{id}_* = \mathrm{id}. \qquad (23)$$

Therefore, the mappings $d_j^n: \Delta^{n-1} \to \Delta^n$ and $s_j^n: \Delta^{n+1} \to \Delta^n$ satisfy the equations obtained from (16)–(18), by replacing δ_j and σ_j by d_j and s_j, respectively.

The following lemma is an easy consequence of (9) and (8).

LEMMA 1.10. *If $\boldsymbol{f} = (f, f_\mu)$ is a coherent mapping and $u: [k] \to [n]$ is an increasing function, then for every $\boldsymbol{\mu} \in M_n$ one has*

$$f_\mu(x, u_*(t)) = q f_{u^*(\mu)}(p(x), t), \qquad (24)$$

i.e., the following diagram commutes.

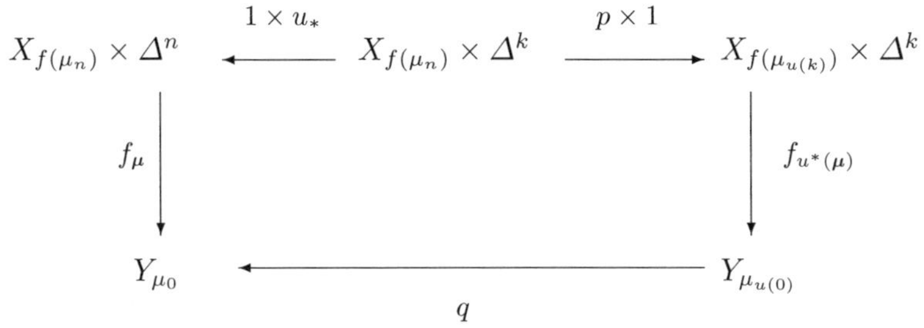

$$
\begin{array}{ccc}
X_{f(\mu_n)} \times \Delta^n & \xleftarrow{\;\;1 \times u_*\;\;} X_{f(\mu_n)} \times \Delta^k & \xrightarrow{\;\;p \times 1\;\;} X_{f(\mu_{u(k)})} \times \Delta^k \\
\downarrow f_\mu & & \downarrow f_{u^*(\mu)} \\
Y_{\mu_0} & \xleftarrow{\qquad\qquad q \qquad\qquad} & Y_{\mu_{u(0)}}
\end{array}
$$
$$(25)$$

Proof. In the special cases, when $u = \delta_j$ or $u = \sigma_j$, (25) holds, because it reduces to (9) and (8), respectively. However, composing commutative diagrams of type (25) for two increasing functions $u: [k] \to [m]$ and $v: [m] \to [n]$, one obtains the commutative diagram (25) for the composition $vu: [k] \to [n]$. Indeed, using (23), (24) and (15), one has

$$f_\mu(x, (vu)_*(t)) = f_\mu(x, v_* u_*(t)) = q f_{v^*(\mu)}(p(x), u_*(t)) \\ = q f_{u^* v^*(\mu)}(p(x), t) = q f_{(vu)^*(\mu)}(p(x), t). \qquad (26)$$

Since all increasing functions can be obtained by composing functions of type δ_i^n and σ_j^n, it follows that (25) holds in general. $\square$

Recall that a *simplicial set* S consists of a sequence of sets $S_n, n = 0, 1, \ldots$, and two sequences of functions $d_n^j : S_n \to S_{n-1}$, $n \geq 1$, and $s_n^j : S_n \to S_{n+1}$, $n \geq 0$, $j = 0, \ldots, n$, which satisfy conditions (19)–(21). The elements of S_n are called *n - simplices*, while d_n^j and s_n^j are called the *face operators* and the *degeneracy operators*, respectively. A *simplicial mapping* $\phi : S \to T$ between simplicial sets is a sequence of mappings $\phi_n : S_n \to T_n$, i.e., a *graded mapping* of degree 0, which commutes with the face and degeneracy operators,

$$d_n^j \phi_n = \phi_{n-1} d_n^j, \ s_n^j \phi_n = \phi_{n+1} s_n^j. \tag{27}$$

Alternatively, a simplicial set can be defined as a contravariant functor $[\mathbb{N}] \to \mathsf{Set}$, where $[\mathbb{N}]$ is the *category of finite ordinals*, whose objects are the sets $[n]$, $n \in \{0, 1, \ldots, \}$, and whose morphisms are all increasing functions $u : [k] \to [n]$.

Clearly, $M_* = (M_n, n = 0, 1, \ldots)$, together with the face and degeneracy operators, is a simplicial set, called the *simplicial set of multiindices* in M. Similarly, $\Delta^* = (\Delta^n, n = 0, 1, \ldots)$, together with the face and degeneracy operators, is a *cosimplicial set*, referred to as the *cosimplicial set of standard simplices.*

REMARK 1.11. According to Remark 1.1, if $\boldsymbol{f} = (f, f_\mu) : \boldsymbol{X} \to \boldsymbol{Y}$ is a mapping between inverse systems of topological spaces, then (f_μ) can be viewed as a natural transformation between functors $\boldsymbol{X} f$ and $\boldsymbol{Y}$. Recently it was shown in (Fritsch to appear) that an analogous statement holds also for coherent mappings $\boldsymbol{f} = (f, f_{\boldsymbol{\mu}}) : \boldsymbol{X} \to \boldsymbol{Y}$, i.e., the collection of mappings $(f_{\boldsymbol{\mu}})$ can be interpreted as a natural transformation between suitable functors P and Q.

The source category on which the functors P and Q are defined is the category ΓM_* of simplices of the simplicial set M_* of multiindices (see (Fritsch, Piccinini 1990), p. 141). Its objects are the simplices of M_*, i.e., multiindices $\boldsymbol{\mu} = (\mu_0, \ldots, \mu_n) \in M_n$, $n \geq 0$. Morphisms $\boldsymbol{\nu} \to \boldsymbol{\mu}$ are pairs $(\boldsymbol{\mu}, u)$, where $u : [k] \to [n]$ is an increasing function and $\boldsymbol{\nu} = \boldsymbol{\mu} u$. Furthermore, $\mathrm{id} : \boldsymbol{\mu} \to \boldsymbol{\mu}$ is the pair $(\boldsymbol{\mu}, \mathrm{id})$ and the composition $(\boldsymbol{\mu}, u) \circ (\boldsymbol{\nu}, v) = (\boldsymbol{\mu}, uv)$, where $v : [l] \to [k]$.

The target category is an extension of the category Top, here denoted by $\overline{\mathsf{Top}}$. Its objects are topological spaces. To define morphisms $X \to Y$ one considers pairs (φ, φ') of mappings $\varphi : S \to X$ and $\varphi' : S \to Y$, where S is a topological space. Such a pair is considered to be equivalent to a pair $(\tilde{\varphi}, \tilde{\varphi}')$, where $\tilde{\varphi} : \tilde{S} \to X$ and $\tilde{\varphi}' : \tilde{S} \to Y$, provided there exists a homeomorphism $\chi : S \to \tilde{S}$ such that $\tilde{\varphi}\chi = \varphi$ and $\tilde{\varphi}'\chi = \varphi'$. Morphisms $X \to Y$ of $\overline{\mathsf{Top}}$ are equivalence classes $[\varphi, \varphi']$ of pairs (φ, φ'). To compose two morphisms $[\varphi, \varphi'] : X \to Y$ and $[\psi, \psi'] : Y \to Z$, one takes the pull-back (η, η') of the mappings φ' and ψ. Then the composition is the class $[\varphi\eta, \psi'\eta'] : X \to Z$. The identity morphism $X \to X$ is the class $[\mathrm{id}, \mathrm{id}]$. The category Top can be considered a subcategory of $\overline{\mathsf{Top}}$ by identifying every mapping $\varphi : X \to Y$ with the class $[\mathrm{id}, \varphi]$.

The functor $P: \Gamma M_* \to \overline{\mathsf{Top}}$ assigns to every $\boldsymbol{\mu} \in M_n$ the space $X_{f(\mu_n)} \times \Delta^n$ and to every morphism $(\boldsymbol{\mu}, u): \boldsymbol{\nu} \to \boldsymbol{\mu}$ the class $[\varphi, \varphi']: X_{f(\mu_{u(k)})} \times \Delta^k \to X_{f(\mu_n)} \times \Delta^n$, where

$$\varphi = p \times 1: X_{f(\mu_n)} \times \Delta^k \to X_{f(\mu_{u(k)})} \times \Delta^k, \tag{28}$$

$$\varphi' = 1 \times u_*: X_{f(\mu_n)} \times \Delta^k \to X_{f(\mu_n)} \times \Delta^n. \tag{29}$$

Verification of the functoriality reduces to showing that the pull-back of the mappings $1 \times v_*: X_{f(\mu_{u(k)})} \times \Delta^l \to X_{f(\mu_{u(k)})} \times \Delta^k$ and $p \times 1: X_{f(\mu_n)} \times \Delta^k \to X_{f(\mu_{u(k)})} \times \Delta^k$ consists of the mappings $p \times 1: X_{f(\mu_n)} \times \Delta^l \to X_{f(\mu_{u(k)})} \times \Delta^l$ and $1 \times v_*: X_{f(\mu_n)} \times \Delta^l \to X_{f(\mu_n)} \times \Delta^k$. The functor $Q: \Gamma M_* \to \overline{\mathsf{Top}}$ is actually a functor $Q: \Gamma M_* \to \mathsf{Top}$ to the subcategory $\mathsf{Top} \subseteq \overline{\mathsf{Top}}$. It assigns to $\boldsymbol{\mu}$ the space Y_{μ_0} and to the morphism $(\boldsymbol{\mu}, u)$ the mapping $q_{\mu_0 \mu_{u(0)}}: Y_{\mu_{u(0)}} \to Y_{\mu_0}$. The condition which makes $(f_{\boldsymbol{\mu}})$ a natural transformation $P \to Q$ is the equality $f_{\boldsymbol{\mu}} P(\boldsymbol{\mu}, u) = Q(\boldsymbol{\mu}, u) f_{\boldsymbol{\nu}}$, which must hold for every morphism $(\boldsymbol{\mu}, u): \boldsymbol{\nu} \to \boldsymbol{\mu}$. However, this condition is just the requirement that diagram (25) be commutative.

REMARK 1.12. There are situations where one needs to consider coherent mappings $\boldsymbol{f}: X \to \boldsymbol{Y}$ from a space X to an inverse system $\boldsymbol{Y}$. In such cases, one views X as a rudimentary inverse system (see 1.1). Then $\boldsymbol{f} = (f_{\boldsymbol{\mu}})$ is given by mappings $f_{\boldsymbol{\mu}}: X \times \Delta^n \to Y_{\mu_0}$, $\boldsymbol{\mu} \in M_n$, and formula (9), assumes the simpler form $f_{\boldsymbol{\mu}}(x, d_j t) = q f_{d^j \boldsymbol{\mu}}(x, t)$.

The next lemma shows that, in the case of ordered indexing sets M, in the definition of a coherent mapping non-degenerate multiindices suffice.

LEMMA 1.13. *Let* $\boldsymbol{X}$ *and* $\boldsymbol{Y}$ *be inverse systems over* Λ *and* M, *respectively. Let* $f: M \to \Lambda$ *be an increasing function and let* $f_{\boldsymbol{\mu}}: X_{f(\mu_n)} \times \Delta^n \to Y_{\mu_0}$ *be mappings, defined for non-degenerate* $\boldsymbol{\mu} \in M_n$, *which satisfy* (7). *If* M *is ordered, one can extend these data in a unique way to a coherent mapping* $\boldsymbol{f} = (f, f_{\boldsymbol{\mu}}): \boldsymbol{X} \to \boldsymbol{Y}$.

Proof. First note that with every multiindex $\boldsymbol{\mu} \in M_n$ one can associate a unique non-degenerate multiindex $\boldsymbol{\nu} = (\nu_0, \dots, \nu_k) \in M_k$, $0 \le k \le n$, and a unique increasing surjection $u: [n] \to [k]$ such that

$$\boldsymbol{\mu} = \boldsymbol{\nu} u = u^*(\boldsymbol{\nu}). \tag{30}$$

Indeed, $[n]$ decomposes into $k+1$ disjoint subset, the counter-images of different points of the set $\mu[n] \subseteq M$ under the mapping $\boldsymbol{\mu}: [n] \to M$. If $i \le i''$ belong to the same class of this decomposition and $i \le i' \le i''$, then i' also belongs to that class. Indeed, $\boldsymbol{\mu}(i) \le \boldsymbol{\mu}(i') \le \boldsymbol{\mu}(i'')$ and $\boldsymbol{\mu}(i) = \boldsymbol{\mu}(i'')$ imply $\boldsymbol{\mu}(i) = \boldsymbol{\mu}(i')$, because $\le$ is anti-symmetric. Therefore, one can identify the corresponding quotient set of $[n]$ with $[k]$ and the quotient mapping with an increasing surjection $u: [n] \to [k]$. Moreover, $\boldsymbol{\mu}$ induces an increasing injection $\boldsymbol{\nu}: [k] \to M$ such that $\boldsymbol{\mu} = \boldsymbol{\nu} u$. Clearly, $\boldsymbol{\nu}$ is a non-degenerate multiindex $\boldsymbol{\nu} = (\nu_0, \ldots, \nu_k)$, $\nu_0 < \ldots < \nu_k$.

For a degenerate multiindex $\boldsymbol{\mu} \in M_n$, we define $f_{\boldsymbol{\mu}}: X_{f(\mu_n)} \times \Delta^n \to Y_{\mu_0}$ by putting

$$f_{\boldsymbol{\mu}}(x,t) = f_{\boldsymbol{\nu}}(x, u_*(t)), \;\; x \in X_{f(\mu_n)}, t \in \Delta^n, \tag{31}$$

where $\boldsymbol{\nu} = (\nu_0, \ldots, \nu_k)$ is non-degenerate, $u: [n] \to [k]$ is an increasing surjection and (30) holds. Note that $u(0) = 0$, $u(n) = k$ and therefore, $\mu_0 = \nu_0$, $\mu_n = \nu_k$. We will now show that the mappings $f_{\boldsymbol{\mu}}$ satisfy the coherence conditions.

First consider the case when $\boldsymbol{\mu}$ is non-degenerate. Then $d^j\boldsymbol{\mu}$ is also non-degenerate and (7) holds by assumption. To verify (8), note that $s^j\boldsymbol{\mu} = \sigma_j^*(\boldsymbol{\mu}) = \boldsymbol{\mu}\sigma_j$ is the factorization (30) of $s^j\boldsymbol{\mu}$. Therefore, by (31),

$$f_{s^j\boldsymbol{\mu}}(x,t) = f_{\boldsymbol{\mu}}(x, \sigma_{j*}(t)) = f_{\boldsymbol{\mu}}(x, s_j t), \tag{32}$$

which is the desired relation (8).

Now assume that $\boldsymbol{\mu} = \boldsymbol{\nu} u$ is degenerate. The factorization (14) of u must be of the form

$$u = \sigma_{j_1} \cdots \sigma_{j_r}, \tag{33}$$

because u is surjective and the functions δ_i are not. By (18), for any i, j, the composition $\sigma_i \delta_j$ either equals $\delta_{j'}\sigma_{i'}$, for an appropriate choice of indices i', j', or equals the identity. Therefore, for $u\delta_j = \sigma_{j_1} \cdots \sigma_{j_r}\delta_j$, there are three possibilities:

$$u\delta_j = \delta_{j'}\sigma_{j'_1} \cdots \sigma_{j'_{r'}}, \tag{34}$$

$$u\delta_j = \sigma_{j'_1} \cdots \sigma_{j'_{r'}} \tag{35}$$

or $u\delta_j = \mathrm{id}$. We will verify (7) in all three cases.

In the first case,

$$u_* d_j = (u\delta_j)_* = d_{j'} s_{j'_1} \cdots s_{j'_{r'}}. \tag{36}$$

Since $\boldsymbol{\nu}$ is non-degenerate, (31) and the assumed coherence conditions for non-degenerate multiindices imply

$$\begin{aligned} f_{\boldsymbol{\mu}}(x, d_j t) &= f_{\boldsymbol{\nu}}(x, u_* d_j t) = f_{\boldsymbol{\nu}}(x, d_{j'} s_{j'_1} \cdots s_{j'_{r'}} t) \\ &= q f_{d^{j'}\boldsymbol{\nu}}(p(x), s_{j'_1} \cdots s_{j'_{r'}} t). \end{aligned} \tag{37}$$

On the other hand,

$$d^j\boldsymbol{\mu} = \delta_j^*(\boldsymbol{\nu} u) = \boldsymbol{\nu} u \delta_j = \boldsymbol{\nu}\delta_{j'}\sigma_{j'_1} \cdots \sigma_{j'_{r'}}. \tag{38}$$

Note that $\boldsymbol{\nu}' = \boldsymbol{\nu}\delta_{j'} = d^{j'}\boldsymbol{\nu}$ is non-degenerate, while $u' = \sigma_{j'_1} \cdots \sigma_{j'_{r'}}$ is an increasing surjection. Therefore, (31) and (38) yield

$$q f_{d^j\boldsymbol{\mu}}(p(x), t) = q f_{d^{j'}\boldsymbol{\nu}}(p(x), s_{j'_1} \cdots s_{j'_{r'}} t). \tag{39}$$

Comparison of (37) with (39) yields (7).

In the second case, the argument is similar and somewhat simpler. Formula (35) implies

$$u_* d_j = s_{j'_1} \cdots s_{j'_{r'}}, \tag{40}$$

Therefore, (31) yields

$$f_{\mu}(x, d_j t) = f_{\nu}(x, u_* d_j t) = f_{\nu}(x, s_{j'_1} \cdots s_{j'_{r'}} t). \tag{41}$$

On the other hand,

$$d^j \boldsymbol{\mu} = \boldsymbol{\nu} u \delta_j = \boldsymbol{\nu} \sigma_{j'_1} \cdots \sigma_{j'_{r'}} \tag{42}$$

Consequently, by (31),

$$f_{d^j \boldsymbol{\mu}}(x, t) = f_{\nu}(x, s_{j'_1} \cdots s_{j'_{r'}} t). \tag{43}$$

Comparison of (41) with (43) again yields (7). Notice that in this case p and q equal identity, because (42) shows that $d^j \boldsymbol{\mu}$ and $\boldsymbol{\nu}$ have the same initial and terminal element, and the same holds for $\boldsymbol{\nu}$ and $\boldsymbol{\mu}$.

Finally, if $u\delta_j = \text{id}$, then $u_* d_j = (u\delta_j)_* = \text{id}$ and thus, by (31),

$$f_{\mu}(x, d_j t) = f_{\nu}(x, t). \tag{44}$$

On the other hand, $d^j \boldsymbol{\mu} = \boldsymbol{\nu} u \delta_j = \boldsymbol{\nu}$ and thus,

$$f_{d^j \boldsymbol{\mu}}(x, t) = f_{\nu}(x, t). \tag{45}$$

Comparison of (44) with (45) completes the verification of (7) in all cases.

To verify condition (8), note that

$$u_* s_j = u_* \sigma_{j*} = \sigma_{j_1 *} \cdots \sigma_{j_r *} \sigma_{j*} = s_{j_1} \cdots s_{j_r} s_j \tag{46}$$

and thus,

$$f_{\mu}(x, s_j t) = f_{\nu}(x, u_* s_j t) = f_{\nu}(x, s_{j_1} \cdots s_{j_r} s_j t). \tag{47}$$

On the other hand,

$$s_j \boldsymbol{\mu} = \sigma_j^*(\boldsymbol{\nu} u) = \boldsymbol{\nu} u \sigma_j = \boldsymbol{\nu} \sigma_{j_1} \cdots \sigma_{j_r} \sigma_j \tag{48}$$

and thus, by (31),

$$f_{s_j \boldsymbol{\mu}}(x, t) = f_{\nu}(x, s_{j_1} \cdots s_{j_r} s_j t). \tag{49}$$

Comparison of (47) with (49) establishes the desired condition (8). □

We will now prove a technical lemma which in some situations facilitates the construction of coherent mappings.

LEMMA 1.14. *Let* $\boldsymbol{X}$ *and* $\boldsymbol{Y}$ *be inverse systems over a directed set* Λ *and over a cofinite set* M*, respectively. Let* f *be a function which to every non-degenerate multiindex* $\boldsymbol{\mu} \in M_n$ *assigns an element* $f(\boldsymbol{\mu}) \in \Lambda$ *such that*

$$f(\boldsymbol{\mu}) \geq f(d^j \boldsymbol{\mu}), \; 0 \leq j \leq n, n > 0; \tag{50}$$

Moreover, let $f_{\boldsymbol{\mu}}: X_{f(\boldsymbol{\mu})} \times \Delta^n \to Y_{\mu_0}$ *be mappings such that, for* $n > 0$,

$$f_{\boldsymbol{\mu}}(x, d_j t) = \begin{cases} q_{\mu_0 \mu_1} f_{d^0 \boldsymbol{\mu}}(p_{f(d^0 \boldsymbol{\mu}) f(\boldsymbol{\mu})}(x), t), & j = 0, \\ f_{d^j \boldsymbol{\mu}}(p_{f(d^j \boldsymbol{\mu}) f(\boldsymbol{\mu})}(x), t), & 0 < j \leq n. \end{cases} \tag{51}$$

Then there exists a coherent mapping $\boldsymbol{f}' = (f', f'_{\boldsymbol{\mu}}): \boldsymbol{X} \to \boldsymbol{Y}$ *such that*

$$f(\boldsymbol{\mu}) \leq f'(\mu_n), \tag{52}$$

$$f'_{\boldsymbol{\mu}}(x, t) = f_{\boldsymbol{\mu}}(p_{f(\boldsymbol{\mu}) f'(\mu_n)}(x), t). \tag{53}$$

Proof. By assumption, every index $\mu \in M$ has only finitely many predecessors. Therefore, there are only finitely many non-degenerate multiindices $\boldsymbol{\mu}$, whose terminal element $\mu_n = \mu$. The function f maps these multiindices into a finite subset of Λ. Since Λ is directed, it is possible to define $f'(\mu) \in \Lambda$ so that $f(\boldsymbol{\mu}) \leq f'(\mu)$, for all these multiindices $\boldsymbol{\mu}$. Clearly, $f(\boldsymbol{\mu}) \leq f'(\mu_n)$, for every non-degenerate multiindex $\boldsymbol{\mu}$ of length n. In view of Lemma 1.2, one can assume that $f': M \to \Lambda$ is an increasing function. For every non-degenerate $\boldsymbol{\mu}$, we now define $f'_{\boldsymbol{\mu}}: X_{f'(\mu_n)} \to Y_{\mu_0}$, by (53). By (51), it readily follows that the mappings $f'_{\boldsymbol{\mu}}$ satisfy the coherence conditions (7), for non-degenerate $\boldsymbol{\mu}$. To extend $(f', f'_{\boldsymbol{\mu}})$, to a coherent mapping, it suffices to apply Lemma 1.13. □

We conclude this subsection by defining the *shift* of a coherent mapping $\boldsymbol{f} = (f, f_{\boldsymbol{\mu}}): \boldsymbol{X} \to \boldsymbol{Y}$ by an increasing function $f': M \to \Lambda$, such that $f \leq f'$. It is the coherent mapping $\boldsymbol{f}' = (f', f'_{\boldsymbol{\mu}}): \boldsymbol{X} \to \boldsymbol{Y}$, where

$$f'_{\boldsymbol{\mu}}(x, t) = f_{\boldsymbol{\mu}}(p_{f(\mu_n) f'(\mu_n)}(x), t). \tag{54}$$

Two coherent mappings $\boldsymbol{f}', \boldsymbol{f}'': \boldsymbol{X} \to \boldsymbol{Y}$ are said to be *congruent*, denoted by $\boldsymbol{f}' \equiv \boldsymbol{f}''$, provided they admit a common shift $\boldsymbol{f}$. The following lemma is proved just as Lemma 1.4.

LEMMA 1.15. *If* M *is cofinite and* Λ *is directed, then, on the set of coherent mappings* $\boldsymbol{f}: \boldsymbol{X} \to \boldsymbol{Y}$*, the congruence* $\equiv$ *is an equivalence relation.* □

1.3 Composition of coherent mappings

In this subsection we will define composition of coherent mappings. Let $\boldsymbol{X} = (X_\lambda, p_{\lambda\lambda'}, \Lambda)$, $\boldsymbol{Y} = (Y_\mu, q_{\mu\mu'}, M)$ and $\boldsymbol{Z} = (Z_\nu, r_{\nu\nu'}, N)$ be inverse systems and let $\boldsymbol{f} = (f, f_\mu): \boldsymbol{X} \to \boldsymbol{Y}$ and $\boldsymbol{g} = (g, g_\nu): \boldsymbol{Y} \to \boldsymbol{Z}$ be coherent mappings. Their composition $\boldsymbol{h} = \boldsymbol{gf}: \boldsymbol{X} \to \boldsymbol{Z}$ will be given by an increasing function $h: N \to \Lambda$ and by mappings $h_{\boldsymbol{\nu}}: X_{h(\nu_n)} \times \Delta^n \to Z_{\nu_0}$, for $\boldsymbol{\nu} = (\nu_0, \dots, \nu_n) \in N_n$.

It is natural to define $h: N \to \Lambda$ by $h = fg$. If $\boldsymbol{\nu} = (\nu_0)$ is of length 0, we put $h_{\nu_0} = g_{\nu_0} f_{g(\nu_0)}$. If $\boldsymbol{\nu} = (\nu_0, \nu_1)$ is of length 1, we have the homotopies $f_{g(\nu_0)g(\nu_1)}: X_{fg(\nu_1)} \times I \to Y_{g(\nu_0)}$ and $g_{\nu_0\nu_1}: Y_{g(\nu_1)} \times I \to Z_{\nu_0}$. They yield homotopies $g_{\nu_0} f_{g(\nu_0)g(\nu_1)}$ and $g_{\nu_0\nu_1}(f_{g(\nu_1)} \times 1)$ from $X_{fg(\nu_1)} \times I$ to Z_{ν_0}, which connect $g_{\nu_0} f_{g(\nu_0)} p_{fg(\nu_0), fg(\nu_1)}$ to $g_{\nu_0} q_{g(\nu_0), g(\nu_1)} f_{g(\nu_1)}$ and $g_{\nu_0} q_{g(\nu_0), g(\nu_1)} f_{g(\nu_1)}$ to $r_{\nu_0\nu_1} g_{\nu_1} f_{g(\nu_1)}$, respectively. It is natural to juxtapose these two homotopies to produce the desired homotopy $h_{\nu_0\nu_1}: X_{fg(\nu_1)} \times I \to Z_{\nu_0}$, which connects $g_{\nu_0} f_{g(\nu_0)} p_{fg(\nu_0), fg(\nu_1)}$ to $r_{\nu_0\nu_1} g_{\nu_1} f_{g(\nu_1)}$. This is achieved by decomposing the segment $I = [0,1]$ in two subsegments $P_0^1 = [0, \frac{1}{2}]$, $P_1^1 = [\frac{1}{2}, 1]$ and putting

$$h_{\nu_0\nu_1}(x,t) = \begin{cases} g_{\nu_0}(f_{g(\nu_0)g(\nu_1)}(x, 2t)), & t \in P_0^1, \\ g_{\nu_0\nu_1}(f_{g(\nu_1)}(x), 2t-1), & t \in P_1^1. \end{cases} \tag{1}$$

To generalize the described construction to indices $\boldsymbol{\nu} = (\nu_0, \dots, \nu_n) \in N_n$ of an arbitrary length n, we shall define a decomposition of Δ^n into $n+1$ subpolyhedra P_i^n, $0 \le i \le n$, which are direct products of simplices of dimension $n-i$ and i respectively. More precisely, there exist affine homeomorphisms $c_i^n: P_i^n \to \Delta^{n-i} \times \Delta^i$, given by mappings $a_i^n: P_i^n \to \Delta^{n-i}$ and $b_i^n: P_i^n \to \Delta^i$. For $n = 2$ and $n = 3$, the decomposition of Δ^n is shown on Fig. 1.2 and Fig. 1.3, respectively. The mappings a_i^n and b_i^n are compositions of obvious projections and similarities with ratio 2.

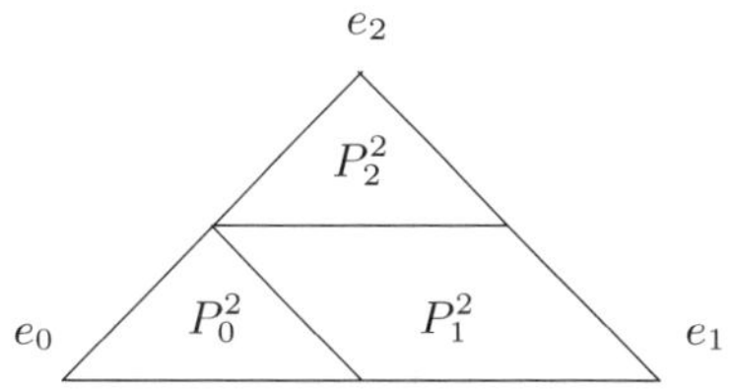

Fig. 1.2. Decomposition of Δ^2

Using barycentric coordinates, P_i^n is defined precisely as the set of all points $t = (t_0, \dots, t_n) \in \Delta^n$ such that

$$t_0 + \ldots + t_{i-1} \le \frac{1}{2} \le t_0 + \ldots + t_i. \tag{2}$$

For $i = 0$, (2) should be interpreted as the condition $1/2 \le t_0$. Note that the inequality $t_0 + \ldots + t_{i-1} \le 1/2$ defines the half-space to which belong the barycenters corresponding to those distributions of mass 1 to the points

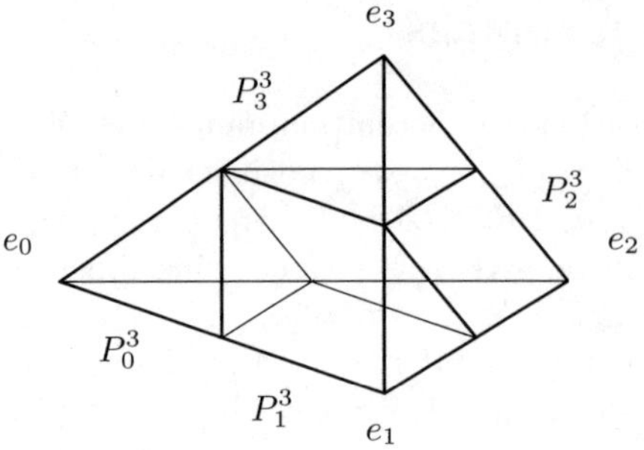

Fig. 1.3. Decomposition of Δ^3

$e_0, \ldots, e_n$, for which the total mass at the points $e_0, \ldots, e_{i-1}$ does not exceed the total mass at the remaining points $e_i, \ldots, e_n$. The inequality $t_0 + \ldots + t_i \geq 1/2$ admits an analogous interpretation. In barycentric coordinates, the mappings $a_i^n\colon P_i^n \to \Delta^{n-i}$ and $b_i^n\colon P_i^n \to \Delta^i$ are given by

$$a_i^n(t) = (\#, 2t_{i+1}, \ldots, 2t_n), \tag{3}$$

$$b_i^n(t) = (2t_0, \ldots, 2t_{i-1}, \#), \tag{4}$$

where # stands for the difference between 1 and the sum of the remaining barycentric coordinates, i.e., in (3), $\# = 1 - 2(t_{i+1} + \ldots + t_n)$ and in (4), $\# = 1 - 2(t_0 + \ldots + t_{i-1})$. Sometimes we omit the dimensional index n.

We now define $h_{\boldsymbol{\nu}}(x,t)$, for $\boldsymbol{\nu} = (\nu_0, \ldots, \nu_n)$, $x \in X_{h(\nu_n)}$ and $t \in P_i^n$, by putting

$$h_{\boldsymbol{\nu}}(x,t) = g_{\nu_0 \ldots \nu_i}(f_{g(\nu_i)\ldots g(\nu_n)}(x, a_i^n(t)), b_i^n(t)). \tag{5}$$

In other words, $h_{\boldsymbol{\nu}}$ on $X_{h(\nu_n)} \times P_i^n$ is the composition

$$X_{h(\nu_n)} \times P_i^n \xrightarrow{1\times c} X_{h(\nu_n)} \times \Delta^{n-i} \times \Delta^i \xrightarrow{f\times 1} Y_{g(\nu_i)} \times \Delta^i \xrightarrow{g} Z_{\nu_0} \tag{6}$$

(we left out the indices of the mappings $c_i^n, f_{g(\nu_i)\ldots g(\nu_n)}$ and $g_{\nu_0\ldots\nu_i}$).

Taking into account the construction of $h_{\boldsymbol{\nu}}$ and the coherence properties of $f_{\boldsymbol{\mu}}$ and $g_{\boldsymbol{\nu}}$, the next lemma is almost obvious. Nevertheless, we give a formal proof.

LEMMA 1.16. *For every multiindex* $\boldsymbol{\nu} = (\nu_0, \ldots, \nu_n)$*, there is a unique mapping* $h_{\boldsymbol{\nu}}\colon X_{h(\nu_n)} \times \Delta^n \to Z_{\nu_0}$ *such that* $h_{\boldsymbol{\nu}}|X_{h(\nu_n)} \times P_i^n$ *is given by* (5). *Moreover,* $h = fg$ *and the mappings* $h_{\boldsymbol{\nu}}$ *form a* coherent mapping $\boldsymbol{h}\colon \boldsymbol{X} \to \boldsymbol{Z}$. *By definition,* $\boldsymbol{h} = \boldsymbol{gf}$ *is the composition of the coherent mappings* $\boldsymbol{f}$ *and* $\boldsymbol{g}$.

Proof. Assume that $h_{\boldsymbol{\nu}}$ is well defined on $X_{h(\nu_n)} \times (P_0^n \cup \ldots \cup P_j^n)$. In order to prove that this mapping extends to $X_{h(\nu_n)} \times (P_0^n \cup \ldots \cup P_{j+1}^n)$, note that $(P_0^n \cup \ldots \cup P_j^n) \cap P_{j+1}^n = P_j^n \cap P_{j+1}^n$. Consequently, it suffices to show that on $X_{h(\nu_n)} \times (P_j^n \cap P_{j+1}^n)$, (5) yields the same values, for $i = j$ and $i = j+1$. Since

$$P_j^n \cap P_{j+1}^n = \{t \in \Delta^n | t_0 + \ldots + t_j = \frac{1}{2}\}, \tag{7}$$

(3) and (4) imply that, for $t \in P_j^n \cap P_{j+1}^n$,

$$a_j^n(t) = (0, 2t_{j+1}, \ldots, 2t_n) = d_0 a_{j+1}^n(t). \tag{8}$$

$$b_{j+1}^n(t) = (2t_0, \ldots, 2t_j, 0) = d_{j+1} b_j^n(t), \tag{9}$$

Using (8) and the coherence condition (1.2.7) for f_μ, we conclude that, for $i = j$, (5) yields the value

$$h_\nu(x,t) = g_{\nu_0 \ldots \nu_j}(q f_{g(\nu_{j+1}) \ldots g(\nu_n)}(x, a_{j+1}^n(t)), b_j^n(t)). \tag{10}$$

However, using (9) and the coherence condition for g_ν, we conclude that, for $i = j+1$, (5) yields the same value.

In view of Lemma 1.13, it suffices to show that the mappings h_ν satisfy the coherence conditions (1.2.7). By the definitions of d_j^n, a_i^n and b_i^n, one easily verifies that, for $j \le i$,

$$d_j(P_i^{n-1}) \subseteq P_{i+1}^n, \tag{11}$$

$$a_{i+1}^n(d_j t) = a_i^{n-1}(t), \tag{12}$$

$$b_{i+1}^n(d_j t) = d_j b_i^{n-1}(t). \tag{13}$$

Similarly, for $i < j$, one has

$$d_j(P_i^{n-1}) \subseteq P_i^n, \tag{14}$$

$$a_i^n(d_j t) = d_{j-i} a_i^{n-1}(t), \tag{15}$$

$$b_i^n(d_j t) = b_i^{n-1}(t). \tag{16}$$

Therefore, for $j \le i$ and $t \in P_i^{n-1}$, (11)–(13) yield

$$\begin{aligned} h_\nu(x, d_j t) &= g_{\nu_0 \ldots \nu_{i+1}}(f_{g(\nu_{i+1}) \ldots g(\nu_n)}(x, a_{i+1}^n(d_j t)), b_{i+1}^n(d_j t)) \\ &= g_{\nu_0 \ldots \nu_{i+1}}(f_{g(\nu_{i+1}) \ldots g(\nu_n)}(x, a_i^{n-1}(t)), d_j b_i^{n-1}(t)) \\ &= r g_{d^j(\nu_0 \ldots \nu_{i+1})}(f_{g(\nu_{i+1}) \ldots g(\nu_n)}(x, a_i^{n-1}(t)), b_i^{n-1}(t)) \\ &= r h_{d^j \nu}(x,t). \end{aligned} \tag{17}$$

Similarly, for $i < j$ and $t \in P_i^{n-1}$, (14)–(16) yield

$$h_\nu(x, d_j t) = h_{d^j \nu}(p(x), t). \;\square \tag{18}$$

Note that the composition of two level coherent mappings is again a level coherent mapping.

1.4 The coherence operator C

With every mapping $\boldsymbol{f} = (f, f_\mu): \boldsymbol{X} \to \boldsymbol{Y}$ from inv-Top one can associate a coherent mapping $C(\boldsymbol{f}): \boldsymbol{X} \to \boldsymbol{Y}$. It consists of the function $f: M \to \Lambda$ and of mappings $f_{\boldsymbol{\mu}}: X_{f(\mu_n)} \times \Delta^n \to Y_{\mu_0}$, $\boldsymbol{\mu} = (\mu_0, \dots, \mu_n) \in M_n$, defined by

$$f_{\boldsymbol{\mu}}(x, t) = f_{\mu_0} p_{f(\mu_0) f(\mu_n)}(x). \tag{1}$$

Since $f_{\mu_0} p_{f(\mu_0) f(\mu_1)} = q_{\mu_0 \mu_1} f_{\mu_1}$, one has

$$f_{\boldsymbol{\mu}}(x, d_0 t) = q_{\mu_0 \mu_1} f_{\mu_1} p_{f(\mu_1) f(\mu_n)}(x) = q_{\mu_0 \mu_1} f_{d^0 \boldsymbol{\mu}}(x, t), \tag{2}$$

which is (1.2.7), for $i = 0$. Similar arguments establish (1.2.7), for $0 < i \le n$.

Using the *coherence operator* C, one defines the *coherent identity mapping* as $C(\boldsymbol{1_X}): \boldsymbol{X} \to \boldsymbol{X}$. Note that $C(\boldsymbol{1_X})$ is a level coherent mapping. Moreover, the following lemma holds.

LEMMA 1.17. *For mappings $\boldsymbol{f}: \boldsymbol{X} \to \boldsymbol{Y}$, $\boldsymbol{g}: \boldsymbol{Y} \to \boldsymbol{Z}$,*

$$C(\boldsymbol{gf}) = C(\boldsymbol{g})\, C(\boldsymbol{f}). \tag{3}$$

If $\boldsymbol{f}, \boldsymbol{f}': \boldsymbol{X} \to \boldsymbol{Y}$ are congruent mappings, M is cofinite and Λ is directed, then

$$C(\boldsymbol{f}) \equiv C(\boldsymbol{f}'). \tag{4}$$

Proof. If $\boldsymbol{f} = (f, f_{\mu_0}), \boldsymbol{g} = (g, g_{\nu_0})$, then the composition $\boldsymbol{h} = \boldsymbol{gf} = (fg, g_{\nu_0} f_{g(\nu_0)})$. Therefore, $C(\boldsymbol{h})$ is the coherent mapping, given by $h = fg$ and by $h_{\boldsymbol{\nu}}$, where

$$h_{\boldsymbol{\nu}}(x, t) = g_{\nu_0} f_{g(\nu_0)} p_{h(\nu_0) h(\nu_n)}(x). \tag{5}$$

On the other hand, $C(\boldsymbol{g})C(\boldsymbol{f}) = \boldsymbol{k}$ is given by $k = fg$ and by $k_{\boldsymbol{\nu}}$, where for $t \in P_i^n$,

$$\begin{aligned} k_{\boldsymbol{\nu}}(x, t) &= g_{\nu_0} q_{g(\nu_0) g(\nu_i)} f_{g(\nu_i)} p_{fg(\nu_i) fg(\nu_n)}(x) \\ &= g_{\nu_0} f_{g(\nu_0)} p_{fg(\nu_0) fg(\nu_n)}(x). \end{aligned} \tag{6}$$

Formulae (5) and (6) show that $C(\boldsymbol{h}) = \boldsymbol{k}$, i.e., (3) holds.

In proving the second assertion, there is no loss of generality in assuming that $\boldsymbol{f}'$ is the shift of $\boldsymbol{f}$ by an increasing function f' and thus,

$$f'_{\mu_0}(x) = f_{\mu_0} p_{f(\mu_0) f'(\mu_0)}(x). \tag{7}$$

It then suffices to show that $C(\boldsymbol{f}')$ is a shift of $C(\boldsymbol{f})$ by f'. Note that $C(\boldsymbol{f}')$ is given by f' and by the mappings $f'_{\boldsymbol{\mu}}$, which satisfy

$$f'_{\boldsymbol{\mu}}(x, t) = f'_{\mu_0} p_{f'(\mu_0) f'(\mu_n)}(x) = f_{\mu_0} p_{f(\mu_0) f'(\mu_n)}(x). \tag{8}$$

On the other hand, by (1.2.54) and (1), the shift of $C(\boldsymbol{f})$ is also given by f' and by the mappings $f'_{\boldsymbol{\mu}}$ from (8). □

We will now define an operator E, called the *forgetful operator*. It assigns to every coherent mapping $\boldsymbol{f} = (f, f_\mu): \boldsymbol{X} \to \boldsymbol{Y}$ the homotopy mapping $E(\boldsymbol{f}) = (f, f_{\mu_0}): \boldsymbol{X} \to \boldsymbol{Y}$, obtained by forgetting part of the structure of $\boldsymbol{f}$. Note that $f_{\mu_0\mu_1}$ is a homotopy, which insures validity of condition (1.2.2). Putting $[\boldsymbol{X}] = (X_\lambda, [p_{\lambda\lambda'}], \Lambda)$ and $[\boldsymbol{Y}] = (Y_\mu, [q_{\mu\mu'}], M)$, we see that $(f, [f_{\mu_0}]): [\boldsymbol{X}] \to [\boldsymbol{Y}]$ is a morphism of inv - H(Top), which we also denote by $E(\boldsymbol{f})$.

LEMMA 1.18. *For coherent mappings* $\boldsymbol{f}: \boldsymbol{X} \to \boldsymbol{Y}$, $\boldsymbol{g}: \boldsymbol{Y} \to \boldsymbol{Z}$,

$$E(\boldsymbol{g}\boldsymbol{f}) = E(\boldsymbol{g})\, E(\boldsymbol{f}). \tag{9}$$

Proof. Let $\boldsymbol{f} = (f, f_\mu)$, $\boldsymbol{g} = (g, g_\nu)$. Then $\boldsymbol{h} = \boldsymbol{g}\boldsymbol{f}$ is given by $h = fg$ and by mappings h_ν, where $h_{\nu_0} = g_{\nu_0} f_{g(\nu_0)}$. Therefore,

$$E(\boldsymbol{h}) = (fg, [g_{\nu_0} f_{g(\nu_0)}]). \tag{10}$$

On the other hand,

$$E(\boldsymbol{f}) = (f, [f_{\mu_0}]),\ E(\boldsymbol{g}) = (g, [g_{\nu_0}]), \tag{11}$$

and thus,

$$E(\boldsymbol{g})E(\boldsymbol{f}) = (fg, [g_{\nu_0}][f_{g(\nu_0)}]) = (fg, [g_{\nu_0} f_{g(\nu_0)}]). \tag{12}$$

Comparison of (10) and (12) establishes (9). □

REMARK 1.19. If $\boldsymbol{f} = (f, f_{\mu_0}): \boldsymbol{X} \to \boldsymbol{Y}$ is a mapping, then $EC(\boldsymbol{f})$ is the morphism $(f, [f_{\mu_0}]): [\boldsymbol{X}] \to [\boldsymbol{Y}]$ of inv - H(Top). Note that to the coherent identity mapping $C(\boldsymbol{1}_{\boldsymbol{X}})$, E assigns the identity morphism of inv - H(Top).

Bibliographic notes

The present definition of an inverse system was first used in (Lefschetz 1931, 1942). Level morphisms were considered in (Freudenthal 1937). The more general morphisms of systems and the category inv - $\mathcal{C}$ were defined in (Eilenberg, Steenrod 1952). The category pro - $\mathcal{C}$ was introduced in (Grothendieck 1959-60). Inverse systems indexed by cofinite directed sets are used systematically in shape theory since (Mardešić, Segal 1970, 1971). More information on the bibliography concerning 1.1 can be found in (Mardešić, Segal 1982). The remaining subsections of 1 are based on (Lisica, Mardešić 1983, 1984a, 1984b). A version of coherent mappings, like the one in 1.2, was developed in (Miminoshvili 1980, 1984) and (Lisitsa 1982a). The composition formula of 1.3 first appears in the above mentioned papers of Lisica and Mardešić.

2. Coherent homotopy

On the set of coherent mappings between two inverse systems one defines the equivalence relation of homotopy. Inverse systems as objects and homotopy classes of coherent mappings as morphisms form the coherent homotopy category CH(pro-Top), which is the main object of study in Chapter I. Composition in CH(pro-Top) is defined by composing coherent mappings belonging to the corresponding homotopy classes. The identity morphisms are the homotopy classes of the identity coherent mappings. We also define two important functors, the coherence functor C : pro-Top $\to$ CH(pro-Top) and the forgetful functor E : CH(pro-Top) $\to$ pro-H(Top).

2.1 The coherent homotopy category CH(pro-Top)

Let $\boldsymbol{X} = (X_\lambda, p_{\lambda\lambda'}, \Lambda)$ and $\boldsymbol{Y} = (Y_\mu, q_{\mu\mu'}, M)$ be inverse systems in Top. A *coherent homotopy* from $\boldsymbol{X}$ to $\boldsymbol{Y}$ is a coherent mapping $\boldsymbol{F} = (F, F_{\boldsymbol{\mu}}) : \boldsymbol{X} \times I \to \boldsymbol{Y}$, where $\boldsymbol{X} \times I = (X_\lambda \times I, p_{\lambda\lambda'} \times 1, \Lambda)$. We say that $\boldsymbol{F}$ *connects* coherent mappings $\boldsymbol{f} = (f, f_{\boldsymbol{\mu}})$ and $\boldsymbol{f}' = (f', f'_{\boldsymbol{\mu}})$ provided $F \geq f, f'$ and for $x \in X_{F(\mu_n)}$, $t \in \Delta^n$,

$$F_{\boldsymbol{\mu}}(x, 0, t) = f_{\boldsymbol{\mu}}(p_{f(\mu_n)F(\mu_n)}(x), t), \;\; F_{\boldsymbol{\mu}}(x, 1, t) = f'_{\boldsymbol{\mu}}(p_{f'(\mu_n)F(\mu_n)}(x), t), \quad (1)$$

i.e., the restrictions of $\boldsymbol{F}$ to $\boldsymbol{X} \times 0$ and $\boldsymbol{X} \times 1$ are shifts of $\boldsymbol{f}$, respectively $\boldsymbol{f}'$, by F. We say that the coherent mappings $\boldsymbol{f}$ and $\boldsymbol{f}'$ are *coherently homotopic* or just *homotopic* and we write $\boldsymbol{f} \simeq \boldsymbol{f}'$, provided such a coherent homotopy $\boldsymbol{F}$ exists. If the function $F = \mathrm{id}$, we speak of *level coherent homotopy* and of *level homotopic* coherent mappings.

In the remaining part of this chapter, we will consider only inverse systems indexed by cofinite directed sets. This restriction is not needed if one considers level homotopy.

LEMMA 2.1. *Homotopy of coherent mappings is an equivalence relation.*

Proof. Reflexivity and symmetry are obvious. To prove transitivity, assume that $\boldsymbol{F}'$ connects $\boldsymbol{f}$ to $\boldsymbol{f}'$ and $\boldsymbol{F}''$ connects $\boldsymbol{f}'$ to $\boldsymbol{f}''$. Let $F : M \to \Lambda$ be an increasing function such that $F \geq F', F''$. Define $F_{\boldsymbol{\mu}} : X_{F(\mu_n)} \times I \times \Delta^n \to Y_{\mu_0}$, $\boldsymbol{\mu} \in M_n$, by

$$F_\mu(x,s,t) = \begin{cases} F'_\mu(p_{F'(\mu_n),F(\mu_n)}(x), 2s, t), & 0 \le s \le 1/2, \\ F''_\mu(p_{F''(\mu_n),F(\mu_n)}(x), 2s-1, t), & 1/2 \le s \le 1. \end{cases} \tag{2}$$

Then $\boldsymbol{F} = (F, F_\mu)$ is a coherent homotopy, which connects $\boldsymbol{f}$ to $\boldsymbol{f}''$. □

We will denote the homotopy class of $\boldsymbol{f}$ by $[\boldsymbol{f}]$.

REMARK 2.2. By obvious simpifications in the above proof, one concludes that level homotopy is also an equivalence relation.

Homotopy generalizes congruence because of the following lemma.

LEMMA 2.3. *Congruent coherent mappings $\boldsymbol{f}, \boldsymbol{f}' : \boldsymbol{X} \to \boldsymbol{Y}$ are homotopic.*

Proof. In view of Lemma 2.1, it suffices to consider the case when $\boldsymbol{f}'$ is a shift of $\boldsymbol{f}$ by f'. Put $F = f'$ and $F_\mu(x,s,t) = f'_\mu(x,t)$. Then $\boldsymbol{F} = (F, F_\mu)$ is a coherent homotopy, which connects $\boldsymbol{f}$ to $\boldsymbol{f}'$. □

The next lemma shows that the homotopy class of the composition of two coherent mappings depends only on the homotopy classes of these mappings.

LEMMA 2.4. *Let $\boldsymbol{f}, \boldsymbol{f}' : \boldsymbol{X} \to \boldsymbol{Y}$ and $\boldsymbol{g}, \boldsymbol{g}' : \boldsymbol{Y} \to \boldsymbol{Z}$ be coherent mappings. If $\boldsymbol{f} \simeq \boldsymbol{f}'$ and $\boldsymbol{g} \simeq \boldsymbol{g}'$, then also $\boldsymbol{g}\boldsymbol{f} \simeq \boldsymbol{g}'\boldsymbol{f}'$.*

In the proof of Lemma 2.4 we will use the following lemma, whose proof we postpone.

LEMMA 2.5. *If $\boldsymbol{f} : \boldsymbol{X} \to \boldsymbol{Y}$ and $\boldsymbol{g}, \boldsymbol{g}' : \boldsymbol{Y} \to \boldsymbol{Z}$ are coherent mappings, then $\boldsymbol{g} \equiv \boldsymbol{g}'$ implies $\boldsymbol{g}\boldsymbol{f} \simeq \boldsymbol{g}'\boldsymbol{f}$.*

Proof of Lemma 2.4. In view of Lemma 2.1, it suffices to prove the following two assertions:

(*i*) $\boldsymbol{f} \simeq \boldsymbol{f}'$ implies $\boldsymbol{g}\boldsymbol{f} \simeq \boldsymbol{g}\boldsymbol{f}'$,

(*ii*) $\boldsymbol{g} \simeq \boldsymbol{g}'$ implies $\boldsymbol{g}\boldsymbol{f} \simeq \boldsymbol{g}'\boldsymbol{f}$.

Proof of (*i*). Let $\boldsymbol{F} = (F, F_\mu) : \boldsymbol{X} \times I \to \boldsymbol{Y}$ be a coherent homotopy, which connects $\boldsymbol{f}$ to $\boldsymbol{f}'$. Then the composition $\boldsymbol{H} = \boldsymbol{g}\boldsymbol{F}$, is a coherent homotopy $\boldsymbol{H} = (H, H_\nu) : \boldsymbol{X} \times I \to \boldsymbol{Z}$, which connects the coherent mappings $\boldsymbol{h} = \boldsymbol{g}\boldsymbol{f}$ and $\boldsymbol{h}' = \boldsymbol{g}\boldsymbol{f}'$. Indeed, for $x \in X_{Fg(\nu_n)}, t \in P^n_i$, the composition rule yields

$$\begin{aligned} &H_\nu(x,0,t) = g_{\nu_0\ldots\nu_i}(F_{g(\nu_i)\ldots g(\nu_n)}(x,0,a_i(t)), b_i(t)) = \\ &g_{\nu_0\ldots\nu_i}(f_{g(\nu_i)\ldots g(\nu_n)}(p_{fg(\nu_n)Fg(\nu_n)}(x), a_i(t)), b_i(t)) = \\ &h_\nu(p_{fg(\nu_n)Fg(\nu_n)}(x), t). \end{aligned} \tag{3}$$

Similarly

$$H_\nu(x,1,t) = h'_\nu(p_{f'g(\nu_n)Fg(\nu_n)}(x), t). \tag{4}$$

Proof of (*ii*). Let $\boldsymbol{G} = (G, G_\nu) : \boldsymbol{Y} \times I \to \boldsymbol{Z}$ be a coherent homotopy, which connects $\boldsymbol{g}$ to $\boldsymbol{g}'$. Denote by $\boldsymbol{g}^*$ and $\boldsymbol{g}'^*$ the shifts of $\boldsymbol{g}$ and $\boldsymbol{g}'$ by G, respectively. The composition $\boldsymbol{K} = \boldsymbol{G}(\boldsymbol{f} \times 1)$ is a coherent homotopy which connects $\boldsymbol{h}^* = \boldsymbol{g}^*\boldsymbol{f}$ to $\boldsymbol{h}'^* = g'^*\boldsymbol{f}$ and thus, $\boldsymbol{g}^*\boldsymbol{f} \simeq \boldsymbol{g}'^*\boldsymbol{f}$. Indeed,

$$K_{\boldsymbol{\nu}}(x,0,t) = g_{\nu_0\ldots\nu_i}(q_{g(\nu_i)G(\nu_i)}f_{G(\nu_i)\ldots G(\nu_n)}(x,a_i(t)),b_i(t)) = h^*_{\boldsymbol{\nu}}(x,t), \quad (5)$$

for $t \in P_i^n$. Similarly,

$$K_{\boldsymbol{\nu}}(x,1,t) = h'^*_{\boldsymbol{\nu}}(x,t). \quad (6)$$

Since $\boldsymbol{g} \equiv \boldsymbol{g}^*$ and $\boldsymbol{g}' \equiv \boldsymbol{g}'^*$, Lemma 2.5, implies that $\boldsymbol{gf} \simeq \boldsymbol{g}^*\boldsymbol{f}$ and $\boldsymbol{g}'\boldsymbol{f} \simeq \boldsymbol{g}'^*\boldsymbol{f}$. Consequently, $\boldsymbol{gf} \simeq \boldsymbol{g}'\boldsymbol{f}$. □

Proof of Lemma 2.5. It suffices to prove the assertion in the case when $\boldsymbol{g}' = (g', g'_{\boldsymbol{\nu}}) : \boldsymbol{Y} \to \boldsymbol{Z}$ is a shift of $\boldsymbol{g} = (g, g_{\boldsymbol{\nu}}) : \boldsymbol{Y} \to \boldsymbol{Z}$, i.e, $g' \geq g$ and $g'_{\boldsymbol{\nu}}(x,t) = g_{\boldsymbol{\nu}}(q_{g(\nu_n)g'(\nu_n)}(x),t)$. In order to define a coherent homotopy $\boldsymbol{H} = (H, H_{\boldsymbol{\nu}}) : \boldsymbol{X} \times I \to \boldsymbol{Z}$, connecting $\boldsymbol{h} = \boldsymbol{gf}$ to $\boldsymbol{h}' = \boldsymbol{g}'\boldsymbol{f}$, we will first define mappings $F_{\boldsymbol{\nu}} : X_{fg'(\nu_n)} \times I \times \Delta^n \to Y_{g(\nu_0)}$, $\boldsymbol{\nu} \in N_n$, such that

$$F_{\boldsymbol{\nu}}(x,s,d_jt) = qF_{d^j\boldsymbol{\nu}}(p(x),s,t), \quad (7)$$

$$F_{\boldsymbol{\nu}}(x,0,t) = f_{g(\nu_0)\ldots g(\nu_n)}(p_{fg(\nu_n)fg'(\nu_n)}(x),t), \quad (8)$$

$$F_{\boldsymbol{\nu}}(x,1,t) = q_{g(\nu_0)g'(\nu_0)}f_{g'(\nu_0)\ldots g'(\nu_n)}(x,t). \quad (9)$$

Once the mappings $F_{\boldsymbol{\nu}}$ are defined, one defines $\boldsymbol{H}$ by putting $H = fg'$ and

$$H_{\boldsymbol{\nu}}(x,s,t) = g_{\nu_0\ldots\nu_i}(F_{\nu_i\ldots\nu_n}(x,s,a_i(t)),b_i(t)), \quad (10)$$

for $t \in P_i^n$.

To see that the mappings $H_{\boldsymbol{\nu}}$ are well defined on $X_{fg'(\nu_n)} \times I \times \Delta^n$ and satisfy the coherence conditions, one argues as in 1.3, where the analogous assertions for the composition $\boldsymbol{gf}$ were verified. Of course, one must use (7) instead of the coherence conditions for $\boldsymbol{f}$. Moreover, since

$$\begin{gathered} h_{\boldsymbol{\nu}}(p_{fg(\nu_n)fg'(\nu_n)}(x),t) = \\ g_{\nu_0\ldots\nu_i}(f_{g(\nu_i)\ldots g(\nu_n)}(p_{fg(\nu_n)fg'(\nu_n)}(x),a_i(t)),b_i(t)), \end{gathered} \quad (11)$$

$$(h')_{\boldsymbol{\nu}}(x,t) = g'_{\nu_0\ldots\nu_i}(f_{g'(\nu_i)\ldots g'(\nu_n)}(x,a_i(t)),b_i(t)), \quad (12)$$

one concludes that (10), (8) and (9) imply

$$H_{\boldsymbol{\nu}}(x,0,t) = h_{\boldsymbol{\nu}}(p_{fg(\nu_n)fg'(\nu_n)}(x),t), \quad (13)$$

$$H_{\boldsymbol{\nu}}(x,1,t) = h'_{\boldsymbol{\nu}}(x,t) \quad (14)$$

which proves that indeed, $\boldsymbol{H}$ connects $\boldsymbol{h}$ to $\boldsymbol{h}'$.

We will now define $F_{\boldsymbol{\nu}}$. First consider the standard triangulation of $I \times \Delta^n$. It is given by $(n+1)$-simplices T_i^{n+1}, $0 \leq i \leq n$, spanned by the vertices

$$(0,e_0),\ldots,(0,e_i),(1,e_i),\ldots,(1,e_n). \quad (15)$$

Also consider the simplicial mapping $\varepsilon^{n+1} : I \times \Delta^n \to \Delta^{n+1}$, given by

$$\varepsilon^{n+1}(0,e_j) = e_j,\ \varepsilon^{n+1}(1,e_j) = e_{j+1}. \quad (16)$$

Then put

$$F_{\boldsymbol{\nu}}(x,s,t) = f_{g(\nu_0)\ldots g(\nu_i)g'(\nu_i)\ldots g'(\nu_n)}(x,\varepsilon^{n+1}(s,t)), \tag{17}$$

for $x \in X_{fg'(\nu_n)}$, $(s,t) \in T_i^{n+1}$. To see that $F_{\boldsymbol{\nu}}$ is well defined, note that

$$(T_0^{n+1} \cup \ldots \cup T_i^{n+1}) \cap T_{i+1}^{n+1} = T_i^{n+1} \cap T_{i+1}^{n+1}, \tag{18}$$

$$T_i^{n+1} \cap T_{i+1}^{n+1} = [(0,e_0),\ldots,(0,e_i),(1,e_{i+1}),\ldots,(1,e_n)], \tag{19}$$

$$\varepsilon^{n+1}(T_i^{n+1} \cap T_{i+1}^{n+1}) = [e_0,\ldots,e_i,e_{i+2},\ldots,e_n] = d_{i+1}(\Delta^n). \tag{20}$$

It now suffices to see that the expressions $f_{g(\nu_0)\ldots g(\nu_i)g'(\nu_i)\ldots g'(\nu_n)}(x,d_{i+1}t')$ and $f_{g(\nu_0)\ldots g(\nu_{i+1})g'(\nu_{i+1})\ldots g'(\nu_n)}(x,d_{i+1}t')$, for $x \in X_{fg'(\nu_n)}$, $t' \in \Delta^n$, assume the same value $f_{g(\nu_0)\ldots g(\nu_i)g'(\nu_{i+1})\ldots g'(\nu_n)}(x,t')$. Indeed, this is an immediate consequence of the coherence conditions for $\boldsymbol{f}$.

Let us now show that $F_{\boldsymbol{\nu}}$ satisfies condition (7). First note that

$$(1 \times d_j)(T_i^n) \subseteq \begin{cases} T_{i+1}^{n+1}, & j \le i, \\ T_i^{n+1}, & i < j. \end{cases} \tag{21}$$

$$\varepsilon^{n+1}(1 \times d_j)\,|\,T_i^n = \begin{cases} d_j\varepsilon^n |\, T_i^n, & j \le i, \\ d_{j+1}\varepsilon^n |\, T_i^n, & i < j. \end{cases} \tag{22}$$

Since the mappings involved are simplicial, formulae (21) and (22) are readily verified by checking their validity at the vertices.

If $x \in X_{fg'(\nu_n)}$ and $(s,t) \in T_i^n$, then $(s,d_jt) \in T_{i+1}^{n+1}$, for $j \le i$, and $(s,d_jt) \in T_i^{n+1}$, for $i < j$. Therefore, in the first case,

$$\begin{aligned} F_{\boldsymbol{\nu}}(x,s,d_jt) &= f_{g(\nu_0)\ldots g(\nu_{i+1})g'(\nu_{i+1})\ldots g'(\nu_n)}(x,\varepsilon(s,d_jt)) \\ &= f_{g(\nu_0)\ldots g(\nu_{i+1})g'(\nu_{i+1})\ldots g'(\nu_n)}(x,d_j\varepsilon(s,t)) \\ &= qf_{d^j(g(\nu_0)\ldots g(\nu_{i+1})g'(\nu_{i+1})\ldots g'(\nu_n))}(p(x),\varepsilon(s,t)). \end{aligned} \tag{23}$$

On the other hand,

$$qF_{d^j\boldsymbol{\nu}}(p(x),s,t) = qf_{g(\nu'_0)\ldots g(\nu'_i)g'(\nu'_i)\ldots g'(\nu'_{n-1})}(p(x),\varepsilon(s,t)), \tag{24}$$

where $\boldsymbol{\nu}' = (\nu'_0,\ldots,\nu'_{n-1}) = d^j\boldsymbol{\nu}$. Since $j \le i$, it follows that

$$\begin{aligned} &d^j(g(\nu_0),\ldots,g(\nu_{i+1}),g'(\nu_{i+1}),\ldots,g'(\nu_n)) \\ &= (g(\nu_0),\ldots,\widehat{g(\nu_j)},\ldots,g(\nu_{i+1}),g'(\nu_{i+1}),\ldots,g'(\nu_n)). \end{aligned} \tag{25}$$

However, $\boldsymbol{\nu}' = (\nu_0,\ldots,\widehat{\nu_j},\ldots,\nu_{i+1},\ldots,\nu_n)$ and thus,

$$\begin{aligned} &(g(\nu'_0),\ldots,g(\nu'_i),g'(\nu'_i),\ldots,g'(\nu'_{n-1})) \\ &= (g(\nu_0),\ldots,\widehat{g(\nu_j)},\ldots,g(\nu_{i+1}),g'(\nu_{i+1}),\ldots,g'(\nu_n)). \end{aligned} \tag{26}$$

Consequently, the right sides of (23) and (24) coincide.

In the second case, the verification of (7) is similar.

$$\begin{aligned} F_{\boldsymbol{\nu}}(x,s,d_jt) &= f_{g(\nu_0)\dots g(\nu_i)g'(\nu_i)\dots g'(\nu_n)}(x,\varepsilon(s,d_jt)) \\ &= f_{g(\nu_0)\dots g(\nu_i)g'(\nu_i)\dots g'(\nu_n)}(x,d_{j+1}(\varepsilon(s,t))) \\ &= qf_{d^{j+1}(g(\nu_0)\dots g(\nu_i)g'(\nu_i)\dots g'(\nu_n))}(p(x),\varepsilon(s,t)). \end{aligned} \tag{27}$$

Since now $i < j$,

$$\begin{aligned} & d^{j+1}(g(\nu_0),...,g(\nu_i),g'(\nu_i),...,g'(\nu_n)) \\ &= (g(\nu_0),...,g(\nu_i),g'(\nu_i),...,\widehat{g'(\nu_j)},...,g'(\nu_n)). \end{aligned} \tag{28}$$

Moreover, $\boldsymbol{\nu}' = d^j\boldsymbol{\nu} = (\nu_0,...,\nu_{i+1},...,\widehat{\nu_j},...,\nu_n)$ and thus,

$$\begin{aligned} & (g(\nu_0'),...,g(\nu_i'),g'(\nu_i'),...,g'(\nu_{n-1}')) \\ &= (g(\nu_0),...,g(\nu_i),g'(\nu_i),...,\widehat{g'(\nu_j)},...,...,g'(\nu_n)). \end{aligned} \tag{29}$$

Since the right sides of (28) and (29) are equal, the left sides of (27) and (24) also coincide.

We will now verify conditions (8) and (9). For $t \in \Delta^n$, we have $(0,t) \in [(0,e_0),\dots,(0,e_n)] \subseteq T_n^{n+1}$ and $\varepsilon^{n+1}(0,t) = d_{n+1}(t)$. Consequently, for $x \in X_{fg'(\nu_n)}$,

$$\begin{aligned} F_{\boldsymbol{\nu}}(x,0,t) &= f_{g(\nu_0)\dots g(\nu_n)g'(\nu_n)}(x,d_{n+1}t) \\ &= f_{g(\nu_0)\dots g(\nu_n)}(p_{fg(\nu_n)fg'(\nu_n)}(x),t). \end{aligned} \tag{30}$$

Similarly, $(1,t) \in [(1,e_0),\dots,(1,e_n)] \subseteq T_0^{n+1}$ and $\varepsilon^{n+1}(1,t)) = d_0(t)$ and thus,

$$\begin{aligned} F_{\boldsymbol{\nu}}(x,1,t) &= f_{g(\nu_0)g'(\nu_0)\dots g'(\nu_n)}(x,d_0t) \\ &= q_{g(\nu_0)g'(\nu_0)}f_{g'(\nu_0)\dots g'(\nu_n)}(x,t). \;\square \end{aligned} \tag{31}$$

REMARK 2.6. The analogue of Lemma 2.4 for level coherent mappings and level coherent homotopies also holds. The proof is much simpler, because in this case all index functions are identities. In particular, no shifts appear and the assertion $\boldsymbol{g} \equiv \boldsymbol{g}'$ is replaced by the assertion $\boldsymbol{g} = \boldsymbol{g}'$. Hence, there is no need for the analogue of Lemma 2.5.

We will now define the *coherent homotopy category* CH(pro-Top), which is the main object of study in this chapter. The objects of this category are inverse systems of spaces and mappings, indexed by cofinite directed sets. Morphisms are homotopy classes of coherent mappings. The *identity morphism* $[\boldsymbol{1_X}]$ is the homotopy class of the coherent identity mapping $C(\boldsymbol{1_X}) : \boldsymbol{X} \to \boldsymbol{X}$. *Composition of homotopy classes of coherent mappings* $[\boldsymbol{g}][\boldsymbol{f}]$ is defined by composing their representatives, i.e., by the formula

$$[\boldsymbol{g}][\boldsymbol{f}] = [\boldsymbol{g}\boldsymbol{f}]. \tag{32}$$

By Lemma 2.4, composition is well defined. In order to prove that CH(pro Top) is indeed a category, it remains to establish associativity of the composition, $[\boldsymbol{h}]([\boldsymbol{g}][\boldsymbol{f}]) = ([\boldsymbol{h}][\boldsymbol{g}])[\boldsymbol{f}]$, as well as the equalities $[\boldsymbol{g}] = [\boldsymbol{g}][\boldsymbol{1_X}]$ and $[\boldsymbol{f}] = [\boldsymbol{1_Y}][\boldsymbol{f}]$. The rather lengthy proofs are carried out in the next two subsections.

REMARK 2.7. By means of adequate reindexing procedures (see (Mardešić, Segal 1982), I.1.2), one can extend CH(pro-Top) also to non-cofinite inverse systems. We renounce doing this because we are intersted in coherent homotopy primarily for its use in strong shape theory of spaces. However, since arbitrary spaces admit adequate cofinite expansions, we do not need this additional generality.

The coherent homotopy category $\mathrm{CH}(\mathsf{Top}^{\Lambda})$ has as objects inverse systems of spaces, indexed by a fixed preordered set Λ. Morphisms are level homotopy classes of level coherent mappings.

2.2 Associativity of the composition

In this subsection we prove the following theorem, which shows that the composition of homotopy classes of coherent mappings is associative.

THEOREM 2.8. *Let $\boldsymbol{f} : \boldsymbol{X} \to \boldsymbol{Y}, \boldsymbol{g} : \boldsymbol{Y} \to \boldsymbol{Z}$ and $\boldsymbol{h} : \boldsymbol{Z} \to \boldsymbol{W}$ be coherent mappings. Then $\boldsymbol{h}(\boldsymbol{g}\boldsymbol{f}) \simeq (\boldsymbol{h}\boldsymbol{g})\boldsymbol{f}$.*

The explicit formula for the composition of coherent mappings required decomposing Δ^n into subpolyhedra P_i^n, $0 \le i \le n$ (see 1.3). To obtain explicit expressions for the compositions $\boldsymbol{k} = \boldsymbol{h}(\boldsymbol{g}\boldsymbol{f})$ and $\boldsymbol{k}' = (\boldsymbol{h}\boldsymbol{g})\boldsymbol{f}$, we must further decompose P_i^n, $0 \le i \le n$, into subpolyhedra P_{ik}^n, $i \le k \le n$, and P_k^n, $0 \le k \le n$, into subpolyhedra Q_{ik}^n, $0 \le i \le k$. By definition,

$$P_{ik}^n = (a_i^n)^{-1}(P_{k-i}^{n-i}) = (c_i^n)^{-1}(P_{k-i}^{n-i} \times \Delta^i) \subseteq P_i^n, \tag{1}$$

$$Q_{ik}^n = (b_k^n)^{-1}(P_i^k) = (c_k^n)^{-1}(\Delta^{n-k} \times P_i^k) \subseteq P_k^n. \tag{2}$$

Figure 2.1 shows these decompositions for $n = 2$.

In order to define a homotopy $\boldsymbol{H} : \boldsymbol{X} \times I \to \boldsymbol{W}$, which connects $\boldsymbol{k}$ to $\boldsymbol{k}'$, we will use the natural piecewise linear mapping $\theta^n : \Delta^n \to \Delta^n$, which maps Q_{ik}^n homeomorphically to P_{ik}^n, and is suggested by Figure 2.1. More precisely, we will use the following lemma.

LEMMA 2.9. *There exists a unique mapping $\theta^n : \Delta^n \to \Delta^n$ such that*

$$\theta^n(Q_{ik}^n) = P_{ik}^n, \ 0 \le i \le k \le n, \tag{3}$$

$$a_{k-i}^{n-i} a_i^n \theta^n(t) = a_k^n(t), \ t \in Q_{ik}^n, \tag{4}$$

$$b_{k-i}^{n-i} a_i^n \theta^n(t) = a_i^k b_k^n(t), \ t \in Q_{ik}^n, \tag{5}$$

$$b_i^n \theta^n(t) = b_i^k b_k^n(t), \ t \in Q_{ik}^n. \tag{6}$$

The mappings θ^n commute with the boundary operators, i.e.,

$$d_j \theta^{n-1} = \theta^n d_j. \tag{7}$$

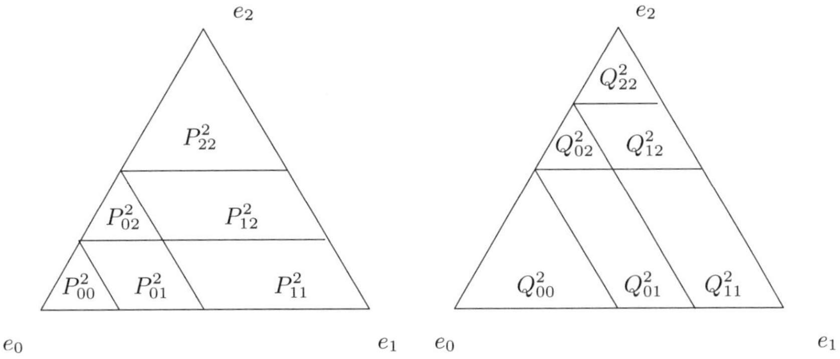

Fig. 2.1. Decompositions of Δ^2

We postpone the somewhat lengthy formal proof of the lemma and proceed to prove the theorem.

Proof of Theorem 2.8. Let $\boldsymbol{W} = (W_\pi, w_{\pi\pi'}, \Pi)$, $\boldsymbol{h} = (h, h_{\boldsymbol{\pi}})$, $\boldsymbol{k} = (k, k_{\boldsymbol{\pi}})$ and $\boldsymbol{k}' = (k', k'_{\boldsymbol{\pi}})$. By definition, for $\boldsymbol{\pi} \in \Pi_n$ and $x \in X_{fgh(\pi_n)}, t \in P^n_i, 0 \leq i \leq n$, we have

$$k_{\boldsymbol{\pi}}(x,t) = h_{\pi_0\ldots\pi_i}((gf)_{h(\pi_i)\ldots h(\pi_n)}(x, a^n_i(t)), b^n_i(t)). \tag{8}$$

Therefore, for $t \in P^n_{ik}$, $k_{\boldsymbol{\pi}}(x,t)$ assumes the value

$$h_{\pi_0\ldots\pi_i}(g_{h(\pi_i)\ldots h(\pi_k)}(f_{gh(\pi_k)\ldots gh(\pi_n)}(x, a^{n-i}_{k-i}a^n_i(t)), b^{n-i}_{k-i}a^n_i(t)), b^n_i(t)). \tag{9}$$

Similarly, for $x \in X_{fgh(\pi_n)}$ and $t \in Q^n_{ik}$, $k'_{\boldsymbol{\pi}}(x,t)$ assumes the value

$$h_{\pi_0\ldots\pi_i}(g_{h(\pi_i)\ldots h(\pi_k)}(f_{gh(\pi_k)\ldots gh(\pi_n)}(x, a^n_k(t)), a^k_i b^n_k(t)), b^k_i b^n_k(t)). \tag{10}$$

Using the mapping θ^n from Lemma 2.9, we define a linear homotopy $\Theta^n : I \times \Delta^n \to \Delta^n$,

$$\Theta^n(s,t) = (1-s)t + s\theta^n(t), \tag{11}$$

We then define $\boldsymbol{H} = (H, H_{\boldsymbol{\pi}})$, by putting $H = fgh$ and

$$H_{\boldsymbol{\pi}}(x,s,t) = k_{\boldsymbol{\pi}}(x, \Theta^n(s,t)). \tag{12}$$

Comparing (9) and (10) and taking into account (3)–(6), we see that

$$k_{\boldsymbol{\pi}}(x, \theta^n(t)) = k'_{\boldsymbol{\pi}}(x,t), \ t \in \Delta^n. \tag{13}$$

Note that (12), (11) and (13) imply

$$H_{\boldsymbol{\pi}}(x,0,t) = k_{\boldsymbol{\pi}}(x,t), \tag{14}$$

$$H_\pi(x,1,t) = k'_\pi(x,t). \tag{15}$$

It is easy to verify that the mappings H_π satisfy the coherence condition (1.2.7). Indeed, since d_j is an affine mapping, (11) and (7) imply

$$\begin{aligned} d_j(\Theta^{n-1}(s,t)) &= (1-s)d_j t + s d_j(\theta^{n-1}(t)) \\ &= (1-s)d_j t + s\theta^n d_j t = \Theta^n(s, d_j t). \end{aligned} \tag{16}$$

Therefore,

$$\begin{array}{l} H_\pi(x,s,d_j(t)) = \\ k_\pi(x,\Theta^n(s,d_j t)) = k_\pi(x, d_j(\Theta^{n-1}(s,t))) = \\ w k_{d^j\pi}(p(x),\Theta^{n-1}(s,t)) = w H_{d^j\pi}(p(x),s,t). \ \Box \end{array} \tag{17}$$

REMARK 2.10. If $\boldsymbol{f},\boldsymbol{g}$ and $\boldsymbol{h}$ are level coherent mappings, i.e., f,g and h are identity mappings, then $H = fgh = \mathrm{id}$ and thus, $\boldsymbol{H}$ is a level homotopy.

Proof of Lemma 2.9. First notice that the horizontal arrows in the diagram which follows are homeomorphisms. Therefore, there exists a unique homeomorphism $\theta^n_{ik} : Q^n_{ik} \to P^n_{ik}$, which makes the diagram commutative.

$$\begin{array}{ccccc} Q^n_{ik} & \xrightarrow{c^n_k} & \Delta^{n-k}\times P^k_i & \xrightarrow{1\times c^k_i} & \Delta^{n-k}\times\Delta^{k-i}\times\Delta^i \\ \downarrow{\scriptstyle \theta^n_{ik}} & & \downarrow & & \downarrow{\scriptstyle \mathrm{id}} \\ P^n_{ik} & \xrightarrow[c^n_i]{} & P^{n-i}_{k-i}\times\Delta^i & \xrightarrow[c^{n-i}_{k-i}\times 1]{} & \Delta^{n-k}\times\Delta^{k-i}\times\Delta^i \end{array} \tag{18}$$

We will show that the mappings θ^n_{ik}, defined by (18), are restrictions to Q^n_{ik} of a well-defined mapping $\theta^n : \Delta^n \to \Delta^n$. To prove this assertion, it suffices to show that

$$\theta^n_{ik}|Q^n_{ik}\cap Q^n_{i+1,k} = \theta^n_{i+1,k}|Q^n_{ik}\cap Q^n_{i+1,k}, \tag{19}$$

$$\theta^n_{ik}|Q^n_{ik}\cap Q^n_{i,k+1} = \theta^n_{i,k+1}|Q^n_{ik}\cap Q^n_{i,k+1}. \tag{20}$$

Note that $c^n_i(t) = (a^n_i(t), b^n_i(t))$, $t\in P^n_i$, implies

$$(1\times c^k_i)\, c^n_k(t) = (a^n_k(t), a^k_i b^n_k(t), b^k_i b^n_k(t)),\ t\in Q^n_{ik}, \tag{21}$$

$$(c^{n-i}_{k-i}\times 1)\, c^n_i(t') = (a^{n-i}_{k-i}a^n_i(t'), b^{n-i}_{k-i}a^n_i(t'), b^n_i(t')),\ t'\in P^n_{ik}. \tag{22}$$

Using formulae (1.3.3) and (1.3.4), it is readily seen that

$$a^k_i b^n_k(t) = (4(t_0+\ldots+t_i)-1, 4t_{i+1},\ldots,4t_{k-1},\#),\ t\in Q^n_{ik}, \tag{23}$$

$$b^k_i b^n_k(t) = (4t_0,\ldots,4t_{i-1},\#),\ t\in Q^n_{ik}. \tag{24}$$

$$a^{n-i}_{k-i}a^n_i(t) = (\#, 4t_{k+1},\ldots,4t_n),\ t\in P^n_{ik}, i<k, \tag{25}$$

$$b^{n-i}_{k-i}a^n_i(t) = (\#, 4t_{i+1},\ldots,4t_{k-1}, 4(t_k+\ldots+t_n)-1),\ t\in P^n_{ik}, i<k, \tag{26}$$

To establish (19), assume that $t \in Q^n_{ik} \cap Q^n_{i+1,k}$ and put $t' = \theta^n_{ik}(t)$, $t'' = \theta^n_{i+1,k}(t)$. Commutativity of (18), applied to (i,k) and $(i+1,k)$, yields

$$(1 \times c^k_i)\, c^n_k(t) = (c^{n-i}_{k-i} \times 1)\, c^n_i(t'), \tag{27}$$

$$(1 \times c^k_{i+1})\, c^n_k(t) = (c^{n-i-1}_{k-i-1} \times 1)\, c^n_{i+1}(t''). \tag{28}$$

In view of (1.3.3), (1.3.4), (21)–(26), formula (27) yields

$$\begin{gathered} (\#, 2t_{k+1}, \ldots, 2t_n) = (\#, 4t'_{k+1}, \ldots, 4t'_n), \\ (4(t_0 + \ldots + t_i) - 1, 4t_{i+1}, \ldots, 4t_{k-1}, \#) = \\ (\#, 4t'_{i+1}, \ldots, 4t'_{k-1}, 4(t'_k + \ldots + t'_n) - 1), \\ (4t_0, \ldots, 4t_{i-1}, \#) = (2t'_0, \ldots, 2t'_{i-1}, \#). \end{gathered} \tag{29}$$

Similarly, formula (28) yields

$$\begin{gathered} (\#, 2t_{k+1}, \ldots, 2t_n) = (\#, 4t''_{k+1}, \ldots, 4t''_n), \\ (4(t_0 + \ldots + t_{i+1}) - 1, 4t_{i+2}, \ldots, 4t_{k-1}, \#) = \\ (\#, 4t''_{i+2}, \ldots, 4t''_{k-1}, 4(t''_k + \ldots + t''_n) - 1), \\ (4t_0, \ldots, 4t_i, \#) = (2t''_0, \ldots, 2t''_i, \#). \end{gathered} \tag{30}$$

Comparing (29) with (30), we see immediately that

$$t'_l = t''_l = \begin{cases} 2t_l, & 0 \le l \le i-1, \\ t_l, & i+2 \le l \le k-1, \\ \frac{1}{2}t_l, & k+1 \le l \le n. \end{cases} \tag{31}$$

Moreover,

$$t'_{i+1} = t_{i+1}, \; t''_i = 2t_i. \tag{32}$$

To complete the proof that $t' = t''$, it suffices to show that also $t'_i = t''_i$, $t'_{i+1} = t''_{i+1}$. Since Q^n_{ik} is the set of all points $t = (t_0, \ldots, t_n) \in P^n_k$, which satisfy the inequality

$$t_0 + \ldots + t_{i-1} \le \frac{1}{4} \le t_0 + \ldots + t_i, \tag{33}$$

we see that $t \in Q^n_{ik} \cap Q^n_{i+1,k}$ implies

$$4(t_0 + \ldots + t_i) = 1. \tag{34}$$

Therefore, a comparison of the first terms in the second equation of (29) yields the equality $0 = 2 - 4(t'_{i+1} + \ldots + t'_n)$, or equivalently, $2(t'_0 + \ldots + t'_i) = 1$. Morever, by (31) and (32), $2(t''_0 + \ldots + t''_i) = 4(t_0 + \ldots + t_i) = 1$. Since, $t'_0 = t''_0, \ldots, t'_{i-1} = t''_{i-1}$, one concudes that indeed, $t'_i = t''_i$. Furthermore, the first terms in the second equation of (30) equal $4t_{i+1}$ and $2 - 4(t''_{i+2} + \ldots + t''_n) = 4(t''_0 + \ldots + t''_i + t''_{i+1}) - 2$, respectively. Since $4(t''_0 + \ldots + t''_i) = 8(t_0 + \ldots + t_i) = 2$, one concludes that $4t_{i+1} = 4t''_{i+1}$. Therefore, (32) implies $t'_{i+1} = t''_{i+1}$.

To establish (20), assume that $t \in Q^n_{ik} \cap Q^n_{i,k+1}$ and thus, $t \in P^n_k \cap P^n_{k+1}$. Put $t' = \theta^n_{ik}(t)$, $t^* = \theta^n_{i,k+1}(t)$. Commutativity of (18), applied to $(i, k+1)$, yields

$$(1 \times c_i^{k+1})\, c^n_{k+1}(t) = (c^{n-i}_{k+1-i} \times 1)\, c^n_i(t^*), \tag{35}$$

which in turn yields

$$\begin{array}{c} (\#, 2t_{k+2}, \dots, 2t_n) = (\#, 4t^*_{k+2}, \dots, 4t^*_n), \\ (4(t_0 + \ldots + t_i) - 1, 4t_{i+1}, \dots, 4t_k, \#) = \\ (\#, 4t^*_{i+1}, \dots, 4t^*_k, 4(t^*_{k+1} + \ldots + t^*_n) - 1), \\ (4t_0, \dots, 4t_{i-1}, \#) = (2t^*_0, \dots, 2t^*_{i-1}, \#). \end{array} \tag{36}$$

By comparison of (29) with (36), one concludes immediately that

$$t'_l = t^*_l = \begin{cases} 2t_l, & 0 \le l \le i-1, \\ t_l, & i+1 \le l \le k-1, \\ \frac{1}{2} t_l, & k+2 \le l \le n. \end{cases} \tag{37}$$

Moreover,

$$t'_{k+1} = \frac{1}{2} t_{k+1}, \; t^*_k = t_k. \tag{38}$$

To complete the proof that $t' = t^*$, it suffices to show that also $t'_k = t^*_k$, $t'_{k+1} = t^*_{k+1}$. Note that $t \in P^n_k \cap P^n_{k+1}$ and (1.3.2) imply

$$2(t_{k+1} + \ldots + t_n) = 1. \tag{39}$$

Therefore, the last terms in the second equation of (29) equal $4t_k$ and $4t'_k$ and thus, $t_k = t'_k$. Indeed, by (39), $2(t_0 + \ldots + t_k) = 1$. However, this implies $2 - 4(t_0 + \ldots + t_{k-1}) = 4t_k$. Moreover, (37) and (38) yield $4(t'_k + \ldots + t'_n) - 1 = 4t'_k + 2(t_{k+1} + \ldots + t_n) - 1 = 4t'_k$. However, by (38), $t^*_k = t_k$. Now consider the last terms in the second equation of (36). By (39), the first of these terms equals $2 - 4(t_0 + \ldots + t_k) = 0$. By (37), the second term equals $4(t^*_{k+1} + \ldots + t^*_n) - 1 = 4t^*_{k+1} + 2(t_{k+2} + \ldots + t_n) - 1 = 4t^*_{k+1} - 2t_{k+1}$. Consequently, $2t^*_{k+1} = t_{k+1}$. However, by (38), $2t'_{k+1} = t_{k+1}$. Hence, $t'_{k+1} = t^*_{k+1}$.

Since formulae (4)–(6) follow from (21), (22) and from the commutativity of (18), to complete the proof of Lemma 2.9, it only remains to prove (7).

First note that formulae (1), (2) and (1.3.11)–(1.3.16) imply

$$d_j(Q^{n-1}_{ik}) \subseteq Q^n_{i'k'}, \tag{40}$$

where

$$(i', k') = \begin{cases} (i+1, k+1), & j \le i, \\ (i, k+1), & i < j \le k, \\ (i, k), & k < j. \end{cases} \tag{41}$$

Indeed, let $t \in Q^{n-1}_{ik}$ and thus, $b^{n-1}_k(t) \in P^k_i$. If $k < j$, then $b^n_k(d_j t) = b^{n-1}_k(t)$ and thus, $d_j t \in Q^n_{ik}$. If $j \le k$, then $b^n_{k+1}(d_j t) = d_j b^{n-1}_k(t) \in d_j(P^k_i)$. However, $d_j(P^k_i) \subseteq P^{k+1}_{i+1}$, if $j \le i$, while $d_j(P^k_i) \subseteq P^{k+1}_i$, if $i < j$. Consequently, $d_j t \in Q^n_{i+1,k+1}$, if $j \le i$, while $d_j t \in Q^n_{i,k+1}$, if $i < j$.

Similarly,

$$d_j(P_{ik}^{n-1}) \subseteq P_{i'k'}^{n}. \tag{42}$$

Indeed, let $t \in P_{ik}^{n-1}$ and thus, $a_i^{n-1}(t) \in P_{k-i}^{n-1-i}$. If $j \leq i$, then $a_{i+1}^n(d_j t) = a_i^{n-1}(t) \in P_{k-i}^{n-1-i}$ and thus, $d_j t \in P_{i+1,k+1}^n$. Now assume that $i < j$ and thus, $a_i^n(d_j t) = d_{j-i}a_i^{n-1}(t) \subseteq d_{j-i}(P_{k-i}^{n-1-i})$. However, $d_{j-i}(P_{k-i}^{n-1-i}) \subseteq P_{k-i+1}^{n-i}$, for $j-i \leq k-i$ and $d_{j-i}(P_{k-i}^{n-1-i}) \subseteq P_{k-i}^{n-i}$, for $k-i < j-i$. Consequently, $d_j t \in P_{i,k+1}^n$, for $j \leq k$ and $d_j t \in P_{ik}^n$, for $k < j$.

By (40), we need to show that

$$\theta_{i'k'}^n d_j | Q_{ik}^{n-1} = d_j \theta_{ik}^{n-1}. \tag{43}$$

However, since $(c_{k'-i'}^{n-i'} \times 1)\, c_{i'}^n : P_{i'k'}^n \to \Delta^{n-k'} \times \Delta^{k'-i'} \times \Delta^{i'}$ is a homeomorphism, it suffices to show that

$$(c_{k'-i'}^{n-i'} \times 1)\, c_{i'}^n \theta_{i'k'}^n d_j | Q_{ik}^{n-1} = (c_{k'-i'}^{n-i'} \times 1)\, c_{i'}^n d_j \theta_{ik}^{n-1}. \tag{44}$$

In order to establish (44), consider the natural mapping $d_j' : \Delta^{n-1-k} \times \Delta^{k-i} \times \Delta^i \to \Delta^{n-k'} \times \Delta^{k'-i'} \times \Delta^{i'}$, induced by the appropriate boundary operators. More precisely, let d_j' be given by

$$d_j' = \begin{cases} 1 \times 1 \times d_j, & j \leq i, \\ 1 \times d_{j-i} \times 1, & i < j \leq k, \\ d_{j-k} \times 1 \times 1, & k < j. \end{cases} \tag{45}$$

Using formulae (1.3.12), (1.3.13), (1.3.15), (1.3.16), a straightforward computation yields the following formulae.

$$(1 \times c_{i'}^{k'})\, c_{k'}^n d_j(t) = d_j'(1 \times c_i^k)\, c_k^{n-1}(t),\ t \in Q_{ik}^{n-1}, \tag{46}$$

$$(c_{k'-i'}^{n-i'} \times 1)\, c_{i'}^n d_j(t') = d_j'(c_{k-i}^{n-1-i} \times 1)\, c_i^{n-1}(t'),\ t' \in P_{ik}^{n-1}. \tag{47}$$

By definition of $\theta_{i'k'}^n$ and θ_{ik}^{n-1}, we also have

$$(c_{k'-i'}^{n-i'} \times 1)\, c_{i'}^n \theta_{i'k'}^n = (1 \times c_{i'}^{k'})\, c_{k'}^n, \tag{48}$$

$$(c_{k-i}^{n-1-i} \times 1)\, c_i^{n-1} \theta_{ik}^{n-1} = (1 \times c_i^k)\, c_k^{n-1}. \tag{49}$$

Clearly, (46)–(49) imply the desired relation (44).

For the sake of completness, we will now formally verify (46) and (47). First assume that $k < j$ and thus, $i' = i$, $k' = k$. Then, by (1.3.15) and (1.3.16),

$$c_{k'}^n(d_j t) = (a_k^n(d_j t), b_k^n(d_j t)) = (d_{j-k} a_k^{n-1}(t), b_k^{n-1}(t)). \tag{50}$$

Consequently,

$$(1 \times c_{i'}^{k'})\, c_{k'}^n(d_j t) = (d_{j-k} a_k^{n-1}(t), a_i^k b_k^{n-1}(t), b_i^k b_k^{n-1}(t)), \tag{51}$$

which coincides with the right side of (46).

Now assume that $j \leq k$ and thus, $k' = k+1$. In this case (1.3.12) and (1.3.13) yield

$$c^n_{k'}(d_j t) = (a^n_{k+1}(d_j t), b^n_{k+1}(d_j t)) = (a^{n-1}_k(t), d_j b^{n-1}_k(t)). \tag{52}$$

If $i < j$, one has $i' = i$ and thus, the left side of (46) equals

$$(1 \times c^{k+1}_i)(a^{n-1}_k(t), d_j b^{n-1}_k(t)) = (a^{n-1}_k(t), a^{k+1}_i d_j b^{n-1}_k(t), b^{k+1}_i d_j b^{n-1}_k(t)). \tag{53}$$

However, by (1.3.15) and (1.3.16),

$$(a^{k+1}_i d_j b^{n-1}_k(t), b^{k+1}_i d_j b^{n-1}_k(t)) = (d_{j-i} a^k_i b^{n-1}_k(t), b^k_i b^{n-1}_k(t)). \tag{54}$$

This shows that, for $i < j$, the left side of (46) coincides with its right side.

Finally, assume that $j \leq i$ and thus, $i' = i+1$. Then the left side of (46) equals

$$(1 \times c^{k+1}_{i+1})(a^{n-1}_k(t), d_j b^{n-1}_k(t)) = (a^{n-1}_k(t), a^{k+1}_{i+1} d_j b^{n-1}_k(t), b^{k+1}_{i+1} d_j b^{n-1}_k(t)). \tag{55}$$

However,

$$(a^{k+1}_{i+1} d_j b^{n-1}_k(t), b^{k+1}_{i+1} d_j b^{n-1}_k(t)) = (a^k_i b^{n-1}_k(t), d_j b^k_i b^{n-1}_k(t)), \tag{56}$$

which shows that in this case too, the left side of (46) coincides with its right side.

To verify formula (47), first assume that $k < j$ and thus, $i' = i$, $k' = k$. Then,

$$c^n_{i'}(d_j t') = (d_{j-i}\, a^{n-1}_i(t'), b^{n-1}_i(t')). \tag{57}$$

Consequently, the left side of (47) equals

$$(a^{n-i}_{k-i} d_{j-i}\, a^{n-1}_i(t'), b^{n-i}_{k-i} d_{j-i}\, a^{n-1}_i(t'), b^{n-1}_i(t')). \tag{58}$$

Furthermore, since $k < j$, (58) coincides with

$$(d_{j-k} \times 1 \times 1)(a^{n-i-1}_{k-i} a^{n-1}_i(t'), b^{n-i-1}_{k-i} a^{n-1}_i(t'), b^{n-1}_i(t')), \tag{59}$$

i.e., with $(d_{j-k} \times 1 \times 1)(c^{n-1-i}_{k-i} \times 1)\, c^{n-1}_i(t')$.

Now assume that $j \leq k$, $i < j$ and thus, $i' = i$, $k' = k+1$. Then

$$c^n_{i'}(d_j t') = (d_{j-i} a^{n-1}_i(t'), b^{n-1}_i(t')), \tag{60}$$

and the left side of (47) equals

$$(a^{n-i}_{k+1-i} d_{j-i} a^{n-1}_i(t'), b^{n-i}_{k+1-i} d_{j-i} a^{n-1}_i(t'), b^{n-1}_i(t')). \tag{61}$$

However, since $j - i \leq k - i$, (61) coincides with

$$(1 \times d_{j-i} \times 1)(a^{n-i-1}_{k-i} a^{n-1}_i(t'), b^{n-i-1}_{k-i} a^{n-1}_i(t'), b^{n-1}_i(t')). \tag{62}$$

Finally, let $j \leq i$, and thus, $i' = i+1$, $k' = k+1$. Then

$$c^n_{i'}(d_j t') = (a^n_{i+1}(d_j t'), b^n_{i+1}(d_j t')). \tag{63}$$

Hence, the application of $(c^{n-i-1}_{k-i} \times 1)$ to (63) yields

$$(1 \times 1 \times d_j)(a^{n-i-1}_{k-i} a^{n-1}_i(t'), b^{n-i-1}_{k-i} a^{n-1}_i(t'), b^{n-1}_i(t')). \;\square \tag{64}$$

2.3 The identity morphism

In this subsection we prove that the homotopy class $[C(\mathbf{1_X})]$ of the coherent identity mapping $C(\mathbf{1_X})$ is indeed the identity morphism of the coherent homotopy category CH(pro-Top), i.e., the following theorem holds.

THEOREM 2.11. *Let $\boldsymbol{f}: \boldsymbol{X} \to \boldsymbol{Y}$ be a coherent mapping. Then $\boldsymbol{f}C(\mathbf{1_X}) \simeq \boldsymbol{f}$ and $C(\mathbf{1_Y})\boldsymbol{f} \simeq \boldsymbol{f}$.*

For later use, we will prove a more general statement.

LEMMA 2.12. *Let $\boldsymbol{f} = (f, f_\mu): \boldsymbol{X} \to \boldsymbol{Y}$ be a mapping of systems and let $\boldsymbol{g} = (g, g_\nu): \boldsymbol{Y} \to \boldsymbol{Z}$ be a coherent mapping. Then the composition $\boldsymbol{h} = \boldsymbol{g}C(\boldsymbol{f})$ is homotopic to the coherent mapping $\boldsymbol{h}': \boldsymbol{X} \to \boldsymbol{Z}$, given by $h' = fg$ and by the mappings $h'_\nu: X_{h'(\nu_n)} \times \Delta^n \to Z_{\nu_0}$, where*

$$h'_\nu(x,t) = g_\nu(f_{g(\nu_n)}(x), t). \tag{1}$$

Similarly, if $\boldsymbol{f} = (f, f_\mu): \boldsymbol{X} \to \boldsymbol{Y}$ is a coherent mapping and $\boldsymbol{g} = (g, g_\nu): \boldsymbol{Y} \to \boldsymbol{Z}$ is a mapping of systems, then the composition $\boldsymbol{k} = C(\boldsymbol{g})\boldsymbol{f}$ is homotopic to the coherent mapping $\boldsymbol{k}': \boldsymbol{X} \to \boldsymbol{Z}$, given by $k' = fg$ and by the mappings $k'_\nu: X_{k'(\nu_n)} \times \Delta^n \to Z_{\nu_0}$, where

$$k'_\nu(x,t) = g_{\nu_0} f_{g(\nu)}(x,t). \tag{2}$$

If $\boldsymbol{f}$ and $\boldsymbol{g}$ are level preserving, then so are $\boldsymbol{h}$, $\boldsymbol{h}'$, $\boldsymbol{k}$, $\boldsymbol{k}'$ as well as the corresponding homotopies.

Proof of Lemma 2.12. To prove the first assertion, notice that the coherence conditions for $\boldsymbol{g}$ imply the coherence conditions for $\boldsymbol{h}'$, so that $\boldsymbol{h}'$ is indeed a coherent mapping. In order to define a coherent homotopy $\boldsymbol{H} = (H, H_\nu): \boldsymbol{X} \times I \to \boldsymbol{Z}$, which connects $\boldsymbol{h}'$ to $\boldsymbol{h}$, note that the composition rule yields

$$h_\nu(x,t) = g_{\nu_0\ldots\nu_i}(f_{g(\nu_i)} p_{h(\nu_i)h(\nu_n)}(x), b_i^n(t)), \tag{3}$$

for $t \in P_i^n$, $0 \le i \le n$. By (1.3.9), there is a well-defined mapping $\varphi^n: \Delta^n \to \Delta^n$, such that

$$\varphi^n(t) = d_n^n \ldots d_{i+1}^{i+1} b_i^n(t),\ t \in P_i^n. \tag{4}$$

Note that the mapping φ^n satisfies

$$h'_\nu(x, \varphi^n(t)) = h_\nu(x,t),\ t \in \Delta^n. \tag{5}$$

Indeed, if $t \in P_i^n$, (4) and the coherence conditions for $\boldsymbol{g}$ yield

$$\begin{aligned} h'_\nu(x, \varphi^n(t)) &= g_\nu(f_{g(\nu_n)}(x), d_n \ldots d_{i+1} b_i^n(t)) \\ &= g_{\nu_0\ldots\nu_{n-1}}(q_{g(\nu_{n-1})g(\nu_n)} f_{g(\nu_n)}(x), d_{n-1} \ldots d_{i+1} b_i^n(t)) \\ &= \ldots = g_{\nu_0\ldots\nu_i}(q_{g(\nu_i)g(\nu_n)} f_{g(\nu_n)}(x), b_i^n(t)), \end{aligned} \tag{6}$$

which, together with (3), proves (5). We now define a mapping $\Phi^n: I \times \Delta^n \to \Delta^n$, by putting

$$\Phi^n(s,t) = (1-s)t + s\varphi^n(t). \tag{7}$$

Then, put $H = fg = h$ and define the mappings $H_\nu: X_{h(\nu_n)} \times I \times \Delta^n \to Y_{\nu_0}$, $\boldsymbol{\nu} = (\nu_0, \dots, \nu_n) \in N_n$, by putting

$$H_\nu(x,s,t) = h'_\nu(x, \Phi^n(s,t)). \tag{8}$$

Clearly,

$$H_\nu(x,0,t) = h'_\nu(x,t),\ t \in \Delta^n, \tag{9}$$

$$H_\nu(x,1,t) = h'_\nu(x,\varphi^n(t)) = h_\nu(x,t),\ t \in \Delta^n. \tag{10}$$

In order to show that $\boldsymbol{H}$ satisfies the coherence conditions, it suffices to show that, for any $0 \le j \le n$,

$$d_j^n \varphi^{n-1} = \varphi^n d_j^n. \tag{11}$$

Indeed, this relation implies

$$d_j^n \Phi^{n-1} = \Phi^n(1 \times d_j^n). \tag{12}$$

Consequently, for $t \in \Delta^{n-1}$, one has

$$\begin{gathered} H_\nu(x,s,d_jt) = h'_\nu(x,\Phi^n(s,d_jt)) = \\ h'_\nu(x, d_j\Phi^{n-1}(s,t)) = rh'_{d_j\nu}(p(x),\Phi^{n-1}(s,t)) = rH_{d_j\nu}(p(x),s,t). \end{gathered} \tag{13}$$

To establish (11), note that for $t \in P_i^{n-1}$ and $j \le i$, (1.3.11) and (1.3.13) imply $d_j(t) \in P_{i+1}^n$ and $b_{i+1}^n(d_jt) = d_j b_i^{n-1}(t)$. Therefore,

$$\varphi^n(d_j^n t) = d_n^n \dots d_{i+2}^{i+2} b_{i+1}^n(d_jt) = d_n^n \dots d_{i+2}^{i+2} d_j b_i^{n-1}(t). \tag{14}$$

On the other hand, by repeated application of (1.2.16), one obtains

$$d_j\varphi^{n-1}(t) = d_j d_{n-1}^{n-1} \dots d_{i+1}^{i+1} b_i^{n-1}(t) = d_n^n \dots d_{i+2}^{i+2} d_j b_i^{n-1}(t). \tag{15}$$

Comparison of (14) with (15) yields (11). Now assume that $i < j$. Then (1.3.14) and (1.3.16) imply $d_j^n(t) \in P_i^n$ and $b_i^n(d_j^n t) = b_i^{n-1}(t)$. Therefore,

$$\varphi^n(d_j^n t) = d_n^n d_{n-1}^{n-1} \dots d_{i+1}^{i+1} b_i^{n-1}(t). \tag{16}$$

On the other hand,

$$d_j\varphi^{n-1}(t) = d_j d_{n-1}^{n-1} \dots d_{i+1}^{i+1} b_i^{n-1}(t). \tag{17}$$

However, since $i < j$, (1.2.5) implies $d_n^n d_{n-1}^{n-1} \dots d_{i+1}^{i+1} = d_j d_{n-1}^{n-1} \dots d_{i+1}^{i+1}$ and thus, the right sides of (16) and (17) coincide.

The proof of the second assertion is similar. We need a homotopy $\boldsymbol{K} = (K, K_\nu): \boldsymbol{X} \times I \to \boldsymbol{Z}$, which connects $\boldsymbol{k}'$ to $\boldsymbol{k}$. Put $K = fg = k$. In order to define the mappings $K_{\boldsymbol{\nu}}: X_{k(\nu_n)} \times I \times \Delta^n \to Y_{\mu_0}$, $\boldsymbol{\nu} = (\nu_0, \ldots, \nu_n) \in N_n$, first note that, by the composition rule,

$$k_{\boldsymbol{\nu}}(x,t) = g_{\nu_0} q_{g(\nu_0)g(\nu_i)} f_{g(\nu_i)\ldots g(\nu_n)}(x, a_i^n(t)),\ t \in P_i^n. \tag{18}$$

Then note that, by (1.3.8), there is a well-defined mapping $\psi^n : \Delta^n \to \Delta^n$, such that

$$\psi^n(t) = d_0^n \ldots d_0^{n-i+1} a_i^n(t),\ t \in P_i^n. \tag{19}$$

Note that the mapping ψ^n satisfies

$$k'_{\boldsymbol{\nu}}(x, \psi^n(t)) = k_{\boldsymbol{\nu}}(x,t), t \in \Delta^n. \tag{20}$$

Indeed, if $t \in P_i^n$, (19) and the coherence conditions for $\boldsymbol{f}$ yield

$$\begin{aligned} k'_{\boldsymbol{\nu}}(x, \psi^n(t)) &= g_{\nu_0}(f_{g(\boldsymbol{\nu})}(x), d_0^n \ldots d_0^{n-i+1} a_i^n(t)) \\ &= g_{\nu_0} q_{g(\nu_0)g(\nu_i)} f_{g(\nu_i)\ldots g(\nu_n)}(x, a_i^n(t)), \end{aligned} \tag{21}$$

which, together with (18), proves (20). Define a mapping $\Psi^n: I \times \Delta^n \to \Delta^n$ by

$$\Psi^n(s,t) = (1-s)t + s\psi^n(t), \tag{22}$$

and put

$$K_{\boldsymbol{\nu}}(x,s,t) = k'_{\boldsymbol{\nu}}(x, \Psi^n(s,t)). \tag{23}$$

Clearly,

$$K_{\boldsymbol{\nu}}(x,0,t) = k'_{\boldsymbol{\nu}}(x,t),\ t \in \Delta^n, \tag{24}$$

$$K_{\boldsymbol{\nu}}(x,1,t) = k'_{\boldsymbol{\nu}}(x, \psi^n(t)) = k_{\boldsymbol{\nu}}(x,t),\ t \in \Delta^n. \tag{25}$$

In order to show that $\boldsymbol{K}$ satisfies the coherence conditions, it suffices to show that, for any $0 \le j \le n$, one has

$$d_j^n \psi^{n-1} = \psi^n d_j^n, \tag{26}$$

and thus also

$$d_j^n \Psi^{n-1} = \Psi^n(1 \times d_j^n). \tag{27}$$

To establish (26), first assume that $t \in P_i^{n-1}$ and $j \le i$. Then (1.3.11) and (1.3.12) imply $a_{i+1}^n(d_j t) = a_i^{n-1}(t)$. Therefore,

$$\psi^n(d_j t) = d_0^n \ldots d_0^{n-i} a_{i+1}^n(d_j t) = d_0^n \ldots d_0^{n-i} a_i^{n-1}(t). \tag{28}$$

On the other hand,

$$d_j \psi^{n-1}(t) = d_j d_0^{n-1} \ldots d_0^{n-i} a_i^{n-1}(t). \tag{29}$$

However, since $j \le i$, (1.2.5) implies $d_j d_0^{n-1} \ldots d_0^{n-i} = d_0 d_0^{n-1} \ldots d_0^{n-i}$ and thus, the right sides of (28) and (29) coincide. Now assume that $i < j$. Then (1.3.14) and (1.3.15) imply $d_j t \in P_i^n$ and $a_i^n(d_j t) = d_{j-i} a_i^{n-1}(t)$. Therefore,

$$\psi^n(d_j t) = d_0^n \ldots d_0^{n-i+1} a_i^n(d_j t) = d_0^n \ldots d_0^{n-i+1} d_{j-i} a_i^{n-1}(t). \tag{30}$$

On the other hand, by (1.2,16),

$$d_j^n \psi^{n-1}(t) = d_j^n d_0^{n-1} \dots d_0^{n-i+1} a_i^{n-1}(t) = d_0^n \dots d_0^{n-i+1} d_{j-i} a_i^{n-1}(t). \quad (31)$$

Comparison of (30) with (31) yields (26). □

Proof of Theorem 2.11. If $\boldsymbol{g} = \mathbf{1}_{\boldsymbol{Y}}$, then $g = \mathrm{id}$ and $g_\nu = \mathrm{id}$. Consequently, $k' = f$ and, by (2), $k'_\nu(x,t) = f_\nu(x,t)$, i.e., $\boldsymbol{k}' = \boldsymbol{f}$. Therefore, Lemma 2.12 implies that $C(\mathbf{1}_{\boldsymbol{Y}})\boldsymbol{f} \simeq \boldsymbol{f}$. The proof of $\boldsymbol{f}C(\mathbf{1}_{\boldsymbol{X}}) \simeq \boldsymbol{f}$ is analogous. □

Another special case of Lemma 2.12 deserves to be mentioned.

LEMMA 2.13. *Let $\boldsymbol{p} = (p_\lambda)\colon X \to \boldsymbol{X}$ be a mapping and $\boldsymbol{f} = (f, f_\mu)\colon \boldsymbol{X} \to \boldsymbol{Y}$ a coherent mapping. Then the composition $\boldsymbol{f}C(\boldsymbol{p})$ is homotopic to the coherent mapping $\boldsymbol{h} = (h_\mu)\colon X \to \boldsymbol{Y}$, where*

$$h_\mu(x,t) = f_\mu(p_{f(\mu_n)}(x), t). \ \square \quad (32)$$

The coherence operator C, defined in 1.4, enables us to define the *coherence functors* C: inv-Top $\to$ CH(pro-Top), C:Top$^\Lambda \to$ CH(Top$^\Lambda$) and C: pro-Top $\to$ CH(pro-Top). By definition, $C(\boldsymbol{X}) = \boldsymbol{X}$. If $\boldsymbol{f}$ is a mapping, then

$$C(\boldsymbol{f}) = [C(\boldsymbol{f})] \quad (33)$$

is the homotopy class of the coherent mapping $C(\boldsymbol{f})$. In the special case of level mappings, $C(\boldsymbol{f})$ is a level coherent mapping. Note that, by Lemma 1.17, one has

$$C(\boldsymbol{gf}) = [C(\boldsymbol{gf})] = [C(\boldsymbol{g})C(\boldsymbol{f})] = [C(\boldsymbol{g})][C(\boldsymbol{f})] = C(\boldsymbol{g})C(\boldsymbol{f}). \quad (34)$$

Moreover, $C(\mathbf{1}_{\boldsymbol{X}}) = [C(\mathbf{1}_{\boldsymbol{X}})]$. If $[\boldsymbol{f}]$ is a congruence class of mappings, put

$$C[\boldsymbol{f}] = [C(\boldsymbol{f})]. \quad (35)$$

That $C[\boldsymbol{f}]$ is well defined by (35), is a consequence of the following lemma.

LEMMA 2.14. *If $\boldsymbol{f} = (f, f_{\mu_0}), \boldsymbol{f}' = (f', f'_{\mu_0})\colon \boldsymbol{X} \to \boldsymbol{Y}$ are congruent mappings, $\boldsymbol{f} \equiv \boldsymbol{f}'$, then the coherent mappings $C(\boldsymbol{f}), C(\boldsymbol{f}')$ are homotopic, i.e., $[C(\boldsymbol{f})] = [C(\boldsymbol{f}')]$.*

Proof. By Lemma 1.17, $\boldsymbol{f} \equiv \boldsymbol{f}'$ implies $C(\boldsymbol{f}) \equiv C(\boldsymbol{f}')$. However, by Lemma 2.3, congruent coherent mappings are homotopic. □

Note that

$$\begin{aligned} C([\boldsymbol{g}][\boldsymbol{f}]) = C[\boldsymbol{gf}] = [C(\boldsymbol{gf})] = \\ [C(\boldsymbol{g})C(\boldsymbol{f})] = [C(\boldsymbol{g})][C(\boldsymbol{f})] = C[\boldsymbol{g}]C[\boldsymbol{f}]. \end{aligned} \quad (36)$$

REMARK 2.15. If X and Y are spaces, inv-Top (X, Y) and pro-Top (X, Y) coincide and C induces a bijection $\mathrm{H}(\mathsf{Top})\,(X, Y) \to \mathrm{CH}(\text{pro-}\mathsf{Top})\,(X, Y)$. Indeed, a mapping $f: X \to Y$ between spaces can be interpreted as a morphism of inv-Top (or equivalently, of pro-Top), between rudimentary systems, indexed by a singleton $\{\mu_0\}$. Then $C(f)$ is the homotopy class of the coherent mapping $C(f): X \to Y$, given by the mappings $f_{\boldsymbol{\mu}}: X \times \Delta^n \to Y$, where $f_{\boldsymbol{\mu}}(x, t) = f(x)$, for $\boldsymbol{\mu} = (\mu_0, \ldots, \mu_0) \in M_n$. If $f, f': X \to Y$ are mappings connected by a homotopy $F: X \times I \to Y$, then the mappings $F_{\boldsymbol{\mu}}: X \times I \times \Delta^n \to Y$, defined by $F_{\boldsymbol{\mu}}(x, s, t) = F(x, s)$, define a coherent homotopy, which connects $C(f)$ to $C(f')$. Therefore, $C([f]) = [C(f)]$, yields a well-defined function of $\mathrm{H}(\mathsf{Top})(X, Y)$ to $\mathrm{CH}(\text{pro-}\mathsf{Top})(X, Y)$. This function is injective. Indeed, if $C([f]) = C([f'])$, there is a coherent homotopy $\boldsymbol{H} = (H_{\boldsymbol{\mu}}): X \times I \to Y$, which connects $C(f)$ to $C(f')$, i.e., $H_{\boldsymbol{\mu}}(x, 0, t) = f(x)$, $H_{\boldsymbol{\mu}}(x, 1, t) = f'(x)$. Putting $F(x, s) = H_{\mu_0}(x, s, 0)$, we obtain a homotopy $F: X \times I \to Y$, which connects f to f' and thus, proves that $[f] = [f']$. Finally, the considered function is also surjective. Indeed, if $\boldsymbol{f}: X \to Y$ is a coherent mapping, given by mappings $f_{\boldsymbol{\mu}}: X \times \Delta^n \to Y$, then the coherence condition (1.2.8) shows that $f_{\boldsymbol{\mu}}(x, t) = f_{\mu_0}(x)$, $t \in \Delta^n$. Consequently, the homotopy class $[f]$ of the mapping $f = f_{\mu_0}: X \to Y$ is transformed into $C([f]) = [\boldsymbol{f}]$.

Using the forgetful operator E from 1.4, we now define the *forgetful functor* E: $\mathrm{CH}(\text{pro-}\mathsf{Top}) \to \text{pro-}\mathrm{H}(\mathsf{Top})$. By definition, $E(\boldsymbol{X}) = [\boldsymbol{X}]$. Moreover, $E(\boldsymbol{f}) = (f, [f_{\mu_0}])$ is a morphism of inv-$\mathrm{H}(\mathsf{Top})$. Put

$$E([\boldsymbol{f}]) = [E(\boldsymbol{f})], \tag{37}$$

The congruence class $[E(\boldsymbol{f})]$ of $E(\boldsymbol{f})$ depends only on the homotopy class of $\boldsymbol{f}$. Indeed, if $\boldsymbol{F} = (F, F_{\boldsymbol{\mu}}): \boldsymbol{X} \times I \to \boldsymbol{Y}$ is a coherent homotopy, which connects $\boldsymbol{f}$ to $\boldsymbol{f}'$, then $F \geq f, f'$ and F_{μ_0} is a homotopy, which connects $f_{\mu_0} p_{f(\mu_0) F(\mu_0)}$ to $f'_{\mu_0} p_{f'(\mu_0) F(\mu_0)}$. Consequently,

$$[f_{\mu_0}][p_{f(\mu_0) F(\mu_0)}] = [f'_{\mu_0}][p_{f'(\mu_0) F(\mu_0)}]. \tag{38}$$

However, (38) shows that $E(\boldsymbol{f})$ and $E(\boldsymbol{f}')$ are congruent mappings, i.e., $[E(\boldsymbol{f})] = [E(\boldsymbol{f}')]$. Note that, $E([\mathbf{1}_{\boldsymbol{X}}]) = [\mathbf{1}_{\boldsymbol{X}}]$. Moreover, by Lemma 1.18,

$$E([\boldsymbol{g}][\boldsymbol{f}]) = E[\boldsymbol{g}\boldsymbol{f}] = [E(\boldsymbol{g}\boldsymbol{f})] = [E(\boldsymbol{g})E(\boldsymbol{f})] = [E(\boldsymbol{g})][E(\boldsymbol{f})]. \tag{39}$$

REMARK 2.16. If $\boldsymbol{f} = (f, f_{\mu_0}): \boldsymbol{X} \to \boldsymbol{Y}$ is a mapping, then $EC(\boldsymbol{f})$ is the congruence class of the morphism $[\boldsymbol{f}] = (f, [f_{\mu_0}])$ of inv-$\mathrm{H}(\mathsf{Top})$, obtained by applying the homotopy functor H: $\mathsf{Top} \to \mathrm{H}(\mathsf{Top})$ to the mapping $\boldsymbol{f}$. Indeed, $C(\boldsymbol{f})$ is the homotopy class of the coherent mapping $(f, f_{\boldsymbol{\mu}})$, where $f_{\boldsymbol{\mu}}(x, t) = f_{\mu_0} p_{f(\mu_0) f(\mu_n)}(x)$. Therefore, $EC(\boldsymbol{f})$ is the congruence class of the morphism $E(f, f_{\boldsymbol{\mu}})) = (f, [f_{\mu_0}])$.

Up to now we considered several categories of inverse systems of spaces $\boldsymbol{X} = (X_\lambda, p_{\lambda\lambda'}, \Lambda)$. The same notions and results have their analogues for inverse systems of pairs of spaces $(\boldsymbol{X},\boldsymbol{X}^0) = ((X,X^0)_\lambda, p_{\lambda\lambda'}, \Lambda)$, where $(X,X^0)_\lambda$ denotes the pair of spaces (X_λ, X^0_λ), $X^0_\lambda \subseteq X_\lambda$, and $p_{\lambda\lambda'}\colon (X,X^0)_{\lambda'} \to (X,X^0)_\lambda$ is a mapping of pairs, i.e., it is a mapping $p_{\lambda\lambda'}\colon X_{\lambda'} \to X_\lambda$ such that $p_{\lambda\lambda'}(X^0_{\lambda'}) \subseteq X^0_\lambda$. In the definition of a mapping of systems $\boldsymbol{f}\colon (\boldsymbol{X},\boldsymbol{X}^0) \to (\boldsymbol{Y},\boldsymbol{Y}^0)$ we require that $f_\mu\colon (X,X^0)_{f(\mu)} \to (Y,Y^0)_\mu$ be a mapping of pairs. In this way we obtain categories $(\mathsf{Top}_2)^\Lambda$ and inv-Top_2. Congruence of mappings is defined as before and it yields the category pro-Top_2. In the definition of pro-H(Top_2), one requires that all homotopies be homotopies of pairs.

In order to define a coherent mapping $\boldsymbol{f}\colon (\boldsymbol{X},\boldsymbol{X}^0) \to (\boldsymbol{Y},\boldsymbol{Y}^0)$, we require that all $f_{\boldsymbol{\mu}}\colon (X,X^0)_{f(\mu_n)} \times \Delta^n \to (Y,Y^0)_{\mu_0}$ be mappings of pairs, where $(X,X^0)_\lambda \times \Delta^n$ denotes the pair $(X_\lambda \times \Delta^n, X^0_\lambda \times \Delta^n)$. The coherence conditions remain unchanged. Composition of coherent mappings is defined using the same formulae as before. One must only verify that $h_{\boldsymbol{\nu}}(X^0_{h(\nu_n)} \times P^n_i) \subseteq Z^0_{\nu_0}$. As in 2.1, a coherent homotopy is a coherent mapping of systems of pairs $\boldsymbol{F}\colon (\boldsymbol{X},\boldsymbol{X}^0) \times I \to (\boldsymbol{Y},\boldsymbol{Y}^0)$. Theorems 2.8 and 2.11 remain valid because all homotopies constructed in the course of the proofs in the case of systems of spaces are actually homotopies of systems of pairs. The reason for this lies in the fact that all the homotopies, which appear in these proofs, either use operations on Δ^n or $I \times \Delta^n$, which remain unchanged, or use already given coherent mappings and coherent homotopies, which are, by assumption, coherent mappings and homotopies of pairs. In this way we obtain the category CH(pro-Top_2). The same applies to the functors C: inv-$\mathsf{Top}_2 \to$ CH(pro-Top_2), C: pro-$\mathsf{Top}_2 \to$ CH(pro-Top_2) and E: CH(pro-Top_2) $\to$ pro-H(Top_2).

Bibliographic notes

The category CH(pro-Top) has been introduced in (Lisica, Mardešić 1983, 1984a, 1984b) (under the name ***CPHTop***). It appears that coherent homotopy categories were first defined in (Boardman, Vogt 1973) and (Vogt 1973), for the more general coherent inverse systems (see 10.4). Other versions of coherent homotopy can be found in (Porter 1978), (Miminoshvili 1980, 1982), (Lisitsa 1982a, 1983a), (Cordier 1982, 1989) and (Šekutkovski 1988, 1997). The special feature of our construction is the fact that it refers to usual inverse systems and not to coherent systems. This restriction does not affect applications to strong shape, but brings considerable simplification and geometric transparency.

3. Coherent homotopy of sequences

This section is devoted to coherent homotopy of inverse sequences (also called *towers*) and is not needed in the remaining part of Chapter I. Restriction to inverse sequences greatly simplifies the theory, because in this case the use of homotopies of orders higher than 2 can be avoided. On the other hand, inverse sequences suffice to develop strong shape theory of metric compact, which is the most useful part of strong shape theory. In the introductory subsection, we define coherent homotopy theories, which use homotopies up to a given finite order $r \geq 0$. However, in the subsection which follows, we use only the case $r = 1$.

3.1 Coherent homotopy of finite height

For every integer $r \geq 0$, we will define a category $\mathrm{CH}^{(r)}$ (pro-Top), called the *coherent homotopy category of height* r and a *forgetful functor* $E^{(r)}$: CH(pro-Top)→ $\mathrm{CH}^{(r)}$(pro-Top). Moreover, for $r < r'$, we will define a *forgetful functor* $E^{(rr')}$: $\mathrm{CH}^{(r')}$(pro-Top)→ $\mathrm{CH}^{(r)}$(pro-Top) such that $E^{(rr')}E^{(r'r'')} = E^{(rr'')}$, for $r < r' < r''$, and $E^{(rr')}E^{(r')} = E^{(r)}$, for $r < r'$. We will show that $\mathrm{CH}^{(0)}$(pro-Top) coincides with the category pro-H(Top). On the other hand, the category CH(pro-Top) can be viewed as the *coherent homotopy category of height* ∞.

We first define *coherent mappings of height* r, shorter called r-*coherent mappings* $\boldsymbol{f}\colon \boldsymbol{X} \to \boldsymbol{Y}$. If $\boldsymbol{X} = (X_\lambda, p_{\lambda\lambda'}, \Lambda)$ and $\boldsymbol{Y} = (Y_\mu, q_{\mu\mu'}, M)$, $\boldsymbol{f}$ consists of an increasing function $f\colon M \to \Lambda$ and of mappings $f_{\boldsymbol{\mu}}\colon X_{f(\mu_n)} \times \Delta^n \to Y_{\mu_0}$, defined for multiindices $\boldsymbol{\mu}$ of length $n \leq r$. One requires that the following r-*coherence conditions* be satisfied:

$$f_{\boldsymbol{\mu}}(x, d_j t) = q f_{d^j \boldsymbol{\mu}}(p(x), t),\ 0 < n \leq r, \tag{1}$$

$$f_{\boldsymbol{\mu}}(x, s_j(t)) = f_{s^j(\boldsymbol{\mu})}(x, t),\ 0 \leq n \leq r - 1. \tag{2}$$

One imposes the *additional condition*: for any multiindex $(\mu_0, \ldots, \mu_{r+1})$ on M, there exists a mapping $f_{\mu_0 \ldots \mu_{r+1}}\colon X_{f(\mu_{r+1})} \times \Delta^{r+1} \to Y_{\mu_0}$, which also satisfies (1). This mapping is not part of the structure of $\boldsymbol{f}$. However, the existence of such a mapping is an additional condition on the mappings $f_{\boldsymbol{\mu}}$, where the length of $\boldsymbol{\mu}$ equals r. The *composition of* r-*coherent mappings*

$\boldsymbol{f}: \boldsymbol{X} \to \boldsymbol{Y}$ and $\boldsymbol{g}: \boldsymbol{Y} \to \boldsymbol{Z}$ is the r-coherent mapping $\boldsymbol{h} = \boldsymbol{gf}$, defined by formula (1.3.5), applied to multiindices $\boldsymbol{\nu}$ of length $n \leq r$. The same formula insures existence of the mapping $h_{\mu_0 \ldots \mu_{r+1}}$, required by the additional condition.

Like in 2.1, *coherent homotopy of height* r, shorter, r-*coherent homotopy*, is an r-coherent mapping $\boldsymbol{F}: \boldsymbol{X} \times I \to \boldsymbol{Y}$. We say that $\boldsymbol{F}$ connects r-coherent mappings $\boldsymbol{f}, \boldsymbol{f}': \boldsymbol{X} \to \boldsymbol{Y}$, provided $F \geq f, f'$ and (2.1.1) holds, for multiindices $\boldsymbol{\mu}$ of length $n \leq r$. If this is the case, we say that $\boldsymbol{f}$ and $\boldsymbol{f}'$ are homotopic, $\boldsymbol{f} \simeq \boldsymbol{f}'$. Homotopy is an equivalence relation of r-coherent mappings. As usual, the homotopy class of an r-coherent mapping $\boldsymbol{f}$ is denoted by $[\boldsymbol{f}]$.

All the results established for coherent mappings and their homotopies in 1 and 2 have valid analogues for r-coherent mappings and their homotopies. The only change in the proofs is that the lengths of all the multiindices involved is restricted to the interval $[r]$.

For $r' > r$, an r'-coherent mapping $\boldsymbol{f} = (f, f_{\mu_0}, \ldots, f_{\mu_0 \ldots \mu_{r'}}): \boldsymbol{X} \to \boldsymbol{Y}$, determines an r-coherent mapping $E^{(rr')}(\boldsymbol{f}) = (f, f_{\mu_0}, \ldots, f_{\mu_0 \ldots \mu_r})$. Clearly, $E^{(rr')}(\boldsymbol{1_X}) = \boldsymbol{1_X}$ and $E^{(rr')}(\boldsymbol{gf}) = E^{(rr')}(\boldsymbol{g}) E^{(rr')}(\boldsymbol{f})$. Moreover, $\boldsymbol{f} \simeq \boldsymbol{f}'$ implies $E^{(rr')}(\boldsymbol{f}) \simeq E^{(rr')}(\boldsymbol{f}')$. Therefore, putting $E^{(rr')}[\boldsymbol{f}] = [E^{(rr')}(\boldsymbol{f})]$, one obtains a functor $E^{(rr')}$: $\mathrm{CH}^{(r')}(\text{pro-}\mathsf{Top}) \to \mathrm{CH}^{(r)}(\text{pro-}\mathsf{Top})$. The functor $E^{(r)}$: $\mathrm{CH}(\text{pro-}\mathsf{Top}) \to \mathrm{CH}^{(r)}(\text{pro-}\mathsf{Top})$ is defined analogously, i.e., by restricting $\boldsymbol{f} = (f, f_{\mu})$ to $E^{(r)}(\boldsymbol{f}) = (f, f_{\mu_0}, \ldots, f_{\mu_0 \ldots \mu_r})$.

The following technical lemma shows that in the definition of r-coherent homotopies, one can omit the additional condition, i.e., the requirement concerning the existence of $F_{\mu_0 \ldots \mu_{r+1}}$.

LEMMA 3.1. *Let $\boldsymbol{f}, \boldsymbol{f}': \boldsymbol{X} \to \boldsymbol{Y}$ be r-coherent mappings. Let $F \geq f, f'$ be an increasing function and let $F_{\mu}: X_{F(\mu_n)} \times I \times \Delta^n \to Y_{\mu_0}$, $\boldsymbol{\mu} \in M_n$, $n \leq r$, be mappings, which satisfy the r-coherence conditions* (1), (2) *and the conditions*

$$F_{\mu}(x, 0, t) = f_{\mu}(p_{f(\mu_n)F(\mu_n)}(x), t), \; F_{\mu}(x, 1, t) = f'_{\mu}(p_{f'(\mu_n)F(\mu_n)}(x), t), \quad (3)$$

for $n \leq r$. Then $\boldsymbol{F} = (F, F_{\mu_0}, \ldots, F_{\mu_0 \ldots \mu_r})$ is an r-coherent homotopy, which connects $\boldsymbol{f}$ to $\boldsymbol{f}'$.

Proof. We only need to verify the additional condition for $\boldsymbol{F}$, i.e., for a given multiindex $(\mu_0, \ldots, \mu_{r+1})$, we must exhibit an appropriate mapping $F_{\mu_0 \ldots \mu_{r+1}}: X_{F(\mu_{r+1})} \times I \times \Delta^{r+1} \to Y_{\mu_0}$. For $x \in X_{F(\mu_{r+1})}$, we put

$$F_{\mu_0 \ldots \mu_{r+1}}(x, s, d_j(t')) = q F_{d^j(\mu_0, \ldots, \mu_{r+1})}(p(x), s, t'), \; t' \in \Delta^r, \quad (4)$$

$$F_{\mu_0 \ldots \mu_{r+1}}(x, 0, t) = f_{\mu_0 \ldots \mu_{r+1}}(p_{f(\mu_{r+1})F(\mu_{r+1})}(x), t), \; t \in \Delta^{r+1}, \quad (5)$$

where $f_{\mu_0 \ldots \mu_{r+1}}$ is obtained from the additional condition for $\boldsymbol{f}$. Since

$$f_{\mu_0 \ldots \mu_{r+1}}(x, d_j(t)) = q f_{d^j(\mu_0, \ldots, \mu_{r+1})}(p(x), t), \quad (6)$$

(3) implies that (4) and (5) assume the same values at points $(x, 0, d_j t')$. Consequently, the mapping $F_{\mu_0 \ldots \mu_{r+1}}$ is well defined on the set

$$X_{F(\mu_{r+1})} \times ((I \times \partial \Delta^{r+1}) \cup (0 \times \Delta^{r+1})). \tag{7}$$

The mapping is readily extended to all of $X_{F(\mu_{r+1})} \times I \times \Delta^{r+1}$, using any retraction $\rho: I \times \Delta^{r+1} \to (I \times \partial \Delta^{r+1}) \cup (0 \times \Delta^{r+1})$. The obtained mapping $F_{\mu_0 \ldots \mu_{r+1}}$ may not satisfy the second equality in (3), for $\boldsymbol{\mu} = (\mu_0, \ldots, \mu_{r+1})$. However, by the definition of an r-coherent homotopy, that condition is not required. □

We will now define a functor U: $\mathrm{CH}^{(0)}(\text{pro-}\mathsf{Top}) \to \text{pro-}\mathrm{H}(\mathsf{Top})$. For objects, we put $U(\boldsymbol{X}) = \boldsymbol{X}$. Let $[\boldsymbol{f}]: \boldsymbol{X} \to \boldsymbol{Y}$ be a morphism of $\mathrm{CH}^{(0)}(\text{pro-}\mathsf{Top})$, where $\boldsymbol{f} = (f, f_{\boldsymbol{\mu}})$ is a coherent mapping of height $r = 0$. The coherence conditions (1) and (2) do not apply, but the additional condition insures the existence of homotopies $f_{\mu_0\mu_1}: X_{f(\mu_1)} \times \Delta^1 \to Y_{\mu_0}$ such that

$$f_{\mu_0\mu_1}(x, e_1) = f_{\mu_0} p_{f(\mu_0)f(\mu_1)}(x), \; f_{\mu_0\mu_1}(x, e_0) = q_{\mu_0\mu_1} f_{\mu_1}(x). \tag{8}$$

Consequently,

$$[f_{\mu_0}][p_{f(\mu_0)f(\mu_1)}] = [q_{\mu_0\mu_1}][f_{\mu_1}], \tag{9}$$

which shows that $(f, [f_{\mu_0}]): \boldsymbol{X} \to \boldsymbol{Y}$ is a morphism of inv-$\mathrm{H}(\mathsf{Top})$. If $\boldsymbol{f}$ and $\boldsymbol{f}'$ are 0-coherent mappings, which belong to the same homotopy class $[\boldsymbol{f}] = [\boldsymbol{f}']$, then they are connected by a 0-coherent homotopy $\boldsymbol{F} = (F, F_{\mu_0})$ and thus, by (3),

$$f_{\mu_0} p_{f(\mu_0)F(\mu_0)} = F_{\mu_0} i_0, \; f'_{\mu_0} p_{f'(\mu_0)F(\mu_0)} = F_{\mu_0} i_1. \tag{10}$$

Here i_0, i_1 denote the standard embeddings, given by $i_0(x) = (x, 0)$, $i_1(x) = (x, 1)$. Since the homotopy classes $[F_{\mu_0} i_0] = [F_{\mu_0} i_1]$, (10) implies

$$[f_{\mu_0}][p_{f(\mu_0)F(\mu_0)}] = [f'_{\mu_0}][p_{f'(\mu_0)F(\mu_0)}], \tag{11}$$

which shows that the morphisms $(f, [f_{\mu_0}])$ and $(f', [f'_{\mu_0}])$ are congruent and therefore, determine the same morphism $[(f, [f_{\mu_0}])]$ of pro-$\mathrm{H}(\mathsf{Top})$. By definition, this morphism is $U[\boldsymbol{f}]$. It is readily seen that

$$\begin{aligned} &U([\boldsymbol{g}][\boldsymbol{f}]) = U[\boldsymbol{g}\boldsymbol{f}] = [(fg, [g_{\nu_0}][f_{g(\nu_0)}])] = \\ &[(g, [g_{\nu_0}])][(f, [f_{\mu_0}])] = U[\boldsymbol{g}]U[\boldsymbol{f}]. \end{aligned} \tag{12}$$

The functor U identifies the categories $\mathrm{CH}^{(0)}(\text{pro-}\mathsf{Top})$ and pro-$\mathrm{H}(\mathsf{Top})$. More precisely, the following theorem holds.

THEOREM 3.2. *The functor* U: $\mathrm{CH}^{(0)}(\text{pro-}\mathsf{Top}) \to \text{pro-}\mathrm{H}(\mathsf{Top})$ *is an isomorphism of categories.*

Proof. First assume that $U[\boldsymbol{f}] = U[\boldsymbol{f}']$, i.e., that $(f, [f_{\mu_0}])$ and $(f', [f'_{\mu_0}])$ are congruent morphisms of inv-H(Top). There is no loss of generality in assuming that $(f', [f'_{\mu_0}])$ is a shift of $(f, [f_{\mu_0}])$, i.e., $[f'_{\mu_0}] = [f_{\mu_0}][p_{f(\mu_0)f'(\mu_0)}]$. Then, there exists a homotopy $F_{\mu_0}: X_{f'(\mu_0)} \times I \to Y_{\mu_0}$, which connects $f_{\mu_0} p_{f(\mu_0)f'(\mu_0)}$ to f'_{μ_0}. By Lemma 3.1, (f', F_{μ_0}) is a 0-coherent homotopy, which connects the 0-coherent mappings $\boldsymbol{f} = (f, f_{\mu_0})$, and $\boldsymbol{f}' = (f', f'_{\mu_0})$. Consequently, $[\boldsymbol{f}] = [\boldsymbol{f}']$.

Now consider a morphism $[(f, [f_{\mu_0}])]: \boldsymbol{X} \to \boldsymbol{Y}$ of pro-H(Top). Then $\boldsymbol{f} = (f, f_{\mu_0})$ is a 0-coherent mapping. Indeed, for any $(\mu_0, \mu_1) \in M_1$, (9) holds and any homotopy $f_{\mu_0\mu_1}: X_{f(\mu_1)} \times \Delta^1 \to Y_{\mu_0}$, which realizes (9), shows that the additional condition for $\boldsymbol{f} = (f, f_{\mu_0})$ is fulfilled. Clearly, $U[\boldsymbol{f}] = [(f, [f_{\mu_0}])]$. □

We will now analyze the category $\mathrm{CH}^{(1)}$(pro-Top) and see that it differs from the category $\mathrm{CH}^{(0)}$(pro-Top). We will first characterize 1-coherent mappings.

Let $\Phi, \Psi: X \times I \to Y$ be homotopies such that $\Phi(x,1) = \Psi(x,0)$. Recall that their *juxtaposition* is the homotopy $\Phi * \Psi: X \times I \to Y$, defined by

$$(\Phi * \Psi)(x,t) = \begin{cases} \Phi(x,2t), & 0 \le t \le \frac{1}{2}, \\ \Psi(x,2t-1), & \frac{1}{2} \le t \le 1. \end{cases} \tag{13}$$

LEMMA 3.3. *Let $\boldsymbol{f} = (f, f_{\mu_0}, f_{\mu_0\mu_1}): \boldsymbol{X} \to \boldsymbol{Y}$ be a 1-coherent mapping. Then the 1-coherence conditions consist of (8) and of the following condition:*

$$f_{\mu_0\mu_0}(x,t) = f_{\mu_0}(x), \ t \in \Delta^1. \tag{14}$$

Moreover, the additional condition implies

$$f_{\mu_0\mu_1}(p_{f(\mu_1)f(\mu_2)} \times 1) * q_{\mu_0\mu_1} f_{\mu_1\mu_2} \simeq f_{\mu_0\mu_2} (\mathrm{rel}\,\{0,1\}). \tag{15}$$

Conversely, if $f: M \to \Lambda$ is an increasing function and $f_{\mu_0}: X_{f(\mu_0)} \to Y_{\mu_0}$, $\mu_0 \in M_0$, $f_{\mu_0\mu_1}: X_{f(\mu_1)} \times \Delta^1 \to Y_{\mu_0}$, $(\mu_0, \mu_1) \in M_1$, are mappings such that (8), (14) and (15) hold, then the collection $\boldsymbol{f} = (f, f_{\mu_0}, f_{\mu_0\mu_1})$ is a 1-coherent mapping $\boldsymbol{f}: \boldsymbol{X} \to \boldsymbol{Y}$.

Proof. First assume that $\boldsymbol{f}$ is a 1-coherent mapping. Then (8) is a consequence of (1), for $\boldsymbol{\mu} = (\mu_0, \mu_1) \in M_1$, $j = 0,1$, and (14) is a consequence of (2), for $\boldsymbol{\mu} = (\mu_0)$, $j = 1$, $t = 1 \in \Delta^0$. Moreover, for $\boldsymbol{\mu} = (\mu_0, \mu_1, \mu_2) \in M_2$, the additional condition yields a mapping $f_{\boldsymbol{\mu}}: X_{f(\mu_2)} \times \Delta^2 \to Y_{\mu_0}$, which satisfies (1), for $t \in \Delta^1$. The homotopies

$$f_{\mu_0\mu_1}(p_{f(\mu_1)f(\mu_2)} \times 1),\ q_{\mu_0\mu_1} f_{\mu_1\mu_2}: X_{f(\mu_2)} \times I \to Y_{\mu_0}, \tag{16}$$

connect the mappings $f_{\mu_0} p_{f(\mu_0)f(\mu_2)}$ to $q_{\mu_0\mu_1} f_{\mu_1} p_{f(\mu_1)f(\mu_2)}$ and the mappings $q_{\mu_0\mu_1} f_{\mu_1} p_{f(\mu_1)f(\mu_2)}$ to $q_{\mu_0\mu_2} f_{\mu_2}$, respectively. Therefore, the left side of (15) is well defined. To establish the existence of a homotopy $\mathrm{rel}\,\{0,1\}$, which realizes (15), subdivide the square $I \times I$ into three 2-simplices, by joining the points (0,1) and (1,1) with the point $(\frac{1}{2},0)$. Let $\phi: I \times I \to \Delta^2$ be the simplicial mapping, which sends the vertices (0,0) and (0,1) to e_0, sends the vertex $(\frac{1}{2},0)$ to e_1 and sends the vertices (1,0) and (1,1) to e_2. Choose a mapping $f_{\mu_0\mu_1\mu_2}: X_{f(\mu_2)} \times \Delta^2 \to Y_{\mu_0}$, which satisfies the additional condition for $\boldsymbol{f}$. Then define $H: X_{f(\mu_2)} \times I \times I \to Y_{\mu_0}$, by putting

$$H(x,t,s) = f_{\mu_0\mu_1\mu_2}(x, \phi(t,s)). \tag{17}$$

Using (1), for $\boldsymbol{\mu} = (\mu_0, \mu_1, \mu_2)$, it is readily seen that indeed, H realizes (15).

Conversely, if H is a homotopy rel $\{0,1\}$, which realizes (15), then there is a unique mapping $f_{\mu_0\mu_1\mu_2}$ such that $f_{\mu_0\mu_1\mu_2}\phi = H$. Clearly, this mapping satisfies (1), for $\boldsymbol{\mu} = (\mu_0, \mu_1, \mu_2)$, which proves that $\boldsymbol{f} = (f, f_{\mu_0}, f_{\mu_0\mu_1})$ satisfies also the additional condition and thus, is a 1 - coherent mapping. □

For later use we now state a lemma characterizing 1 - coherent homotopies. It is an immediate consequence of Lemma 3.1.

LEMMA 3.4. *Let $\boldsymbol{F} = (F, F_{\mu_0}, F_{\mu_0\mu_1}): \boldsymbol{X} \times I \to \boldsymbol{Y}$ be a 1 - coherent homotopy, which connects 1 - coherent mappings $\boldsymbol{f}, \boldsymbol{f}': \boldsymbol{X} \to \boldsymbol{Y}$. Then the following equalities hold.*

$$F_{\mu_0}(x,0) = f_{\mu_0} p_{f(\mu_0)F(\mu_0)}(x), \quad F_{\mu_0}(x,1) = f'_{\mu_0} p_{f'(\mu_0)F(\mu_0)}(x), \tag{18}$$

$$F_{\mu_0\mu_1}(x,0,t) = f_{\mu_0\mu_1}(p_{f(\mu_1)F(\mu_1)}(x), t), \tag{19}$$

$$F_{\mu_0\mu_1}(x,1,t) = f'_{\mu_0\mu_1}(p_{f'(\mu_1)F(\mu_1)}(x), t). \tag{20}$$

$$F_{\mu_0\mu_1}(x,s,e_1) = q_{\mu_0\mu_1} F_{\mu_1}(x,s), \tag{21}$$

$$F_{\mu_0\mu_1}(x,s,e_0) = F_{\mu_0}(p_{F(\mu_0)F(\mu_1)}(x), s), \tag{22}$$

$$F_{\mu_0\mu_0}(x,s,t) = F_{\mu_0}(x,s). \tag{23}$$

Conversely, let $\boldsymbol{f}, \boldsymbol{f}'$ be 1 - coherent mappings. If $F: M \to \Lambda$ is an increasing function, $F \geq f, f'$, and $F_{\mu_0}: X_{f(\mu_0)} \to Y_{\mu_0}$, $F_{\mu_0\mu_1}: X_{f(\mu_1)} \times \Delta^1 \to Y_{\mu_0}$ are mappings such that (18)–(23) hold, then the collection $\boldsymbol{F} = (F, F_{\mu_0}, F_{\mu_0\mu_1})$ is a 1 - coherent homotopy $\boldsymbol{F}: \boldsymbol{X} \to \boldsymbol{Y}$, which connects $\boldsymbol{f}$ to $\boldsymbol{f}'$. □

EXAMPLE 3.5. We will now exhibit an example of two inverse systems $\boldsymbol{X}$ and $\boldsymbol{Y}$ such that the set of morphisms $\mathrm{CH}^{(0)}(\text{pro-}\mathsf{Top})(\boldsymbol{X}, \boldsymbol{Y})$ consists of a single morphism, while the set of morphisms $\mathrm{CH}^{(1)}(\text{pro-}\mathsf{Top})$ $(\boldsymbol{X}, \boldsymbol{Y})$ is uncountable. This proves that the categories $\mathrm{CH}^{(0)}(\text{pro-}\mathsf{Top})$ and $\mathrm{CH}^{(1)}(\text{pro-}\mathsf{Top})$ are different.

We take as $\boldsymbol{X}$ the rudimentary system, whose only space is a singleton $X = \{*\}$ and we take as $\boldsymbol{Y}$ the inverse sequence $(Y_{m_0}, q_{m_0m_1}, \mathbb{N})$, where $Y_{m_0} = S^1 = \{z \in \mathbb{C} : |z| = 1\}$ is the 1 - sphere, for all $m_0 \in \mathbb{N}$, and $q_{m_0m_1}(z) = z^{3^{m_1 - m_0}}$.

A 0 - coherent mapping $\boldsymbol{f} = (f_{m_0}): \{*\} \to \boldsymbol{Y}$ is given by the points $y_{m_0} = f_{m_0}(*) \in Y_{m_0}$, $m_0 \in \mathbb{N}$. The additional condition is a consequence of the path-connectedness of S^1. Let $\boldsymbol{g} = (g_{m_0}): \{*\} \to \boldsymbol{Y}$ be the 0 - coherent mapping, given by $g_{m_0}(*) = 1$, $m_0 \in \mathbb{N}$. If $F_{m_0}: \{*\} \times I \to Y_{m_0}$ is a path connecting the points y_{m_0} and 1, then $\boldsymbol{F} = (F_{m_0})$ is a 0 - coherent homotopy, which connects $\boldsymbol{f}$ to $\boldsymbol{g}$. Hence, $[\boldsymbol{g}]$ is the only morphism of $\mathrm{CH}^{(0)}(\text{pro-}\mathsf{Top})(\boldsymbol{X}, \boldsymbol{Y})$.

We now define an uncountable collection of 1-coherent mappings $\boldsymbol{f}\colon \{*\} \to \boldsymbol{Y}$, which belong to distinct elements of $\mathrm{CH}^{(1)}(\text{pro-}\mathsf{Top})$. First consider the set $\boldsymbol{A}$ of all sequences $\alpha = (\alpha_1, \alpha_2, \ldots)$ of zeros and ones. Define an equivalence relation $\sim$ on A by putting $\alpha \sim \alpha'$, provided α and α' coincide eventually, i.e., $\alpha_j = \alpha'_j$, for all sufficiently large j. Clearly, each equivalence class contains only $\aleph_0$ sequences α. Since A is a set of cardinality $2^{\aleph_0}$, there are $2^{\aleph_0}$ equivalence classes on A. Now assign to every $\alpha \in A$ a 1-coherent mapping $\boldsymbol{f}^{\,\alpha}$, given by the points $f^\alpha_{m_0}(*) = 1$, $m_0 \in \mathbb{N}$, and by loops $f^\alpha_{m_0 m_1}\colon \{*\} \times \Delta^1 \to S^1$, based at 1 and such that

$$\deg f^\alpha_{m_0 m_1} = \alpha_{m_0} + 3\alpha_{m_0+1} + \ldots + 3^{m_1-m_0-1}\alpha_{m_1-1}, \tag{24}$$

if $m_0 < m_1$; for $m_0 = m_1$, put

$$f^\alpha_{m_0 m_0}(x,t) = 1. \tag{25}$$

These conditions insure the validity of (8), (14), (15) and one obtains a 1-coherent mapping $\boldsymbol{f}^{\,\alpha} = (f^\alpha_{m_0}, f^\alpha_{m_0 m_1})$. Note that (15) holds because, for $m_0 < m_1 < m_2$, one has

$$\begin{aligned} &\deg(f^\alpha_{m_0 m_1} * q_{m_0 m_1} f^\alpha_{m_1 m_2}) = \\ &\deg f^\alpha_{m_0 m_1} + \deg q_{m_0 m_1} \deg f^\alpha_{m_1 m_2} = \deg f^\alpha_{m_0 m_2}. \end{aligned} \tag{26}$$

We will show that, whenever two 1-coherent mappings $\boldsymbol{f}^{\,\alpha}$, $\boldsymbol{f}^{\,\alpha'}$ are connected by a 1-coherent homotopy $\boldsymbol{F}$, then necessarily $\alpha \sim \alpha'$. Indeed, if $\boldsymbol{F} = (F_{m_0}, F_{m_0 m_1})$, then (18) implies that F_{m_0} and F_{m_1} are loops based at 1. Moreover, if we take into account (19)–(22) and identify $I \times \Delta^1 = I \times I$ with a disc D^2 in such a way that $(I \times 0) \cup (1 \times I)$ is mapped onto the lower half S^- of ∂D^2, while $(0 \times I) \cup (I \times 1)$ is mapped onto the upper half S^+ of ∂D^2, then $F_{m_0 m_1}$ can be viewed as a homotopy rel $\{0,1\}$, which connects the homotopy $f^{\alpha'}_{m_0 m_1} * F_{m_0}$ to $f^\alpha_{m_0 m_1} * (q_{m_0 m_1} F_{m_1})$. Consequently,

$$\deg f^{\alpha'}_{m_0 m_1} + \deg F_{m_0} = \deg f^\alpha_{m_0 m_1} + 3^{m_1 - m_0} \deg F_{m_1},\ m_0 < m_1. \tag{27}$$

Now note that $|\alpha_j - \alpha'_j| \le 1$ and therefore, (24) implies

$$\begin{aligned} &|\deg f^{\alpha'}_{m_0 m_1} - \deg f^\alpha_{m_0 m_1}| \\ \le\ &|\alpha'_{m_0} - \alpha_{m_0}| + 3|\alpha'_{m_0+1} - \alpha_{m_0+1}| + \ldots + 3^{m_1-m_0-1}|\alpha'_{m_1-1} - \alpha_{m_1-1}| \\ &\le 1 + 3 + \ldots + 3^{m_1-m_0-1} = \tfrac{1}{2}(3^{m_1-m_0} - 1). \end{aligned} \tag{28}$$

Consequently,

$$|\deg F_{m_0} + \deg f^{\alpha'}_{m_0 m_1} - \deg f^\alpha_{m_0 m_1}| \le |\deg F_{m_0}| + \frac{1}{2}(3^{m_1-m_0} - 1). \tag{29}$$

On the other hand, $\deg F_{m_1} \neq 0$ implies

$$3^{m_1-m_0}|\deg F_{m_1}| \ge 3^{m_1-m_0}. \tag{30}$$

By (27), (29) and (30), it is clear that, for a fixed m_0 and for sufficiently large $m_1 > m_0$, one must have $\deg F_{m_1} = 0$. In other words, the members of the sequence $(\deg F_1, \deg F_2, \ldots)$ vanish eventually. Assume that they vanish for all $m \geq m^*$. Then by (27),

$$\deg f^{\alpha}_{m_0 m_1} = \deg f^{\alpha'}_{m_0 m_1}, \tag{31}$$

whenever $m_1 > m_0 \geq m^*$. Using (24) and its analogue for $\deg f^{\alpha'}_{m_0 m_1}$, as well as the fact that $\alpha_{m_0}, \alpha'_{m_0} \in \{0, 1\}$, we conclude from (31) that $\alpha_{m_0} = \alpha'_{m_0}$, for all $m_0 \geq m^*$, i.e., $\alpha \sim \alpha'$. Since A contains $2^{\aleph_0}$ equivalence classes, we obtain $2^{\aleph_0}$ distinct elements $[\boldsymbol{f}^{\alpha}]$ of $\mathrm{CH}^{(1)}(\text{pro-}\mathsf{Top})(\{*\}, \boldsymbol{Y})$ by choosing $2^{\aleph_0}$ elements α of A, representing different classes.

REMARK 3.6. Notions and statements considered in 2.3 for pairs of spaces are readily transposed into notions and statements concerning the categories $\mathrm{CH}^{(r)}(\text{pro-}\mathsf{Top}_2)$, as well as the functors $E^{(rr')}$ and $E^{(r)}$.

3.2 Coherent homotopy of inverse sequences

The relationship between the categories CH(pro-Top) and $\mathrm{CH}^{(r)}$(pro-Top), for various $r \geq 0$, has not yet been sufficiently clarified. However, in the special case of *inverse sequences* (*towers*), i.e., inverse systems indexed by the set of integers $\mathbb{N} = \{1, 2, \ldots\}$, the restrictions of the categories CH(pro-Top) and $\mathrm{CH}^{(r)}$(pro-Top) coincide, for all $r \geq 1$. To state this result precisely, denote by CH(tow-Top) the full subcategory of CH(pro-Top), whose objects are inverse sequences and denote by $\mathrm{CH}^{(1)}$(tow-Top) the corresponding full subcategory of $\mathrm{CH}^{(1)}$(pro-Top). Denote the restriction of the functor $E^{(1)}$: CH(pro-Top) $\to$ $\mathrm{CH}^{(1)}$(pro-Top) to the corresponding subcategories also by $E^{(1)}$. Then the following theorem holds.

THEOREM 3.7. *The functor* $E^{(1)}$: CH(tow-Top) $\to$ $\mathrm{CH}^{(1)}$(tow-Top) *is an isomorphism of categories.*

In order to prove Theorem 3.7, it suffices to prove the following two lemmas.

LEMMA 3.8. *Let* $\boldsymbol{X}$ *and* $\boldsymbol{Y}$ *be inverse sequences and let* $\boldsymbol{h}: \boldsymbol{X} \to \boldsymbol{Y}$ *be a* 1*-coherent mapping. Then there exists a coherent mapping* $\boldsymbol{f}: \boldsymbol{X} \to \boldsymbol{Y}$ *such that* $E^{(1)}(\boldsymbol{f}) \simeq \boldsymbol{h}$.

LEMMA 3.9. *Let* $\boldsymbol{X}$ *and* $\boldsymbol{Y}$ *be inverse sequences and let* $\boldsymbol{f}, \boldsymbol{f}': \boldsymbol{X} \to \boldsymbol{Y}$ *be coherent mappings, such that* $\boldsymbol{h} = E^{(1)}(\boldsymbol{f})$ *and* $\boldsymbol{h}' = E^{(1)}(\boldsymbol{f}')$ *are connected by a* 1*-coherent homotopy* $\boldsymbol{H}$. *Then,* $\boldsymbol{f}$ *and* $\boldsymbol{f}'$ *are connected by a coherent homotopy.*

In the proofs of Lemmas 3.8 and 3.9, we will use the following simple lemma, concerning retractions of the standard simplex Δ^n, $n \geq 1$, to the polygonal line $L^n = [e_0, e_1] \cup [e_1, e_2] \cup \ldots \cup [e_{n-1}, e_n] \subseteq \Delta^n$.

LEMMA 3.10. *For every $n \geq 1$, there exists a retraction $\rho^n \colon \Delta^n \to L^n$ such that, for $n \geq 2$,*

$$\rho^n d_0 = d_0 \rho^{n-1}, \; \rho^n d_n = d_n \rho^{n-1}. \tag{1}$$

Proof of Lemma 3.10. Existence of the retractions ρ^n is proved by induction on n. If $n = 1$, we take for $\rho^1 \colon \Delta^1 \to L^1 = [e_0, e_1]$ the identity mapping, which is a retraction. Now assume that $n \geq 2$ and that ρ^{n-1} satisfies the induction hypothesis. Since $d_0 \colon \Delta^{n-1} \to [e_1, \ldots, e_n] \subseteq \Delta^n$ is a homeomorphism, every point $t \in [e_1, \ldots, e_n]$ admits a unique $t' \in \Delta^{n-1}$ such that $t = d_0(t')$. Putting

$$\rho^n(t) = d_0 \rho^{n-1}(t') \in d_0(L^{n-1}) \subseteq L^n, \tag{2}$$

for $t \in [e_1, \ldots, e_n]$, we insure the validity of the first of the equalities in (1). Moreover, if $t \in [e_1, \ldots, e_n] \cap L^n$, then $t' \in L^{n-1}$. Consequently, $\rho^{n-1}(t') = t'$ and thus, $\rho^n(t) = d_0(t') = t$. Similarly, since $d_n \colon \Delta^{n-1} \to [e_0, \ldots, e_{n-1}] \subseteq \Delta^n$ is a homeomorphism, every $t \in [e_0, \ldots, e_{n-1}]$ admits a unique $t'' \in \Delta^{n-1}$ such that $t = d_n(t'')$. Putting

$$\rho^n(t) = d_n \rho^{n-1}(t'') \in d_n(L^{n-1}) \subseteq L^n, \tag{3}$$

for $t \in [e_0 \ldots e_{n-1}]$, we insure the validity of the second of the equalities in (1). Moreover, $t \in [e_0, \ldots, e_{n-1}] \cap L^n$ implies $t'' \in L^{n-1}$ and thus, $\rho^n(t) = d_n \rho^{n-1}(t'') = t$.

Let us verify that (2) and (3) yield the same values, for $t \in [e_1, \ldots, e_n] \cap [e_0 \ldots e_{n-1}] = [e_1 \ldots e_{n-1}]$ and thus,

$$\rho^n \colon [e_1, \ldots, e_n] \cup [e_0, \ldots, e_{n-1}] \to L^n \tag{4}$$

is a well-defined mapping. Indeed, the affine mapping

$$u = d_0 d_{n-1} \colon \Delta^{n-2} \to [e_1, \ldots, e_{n-1}] \subseteq \Delta^n \tag{5}$$

maps the vertex e_i, to the vertex e_{i+1}, $0 \leq i \leq n-2$, and is therefore, a homeomorphism. Consequently, there is a unique point $t^* \in \Delta^{n-2}$ such that $t = u(t^*)$. Putting $t' = d_{n-1}(t^*) \in \Delta^{n-1}$, we see that $d_0(t') = u(t^*) = t$, and thus, (2) yields the value

$$\rho^n(t) = d_0 \rho^{n-1}(t') = d_0 \rho^{n-1} d_{n-1}(t^*) = d_0 d_{n-1} \rho^{n-2}(t^*) = u \rho^{n-2}(t^*). \tag{6}$$

However, by (1.2.16),

$$d_n d_0 = d_0 d_{n-1} = u. \tag{7}$$

Therefore, the point $t'' = d_0(t^*) \in \Delta^{n-1}$ has the property that $d_n(t'') = d_n d_0(t^*) = u(t^*) = t$. Consequently, (3) yields the same value

$$\rho^n(t) = d_n\rho^{n-1}(t'') = d_n\rho^{n-1}d_0(t^*) = d_nd_0\rho^{n-2}(t^*) = u\rho^{n-2}(t^*). \tag{8}$$

To complete the proof of Lemma 3.10, it now suffices to extend ρ^n to all of $\Delta^n = [e_0, \dots, e_n]$, which is possible because L is an absolute retract. □

Proof of Lemma 3.8. Let $\boldsymbol{X} = (X_{m_0}, p_{m_0m_1}, \mathbb{N})$, $\boldsymbol{Y} = (Y_{m_0}, q_{m_0m_1}\mathbb{N})$ and $\boldsymbol{h} = (h, h_{m_0}, h_{m_0m_1})$. The coherent mapping $\boldsymbol{f}$, which we shall exhibit, will be given by $f = h$ and by mappings $f_{\boldsymbol{m}}: X_{h(m_n)} \times \Delta^n \to Y_{m_0}$, for multiindices $\boldsymbol{m} = (m_0, \dots, m_n)$ of all lengths $n \geq 0$. By Lemma 1.13, it suffices to consider non-degenerate multiindices. For $n = 0$, put $f_{m_0} = h_{m_0}$. In order to define $f_{\boldsymbol{m}}$, for $n \geq 1$, first put $l(\boldsymbol{m}) = m_n - m_0$ and consider the mapping $g_{\boldsymbol{m}}: X_{h(m_n)} \times L^{l(\boldsymbol{m})} \to Y_{m_0}$, which on $X_{h(m_n)} \times [e_{i-1}, e_i]$, is given by h_{m_0+i-1,m_0+i}, $1 \leq i \leq l(\boldsymbol{m})$. More precisely, denote by $w_i^l: \Delta^1 \to \Delta^l$, $1 \leq i \leq l$, the simplicial mapping, given by $w_i^l(e_0) = e_{i-1}$, $w_i^l(e_1) = e_i$. For $t \in [e_{i-1}, e_i]$, there is a unique $t' \in \Delta^1$ such that $w_i^{l(\boldsymbol{m})}(t') = t$. Then put

$$g_{\boldsymbol{m}}(x,t) = q_{m_0,m_0+i-1}h_{m_0+i-1,m_0+i}(p_{h(m_0+i)h(m_n)}(x), t'). \tag{9}$$

The mapping $g_{\boldsymbol{m}}$ is well defined. Indeed, if $t \in [e_{i-1}, e_i] \cap [e_i, e_{i+1}]$, then $t = e_i$. Therefore, $w_i^{l(\boldsymbol{m})}(t') = e_i$ implies $t' = e_1 = d_0(e_0)$. Taking into account the coherence conditions for $\boldsymbol{h}$, we see that (9) assumes the value

$$h_{m_0+i}(p_{h(m_0+i)h(m_n)}(x), e_0). \tag{10}$$

On the other hand, $w_{i+1}^{l(\boldsymbol{m})}(t'') = e_i$ implies $t'' = e_0 = d_1^1(e_0)$ and (9) (for $i+1$) yields again the value (10).

Next consider the simplicial mapping $v_{\boldsymbol{m}}: \Delta^n \to \Delta^{l(\boldsymbol{m})}$, given by

$$v_{\boldsymbol{m}}(e_i) = e_{m_i - m_0}, \quad 0 \leq i \leq n. \tag{11}$$

Then define the mapping $f_{\boldsymbol{m}}: X_{h(m_n)} \times \Delta^n \to Y_{m_0}$, $n \geq 1$, by putting

$$f_{\boldsymbol{m}}(x,t) = g_{\boldsymbol{m}}(x, \rho^{l(\boldsymbol{m})}v_{\boldsymbol{m}}(t)). \tag{12}$$

We need to prove the following two assertions:

(*i*) The mappings $f_{\boldsymbol{m}}$ satisfy the coherence conditions and thus make $\boldsymbol{f} = (f, f_{\boldsymbol{m}})$ a coherent mapping;

(*ii*) $\boldsymbol{h}$ is homotopic to $E^{(1)}(\boldsymbol{f})$.

In order to prove (*i*), let us first show that

$$v_{\boldsymbol{m}}d_j^n = \begin{cases} d_0^{l(\boldsymbol{m})} \dots d_0^{l(d^0\boldsymbol{m})+1} v_{d^0\boldsymbol{m}}, & j = 0, \\ v_{d^j\boldsymbol{m}}, & 0 < j < n, \\ d_{l(\boldsymbol{m})}^{l(\boldsymbol{m})} \dots d_{l(d^n\boldsymbol{m})+1}^{l(d^n\boldsymbol{m})+1} v_{d^n\boldsymbol{m}}, & j = n. \end{cases} \tag{13}$$

Since all the mappings involved are simplicial, it suffices to verify the validity of (13) at the vertices $e_0, \dots, e_{n-1}$ od Δ^{n-1}. For $j = 0$, we have $d_0(e_i) = e_{i+1}$ and therefore, $v_{\boldsymbol{m}}d_0(e_i) = e_{m_{i+1}-m_0}$. On the other hand, $d^0\boldsymbol{m} = (m_1, \dots, m_n)$ and thus, $v_{d^0\boldsymbol{m}}(e_i) = e_{m_{i+1}-m_1}$. Since $d_0 \dots d_0$ shifts indices of

e_k by $l(\boldsymbol{m}) - l(d^0\boldsymbol{m}) = m_1 - m_0$ units, it maps $e_{m_{i+1}-m_1}$ also into $e_{m_{i+1}-m_0}$. Now assume that $0 < j < n$. Then $d_j(e_i) = e_i$, for $i < j$ and $d_j(e_i) = e_{i+1}$, for $j \le i$. Therefore, $v_{\boldsymbol{m}} d_j(e_i) = e_{m_i - m_0}$, for $i < j$ and $v_{\boldsymbol{m}} d_j(e_i) = e_{m_{i+1}-m_0}$, for $j \le i$. Since $d^j \boldsymbol{m} = (m_0, \ldots, \widehat{m_j}, \ldots, m_n)$, $v_{d^j \boldsymbol{m}}(e_i)$ assume the same values. Finally, for $j = n$, $v_{\boldsymbol{m}} d_n(e_i) = e_{m_i - m_0}$ and $v_{d^n \boldsymbol{m}}(e_i)$ assumes the same value. Since $d_{l(\boldsymbol{m})} \ldots d_{l(d^n \boldsymbol{m})+1}$ is an inclusion, (12) holds in this case too.

In the proof of (i) we also need the following facts.

$$q_{m_0 m_1} g_{d^0 \boldsymbol{m}}(x,t) = g_{\boldsymbol{m}}(x, d_0^{l(\boldsymbol{m})} \ldots d_0^{l(d^0 \boldsymbol{m})+1} t), \tag{14}$$

$$g_{d^j \boldsymbol{m}}(x,t) = g_{\boldsymbol{m}}(x,t),\ 0 < j < n, \tag{15}$$

$$g_{d^n \boldsymbol{m}}(p(x),t) = g_{\boldsymbol{m}}(x, d_{l(\boldsymbol{m})}^{l(\boldsymbol{m})} \ldots d_{l(d^n \boldsymbol{m})+1}^{l(d^n \boldsymbol{m})+1} t). \tag{16}$$

For $t \in [e_{i-1}, e_i]$, $1 \le i \le l(d^0\boldsymbol{m})$, let $t' \in \Delta^1$ be such that $w_i^{l(d^0\boldsymbol{m})}(t') = t$. Since $l(d^0\boldsymbol{m}) + (m_1 - m_0) = l(\boldsymbol{m})$ and $d_0^{l(\boldsymbol{m})} \ldots d_0^{l(d^0 \boldsymbol{m})+1}$ shifts the indices of e_k by $(m_1 - m_0)$ units, we see that $w_{i+m_1-m_0}^{l(\boldsymbol{m})}(t') = d_0^{l(\boldsymbol{m})} \ldots d_0^{l(d^0\boldsymbol{m})+1}(t)$. Therefore,

$$g_{d^0 \boldsymbol{m}}(x,t) = q_{m_1, m_1+i-1} h_{m_1+i-1, m_1+i}(p_{h(m_1+i)h(m_n)}(x), t'). \tag{17}$$

$$g_{\boldsymbol{m}}(x, d_0 \ldots d_0 t) = q_{m_0, m_1+i-1} h_{m_1+i-1, m_1+i}(p_{h(m_1+i)h(m_n)}(x), t'). \tag{18}$$

Comparison of (17) and (18) establishes (14).

The validity of (15) is an immediate consequence of the fact that $l(d^0\boldsymbol{m}) = l(\boldsymbol{m})$. To prove (16), for $t \in [e_{i-1}, e_i]$, $1 \le i \le l(d^n \boldsymbol{m})$, choose $t' \in \Delta^1$ so that $w_i^{l(d^n \boldsymbol{m})}(t') = t$. Note that $d_{l(\boldsymbol{m})}^{l(\boldsymbol{m})} \ldots d_{l(d^n\boldsymbol{m})+1}^{l(d^n \boldsymbol{m})+1}$ is an inclusion. Therefore, $w_i^{l(\boldsymbol{m})}(t') = d_{l(\boldsymbol{m})} \ldots d_{l(d^n \boldsymbol{m})+1} t$ and thus,

$$g_{d^n \boldsymbol{m}}(p_{h(m_{n-1})h(m_n)}(x), t) = q_{m_0, m_0+i-1} h_{m_0+i-1, m_0+i}(p(x), t'). \tag{19}$$

$$g_{\boldsymbol{m}}(x, d_{l(\boldsymbol{m})} \ldots d_{l(d^n \boldsymbol{m})+1} t) = q_{m_0, m_0+i-1} h_{m_0+i-1, m_0+i}(p(x), t'). \tag{20}$$

Comparison of (19) and (20) establishes (16).

Using (13)–(16) and (1), it is easy to verify the desired coherence conditions for $f_{\boldsymbol{m}}$. Indeed, for $j = 0$, we have

$$\begin{aligned} &f_{\boldsymbol{m}}(x, d_0 t) = g_{\boldsymbol{m}}(x, \rho^{l(\boldsymbol{m})} v_{\boldsymbol{m}}(d_0 t)) = \\ &g_{\boldsymbol{m}}(x, \rho^{l(\boldsymbol{m})} d_0 \ldots d_0 v_{d^0 \boldsymbol{m}}(t)) = g_{\boldsymbol{m}}(x, d_0 \ldots d_0 \rho^{l(d^0 \boldsymbol{m})} v_{d^0 \boldsymbol{m}}(t)) = \\ &q_{m_0 m_1} g_{d^0 \boldsymbol{m}}(x, \rho^{l(d^0 \boldsymbol{m})} v_{d^0 \boldsymbol{m}}(t)) = q_{m_0 m_1} f_{d^0 \boldsymbol{m}}(x,t). \end{aligned} \tag{21}$$

Analogous arguments apply to the remaining cases.

In order to prove (ii), we must exhibit a 1-coherent homotopy $\boldsymbol{F}$, which connects $\boldsymbol{h}$ to $\boldsymbol{h}' = E^{(1)}(\boldsymbol{f})$. By Lemma 3.4, it suffices to define $F, F_{m_0}, F_{m_0 m_1}$ so that (3.1.18)–(3.1.23) (for $\boldsymbol{h}, \boldsymbol{h}'$) hold. First put $F = h$ and define F_{m_0}, by $F_{m_0}(x,s) = h_{m_0}(x)$. Since, $\boldsymbol{h}' = (f, f_{m_0}, f_{m_0 m_1})$, where $f = h$, $f_{m_0} = h_{m_0}$, condition (3.1.18) is satisfied. We next define $F_{\boldsymbol{m}}$, for $\boldsymbol{m} = (m_0, m_1)$. If $l(\boldsymbol{m}) = 0$, i.e., $\boldsymbol{m} = (m_0, m_0)$, we put $F_{m_0 m_0}(x,s,t) = h_{m_0}(x)$, so that

(3.1.23) holds. Now assume that $l(\boldsymbol{m}) = 1$, i.e., $\boldsymbol{m} = (m_0, m_0 + 1)$. In this case $w_1^{l(\boldsymbol{m})} = \mathrm{id}$ and therefore, by (9), $g_{m_0,m_0+1} = h_{m_0,m_0+1}$. Since also $v_{\boldsymbol{m}} = \mathrm{id}$ and $\rho^1 = \mathrm{id}$, (12) shows that

$$f_{m_0,m_0+1} = h_{m_0,m_0+1}. \tag{22}$$

Putting $F_{m_0,m_0+1}(x, s, t) = h_{m_0,m_0+1}(x, t)$, we see that (3.1.19)–(3.1.23) hold, for $l(\boldsymbol{m}) = 1$.

Now assume that $l(\boldsymbol{m}) = 2$, i.e., $\boldsymbol{m} = (m_0, m_0 + 2)$. Then, by (3.1.15) and (22),

$$\begin{gathered} f_{m_0,m_0+2} \simeq f_{m_0,m_0+1}(p \times 1) * q f_{m_0+1,m_0+2} = \\ h_{m_0,m_0+1}(p \times 1) * q h_{m_0+1,m_0+2} \simeq h_{m_0,m_0+2} \, (\mathrm{rel}\,\{0,1\}). \end{gathered} \tag{23}$$

However, this relation yields a homotopy $F_{m_0m_0+2}\colon X_{f(m_0+2)} \times I \times \Delta^1$, which satisfies (3.1.19)–(3.1.23).

An analogous argument leads to the definition of $F_{m_0m_1}$, when $l(\boldsymbol{m}) \geq 3$. We first define the juxtaposition of l homotopies $\Phi_i, i = 1, \dots, l$, by putting $\Phi_1 * \dots * \Phi_l = (\Phi_1 * \dots * \Phi_{l-1}) * \Phi_l$. Clearly, juxtaposition is associative up to homotopy rel $\{0, 1\}$ and formula (3.1.15) generalizes to l terms. We thus obtain

$$\begin{gathered} f_{m_0,m_0+l(\boldsymbol{m})} \simeq \\ f_{m_0,m_0+1}(p \times 1) * q f_{m_0+1,m_0+2}(p \times 1) * \dots * q f_{m_0+l(\boldsymbol{m})-1,m_0+l(\boldsymbol{m})} = \\ h_{m_0,m_0+1}(p \times 1) * q h_{m_0+1,m_0+2}(p \times 1) * \dots * q h_{m_0+l(\boldsymbol{m})-1,m_0+l(\boldsymbol{m})} \\ \simeq h_{m_0,m_0+l(\boldsymbol{m})} \, (\mathrm{rel}\,\{0,1\}). \end{gathered} \tag{24}$$

However, this relation yields a homotopy $F_{m_0,m_0+l(\boldsymbol{m})}\colon X_{f(m_0+l(\boldsymbol{m}))} \times I \times \Delta^1 \to Y_{m_0}$, which satisfies (3.1.19)–(3.1.23) and thus, establishes (*ii*). □

Proof of Lemma 3.9. In the proof of Lemma 3.8, we associated with every 1-coherent mapping $\boldsymbol{h}$ a coherent mapping $\boldsymbol{f}$. Applying the same construction to $\boldsymbol{H} = (H, H_{m_0}, H_{m_0m_1})$, we obtain a coherent homotopy $\boldsymbol{F} = (F, F_{\boldsymbol{m}})\colon \boldsymbol{X} \times I \to \boldsymbol{Y}$. Note that $F = H$, $F_{m_0} = H_{m_0}$ and for multiindices of length $n \geq 1$, $F_{\boldsymbol{m}}$ is defined by (see (12))

$$F_{\boldsymbol{m}}(x, s, t) = G_{\boldsymbol{m}}(x, s, \rho^{l(\boldsymbol{m})} v_{\boldsymbol{m}}(t)), \tag{25}$$

where $G_{\boldsymbol{m}}\colon X_{H(m_n)} \times I \times L^{l('boldm)} \to Y_{m_0}$ is given by (see (9))

$$G_{\boldsymbol{m}}(x, s, t) = q_{m_0,m_0+i-1} H_{m_0+i-1,m_0+i}(p_{H(m_0+i)H(m_n)}(x), s, t'). \tag{26}$$

Here $w_i^{l(\boldsymbol{m})}(t') = t \in [e_{i-1}, e_i] \subseteq L^{l(\boldsymbol{m})}$.

At the levels $s = 0, 1$, $\boldsymbol{F}$, reduces to coherent mappings, which we denote by $\boldsymbol{f}^*$ and $\boldsymbol{f}'^*$ respectively. Since $\boldsymbol{f}^* \simeq \boldsymbol{f}'^*$, the proof of Lemma 3.9 will be completed if we show that $\boldsymbol{f} \simeq \boldsymbol{f}^*$ and $\boldsymbol{f}' \simeq \boldsymbol{f}'^*$.

The mapping $\boldsymbol{f}^*$ is given by $F = H$ and by mappings $f^*_{\boldsymbol{m}}$, where

$$f^*_{\boldsymbol{m}}(x,t) = F_{\boldsymbol{m}}(x,0,t). \tag{27}$$

Since

$$H_{m_0m_1}(x,0,t) = h_{m_0m_1}(p_{h(m_1)H(m_1)}(x),t), \tag{28}$$

we see that (26) and (9) imply

$$G_{\boldsymbol{m}}(x,0,t) = g_{\boldsymbol{m}}(p_{h(m_n)H(m_n)}(x),t). \tag{29}$$

Now (27), (25) and (29) imply

$$f^*_{\boldsymbol{m}}(x,t) = g_{\boldsymbol{m}}(p_{h(m_n)H(m_n)}(x),\rho^{l(\boldsymbol{m})}v_{\boldsymbol{m}}(t)). \tag{30}$$

We will now define a coherent homotopy $\boldsymbol{F}^* = (F^*, F^*_{\boldsymbol{m}})$, which connects $\boldsymbol{f}$ to $\boldsymbol{f}^*$. Put $F^* = H$ and

$$F^*_{m_0}(x,s) = f_{m_0}p_{f(m_0)H(m_0)}(x). \tag{31}$$

In order to define $F^*_{\boldsymbol{m}}$, for $n \geq 1$, consider the linear homotopy $P^n\colon I \times \Delta^n \to \Delta^n$, given by

$$P^n(s,t) = (1-s)t + s\rho^n(t). \tag{32}$$

Note that (1) implies

$$P^n(1 \times d_0) = d_0P^{n-1},\ \ P^n(1 \times d_n) = d_nP^{n-1}. \tag{33}$$

Now put

$$F^*_{\boldsymbol{m}}(x,s,t) = f_{\boldsymbol{m}^*}(p_{f(m_n)H(m_n)}(x), P^{l(\boldsymbol{m})}(s,v_{\boldsymbol{m}}(t))), \tag{34}$$

where $\boldsymbol{m}^* = (m_0, m_0+1, \dots, m_n)$. Note that the length of $\boldsymbol{m}^*$ equals $l(\boldsymbol{m})$. Moreover, the following equalities are readily verified.

$$(d_j\boldsymbol{m})^* = \begin{cases} d_0^{l(d^0\boldsymbol{m})+1}\dots d_0^{l(\boldsymbol{m})}\boldsymbol{m}^*, & j=0,\\ \boldsymbol{m}^*, & 0<j<n,\\ d_{l(d^n\boldsymbol{m})+1}^{l(d^n\boldsymbol{m})+1}\dots d_{l(\boldsymbol{m})}^{l(\boldsymbol{m})}\boldsymbol{m}^*, & j=n. \end{cases} \tag{35}$$

The following formulae are an immediate consequence of (35) and of the coherence conditions for $\boldsymbol{f}$.

$$q_{m_0m_1}f_{(d_0\boldsymbol{m})^*}(x,t) = f_{\boldsymbol{m}^*}(x, d_0^{l(\boldsymbol{m})}\dots d_0^{l(d^0\boldsymbol{m})+1}t), \tag{36}$$

$$f_{(d^j\boldsymbol{m})^*}(x,t) = f_{\boldsymbol{m}^*}(x,t),\ 0<j<n, \tag{37}$$

$$f_{(d^n\boldsymbol{m})^*}(p_{h(m_{n-1})h(m_n)}(x),t) = f_{\boldsymbol{m}^*}(x, d_{l(\boldsymbol{m})}^{l(\boldsymbol{m})}\dots d_{l(d^n\boldsymbol{m})+1}^{l(d^n\boldsymbol{m})+1}t). \tag{38}$$

The coherence conditions for $\boldsymbol{F}^*$ are an easy consequence of (13), (33) and (36)–(38). Indeed, for $j=0$, we have

$$\begin{aligned} &F^*_{\boldsymbol{m}}(x,s,d_0t) = f_{\boldsymbol{m}^*}(p(x), P^{l(\boldsymbol{m})}(s, d_0\dots d_0v_{d^0\boldsymbol{m}}(t)) =\\ &f_{\boldsymbol{m}^*}(p(x), d_0\dots d_0P^{l(d^0\boldsymbol{m})}(s, v_{d^0\boldsymbol{m}}(t))) =\\ &qf_{d^0\boldsymbol{m}}(p(x), P^{l(d^0\boldsymbol{m})}(s, v_{d^0\boldsymbol{m}}(t))) = qF^*_{d^0\boldsymbol{m}}(p(x),s,t). \end{aligned} \tag{39}$$

Analogous arguments apply to the remaining cases.

Let us now show that

$$F^*_{\boldsymbol{m}}(x,0,t) = f_{\boldsymbol{m}}(p_{f(m_n)H(m_n)}(x),t), \tag{40}$$

$$F^*_{\boldsymbol{m}}(x,1,t) = f^*_{\boldsymbol{m}}(x,t). \tag{41}$$

By (34) and (32),

$$F^*_{\boldsymbol{m}}(x,0,t) = f_{\boldsymbol{m}^*}(p_{f(m_n)H(m_n)}(x), v_{\boldsymbol{m}}(t)). \tag{42}$$

Note that $\boldsymbol{m}$ is obtained from $\boldsymbol{m}^*$ by deleting the terms $m_0+1,\dots,m_1-1,\dots,m_{n-1}+1,\dots,m_n-1$, i.e.,

$$\boldsymbol{m} = (d_{m_0+1}\dots d_{m_1-1})\dots(d_{m_{n-1}+1}\dots d_{m_n-1})\boldsymbol{m}^*. \tag{43}$$

On the other hand, it is readily seen that

$$(d_{m_n-1}\dots d_{m_{n-1}+1})\dots(d_{m_1-1}\dots d_{m_0+1})t = v_{\boldsymbol{m}}(t). \tag{44}$$

Therefore, the coherence conditions for $\boldsymbol{f}$ show that

$$f_{\boldsymbol{m}}(p(x),t) = f_{\boldsymbol{m}^*}(p(x), v_{\boldsymbol{m}}(t)). \tag{45}$$

Now (42) and (45) establish (40).

To prove (41), note that (34) yields

$$F^*_{\boldsymbol{m}}(x,1,t) = f_{\boldsymbol{m}^*}(p_{f(m_n)H(m_n)}(x), \rho^{l(\boldsymbol{m})}v_{\boldsymbol{m}}(t)). \tag{46}$$

Since $\rho^{l(\boldsymbol{m})}v_{\boldsymbol{m}}(t) \in L^{l(\boldsymbol{m})}$, there exist an index i, $1 \le i \le l(\boldsymbol{m})$, and a $t' \in \Delta^1$ such that $\rho^{l(\boldsymbol{m})}v_{\boldsymbol{m}}(t) = w_i^{l(\boldsymbol{m})}(t')$. Taking into account the fact that

$$w_i^{l(\boldsymbol{m})} = d^{\,l(\boldsymbol{m})}_{l(\boldsymbol{m})}\dots d^{\,i+1}_{i+1}d^i_0\dots d^2_0, \tag{47}$$

we see that the coherence conditions for $\boldsymbol{f}$ imply

$$\begin{aligned} &f_{\boldsymbol{m}^*}(p(x), \rho^{l(\boldsymbol{m})}v_{\boldsymbol{m}}(t)) = \\ &q_{m_0,m_0+i-1}f_{m_0+i-1,m_0+i}(p_{f(m_0+i)H(m_n)}(x),t'). \end{aligned} \tag{48}$$

However, since $E^{(1)}(\boldsymbol{f}) = \boldsymbol{h}$, one has $h = f$ and $h_{m_0m_1} = f_{m_0m_1}$. Therefore, (28) yields

$$f_{m_0+i-1,m_0+i}(p(x),t') = H_{m_0+i-1,m_0+i}(p_{H(m_0+i)H(m_n)}(x),0,t'). \tag{49}$$

Now, (46), (48), (49), (26), (25), (29) and (30) imply

$$\begin{aligned} &F^*_{\boldsymbol{m}}(x,1,t) = \\ &q_{m_0,m_0+i-1}H_{m_0+i-1,m_0+i}(p_{H(m_0+i)H(m_n)}(x),0,t') = \\ &G_{\boldsymbol{m}}(x,0,\rho^{l(\boldsymbol{m})}v_{\boldsymbol{m}}(t)) = f^*_{\boldsymbol{m}}(x,t), \end{aligned} \tag{50}$$

which completes the proof of (41). Formulae (40) and (41) show that $\boldsymbol{f} \simeq \boldsymbol{f}^*$. An analogous proof yields $\boldsymbol{f}' \simeq \boldsymbol{f}'^*$. □

REMARK 3.11. Similar arguments show that, for any $r \geq 1$, the restriction of the functor $E^{(1r)}$ is an ismorphism of categories $\mathrm{CH}^{(r)}(\text{tow-}\mathsf{Top}) \to \mathrm{CH}^{(1)}(\text{tow-}\mathsf{Top})$. Since $E^{(1r)}E^{(r)} = E^{(1)}$ and $E^{(1r)}E^{(rr')} = E^{(1r')}$, $r \leq r'$, it follows that the restrictions of $E^{(r)}$ and $E^{(rr')}$ are isomorphisms of categories $\mathrm{CH}(\text{tow-}\mathsf{Top}) \to \mathrm{CH}^{(r)}(\text{tow-}\mathsf{Top})$ and $\mathrm{CH}^{(r')}(\text{tow-}\mathsf{Top}) \to \mathrm{CH}^{(r)}(\text{tow-}\mathsf{Top})$, respectively.

Bibliographic notes

The coherent homotopy category for inverse sequences $\mathrm{CH}^{(1)}(\text{tow-}\mathsf{Top})$ was first defined in (Lisitsa 1977) with the purpose of obtaining an alternate description of Bauer's strong shape (for metric compacta) (Bauer 1976). Lisitsa's definition was suggested by Sitnikov's description of the Steenrod homology groups (Sitnikov 1951, 1954). The proof that the category $\mathrm{CH}^{(1)}(\text{tow-}\mathsf{Top})$ is isomorhic to the coherent homotopy category $\mathrm{CH}(\text{tow-}\mathsf{Top})$ (Theorem 3.7) is from (Lisica, Mardešić 1985a). The papers (Miminoshvili 1982) and (Lisitsa 1983a) contain variants of $\mathrm{CH}^{(1)}(\text{pro-}\mathsf{Top})$ for the more general setting of 1-coherent inverse systems.

4. Coherent homotopy and localization

It is well known that localization of the category Top at homotopy equivalences yields the homotopy category H(Top) (see Theorem 4.35). The analogous procedure for inverse systems, indexed by Λ, is the localization of the category Top^{Λ} at level homotopy equivalences. The resulting category will be denoted by $\mathrm{Ho}(\mathsf{Top}^{\Lambda})$. One also considers the corresponding localization of pro-Top and denotes it by Ho(pro-Top). The main results of this section show that the localized categories, obtained in this manner, are isomorphic to the corresponding coherent homotopy categories $\mathrm{CH}(\mathsf{Top}^{\Lambda})$ and CH(pro-Top), respectively. Therefore, the coherent homotopy categories can be viewed as concrete realizations of the rather abstract localized categories. The first subsection is devoted to an isomorphism theorem in coherent homotopy, which enables one to obtain functors between the categories in question. The second subsection defines cotelescopes (homotopy limits), a tool needed to prove that these functors are isomorphisms of categories.

4.1 An isomorphism theorem in CH(pro-Top)

A mapping $\boldsymbol{f}\colon \boldsymbol{X} \to \boldsymbol{Y}$, i.e., a morphism of inv-Top, is called a *level homotopy equivalence*, provided it is a level mapping $\boldsymbol{f} = (f_\lambda)$ and every $f_\lambda\colon X_\lambda \to Y_\lambda$, $\lambda \in \Lambda$, is a homotopy equivalence. A morphism of pro-Top, i.e., a congruence class of mappings $[\boldsymbol{f}]\colon \boldsymbol{X} \to \boldsymbol{Y}$ is called a *level homotopy equivalence*, provided it has a representative $\boldsymbol{f}$, which is a level homotopy equivalence.

As an introduction to the main result of this subsection, we first prove an elementary theorem on prohomotopy.

THEOREM 4.1. *If $\boldsymbol{f}\colon \boldsymbol{X} \to \boldsymbol{Y}$ is a level homotopy equivalence of inverse systems of spaces, then the homotopy functor H:Top $\to$ H(Top) induces isomorphisms $[\boldsymbol{f}]\colon [\boldsymbol{X}] \to [\boldsymbol{Y}]$ in the categories* $(\mathrm{H}(\mathsf{Top}))^{\Lambda}$ *and* pro-H(Top).

Proof. Let $\boldsymbol{f} = (f_\lambda)$ and let $g_\lambda\colon Y_\lambda \to X_\lambda$ be a homotopy inverse of f_λ, $\lambda \in \Lambda$. Then the homotopy classes $[g_\lambda]$ form a morphism of $(\mathrm{H}(\mathsf{Top}))^{\Lambda}$ and determine a morphism of pro-H(Top). Indeed, for $\lambda \leq \lambda'$, one has

$$p_{\lambda\lambda'}g_{\lambda'} \simeq g_\lambda f_\lambda p_{\lambda\lambda'}g_{\lambda'} = g_\lambda q_{\lambda\lambda'}f_{\lambda'}g_{\lambda'} \simeq g_\lambda q_{\lambda\lambda'} \tag{1}$$

and thus, $[p_{\lambda\lambda'}][g_{\lambda'}] = [g_\lambda][q_{\lambda\lambda'}]$. Moreover, $g_\lambda f_\lambda \simeq \mathrm{id}$ and $f_\lambda g_\lambda \simeq \mathrm{id}$ imply $[g_\lambda][f_\lambda] = \mathrm{id}$ and $[f_\lambda][g_\lambda] = \mathrm{id}$ and thus, $[\boldsymbol{g}][\boldsymbol{f}] = \mathrm{id}$ and $[\boldsymbol{f}][\boldsymbol{g}] = \mathrm{id}$ in both categories. □

REMARK 4.2. The assumptions of Theorem 4.1 do not imply the existence of homotopy inverses g_λ of f_λ, which form a morphism of Top^Λ. More generally, the inverses of $[\boldsymbol{f}]: [\boldsymbol{X}] \to [\boldsymbol{Y}]$ in $(\mathrm{H}(\mathsf{Top}))^\Lambda$ and pro-$\mathrm{H}(\mathsf{Top})$ need not be induced by morphisms of Top^Λ and pro-Top, respectively. This is demonstrated by examples which follow.

EXAMPLE 4.3. Let $\boldsymbol{X}$ be the inverse sequence $(X_m, p_{mm'}, \mathbb{N})$, where $X_m = [m, \infty) \subseteq \mathbb{R}$ and $p_{mm'}$ are inclusions. Let $\boldsymbol{Y} = (Y_m, q_{mm'}, \mathbb{N})$ be the inverse sequence, where $Y_m = \{*\}$ is a single point, for all $m \in \mathbb{N}$. Clearly, the constant mappings $f_m: X_m \to Y_m$ form a level homotopy equivalence $\boldsymbol{f} = (f_m): \boldsymbol{X} \to \boldsymbol{Y}$. Nevertheless, there exists no mapping $\boldsymbol{g}: \boldsymbol{Y} \to \boldsymbol{X}$ whatsoever. Indeed, assume that $\boldsymbol{g} = (g, g_m)$ is such a mapping. Choose an integer $n \in \mathbb{N}$ such that $g_1(*) < n$. Note that $g_1(*) = g_1 q_{g(1)g(n)}(*) = p_{1n} g_n(*) = g_n(*) \in X_n = [n, \infty)$. Consequently, $g_1(*) \geq n$, which is a contradiction.

EXAMPLE 4.4. A popular example in shape theory is the *Warsaw circle* Σ (see Figure 4.1). To describe it one considers the graph G of the function $x \mapsto \sin 1/x$, $0 < x < 1/\pi$, and its closure $\overline{G}$ in the Euclidean plane $\mathbb{R}^2$. By definition, $\Sigma = \overline{G} \cup A$, where A is an arc such that $\overline{G} \cap A = \partial A = \{(0, -1), (1/\pi, 0)\}$.

Fig. 4.1. The Warsaw circle

The Warsaw circle can easily be represented as the limit X of an inverse sequence $\boldsymbol{X} = (X_m, p_{mm'})$ of 1-spheres $X_m = S^1$ with bonding mappings $p_{m,m+1}: X_{m+1} \to X_m$, which are homotopy equivalences and have degree 1. Let $\boldsymbol{Y} = (Y_m, q_{mm'})$ be the inverse sequence, given by $Y_m = S^1$ and $q_{mm'} = \mathrm{id}$. Clearly, the mappings $f_m: X_m \to Y_m$, given by $f_1 = \mathrm{id}$, $f_m = p_{12} \cdots p_{m-1,m}$, for $m \geq 2$, also are homotopy equivalences and they form a morphism $\boldsymbol{f} = (f_m): \boldsymbol{X} \to \boldsymbol{Y}$ of $\mathsf{Top}^{\mathbb{N}}$. Now assume that the class $[\boldsymbol{f}]: [\boldsymbol{X}] \to [\boldsymbol{Y}]$ has an inverse in pro-$\mathrm{H}(\mathsf{Top})$, given by a mapping $\boldsymbol{g} = (g, g_m): \boldsymbol{Y} \to \boldsymbol{X}$. Then $\boldsymbol{g}$ would induce the mapping $g = \lim \boldsymbol{g}: Y \to X$, where $Y = \lim \boldsymbol{Y} = S^1$, such that the induced shape morphism $S(g): Y \to X$ is the inverse of $S(f): X \to Y$, where $f = \lim \boldsymbol{f}$. Consequently, f would induce an isomorphism of Čech homology groups $f_*: \check{H}_1(X; \mathbb{Z}) \to \check{H}_1(Y; \mathbb{Z})$ with inverse g_*. However, $g_* = 0$, because every mapping $S^1 \to \Sigma$ is homotopically trivial. This is in contradiction with $g_* f_* = \mathrm{id}$ and $\check{H}_1(X; \mathbb{Z}) \approx \mathbb{Z} \neq 0$.

In 2.3 we defined the coherence functors C: $\mathsf{Top}^{\Lambda} \to \mathrm{CH}(\mathsf{Top}^{\Lambda})$ and C: pro-Top $\to$ CH(pro-Top). The main result of this subsection is the next theorem, which considerably strengthens Theorem 4.1 and alone suffices to justify introduction of the categories $\mathrm{CH}(\mathsf{Top}^{\Lambda})$ and CH(pro-Top).

THEOREM 4.5. *If* $\boldsymbol{f}: \boldsymbol{X} \to \boldsymbol{Y}$ *is a level homotopy equivalence from* Top^{Λ}, *then* $C(\boldsymbol{f})$ *is an isomorphism of* $\mathrm{CH}(\mathsf{Top}^{\Lambda})$.

COROLLARY 4.6. *If* $[\boldsymbol{f}]: \boldsymbol{X} \to \boldsymbol{Y}$ *is a level homotopy equivalence from* pro-Top, *then* $C[\boldsymbol{f}]$ *is an isomorphism of* CH(pro-Top).

Proof. By assumption $[\boldsymbol{f}]$ has a representative $\boldsymbol{f}$, which is a level homotopy equivalence. Therefore, by Theorem 4.5, $C(\boldsymbol{f})$ is an isomorphism of $\mathrm{CH}(\mathsf{Top}^{\Lambda})$. Since, $C[\boldsymbol{f}] = C(\boldsymbol{f})$, we conclude that $C[\boldsymbol{f}]$ is an isomorphism of CH(pro Top). □

The proof of Theorem 4.5 is non-trivial. An important ingredient in the proof is the following lemma on homotopy equivalences, due to R. Vogt (Vogt 1972).

LEMMA 4.7. *Let X, Y be arbitrary spaces and let $f: X \to Y$ be a homotopy equivalence with a homotopy inverse $g: Y \to X$. Let $H: X \times I \to X$ be a homotopy which connects 1_X with gf. Then there exists a homotopy $K: Y \times I \to Y$, which connects 1_Y with fg and is such that*

$$K(f \times 1) \simeq fH\,(\mathrm{rel}(X \times \partial I)), \tag{2}$$

$$H(g \times 1) \simeq gK\,(\mathrm{rel}(Y \times \partial I)). \tag{3}$$

In order to prove Lemma 4.7, we first observe that the relation $\simeq$ rel $(X \times \partial I)$ is an equivalence relation on the set of all homotopies $U: X \times I \to Y$, which connect two given mappings $p, q: X \to Y$. Indeed, the constant homotopy $H(x,s,t) = U(x,s)$ realizes $U \simeq U\,(\mathrm{rel}\,(X \times \partial I))$. If $H: X \times I \times I \to Y$ realizes $U \simeq U'\,(\mathrm{rel}\,(X \times \partial I))$, i.e., if

$$H(x,s,0) = U(x,s),\ \ H(x,s,1) = U'(x,s), \tag{4}$$

$$H(x,0,t) = p(x),\ \ H(x,1,t) = q(x), \tag{5}$$

then $H^-: X \times I \times I \to Y$, given by $H^-(x,s,t) = H(x,s,1-t)$, realizes $U' \simeq U\,(\mathrm{rel}\,(X \times \partial I))$. Finally, if $H': X \times I \times I \to Y$ realizes $U' \simeq U''\,(\mathrm{rel}\,(X \times \partial I))$, then the juxtaposition $K = H * H'$ (with respect to the variable t) realizes $U \simeq U''\,(\mathrm{rel}\,(X \times \partial I))$. We will denote the class of U by $[U]$ and the set of all classes $[U]$ by $[X,Y]_p^q$.

If homotopies $U, V: X \times I \to Y$ connect mappings p, q and q, r, respectively, then their juxtaposition $U * V: X \times I \to Y$ connects the mappings p, r. The class $[U * V] \in [X,Y]_p^r$ depends only on the classes $[U] \in [X,Y]_p^q$ and $[V] \in [X,Y]_q^r$. Indeed, if $H: X \times I \times I \to Y$, $K: X \times I \times I \to Y$ are homotopies rel $(X \times \partial I)$, which realize relations $U \simeq U'$ (rel $(X \times \partial I)$) and $V \simeq V'$ (rel $(X \times \partial I)$), respectively, then the juxtaposition $L = H * K$ (with respect to the variable s) realizes $U * V \simeq U' * V'$ (rel $(X \times \partial I)$). Therefore, one can define juxtaposition of classes by putting $[U] * [V] = [U * V]$. It is readily seen that $([U] * [V]) * [W] = [U] * ([V] * [W])$. Furthermore, $[U^-] * [U]$ and $[U] * [U^-]$ are classes of the constant homotopies $p: X \times I \to Y$ and $q: X \times I \to Y$, given by mappings p and q, respectively. Finally, $[p] * [U] = [U]$, $[U] * [q] = [U]$.

LEMMA 4.8. *Let $U: X \times I \to Y$, $V: Y \times I \to Z$ be homotopies which connect mappings $u, u': X \to Y$ and $v, v': Y \to Z$, respectively. Then*

$$[vU] * [V(u' \times 1)] = [V(u \times 1)] * [v'U]. \tag{6}$$

Proof. Consider the mapping $H: X \times I \times I \to Z$, given by

$$H(x, s, t) = V(U(x, t), s) \tag{7}$$

(see Fig. 4.2).

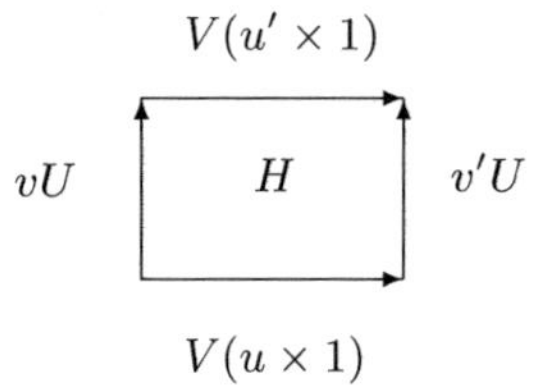

Fig. 4.2.

Note that

$$H(x, 0, t) = vU(x, t),\ H(x, 1, t) = v'U(x, t), \tag{8}$$

$$H(x, s, 0) = V(u(x), s),\ H(x, s, 1) = V(u'(x), s). \tag{9}$$

Let $\Phi: I \times I \to I \times I$ be a mapping, which sends $0 \times I$ to $(0, 0)$ and $1 \times I$ to $(1, 1)$ and which maps the segments $[0, 1/2] \times 0, [1/2, 1] \times 0, [0, 1/2] \times 1$ and $[1/2, 1] \times 1$ linearly onto $I \times 0, 1 \times I, 0 \times I$ and $I \times 1$, respectively. Then $H(1 \times \Phi): X \times I \times I \to Z$ is a homotopy rel $(X \times \partial I)$, which realizes (6). □

Every mapping $f: X \to Y$ induces a function $f^*: [Y, Z]_p^q \to [X, Z]_{pf}^{qf}$, where $f^*[V] = [V(f \times 1)]$. To see that f^* is well defined, consider a homotopy K, which realizes the relation $V \simeq V'$ (rel$(Y \times \partial I)$). Then $K(f \times 1 \times 1)$ realizes the relation $V(f \times 1) \simeq V'(f \times 1)$ (rel$(X \times \partial I)$). Similarly, f induces a function $f_*: [Z, X]_u^v \to [Z, Y]_{fu}^{fv}$, given by $f_*[W] = [fW]$.

LEMMA 4.9. *If $p, q: Y \to Z$ are mappings and $f: X \to Y$ is a homotopy equivalence, then $f^*: [Y, Z]_p^q \to [X, Z]_{pf}^{qf}$ is a bijection.*

Proof. Let $g: Y \to X$ be a homotopy inverse of $f: X \to Y$. We will first show that the composition $g^* f^*$ of $f^*: [Y, Z]_p^q \to [X, Z]_{pf}^{qf}$ and $g^*: [X, Z]_{pf}^{qf} \to [Y, Z]_{pfg}^{qfg}$ is a bijection and thus, f^* is an injection, while g^* is a surjection. Indeed, for $[W] \in [Y, Z]_{pfg}^{qfg}$, put

$$\phi[W] = [pF^-] * [W] * [qF], \tag{10}$$

where $F: Y \times I \to Y$ is a homotopy which realizes $fg \simeq 1_Y$. Since $pF^- = (pF)^-$, $qF^- = (qF)^-$, the above described properties of the operation $*$ imply that $\phi: [Y, Z]_{pfg}^{qfg} \to [Y, Z]_p^q$ is a bijection with inverse

$$\phi^{-1}[V] = [pF] * [V] * [qF^-]. \tag{11}$$

Consequently, to show that $g^* f^*$ is a bijection, it suffices to show that $\phi g^* f^*$ is a bijection. Actually, $\phi g^* f^* = \mathrm{id}$.

Indeed, for $[V] \in [Y, Z]_p^q$,

$$\phi g^* f^* [V] = [pF^-] * [V(fg \times 1)] * [qF]. \tag{12}$$

However, by Lemma 4.8, $[V(fg \times 1)] * [qF] = [pF] * [V(1_Y \times 1)] = [pF] * [V]$ and thus,

$$\phi g^* f^* [V] = [pF^-] * [pF] * [V] = [V]. \tag{13}$$

The same argument, applied to mappings $u, v: X \to Z$ and $g: Y \to X$ shows that $g^*: [X, Z]_u^v \to [Y, Z]_{ug}^{vg}$ is an injection and $f^*: [Y, Z]_{ug}^{vg} \to [X, Z]_{ugf}^{vgf}$ is a surjection. In particular, for $u = pf$ and $v = qf$, we conclude that $g^*: [X, Z]_{pf}^{qf} \to [Y, Z]_{pfg}^{qfg}$ is an injection. Since we already proved that this mapping is a surjection, one concludes that it is a bijection. It then follows that $f^*: [Y, Z]_p^q \to [X, Z]_{pf}^{qf}$ also is a bijection. □

Proof of Lemma 4.7. By Lemma 4.9, $f^*: [Y, Y]_1^{fg} \to [X, Y]_f^{fgf}$ is a bijection. Since $[fH] \in [X, Y]_f^{fgf}$, there exists a homotopy $K: Y \times I \to Y$, which connects 1_Y with fg and is such that $f^*[K] = [fH]$, i.e., (2) holds. To prove that also (3) holds, apply Lemma 4.9 to the mappings $1_X, gf: X \to X$ and the homotopy equivalence $g: Y \to X$. One concludes that $g^*: [X, X]_1^{gf} \to [Y, X]_g^{gfg}$ is a bijection. Since $[gK] \in [Y, X]_g^{gfg}$, there exists a homotopy $H': X \times I \to X$, which connects 1_X with gf and is such that $g^*[H'] = [gK]$, i.e.,

$$H'(g \times 1) \simeq gK \, (\mathrm{rel}(Y \times \partial I)). \tag{14}$$

Now consider the function $\psi: [Y, Y]_{fg}^{fgfg} \to [Y, Y]_1^{fg}$, given by

$$\psi[W] = [K] * [W] * [K^-(fg \times 1)]. \tag{15}$$

Observe that the composition $g_* \psi f_* = \mathrm{id}$, where $f_*: [Y, X]_g^{gfg} \to [Y, Y]_{fg}^{fgfg}$ and $g_*: [Y, Y]_1^{fg} \to [Y, X]_g^{gfg}$. Indeed,

$$\begin{aligned} g_*\psi f_*[U] &= [gK] * [gfU] * [gK^-(fg \times 1)] \\ &= [H'(g \times 1)] * [gfU] * [H'^-(gfg \times 1)], \end{aligned} \tag{16}$$

because of $g(U * V) = gU * gV$ and (14). However, by Lemma 4.8,

$$[H'(g \times 1)] * [gfU] = [U] * [H'(gfg \times 1)] \tag{17}$$

and thus, (16) yields

$$g_*\psi f_*[U] = [U]. \tag{18}$$

Now note that $[H(g \times 1)] \in [Y, X]_g^{gfg}$. Consequently, (18) implies

$$[H(g \times 1)] = g_*\psi f_*[H(g \times 1)] = [gK] * [gfH(g \times 1)] * [gK^-(fg \times 1)]. \tag{19}$$

However, by(2), $[fH] = [K(f \times 1)]$ and thus, $[gfH(g \times 1)] = [gK(fg \times 1)]$. Therefore, (19) becomes $[H(g \times 1)] = [gK]$, which is equivalent to (3). □

The following lemma is an essential part of the proof of Theorem 4.5.

LEMMA 4.10. *Let* $\boldsymbol{f} = (f_\lambda): \boldsymbol{X} \to \boldsymbol{Y}$ *be a level homotopy equivalence. Then there exist a level coherent mapping* $\boldsymbol{g} = (g_{\boldsymbol{\lambda}}): \boldsymbol{Y} \to \boldsymbol{X}$ *and a level coherent homotopy* $\boldsymbol{H} = (H_{\boldsymbol{\lambda}}): \boldsymbol{X} \times I \to \boldsymbol{X}$ *such that*

$$H_{\boldsymbol{\lambda}}(x, 0, t) = p_{\lambda_0\lambda_n}(x), \tag{20}$$

$$H_{\boldsymbol{\lambda}}(x, 1, t) = g_{\boldsymbol{\lambda}}(f_{\lambda_n}(x), t). \tag{21}$$

In the proof of this lemma (and later in the book) we will repeatedly use the following simple *homotopy extension property* (HEP).

(HEP) *Let X and Y be arbitrary topological spaces and let (P, Q) be a polyhedral pair. Then every mapping $f: X \times (P \times 0 \cup Q \times I) \to Y$ admits an extension $\tilde{f}: X \times P \times I \to Y$.*

Indeed, since the inclusion $i: Q \to P$ is a cofibration, the identity mapping on $P \times 0 \cup Q \times I$ admits a mapping $r: P \times I \to P \times 0 \cup Q \times I$, which is a retraction. Therefore, $\tilde{f} = f(1 \times r)$ is an extension of f.

Proof of Lemma 4.10. For every $\boldsymbol{\lambda} = (\lambda_0, \dots, \lambda_n)$, $n \geq 0$, we must exhibit mappings $g_{\boldsymbol{\lambda}}: Y_{\lambda_n} \times \Delta^n \to X_{\lambda_0}$, $H_{\boldsymbol{\lambda}}: X_{\lambda_n} \times I \times \Delta^n \to X_{\lambda_0}$, which satisfy (20), (21) and the corresponding coherence conditions. We will construct these mappings by induction on n. For $n = 0$, we choose as $g_{\lambda_0}: Y_{\lambda_0} \to X_{\lambda_0}$ an arbitrary homotopy inverse of $f_{\lambda_0}: X_{\lambda_0} \to Y_{\lambda_0}$ and we choose as $H_{\lambda_0}: X_{\lambda_0} \times I \to X_{\lambda_0}$ an arbitrary homotopy, which realizes the relation $1 \simeq g_{\lambda_0} f_{\lambda_0}$.

In order to proceed with the induction, we choose homotopies $K_{\lambda_0}: Y_{\lambda_0} \times I \to Y_{\lambda_0}$, according to Lemma 4.7, i.e., such that K_{λ_0} realizes the relation $1 \simeq f_{\lambda_0} g_{\lambda_0}$ and

$$K_{\lambda_0}(f_{\lambda_0} \times 1) \simeq f_{\lambda_0} H_{\lambda_0} \ (\mathrm{rel}(X_{\lambda_0} \times \partial I)). \tag{22}$$

Now assume that we have already defined $g_{\boldsymbol{\lambda}'}$ and $H_{\boldsymbol{\lambda}'}$, for $\boldsymbol{\lambda}'$ of length $n' < n$, $n \geq 1$, and that they satisfy all the required conditions. Let $\boldsymbol{\lambda}$ be a multiindex of length n. By Lemma 1.13, it suffices to consider non-degenerate multiindices. Using repeatedly the homotopy extension property (HEP), we will define certain mappings $\psi_{\boldsymbol{\lambda}}: Y_{\lambda_n} \times I \times \Delta^n \to X_{\lambda_0}$ and $\Phi_{\boldsymbol{\lambda}}: X_{\lambda_n} \times I \times I \times \Delta^n \to X_{\lambda_0}$. Then $g_{\boldsymbol{\lambda}}$ and $H_{\boldsymbol{\lambda}}$ will be defined by

$$g_{\boldsymbol{\lambda}}(y,t) = \psi_{\boldsymbol{\lambda}}(y,0,t), \tag{23}$$

$$H_{\boldsymbol{\lambda}}(x,s,t) = \Phi_{\boldsymbol{\lambda}}(x,0,s,t). \tag{24}$$

We first define $\Phi_{\boldsymbol{\lambda}}$ on $X_{\lambda_n} \times (I \times 0 \times \partial\Delta^n)$, by putting

$$\Phi_{\boldsymbol{\lambda}}(x,r,0,d_j t') = \begin{cases} p_{\lambda_0\lambda_1} H_{d^0\boldsymbol{\lambda}}(x,r,t'), & j=0, \\ H_{d^j\boldsymbol{\lambda}}(x,r,t'), & 0<j<n, \\ H_{d^n\boldsymbol{\lambda}}(p_{\lambda_{n-1}\lambda_n}(x),r,t'), & j=n, \end{cases} \tag{25}$$

where $t' \in \Delta^{n-1}$. The mapping is well defined, because $H_{d^j\boldsymbol{\lambda}}$ already satisfies the corresponding coherence conditions. We then put

$$\Phi_{\boldsymbol{\lambda}}(x,0,0,t) = p_{\lambda_0\lambda_n}(x),\ t \in \Delta^n. \tag{26}$$

Using (20) for $H_{d^j\boldsymbol{\lambda}}$, one verifies that $\Phi_{\boldsymbol{\lambda}}$ is well defined on

$$X_{\lambda_n} \times ((I \times 0 \times \partial\Delta^n) \cup (0 \times 0 \times \Delta^n)). \tag{27}$$

One now extends $\Phi_{\boldsymbol{\lambda}}$ to all of $X_{\lambda_n} \times I \times 0 \times \Delta^n$, using property (HEP).

Next, one considers the mapping $\theta_{\boldsymbol{\lambda}}: X_{\lambda_n} \times \Delta^n \to X_{\lambda_0}$, given by

$$\theta_{\boldsymbol{\lambda}}(x,t) = \Phi_{\boldsymbol{\lambda}}(x,1,0,t). \tag{28}$$

Using (25), (21) for $H_{d^j\boldsymbol{\lambda}}$ and the fact that $f_{\lambda_{n-1}} p_{\lambda_{n-1}\lambda_n} = q_{\lambda_{n-1}\lambda_n} f_{\lambda_n}$, one readily concludes that

$$\theta_{\boldsymbol{\lambda}}(x,d_j t') = \begin{cases} p_{\lambda_0\lambda_1} g_{d^0\boldsymbol{\lambda}}(f_{\lambda_n}(x),t'), & j=0, \\ g_{d^j\boldsymbol{\lambda}}(f_{\lambda_n}(x),t'), & 0<j<n, \\ g_{d^n\boldsymbol{\lambda}}(q_{\lambda_{n-1}\lambda_n} f_{\lambda_n}(x),t'), & j=n. \end{cases} \tag{29}$$

Now extend $\Phi_{\boldsymbol{\lambda}}$ to $X_{\lambda_n} \times 1 \times I \times \Delta^n$, by putting

$$\Phi_{\boldsymbol{\lambda}}(x,1,s,t) = \theta_{\boldsymbol{\lambda}}(H_{\lambda_n}(x,s),t). \tag{30}$$

Note that (30) yields $\Phi_{\boldsymbol{\lambda}}(x,1,0,t) = \theta_{\boldsymbol{\lambda}}(H_{\lambda_n}(x,0),t) = \theta_{\boldsymbol{\lambda}}(x,t)$, which agrees with the value obtained from (28). Also note that (29) implies

$$\Phi_{\boldsymbol{\lambda}}(x,1,s,d_j t') = \begin{cases} p_{\lambda_0\lambda_1} g_{d^0\boldsymbol{\lambda}}(f_{\lambda_n} H_{\lambda_n}(x,s),t'), & j=0, \\ g_{d^j\boldsymbol{\lambda}}(f_{\lambda_n} H_{\lambda_n}(x,s),t'), & 0<j<n, \\ g_{d^n\boldsymbol{\lambda}}(q_{\lambda_{n-1}\lambda_n} f_{\lambda_n} H_{\lambda_n}(x,s),t'), & j=n, \end{cases} \tag{31}$$

Moreover,

$$\Phi_{\boldsymbol{\lambda}}(x,1,1,t) = \theta_{\boldsymbol{\lambda}}(H_{\lambda_n}(x,1),t) = \theta_{\boldsymbol{\lambda}}(g_{\lambda_n} f_{\lambda_n}(x),t). \tag{32}$$

We will now extend $\Phi_{\boldsymbol{\lambda}}$ to $X_{\lambda_n} \times I \times I \times d_j(\Delta^{n-1})$, for each $0 \leq j \leq n$, and will thus, obtain an extension to $X_{\lambda_n} \times I \times I \times \partial\Delta^n$. Note that $\Phi_{\boldsymbol{\lambda}}$ is already defined on $X_{\lambda_0} \times (I \times 0 \cup 1 \times I) \times d_j(\Delta^{n-1})$, by (25) and (30). To extend it further, we decompose the square $I \times I$ in two triangles $T^- = \{(r,s) | r+s \leq 1\}$ and $T^+ = \{(r,s) | r+s \geq 1\}$. We first extend $\Phi_{\boldsymbol{\lambda}}$ to $X_{\lambda_n} \times T^- \times d_j(\Delta^{n-1})$, by putting

$$\Phi_{\boldsymbol{\lambda}}(x,r,s,d_jt') = \begin{cases} p_{\lambda_0\lambda_1} H_{d^0\boldsymbol{\lambda}}(x, r+s, t'), & j = 0, \\ H_{d^j\boldsymbol{\lambda}}(x, r+s, t'), & 0 < j < n, \\ H_{d^n\boldsymbol{\lambda}}(p_{\lambda_{n-1}\lambda_n}(x), r+s, t'), & j = n. \end{cases} \tag{33}$$

Note that on the set $\{(r,s) | r+s = 1\}$, (33) becomes

$$\Phi_{\boldsymbol{\lambda}}(x,r,s,d_jt') = \begin{cases} p_{\lambda_0\lambda_1} g_{d^0\boldsymbol{\lambda}}(f_{\lambda_n}(x), t'), & j = 0, \\ g_{d^j\boldsymbol{\lambda}}(f_{\lambda_n}(x), t'), & 0 < j < n, \\ g_{d^n\boldsymbol{\lambda}}(q_{\lambda_{n-1}\lambda_n} f_{\lambda_n}(x), t'), & j = n. \end{cases} \tag{34}$$

This is seen by applying (21) to $H_{d^j\boldsymbol{\lambda}}$.

To extend $\Phi_{\boldsymbol{\lambda}}$ to $X_{\lambda_n} \times T^+ \times d_j(\Delta^{n-1})$, note that (22) yields a homotopy $L_{\lambda_n}: X_{\lambda_n} \times I \times I \to Y_{\lambda_n}$ (rel $(X_{\lambda_n} \times \partial I)$), such that

$$L_{\lambda_n}(x,r,0) = f_{\lambda_n} H_{\lambda_n}(x,r), \tag{35}$$

$$L_{\lambda_n}(x,r,1) = K_{\lambda_n}(f_{\lambda_n}(x), r). \tag{36}$$

Let $\omega: I \times I \to T^+$ be a mapping onto T^+, such that

$$\omega(r,0) = (1,r),\ \omega(r,1) = (r,1), \tag{37}$$

$$\omega(0,s) = (1-s,s),\ \omega(1,s) = (1,1). \tag{38}$$

Moreover, we require that $\omega^{-1}(1,1) = 1 \times I$ and that $(1,1)$ is the only point $(r',s') \in T^+$, for which $\omega^{-1}(r',s')$ is not a singleton. An example of such a mapping is given by $\omega(r,s) = (1 + s(r-1), r + s(1-r))$. Since L_{λ_n} is a homotopy rel $(X_{\lambda_n} \times \partial I)$, $L_{\lambda_n}(X_{\lambda_n} \times 1 \times I)$ is a singleton. Therefore, there exists a mapping $M_{\lambda_n}: X_{\lambda_n} \times T^+ \to Y_{\lambda_n}$, such that

$$L_{\lambda_n}(x,r,s) = M_{\lambda_n}(x, \omega(r,s)). \tag{39}$$

For $(r,s) \in T^+$, we now put

$$\Phi_{\boldsymbol{\lambda}}(x,r,s,d_jt') = \begin{cases} p_{\lambda_0\lambda_1} g_{d^0\boldsymbol{\lambda}}(M_{\lambda_n}(x,r,s), t'), & j = 0, \\ g_{d^j\boldsymbol{\lambda}}(M_{\lambda_n}(x,r,s), t'), & 0 < j < n, \\ g_{d^n\boldsymbol{\lambda}}(q_{\lambda_{n-1}\lambda_n} M_{\lambda_n}(x,r,s), t'), & j = n. \end{cases} \tag{40}$$

If $r+s = 1$, (38) implies $(r,s) = \omega(0,s)$ and thus, by (35), $M_{\lambda_n}(x,r,s) = L_{\lambda_n}(x,0,s)$. Since L_{λ_n} is a homotopy rel $(X_{\lambda_n} \times \partial I)$, one has $L_{\lambda_n}(x,0,s) = L_{\lambda_n}(x,0,0) = f_{\lambda_n} H_{\lambda_n}(x,0) = f_{\lambda_n}(x)$. Consequently, (40) yields the same values as (34). Furthermore, by (37), $(1,s) = \omega(s,0)$ and thus, $M_{\lambda_n}(x,1,s) = L_{\lambda_n}(x,s,0)$. Therefore, (40) and (35) yield

$$\Phi_{\boldsymbol{\lambda}}(x,1,s,d_jt') = \begin{cases} p_{\lambda_0\lambda_1}g_{d^0\boldsymbol{\lambda}}(f_{\lambda_n}H_{\lambda_n}(x,s),t'), & j=0, \\ g_{d^j\boldsymbol{\lambda}}(f_{\lambda_n}H_{\lambda_n}(x,s),t'), & 0<j<n, \\ g_{d^n\boldsymbol{\lambda}}(q_{\lambda_{n-1}\lambda_n}f_{\lambda_n}H_{\lambda_n}(x,s),t'), & j=n. \end{cases} \tag{41}$$

Now (29) shows that this value coincides with the value given by (30). Thus, $\Phi_{\boldsymbol{\lambda}}$ has been extended also to $X_{\boldsymbol{\lambda}} \times I \times I \times d_j(\Delta^{n-1})$.

Notice that $(r,1) = \omega(r,1)$, $M_{\lambda_n}(x,r,1) = L_{\lambda_n}(x,r,1)$. Therefore, (40) and (36) yield

$$\Phi_{\boldsymbol{\lambda}}(x,r,1,d_jt') = \begin{cases} p_{\lambda_0\lambda_1}g_{d^0\boldsymbol{\lambda}}(K_{\lambda_n}(f_{\lambda_n}(x),r),t'), & j=0, \\ g_{d^j\boldsymbol{\lambda}}(K_{\lambda_n}(f_{\lambda_n}(x),r),t'), & 0<j<n, \\ g_{d^n\boldsymbol{\lambda}}(q_{\lambda_{n-1}\lambda_n}K_{\lambda_n}(f_{\lambda_n}(x),r),t'), & j=n. \end{cases} \tag{42}$$

The next goal is to extend $\Phi_{\boldsymbol{\lambda}}$ to $X_{\lambda_n} \times I \times 1 \times \Delta^n$ and to define $\psi_{\boldsymbol{\lambda}}: Y_{\lambda_n} \times I \times \Delta^n \to X_{\lambda_0}$. Note that $\Phi_{\boldsymbol{\lambda}}$ is already defined on $X_{\lambda_n} \times 1 \times 1 \times \Delta^n$, by (32), and on $X_{\lambda_n} \times I \times 1 \times \partial\Delta^n$, by (42). Let $\psi_{\boldsymbol{\lambda}}: Y_{\lambda_n} \times (1 \times \Delta^n \cup I \times \partial\Delta^n) \to X_{\lambda_0}$ be the mapping obtained from (32) and (42), by omitting the factor f_{λ_n}, i.e., let

$$\psi_{\boldsymbol{\lambda}}(y,1,t) = \theta_{\boldsymbol{\lambda}}(g_{\lambda_n}(y),t), \tag{43}$$

$$\psi_{\boldsymbol{\lambda}}(y,r,d_jt') = \begin{cases} p_{\lambda_0\lambda_1}g_{d^0\boldsymbol{\lambda}}(K_{\lambda_n}(y,r),t'), & j=0, \\ g_{d^j\boldsymbol{\lambda}}(K_{\lambda_n}(y,r),t'), & 0<j<n, \\ g_{d^n\boldsymbol{\lambda}}(q_{\lambda_{n-1}\lambda_n}K_{\lambda_n}(y,r),t'), & j=n. \end{cases} \tag{44}$$

The mapping $\psi_{\boldsymbol{\lambda}}$ is well defined because of (29) and the fact that $K_{\lambda_n}(y,1) = f_{\lambda_n}g_{\lambda_n}(y)$. By property (HEP), this mapping admits an extension $\psi_{\boldsymbol{\lambda}}: Y_{\lambda_n} \times I \times \Delta^n \to X_{\lambda_0}$. Using this extension, we define the desired mapping $g_{\boldsymbol{\lambda}}: Y_{\lambda_n} \times \Delta^n \to X_{\lambda_0}$, by (23). Since $K_{\lambda_n}(y,0) = y$, (44) and (23) show that $g_{\boldsymbol{\lambda}}$ satisfies the coherence conditions.

We now define $\Phi_{\boldsymbol{\lambda}}$ on $X_{\lambda_n} \times I \times 1 \times \Delta^n$, by putting

$$\Phi_{\boldsymbol{\lambda}}(x,r,1,t) = \psi_{\boldsymbol{\lambda}}(f_{\lambda_n}(x),r,t). \tag{45}$$

By (44), $\Phi_{\boldsymbol{\lambda}}(x,r,1,d_jt')$ coincides with (42). Moreover, by (43), $\Phi_{\boldsymbol{\lambda}}(x,1,1,t)$ coincides with (32). Therefore, $\Phi_{\boldsymbol{\lambda}}$ is extended to

$$X_{\lambda_n} \times (\,\partial(I \times I \times \Delta^n) \setminus \mathrm{Int}(0 \times I \times \Delta^n)) \tag{46}$$

and property (HEP) yields the desired mapping $\Phi_{\boldsymbol{\lambda}}: X_{\lambda_n} \times I \times I \times \Delta^n \to X_{\lambda_0}$.

The mapping $H_{\boldsymbol{\lambda}}$ is now defined by (24). Condition (20) follows from (26). One derives (21) by applying (45) and (23). Finally, the coherence conditions for $H_{\boldsymbol{\lambda}}$ follow from (33). □

LEMMA 4.11. *Let* $\boldsymbol{f} = (f_\lambda): \boldsymbol{X} \to \boldsymbol{Y}$ *be a level homotopy equivalence and let* $\boldsymbol{g}: \boldsymbol{Y} \to \boldsymbol{X}$ *and* $\boldsymbol{H}: \boldsymbol{X} \times I \to \boldsymbol{X}$ *be level coherent mappings, which have the properties stated in Lemma* 4.10. *Then the composition* $\boldsymbol{F} = \boldsymbol{H}C(\mathbf{1}_{\boldsymbol{X}\times I}): \boldsymbol{X} \times I \to \boldsymbol{X}$ *is a level coherent homotopy* $\boldsymbol{F} = (F_{\boldsymbol{\lambda}})$, *which connects* $C(\mathbf{1}_{\boldsymbol{X}})$ *to* $\boldsymbol{g}C(\boldsymbol{f})$. *Consequently, the induced morphism* $[\boldsymbol{g}]: \boldsymbol{Y} \to \boldsymbol{X}$ *in* CH(Top^Λ) *is a left inverse of* $[C(\boldsymbol{f})]: \boldsymbol{X} \to \boldsymbol{Y}$, *i.e.,* $[\boldsymbol{g}][C(\boldsymbol{f})] = \mathrm{id}$.

Proof. First recall that $C(\mathbf{1_X})$ is a level coherent mapping, given by the mappings $1_{\boldsymbol{\lambda}}(x,t) = p_{\lambda_0\lambda_n}(x)$. The composition $\boldsymbol{h} = \boldsymbol{g}C(\boldsymbol{f})\colon \boldsymbol{X} \to \boldsymbol{Y}$ is also a level coherent mapping $\boldsymbol{h} = (h_{\boldsymbol{\lambda}})$. By the composition formula (1.3.5), one has

$$h_{\boldsymbol{\lambda}}(x,t) = g_{\lambda_0\ldots\lambda_i}(f_{\lambda_i}p_{\lambda_i\lambda_n}(x), b_i^n(t)),\ t \in P_i^n. \tag{47}$$

By the same formula,

$$F_{\boldsymbol{\lambda}}(x,s,t) = H_{\lambda_0\ldots\lambda_i}(p_{\lambda_i\lambda_n}(x), s, b_i^n(t)),\ t \in P_i^n. \tag{48}$$

Now (20), (21) and (47) imply

$$F_{\boldsymbol{\lambda}}(x,0,t) = p_{\lambda_0\lambda_n}(x),\ t \in \Delta^n, \tag{49}$$

$$F_{\boldsymbol{\lambda}}(x,1,t) = h_{\boldsymbol{\lambda}}(x,t),\ t \in \Delta^n.\ \square \tag{50}$$

The next lemma and its proof are analogous to Lemma 4.10 and its proof. For easier comparison, we use the same notation, distinguished only by an apostrophe. To simplify notation, we suppress some of the indices.

LEMMA 4.12. *Let* $\boldsymbol{f} = (f_\lambda)\colon \boldsymbol{X} \to \boldsymbol{Y}$ *be a level homotopy equivalence. Then there exist a level coherent mapping* $\boldsymbol{g}' = (g'_{\boldsymbol{\lambda}})\colon \boldsymbol{Y} \to \boldsymbol{X}$ *and a level coherent homotopy* $\boldsymbol{H}' = (H'_{\boldsymbol{\lambda}})\colon \boldsymbol{Y} \times I \to \boldsymbol{Y}$ *such that*

$$H'_{\boldsymbol{\lambda}}(y,0,t) = q_{\lambda_0\lambda_n}(y), \tag{51}$$

$$H'_{\boldsymbol{\lambda}}(y,1,t) = f_{\lambda_0}(g'_{\boldsymbol{\lambda}}(y),t). \tag{52}$$

Proof. For every $\boldsymbol{\lambda}$, $n \geq 0$, we must exhibit mappings $g'_{\boldsymbol{\lambda}}\colon Y_{\lambda_n} \times \Delta^n \to X_{\lambda_0}$, $H'_{\boldsymbol{\lambda}}\colon Y_{\lambda_n} \times I \times \Delta^n \to Y_{\lambda_0}$, which satisfy (51), (52) and the corresponding coherence conditions. It suffices to consider the case of non-degenerate $\boldsymbol{\lambda}$. We will define these mappings by (23) and (24), using mappings $\psi'_{\boldsymbol{\lambda}}\colon Y_{\lambda_n} \times I \times \Delta^n \to X_{\lambda_0}$ and $\Phi'_{\boldsymbol{\lambda}}\colon Y_{\lambda_n} \times I \times I \times \Delta^n \to Y_{\lambda_0}$, instead of $\psi_{\boldsymbol{\lambda}}$ and $\Phi_{\boldsymbol{\lambda}}$. In the constructions of the mappings $\psi'_{\boldsymbol{\lambda}}, \Phi'_{\boldsymbol{\lambda}}$ which follow, the mappings g_{λ_0}, H_{λ_0}, K_{λ_0}, L_{λ_0} are those already used in the proof of Lemma 4.10.

To begin the induction, we put

$$g'_{\lambda_0} = g_{\lambda_0}, \tag{53}$$

$$H'_{\lambda_0} = K_{\lambda_0}. \tag{54}$$

In formulae (25)–(28), we replace Φ, θ, x, X, p by Φ', θ', y, Y, q. Therefore, (using the shorter notation which omits appropriate indices of p and q), (29) becomes

$$\theta'_{\boldsymbol{\lambda}}(y, d_j t') = f_{\lambda_0} p g'_{d^j\boldsymbol{\lambda}}(q(y), t'),\ 0 \leq j \leq n, \tag{55}$$

The definition of $\Phi'_{\boldsymbol{\lambda}}$ on $Y_{\lambda_n} \times 1 \times I \times \Delta^n$ is given by

$$\Phi'_{\boldsymbol{\lambda}}(y,1,s,t) = K_{\lambda_0}(\theta'_{\boldsymbol{\lambda}}(y,t),s). \tag{56}$$

By (55), one obtains

$$\Phi'_{\boldsymbol\lambda}(y,1,s,d_jt') = K_{\lambda_0}(f_{\lambda_0}pg'_{d^j\boldsymbol\lambda}(q(y),t'),s),\ 0 \le j \le n. \tag{57}$$

Moreover,

$$\Phi'_{\boldsymbol\lambda}(y,1,1,t) = f_{\lambda_0}g'_{\lambda_0}\theta'(y,t). \tag{58}$$

$\Phi'_{\boldsymbol\lambda}$ is extended to $Y_{\lambda_n} \times T^- \times d_j(\Delta^{n-1})$, by

$$\Phi'_{\boldsymbol\lambda}(y,r,s,d_jt') = K_{\lambda_0}(pg'_{d^j\boldsymbol\lambda}(q(y),t'),r+s),\ 0 \le j \le n. \tag{59}$$

To extend $\Phi'_{\boldsymbol\lambda}$ to $Y_{\lambda_n} \times T^+ \times d_j(\Delta^{n-1})$, we apply again (22), and use the opposite homotopy $L^-_{\lambda_0}$. Clearly, $L^-_{\lambda_0}$ connects $K_{\lambda_0}(f_{\lambda_0} \times 1)$ to $f_{\lambda_0}H_{\lambda_0}$. Using the same mapping ω as before, we define $M'_{\lambda_0}: X_{\lambda_0} \times T^+ \times d_j(\Delta^{n-1}) \to Y_{\lambda_0}$ by $L^-_{\lambda_0}(y,r,s) = M'_{\lambda_0}(y,\omega(r,s))$. For $(r,s) \in T^+$, we now put

$$\Phi'_{\boldsymbol\lambda}(y,r,s,d_jt') = M'_{\lambda_0}(pg'_{d^j\boldsymbol\lambda}(q(y),t'),r,s),\ 0 \le j \le n. \tag{60}$$

Note that (60) and (35) yield

$$\Phi'_{\boldsymbol\lambda}(y,r,1,d_jt') = f_{\lambda_0}H_{\lambda_0}(pg'_{d^j\boldsymbol\lambda}(q(y),t'),r),\ 0 \le j \le n. \tag{61}$$

We now define the mapping $\psi'_{\boldsymbol\lambda}$ on $Y_{\lambda_n} \times (I \times \partial\Delta^n \cup 1 \times \Delta^n)$, by omitting the factor f_{λ_0} in (61) and (58). Then we extend $\psi'_{\boldsymbol\lambda}$ to a mapping $\psi'_{\boldsymbol\lambda}: Y_{\lambda_n}\ (I \times \partial\Delta^n) \to X_{\lambda_0}$. Putting $\Phi'_{\boldsymbol\lambda}(y,r,1,t) = f_{\lambda_0}\psi'_{\boldsymbol\lambda}(y,r,t)$, we obtain an extension of $\Phi'_{\boldsymbol\lambda}$ to $Y_{\lambda_n} \times (\partial(I \times I \times \Delta^n) \setminus \mathrm{Int}(0 \times I \times \Delta^n))$. Finally, this mapping extends further to all of $Y_{\lambda_n} \times I \times I \times \Delta^n$. One readily verifies that the mappings $g'_{\boldsymbol\lambda}$ and $H'_{\boldsymbol\lambda}$ have all the desired properties. □

LEMMA 4.13. *Let $\boldsymbol f = (f_\lambda): \boldsymbol X \to \boldsymbol Y$ be a level homotopy equivalence and let $\boldsymbol g': \boldsymbol Y \to \boldsymbol X$ and $\boldsymbol H': \boldsymbol Y \times I \to \boldsymbol Y$ be level coherent mappings, which have the properties stated in Lemma 4.12. Then the composition $\boldsymbol F' = C(\mathbf{1_Y})\boldsymbol H': \boldsymbol Y \times I \to \boldsymbol Y$ is a level coherent homotopy $\boldsymbol F' = (F'_{\boldsymbol\lambda})$, which connects $C(\mathbf{1_Y})$ with $C(\boldsymbol f)\boldsymbol g'$. Consequently, the induced morphism $[\boldsymbol g']: [\boldsymbol Y] \to [\boldsymbol X]$ in* CH(Top$^\Lambda$) *is a right inverse of $[C(\boldsymbol f)]: [\boldsymbol X] \to [\boldsymbol Y]$, i.e., $[C(\boldsymbol f)][\boldsymbol g'] = \mathrm{id}$.*

Proof. The composition $\boldsymbol h' = C(\boldsymbol f)\boldsymbol g': \boldsymbol Y \to \boldsymbol X$ is a level coherent mapping $\boldsymbol h' = (h'_{\boldsymbol\lambda})$. By the composition formula (1.3.5),

$$h'_{\boldsymbol\lambda}(y,t) = f_{\lambda_0}p_{\lambda_0\lambda_i}g'_{\lambda_i\ldots\lambda_n}(y,a^n_i(t)),\ t \in P^n_i. \tag{62}$$

By the same formula

$$F'_{\boldsymbol\lambda}(y,s,t) = q_{\lambda_0\lambda_i}H'_{\lambda_i\ldots\lambda_n}(y,s,a^n_i(t)),\ t \in P^n_i. \tag{63}$$

Consequently,

$$F'_{\boldsymbol\lambda}(y,0,t) = q_{\lambda_0\lambda_n}(y),\ t \in \Delta^n. \tag{64}$$

$$F'_{\boldsymbol\lambda}(y,1,t) = h'_{\boldsymbol\lambda}(y,t),\ t \in \Delta^n. \ \square \tag{65}$$

Proof of Theorem 4.5. By Lemmas 4.11 and 4.13, $[C(\boldsymbol{f})]$ has a left inverse $[\boldsymbol{g}]$ and a right inverse $[\boldsymbol{g}]'$. However, $[\boldsymbol{g}] = [\boldsymbol{g}']$, because

$$[\boldsymbol{g}] = [\boldsymbol{g}][C(\mathbf{1}_{\boldsymbol{Y}})] = [\boldsymbol{g}][C(\boldsymbol{f})][\boldsymbol{g}'] = [C(\mathbf{1}_{\boldsymbol{X}})][\boldsymbol{g}'] = [\boldsymbol{g}']. \square \tag{66}$$

REMARK 4.14. In the proof of Theorem 4.5 (in proving Lemmas 4.9 and 4.11), we did not use the full strength of Lemma 4.7. Indeed, we used (2) twice, but we did not use (3).

4.2 Cotelescopes (homotopy limits)

In this subsection we consider an important construction, which with every inverse system $\boldsymbol{X}$ associates a space $T(\boldsymbol{X})$, called the *cotelescope* of $\boldsymbol{X}$. It proves the existence of *homotopy limits* holim$\boldsymbol{X}$ (see Remark 4.20). Generalizing the construction to the *cotelescope system* $\boldsymbol{T}(\boldsymbol{X})$, we will obtain the tool, which enables us to replace coherent mappings $\boldsymbol{f}: \boldsymbol{X} \to \boldsymbol{Y}$ by mappings of systems $\boldsymbol{T}(\boldsymbol{f}): \boldsymbol{T}(\boldsymbol{X}) \to \boldsymbol{T}(\boldsymbol{Y})$.

For a space X and $n \geq 0$, let $(X)^{\Delta^n}$ denote the space of mappings $\eta: \Delta^n \to X$ (i.e., singular n-simplexes of X), endowed with the compact-open topology. Let $e: (X)^{\Delta^n} \times \Delta^n \to X$ denote the evaluation mapping, given by $e(\eta, t) = \eta(t)$. Since Δ^n is compact, e is continuous (see e.g., (Engelking 1977), 3.4.3 and 2.6.11). Given an inverse system $\boldsymbol{X} = (X_\lambda, p_{\lambda\lambda'}, \Lambda)$, consider the direct product

$$\prod_{\boldsymbol{\lambda}} (X_{\lambda_0})^{\Delta^n}, \tag{1}$$

taken over all multiindices $\boldsymbol{\lambda} = (\lambda_0, \ldots, \lambda_n) \in \Lambda_n$ and all lengths $n \geq 0$. By definition, $T(\boldsymbol{X})$ is the subspace of (1), which consists of all $\omega = (\omega_{\boldsymbol{\lambda}})$ satisfying the following conditions.

$$\omega_{\boldsymbol{\lambda}}(d_j t') = \begin{cases} p_{\lambda_0\lambda_1}\omega_{d^0\boldsymbol{\lambda}}(t'), & j = 0, \\ \omega_{d^j\boldsymbol{\lambda}}(t'), & 0 < j \leq n, \end{cases} \tag{2}$$

for $t' \in \Delta^{n-1}, n > 0$;

$$\omega_{\boldsymbol{\lambda}}(s_j t'') = \omega_{s^j\boldsymbol{\lambda}}(t''),\ 0 \leq j \leq n, \tag{3}$$

for $t'' \in \Delta^{n+1}, n \geq 0$. Note that $\omega_{\lambda_0}(1)$ is a point of X_{λ_0}, while $\omega_{\lambda_0\lambda_1}$ is a path in X_{λ_0}, which connects the points $\omega_{\lambda_0}(1)$ and $p_{\lambda_0\lambda_1}(\omega_{\lambda_1}(1))$.

We also define a coherent mapping $\pi_{\boldsymbol{X}}: T(\boldsymbol{X}) \to \boldsymbol{X}$. It consists of mappings $(\pi_{\boldsymbol{X}})_{\boldsymbol{\lambda}}: T(\boldsymbol{X}) \times \Delta^n \to X_{\lambda_0}$, given by

$$(\pi_{\boldsymbol{X}})_{\boldsymbol{\lambda}}(\omega, t) = \omega_{\boldsymbol{\lambda}}(t). \tag{4}$$

The mapping $(\pi_{\boldsymbol{X}})_{\boldsymbol{\lambda}}$ is continuous, because it is the composition of two continuous mappings. The first one is $p_{\boldsymbol{\lambda}} \times 1$, where $p_{\boldsymbol{\lambda}}$ is the projection, $p_{\boldsymbol{\lambda}}(\omega) = \omega_{\boldsymbol{\lambda}}$.

The second one is the evaluation mapping $e_{\lambda_0}\colon (X_{\lambda_0})^{\Delta^n} \times \Delta^n \to X_{\lambda_0}$. Using (2) and (3), one readily verifies the coherence conditions.

$$(\pi_{\boldsymbol{X}})_{\boldsymbol{\lambda}}(\omega, d_j t') = \begin{cases} p_{\lambda_0\lambda_1}(\pi_{\boldsymbol{X}})_{d^0\boldsymbol{\lambda}}(\omega, t'), & j = 0, \\ (\pi_{\boldsymbol{X}})_{d^j\boldsymbol{\lambda}}(\omega, t'), & 0 < j \leq n, \end{cases} \tag{5}$$

for $t' \in \Delta^{n-1}, n > 0$;

$$(\pi_{\boldsymbol{X}})_{\boldsymbol{\lambda}}(\omega, s_j t'') = (\pi_{\boldsymbol{X}})_{s^j\boldsymbol{\lambda}}(\omega, t''),\ 0 \leq j \leq n, \tag{6}$$

for $t'' \in \Delta^{n+1}, n \geq 0$.

REMARK 4.15. The name cotelescope is justified by the fact that the construction of $T(\boldsymbol{X})$ is dual to the construction of the telescope of an inverse sequence (see (Edwards, Hastings 1976a), Definition 3.7.2).

LEMMA 4.16. *Let X be a space and let $\boldsymbol{f} = (f_\mu)\colon X \to \boldsymbol{Y} = (Y_\mu, q_{\mu\mu'}, M)$ be a coherent mapping. Then there exists a unique mapping $g = R(\boldsymbol{f})\colon X \to T(\boldsymbol{Y})$ such that*

$$f_{\boldsymbol{\mu}}(x, t) = (\pi_{\boldsymbol{Y}})_{\boldsymbol{\mu}}(g(x), t). \tag{7}$$

If $g(x) = \omega = (\omega_{\boldsymbol{\mu}})$, then

$$\omega_{\boldsymbol{\mu}}(t) = f_{\boldsymbol{\mu}}(x, t). \tag{8}$$

Proof. For $x \in X$, put $g(x) = \omega = (\omega_{\boldsymbol{\mu}})$, where $\omega_{\boldsymbol{\mu}} \in (Y_{\mu_0})^{\Delta^n}$ is given by (8). By the coherence conditions for $f_{\boldsymbol{\mu}}$, $\omega = (\omega_{\boldsymbol{\mu}})$ satisfies conditions (2) and (3) and thus, $\omega \in T(\boldsymbol{Y})$. Since

$$(\pi_{\boldsymbol{Y}})_{\boldsymbol{\mu}}(\omega, t) = \omega_{\boldsymbol{\mu}}(t), \tag{9}$$

(8) implies (7). Conversely, if g is a mapping such that (7) holds and we put $g(x) = \omega$, then (9) implies (8). □

LEMMA 4.17. *For a coherent mapping $\boldsymbol{f}\colon X \to \boldsymbol{Y}$, $[R(\boldsymbol{f})]$ is the only class $[g]\colon X \to T(\boldsymbol{Y})$ in* H(Top) *such that in* CH(pro-Top) *one has*

$$[\boldsymbol{f}] = [\pi_{\boldsymbol{Y}}]C[g]. \tag{10}$$

Proof. By Lemma 4.16, $g = R(\boldsymbol{f})\colon X \to T(\boldsymbol{Y})$ satisfies (7). Then, Lemma 2.12 shows that $\boldsymbol{f} \simeq \pi_{\boldsymbol{Y}} C(g)$, i.e., (10) holds. Now assume that $g'\colon X \to T(\boldsymbol{Y})$ is another mapping such that $[\boldsymbol{f}] = [\pi_{\boldsymbol{Y}}]C[g']$. Let $\boldsymbol{f}'\colon X \to \boldsymbol{Y}$ be the coherent mapping given by $f'_{\boldsymbol{\mu}}(x, t) = (\pi_{\boldsymbol{Y}})_{\boldsymbol{\mu}}(g'(x), t)$. By Lemma 2.12, $[\boldsymbol{f}'] = [\pi_{\boldsymbol{Y}}]C[g'] = [\boldsymbol{f}]$. Consequently, there exists a coherent homotopy $\boldsymbol{F} = (F_{\boldsymbol{\mu}})\colon X \times I \to \boldsymbol{Y}$, which connects $\boldsymbol{f}$ to $\boldsymbol{f}'$. Lemma 4.16, applied to $\boldsymbol{F}$, yields a unique homotopy $G\colon X \times I \to T(\boldsymbol{Y})$ such that

$$F_{\boldsymbol{\mu}}(x, s, t) = (\pi_{\boldsymbol{Y}})_{\boldsymbol{\mu}}(G(x, s), t). \tag{11}$$

Since $f_\mu(x,t) = F_\mu(x,0,t) = (\pi_{\boldsymbol{Y}})_\mu(G(x,0),t)$, the uniqueness in Lemma 4.16 and (7) imply

$$G(x,0) = g(x), \; x \in X. \tag{12}$$

An analogous argument shows that

$$G(x,1) = g'(x), \; x \in X. \tag{13}$$

Consequently, G is a homotopy, which connects g and g', i.e., $[g] = [g']$. □

Clearly, if $[\boldsymbol{f}] = [\boldsymbol{f}']$, Lemma 4.17 implies that $[R(\boldsymbol{f})] = [R(\boldsymbol{f}')]$. Consequently, one can also define $R[\boldsymbol{f}]$ by putting $R[\boldsymbol{f}] = [R(\boldsymbol{f})]$.

In the more general case, when the domain of the coherent mapping $\boldsymbol{f}: \boldsymbol{X} \to \boldsymbol{Y}$ is a system, one defines $T(\boldsymbol{f}): T(\boldsymbol{X}) \to T(\boldsymbol{Y})$ by putting

$$T(\boldsymbol{f}) = R(\boldsymbol{f}\pi_{\boldsymbol{X}}). \tag{14}$$

Note that the domain of $\boldsymbol{f}\pi_{\boldsymbol{X}}$ is a space and therefore, by Lemma 4.16, the right side of (14) was defined before. The homotopy class $[T(\boldsymbol{f})]$ depends only on the homotopy class $[\boldsymbol{f}]$ in CH(pro-Top). Indeed, $[\boldsymbol{f}] = [\boldsymbol{f}']$ implies $[\boldsymbol{f}\pi_{\boldsymbol{X}}] = [\boldsymbol{f}][\pi_{\boldsymbol{X}}] = [\boldsymbol{f}'][\pi_{\boldsymbol{X}}] = [\boldsymbol{f}'\pi_{\boldsymbol{X}}]$ and thus, by Lemma 4.17, $[T(\boldsymbol{f})] = [T(\boldsymbol{f}')]$. Consequently, one can also define $T[\boldsymbol{f}]$ by putting $T[\boldsymbol{f}] = [T(\boldsymbol{f})]$.

Clearly, $T[\boldsymbol{f}]$ is the only class $[g]: T(\boldsymbol{X}) \to T(\boldsymbol{Y})$ in H(Top), for which $[\boldsymbol{f}][\pi_{\boldsymbol{X}}] = [\pi_{\boldsymbol{Y}}]C[g]$ in CH(pro-Top). In particular, the following diagram commutes.

$$\begin{array}{ccc} \boldsymbol{X} & \xleftarrow{[\pi_{\boldsymbol{X}}]} & T(\boldsymbol{X}) \\ {\scriptstyle [\boldsymbol{f}]}\downarrow & & \downarrow{\scriptstyle CT[\boldsymbol{f}]} \\ \boldsymbol{Y} & \xleftarrow[{[\pi_{\boldsymbol{Y}}]}]{} & T(\boldsymbol{Y}) \end{array} \tag{15}$$

LEMMA 4.18. *The function T, which to a system $\boldsymbol{X}$ assigns the space $T(\boldsymbol{X})$ and to a morphism $[\boldsymbol{f}]$ assigns $T[\boldsymbol{f}]$, is a functor T:* CH(pro-Top) $\to$ H(Top)*. The morphisms $[\pi_{\boldsymbol{X}}]: T(\boldsymbol{X}) \to \boldsymbol{X}$ define a natural transformation between the functors CT,* id: CH(pro-Top) $\to$ CH(pro-Top).

Proof. If $\boldsymbol{f}: \boldsymbol{X} \to \boldsymbol{Y}$ and $\boldsymbol{g}: \boldsymbol{Y} \to \boldsymbol{Z}$ are coherent mappings, an application of (15), for $\boldsymbol{f}$, $\boldsymbol{g}$ and $\boldsymbol{gf}$, shows that

$$[\boldsymbol{gf}][\pi_{\boldsymbol{X}}] = [\boldsymbol{g}][\boldsymbol{f}][\pi_{\boldsymbol{X}}] = [\pi_{\boldsymbol{Z}}]C(T[\boldsymbol{g}]T[\boldsymbol{f}]), \tag{16}$$

$$[\boldsymbol{gf}][\pi_{\boldsymbol{X}}] = [\pi_{\boldsymbol{Z}}]CT[\boldsymbol{gf}]. \tag{17}$$

By the uniqueness part of Lemma 4.17, one concludes that $T[\boldsymbol{g}]T[\boldsymbol{f}] = T[\boldsymbol{gf}] = T([\boldsymbol{g}][\boldsymbol{f}])$. Furthermore, $C[\mathbf{1}_{T(\boldsymbol{X})}] = [\mathbf{1}]_{T(\boldsymbol{X})}$ is the identity morphism in CH(pro-Top) and thus,

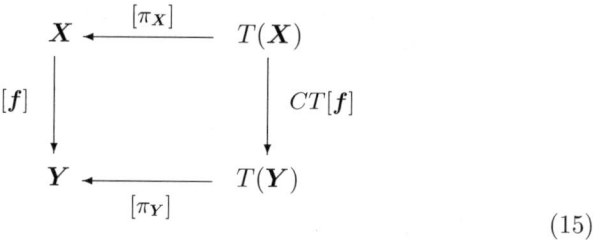

$$[\mathbf{1}_{\boldsymbol{X}}][\pi_{\boldsymbol{X}}] = [\pi_{\boldsymbol{X}}] = [\pi_{\boldsymbol{X}}]C[\mathbf{1}_{T(\boldsymbol{X})}]. \tag{18}$$

On the other hand, by (15),

$$[\mathbf{1}_{\boldsymbol{X}}][\pi_{\boldsymbol{X}}] = [\pi_{\boldsymbol{X}}] = [\pi_{\boldsymbol{X}}]CT[\mathbf{1}_{\boldsymbol{X}}]. \tag{19}$$

Hence, the uniqueness part of Lemma 4.17 implies that $T[\mathbf{1}_{\boldsymbol{X}}] = [\mathbf{1}_{T(\boldsymbol{X})}]$. This completes the proof of the assertion that T is a functor. That the morphisms $[\pi_{\boldsymbol{X}}]$ define a natural transformation follows immediately from the commutativity of diagram (15). Note that $CT(\boldsymbol{X}) = T(\boldsymbol{X})$. □

COROLLARY 4.19. *The cotelescope functor* T: CH(pro-Top) $\to$ H(Top) *is a right adjoint of the coherence functor* C: H(Top) $\to$ CH(pro-Top). *More precisely, the functions* $\eta_{X\boldsymbol{Y}}$: CH(pro-Top)$(C(X), \boldsymbol{Y}) \to$ H(Top)$(X, T(\boldsymbol{Y}))$, *given by* $\eta_{X\boldsymbol{Y}}[\boldsymbol{f}] = R[\boldsymbol{f}]$, *form a natural equivalence.*

Proof. That $\eta_{X\boldsymbol{Y}}$ is a bijection is an immediate consequence of Lemma 4.17. To prove naturality in the first variable, note that $C(X) = X$ and consider a homotopy class $[f']: X' \to X$. The function CH(pro-Top)$(X, \boldsymbol{Y}) \to$ CH(pro-Top)$(X', \boldsymbol{Y})$, induced by $C[f'] = [C(f')]: X' \to X$ maps a class $[\boldsymbol{f}]: X \to \boldsymbol{Y}$ to $[\boldsymbol{f}][C(f')]$. However, by Lemma 4.17,

$$[\boldsymbol{f}C(f')] = [\boldsymbol{f}][C(f')] = [\pi_{\boldsymbol{Y}}]CR[\boldsymbol{f}]C[f']. \tag{20}$$

Applying again Lemma 4.17, one concludes that $R[\boldsymbol{f}C(f')] = R[\boldsymbol{f}][f']$, which is the desired naturality condition, for the first variable. Concerning the second variable, consider a coherent mapping $\boldsymbol{g}: \boldsymbol{Y} \to \boldsymbol{Y}'$ and note that the function CH(pro-Top)$(X, \boldsymbol{Y}) \to$ CH(pro-Top)$(X, \boldsymbol{Y}')$, induced by $[\boldsymbol{g}]$, maps a class $[\boldsymbol{f}]: X \to \boldsymbol{Y}$ to $[\boldsymbol{g}][\boldsymbol{f}]$. By Lemma 4.17,

$$[\boldsymbol{g}][\boldsymbol{f}] = [\pi_{\boldsymbol{Y}'}]C(T[\boldsymbol{g}][R(\boldsymbol{f})]). \tag{21}$$

By uniqueness in Lemma 4.17, one concludes that $R[\boldsymbol{g}\boldsymbol{f}] = T[\boldsymbol{g}][R(\boldsymbol{f})]$, which is the desired naturality condition, for the second variable. □

REMARK 4.20. Recall that the inverse limit lim: inv-Top $\to$ Top is a right adjoint of the inclusion functor Top $\to$ inv-Top, i.e., there is a natural equivalence $\eta_{X\boldsymbol{Y}}$: inv-Top$(X, \boldsymbol{Y}) \to$ Top$(X, \lim \boldsymbol{Y})$, defined by $\eta_{X\boldsymbol{Y}}(\boldsymbol{f}) = \lim \boldsymbol{f}$. This property determines lim because of the following general fact. If $F: \mathcal{A} \to \mathcal{B}$ and $G: \mathcal{B} \to \mathcal{A}$ are adjoint functors, then F determines G up to isomorphism and vice versa (see e.g., Corollary 16.4.4 of (Schubert 1970)). It is therefore, natural to define *homotopy limits* as right adjoint functors of the coherence functor C: H(Top) $\to$ CH(pro-Top). Consequently, Corollary 4.19 shows that the cotelescope $T(\boldsymbol{X})$ is a homotopy limit of $\boldsymbol{X}$. The construction of homotopy limits described in (Vogt 1973) is more involved than the construction of the cotelescope, because it applies to the more general coherent systems.

EXAMPLE 4.21. In general the morphism $[\pi_{\boldsymbol{X}}]\colon T(\boldsymbol{X}) \to \boldsymbol{X}$ is not an isomorphism of CH(pro-Top). An easy example is given by the inverse sequence $\boldsymbol{X} = (X_m, p_{mm'}, \mathbb{N})$, where $X_m = S^1 = \{z \in \mathbb{C}\colon |z| = 1\}$, $p_{m,m+1}(z) = z^2$. Indeed, assume that $[\pi_{\boldsymbol{X}}]$ is an isomorphism. Application of the forgetful functor E (see 2.3) yields an isomorphism $E[\pi_{\boldsymbol{X}}]\colon T(\boldsymbol{X}) \to [\boldsymbol{X}] = (X_m, [p_{mm'}], \mathbb{N})$. Application of the homology functor $H_1(\,.\,;\mathbb{Z})$ to $E[\pi_{\boldsymbol{X}}]$ yields an isomorphism of pro-groups $H_1(T(\boldsymbol{X});\mathbb{Z}) \to H_1([\boldsymbol{X}];\mathbb{Z})$. However, $H_1(T(\boldsymbol{X});\mathbb{Z})$ is a group, while $H_1([\boldsymbol{X}];\mathbb{Z})$ is the pro-group $\mathbb{Z} \xleftarrow{2} \mathbb{Z} \xleftarrow{2} \mathbb{Z} \leftarrow \ldots$, which is non-movable (see e.g., (Mardešić, Segal 1982), II.6.1, Example 1) and thus, cannot be isomorphic to a group.

We will now generalize the previous constructions and assign to every inverse system $\boldsymbol{X}$, indexed by Λ, a new system $\boldsymbol{T}(\boldsymbol{X})$, also indexed by Λ, and called the *cotelescope system* of $\boldsymbol{X}$. For every $\lambda \in \Lambda$, let Λ_λ be the set $\{\nu \in \Lambda | \nu \le \lambda\}$. Since the restriction $\boldsymbol{X}_\lambda = \boldsymbol{X}|\Lambda_\lambda = (X_\nu, p_{\nu\nu'}, \Lambda_\lambda)$ is also an inverse system, the cotelescope $T(\boldsymbol{X}_\lambda)$ is a well-defined subspace of $\prod_{\boldsymbol{\lambda}\le\lambda}(X_{\lambda_0})^{\Delta^n}$, where $\boldsymbol{\lambda} \le \lambda$ means that the last term λ_n of $\boldsymbol{\lambda} = (\lambda_0, \ldots, \lambda_n)$ satisfies $\lambda_n \le \lambda$. For $\lambda \le \lambda'$, there is a natural projection

$$u_{\lambda\lambda'}\colon \prod_{\boldsymbol{\lambda}\le\lambda'} (X_{\lambda_0})^{\Delta^n} \to \prod_{\boldsymbol{\lambda}\le\lambda} (X_{\lambda_0})^{\Delta^n}. \tag{22}$$

If $\omega = (\omega_{\boldsymbol{\lambda}})$, $\boldsymbol{\lambda} \le \lambda'$, belongs to the domain of $u_{\lambda\lambda'}$, then by definition, $u_{\lambda\lambda'}(\omega)$ is the restriction of ω to the coordinates $\omega_{\boldsymbol{\lambda}}$, where $\boldsymbol{\lambda} \le \lambda$. Clearly, the restriction of the projection $u_{\lambda\lambda'}$ to $T(\boldsymbol{X}_{\lambda'})$ yields a mapping $u_{\lambda\lambda'}\colon T(\boldsymbol{X}_{\lambda'}) \to T(\boldsymbol{X}_\lambda)$. It is readily seen that $\boldsymbol{T}(\boldsymbol{X}) = (T(\boldsymbol{X}_\lambda), u_{\lambda\lambda'}, \Lambda)$ is an inverse system.

We will also define a level coherent mapping $\boldsymbol{\pi}_{\boldsymbol{X}}\colon \boldsymbol{T}(\boldsymbol{X}) \to \boldsymbol{X}$, consisting of mappings $(\boldsymbol{\pi}_{\boldsymbol{X}})_{\boldsymbol{\lambda}}\colon T(\boldsymbol{X}_{\lambda_n}) \times \Delta^n \to X_{\lambda_0}$, $\boldsymbol{\lambda} = (\lambda_0, \ldots, \lambda_n)$, defined by

$$(\boldsymbol{\pi}_{\boldsymbol{X}})_{\boldsymbol{\lambda}}(\omega, t) = \omega_{\boldsymbol{\lambda}}(t). \tag{23}$$

Note that every element $\omega \in T(\boldsymbol{X}_{\lambda_n}) \subseteq \prod (X_{\lambda_0})^{\Delta^n}$ is a collection $\omega = (\omega_{\boldsymbol{\lambda}})$, indexed by all $\boldsymbol{\lambda} \le \lambda_n$. Since $\boldsymbol{\lambda} \le \lambda_n$ is such an index, the coordinate $\omega_{\boldsymbol{\lambda}} \in (X_{\lambda_0})^{\Delta^n}$ is well defined.

It is easy to see that the mappings $\pi_{\boldsymbol{\lambda}} = (\boldsymbol{\pi}_{\boldsymbol{X}})_{\boldsymbol{\lambda}}$ satisfy the coherence conditions. Let us verify this in the case of the face operator d_n. By definition,

$$\pi_{d^n\boldsymbol{\lambda}}(u_{\lambda_{n-1}\lambda_n}(\omega), t') = (u_{\lambda_{n-1}\lambda_n}(\omega))_{d^n\boldsymbol{\lambda}}(t') = \omega_{d^n\boldsymbol{\lambda}}(t'), \tag{24}$$

because $d^n\boldsymbol{\lambda} \le \lambda_{n-1}$. However, also $\pi_{\boldsymbol{\lambda}}(\omega, d_n t') = \omega_{\boldsymbol{\lambda}}(d_n t') = \omega_{d^n\boldsymbol{\lambda}}(t')$. The verification of the coherence conditions in the remaining cases is similar.

The next lemma is analogous to Lemma 4.16.

LEMMA 4.22. *Let $\boldsymbol{f} = (f, f_{\boldsymbol{\mu}})\colon \boldsymbol{X} \to \boldsymbol{Y} = (Y_\mu, q_{\mu\mu'}, M)$ be a coherent mapping. Then there exists a unique mapping $\boldsymbol{g} = (g, g_\mu) = R(\boldsymbol{f})\colon \boldsymbol{X} \to \boldsymbol{T}(\boldsymbol{Y})$ such that $g = f$ and the mappings g_μ satisfy the following condition*

$$f_{\boldsymbol{\mu}}(p_{f(\mu_n)f(\mu)}(x), t) = (\boldsymbol{\pi}_{\boldsymbol{Y}_\mu})_{\boldsymbol{\mu}}(g_\mu(x), t), \;\; \boldsymbol{\mu} \le \mu. \tag{25}$$

Proof. For every $\mu \in M$, the coherent mapping $\boldsymbol{f}: \boldsymbol{X} \to \boldsymbol{Y}$ induces a coherent mapping $\boldsymbol{f}_\mu: X_{f(\mu)} \to \boldsymbol{Y}_\mu$, given by mappings $h_{\boldsymbol{\mu}}: X_{f(\mu)} \times \Delta^n \to Y_{\mu_0}$, where

$$h_{\boldsymbol{\mu}}(x,t) = f_{\boldsymbol{\mu}}(p_{f(\mu_n)f(\mu)}(x),t), \ \boldsymbol{\mu} \le \mu. \tag{26}$$

Application of Lemma 4.16 to $\boldsymbol{f}_\mu$ yields a unique mapping $g_\mu: X_{f(\mu)} \to T(\boldsymbol{Y}_\mu)$ such that (25) holds. The function $g = f$ and the mappings g_μ form a mapping $\boldsymbol{g} = (g, g_\mu): \boldsymbol{X} \to \boldsymbol{T}(\boldsymbol{Y})$, i.e.,

$$g_\mu p_{f(\mu)f(\mu')} = v_{\mu\mu'} g_{\mu'}, \ \mu \le \mu', \tag{27}$$

where $v_{\mu\mu'}: T(\boldsymbol{Y}_{\mu'}) \to T(\boldsymbol{Y}_\mu)$ is the natural projection (see (22)). Indeed, if we put $g_{\mu'}(x) = \omega'$ and $g_\mu p_{f(\mu)f(\mu')}(x) = \omega$, then, for $\boldsymbol{\mu} \le \mu$,

$$(v_{\mu\mu'} g_{\mu'}(x))_{\boldsymbol{\mu}} = (g_{\mu'}(x))_{\boldsymbol{\mu}} = \omega'_{\boldsymbol{\mu}} \tag{28}$$

$$(g_\mu p_{f(\mu)f(\mu')}(x))_{\boldsymbol{\mu}} = \omega_{\boldsymbol{\mu}}, \tag{29}$$

and we must prove that $\omega'_{\boldsymbol{\mu}}(t) = \omega_{\boldsymbol{\mu}}(t)$, for all $t \in \Delta^n$. However, (25) shows that

$$\begin{aligned} f_{\boldsymbol{\mu}}(p_{f(\mu_n)f(\mu')}(x),t) &= f_{\boldsymbol{\mu}}(p_{f(\mu_n)f(\mu)} p_{f(\mu)f(\mu')}(x),t) = \\ (\pi_{\boldsymbol{Y}_\mu})_{\boldsymbol{\mu}}\,(g_\mu p_{f(\mu)f(\mu')}(x),t) &= (\pi_{\boldsymbol{Y}_\mu})_{\boldsymbol{\mu}}\,(\omega,t) = \omega_{\boldsymbol{\mu}}(t). \end{aligned} \tag{30}$$

Also by (25), we have

$$f_{\boldsymbol{\mu}}(p_{f(\mu_n)f(\mu')}(x),t) = (\pi_{\boldsymbol{Y}_{\mu'}})_{\boldsymbol{\mu}}\,(g_{\mu'}(x),t) = (\pi_{\boldsymbol{Y}_{\mu'}})_{\boldsymbol{\mu}}\,(\omega',t) = \omega'_{\boldsymbol{\mu}}(t). \tag{31}$$

Comparison of (30) with (31) shows that indeed, $\omega'_{\boldsymbol{\mu}}(t) = \omega_{\boldsymbol{\mu}}(t)$. □

To state the results which follow, we need yet another homotopy category of inverse systems of spaces, which we denote by H(pro-Top). We call this category the *naïve homotopy category of inverse systems.* Its objects are systems, indexed by cofinite directed sets and its morphisms are equivalence classes $[\boldsymbol{f}]$ of mappings $\boldsymbol{f}$ with respect to the equivalence relation of *homotopy* $\simeq$, defined as follows. Two mappings $\boldsymbol{f} = (f, f_\mu)$, $\boldsymbol{f}' = (f', f'_\mu): \boldsymbol{X} \to \boldsymbol{Y}$ are considered *homotopic*, $\boldsymbol{f} \simeq \boldsymbol{f}'$, provided there exists a mapping $\boldsymbol{F} = (F, F_\mu): \boldsymbol{X} \times I \to \boldsymbol{Y}$, i.e., a morphism of inv-Top such that $F \ge f, f'$ and

$$F_\mu(x,0) = f_\mu p_{f(\mu)F(\mu)}(x), \ F_\mu(x,1) = f'_\mu p_{f'(\mu)F(\mu)}(x). \tag{32}$$

The relation $\simeq$ is indeed an equivalence relation on the set inv-Top $(\boldsymbol{X}, \boldsymbol{Y})$. Reflexivity and anti-symmetry are obvious. To prove transitivity, consider mappings $\boldsymbol{F}' = (F', F'_\mu)$ and $\boldsymbol{F}'' = (F'', F''_\mu)$, which connect $\boldsymbol{f}$ with $\boldsymbol{f}'$ and $\boldsymbol{f}'$ with $\boldsymbol{f}''$, respectively. There is no loss of generality in assuming that $F' = F'' = F$. If this is not the case, one chooses an increasing function $F \ge F', F''$ and then replaces $\boldsymbol{F}'$, $\boldsymbol{F}''$ by the respective shifts by F. Consider $F_\mu = F'_\mu * F''_\mu$ and note that $\boldsymbol{F} = (F, F_\mu)$ is a mapping, which connects $\boldsymbol{f}$ with $\boldsymbol{f}''$. Note that congruent mappings always belong to the same homotopy class of H(pro-Top), which justifies its notation. Indeed, if $\boldsymbol{f}' = (f', f'_\mu)$ is a shift of $\boldsymbol{f} = (f, f_\mu)$, then $F = f'$ and $F_\mu(x,t) = f'_\mu(x)$ define a homotopy $\boldsymbol{F}: \boldsymbol{X} \times I \to \boldsymbol{Y}$, which connects $\boldsymbol{f}$ and $\boldsymbol{f}'$.

If $\boldsymbol{F}$ and $\boldsymbol{G}$ are homotopies, which connect $\boldsymbol{f}$ with $\boldsymbol{f}'$ and $\boldsymbol{g}$ with $\boldsymbol{g}'$, then $\boldsymbol{gF}$ and $\boldsymbol{G}(\boldsymbol{f}' \times 1)$ are homotopies, which connect $\boldsymbol{gf}$ with $\boldsymbol{gf}'$ and $\boldsymbol{gf}'$ with $\boldsymbol{g}'\boldsymbol{f}'$. Therefore, composition of morphisms is well defined by the formula $[\boldsymbol{g}][\boldsymbol{f}] = [\boldsymbol{gf}]$. It is associative, because composition in inv-Top is associative. Moreover, $[\boldsymbol{f}][\boldsymbol{1_X}] = [\boldsymbol{f}]$ and $[\boldsymbol{1_Y}][\boldsymbol{f}] = [\boldsymbol{f}]$.

LEMMA 4.23. *Let $\boldsymbol{f} = (f, f_\mu), \boldsymbol{f}' = (f', f'_\mu): \boldsymbol{X} \to \boldsymbol{Y}$ be coherent mappings. $[\boldsymbol{f}] = [\boldsymbol{f}']$ in* CH(pro-Top) *if and only if $[R(\boldsymbol{f})] = [R(\boldsymbol{f}')]$ in* H(pro-Top).

Proof. Let $\boldsymbol{F} = (F, F_\mu): \boldsymbol{X} \times I \to \boldsymbol{Y}$ be a coherent homotopy, which connects $\boldsymbol{f}$ to $\boldsymbol{f}'$. Put $\boldsymbol{G} = R(\boldsymbol{F})$. Then $\boldsymbol{G} = (G, G_\mu): \boldsymbol{X} \times I \to \boldsymbol{T}(\boldsymbol{Y})$ is a mapping such that $G = F$ and

$$F_\mu(p_{F(\mu_n)F(\mu)}(x), s, t) = (\pi_{\boldsymbol{Y}_\mu})_\mu (G_\mu(x,s), t), \ \boldsymbol{\mu} \le \mu. \tag{33}$$

For $s = 0$, one obtains

$$f_\mu(p_{f(\mu_n)F(\mu)}(x), t) = (\pi_{\boldsymbol{Y}_\mu})_\mu (G_\mu(x,0), t), \ \boldsymbol{\mu} \le \mu, \tag{34}$$

If we compare this with (25) and apply uniqueness from Lemma 4.16, we see that $G_\mu(x,0) = g_\mu p_{f(\mu)F(\mu)}(x)$, where $\boldsymbol{g} = (g, g_\mu) = R(\boldsymbol{f})$. Similarly we obtain $G_\mu(x,1) = g'_\mu p_{f'(\mu)F(\mu)}(x)$, where $\boldsymbol{g}' = (g', g'_\mu) = R(\boldsymbol{f}')$. Consequently, $\boldsymbol{G}$ is a homotopy, which connects $R(\boldsymbol{f})$ to $R(\boldsymbol{f}')$.

To prove the converse, assume that $\boldsymbol{G} = (G, G_\mu)$ is such a homotopy. Consider the coherent mapping $\boldsymbol{F} = (F, F_\mu): \boldsymbol{X} \times I \to \boldsymbol{Y}$, given by $F = G$ and

$$F_\mu(x,s,t) = (\pi_{\boldsymbol{Y}_\rho})_\mu (G_\rho(x,s), t), \tag{35}$$

where ρ stands for μ_n. Note that the definition of $\boldsymbol{g} = R(\boldsymbol{f})$ implies

$$\begin{aligned} F_\mu(x,0,t) &= (\pi_{\boldsymbol{Y}_\rho})_\mu (G_{\mu_n}(x,0), t) = (\pi_{\boldsymbol{Y}_\rho})_\mu (g_{\mu_n} p_{f(\mu_n)F(\mu_n)}(x), t) \\ &= f_\mu(p_{f(\mu_n)F(\mu_n)}(x), t). \end{aligned} \tag{36}$$

Analogously,

$$F_\mu(x,1,t) = f'_\mu(p_{f'(\mu_n)F(\mu_n)}(x), t). \tag{37}$$

This proves that $\boldsymbol{F}$ is a coherent homotopy, which connects coherent mappings $\boldsymbol{f}^*, \boldsymbol{f}'^*: \boldsymbol{X} \to \boldsymbol{Y}$, where $\boldsymbol{f}^*$ is the shift of $\boldsymbol{f}$ by F and $\boldsymbol{f}'^*$ is the shift of $\boldsymbol{f}'$ by F. Consequently, $[\boldsymbol{f}] = [\boldsymbol{f}^*] = [\boldsymbol{f}'^*] = [\boldsymbol{f}']$. □

Before stating the next lemma, notice that, for mappings $\boldsymbol{f}, \boldsymbol{f}': \boldsymbol{X} \to \boldsymbol{Y}$ belonging to the same class of H(pro-Top), one has $[C(\boldsymbol{f})] = [C(\boldsymbol{f}')]$ in CH(pro-Top). Indeed, if $\boldsymbol{F}: \boldsymbol{X} \times I \to \boldsymbol{Y}$ is a homotopy connecting $\boldsymbol{f}$ to $\boldsymbol{f}'$, then $C(\boldsymbol{F}): \boldsymbol{X} \times I \to \boldsymbol{Y}$ satisfies

$$\begin{aligned} C(\boldsymbol{F})_\mu(x,0,t) &= F_{\mu_0}(p_{F(\mu_0)F(\mu_n)}(x), 0) = f_{\mu_0} p_{f(\mu_0)F(\mu_n)}(x) \\ &= (C\boldsymbol{f})_\mu(p_{f(\mu_n)F(\mu_n)}(x), t). \end{aligned} \tag{38}$$

Similarly,

$$C(\boldsymbol{F})_{\boldsymbol{\mu}}(x,1,t) = (C\boldsymbol{f}')_{\boldsymbol{\mu}}(p_{f'(\mu_n)F(\mu_n)}(x),t). \tag{39}$$

Formulae (38) and (39) show that the coherent mappings $C(\boldsymbol{f})$ and $C(\boldsymbol{f}')$ have homotopic shifts and thus, are homotopic. Consequently, by putting $C[\boldsymbol{f}] = [C(\boldsymbol{f})]$, one obtains a functor C: H(pro-Top) $\to$ CH(pro-Top), also called the *coherence functor*. The following lemma is analogous to Lemma 4.17.

LEMMA 4.24. *For every coherent mapping* $\boldsymbol{f}\colon \boldsymbol{X} \to \boldsymbol{Y}$, $[R(\boldsymbol{f})]$ *is the only class* $[\boldsymbol{g}]\colon \boldsymbol{X} \to \boldsymbol{T}(\boldsymbol{Y})$ *in* H(pro-Top) *such that in* CH(pro-Top),

$$[\boldsymbol{f}] = [\boldsymbol{\pi_Y}]C[\boldsymbol{g}]. \tag{40}$$

Proof. Put $\boldsymbol{g} = (g, g_\mu) = R(\boldsymbol{f})$. Since $\boldsymbol{g}$ is a mapping and $\boldsymbol{\pi_Y}$ is a level coherent mapping, Lemma 2.12 shows that the composition $\boldsymbol{\pi_Y}C(\boldsymbol{g})$ is homotopic to the coherent mapping, given by the function $g = f$ and the mappings $(\boldsymbol{\pi_Y})_{\boldsymbol{\mu}}(g_{\mu_n} \times 1)$. However, for $x \in X_{f(\mu_n)}$, one has $g_{\mu_n}(x) = \omega \in T(\boldsymbol{Y}_{\mu_n})$. Therefore, $(\boldsymbol{\pi_Y})_{\boldsymbol{\mu}}(\omega,t) = \omega_{\boldsymbol{\mu}}(t) = (\pi_{\boldsymbol{Y}_\rho})_{\boldsymbol{\mu}}(\omega,t)$, where ρ stands for μ_n. Now formula (25) shows that

$$(\boldsymbol{\pi_Y})_{\boldsymbol{\mu}}(g_{\mu_n} \times 1) = f_{\boldsymbol{\mu}}. \tag{41}$$

Consequently, $\boldsymbol{\pi_Y}C(\boldsymbol{g}) \simeq \boldsymbol{f}$.

To prove uniqueness, assume that $\boldsymbol{g}' = (g', g'_\mu)\colon \boldsymbol{X} \to \boldsymbol{T}(\boldsymbol{Y})$ is another mapping such that $\boldsymbol{f} \simeq \boldsymbol{\pi_Y}C(\boldsymbol{g}')$. Define a coherent mapping $\boldsymbol{f}' = (f', f'_{\boldsymbol{\mu}})\colon \boldsymbol{X} \to \boldsymbol{Y}$, by putting $f' = g'$ and

$$f'_{\boldsymbol{\mu}}(x,t) = (\boldsymbol{\pi_Y})_{\boldsymbol{\mu}}(g'_{\mu_n}(x),t). \tag{42}$$

Then, by Lemma 2.12, $\boldsymbol{f}' \simeq \boldsymbol{\pi_Y}C(\boldsymbol{g}')$ and thus, $\boldsymbol{f} \simeq \boldsymbol{f}'$.

Let us show that $R(\boldsymbol{f}') = \boldsymbol{g}'$. In view of Lemma 4.22, it suffices to show that $\boldsymbol{f}'$ and $\boldsymbol{g}'$ satisfy (25). For any $\mu \in M$ and $\boldsymbol{\mu} \le \mu$, (42) yields

$$\begin{gathered} f'_{\boldsymbol{\mu}}(p_{f'(\mu_n)f'(\mu)}(x),t) = (\boldsymbol{\pi_Y})_{\boldsymbol{\mu}}(g'_{\mu_n}p_{f'(\mu_n)f'(\mu)}(x),t) = \\ (\boldsymbol{\pi_Y})_{\boldsymbol{\mu}}(v_{\mu_n\mu}g'_\mu(x),t). \end{gathered} \tag{43}$$

Now put $g'_\mu(x) = \omega'$, $v_{\mu_n\mu}g'_\mu(x) = \omega$ and note that the definition of $v_{\mu_n\mu}$ implies $\omega_{\boldsymbol{\mu}}(t) = \omega'_{\boldsymbol{\mu}}(t)$. Moreover, note that

$$(\boldsymbol{\pi_Y})_{\boldsymbol{\mu}}(v_{\mu_n\mu}g'_\mu(x),t) = (\boldsymbol{\pi_Y})_{\boldsymbol{\mu}}(\omega,t) = \omega_{\boldsymbol{\mu}}(t), \tag{44}$$

$$(\pi_{\boldsymbol{Y}_\mu})_{\boldsymbol{\mu}}(g'_\mu(x),t) = (\pi_{\boldsymbol{Y}_\mu})_{\boldsymbol{\mu}}(\omega',t) = \omega'_{\boldsymbol{\mu}}(t). \tag{45}$$

Consequently, the left sides of (44) and (45) coincide and (43) yields

$$f'_{\boldsymbol{\mu}}(p_{f'(\mu_n)f'(\mu)}(x),t) = (\boldsymbol{\pi}_{\boldsymbol{Y}_\mu})_{\boldsymbol{\mu}}(g'_\mu(x),t). \tag{46}$$

Finally, by Lemma 4.23, $\boldsymbol{f} \simeq \boldsymbol{f}'$ implies the desired conclusion $\boldsymbol{g}' = R(\boldsymbol{f}') \simeq R(\boldsymbol{f}) = \boldsymbol{g}$. □

With every coherent mapping $\boldsymbol{f}: \boldsymbol{X} \to \boldsymbol{Y}$ we now associate a morphism $\boldsymbol{T}(\boldsymbol{f}): \boldsymbol{T}(\boldsymbol{X}) \to \boldsymbol{T}(\boldsymbol{Y})$ of H(pro-Top). By definition,

$$\boldsymbol{T}(\boldsymbol{f}) = R(\boldsymbol{f}\boldsymbol{\pi}_{\boldsymbol{X}}). \tag{47}$$

It follows from Lemma 4.23 that the homotopy class $[\boldsymbol{T}(\boldsymbol{f})]$ in H(pro-Top) depends only on the homotopy class $[\boldsymbol{f}]$ in CH(pro-Top). Therefore, we can also define $\boldsymbol{T}[\boldsymbol{f}]$ by putting $\boldsymbol{T}[\boldsymbol{f}] = [\boldsymbol{T}(\boldsymbol{f})]$. An application of Lemma 4.24 to $\boldsymbol{f}\boldsymbol{\pi}_{\boldsymbol{X}}$ shows that $\boldsymbol{T}[\boldsymbol{f}]$ is the only morphism $[\boldsymbol{g}]: \boldsymbol{T}(\boldsymbol{X}) \to \boldsymbol{T}(\boldsymbol{Y})$ of H(pro-Top), for which $[\boldsymbol{f}][\boldsymbol{\pi}_{\boldsymbol{X}}] = [\boldsymbol{\pi}_{\boldsymbol{Y}}]C[\boldsymbol{g}]$ in CH(pro-Top). In particular,

$$[\boldsymbol{f}][\boldsymbol{\pi}_{\boldsymbol{X}}] = [\boldsymbol{\pi}_{\boldsymbol{Y}}]C\boldsymbol{T}[\boldsymbol{f}], \tag{48}$$

which expresses commutativity of the analogue of diagram (15).

All essential properties of the cotelescope system are stated in the following theorem.

THEOREM 4.25. *The function $\boldsymbol{T}$, which to a system $\boldsymbol{X}$ assigns the system $\boldsymbol{T}(\boldsymbol{X})$ and to a morphism $[\boldsymbol{f}]$ assigns the morphism $\boldsymbol{T}[\boldsymbol{f}]$, is a functor $\boldsymbol{T}$:* CH(pro-Top) $\to$ H(pro-Top). *The morphisms $[\boldsymbol{\pi}_{\boldsymbol{X}}]: \boldsymbol{T}(\boldsymbol{X}) \to \boldsymbol{X}$ define a natural equivalence between the functors $C\boldsymbol{T}$,* id: CH(pro-Top) $\to$ CH(pro-Top).

Proof. That $\boldsymbol{T}$ is a functor and $\boldsymbol{\pi}_{\boldsymbol{X}}$ is a natural transformation is an immediate consequence of Lemma 4.24 and is proved like the corresponding part of Lemma 4.18. It remains to prove that $[\boldsymbol{\pi}_{\boldsymbol{X}}]: \boldsymbol{T}(\boldsymbol{X}) \to \boldsymbol{X}$ is an isomorphism in CH(pro-Top).

According to Lemma 4.24,

$$\boldsymbol{\rho}_{\boldsymbol{X}} = RC(\mathbf{1}_{\boldsymbol{X}}) \tag{49}$$

is a well-defined mapping $\boldsymbol{\rho}_{\boldsymbol{X}}: \boldsymbol{X} \to \boldsymbol{T}(\boldsymbol{X})$, for which

$$[\boldsymbol{\pi}_{\boldsymbol{X}}]C[\boldsymbol{\rho}_{\boldsymbol{X}}] = [\mathbf{1}]_{\boldsymbol{X}}. \tag{50}$$

To prove that also

$$C[\boldsymbol{\rho}_{\boldsymbol{X}}][\boldsymbol{\pi}_{\boldsymbol{X}}] = [\mathbf{1}]_{\boldsymbol{T}(\boldsymbol{X})}, \tag{51}$$

note that (50) implies

$$[\boldsymbol{\pi}_{\boldsymbol{X}}]C[\boldsymbol{\rho}_{\boldsymbol{X}}][\boldsymbol{\pi}_{\boldsymbol{X}}] = [\boldsymbol{\pi}_{\boldsymbol{X}}]. \tag{52}$$

By the uniqueness in Lemma 4.24, a comparison of (52) with the trivial relation

$$[\boldsymbol{\pi}_{\boldsymbol{X}}]C[\mathbf{1}_{\boldsymbol{T}(\boldsymbol{X})}] = [\boldsymbol{\pi}_{\boldsymbol{X}}] \tag{53}$$

yields the desired relation (51). Hence, $C[\boldsymbol{\rho}_{\boldsymbol{X}}]$ is the inverse of $[\boldsymbol{\pi}_{\boldsymbol{X}}]$. □

REMARK 4.26. Constructions, which to a coherent (non-commutative) mapping $\boldsymbol{f}$ assign a (commutative) mapping like $\boldsymbol{T}(\boldsymbol{f})$, are referred to as *rectifications* or *rigidifications* of $\boldsymbol{f}$.

There is an analogue of Theorem 4.25, for systems indexed by a fixed ordered set Λ. The role of H(pro-Top) is taken up by the category H(Top^{Λ}). Its objects are systems indexed by Λ. Two level mappings $\boldsymbol{f} = (f_\lambda)$, $\boldsymbol{f}' = (f'_\lambda): \boldsymbol{X} \to \boldsymbol{Y}$ are considered *homotopic*, $\boldsymbol{f} \simeq \boldsymbol{f}'$, provided there exists a level mapping $\boldsymbol{F} = (F_\lambda): \boldsymbol{X} \times I \to \boldsymbol{Y}$ such that $F_\lambda(x,0) = f_\lambda(x)$, $F_\lambda(x,1) = f'_\lambda(x)$. Morphisms of H($\mathsf{Top}^{\Lambda}$) are homotopy classes $[\boldsymbol{f}]$ of level mappings.

THEOREM 4.27. *The function $\boldsymbol{T}$, which to a system $\boldsymbol{X}$ assigns the system $\boldsymbol{T}(\boldsymbol{X})$ and to a morphism $[\boldsymbol{f}]$ assigns the morphism $\boldsymbol{T}[\boldsymbol{f}]$ is a functor $\boldsymbol{T}: \mathrm{CH}(\mathsf{Top}^{\Lambda}) \to \mathrm{H}(\mathsf{Top}^{\Lambda})$. The morphisms $[\boldsymbol{\pi}_{\boldsymbol{X}}]: \boldsymbol{T}(\boldsymbol{X}) \to \boldsymbol{X}$ define a natural equivalence between the functors $C\boldsymbol{T}, \mathrm{id}: \mathrm{CH}(\mathsf{Top}^{\Lambda}) \to \mathrm{CH}(\mathsf{Top}^{\Lambda})$.*

Proof. It is obtained by obvious changes in the proof of Theorem 4.25. □

In the special case when a coherent mapping is induced by a mapping $\boldsymbol{f}: \boldsymbol{X} \to \boldsymbol{Y}$ (commutative), the class $[\boldsymbol{T}C(\boldsymbol{f})]$ has a representative also induced by a mapping, denoted by $\tau(\boldsymbol{f}): \boldsymbol{T}(\boldsymbol{X}) \to \boldsymbol{T}(\boldsymbol{Y})$ (see Lemma 4.29). If $\boldsymbol{f} = (f, f_\mu)$, we define $\tau(\boldsymbol{f})$ as the mapping consisting of the function f and of the mappings $(\tau(\boldsymbol{f}))_\mu = \tau_\mu: T(\boldsymbol{X}_{f(\mu)}) \to T(\boldsymbol{Y}_\mu)$, where

$$\tau_\mu(\omega) = \eta = (\eta_{\boldsymbol{\mu}}),\ \boldsymbol{\mu} \le \mu, \tag{54}$$

$$\eta_{\boldsymbol{\mu}} = f_{\mu_0} \omega_{f(\boldsymbol{\mu})}. \tag{55}$$

Note that, for $\boldsymbol{\mu} = (\mu_0, \dots, \mu_n) \le \mu$, one has $f(\boldsymbol{\mu}) = (f(\mu_0), \dots, f(\mu_n)) \le f(\mu)$ and thus, $\omega_{f(\boldsymbol{\mu})} \in (X_{f(\mu_0)})^{\Delta^n}$ and $\eta_{\boldsymbol{\mu}} \in (Y_{\mu_0})^{\Delta^n}$ are well defined. It is readily seen that $\omega \in T(\boldsymbol{X}_{f(\mu)})$ implies $\eta \in T(\boldsymbol{Y}_\mu)$. Indeed,

$$\begin{aligned}\eta_{\boldsymbol{\mu}}(d_0 t') = f_{\mu_0} \omega_{f(\boldsymbol{\mu})}(d_0 t') = f_{\mu_0} p_{f(\mu_0) f(\mu_1)} \omega_{d^0 f(\boldsymbol{\mu})}(t') = \\ q_{\mu_0 \mu_1} f_{\mu_1} \omega_{d^0 f(\boldsymbol{\mu})}(t') = q_{\mu_0 \mu_1} \eta_{d^0 \boldsymbol{\mu}}(t').\end{aligned} \tag{56}$$

Analogous formulae hold for d_i, $i = 1, \dots, n$ and s_j, $j = 0, \dots, n$.

Clearly, for $\mu \le \mu'$,

$$\tau_\mu u_{f(\mu) f(\mu')} = v_{\mu\mu'} \tau_{\mu'}, \tag{57}$$

where $u_{\lambda\lambda'}$ and $v_{\mu\mu'}$ are the bonding mappings in $\boldsymbol{T}(\boldsymbol{X})$ and $\boldsymbol{T}(\boldsymbol{Y})$, respectively. Therefore, $\tau(\boldsymbol{f})$ is indeed a mapping.

We now define a level mapping $\phi_{\boldsymbol{X}}: \boldsymbol{X} \to \boldsymbol{T}(\boldsymbol{X})$. It consists of mappings $(\phi_{\boldsymbol{X}})_\lambda = \phi_\lambda: X_\lambda \to T(\boldsymbol{X}_\lambda)$, given by $\phi_\lambda(x) = \omega = (\omega_{\boldsymbol{\lambda}})$, $\boldsymbol{\lambda} \le \lambda$, where

$$\omega_{\boldsymbol{\lambda}}(t) = p_{\lambda_0 \lambda}(x). \tag{58}$$

One readily verifies the equality

$$u_{\lambda\lambda'} \phi_{\lambda'} = \phi_\lambda p_{\lambda\lambda'}, \lambda \le \lambda'. \tag{59}$$

REMARK 4.28. If $\boldsymbol{X}$ is a rudimentary system, i.e., $\Lambda = \{\lambda_0\}$, $X_{\lambda_0} = X$, then the system $\boldsymbol{T}(\boldsymbol{X})$ is also rudimentary and consists of the single space $T(X)$. Any element $\omega = (\omega_{\boldsymbol{\nu}})$ of this space is determined by ω_{λ_0}, i.e., by the point $\omega_{\lambda_0}(1) \in X$, because all other multiindices $\boldsymbol{\nu}$ are degenerate. Consequently, one can identify X with $T(X)$, by identifying $x \in X$ with $\omega \in T(X)$, where $\omega_{\lambda_0}(1) = x$. Note that in this case, $\phi_X = \phi_{\lambda_0}: X \to T(X)$ is the mapping, which sends $x \in X$ precisely to this ω. In other words, it is the mapping ϕ_X which identifies X with $T(X)$.

LEMMA 4.29. *For every mapping $\boldsymbol{f}: \boldsymbol{X} \to \boldsymbol{Y}$, $[\tau(\boldsymbol{f})] = [\boldsymbol{T}C(\boldsymbol{f})]$ in* H(pro-Top). *If $\boldsymbol{f}$ belongs to* Top^{Λ}, *then $[\tau(\boldsymbol{f})] = [\boldsymbol{T}C(\boldsymbol{f})]$ in* $\mathrm{H}(\mathsf{Top}^{\Lambda})$.

Proof. By the first part of Lemma 2.12, the coherent mapping $\boldsymbol{\pi_Y} C\tau(\boldsymbol{f})$ is homotopic to the coherent mapping $\boldsymbol{h}: \boldsymbol{T}(\boldsymbol{X}) \to \boldsymbol{Y}$, given by the function f and by the mappings $h_{\boldsymbol{\mu}}: T(\boldsymbol{X}_{f(\mu_n)}) \times \Delta^n \to Y_{\mu_0}$, where

$$h_{\boldsymbol{\mu}}(\omega, t) = (\boldsymbol{\pi_Y})_{\boldsymbol{\mu}}(\tau_{\mu_n}(\omega), t) = (\tau_{\mu_n}\omega)_{\boldsymbol{\mu}}(t) = f_{\mu_0}\omega_{f(\boldsymbol{\mu})}(t). \tag{60}$$

By the second part of the same lemma, the coherent mapping $C(\boldsymbol{f})\boldsymbol{\pi_X}$ is homotopic to the coherent mapping $\boldsymbol{k}: \boldsymbol{T}(\boldsymbol{X}) \to \boldsymbol{Y}$, given by the function f and by the mappings $k_{\boldsymbol{\mu}}: T(\boldsymbol{X}_{f(\mu_n)}) \times \Delta^n \to Y_{\mu_0}$, $\boldsymbol{\mu} \leq \lambda$, where

$$k_{\boldsymbol{\mu}}(\omega, t) = f_{\mu_0}(\boldsymbol{\pi_X})_{f(\boldsymbol{\mu})}(\omega, t) = f_{\mu_0}\omega_{f(\boldsymbol{\mu})}(t). \tag{61}$$

Comparing (60) and (61), one concludes that

$$C(\boldsymbol{f})\boldsymbol{\pi_X} \simeq \boldsymbol{\pi_Y} C\tau(\boldsymbol{f}). \tag{62}$$

On the other hand, by Lemma 4.24,

$$C(\boldsymbol{f})\boldsymbol{\pi_X} \simeq \boldsymbol{\pi_Y} C\boldsymbol{T}C(\boldsymbol{f}). \tag{63}$$

Consequently, by the uniqueness part of Lemma 4.24, it follows that

$$\tau(\boldsymbol{f}) \simeq \boldsymbol{T}C(\boldsymbol{f}). \ \square \tag{64}$$

THEOREM 4.30. *The function τ, which to a system $\boldsymbol{X}$ assigns the system $\tau(\boldsymbol{X}) = \boldsymbol{T}(\boldsymbol{X})$ and to a mapping $\boldsymbol{f}: \boldsymbol{X} \to \boldsymbol{Y}$ assigns the mapping $\tau(\boldsymbol{f})$, is a functor τ:* inv-Top $\to$ inv-Top. *The mappings $\boldsymbol{\phi_X}: \boldsymbol{X} \to \boldsymbol{T}(\boldsymbol{X})$ are level homotopy equivalences and define a natural transformation between the functors* id *and τ.*

Proof. For mappings $\boldsymbol{f} = (f, f_\mu): \boldsymbol{X} \to \boldsymbol{Y}$ and $\boldsymbol{g} = (g, g_\nu): \boldsymbol{Y} \to \boldsymbol{Z}$, one has $\tau(\boldsymbol{gf}) = \tau(\boldsymbol{g})\tau(\boldsymbol{f})$. Indeed, put $\boldsymbol{gf} = \boldsymbol{h}$ and put $\tau(\boldsymbol{f}) = (f, \tau_\mu)$, $\tau(\boldsymbol{g}) = (g, \tau'_\nu)$ and $\tau(\boldsymbol{h}) = (h, \tau''_\nu)$, respectively. We must verify that $\tau''_\nu = \tau'_\nu \tau_{g(\nu)}$, for $\nu \in \mathrm{N}$. Indeed, by (54) and (55), $(\tau_\mu(\omega))_{\boldsymbol{\mu}} = f_{\mu_0}\omega_{f(\boldsymbol{\mu})}$ and $(\tau'_\nu(\eta))_{\boldsymbol{\nu}} = g_{\nu_0}\eta_{g(\boldsymbol{\nu})}$. Therefore,

$$(\tau'_\nu \tau_{g(\nu)}(\omega))_{\boldsymbol{\nu}} = g_{\nu_0}(\tau_{g(\nu)}(\omega))_{g(\boldsymbol{\nu})} = g_{\nu_0} f_{g(\nu_0)}\omega_{fg(\boldsymbol{\nu})}. \tag{65}$$

On the other hand,

$$(\tau''_{\nu}(\omega))_{\boldsymbol{\nu}} = h_{\nu_0}\omega_{h(\omega)} = g_{\nu_0}f_{g(\nu_0)}\omega_{fg(\boldsymbol{\nu})}. \tag{66}$$

Moreover, $\tau(\mathbf{1_X}) = \mathrm{id}$, because, for the corresponding mappings τ_λ, one has $(\tau_\lambda(\omega))_{\boldsymbol{\lambda}} = \omega_{\boldsymbol{\lambda}}$.

To show that the mappings $\phi_{\boldsymbol{X}}$ form a natural transformation, we must show that, for every mapping $\boldsymbol{f}: \boldsymbol{X} \to \boldsymbol{Y}$,

$$\tau(\boldsymbol{f})\phi_{\boldsymbol{X}} = \phi_{\boldsymbol{Y}}\boldsymbol{f}, \tag{67}$$

i.e., $\tau_\mu\phi_{f(\mu)}(x) = \phi_\mu f_\mu(x)$, for every $x \in X_{f(\mu)}$ and $\mu \in M$. Indeed, putting $\phi_{f(\mu)}(x) = \omega$ and $\tau_\mu(\omega) = \eta$, we have $(\eta_{\boldsymbol{\mu}})(t) = f_{\mu_0}\omega_{f(\boldsymbol{\mu})}$, $\omega_{f(\boldsymbol{\mu})} = p_{f(\mu_0)f(\mu)}(x)$ and thus, $(\eta_{\boldsymbol{\mu}})(t) = f_{\mu_0}p_{f(\mu_0)f(\mu)}(x)$. On the other hand, putting $\phi_\mu f_\mu(x) = \eta'$, we have $(\eta'_{\boldsymbol{\mu}})(t) = q_{\mu_0\mu}f_\mu(x)$. However, $f_{\mu_0}p_{f(\mu_0)f(\mu)} = q_{\mu_0\mu}f_\mu$, which shows that (67) holds.

It remains to prove that every mapping $(\phi_{\boldsymbol{X}})_\lambda = \phi_\lambda: X_\lambda \to T(\boldsymbol{X}_\lambda)$ is a homotopy equivalence. To prove this assertion, we define a mapping $\psi_\lambda: T(\boldsymbol{X}_\lambda) \to X_\lambda$, by putting

$$\psi_\lambda(\omega) = \omega_\lambda(1). \tag{68}$$

Note that λ is a multiindex $\boldsymbol{\lambda}$ of length 0, which satisfies the condition $\boldsymbol{\lambda} \leq \lambda$. Therefore, ω_λ is a coordinate of ω and $\omega_\lambda(1) \in X_\lambda$ is well defined. It is readily seen that

$$\psi_\lambda\phi_\lambda = \mathrm{id}. \tag{69}$$

Indeed, if we put $\phi_\lambda(x) = \omega$, then $\psi_\lambda(\omega) = \omega_\lambda(1)$ and $\omega_\lambda(1) = p_{\lambda\lambda}(x) = x$.

We will now prove that

$$\phi_\lambda\psi_\lambda \simeq \mathrm{id}. \tag{70}$$

First note that

$$\phi_\lambda\psi_\lambda(\omega) = \phi_\lambda(\omega_\lambda(1)) = \zeta = (\zeta_{\boldsymbol{\lambda}}),\ \boldsymbol{\lambda} \leq \lambda, \tag{71}$$

where $\zeta_{\boldsymbol{\lambda}}(t) = p_{\lambda_0\lambda}\omega_\lambda(1)$. Now consider the homotopy $F: T(\boldsymbol{X}_\lambda) \times I \to T(\boldsymbol{X}_\lambda)$, defined by $F(\omega, s) = (\eta_{\boldsymbol{\lambda}})$, where $\eta_{\boldsymbol{\lambda}}: \Delta^n \to X_{\lambda_0}$ is given by the formula

$$\eta_{\boldsymbol{\lambda}}(t) = \omega_{\boldsymbol{\lambda}\lambda}((1-s)t, s),\ \boldsymbol{\lambda} \leq \lambda; \tag{72}$$

here $((1-s)t, s)$ denotes the point

$$((1-s)t_0, \dots, (1-s)t_n, s) \in \Delta^{n+1}, \tag{73}$$

for $s \in I$, $t = (t_0, \dots, t_n) \in \Delta^n$, and $\boldsymbol{\lambda}\lambda$ is the multiindex $(\lambda_0, \dots, \lambda_n, \lambda) \in \Lambda_{n+1}$, where $\boldsymbol{\lambda} = (\lambda_0, \dots, \lambda_n)$. Therefore, $\eta_{\boldsymbol{\lambda}}$ is a well-defined element of $(X_{\lambda_0})^{\Delta^n}$.

To see that $(\omega, s) \mapsto \eta_{\boldsymbol{\lambda}}$ is a continuous function $T(\boldsymbol{X}_\lambda) \times I \to (X_{\lambda_0})^{\Delta^n}$, note that $\eta_{\boldsymbol{\lambda}} = e(v \times 1)(\omega, s)$, where $e: ((X_{\lambda_0})^{\Delta^n})^I \times I \to (X_{\lambda_0})^{\Delta^n}$ is the evaluation mapping and $v: T(\boldsymbol{X}_\lambda) \to ((X_{\lambda_0})^{\Delta^n})^I$ is the composition of the following three mappings: the natural projection $T(\boldsymbol{X}_\lambda) \to X_{\lambda_0}^{\Delta^{n+1}}$, which sends ω to $\omega_{\boldsymbol{\lambda}\lambda}$; the mapping $(X_{\lambda_0})^{\Delta^{n+1}} \to (X_{\lambda_0})^{I\times\Delta^n}$, induced by the mapping $I \times \Delta^n \to \Delta^{n+1}$, which sends (s,t) to $((1-s)t, s)$; the mapping $(X_{\lambda_0})^{I\times\Delta^n} \to ((X_{\lambda_0})^I)^{\Delta^n}$, which sends $H\colon I\times\Delta^n \to X_{\lambda_0}$ to $h\colon \Delta^n \to (X_{\lambda_0})^I$, where $(h(t))(s) = H(s,t)$. Continuity of all of these mappings is well known (see e.g., 3.4.2, 3.4.3, 3.4.8, 2.6.11 of (Engelking 1977)).

In verifying that η satisfies conditions (2) and (3) and thus, $\eta \in T(\boldsymbol{X}_\lambda)$, note that, for $0 \le j \le n$,

$$d_j((1-s)t', s) = ((1-s)d_jt', s),\ t' \in \Delta^{n-1} \tag{74}$$

and $d_j(\boldsymbol{\lambda}\lambda) = (d_j\boldsymbol{\lambda})\lambda$. Therefore, if $0 < j \le n$, one has

$$\eta_{\boldsymbol{\lambda}}(d_jt') = \omega_{\boldsymbol{\lambda}\lambda}((1-s)d_jt', s) = \omega_{(d^j\boldsymbol{\lambda})\lambda}((1-s)t', s) = \eta_{d^j\boldsymbol{\lambda}}(t'). \tag{75}$$

For $j = 0$, one has

$$\begin{gathered}\eta_{\boldsymbol{\lambda}}(d_0t') = \omega_{\boldsymbol{\lambda}\lambda}((1-s)d_0t', s) = \\ p_{\lambda_0\lambda_1}\omega_{(d^0\boldsymbol{\lambda})\lambda}((1-s)t', s) = p_{\lambda_0\lambda_1}\eta_{d^0\boldsymbol{\lambda}}(t').\end{gathered} \tag{76}$$

The verification of condition (3) is similar.

To see that F connects id with $\phi_\lambda\psi_\lambda$, note that, for $s = 0$, (72) yields

$$\eta_{\boldsymbol{\lambda}}(t) = \omega_{\boldsymbol{\lambda}\lambda}(t, 0) = \omega_{\boldsymbol{\lambda}\lambda}(d_{n+1}t) = \omega_{d^{n+1}(\boldsymbol{\lambda}\lambda)}(t) = \omega_{\boldsymbol{\lambda}}(t). \tag{77}$$

For $s = 1$, one has

$$\begin{gathered}\eta_{\boldsymbol{\lambda}}(t) = \omega_{\boldsymbol{\lambda}\lambda}(0,\dots,0,1) = \omega_{\boldsymbol{\lambda}\lambda}(d_0...d_0d_0(1)) = \\ p_{\lambda_0\lambda_1}\cdots p_{\lambda_n\lambda}\omega_{d^0...d^0d^0(\boldsymbol{\lambda}\lambda)}(1) = \ldots = p_{\lambda_0\lambda_1}\cdots p_{\lambda_n\lambda}\omega_\lambda(1) = \\ p_{\lambda_0\lambda}\omega_\lambda(1) = \zeta_{\boldsymbol{\lambda}}(t). \ \square\end{gathered} \tag{78}$$

Note that $\tau(\boldsymbol{f})$ is a level mapping, whenever $\boldsymbol{f}$ is a level mapping. Therefore, the proof of Theorem 4.30 also establishes the following analogue of Theorem 4.25, for systems indexed by a fixed ordered set Λ.

THEOREM 4.31. *The function τ, which to a system $\boldsymbol{X}$ assigns the system $\tau(\boldsymbol{X}) = \boldsymbol{T}(\boldsymbol{X})$ and to a level mapping $\boldsymbol{f}\colon \boldsymbol{X} \to \boldsymbol{Y}$ assigns the level mapping $\tau(\boldsymbol{f})$, is a functor τ: $\mathsf{Top}^\Lambda \to \mathsf{Top}^\Lambda$. The mappings $\phi_{\boldsymbol{X}}\colon \boldsymbol{X} \to \tau(\boldsymbol{X})$ define a natural transformation between the functors* id *and τ.*

REMARK 4.32. The mappings ψ_λ form a homotopy mapping $\psi_{\boldsymbol{X}}\colon \tau(\boldsymbol{X}) \to \boldsymbol{X}$, but not a mapping. Indeed, for $\lambda \le \lambda'$ and $\omega \in T(\boldsymbol{X}_{\lambda'})$, one has $\psi_\lambda u_{\lambda\lambda'}(\omega) = (u_{\lambda\lambda'}(\omega))_\lambda(1) = \omega_\lambda(1)$, while $p_{\lambda\lambda'}\psi_{\lambda'}(\omega) = p_{\lambda\lambda'}\omega_{\lambda'}(1)$.

REMARK 4.33. If $\boldsymbol{A} \subseteq \boldsymbol{X}$ is a subsystem of $\boldsymbol{X}$, i.e., a system, indexed by Λ and formed by subspaces $A_\lambda \subseteq X_\lambda$ such that $p_{\lambda\lambda'}(A_{\lambda'}) \subseteq A_\lambda$, then $\psi_\lambda\tau(i_\lambda) = i_\lambda\psi_\lambda^A$, where $i_\lambda\colon A_\lambda \to X_\lambda$ denotes inclusion and $\psi_{\boldsymbol{A}} = (\psi_\lambda^A)$. Indeed, for $\omega \in T(\boldsymbol{A}_\lambda)$, one has $\psi_\lambda\tau(i_\lambda)(\omega) = \psi_\lambda(\omega) = \omega(1) = i_\lambda\psi_\lambda^A(\omega)$.

REMARK 4.34. $C[\phi_{\boldsymbol{X}}]: \boldsymbol{X} \to \boldsymbol{T}(\boldsymbol{X})$ coincides with $[\underline{\rho}_{\boldsymbol{X}}]: \boldsymbol{X} \to \boldsymbol{T}(\boldsymbol{X})$, the inverse of $[\boldsymbol{\pi}_{\boldsymbol{X}}]: \boldsymbol{T}(\boldsymbol{X}) \to \boldsymbol{X}$. To verify this assertion, it suffices to show that the coherent mapping $\boldsymbol{\pi}_{\boldsymbol{X}} C(\phi_{\boldsymbol{X}})$ is homotopic to $C(\mathbf{1}_{\boldsymbol{X}})$. By Lemma 2.12, $\boldsymbol{\pi}_{\boldsymbol{X}} C(\phi_{\boldsymbol{X}})$ is homotopic to a level coherent mapping $\boldsymbol{h} = (h_{\boldsymbol{\lambda}})$, where $h_{\boldsymbol{\lambda}}(x,t) = (\boldsymbol{\pi}_{\boldsymbol{X}})_{\boldsymbol{\lambda}}(\phi_{\lambda_n}(x), t)$. If $\phi_{\lambda_n}(x) = \omega = (\omega_{\boldsymbol{\nu}})$, $\boldsymbol{\nu} \le \lambda_n$, then $\omega_{\boldsymbol{\nu}}(t) = p_{\nu_0 \lambda_n}(x)$ and thus, $h_{\boldsymbol{\lambda}}(x,t) = p_{\lambda_0 \lambda_n}(x)$. In other words, $\boldsymbol{h} = C(\mathbf{1}_{\boldsymbol{X}})$. The direct verification of the equality $C(\phi_{\boldsymbol{X}})\boldsymbol{\pi}_{\boldsymbol{X}} \simeq \mathbf{1}_{\boldsymbol{T}(\boldsymbol{X})}$ is more involved.

4.3 Localizing pro-Top at level homotopy equivalences

If $\mathcal{A}$ is a category (in a universe $\mathcal{U}$) and Σ is a class of morphisms of $\mathcal{A}$, then (in a higher universe $\mathcal{V}$) there exists a category $\mathcal{A}' = \mathcal{A}(\Sigma^{-1})$, the *localization* of $\mathcal{A}$ at Σ and there exists a functor $P = P_\Sigma: \mathcal{A} \to \mathcal{A}'$, the *localization functor*, such that the following two conditions are satisfied:

(*i*) For every morphism $s \in \Sigma$, $P_\Sigma(s)$ is an isomorphism;

(*ii*) If $F: \mathcal{A} \to \mathcal{B}$ is a functor ($\mathcal{B}$ from an arbitrary universe) such that $F(s)$ is an isomorphism, whenever $s \in \Sigma$, then there exists a unique functor $F': \mathcal{A}' \to \mathcal{B}$ such that $F = F'P$.

Clearly, $\mathcal{A}'$ and P are determined up to a natural isomorphism. The existence of $\mathcal{A}'$ is not so obvious (see (Bauer, Dugundji 1969) or (Schubert 1970)).

To construct the category $\mathcal{A}'$, we first describe an oriented graph $\mathcal{D}'$. Its vertices are the objects of $\mathcal{A}$. There are two types of arrows in $\mathcal{D}'$, namely, the morphisms $u: U \to V$ of $\mathcal{A}$ and arrows $s^-: U \to V$, which are in a bijective correspondence with the morphisms $s: V \to U$, where $s \in \Sigma$. Objects of the category $\mathcal{A}'$ are the vertices of $\mathcal{D}'$, i.e., coincide with the objects of the category $\mathcal{A}$. To define the morphisms $\alpha: A \to B$ of $\mathcal{A}'$, one considers *paths* in $\mathcal{D}'$ from A to B, i.e., sequences of arrows $(u_n, \dots, u_1)$ in $\mathcal{D}'$, where the end of u_i is the origin of u_{i+1}, the origin of u_1 is A and the end of u_n is B. Morphisms $\alpha: A \to B$ of $\mathcal{A}'$ are equivalence classes $[u_n, \dots, u_1]$ of such paths with respect to the equivalence relation $\simeq$, generated by the following relations. First, the relations inherited from $\mathcal{A}$, i.e., $(v, u) \simeq (w)$, whenever u, v, w are morphisms of $\mathcal{A}$ such that $vu = w$. Then, the relations $(s, s^-) \simeq (1_U)$, $(s^-, s) \simeq (1_V)$, $(1_V, s^-) \simeq (s^-)$ and $(s^-, 1_U) \simeq (s^-)$, whenever $s \in \Sigma$. The composition of paths in $\mathcal{D}'$, given by juxtaposition, induces composition of morphisms in $\mathcal{A}'$ and $[1_U]$ is the identity at U. Note that morphisms $\alpha: A \to B$ may not form a set, because of paths of the form (s^-, u), where $u: A \to C$, $s: B \to C$ are morphisms involving a third object C. The functor $P: \mathcal{A} \to \mathcal{A}'$ keeps objects fixed. It maps a morphism $u: A \to B$ of $\mathcal{A}$ to the equivalence class $[u]: A \to B$.

If $F: \mathcal{A} \to \mathcal{B}$ is a functor as in (*ii*), we define $F': \mathcal{A}' \to \mathcal{B}$ as follows. On objects F' coincides with F, i.e., $F'(A) = F(A)$. To define F' on morphisms of $\mathcal{A}'$, we first define a mapping F' on the paths of $\mathcal{D}'$. We begin with paths of length 1. If u is a morphism of $\mathcal{A}$, one puts $F'(u) = F(u)$. If $s \in \Sigma$, one puts $F'(s^-) = (F(s))^{-1}$. For arbitrary paths $(u_n, \dots, u_1)$, one puts $F'(u_n, \ldots, u_1) = F'(u_n) \ldots F'(u_1)$. Let us now show that $F'(u_n, \ldots, u_1)$ depends only on the equivalence class $[u_n, \ldots, u_1]$ of $(u_n, \ldots, u_1)$. Indeed, if $(u_n, \ldots, u_1) \simeq (u'_{n'}, \ldots, u'_1)$, then there exists a chain of paths, beginning with $(u_n, \ldots, u_1)$ and ending with $(u'_{n'}, \ldots, u'_1)$, such that the only difference in two consecutive paths comes from one of the generating relations. Therefore, it suffices to see that in such cases F' assumes the same values. Let the first path be of the form $(u_n, \ldots, u_{i+1}, u_i, \ldots, u_1)$ and let the second path be of the form $(u_n, \ldots, v_i, \ldots, u_1)$. If u_{i+1}, u_i, v_i are morphisms of $\mathcal{A}$ such that $u_{i+1}u_i = v_i$, then $F(u_{i+1})F(u_i) = F(v_i)$ yields the desired conclusion. If $u_{i+1} = s \in \Sigma$, $u_i = s^-$ and $v_i = 1$, then $F'(s)F'(s^-) = F(s)(F(s))^{-1} = 1 = F'(1)$. If $u_{i+1} = 1_V$, $u_i = s^-$ and $v_i = s^-$, then $F'(1_V)F'(s^-) = F(1_V)(F(s))^{-1} = (F(s))^{-1} = F'(s^-)$. Analogous arguments apply in the two remaining cases. Finally, we define F' on morphisms of $\mathcal{A}'$, by putting $F'[u_n, \ldots, u_1] = F'(u_n, \ldots, u_1)$ (see, e.g., [Schubert 1970], §19). One has $F = F'P$, because on morphisms u from $\mathcal{A}$, $F'P(u) = F'[u] = F'(u) = F(u)$.

As an introduction to the main theorem of this section, we will describe a simple, but illuminating example of localization.

Consider the categories $\mathcal{A}$=Top, $\mathcal{B}$=H(Top) and let Σ be the class of all homotopy equivalences in Top. Recall that the homotopy functor $F = H$: Top $\to$ H(Top) maps every mapping $f: X \to Y$ to its homotopy class $H(f) = [f]: X \to Y$. Clearly, for every $s \in \Sigma$, $H(s)$ is an isomorphism in H(Top). Therefore, F induces a well-defined functor F': Top$(\Sigma^{-1}) \to$ H(Top).

THEOREM 4.35. *The functor F': Top$(\Sigma^{-1}) \to$ H(Top), induced by the homotopy functor H: Top $\to$ H(Top), is an isomorphism of categories. In other words, the homotopy category H(Top) is the localization of the category Top at homotopy equivalences.*

To define a functor G: H(Top) $\to$ Top(Σ^{-1}), which is the inverse of the functor F', we need the following lemma.

LEMMA 4.36. *Let P: Top $\to$Top (Σ^{-1}) denote the localization functor. If $f, f': X \to Y$ are homotopic mappings, then $P(f) = P(f')$.*

Proof. By assumption, there exists a homotopy $F: X \times I \to Y$, which connects f to f', i.e., $f = Fi_0$, $f' = Fi_1$, where $i_0, i_1: X \to X \times I$ are the standard embeddings, given by $i_0(x) = (x, 0)$, $i_1(x) = (x, 1)$. If $p: X \times I \to X$ denotes the first projection, given by $p(x,t) = x$, then $pi_0 = \mathrm{id}$, $pi_1 = \mathrm{id}$. Consequently, $P(p)P(i_0) = \mathrm{id}$ and $P(p)P(i_1) = \mathrm{id}$. However, p is a homotopy equivalence (i_0 and i_1 are its homotopy inverses), i.e., $p \in \Sigma$ and thus, $P(p)$ is an isomorphism. Therefore, $P(i_0) = P(i_1)$. Since also, $P(f) = P(F)P(i_0)$, $P(f') = P(F)P(i_1)$, one obtains the desired conclusion, $P(f) = P(f')$. □

Proof of Theorem 4.35. We define the functor G by putting $G(X) = X$ and $G[f] = P(f)$. If f' is another representative of $[f]$, then Lemma 4.36 shows that $P(f) = P(f')$, hence, G is well defined. That G is indeed a functor is an immediate consequence of the fact that P is a functor. For a homotopy class $[f]: X \to Y$, we have $F'G[f] = F'P(f) = H(f) = [f]$ and thus, $F'G = \mathrm{id}$. To prove that also $GF' = \mathrm{id}$, consider any mapping $f: X \to Y$ and note that $GF'P(f) = GH(f) = G[f] = P(f)$, hence, $GF'P = P$. On the other hand, $\mathrm{id} \circ P = P$ and P: $\mathsf{Top} \to \mathsf{Top}(\Sigma^{-1})$ maps every $s \in \Sigma$ into an isomorphism. Applying the uniqueness condition from *(ii)*, one concludes that indeed, $GF' = \mathrm{id}$. □

We now pass to the main results of this section. Consider the category Top^Λ and the class Σ, consisting of level homotopy equivalences. Let $\mathrm{Ho}(\mathsf{Top}^\Lambda)$ be the localization of Top^Λ at Σ. By Theorem 4.5, the coherence functor $C: \mathsf{Top}^\Lambda \to \mathrm{CH}(\mathsf{Top}^\Lambda)$ maps morphisms from Σ into isomorphisms. Consequently, C induces a functor $C': \mathrm{Ho}(\mathsf{Top}^\Lambda) \to \mathrm{CH}(\mathsf{Top}^\Lambda)$.

THEOREM 4.37. *The functor* $C': \mathrm{Ho}(\mathsf{Top}^\Lambda) \to \mathrm{CH}(\mathsf{Top}^\Lambda)$*, induced by the coherence functor* $C: \mathsf{Top}^\Lambda \to \mathrm{CH}(\mathsf{Top}^\Lambda)$*, is an isomorphism of categories.*

Analogously, we define the category Ho(pro-Top) as the localization of the category pro-Top at the class Σ of level homotopy equivalences $[\boldsymbol{f}]$. By Corollary 4.6, the coherence functor C: pro-Top → CH(pro-Top) maps morphisms from Σ into isomorphisms. Consequently, C induces a functor C': Ho(pro-Top) → CH(pro-Top).

THEOREM 4.38. *The functor* C': Ho(pro-Top) → CH(pro Top)*, induced by the coherence functor* C: pro-Top→ CH(pro-Top) *is an isomorphism of categories.*

The proofs of Theorems 4.37 and 4.38 follow the same pattern. We will give the proof of 4.38 in detail and only indicate the necessary changes to obtain the proof of Theorem 4.37.

Recall that a functor $F: \mathcal{A} \to \mathcal{B}$ is called an *equivalence of categories* if there exists a functor $G: \mathcal{B} \to \mathcal{A}$ such that there exist natural equivalences $\pi: FG \to \mathrm{id}$ and $\rho: GF \to \mathrm{id}$. Clearly, an isomorphism of categories is an equivalence. In general, an equivalence of categories need not be an isomorphism. However, the following lemma holds.

LEMMA 4.39. *If a functor* $F: \mathcal{A} \to \mathcal{B}$ *is an equivalence of categories and* F *induces a bijection between the objects of* $\mathcal{A}$ *and the objects of* $\mathcal{B}$*, then* F *is an isomorphism.*

Proof. We must show that, for every pair of objects A, A' from $\mathcal{A}$, the function $F_{AA'}: \mathcal{A}(A, A') \to \mathcal{B}(F(A), F(A'))$, induced by F is a bijection. Let us first show that $F_{AA'}$ is an injection. Indeed, if $f, f': A \to A'$ are morphisms from $\mathcal{A}$ such that $F_{AA'}(f) = F_{AA'}(f')$, i.e., $F(f) = F(f')$. Then also $\rho_{A'}GF(f) = \rho_{A'}GF(f')$. However, by the naturality of ρ, one has $\rho_{A'}GF(f) = f\rho_A$ and $\rho_{A'}GF(f') = f'\rho_A$. Consequently, $f\rho_A = f'\rho_A$. Since ρ_A is an isomorphism, one concludes that $f = f'$. The same argument, for π

shows that also $G_{BB'}: \mathcal{B}(B, B') \to \mathcal{A}(G(B), G(B'))$ is an injection and thus, for $g, g': B \to B'$ from $\mathcal{B}$, $G(g) = G(g')$ implies $g = g'$.

To show that $F_{AA'}$ is also a surjection, consider any morphism $g: F(A) \to F(A')$ from $\mathcal{B}$. Since ρ_A is an isomorphism, there exists a morphism $f: A \to A'$ such that $\rho_{A'} G(g) = f \rho_A$. By the naturality of ρ, one concludes that also $\rho_{A'} GF(f) = f \rho_A$, Consequently, $\rho_{A'} G(g) = \rho_{A'} GF(f)$. Since $\rho_{A'}$ is an isomorphism, one concludes that $G(g) = GF(f)$. However, the latter equality implies the desired conclusion $g = F(f)$. □

Note that the functor C': Ho(pro-Top) $\to$ CH(pro-Top) keeps objects fixed and thus, is a bijection on objects. Therefore, Theorem 4.38 will be proved if we show that C' is an equivalence of categories. For the functor $F = C'$, we will first construct the corresponding functor G: CH(pro-Top) $\to$ Ho(pro-Top). For this we need the following lemma, which is analogous to Lemma 4.36.

LEMMA 4.40. *Let* P: pro-Top$\to$ Ho(pro-Top) *be the localization functor. If* $\boldsymbol{f}, \boldsymbol{f}': \boldsymbol{X} \to \boldsymbol{Y}$ *are mappings which belong to the same class in* H(pro-Top), *then* $P[\boldsymbol{f}] = P[\boldsymbol{f}']$.

Proof. By assumption, there exists a homotopy $\boldsymbol{F} = (F, F_\mu): \boldsymbol{X} \times I \to \boldsymbol{Y}$, which connects $\boldsymbol{f} = (f, f_\mu)$ to $\boldsymbol{f}' = (f', f'_\mu)$ in H(pro-Top). Consider the level mappings $\boldsymbol{i}_0, \boldsymbol{i}_1: \boldsymbol{X} \to \boldsymbol{X} \times I$, which consists of the standard inclusions $i_{0\lambda}, i_{1\lambda}: X_\lambda \to X_\lambda \times I$, where $i_{0\lambda}(x) = (x, 0)$, $i_{1\lambda}(x) = (x, 1)$. Furthermore, let $\boldsymbol{p}: \boldsymbol{X} \times I \to \boldsymbol{X}$ be the level mapping, which consists of the first projections $p_\lambda: X_\lambda \times I \to X_\lambda$. Note that $\boldsymbol{p}\boldsymbol{i}_0 = \mathrm{id}$, $\boldsymbol{p}\boldsymbol{i}_1 = \mathrm{id}$ and thus,

$$P[\boldsymbol{p}]P[\boldsymbol{i}_0] = \mathrm{id}, \ P[\boldsymbol{p}]P[\boldsymbol{i}_1] = \mathrm{id}. \tag{1}$$

Since the mappings p_λ are homotopy equivalences, $[\boldsymbol{p}]$ belongs to Σ. Consequently, the morphism $P[\boldsymbol{p}]$ has an inverse in Ho(pro-Top) and thus, (1) implies

$$P[\boldsymbol{i}_0] = P[\boldsymbol{i}_1]. \tag{2}$$

By the assumption on $\boldsymbol{F}$ (see (4.2.32)), $\boldsymbol{F}\boldsymbol{i}_0$ is a shift of $\boldsymbol{f}$ by F. Therefore,

$$[\boldsymbol{F}][\boldsymbol{i}_0] = [\boldsymbol{F}\boldsymbol{i}_0] = [\boldsymbol{f}]. \tag{3}$$

Applying P to (3), one concludes that

$$P[\boldsymbol{F}]P[\boldsymbol{i}_0] = P[\boldsymbol{f}]. \tag{4}$$

Similarly, one concludes that

$$P[\boldsymbol{F}]P[\boldsymbol{i}_1] = P[\boldsymbol{f}']. \tag{5}$$

Formulae (2), (4) and (5) yield the desired conclusion $P[\boldsymbol{f}] = P[\boldsymbol{f}']$. □

Proof of Theorem 4.38. We define the functor G on objects by putting $G(\boldsymbol{X}) = \boldsymbol{T}(\boldsymbol{X})$. For a morphism $[\boldsymbol{f}]: \boldsymbol{X} \to \boldsymbol{Y}$ of CH(pro-Top), consider the induced mapping $\boldsymbol{T}(\boldsymbol{f}): \boldsymbol{T}(\boldsymbol{X}) \to \boldsymbol{T}(\boldsymbol{Y})$ and put $G[\boldsymbol{f}] = P[\boldsymbol{T}(\boldsymbol{f})]$. If

$\boldsymbol{f}'$ is another representative of $[\boldsymbol{f}]$, then $\boldsymbol{f}$ and $\boldsymbol{f}'$ are homotopic coherent mappings. Therefore, $\boldsymbol{T}(\boldsymbol{f})$ and $\boldsymbol{T}(\boldsymbol{f}')$ belong to the same class in H(pro-Top) and thus, by Lemma 4.40, $P[\boldsymbol{T}(\boldsymbol{f})] = P[\boldsymbol{T}(\boldsymbol{f}')]$. Consequently, G is well defined. Since $\boldsymbol{T}$ and P are functors, so is G.

Note that $C'G[\boldsymbol{f}] = C'P[\boldsymbol{T}(\boldsymbol{f})] = C[\boldsymbol{T}(\boldsymbol{f})] = C\boldsymbol{T}[\boldsymbol{f}]$. Now Theorem 4.25 shows that $[\boldsymbol{\pi}_{\boldsymbol{X}}]$ is a natural equivalence between the functor $C'G$ and the identity functor on CH(pro-Top), i.e., $C'G \approx \mathrm{id}$.

To complete the proof, it remains to show that also id $\approx GC'$. First note that $GC'(\boldsymbol{X}) = \boldsymbol{T}(\boldsymbol{X})$. By Theorem 4.30, $\phi_{\boldsymbol{X}}: \boldsymbol{X} \to \boldsymbol{T}(\boldsymbol{X})$ is a level homotopy equivalence, i.e., $[\phi_{\boldsymbol{X}}] \in \Sigma$. Consequently, $P[\phi_{\boldsymbol{X}}]: \boldsymbol{X} \to \boldsymbol{T}(\boldsymbol{X})$ is an isomorphism of Ho(pro-Top). We will show that the isomorphisms $P[\phi_{\boldsymbol{X}}]$ define a natural equivalence between the functors id and GC'. To achieve this we need to show that, for every morphism $\alpha: \boldsymbol{X} \to \boldsymbol{Y}$ of Ho(pro-Top), one has

$$GC'(\alpha)P[\phi_{\boldsymbol{X}}] = P[\phi_{\boldsymbol{Y}}]\alpha. \tag{6}$$

It suffices to show that (6) holds for α of the form $\alpha = P[\boldsymbol{f}]$, where $\boldsymbol{f}: \boldsymbol{X} \to \boldsymbol{Y}$ is an arbitrary mapping, i.e., that one has

$$GC'(P[\boldsymbol{f}])P[\phi_{\boldsymbol{X}}] = P[\phi_{\boldsymbol{Y}}]P[\boldsymbol{f}]. \tag{7}$$

Indeed, let F', F'': Ho(pro-Top)$\to$ Ho(pro-Top) be functors, where $F'(\boldsymbol{X}) = F''(\boldsymbol{X}) = \boldsymbol{T}(\boldsymbol{X})$, $F'(\alpha) = GC'(\alpha)$, $F''(\alpha) = P[\phi_{\boldsymbol{Y}}]\alpha(P[\phi_{\boldsymbol{X}}])^{-1}$. Then (7) shows that $F'P = F''P$ and therefore, by the universal property of P, $F' = F''$, which yields the desired relation (6).

To prove (7), note that $GC'P[\boldsymbol{f}] = GC[\boldsymbol{f}] = G[C(\boldsymbol{f})] = P[\boldsymbol{T}C(\boldsymbol{f})]$. However, by Lemma 4.29, $\boldsymbol{T}C(\boldsymbol{f})$ and $\tau(\boldsymbol{f})$ are homotopic in H(pro-Top). Therefore, Lemma 4.36 implies that $P[\boldsymbol{T}C(\boldsymbol{f})] = P[\tau(\boldsymbol{f})]$. We thus, have

$$GC'P[\boldsymbol{f}] = P[\tau(\boldsymbol{f})]. \tag{8}$$

By Theorem 4.31, we know that $\phi_{\boldsymbol{Y}}\boldsymbol{f} = \tau(\boldsymbol{f})\phi_{\boldsymbol{X}}$. Consequently,

$$P[\phi_{\boldsymbol{Y}}]P[\boldsymbol{f}] = P[\tau(\boldsymbol{f})]P[\phi_{\boldsymbol{X}}] = GC'P[\boldsymbol{f}]P[\phi_{\boldsymbol{X}}], \tag{9}$$

which is the desired relation (7). □

Proof of Theorem 4.37. As in the proof of Theorem 4.38, we define the functor G: CH(Top$^{\Lambda}$) $\to$ Ho(Top$^{\Lambda}$), by putting $G(\boldsymbol{X}) = \boldsymbol{T}(\boldsymbol{X})$, $G[\boldsymbol{f}] = P[\boldsymbol{T}(\boldsymbol{f})]$. Note that $\boldsymbol{T}(\boldsymbol{f})$ is a level mapping and therefore, $G[\boldsymbol{f}]$ belongs to Ho (Top$^{\Lambda}$). Theorem 4.27 implies that the morphisms $[\boldsymbol{\pi}_{\boldsymbol{X}}]$ form a natural equivalence between $C'G$ and the identity functor on CH(Top$^{\Lambda}$), so that $C'G \approx \mathrm{id}$. To show that also id $\approx GC'$, note that $\phi_{\boldsymbol{X}}$ is a level homotopy equivalence and thus, $P[\phi_{\boldsymbol{X}}]$ is an isomorphism of Ho(Top$^{\Lambda}$). Note that, by Lemma 4.29, $\boldsymbol{T}C(\boldsymbol{f})$ and $\tau(\boldsymbol{f})$ are homotopic in H(Top$^{\Lambda}$), whenever $\boldsymbol{f}$ belongs to Top$^{\Lambda}$. □

Bibliographic notes

This section essentially follows (Mardešić 1999a). Of course the results have a longer history. In the homotopy category H(Top) direct and inverse systems do not always have limits (see e.g., I.5.1, Example 2 in (Mardešić, Segal 1982)). Therefore, homotopy colimits and homotopy limits, which have weaker properties, are used as substitutes. The first appearence of a homotopy colimit is the telescope of a direct sequence with inclusions as bonding morphisms (Milnor 1962). The telescope of an inverse sequence is considered in (Edwards, Hastings 1976a). Constructions of homotopy colimits and limits for arbitrary inverse systems were given by G.B. Segal in (Segal 1968). Further studies of homotopy colimits and limits were carried out in (Bousfield, Kan 1972), (Boardman, Vogt 1973) and (Vogt 1973). The idea of rigidifying systems, using homotopy limits, appears in (Segal 1974). Our definition of the cotelescope $T(\boldsymbol{X})$ follows (Lisitsa 1982b) and (Lisica, Mardešić 1984b). In particular, our Lemmas 4.16, 4.17 and 4.18 follow from Theorems I.5, I.6 of (Lisica, Mardešić 1984b). Construction of the cotelescope system $\boldsymbol{T}(\boldsymbol{X})$ was described in (Thiemann 1995), where some of its properties were stated.

The category Ho(Top^{Λ}) was considered in (Boardman, Vogt 1973), (Vogt 1973). In 1976 D.A. Edwards and H.M. Hastings considered the category Ho(Top^{Λ}), under the additional hypothesis that Λ is directed and cofinal (Edwards, Hastings 1976). They endowed Top^{Λ} with a model category structure in the sense of (Quillen 1967), i.e., they specified morphisms called cofibrations, fibrations and weak equivalences, satisfying certain axioms. The homotopy category of a model category is the localization of that category at weak equivalences. In the case of Top^{Λ}, the weak equivalences are just the level homotopy equivalences.

The category Ho(pro-Top) was defined in (Porter 1974). Edwards and Hastings [loc. cit] used the reindexing theorem to endow pro-Top with a model category structure. Their definition of weak equivalences in pro-Top is rather involved. However, T. Porter analyzed the relationship between level equivalences and weak equivalences of pro-Top and he proved that the corresponding localizations are isomorphic categories (Porter 1988).

Several authors considered coherent homotopy categories, more general than CH(Top^{Λ}) and CH(pro-Top), allowing as objects also coherent systems. In these systems the commutativity relation $p_{\lambda\lambda'}p_{\lambda'\lambda''} = p_{\lambda\lambda''}$, $\lambda \le \lambda' \le \lambda''$, is replaced by coherent homotopies of all orders. However, they also proved that these more general categories are equivalent to the categories Ho(Top^{Λ}) and Ho(pro-Top), respectively. In particular, such results were obtained in (Boardman, Vogt 1973), (Vogt 1973), (Cordier, Porter 1986), (Cordier 1989), (Günther 1991a), (Batanin 1993) and (Šekutkovski 1997). The isomorphism of the categories Ho(pro-Top) and CH(pro-Top), here proved directly, also follows from (Cordier, Porter 1986) and (Cordier 1989). In (Günther 1991a) and (Batanin 1993) different proofs establish isomorphism of both of these categories. The proof of Lemma 4.7, given in 4.1, follows the original proof of (Vogt 1972). A "diagramatical" proof was given in (Hardie, Kamps 1989).

5. Coherent homotopy as a Kleisli category

A well-known construction associates with every monad on a category a new category, called the Kleisli category of the given monad. Following M.A. Batanin, in this section we endow the homotopy category H(pro-Top) with the structure of a monad and show that its Kleisli category is isomorphic to the coherent homotopy category CH(pro-Top). This fact is not used in other sections of the book.

5.1 The Kleisli category of a monad

If $\mathcal{C}$ is a category and $T : \mathcal{C} \to \mathcal{C}$ is a functor of the category $\mathcal{C}$ to itself, then the compositions $T^2 = T \circ T$ and $T^3 = T^2 \circ T$ are well-defined functors $T^2, T^3 : \mathcal{C} \to \mathcal{C}$. A *monad* on a category $\mathcal{C}$ consists of a functor $T : \mathcal{C} \to \mathcal{C}$ and of two natural transformations $\varepsilon : 1_{\mathcal{C}} \to T$, $\mu : T^2 \to T$, which for every object X of $\mathcal{C}$ make the following diagrams commutative.

$$\begin{array}{ccccc} T(X) & \xrightarrow{\varepsilon_{T(X)}} & T^2(X) & \xleftarrow{T(\varepsilon_X)} & T(X) \\ & {\scriptstyle 1_{T(X)}} \searrow & \downarrow {\scriptstyle \mu_X} & \swarrow {\scriptstyle 1_{T(X)}} & \\ & & T(X) & & \end{array} \tag{1}$$

$$\begin{array}{ccc} T^3(X) & \xrightarrow{\mu_{T(X)}} & T^2(X) \\ {\scriptstyle T(\mu_X)} \downarrow & & \downarrow {\scriptstyle \mu_X} \\ T^2(X) & \xrightarrow[\mu_X]{} & T(X) \end{array} \tag{2}$$

The structure of a monad resembles the structure of a monoid, the natural transformations ε and μ corresponding to the unit element and the multiplication, respectively.

EXAMPLE 5.1. Let $T : \mathsf{Set} \to \mathsf{Set}$ be the functor which associates with every set X the set $T(X)$ of all finite sequences $(x_1, \ldots, x_n)$ from X. With a function $f : X \to Y$, T associates the function $T(f) : T(X) \to T(Y)$, where $T(f)(x_1, \ldots, x_n) = (f(x_1), \ldots, f(x_n))$. Let $\varepsilon_X : X \to T(X)$ and $\mu_X : T^2(X) \to T(X)$ be defined by putting

$$\varepsilon_X(x) = (x), \tag{3}$$

$$\mu_X((x_1^1, \ldots, x_{n_1}^1), \ldots, (x_1^k, \ldots, x_{n_k}^k)) = (x_1^1, \ldots, x_{n_1}^1, \ldots, x_1^k, \ldots, x_{n_k}^k). \tag{4}$$

Then (T, ε, μ) is a monad on Set .

With every monad (T, ε, μ) on a category $\mathcal{C}$ is associated a category $\mathcal{K}$, called the *Kleisli category* of (T, ε, μ). Its objects are the objects of $\mathcal{C}$. A morphism $f : X \to Y$ of $\mathcal{K}$ is a morphism $f : X \to T(Y)$ of $\mathcal{C}$. The identity on X in $\mathcal{K}$ is the morphism $\varepsilon_X : X \to T(X)$. The composition gf of morphisms $f : X \to Y$ and $g : Y \to Z$ of $\mathcal{K}$ is given by the following composition in $\mathcal{C}$:

$$X \xrightarrow{f} T(Y) \xrightarrow{T(g)} T^2(Z) \xrightarrow{\mu_Z} T(Z). \tag{5}$$

LEMMA 5.2. *$\mathcal{K}$ is a category.*

Proof. To prove associativity of the composition in $\mathcal{K}$, it suffices to show that morphisms $f : X \to T(Y)$, $g : Y \to T(Z)$ and $h : Z \to T(W)$ from $\mathcal{C}$ satisfy the following equality.

$$\mu_W T(h) \mu_Z T(g) f = \mu_W T(\mu_W) T^2(h) T(g) f. \tag{6}$$

Since μ is natural, the following diagram commutes.

$$\begin{array}{ccc} T^2(Z) & \xrightarrow{T^2(h)} & T^3(W) \\ {\scriptstyle \mu_Z} \downarrow & & \downarrow {\scriptstyle \mu_{T(W)}} \\ T(Z) & \xrightarrow[T(h)]{} & T^2(W) \end{array} \tag{7}$$

and thus,

$$\mu_W T(h) \mu_Z T(g) f = \mu_W \mu_{T(W)} T^2(h) T(g) f. \tag{8}$$

However, the commutativity of (2) yields

$$\mu_W \mu_{T(W)} = \mu_W T(\mu_W) \tag{9}$$

and one obtains (5).

To prove that ε_X is the identity of X in $\mathcal{K}$, one needs to show that, for every morphism $f : X \to T(Y)$ of $\mathcal{C}$, one has

$$\mu_Y T(f)\varepsilon_X = f, \tag{10}$$

$$\mu_Y T(\varepsilon_Y) f = f. \tag{11}$$

To prove (10), observe that, by the naturality of ε, the following diagram commutes

$$\begin{array}{ccc} X & \xrightarrow{\varepsilon_X} & T(X) \\ {\scriptstyle f}\downarrow & & \downarrow{\scriptstyle T(f)} \\ T(Y) & \xrightarrow[\varepsilon_T(Y)]{} & T^2(Y) \end{array} \tag{12}$$

Therefore, by the left part of (1), $\mu_Y T(f)\varepsilon_X = \mu_Y \varepsilon_T(Y) f = 1_{T(Y)} = f$. On the other hand, by the right part of (1) (for Y), $\mu_Y T(\varepsilon_Y) f = 1_{T(Y)} f = f$. □

5.2 CH(pro-Top) is the Kleisli category of a monad

In 4.2 we have defined two functors, $\boldsymbol{T}$: CH(pro-Top) $\to$ H(pro-Top), τ: inv-Top$\to$ inv-Top, and two natural transformations, $[\boldsymbol{\pi}]$ between the functors $C\boldsymbol{T}$, id: CH(pro-Top) $\to$ CH(pro-Top) and ϕ between the functors id, τ: inv-Top$\to$ inv-Top. We will now use these data to endow the category H(pro-Top) with the structure of a monad (T, ε, μ) and then show that the corresponding Kleisli category $\mathcal{K}$ is isomorphic with the category CH(pro-Top).

We first define the functor T: H(pro-Top) $\to$ H(pro-Top). For an object $\boldsymbol{X}$, put $T(\boldsymbol{X}) = \tau(\boldsymbol{X})$. For a morphism $[\boldsymbol{f}]: \boldsymbol{X} \to \boldsymbol{Y}$, put $T[\boldsymbol{f}] = [\tau(\boldsymbol{f})]$, where $\boldsymbol{f} = (f, f_\mu): \boldsymbol{X} \to \boldsymbol{Y}$ is a mapping belonging to the class $[\boldsymbol{f}]$ and $[\tau(\boldsymbol{f})]$ is the class of the mapping $\tau(\boldsymbol{f})$. The morphism $T[\boldsymbol{f}]$ is well defined, because $[\boldsymbol{f}] = [\boldsymbol{f}']$ implies $[\tau(\boldsymbol{f})] = [\tau(\boldsymbol{f}')]$. Indeed, we showed in 4.2 that $[\boldsymbol{f}] = [\boldsymbol{f}']$ in H(pro-Top) implies $[C(\boldsymbol{f})] = [C(\boldsymbol{f}')]$ in CH(pro-Top). Consequently, $\boldsymbol{T}[C(\boldsymbol{f})] = \boldsymbol{T}[C(\boldsymbol{f}')]$ in H(pro-Top). However, by Lemma 4.29, $\boldsymbol{T}[C(\boldsymbol{f})] = [\boldsymbol{T}C(\boldsymbol{f})] = [\tau(\boldsymbol{f})]$ and $\boldsymbol{T}[C(\boldsymbol{f}')] = [\tau(\boldsymbol{f}')]$ in H(pro-Top), and the assertion follows. T is a functor, because τ is a functor.

We define ε by putting $\varepsilon_{\boldsymbol{X}} = [\phi_{\boldsymbol{X}}]$, where $[\phi_{\boldsymbol{X}}]$ is the class of the level mapping $\phi_{\boldsymbol{X}}: \boldsymbol{X} \to \tau(\boldsymbol{X})$ in H(pro-Top). That ε: id $\to T$ is a natural transformation follows from the fact that ϕ is a natural transformation (see Theorem 4.30).

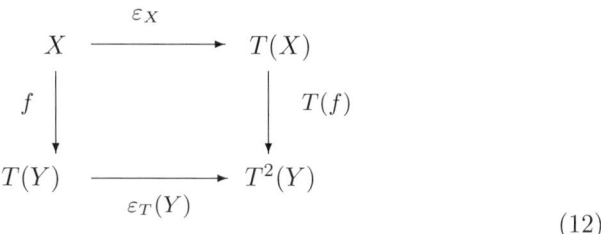

In order to define the natural transformation $\mu: T^2 \to T$, for a system $\boldsymbol{X} = (X_\lambda, p_{\lambda\lambda'}, \Lambda)$, put $\tau(\boldsymbol{X}) = \boldsymbol{X}' = (X'_\lambda, u'_{\lambda\lambda'}, \Lambda)$ and $\tau^2(\boldsymbol{X}) = \boldsymbol{X}'' = (X''_\lambda, u''_{\lambda\lambda'}, \Lambda)$. Clearly, the systems $\boldsymbol{X}'$ and $\boldsymbol{X}''$ consist of mapping spaces and natural projections. It is readily seen that the elements ω of X''_λ can be identified with collections of mappings $\omega_{\boldsymbol{\lambda}'\boldsymbol{\lambda}}: \Delta^{n'} \times \Delta^n \to X_{\lambda_0}$, where $\boldsymbol{\lambda}'$ and $\boldsymbol{\lambda}$ are multiindices of length n' and n, respectively, such that $\boldsymbol{\lambda}' \le \boldsymbol{\lambda} \le \lambda$, i.e., $\lambda'_{n'} \leq \lambda_0$ and $\lambda_n \leq \lambda$. Moreover, the following coherence conditions are satisfied.

$$\omega_{\boldsymbol{\lambda}'\boldsymbol{\lambda}}(d_j t', t) = \begin{cases} p_{\lambda'_0\lambda'_1}\omega_{d^0\boldsymbol{\lambda}',\boldsymbol{\lambda}}(t', t), & j = 0, \\ \omega_{d^0\boldsymbol{\lambda}',\boldsymbol{\lambda}}(t', t), & 0 < j \leq n', \end{cases} \tag{1}$$

$$\omega_{\boldsymbol{\lambda}'\boldsymbol{\lambda}}(t', d_j t) = \omega_{\boldsymbol{\lambda}',d^j\boldsymbol{\lambda}}(t', t),\ 0 \leq j \leq n, \tag{2}$$

$$\omega_{\boldsymbol{\lambda}'\boldsymbol{\lambda}}(s_j t', t) = \omega_{s^j\boldsymbol{\lambda}',\boldsymbol{\lambda}}(t', t)\ 0 \leq j \leq n. \tag{3}$$

$$\omega_{\boldsymbol{\lambda}'\boldsymbol{\lambda}}(t', s_j t) = \omega_{\boldsymbol{\lambda}',s^j\boldsymbol{\lambda}}(t', t)\ 0 \leq j \leq n. \tag{4}$$

Taking into account the definition of $\tau(\boldsymbol{f})$, for a mapping $\boldsymbol{f} = (f, f_\mu): \boldsymbol{X} \to \boldsymbol{Y}$ (see 4.2), it is readily seen that τ^2 maps an element $\omega = (\omega_{\boldsymbol{\lambda}'\boldsymbol{\lambda}}) \in X''_{f(\mu)}$ to the element $\zeta = (\zeta_{\boldsymbol{\mu}'\boldsymbol{\mu}}) \in Y''_\mu$, given by the formula

$$\zeta_{\boldsymbol{\mu}'\boldsymbol{\mu}} = f_{\mu'_0}\omega_{f(\boldsymbol{\mu}')f(\boldsymbol{\mu})},\ \boldsymbol{\mu}' \leq \boldsymbol{\mu} \leq \mu. \tag{5}$$

For a system $\boldsymbol{X}$, we now define a level mapping $\mu_{\boldsymbol{X}}: \tau^2(\boldsymbol{X}) \to \tau(\boldsymbol{X})$, consisting of mappings $\mu_\lambda: X''_\lambda \to X'_\lambda$, $\lambda \in \Lambda$. If $\omega = (\omega_{\boldsymbol{\lambda}'\boldsymbol{\lambda}})$ belongs to X''_λ, we define $\mu_\lambda(\omega) = \eta = (\eta_{\boldsymbol{\nu}}) \in X'_\lambda$ as follows. Let $\boldsymbol{\nu} \leq \lambda$ be a multiindex of length m. Consider the decomposition $\Delta^m = \cup P^m_i$ and the mappings $a^m_i: P^m_i \to \Delta^{m-i}$, $b^m_i: P^m_i \to \Delta^i$, $0 \leq i \leq m$, from 1.3. Then define $\eta_{\boldsymbol{\nu}}: \Delta^m \to X_{\lambda_0}$, $\boldsymbol{\nu} \leq \lambda$, by putting

$$\eta_{\boldsymbol{\nu}}(t) = \omega_{(\nu_0 \dots \nu_i)(\nu_i \dots \nu_m)}(b^m_i(t), a^m_i(t)),\ t \in P^m_i. \tag{6}$$

In order to see that $\eta_{\boldsymbol{\nu}}: \Delta^m \to X_{\nu_0}$ is a well-defined mapping, it suffices to verify that formula (6) yields the same values for $i = j$ and $i = j + 1$. Indeed, for $t \in P^m_j \cap P^m_{j+1}$, formulae (1.3.8), (6) for $i = j$ and (2) yield

$$\begin{aligned} \eta_{\boldsymbol{\nu}}(t) &= \omega_{(\nu_0 \dots \nu_j)(\nu_j \dots \nu_m)}(b_j(t), a_j(t)) \\ &= \omega_{(\nu_0 \dots \nu_j)(\nu_j \dots \nu_m)}(b_j(t), d_0 a_{j+1}(t)) \\ &= \omega_{(\nu_0 \dots \nu_j)(\nu_{j+1} \dots \nu_m)}(b_j(t), a_{j+1}(t)). \end{aligned} \tag{7}$$

Similarly, (1.3.9), (6) for $i = j + 1$ and (1) yield

$$\begin{aligned} \eta_{\boldsymbol{\nu}}(t) &= \omega_{(\nu_0 \dots \nu_{j+1})(\nu_{j+1} \dots \nu_m)}(b_{j+1}(t), a_{j+1}(t)) \\ &= \omega_{(\nu_0 \dots \nu_{j+1})(\nu_{j+1} \dots \nu_m)}(d_{j+1} b_j(t), a_{j+1}(t)) \\ &= \omega_{(\nu_0 \dots \nu_j)(\nu_{j+1} \dots \nu_m)}(b_j(t), a_{j+1}(t)). \end{aligned} \tag{8}$$

Let us now show that $\eta = (\eta_{\boldsymbol{\nu}})$, $\boldsymbol{\nu} \leq \lambda$, satisfies the coherence relations needed to insure that $\eta \in X'_{\nu_0}$. For $t \in P^{m-1}_i$ and $j \leq i$, one has $d_j t \in P^m_{i+1}$ (see (1.3.11). Therefore, by (1.3.12) and (1.3.13),

$$\begin{aligned}\eta_{\boldsymbol{\nu}}(d_j t) &= \omega_{(\nu_0\ldots\nu_{i+1})(\nu_{i+1}\ldots\nu_m)}(b_{i+1}(d_j t), a_{i+1}(d_j t))\\ &= \omega_{(\nu_0\ldots\nu_{i+1})(\nu_{i+1}\ldots\nu_m)}(d_j b_i(t), a_i(t))\\ &= p\omega_{d^j(\nu_0\ldots\nu_{i+1})(\nu_{i+1}\ldots\nu_m)}(b_i(t), a_i(t))\\ &= p\eta_{d^j\boldsymbol{\nu}}(t),\end{aligned} \tag{9}$$

where $p = \mathrm{id}$, if $j > 0$ and $p = p_{\nu_0\nu_1}$, if $j = 0$. Similarly, for $i < j$, (1.3.14)–(1.3.16) yield

$$\begin{aligned}\eta_{\boldsymbol{\nu}}(d_j t) &= \omega_{(\nu_0\ldots\nu_i)(\nu_i\ldots\nu_m)}(b_i(d_j t), a_i(d_j t))\\ &= \omega_{(\nu_0\ldots\nu_i)(\nu_i\ldots\nu_m)}(b_i(t), d_{j-i}a_i(t))\\ &= \omega_{(\nu_0\ldots\nu_i)d^{j-i}(\nu_i\ldots\nu_m)}(b_i(t), a_i(t))\\ &= \eta_{d^j\boldsymbol{\nu}}(t).\end{aligned} \tag{10}$$

That $\eta_{\boldsymbol{\nu}}(s_j t) = \eta_{s^j\boldsymbol{\nu}}(t)$ is verified similarly. Using (5), it is easy to see that $\mu_\lambda u_{\lambda\lambda'} = v_{\lambda\lambda'}\mu_{\lambda'}$, which shows that $\mu_{\boldsymbol{X}}$ is indeed a level mapping. The class of $\mu_{\boldsymbol{X}}$ in H(pro-Top), also denoted by $\mu_{\boldsymbol{X}}$, is the desired function $\mu_{\boldsymbol{X}}: T^2(\boldsymbol{X}) \to T(\boldsymbol{X})$. Using formula (5), it is easy to verify that, for a mapping $\boldsymbol{f}: \boldsymbol{X} \to \boldsymbol{Y}$, one has

$$\tau(\boldsymbol{f})\mu_{\boldsymbol{X}} = \mu_{\boldsymbol{Y}}\tau^2(\boldsymbol{f}). \tag{11}$$

Consequently, the functions $\mu_{\boldsymbol{X}}$ form a natural transformation $\mu: T^2 \to T$.

LEMMA 5.3. *(T, ε, μ) is a monad on* H(pro-Top).

Proof. We must prove commutativity of diagrams (5.1.1) and (5.1.2), i.e., we must show that, for every inverse system $\boldsymbol{X}$, the following relations hold in H(pro-Top).

$$\mu_{\boldsymbol{X}}\phi_{\tau(\boldsymbol{X})} \simeq \mathrm{id}, \tag{12}$$

$$\mu_{\boldsymbol{X}}\tau(\phi_{\boldsymbol{X}}) \simeq \mathrm{id}, \tag{13}$$

$$\mu_{\boldsymbol{X}}\mu_{\tau(\boldsymbol{X})} \simeq \mu_{\boldsymbol{X}}\tau(\mu_{\boldsymbol{X}}), \tag{14}$$

where $\simeq$ denotes the naïve homotopy used in the definition of H(pro-Top).

First note that the left side of (12) is a level mapping $\tau(\boldsymbol{X}) \to \tau(\boldsymbol{X})$, consisting of compositions $\mu_\lambda\chi_\lambda$, where $\chi_\lambda: T(\boldsymbol{X}_\lambda) \to T(\tau(\boldsymbol{X})_\lambda)$ maps $\omega = (\omega_{\boldsymbol{\alpha}})$, $\boldsymbol{\alpha} \le \lambda$, to $\zeta = (\zeta_{\boldsymbol{\nu}})$, $\boldsymbol{\nu} \le \lambda$. Thereby, $\zeta_{\boldsymbol{\nu}}: \Delta^n \to T(\boldsymbol{X}_{\nu_0})$ is given by $\zeta_{\boldsymbol{\nu}}(t) = (\zeta_{\boldsymbol{\nu}'\boldsymbol{\nu}}(t))$, $\boldsymbol{\nu}' \le \boldsymbol{\nu}$, and $\zeta_{\boldsymbol{\nu}'\boldsymbol{\nu}}(t): \Delta^{n'} \to X_{\nu'_0}$ is given by $\zeta_{\boldsymbol{\nu}'\boldsymbol{\nu}}(t', t) = (u_{\nu_0\lambda}(\omega))_{\boldsymbol{\nu}'} = \omega_{\boldsymbol{\nu}'}$. Consequently, $\mu_\lambda\chi_\lambda(\omega) = (\eta_{\boldsymbol{\nu}})$, $\boldsymbol{\nu} \le \lambda$, where for $t \in P_i^n$, $0 \le i \le n$, one has

$$\eta_{\boldsymbol{\nu}}(t) = \zeta_{(\nu_0\ldots\nu_i)(\nu_i\ldots\nu_n)}(b_i^n(t), a_i^n(t)) = \omega_{(\nu_0\ldots\nu_i)}(b_i^n(t)). \tag{15}$$

To complete the proof of (12), we now define a level homotopy $\boldsymbol{H} = (H_\lambda): \tau(\boldsymbol{X}) \times I \to \tau(\boldsymbol{X})$, using the homotopies $\Phi^n: I \times \Delta^n \to \Delta^n$, given by (2.3.4) and (2.3.7). We put $H_\lambda(\omega, s) = (\theta_{\boldsymbol{\nu}})$, $\boldsymbol{\nu} \le \lambda$, where

$$\theta_{\nu}(s,t) = \omega_{\nu}\Phi^n(s,t). \tag{16}$$

Since

$$\Phi^n(0,t) = t,\ \Phi^n(1,t) = d_n ... d_{i+1} b_i^n(t), \tag{17}$$

one has $\theta_{\nu}(0,t) = \omega_{\nu}(t)$ and $\theta_{\nu}(1,t) = \omega_{\nu}(d_n ... d_{i+1} b_i^n(t)) = \omega_{(\nu_0 ... \nu_i)}(b_i^n(t))$. Hence, H_λ connects id with $\mu_\lambda \chi_\lambda$.

Similar arguments show that the left side of (13) is a level mapping $\tau(\boldsymbol{X}) \to \tau(\boldsymbol{X})$, which maps $\omega \in T(\boldsymbol{X}_\lambda)$ to (η_{ν}), $\boldsymbol{\nu} \leq \lambda$, where for $t \in P_i^n$, $0 \leq i \leq n$, one has

$$\eta_{\nu}(t) = p_{\nu_0 \nu_i} \omega_{(\nu_i ... \nu_n)}(a_i^n(t)). \tag{18}$$

In order to define a level homotopy $\boldsymbol{K} = (K_\lambda): \tau(\boldsymbol{X}) \times I \to \tau(\boldsymbol{X})$, which realizes (13), we now use the homotopies $\Psi^n: I \times \Delta^n \to \Delta^n$, given by (2.3.19) and (2.3.22). We put $K_\lambda(\omega, s) = (\theta_{\nu})$, $\boldsymbol{\nu} \leq \lambda$, where

$$\theta_{\nu}(t) = \omega_{\nu}\Psi^n(s,t). \tag{19}$$

Since

$$\Psi^n(0,t) = t,\ \Psi^n(1,t) = d_0 ... d_0 a_i^n(t), \tag{20}$$

the assertion follows.

Before proving (14), note that $\tau^3(\boldsymbol{X})$ is an inverse system of mapping spaces X_λ''' and natural projections $u_{\lambda\lambda'}'''$, also indexed by Λ. The elements ω of X_λ''' are collections of mappings $\omega_{\boldsymbol{\lambda}''\boldsymbol{\lambda}'\boldsymbol{\lambda}}: \Delta^{n''} \times \Delta^{n'} \times \Delta^n \to X_{\lambda_0}$, where $\boldsymbol{\lambda}'' \leq \boldsymbol{\lambda}' \leq \boldsymbol{\lambda} \leq \lambda$. Moreover, coherence conditions like conditions (1)–(4) hold.

To express explicitly the left and the right sides of (14), we must use the decompositions $P_i^n = \cup_k P_{ik}^n$, $P_k^n = \cup_i Q_{ik}^n$, introduced in 2.2. Then a straightforward computation shows that $\mu_{\boldsymbol{X}} \mu_{\tau(\boldsymbol{X})}: \tau^3(\boldsymbol{X}) \to \tau(\boldsymbol{X})$ is a level mapping, consisting of mappings, which with $(\omega_{\boldsymbol{\lambda}''\boldsymbol{\lambda}'\boldsymbol{\lambda}})$, $\boldsymbol{\lambda} \leq \lambda$, associate (η_{ν}), $\boldsymbol{\nu} \leq \lambda$, where, for $t \in P_{ik}^n$,

$$\eta_{\nu}(t) = \omega_{(\nu_0 ... \nu_i)(\nu_i ... \nu_k)(\nu_k ... \nu_n)}(b_i^n(t), b_{k-i}^{n-i} a_i^n(t), a_{k-i}^{n-i} a_i^n(t)). \tag{21}$$

Similarly, $\mu_{\boldsymbol{X}} \tau(\mu_{\boldsymbol{X}})$ is a level mapping, consisting of mappings which with $(\omega_{\boldsymbol{\lambda}''\boldsymbol{\lambda}'\boldsymbol{\lambda}})$ associate (ζ_{ν}), $\boldsymbol{\nu} \leq \lambda$, where, for $t \in Q_{ik}^n$,

$$\zeta_{\nu}(t) = \omega_{(\nu_0 ... \nu_i)(\nu_i ... \nu_k)(\nu_k ... \nu_n)}(b_i^k b_k^n(t), a_i^k b_k^n(t), a_k^n(t)). \tag{22}$$

In order to define a level homotopy $\boldsymbol{L} = (L_\lambda): \tau^3(\boldsymbol{X}) \times I \to \tau(\boldsymbol{X})$, which realizes (14), we use the homotopies $\Theta^n: I \times \Delta^n \to \Delta^n$, given by (2.2.11) and Lemma 2.9. Then, for $t \in Q_{ik}^n$, we put $L_\lambda(\omega, s) = (\xi_{\nu})$, $\boldsymbol{\nu} \leq \lambda$, where

$$\xi_{\nu}(s,t) = \eta_{\nu}(\Theta^n(s,t)). \tag{23}$$

Note that

$$\xi_{\nu}(1,t) = \omega_{(\nu_0 ... \nu_i)(\nu_i ... \nu_k)(\nu_k ... \nu_n)}(b_i^n \theta^n(t), b_{k-i}^{n-i} a_i^n \theta^n(t), a_{k-i}^{n-i} a_i^n \theta^n(t)). \tag{24}$$

However, by formulae (2.2.4)–(2.2.6), the right side of (24) equals $\zeta_{\nu}(t)$. □

We shall now establish the main result of this section.

THEOREM 5.4. *Let* (T, ε, μ) *be the monad considered in Lemma* 5.3. *Then its Kleisli category* $\mathcal{K}$ *is isomorphic to the coherent homotopy category* CH(pro-Top).

Proof. We first define a functor $U: \mathcal{K} \to$ CH(pro-Top). For a system $\boldsymbol{X}$, let $U(\boldsymbol{X}) = \boldsymbol{X}$. By definition of $\mathcal{K}$, a morphism $\boldsymbol{X} \to \boldsymbol{Y}$ is a morphism $[\boldsymbol{f}]: \boldsymbol{X} \to \tau(\boldsymbol{Y})$ of H(pro-Top). Put

$$U[\boldsymbol{f}] = [\boldsymbol{\pi_Y}]C[\boldsymbol{f}], \tag{25}$$

where $\boldsymbol{\pi_Y}: \tau(\boldsymbol{Y}) \to \boldsymbol{Y}$ is the coherent mapping defined in 4.2. Note that $U[\boldsymbol{f}]: \boldsymbol{X} \to \boldsymbol{Y}$ is a morphism of CH(pro-Top).

The identity on $\boldsymbol{X}$ in $\mathcal{K}$ is $[\phi_{\boldsymbol{X}}]: \boldsymbol{X} \to \tau(\boldsymbol{X})$ and U maps it to $U[\phi_{\boldsymbol{X}}] = [\boldsymbol{\pi_X}]C[\phi_{\boldsymbol{X}}]$. However,

$$[\boldsymbol{\pi_X}]C[\phi_{\boldsymbol{X}}] = [\mathrm{id}]. \tag{26}$$

Indeed, $\phi_{\boldsymbol{X}} = (\phi_\lambda): \boldsymbol{X} \to \tau(\boldsymbol{X})$ is a level mapping and therefore, by Lemma 2.12, the composition $\boldsymbol{\pi_X}C(\phi_{\boldsymbol{X}})$ is homotopic to a level coherent mapping $\boldsymbol{h} = (h_{\boldsymbol{\lambda}})$, where

$$h_{\boldsymbol{\lambda}}(x,t) = (\boldsymbol{\pi_X})_{\boldsymbol{\lambda}}(\phi_{\lambda_n}(x), t). \tag{27}$$

However, $\phi_{\lambda_n}(x) = \omega = (\omega_{\boldsymbol{\nu}})$, $\boldsymbol{\nu} \le \lambda$, where $\omega_{\boldsymbol{\nu}}(t) = p_{\nu_0\lambda_n}(x)$ and since $(\boldsymbol{\pi_X})_{\boldsymbol{\lambda}}(\omega, t) = \omega_{\boldsymbol{\lambda}}(t) = p_{\nu_0\lambda_n}(x)$, one concludes that

$$h_{\boldsymbol{\lambda}}(x,t) = p_{\nu_0\lambda_n}(x). \tag{28}$$

Hence, $\boldsymbol{h} \simeq \mathrm{id}$.

In order to show that U preserves composition, consider two morphisms $\boldsymbol{X} \to \boldsymbol{Y}$ and $\boldsymbol{Y} \to \boldsymbol{Z}$ in $\mathcal{K}$, i.e., two morphisms $[\boldsymbol{f}]: \boldsymbol{X} \to \tau(\boldsymbol{Y})$ and $[\boldsymbol{g}]: \boldsymbol{Y} \to \tau(\boldsymbol{Z})$ of H(pro-Top). Their composition in $\mathcal{K}$ is the morphism $[\mu_{\boldsymbol{Z}}\tau(\boldsymbol{g})\boldsymbol{f}]: \boldsymbol{X} \to \boldsymbol{\tau(Z)}$ of H(pro-Top), which U maps to $[\boldsymbol{\pi_Z}]C[\mu_{\boldsymbol{Z}}\tau(\boldsymbol{g})\boldsymbol{f}]$. Since $U[\boldsymbol{g}] = [\boldsymbol{\pi_Z}]C[\boldsymbol{g}]$, we must show that in CH(pro-Top) one has

$$[\boldsymbol{\pi_Z}]C[\mu_{\boldsymbol{Z}}\tau(\boldsymbol{g})\boldsymbol{f}] = [\boldsymbol{\pi_Z}]C[\boldsymbol{g}][\boldsymbol{\pi_Y}]C[\boldsymbol{f}]. \tag{29}$$

Clearly, it suffices to show that

$$C[\mu_{\boldsymbol{Z}}]C[\tau(\boldsymbol{g})] = C[\boldsymbol{g}][\boldsymbol{\pi_Y}]. \tag{30}$$

Since $[\boldsymbol{\pi}]$, is a natural equivalence between the functors $C\boldsymbol{T}$ and id (see Theorem 4.25), one has

$$[\boldsymbol{\pi}_{\tau(\boldsymbol{Z})}]C\boldsymbol{T}C[\boldsymbol{g}] = C[\boldsymbol{g}][\boldsymbol{\pi_Y}]. \tag{31}$$

Moreover, by Lemma 4.29, $\boldsymbol{T}C[\boldsymbol{g}] = [\tau(\boldsymbol{g})]$ and thus, (31) becomes

$$[\boldsymbol{\pi}_{\tau(\boldsymbol{Z})}]C[\tau(\boldsymbol{g})] = C[\boldsymbol{g}][\boldsymbol{\pi_Y}]. \tag{32}$$

Consequently, in order to prove (30), it suffices to show that

$$C[\mu_{\boldsymbol{Z}}] = [\boldsymbol{\pi}_{\tau(\boldsymbol{Z})}]. \tag{33}$$

By the commutativity of the left part of diagram (5.1.1) for $\boldsymbol{Z}$, $\mu_{\boldsymbol{Z}}\varepsilon_{T(\boldsymbol{Z})} =$ id. However, $T(\boldsymbol{Z}) = \tau(\boldsymbol{Z})$ and $\varepsilon_{\tau(\boldsymbol{Z})} = [\phi_{\tau(\boldsymbol{Z})}]$ and thus,

$$C[\mu_{\boldsymbol{Z}}]C[\phi_{\tau(\boldsymbol{Z})}] = C[\mu_{\boldsymbol{Z}}\phi_{\tau(\boldsymbol{Z})}] = [\mathrm{id}]. \tag{34}$$

On the other hand, by (26) for $\tau(\boldsymbol{Z})$, one has

$$[\boldsymbol{\pi}_{\tau(\boldsymbol{Z})}]C[\phi_{\tau(\boldsymbol{Z})}] = [\mathrm{id}]. \tag{35}$$

Since $[\boldsymbol{\pi}_{\tau(\boldsymbol{Z})}]$ is an isomorphism of CH(pro-Top), (33) is a consequence of (34) and (35).

To complete the proof of Theorem 5.2, it only remains to realize that the functor U induces a bijection $\mathcal{K}(\boldsymbol{X}, \boldsymbol{Y}) \to$ CH(pro-Top)$(\boldsymbol{X}, \boldsymbol{Y})$. However, this is precisely the assertion of Lemma 4.24. □

Bibliographic notes

Monads were formally introduced in (Huber 1961) under the name *standard constructions.* The name *monad* was proposed by S. Mac Lane (see (Mac Lane 1971)). Following S. Eilenberg and J.C. Moore, the name *triple* is also in use. The construction of the Kleisli category first appeared in (Kleisli 1965). The construction of the monad (S, ε, μ) on the category H(pro-Top) is due to M.A. Batanin (Batanin 1986), who refers to (Lisica, Mardešić 1984b) for the verification of many of his assertions. A later paper (Batanin 1993) contains a version of his theory for abstract categories.

II. STRONG SHAPE

6. Resolutions

Our approach to the construction of shape categories consists of two steps. In the first step one approximates spaces by polyhedra (or ANR's), i.e., one associates with spaces suitable inverse systems of polyhedra (ANR's). In the second step, one develops a suitable homotopy theory of inverse systems. In order to develop strong shape, one uses coherent homotopy of inverse systems, considered in Chapter I. As associated inverse systems, one uses strong polyhedral (or ANR) expansions, a notion defined in the next section, which insures that different expansions of the same space are naturally isomorphic in the coherent homotopy category CH(pro-Top). A very useful special case of strong expansions are resolutions, defined in this section. In the most important cases (e.g., for paracompact spaces), a resolution $\boldsymbol{p} : X \to \boldsymbol{X}$ is an inverse limit of $\boldsymbol{X}$, satisfying certain additional conditions. For compact spaces, the additional conditions are already fulfilled by limits, hence, in the compact case, resolutions and inverse limits coincide.

6.1 Resolutions of spaces and mappings

Various properties of the inverse limit $X = \lim \boldsymbol{X}$ of an inverse system $\boldsymbol{X} = (X_\lambda, p_{\lambda\lambda'}, \Lambda)$ of compact Hausdorff spaces are inherited from the terms X_λ of $\boldsymbol{X}$. Unfortunately, this is not the case with inverse systems of non-compact spaces. E.g., for compact spaces X_λ, the covering dimension $\dim X_\lambda \leq n$ implies that also $\dim X \leq n$. But, there exist inverse sequences of 0-dimensional paracompact spaces, whose limit X is normal and has covering dimension $\dim X > 0$ (Charalambous 1980).

Similarly, if $\boldsymbol{X}$ and $\boldsymbol{X}'$ are inverse systems of compact polyhedra having the same limit X, then $\boldsymbol{X}$ and $\boldsymbol{X}'$ are isomorphic in CH(pro-Top), hence, also in pro-H(Top). However, it is easy to exhibit non-compact examples, where this property fails.

EXAMPLE 6.1. E.g., let $\boldsymbol{X}$ and $\boldsymbol{X}'$ be inclusion sequences consisting of polyhedra $X_n = \{x \in \mathbb{R}^2 \backslash \{0\} \mid \|x\| < \frac{1}{n}\}$ and $X'_n = \{x \in \mathbb{R}^1 \mid n < x < \infty\}$, respectively. Clearly, $\lim \boldsymbol{X} = \lim \boldsymbol{X}' = \emptyset$. Nevertheless, $\boldsymbol{X}$ and $\boldsymbol{X}'$ differ already in the category pro-H(Top). Indeed, their first homology pro-groups differ, because $H_1(\boldsymbol{X};\mathbb{Z})$ is the inverse sequence $\mathbb{Z} \xleftarrow{1} \mathbb{Z} \xleftarrow{1} \mathbb{Z} \ldots$, whose limit is $\mathbb{Z}$, while $H_1(\boldsymbol{X}';\mathbb{Z})$ is the inverse sequence $0 \xleftarrow{1} 0 \xleftarrow{1} 0 \ldots$, whose limit is 0.

These anomalies are avoided if one imposes suitable additional conditions on the mapping $\boldsymbol{p} = (p_\lambda): X \to \boldsymbol{X}$, i.e., if one replaces the notion of limit by the notion of resolution (see e.g., I.6 of (Mardešić, Segal 1982)).

A *resolution* $\boldsymbol{p}: X \to \boldsymbol{X}$ *of a space* X consists of an inverse system $\boldsymbol{X} = (X_\lambda, p_{\lambda\lambda'}, \Lambda)$ and of a mapping of systems $\boldsymbol{p} = (p_\lambda): X \to \boldsymbol{X}$, which has the property that, for every polyhedron P and every open covering $\mathcal{V}$ of P, the following additional conditions are satisfied:

(R1) *For every mapping* $f: X \to P$, *there exist a* $\lambda \in \Lambda$ *and a mapping* $h: X_\lambda \to P$ *such that the mappings* hp_λ *and* f *are* $\mathcal{V}$ *- near, i.e., every* $x \in X$ *admits a* $V \in \mathcal{V}$ *such that* $hp_\lambda(x), f(x) \in V$.

(R2) *There exists an open covering* $\mathcal{V}'$ *of* P, *such that whenever, for a* $\lambda \in \Lambda$ *and for two mappings* $h, h': X_\lambda \to P$, *the mappings* $hp_\lambda, h'p_\lambda$ *are* $\mathcal{V}'$ *- near, then there exists a* $\lambda' \geq \lambda$, *such that the mappings* $hp_{\lambda\lambda'}, h'p_{\lambda\lambda'}$ *are* $\mathcal{V}$ *- near.*

A resolution $\boldsymbol{p}$ of X is called *cofinite* if its index set Λ is cofinite. If all the terms X_λ belong to a class $\mathcal{C}$ of spaces, then we speak of a $\mathcal{C}$ *- resolution. Polyhedral resolutions* and ANR *- resolutions* are of particular interest.

A *resolution of a mapping* $f: X \to Y$ between two spaces consists of two resolutions $\boldsymbol{p}: X \to \boldsymbol{X}$, $\boldsymbol{q}: Y \to \boldsymbol{Y}$ and of a mapping $\boldsymbol{f}: \boldsymbol{X} \to \boldsymbol{Y}$ such that

$$\boldsymbol{f}\boldsymbol{p} = \boldsymbol{q}f \tag{1}$$

in inv - Top. A resolution $(\boldsymbol{p}, \boldsymbol{q}, \boldsymbol{f})$ of f is called cofinite if both resolutions $\boldsymbol{p}$ and $\boldsymbol{q}$ are cofinite. If $\boldsymbol{p}, \boldsymbol{q}$ are $\mathcal{C}$ - resolutions, we speak of a $\mathcal{C}$ - resolution $(\boldsymbol{p}, \boldsymbol{q}, \boldsymbol{f})$ of the mapping f.

We will often consider a covering $\mathcal{V}$ of a space Y and two $\mathcal{V}$ - near mappings $f, f': X \to Y$. In such cases we will use the abbreviation $(f, f') \prec \mathcal{V}$. We will also use the notation $\mathcal{V}' \prec \mathcal{V}$ to mean that the covering $\mathcal{V}'$ *refines* the covering $\mathcal{V}$ (Note that this differs from the notation used in (Mardešić, Segal 1982)). If $A \subseteq Y$, then the *star* $\mathrm{St}(A, \mathcal{V})$ denotes the union of all members of $\mathcal{V}$, which intersect A. All sets $\mathrm{St}(V, \mathcal{V})$, where $V \in \mathcal{V}$, form an open covering, denoted by $\mathrm{St}(\mathcal{V})$. If for a covering $\mathcal{V}$ the covering $\mathrm{St}(\mathcal{V})$ refines a covering $\mathcal{W}$, we say that $\mathcal{V}$ is a *star-refinement* of $\mathcal{W}$ and we write $\mathcal{V} \prec^* \mathcal{W}$. We denote by $\mathrm{Cov}(X)$ the set of all *normal* (*numerable*) coverings of the space X. These are coverings, which admit a partition of unity subordinated to the covering (see e.g., (Mardešić, Segal 1982), Appendix 1, 3.1).

REMARK 6.2. By a *polyhedron* we always mean the geometric realization of a (possibly not locally finite) simplicial complex, endowed with the CW - topology. Conditions (R1) and (R2) can be stated for any class $\mathcal{C}$ of spaces, in place of the class Pol of polyhedra. However, in the general case one must require that $\mathcal{V}$ be a normal covering. Recall that polyhedra are paracompact spaces and for paracompact spaces open coverings are always normal (see e.g., (Mardešić, Segal 1982), Appendix 1, 1.1, Corollary 3 and 3.1, Corollary 1).

LEMMA 6.3. *Condition* (Ri), i=1,2, *for polyhedra, is equivalent to the corresponding condition, for any of the following classes $\mathcal{C}$: polyhedra endowed with the metric topology* MPol, CW - *complexes* CW, *absolute neighborhood retracts for metric spaces* ANR.

We precede the proof by introducing some useful notions. A class $\mathcal{C}$ of spaces is *approximately dominated* by a class $\mathcal{C}'$ provided every space P from $\mathcal{C}$ and every normal covering $\mathcal{V}$ of P admit a space P' from $\mathcal{C}'$ and admit mappings $\phi: P \to P'$ and $\psi: P' \to P$ such that $(\psi\phi, \mathrm{id}) \prec \mathcal{V}$. If one imposes the stronger condition that $\psi\phi$ and id are connected by a $\mathcal{V}$-homotopy, then one obtains the notion of *approximate homotopy domination.* If also $\mathcal{C}'$ is *approximately (homotopy) dominated* by $\mathcal{C}$, then $\mathcal{C}$ and $\mathcal{C}'$ are *approximately (homotopy) equivalent* classes. It is well known that polyhedra, polyhedra with the metric topology, CW - complexes and ANR's are approximately homotopy equivalent classes. All of these classes are contained in the class AP of *approximate polyhedra,* which consists of all spaces approximately dominated by polyhedra (see (Mardešić 1981b), Theorem 1 and Remark 2). It is now clear that Lemma 6.3 is a consequence of the following Lemma.

LEMMA 6.4. *Let $\mathcal{C}$ be a class of spaces approximately dominated by a class $\mathcal{C}'$. If condition* (Ri), i=1,2, *holds for the class $\mathcal{C}'$, then it also holds for the class $\mathcal{C}$.*

Proof. Let P be a space from $\mathcal{C}$ and let $\mathcal{V}$ be a normal covering of P. In order to verify condition (R1) for $\mathcal{C}$, consider any mapping $f: X \to P$. Let $\mathcal{W}$ be a star-refinement of $\mathcal{V}$. Choose a member P' of $\mathcal{C}'$ and choose mappings $\phi: P \to P', \psi: P' \to P$ such that $(\psi\phi, \mathrm{id}) \prec \mathcal{W}$. By (R1) for $\mathcal{C}'$, there exist a $\lambda \in \Lambda$ and a mapping $h': X_\lambda \to P'$ such that $(h'p_\lambda, \phi f) \prec \psi^{-1}(\mathcal{W})$, and therefore, $(\psi h'p_\lambda, \psi\phi f) \prec \mathcal{W}$. Since also $(\psi\phi f, f) \prec \mathcal{W}$, one concludes that $(hp_\lambda, f) \prec \mathcal{V}$, for $h = \psi h': X_\lambda \to P$.

In order to verify condition (R2) for $\mathcal{C}$, consider $\mathcal{W}, P', \phi$ and ψ as above. Let $\mathcal{W}'$ be the normal covering of P' obtained by applying (R2) for $\mathcal{C}'$ to P' and $\psi^{-1}(\mathcal{W})$. We claim that $\mathcal{V}' = \phi^{-1}(\mathcal{W}')$ has the property required by (R2) for $\mathcal{C}$. Indeed, let $h, h': X_\lambda \to P$ be mappings such that $(hp_\lambda, h'p_\lambda) \prec \mathcal{V}'$. Then $(\phi hp_\lambda, \phi h'p_\lambda) \prec \mathcal{W}'$. Therefore, there exists a $\lambda' \geq \lambda$ such that $(\phi hp_{\lambda\lambda'}, \phi h'p_{\lambda\lambda'}) \prec \psi^{-1}(\mathcal{W})$ and thus, $(\psi\phi hp_{\lambda\lambda'}, \psi\phi h'p_{\lambda\lambda'}) \prec \mathcal{W}$. Since also $(hp_{\lambda\lambda'}, \psi\phi hp_{\lambda\lambda'}) \prec \mathcal{W}$, $(h'p_{\lambda\lambda'}, \psi\phi h'p_{\lambda\lambda'}) \prec \mathcal{W}$, the assumption $\mathcal{W} \prec^* \mathcal{V}$ implies the desired conclusion $(hp_{\lambda\lambda'}, h'p_{\lambda\lambda'}) \prec \mathcal{V}$. □

An inverse system $\boldsymbol{X}$ is said to be a *resolution* provided there exist a topological space X and a resolution $\boldsymbol{p}: X \to \boldsymbol{X}$ of X.

THEOREM 6.5. *If in a resolution $\boldsymbol{X} = (X_\lambda, p_{\lambda\lambda'}, \Lambda)$ all terms X_λ are Tychonoff, topologically complete or compact Hausdorff spaces, then there exists* *a resolution* $\boldsymbol{p} = (p_\lambda): X \to \boldsymbol{X}$*, where the space* X *is Tychonoff, topologically complete or compact Hausdorff, respectively.*

Proof. By definition, there exists a topological space Y and a mapping $\boldsymbol{q} = (q_\lambda): Y \to \boldsymbol{X}$, which is a resolution of Y. We distinguish three cases.

(*i*) All X_λ are Tychonoff spaces. It is well known that the *Tychonoff functor* τ is a *reflector* from the category Top to the full subcategory of Tychonoff spaces, i.e., with every space X is associated a mapping $r_X: X \to \tau(X)$ such that every mapping $\phi: X \to Z$ to a Tychonoff space Z admits a unique mapping $\psi: \tau(X) \to Z$ such that $\phi = \psi r_X$. Moreover, r_X is surjective (Morita 1975a). Put $X = \tau(Y)$ and notice that every mapping $q_\lambda: Y \to X_\lambda$ determines a unique mapping $p_\lambda: X \to X_\lambda$ such that $q_\lambda = p_\lambda r_Y$. If $\lambda \le \lambda'$, one has $p_{\lambda\lambda'} p_{\lambda'} r_Y = p_{\lambda\lambda'} q_{\lambda'} = q_\lambda = p_\lambda r_Y$ and therefore, by uniqueness, $p_{\lambda\lambda'} p_{\lambda'} = p_\lambda$. Let us show that $\boldsymbol{p} = (p_\lambda): X \to \boldsymbol{X}$ is a resolution of the Tychonoff space X, i.e., it has properties (R1),(R2).

Indeed, let P be a polyhedron and let $\mathcal{V} \in \mathrm{Cov(P)}$. For a mapping $f: X \to P$, consider the composition $g = f r_Y$. By (R1) for $\boldsymbol{q}$, there exists a λ and a mapping $h: X_\lambda \to P$ such that

$$(h q_\lambda, g) \prec \mathcal{V}. \tag{2}$$

Since $q_\lambda = p_\lambda r_Y$, (2) becomes

$$(h q_\lambda r_Y, f r_Y) \prec \mathcal{V}. \tag{3}$$

However, r_Y is a surjection and thus, (3) implies $(h q_\lambda, f) \prec \mathcal{V}$, which establishes (R1) for $\boldsymbol{p}$. If $\mathcal{V}' \in \mathrm{Cov}(P)$ satisfies (R2) for $\boldsymbol{q}$, then it also satisfies (R2) for $\boldsymbol{p}$, because $(h p_\lambda, h' p_\lambda) \prec \mathcal{V}'$ implies $(h q_\lambda, h' q_\lambda) \prec \mathcal{V}'$ and therefore, $(h p_{\lambda\lambda'}, h' p_{\lambda\lambda'}) \prec \mathcal{V}$, for some $\lambda' \ge \lambda$.

(*ii*) All X_λ are topologically complete spaces, i.e., they are Tychonoff spaces, complete with respect to their finest uniformity. Recall that this is the uniformity generated by all normal coverings (see 8.1.C and 8.5.13 in (Engelking 1977)). Consider the functor c, which assigns to every Tychonoff space X the completion $c(X)$ of X with respect to the finest uniformity, which induces the topology of X. This functor is a reflector from the category of Tychonoff spaces to the full subcategory of topologically complete spaces. The reflection $r_X: X \to c(X)$ need not be surjective, but it maps X onto a dense subset of $c(X)$. By case (*i*), there is no loss of generality in assuming that Y is a Tychonoff space. Putting $X = c(Y)$ and repeating the argument used in case (*i*), one obtains mappings $p_\lambda: X \to X_\lambda$ such that $q_\lambda = p_\lambda r_Y$. Moreover, for $\lambda \le \lambda'$, one has $p_{\lambda\lambda'} p_{\lambda'} r_Y = p_\lambda r_Y$. Since $r_T(Y)$ is dense in X, it follows that also in this case $p_{\lambda\lambda'} p_{\lambda'} = p_\lambda$.

To verify (R1) for $\boldsymbol{p} = (p_\lambda) : X \to \boldsymbol{X}$, consider a mapping $f : X \to P$, where P is a polyhedron. For $\mathcal{V} \in \mathrm{Cov}(P)$, choose a star-refinement $\mathcal{W}$ of $\mathcal{V}$. Put $g = f r_Y: Y \to P$ and choose λ and $h: X_\lambda \to P$ so that

$$(h q_\lambda, g) \prec \mathcal{W}. \tag{4}$$

It now suffices to show that

$$(hp_\lambda, f) \prec \mathrm{St}(\mathcal{W}). \tag{5}$$

Choose a normal covering $\mathcal{G}$ of X, which refines both coverings $f^{-1}(\mathcal{W})$ and $(hp_\lambda)^{-1}(\mathcal{W})$. For a point $x \in X$, choose a member G of $\mathcal{G}$, which contains x. Since $r_Y(Y)$ is dense in X, there exists a point $y \in Y$, such that $r_Y(y) \in G$. Clearly, there exist members $W', W'' \in \mathcal{W}$ such that $f(x), fr_Y(y) \in W'$ and $hp_\lambda(x), hp_\lambda r_Y(y) \in W''$. Moreover, (4) implies the existence of a member $W \in \mathcal{W}$ such that $hq_\lambda(y), fr_Y(y) \in W$. Consequently, $W' \cup W'' \subseteq \mathrm{St}(W, \mathcal{W})$ and $f(x), hp_\lambda(x) \in W' \cup W''$, which establishes (5). The proof of (R2), used in case (i), applies to the present case as well.

(iii) All X_λ are compact Hausdorff spaces. Consider the functor β, which assigns to every Tychonoff space its Čech – Stone compactification. β is a reflector from the category of Tychonoff spaces to the full subcategory of compact Hausdorff spaces. Moreover, the reflection $r_X: X \to \beta X$ maps X onto a dense subset of βX. By case (i), there is no loss of generality in assuming that Y is a Tychonof space. Repeating the argument described in case (ii), one concludes that the mappings $p_\lambda : X \to X_\lambda$, determined by $q_\lambda = p_\lambda r_Y$, form a resolution $\boldsymbol{p} = (p_\lambda): X \to \boldsymbol{X}$ of the compact Hausdorff space $X = \beta Y$. □

REMARK 6.6. The class of topologically complete spaces includes all paracompact spaces (see 8.5.13 (b) of (Engelking 1977)). In particular, it includes metric spaces, polyhedra, CW - complexes and ANR's.

6.2 Characterization of resolutions

In this subsection we state two conditions, for a mapping $\boldsymbol{p}: X \to \boldsymbol{X}$, which are equivalent to (R1) and (R2) (for polyhedra), but are often easier to verify.

(B1) *For every normal covering $\mathcal{U}$ of X, there exist an index $\lambda \in \Lambda$ and a normal covering $\mathcal{U}_\lambda$ of X_λ, such that $p_\lambda^{-1}(\mathcal{U}_\lambda)$ refines $\mathcal{U}$.*

(B2) *For every $\lambda \in \Lambda$ and every normal covering $\mathcal{U}_\lambda$ of X_λ, there exists a $\lambda' \geq \lambda$ such*

$$p_{\lambda\lambda'}(X_{\lambda'}) \subseteq \mathrm{St}(p_\lambda(X), \mathcal{U}_\lambda). \tag{1}$$

THEOREM 6.7. *A mapping of systems $\boldsymbol{p}: X \to \boldsymbol{X}$ is a resolution if and only if it has properties* (B1) *and* (B2). *More precisely,* (Ri) *for polyhedra implies* (Bi), i=1,2. *Conversely,* (B1)∧(B2) ⇒ (R1) *for polyhedra and* (B2) ⇒ (R2) *for polyhedra.*

Proof. (R1)⇒(B1). For $\mathcal{U} \in \mathrm{Cov}(X)$ let N denote the nerve of $\mathcal{U}$, let $|N|$ be its geometric realization (CW - topology) and let $f: X \to |N|$ be a canonical mapping. Let $\mathcal{W}$ be a star-refinement of the covering $\mathcal{N}$ of $|N|$, formed by all open stars $\mathrm{St}(U,N), U \in \mathcal{U}$, of the vertices of N. Then (R1) yields a $\lambda \in \Lambda$ and a mapping $h\colon X_\lambda \to |N|$ such that $(f, hp_\lambda) \prec \mathcal{W}$. Clearly,

$$p_\lambda^{-1}h^{-1}(W) \subseteq f^{-1}(\mathrm{St}(W,\mathcal{W})),\ W \in \mathcal{W}. \tag{2}$$

Since every $W \in \mathcal{W}$ admits a $U \in \mathcal{U}$ such that $\mathrm{St}(W,\mathcal{W}) \subseteq \mathrm{St}(U,N)$, it follows that

$$p_\lambda^{-1}h^{-1}(W) \subseteq f^{-1}(\mathrm{St}(U,N)) \subseteq U. \tag{3}$$

Consequently, the normal covering $\mathcal{U}_\lambda = h^{-1}(\mathcal{W})$ of X_λ has the desired property $p_\lambda^{-1}(\mathcal{U}_\lambda) \prec \mathcal{U}$.

(R2)$\Rightarrow$(B2). Let $\mathcal{U}_\lambda$ be a normal covering of X_λ. Consider its nerve N and a canonical mapping $f\colon X_\lambda \to |N|$. Then the covering $\mathcal{V}$ of $|N|$, consisting of all open stars of the vertices of N, has the property that $f^{-1}(\mathcal{V})$ refines $\mathcal{U}_\lambda$. Therefore, for any subset $A \subseteq X_\lambda$, one has

$$f^{-1}(\mathrm{St}(f(A)),\mathcal{V}) \subseteq \mathrm{St}(A,\mathcal{U}_\lambda). \tag{4}$$

Notice that, for any subset $B \subseteq |N|$, the closure $\overline{B} \subseteq \mathrm{St}(B,\mathcal{V})$. Therefore, there exists a mapping $g\colon |N| \to I = [0,1]$ such that

$$g|\overline{fp_\lambda(X)} = 0, \tag{5}$$

$$g|(|N| \backslash \mathrm{St}(fp_\lambda(X),\mathcal{V})) = 1. \tag{6}$$

Now consider the covering $\mathcal{W}$ of I, which consists of the sets $[0,1), (0,1]$ and let $\mathcal{W}'$ be the covering of I, obtained by applying (R2) to I and $\mathcal{W}$. The mappings $h = gf\colon X_\lambda \to I$, $h' = 0\colon X_\lambda \to I$ are $\mathcal{W}'$-near, because $hp_\lambda = h'p_\lambda = 0$. Hence, there exists a $\lambda' \geq \lambda$ such that $hp_{\lambda\lambda'}$ and $h'p_{\lambda\lambda'} = 0$ are $\mathcal{W}$-near. However, this implies $hp_{\lambda\lambda'}(X_{\lambda'}) \subseteq [0,1)$. Since $h = gf$, (6) yields

$$fp_{\lambda\lambda'}(X_{\lambda'}) \subseteq \mathrm{St}(fp_\lambda(X),\mathcal{V}). \tag{7}$$

Applying (4) to $A = p_\lambda(X_\lambda)$, one obtains the desired relation (1).

(B1)$\wedge$($B2$) $\Rightarrow$(R1). Let P be a polyhedron, $\mathcal{V} \in \mathrm{Cov}(P)$ and let $f\colon X \to P$ be a mapping. There exists a triangulation K of P such that the open covering $\mathcal{K}$, which consists of open stars $\mathrm{St}(v,K)$ of the vertices v of K, refines $\mathcal{V}$ (see (Mardešić, Segal 1982), Appendix 1, 1.1, Theorem 4). By (B1), there exist a $\lambda \in \Lambda$ and a $\mathcal{U}_\lambda \in \mathrm{Cov}(X_\lambda)$ such that

$$p_\lambda^{-1}(\mathcal{U}_\lambda) \prec f^{-1}(\mathcal{K}). \tag{8}$$

Consider the nerve N of $\mathcal{U}_\lambda$ and a canonical mapping $g\colon X_\lambda \to |N|$. Denote by N' the subcomplex of N, whose simplices $(U_0,\dots,U_n)$ are spanned by vertices $U_0,\dots,U_n$ of N, for which

$$U_0 \cap \ldots \cap U_n \cap p_\lambda(X) \neq \emptyset. \tag{9}$$

Clearly, if $x \in X$ and $(U_0,\dots,U_n)$ is the simplex of N, which contains $gp_\lambda(x)$ in its interior, then

$$p_\lambda(x) \in g^{-1}(\mathrm{St}(U_0, N)) \cap \ldots \cap g^{-1}(\mathrm{St}(U_n, N)) \subseteq U_0 \cap \ldots \cap U_n, \tag{10}$$

which shows that $(U_0, \ldots, U_n)$ belongs to N'. Consequently,

$$gp_\lambda(X) \subseteq |N'|. \tag{11}$$

By (8), there exists a function k, which assigns to each vertex $U \in \mathcal{U}_\lambda$ of N, a vertex $k(U) \in K$ such that

$$p_\lambda^{-1}(U) \subseteq f^{-1}(\mathrm{St}(k(U), K)). \tag{12}$$

Since (9) implies

$$p_\lambda^{-1}(U_0) \cap \ldots \cap p_\lambda^{-1}(U_n) \neq \emptyset, \tag{13}$$

(12) yields

$$\mathrm{St}(k(U_0), K) \cap \ldots \cap \mathrm{St}(k(U_n), K) \neq \emptyset, \tag{14}$$

and therefore, $(k(U_0), \ldots, k(U_n))$ is a simplex of K. This shows that the restriction of k to the vertices of N' defines a simplicial mapping $k'\colon |N'| \to |K| = P$.

By (11), $k'gp_\lambda\colon X \to P$ is a well-defined mapping. Moreover,

$$(k'gp_\lambda, f) \prec \mathcal{V}. \tag{15}$$

Indeed, for $x \in X$ and $(U_0, \ldots, U_n) \in N'$ as above, $k'gp_\lambda(x)$ is an interior point of the simplex $(k'(U_0), \ldots, k'(U_n))$ of K, and thus,

$$k'gp_\lambda(x) \in \mathrm{St}(k'(U_0), K). \tag{16}$$

On the other hand, by (10) and (12)

$$x \in p_\lambda^{-1}(U_0) \subseteq f^{-1}(\mathrm{St}(k'(U_0), K)), \tag{17}$$

and thus,

$$f(x) \in \mathrm{St}(k'(U_0), K). \tag{18}$$

Since $\mathcal{K} \prec \mathcal{V}$, one concludes that there is a $V \in \mathcal{V}$, which contains both points $k'gp_\lambda(x), f(x)$.

To complete the proof, we now use the fact that every subcomplex of a complex is a neighborhood retract (see (Hu 1965), Chapter III, Lemma 10.1). In particular, there exist an open neighborhood H of $|N'|$ in $|N|$ and a retraction $r\colon H \to |N'|$. Since $\{H, |N| \backslash |N'|\}$ is an open covering of $|N|$, $\mathcal{W} = \{g^{-1}(H), g^{-1}(|N| \backslash |N'|)\}$ is a normal covering of X_λ. By (B2), applied to X_λ and $\mathcal{W}$, there exists a $\lambda' \geq \lambda$ such that

$$p_{\lambda\lambda'}(X_{\lambda'}) \subseteq \mathrm{St}(p_\lambda(X), \mathcal{W}). \tag{19}$$

On the other hand, (11) implies

$$p_\lambda(X) \subseteq g^{-1}(|N'|) \subseteq g^{-1}(H), \tag{20}$$

$$p_\lambda(X) \cap g^{-1}(|N| \backslash |N'|)\} = \emptyset, \tag{21}$$

and thus,

$$\mathrm{St}(p_\lambda(X), \mathcal{W}) = g^{-1}(H). \tag{22}$$

Consequently, (19) becomes

$$p_{\lambda\lambda'}(X_{\lambda'}) \subseteq g^{-1}(H). \tag{23}$$

According to (23), $h = k'rgp_{\lambda\lambda'}: X_{\lambda'} \to P$ is a well-defined mapping. Moreover, by (11), $hp_{\lambda'} = k'rgp_\lambda = k'gp_\lambda$ and thus, (15) yields the desired conclusion $(hp_{\lambda'}, f) \prec \mathcal{V}$.

(B2)⇒(R2). If P is a polyhedron and $\mathcal{V} \in \mathrm{Cov}(P)$, choose as $\mathcal{V}'$ any star-refinement of $\mathcal{V}$. Let $h, h': X_\lambda \to P$ be mappings such that

$$(hp_\lambda, h'p_\lambda) \prec \mathcal{V}'. \tag{24}$$

Let $\mathcal{U}_\lambda$ be a normal covering of X_λ, which refines the normal coverings $h^{-1}(\mathcal{V}')$ and $h'^{-1}(\mathcal{V}')$. Consider the set $G = \mathrm{St}(p_\lambda(X), \mathcal{U}_\lambda)$. We will show that

$$(h|G, h'|G) \prec \mathcal{V}. \tag{25}$$

Indeed, for any $y \in G$, there is an $x \in X$ and there is a $U \in \mathcal{U}_\lambda$ such that $y, p_\lambda(x) \in U$. Moreover, there are elements $V', V'' \in \mathcal{V}'$ such that $U \subseteq h^{-1}(V'), U \subseteq h'^{-1}(V'')$. Furthermore, by (24), there is a $V \in \mathcal{V}'$ such that $hp_\lambda(x), h'p_\lambda(x) \in V$. Since $hp_\lambda(x) \in h(U) \subseteq V'$, we see that $V \cap V' \neq \emptyset$. Similarly, $V \cap V'' \neq \emptyset$. Therefore, $V' \cup V'' \subseteq \mathrm{St}(V, \mathcal{V}')$. Finally, $\mathcal{V}' \prec^* \mathcal{V}$ shows that there is an element $W \in \mathcal{V}$, which contains $\mathrm{St}(V, \mathcal{V}')$. However, $h(y) \in h(U) \subseteq V'$ and $h'(y) \in h'(U) \subseteq V''$. Hence, W contains both points $h(y), h'(y)$.

To conclude the proof, we apply (B2) to the covering $\mathcal{U}_\lambda$ and obtain a $\lambda' \geq \lambda$ such that $p_{\lambda\lambda'}(X_{\lambda'}) \subseteq G$. Consequently, (25) yields the desired relation $(hp_{\lambda\lambda'}, h'p_{\lambda\lambda'}) \prec \mathcal{V}$. □

COROLLARY 6.8. *Let M be a topological space, X a subset of M and $(X_\lambda,\ \lambda \in \Lambda)$ a basis of neighborhoods of X in M. Order Λ by putting $\lambda \leq \lambda'$, whenever $X_{\lambda'} \subseteq X_\lambda$, and let $p_{\lambda\lambda'}: X_{\lambda'} \to X_\lambda$, $p_\lambda: X \to X_\lambda$ be inclusion mappings. Then $\boldsymbol{X} = (X_\lambda, p_{\lambda\lambda'}, \Lambda)$ is an inverse system and $\boldsymbol{p} = (p_\lambda): X \to \boldsymbol{X}$ is a mapping of systems. If all X_λ are paracompact spaces, then $\boldsymbol{p}$ is a resolution of X.*

Proof. A normal covering $\mathcal{U}$ of X extends to an open covering $\mathcal{V}$ of some neighborhood G of X. Choose a $\lambda \in \Lambda$ such that $X_\lambda \subseteq G$. The restriction of $\mathcal{V}$ to X_λ is an open covering $\mathcal{U}_\lambda$ of X_λ such that $p_\lambda^{-1}(\mathcal{U}_\lambda) \prec \mathcal{U}$. The covering $\mathcal{U}_\lambda$ is normal, because X_λ is a paracompact space. Hence, (B1) holds. The verification of (B2) is immediate. □

COROLLARY 6.9. *If M is an ANR and X is an arbitrary subset of M, then all open neighborhoods X_λ of X and inclusions $p_{\lambda\lambda'}: X_{\lambda'} \to X_\lambda, p_\lambda: X \to X_\lambda$ form an ANR-resolution of X.*

EXAMPLE 6.10. The following example shows that condition (B1) alone does not imply (R1). Let $\Lambda = \mathbb{N}$, let X_n be the m-ball B^m, let $X = S^{m-1} = \partial B^m$, $p_{nn'} = \mathrm{id}$ and let $p_n: X \to X_n$ be the inclusion mapping. Obviously, $\boldsymbol{p} = (p_n): X \to \boldsymbol{X} = (X_n, p_{nn'}, \mathbb{N})$ has property (B1). To see that it does not have property (R1), consider the polyhedron $P = S^{m-1}$ and any open covering $\mathcal{V}$ of S^{m-1}, so fine that any two $\mathcal{V}$-near mappings into S^{m-1} are homotopic. For $f = \mathrm{id}: X \to P$, condition (R1) would yield an $n \in \mathbb{N}$ and a mapping $h: X_n = B^m \to S^{m-1}$ such that $(hp_n, f) \prec \mathcal{V}$ and therefore, f would be homotopic to the contractible mapping hp_n, which is a contradiction. Clearly, in this example condition (B2) is not satisfied.

EXAMPLE 6.11. We now describe an inverse sequence $\boldsymbol{X} = (X_n, p_{n\,n+1})$, for which (B2) holds, but (B1) does not. Let $X_n \subseteq \mathbb{R}^2$ be the union of the line $R_n = \mathbb{R} \times \{0\}$ and the ray $R_n^+ = \{n\} \times \mathbb{R}^+$, where $\mathbb{R}^+ = \{t \in \mathbb{R}: t \geq 0\}$. Let $p_{n\,n+1}: X_{n+1} \to X_n$ be the mapping, which is identity on $\mathbb{R} \times \{0\}$, it maps the point $(n+1, t)$ to $(n, t-1)$, if $t \geq 1$ and to $(n+1-t, 0)$, if $0 \leq t \leq 1$. Note that the limit $X = \lim \boldsymbol{X}$ is a metric space which has two components U, V, both homeomorphic to $\mathbb{R}$. Moreover, for every n, $\mathbb{R} \times \{0\}$ belongs to the first component U, while $p_n^{-1}(R_n^+)$ belongs to the second component V. Since the mappings $p_{n\,n+1}$ are surjective, so are the projections p_n and thus, (B2) is fulfilled. However, for no n and no open covering $\mathcal{V}_n$ of X_n will $p_n^{-1}(\mathcal{V}_n)$ refine the open covering $\mathcal{V} = \{U, V\}$, because, for every member V_n of $\mathcal{V}_n$ which contains the point $(n, 0) \in X_n$, the set $p_n^{-1}(V_n)$ intersects both components U and V.

REMARK 6.12. Lemma (2.13) of (Mardešić, Watanabe 1989) erroneously asserts that, for approximate mappings into approximate inverse systems, condition (B1) implies (R1). The correct statement is as above, (B1)$\wedge$(B2) $\Rightarrow$ R1. To correct the proof one must use a neighborhood retraction and a new index λ' as in the corresponding proof of above. This gap does not affect the rest of the paper.

REMARK 6.13. The following simple condition (B2)′ implies condition (B2).

(B2)′ *For every $\lambda \in \Lambda$ and every open neighborhood U of $\overline{p_\lambda(X)}$ in X_λ, there exists a $\lambda' \geq \lambda$ such that*

$$p_{\lambda\lambda'}(X_{\lambda'}) \subseteq U. \tag{26}$$

Indeed, for any open covering $\mathcal{U}_\lambda$ of X_λ and any subset $A \subseteq X_\lambda$, $\overline{A} \subseteq \mathrm{St}(A, \mathcal{U}_\lambda)$. Application of (B2)′ to $A = p_\lambda(X)$ and $U = \mathrm{St}(p_\lambda(X), \mathcal{U}_\lambda)$ yields a $\lambda' \geq \lambda$, which satisfies (1). Hence, (B2)′ $\Rightarrow$ (B2).

The converse implication (B2) $\Rightarrow$ (B2)′ holds if all spaces X_λ are normal. Indeed, in a normal space every finite open covering is normal. Therefore, the sets $U, X_\lambda \backslash \overline{p_\lambda(X)}$ form a normal covering $\mathcal{U}_\lambda$ of X_λ. Application of (B2) to $\mathcal{U}_\lambda$ yields an index $\lambda' \geq \lambda$ such that (1) holds. However, $\mathrm{St}(p_\lambda(X), \mathcal{U}_\lambda) = U$ and (1) becomes (26).

REMARK 6.14. If all the projections p_λ in a mapping of systems $\boldsymbol{p}$ are surjective, condition (B2) is fulfilled. On the other hand, condition (B2) cannot be fulfilled if X is empty and all X_λ are non-empty. Consequently, for a resolution $\boldsymbol{p}: X \to \boldsymbol{X}$, $X_\lambda \neq \emptyset$, for all $\lambda \in \Lambda$, implies $X \neq \emptyset$.

REMARK 6.15. In this book, by dimension of a topological space X, we always mean the *covering dimension* $\dim X$, based on normal (open) coverings. Consequently, $\dim X \leq n$ means that every normal covering $\mathcal{U}$ of X admits a normal refinement $\mathcal{V}$ of order $\leq n$. For normal spaces, this is equivalent to the requirement that every finite open covering admits a finite open refinement of order $\leq n$ ((Morita 1975b), Theorem 1.3). The following is an immediate consequence of condition (B1) and the above definition. If $\boldsymbol{p}: X \to \boldsymbol{X}$ is a resolution and $\dim X_\lambda \leq n$, for all λ, then also $\dim X \leq n$.

6.3 Resolutions versus limits

A resolution of a space X and a limit of an inverse system $\boldsymbol{X}$ are both mappings of systems $\boldsymbol{p}: X \to \boldsymbol{X}$. We will now compare these two notions. We first prove a general theorem, which gives sufficient conditions for a resolution to be an inverse limit.

THEOREM 6.16. *Let $\boldsymbol{p}: X \to \boldsymbol{X} = (X_\lambda, p_{\lambda\lambda'}, \Lambda)$ be a resolution. If all X_λ are Tychonoff spaces and X is a topologically complete space, then $\boldsymbol{p}$ is an inverse limit of $\boldsymbol{X}$.*

Proof. We need to verify that $\boldsymbol{p}$ has the *universal property of inverse limits*, i.e., for any mapping $\boldsymbol{q} = (q_\lambda): Y \to \boldsymbol{X}$, there exists a unique mapping $q: Y \to X$ such that $q_\lambda = p_\lambda q$, for all $\lambda \in \Lambda$. The proof consists of several steps.

(*i*) *For any $y \in Y, \lambda \in \Lambda$ and $\mathcal{U}_\lambda \in \mathrm{Cov}(X_\lambda)$, the set*

$$p_\lambda^{-1}(\mathrm{St}(q_\lambda(y), \mathcal{U}_\lambda)) \neq \emptyset. \tag{1}$$

Indeed, by (B2), there is a $\lambda' \geq \lambda$ such that $p_{\lambda\lambda'}(X_{\lambda'}) \subseteq \mathrm{St}(p_\lambda(X), \mathcal{U}_\lambda)$. Since $q_\lambda(y) = p_{\lambda\lambda'}(q_{\lambda'}(y))$ and $q_{\lambda'}(y) \in X_{\lambda'}$, it follows that $q_\lambda(y) \in \mathrm{St}(p_\lambda(X), \mathcal{U}_\lambda)$. However, this is equivalent to $p_\lambda(X) \cap \mathrm{St}(q_\lambda(y), \mathcal{U}_\lambda) \neq \emptyset$ and thus, equivalent to (1).

(*ii*) *For a fixed $y \in Y$, the family $\mathcal{F}(y)$, which consists of all sets of the form*

$$F = p_\lambda^{-1}(\overline{\mathrm{St}(q_\lambda(y), \mathcal{U}_\lambda)}), \tag{2}$$

where $\lambda \in \Lambda$, $\mathcal{U}_\lambda \in \mathrm{Cov}(X_\lambda)$, is a centered system of closed sets.

Indeed, let F_i, $i = 1, \dots, n$, be a finite collection of members from $\mathcal{F}(y)$, given by indices λ_i and coverings $\mathcal{U}_{\lambda_i}$. Choose a $\lambda \geq \lambda_1, \dots, \lambda_n$ and choose a normal covering $\mathcal{U}_\lambda$ of X_λ, which refines all the coverings $p_{\lambda_i\lambda}^{-1}(\mathcal{U}_{\lambda_i})$, $i = 1, \dots, n$. Since $p_{\lambda_i\lambda}(q_\lambda(y)) = q_{\lambda_i}(y)$, one has

$$\mathrm{St}(q_\lambda(y),\mathcal{U}_\lambda) \subseteq p^{-1}_{\lambda_i\lambda}(\mathrm{St}(q_{\lambda_i}(y),\mathcal{U}_{\lambda_i})),\ i=1,\dots,n, \tag{3}$$

which implies

$$p_\lambda^{-1}(\mathrm{St}(q_\lambda(y),\mathcal{U}_\lambda)) \subseteq F_i,\ i=1,\dots,n. \tag{4}$$

However, by (1), the left side of (4) is non-empty and thus, $F_1\cap\ldots\cap F_n \neq \emptyset$.

(iii) The family $\mathcal{F}(y)$ contains arbitrarily small sets with respect to the finest uniformity of X and is therefore, a Cauchy family with respect to that uniformity.

Given a covering $\mathcal{U} \in \mathrm{Cov}(\mathrm{X})$, we must exhibit $\lambda \in \Lambda$, $\mathcal{U}_\lambda \in \mathrm{Cov}(X_\lambda)$ and $U \in \mathcal{U}$ such that F, given by (2), satisfies $F \subseteq U$. To achieve this, apply (B1) to $\mathcal{U}$. One obtains a λ and a $\mathcal{W} \in \mathrm{Cov}(X_\lambda)$ such that

$$p_\lambda^{-1}(\mathcal{W}) \prec \mathcal{U}. \tag{5}$$

Choose coverings $\mathcal{V},\mathcal{U}_\lambda \in \mathrm{Cov}(X_\lambda)$ so that $\mathcal{U}_\lambda \prec^* \mathcal{V} \prec^* \mathcal{W}$. Then there is a $W \in \mathcal{W}$ such that

$$\mathrm{St}(q_\lambda(y),\mathcal{V}) \subseteq W, \tag{6}$$

and there is a $U \in \mathcal{U}$ such that

$$p_\lambda^{-1}(W) \subseteq U. \tag{7}$$

We claim that this choice of λ, $\mathcal{U}_\lambda$ and U yields the desired relation $F \subseteq U$. Indeed, for any $x \in F$ there is a $U'_\lambda \in \mathcal{U}_\lambda$ such that $p_\lambda(x) \in U'_\lambda$ and U'_λ intersects $\mathrm{St}(q_\lambda(y),\mathcal{U}_\lambda)$. Consequently, U'_λ must intersect some member $U''_\lambda \in \mathrm{Cov}(\mathcal{U}_\lambda)$, which contains $q_\lambda(y)$. Since $\mathcal{U}_\lambda \prec^* \mathcal{V}$, one can choose a $V \in \mathcal{V}$ such that $U'_\lambda \cup U''_\lambda \subseteq V$. Then, by (6), $p_\lambda(x) \in V \subseteq \mathrm{St}(q_\lambda(y),\mathcal{V})) \subseteq W$. Finally, (7) yields the desired conclusion $x \in U$.

(iv) There is a unique function $q\colon Y \to X$ such that $p_\lambda q = q_\lambda$, for every $\lambda \in \Lambda$.

For $y \in Y$, consider the Cauchy family $\mathcal{F}(y)$ and define $q(y)$ as the intersection of all members of this family. Since X is complete with respect to its finest uniformity, this intersection is a unique point of X. Note that

$$p_\lambda(q(y)) \in \overline{\mathrm{St}(q_\lambda(y),\mathcal{U}_\lambda)}, \tag{8}$$

for all $\mathcal{U}_\lambda \in \mathrm{Cov}(X_\lambda)$. Since X_λ is a Tychonoff space, the intersection of all sets on the right side of (8) is just the point $q_\lambda(y)$. Consequently, $p_\lambda q(y) = q_\lambda(y)$. In order to prove uniqueness, assume that $q'\colon Y \to X$ is another function, satisfying $p_\lambda q'(y) = q_\lambda(y)$, for all $y \in Y$. Since $q_\lambda(y) \in \overline{\mathrm{St}(q_\lambda(y),\mathcal{U}_\lambda)}$, it follows that $q'(y) \in (p_\lambda)^{-1}\overline{(\mathrm{St}(q_\lambda(y),\mathcal{U}_\lambda))} = F$, i.e., $q'(y)$ belongs to the intersection of all $F \in \mathcal{F}(y)$. Hence, $q'(y) = q(y)$.

(v) $q\colon Y \to X$ is continuous. For $y \in Y$ and an open neighborhood G of $q(y)$ in X, choose $\mathcal{U} \in \mathrm{Cov}(X)$ such that

$$\mathrm{St}(q(y),\mathcal{U}) \subseteq G. \tag{9}$$

Choose $\lambda \in \Lambda, \mathcal{U}_\lambda \in \mathrm{Cov}(X_\lambda), U \in \mathcal{U}$ and $F \in \mathcal{F}(y)$ as in the proof of (iii) and recall that $F \subseteq U$. By the definition of q, the point $q(y) \in F$ and therefore,

$$q(y) \in U \subseteq \mathrm{St}(q(y), \mathcal{U}). \tag{10}$$

On the other hand, by the continuity of q_λ, there exists an open neighborhood H of y in Y such that

$$q_\lambda(H) \subseteq \mathrm{St}(q_\lambda(y), \mathcal{U}_\lambda). \tag{11}$$

To complete the proof, it suffices to show that

$$q(H) \subseteq G. \tag{12}$$

Consider any $y' \in H$. Since $p_\lambda q(y') = q_\lambda(y')$, (11) implies

$$q(y') \in (p_\lambda)^{-1}(\mathrm{St}(q_\lambda(y), \mathcal{U}_\lambda)) \subseteq F \subseteq U. \tag{13}$$

Consequently, (10) and (9) yield the desired conclusion $q(y') \in G$. □

EXAMPLE 6.17. There exist inverse systems $\boldsymbol{X}$, which are not resolutions. E.g., such are the systems described in Example 6.1. Indeed, assume that there exists a space X and a resolution $\boldsymbol{p}: X \to \boldsymbol{X}$. Since the terms X_λ in these systems are polyhedra, Theorem 6.5 enables us to assume that X is topologically complete. Then Theorem 6.16 implies that $X = \lim \boldsymbol{X}$. However, in Example 6.1, $X_\lambda \neq \emptyset$ and $\lim \boldsymbol{X} = \emptyset$, which contradicts Remark 6.14.

EXAMPLE 6.18. We will now exhibit an example of a resolution, which is not a limit. Let $X = [0, \omega_1)$ be the space of all countable ordinals λ. It is well-known that every mapping $f: X \to \mathbb{R}$ is eventually constant, i.e., $f|[\lambda_0, \omega_1)$ is constant, for some countable ordinal λ_0 (see e.g., (Engelking 1977), Example 3.1.27). Let us prove the stronger statement that every mapping $f: X \to P$ to a polyhedron $P = |K|$ is also eventually constant. Let Q be the complex $|K|$ endowed with the metric topology. Then the identity mapping $i: P \to Q$ is continuous (see e.g., (Mardešić, Segal 1982), Appendix 1, 1.3, Corollary 5). Clearly, it suffices to prove that $g = if: X \to Q$ is eventually constant. We first show that, for every $n \in \mathbb{N}$, there is a countable ordinal $\lambda(n)$ such that $\mathrm{diam}\, g([\lambda(n), \omega_1)) \leq \frac{1}{n}$. If this were not the case, one could find an $n_0 \in \mathbb{N}$ and a sequence of countable ordinals $\lambda_1 < \mu_1 < \lambda_2 < \mu_2 < \ldots$ such that the distance $d(\lambda_k, \mu_k) > \frac{1}{n_0}$, for all k. Clearly, both sequences $(\lambda_k), (\mu_k)$ have a common limit point $\nu < \omega_1$. However, g cannot be continuous at the point ν. It is now clear that g is constant on $[\lambda, \omega_1)$, where $\lambda = \sup\{\lambda(n) | n \in \mathbb{N}\}$.

Now let Λ be the set of all countable ordinals $\lambda < \omega_1$ endowed with its natural ordering. Let $X_\lambda = [0, \lambda] \subseteq X$ and let $p_{\lambda\lambda'}: X_{\lambda'} \to X_\lambda$, $\lambda < \lambda'$, and $p_\lambda: X \to X_\lambda$ be retractions such that $p_{\lambda\lambda'}(X_{\lambda'} \backslash X_\lambda) = \{\lambda\}$ and $p_\lambda(X \backslash X_\lambda) = \{\lambda\}$. Clearly, $\boldsymbol{X} = (X_\lambda, p_{\lambda\lambda'}, \Lambda)$ is an inverse system of

0-dimensional compact metric spaces and $\boldsymbol{p}: X \to \boldsymbol{X}$ is a mapping. By the above property of X, every mapping $f: X \to P$ admits an exact factorization $f = hp_\lambda$. Therefore, (R1) is fulfilled. (R2) also holds, because the projections p_λ are surjective. Consequently, $\boldsymbol{p}$ is a resolution of X. However, $\boldsymbol{p}$ cannot be a limit of $\boldsymbol{X}$, because X is not compact.

EXAMPLE 6.19. An inverse system $\boldsymbol{X} = (X_\lambda, p_{\lambda\lambda'}, \Lambda)$ can be the resolution of topologically different spaces. Indeed, consider the resolution $\boldsymbol{p}: X \to \boldsymbol{X}$ described in the preceding example. Let $X' = [0, \omega_1]$ and let $i: X \to X'$ be the inclusion mapping. Then for every $\lambda < \omega_1$, there is a unique extension $q_\lambda: X' \to X_\lambda$ of $p_\lambda: X \to X_\lambda$, because X_λ embeds in $\mathbb{R}$ and p_λ is eventually constant. Clearly, $\boldsymbol{q} = (q_\lambda): X' \to \boldsymbol{X}$ is also a resolution. However, X' is compact, while X is not. This example illustrates the phenomenon considered in Theorem 6.5.

The next theorem gives an important case, when limits are resolutions.

THEOREM 6.20. *Let $\boldsymbol{X} = (X_\lambda, p_{\lambda\lambda'}, \Lambda)$ be an inverse system of compact Hausdorff spaces X_λ with limit $\boldsymbol{p} = (p_\lambda): X \to \boldsymbol{X}$. Then $\boldsymbol{p}$ is a resolution.*

Proof. It suffices to verify conditions (B1) and (B2)$'$.

(B1). An arbitrary covering $\mathcal{U} \in \mathrm{Cov}(X)$ can be refined by a covering consisting of sets of the form $(p_\lambda)^{-1}(V_\lambda)$, where $\lambda \in \Lambda$ and $V_\lambda \subseteq X_\lambda$ is open. By compactness of X, this covering has a finite subcovering

$$\{(p_{\lambda_1})^{-1}(V_{\lambda_1}), \ldots, (p_{\lambda_n})^{-1}(V_{\lambda_n})\}. \tag{14}$$

Choose $\lambda \geq \lambda_1, \ldots, \lambda_n$. Then the collection

$$\mathcal{U}_\lambda = \{(p_{\lambda_1\lambda})^{-1}(V_{\lambda_1}), \ldots, (p_{\lambda_n\lambda})^{-1}(V_{\lambda_n}), X_\lambda \backslash p_\lambda(X)\} \tag{15}$$

is an open covering of X_λ, which has the desired property $(p_\lambda)^{-1}(\mathcal{U}_\lambda) \prec \mathcal{U}$.

(B2)$'$. Assume the contrary, i.e., that there exist a λ and an open neighborhood U of $p_\lambda(X)$ in X_λ such that

$$Y_\mu = X_\mu \backslash (p_{\lambda\mu})^{-1}(U) \neq \emptyset, \tag{16}$$

for every $\mu \geq \lambda$. Putting $M = \{\mu \in \Lambda | \mu \geq \lambda\}$, $q_{\mu\mu'} = p_{\mu\mu'}|Y_{\mu'}: Y_{\mu'} \to Y_\mu$, one obtains an inverse system $\boldsymbol{Y} = (Y_\mu, q_{\mu\mu'}, M)$ of non-empty compact Hausdorff spaces. Therefore, its limit $Y \neq \emptyset$. The inclusion maps $i_\mu: Y_\mu \to X_\mu$, $\mu \in M$, define a level-preserving mapping $\boldsymbol{i}: \boldsymbol{Y} \to \boldsymbol{X}|M$ to the restriction of $\boldsymbol{X}$ to M, which induces a mapping $i: Y \to X' = \lim(\boldsymbol{X}|M)$. Moreover, the identity mappings on X_μ define a mapping $\boldsymbol{j}: \boldsymbol{X} \to \boldsymbol{X}|M$, which induces a mapping $j: X \to X'$. Since M is cofinal in Λ, it follows that j is a homeomorphism. Now choose any point $y \in Y$ and choose a point $x \in X$ such that $j(x) = i(y)$. Then $p_\lambda(x) = p_\lambda j(x) = p_\lambda i(y) = i_\lambda q_\lambda(y) = q_\lambda(y) \in Y_\lambda$. However, by (16), $Y_\lambda = X_\lambda \backslash U$ and thus, $p_\lambda(x) \notin U$, which is a contradiction, because U is a neighborhood of $p_\lambda(x)$. □

If the resolution of a completely regular space X consists of compact Hausdorff spaces, X need not be compact Hausdorff as Example 6.18 shows. One has the following theorem.

THEOREM 6.21. *For a Tychonoff space X the following assertions are equivalent:*

(i) X is pseudocompact.

(ii) X admits a resolution consisting of compact polyhedra.

(iii) X admits a resolution consisting of compact Hausdorff spaces.

Proof. (i) $\Rightarrow$ (ii). For X pseudocompact, consider its Čech – Stone compactification $i: X \to \beta(X)$. Since $\beta(X)$ is a compact Hausdorff space, it admits a limit $\boldsymbol{q}: \beta(X) \to \boldsymbol{X}$, where $\boldsymbol{X} = (X_\lambda, p_{\lambda\lambda'}, \Lambda)$ consists of compact polyhedra X_λ. Clearly, the mappings $p_\lambda = q_\lambda i: X \to X_\lambda$ form a system of mappings $\boldsymbol{p} = X \to \boldsymbol{X}$. It suffices to show that $\boldsymbol{p}$ is a resolution of X. Let $f: X \to P$ be a mapping into a polyhedron $P = |K|$ and let $\mathcal{V} \in \mathrm{Cov}(P)$. Recall that in a space, whose open subsets are paracompact, all subsets are paracompact (see (Lundell, Weingram 1969), Appendix I, Lemma 8). This is the case with polyhedra, because their open subsets are again polyhedra and thus, are paracompact. Consequently, $f(X)$ is paracompact. Obviously, $f(X)$ is also pseudocompact. However, pseudocompact paracompact spaces are always compact (see (Engelking 1977), Theorems 3.10.20 and 5.1.20). It follows that $f(X)$ is compact, and is therefore, contained in a compact subpolyhedron $Q \subseteq P$. This implies that $f: X \to Q$ admits an extension $g: \beta(X) \to Q \subseteq P$, $gi = f$. By Theorem 6.20, $\boldsymbol{q}$ is a resolution. Therefore, property (R1) for $\boldsymbol{q}$, yields $\lambda \in \Lambda$ and $h: X_\lambda \to Q \subseteq P$, such that $(hq_\lambda, g) \prec \mathcal{V}$. However, this implies $(hp_\lambda, f) \prec \mathcal{V}$, which establishes (R1) for $\boldsymbol{p}$. By Theorem 6.7, (R1) $\Rightarrow$ (B1). To prove (B2)$'$ for $\boldsymbol{p}$, note that $i(X)$ is dense in $\beta(X)$ and therefore, $\overline{p_\lambda(X)} = \overline{q_\lambda(\beta(X))}$. Consequently, (B2)$'$ for $\boldsymbol{q}$, yields the desired conclusion.

(ii) $\Rightarrow$ (iii) is obvious. In order to show that (iii) $\Rightarrow$ (i), assume that $\boldsymbol{p}: X \to \boldsymbol{X}$ is a compact resolution of X. Given a mapping $f: X \to \mathbb{R}$, consider a covering $\mathcal{V}$ of $\mathbb{R}$, which consists of open intervals of length 1. Then (R1) yields a λ and a mapping $h: X_\lambda \to \mathbb{R}$, such that $(hp_\lambda, f) \prec \mathcal{V}$. Since $hp_\lambda(X) \subseteq h(X_\lambda)$ is compact and therefore, bounded, it follows that also $f(X)$ is bounded. Hence, X is pseudocompact. □

6.4 Existence of polyhedral and ANR - resolutions

It is well-known that every compact Hausdorff space X is the limit of an inverse system of compact polyhedra ((Mardešić, Segal 1982), I.5.2, Theorem 7). Therefore, Theorem 6.20 shows that compact Hausdorff spaces admit resolutions consisting of compact polyhedra. We will generalize this result to arbitrary spaces and mappings.

THEOREM 6.22. *Every topological space X admits a polyhedral resolution. Every mapping of topological spaces admits a polyhedral resolution.*

THEOREM 6.23. *Every topological space X admits an* ANR-*resolution. Every mapping of topological spaces admits an* ANR-*resolution.*

In the proofs of both theorems, we will use the construction and lemma which follow. Let $\boldsymbol{X} = (X_\lambda, p_{\lambda\lambda'}, \Lambda)$ be an inverse system of spaces and let $\boldsymbol{p} = (p_\lambda): X \to \boldsymbol{X}$ be a mapping of systems. For each $\lambda \in \Lambda$, let $\mathcal{G}_\lambda$ be the set of all open neighborhoods of $\mathrm{Cl}(p_\lambda(X))$ in X_λ. Let Λ^* be the set of all pairs $\lambda^* = (\lambda, G)$, where $\lambda \in \Lambda$, $G \in \mathcal{G}_\lambda$. Let $X^*_{\lambda^*} = G$ and let $p^*_{\lambda^*}: X \to X^*_{\lambda^*}$ be the mapping $p_\lambda: X \to G \subseteq X_\lambda$. Put $\lambda^* \leq \lambda^{*\prime} = (\lambda', G')$, provided $\lambda \leq \lambda'$ and $p_{\lambda\lambda'}(G') \subseteq G$. Let $p^*_{\lambda^*\lambda^{*\prime}}: X^*_{\lambda^{*\prime}} \to X^*_{\lambda^*}$ be the mapping $p_{\lambda\lambda'}|G': G' \to G$. Clearly, $\boldsymbol{X}^* = (X^*_{\lambda^*}, p^*_{\lambda^*\lambda^{*\prime}}, \Lambda^*)$ is an inverse system and $\boldsymbol{p}^* = (p^*_{\lambda^*}): X \to \boldsymbol{X}^*$ is a mapping.

LEMMA 6.24. *If $\boldsymbol{X}$ is an inverse system of normal spaces and $\boldsymbol{p}: X \to \boldsymbol{X}$ is a mapping which has property* (B1), *then $\boldsymbol{p}^*: X \to \boldsymbol{X}^*$ is a resolution.*

Proof. If $\mathcal{U} \in \mathrm{Cov}(X)$, choose a $\lambda \in \Lambda$ and a $\mathcal{U}_\lambda \in \mathrm{Cov}(X_\lambda)$ such that $(p_\lambda)^{-1}(\mathcal{U}_\lambda) \prec \mathcal{U}$. If $\lambda^* = (\lambda, G) \in \Lambda^*$, one has $(p^*_\lambda)^{-1}(\mathcal{U}_\lambda|X^*_{\lambda^*}) \prec \mathcal{U}$. Hence, $\boldsymbol{p}^*$ has property (B1). The mapping $\boldsymbol{p}^*$ also has property (B2)$'$. Indeed, let $\lambda^* = (\lambda, G)$ and let U be an open neighborhood of $\mathrm{Cl}(p^*_{\lambda^*}(X)) = \mathrm{Cl}(p_\lambda(X))$ in $X^*_{\lambda^*} = G$. Then $U \in \mathcal{G}_\lambda$ and $\lambda^{*\prime} = (\lambda, U) \in \Lambda^*$. Since $p_{\lambda\lambda} = \mathrm{id}$ and $U \subseteq G$, one has $\lambda^* \leq \lambda^{*\prime}$. Moreover, $X^*_{\lambda^{*\prime}} = U$ and thus, $p^*_{\lambda^*\lambda^{*\prime}}(X^*_{\lambda^{*\prime}}) = p_{\lambda\lambda}(U) = U$. That $\boldsymbol{p}^*$ is indeed a resolution, it now follows by applying Remark 6.13. □

Proof of Theorem 6.22. With every normal covering $\mathcal{U}$ of X associate a locally finite partition of unity $\Phi_{\mathcal{U}} = (\phi_U, U \in \mathcal{U})$, subordinated to $\mathcal{U}$. Let $N_{\mathcal{U}}$ be the nerve of $\mathcal{U}$ and let $|N_{\mathcal{U}}|$ be its geometric realization (CW-topology). Recall that the vertices of $N_{\mathcal{U}}$ are the members $U \in \mathcal{U}$ and the points y of $|N_{\mathcal{U}}|$ are completely determined by their barycentric coordinates $\alpha_U(y)$, associated with the vertices of $N_{\mathcal{U}}$. Moreover, $\Phi_{\mathcal{U}}$ determines a canonical mapping $p_{\mathcal{U}}: X \to |N_{\mathcal{U}}|$, given by

$$\alpha_U(p_{\mathcal{U}}(x)) = \phi_U(x),\ U \in \mathcal{U}. \tag{1}$$

Let Λ be the set of all finite subsets $\lambda = \{\mathcal{U}_1, \ldots, \mathcal{U}_n\}$, which consist of different normal coverings of X. Let $\leq$ denote the ordering of Λ by inclusion $\subseteq$. With every $\lambda \in \Lambda$ associate the covering $\mathcal{U}_\lambda \in \mathrm{Cov}(X)$, which consists of the sets $U_\lambda = U_1 \cap \ldots \cap U_n$, where $U_i \in \mathcal{U}_i$, $i = 1, \ldots, n$. Let N_λ be the nerve of $\mathcal{U}_\lambda$ and let $X_\lambda = |N_\lambda|$. If $\lambda \leq \lambda' = \{\mathcal{U}_1, \ldots, \mathcal{U}_n, \ldots, \mathcal{U}_{n'}\}$, we consider the function $p_{\lambda\lambda'}$, which maps the vertex $\{U_1, \ldots, U_n, \ldots, U_{n'}\}$ of $N_{\lambda'}$ to the vertex $\{U_1, \ldots, U_n\}$ of N_λ. Since $U_1 \cap \ldots \cap U_n \cap \ldots \cap U_{n'} \subseteq U_1 \cap \ldots \cap U_n$, the function $p_{\lambda\lambda'}$ induces a simplicial mapping $p_{\lambda\lambda'}: X_{\lambda'} \to X_\lambda$. Using barycentric coordinates, this mapping is given by the formula

$$\alpha_{\{U_1,\ldots,U_n\}}(p_{\lambda\lambda'}(y)) = \sum \alpha_{\{U_1,\ldots,U_n,\ldots,U_{n'}\}}(y),\ y \in X_{\lambda'}, \tag{2}$$

where summation is taken over all vertices $\{U_1, \dots, U_n, \dots, U_{n'}\}$ of $N_{\lambda'}$ with prescribed $\{U_1, \dots, U_n\}$. It is readily seen that $p_{\lambda\lambda'} p_{\lambda'\lambda''} = p_{\lambda\lambda''}$, for $\lambda \leq \lambda' \leq \lambda''$. Consequently, $\boldsymbol{X} = (X_\lambda, p_{\lambda\lambda'}, \Lambda)$ is an inverse system of polyhedra.

With every $\lambda = \{\mathcal{U}_1, \dots, \mathcal{U}_n\} \in \Lambda$ we now associate the family Φ_λ of functions

$$\phi_{\{U_1,\dots,U_n\}} = \phi_{U_1} \cdot \dots \cdot \phi_{U_n} : X \to [0,1], \tag{3}$$

where ϕ_{U_k} is the member of $\Phi_{\mathcal{U}_k}$, which belongs to $U_k \in \mathcal{U}_k$, $k = 1, \dots, n$. It is readily seen that Φ_λ is a partition of unity, subordinated to the covering $\mathcal{U}_\lambda$. Let $p_\lambda : X \to X_\lambda$ be the canonical mapping, given by the partition of unity Φ_λ, i.e.,

$$\alpha_{\{U_1,\dots,U_n\}}(p_\lambda(x)) = \phi_{\{U_1,\dots,U_n\}}(x), \ x \in X. \tag{4}$$

Then, by (2),

$$\alpha_{\{U_1,\dots,U_n\}}(p_{\lambda\lambda'} p_{\lambda'}(x)) = \sum \alpha_{\{U_1,\dots,U_n,\dots,U_{n'}\}}(p_{\lambda'}(x)), \ x \in X. \tag{5}$$

However, by (4) and (3),

$$\begin{aligned} \alpha_{\{U_1,\dots,U_n,\dots,U_{n'}\}}(p_{\lambda'}(x)) &= \phi_{\{U_1,\dots,U_n,\dots,U_{n'}\}}(x) \\ &= \phi_{U_1}(x) \cdot \dots \cdot \phi_{U_n}(x) \cdot \dots \cdot \phi_{U_{n'}}(x). \end{aligned} \tag{6}$$

Since

$$\sum_{U_{n+1}} \phi_{U_{n+1}}(x) = 1, \dots, \sum_{U_{n'}} \phi_{U_{n'}}(x) = 1, \tag{7}$$

(5) becomes

$$\begin{aligned} \alpha_{\{U_1,\dots,U_n\}}(p_{\lambda\lambda'} p_{\lambda'}(x)) &= \phi_{U_1}(x) \cdot \dots \cdot \phi_{U_n}(x) \\ &= \phi_{\{U_1,\dots,U_n\}}(x) = \alpha_{\{U_1,\dots,U_n\}}(p_\lambda(x)), \end{aligned} \tag{8}$$

which shows that $p_{\lambda\lambda'} p_{\lambda'} = p_\lambda$. Consequently, $\boldsymbol{p} = (p_\lambda) : X \to \boldsymbol{X}$ is a mapping of systems.

Furthermore, $\boldsymbol{p}$ has property (B1), because, for any normal covering $\mathcal{U}$ of X and $\lambda = \{\mathcal{U}\}$, $p_\lambda = p_\mathcal{U}$ is the canonical mapping, belonging to $\mathcal{U}$. Therefore, the open covering $\mathcal{K}$ of X_λ, formed by the open stars of the vertices of $N_\mathcal{U}$, satisfies the condition $(p_\lambda)^{-1}(\mathcal{K}) \prec \mathcal{U}$. Now Lemma 6.24 shows that $\boldsymbol{p}^* : X \to \boldsymbol{X}^*$ is a resolution of X. Since the terms of $\boldsymbol{X}^*$ are open subsets of polyhedra, they are themselves polyhedra. Hence, $\boldsymbol{p}^*$ is a polyhedral resolution. This completes the proof of the first assertion.

Now consider a mapping $f : X \to Y$. In order to obtain a polyhedral resolution of f, we first apply the above described constructions to Y. This yields a mapping $\boldsymbol{q} = (q_\mu) : Y \to \boldsymbol{Y} = (Y_\mu, q_{\mu\mu'}, M)$ and a polyhedral resolution $\boldsymbol{q}^* = (q^*_{\mu^*}) : Y \to \boldsymbol{Y}^* = (Y^*_{\mu^*}, q_{\mu^*\mu^{*\prime}}, M)$. Analogously, we define a mapping $\boldsymbol{p} : X \to \boldsymbol{X}$ and a polyhedral resolution $\boldsymbol{p}^* : X \to \boldsymbol{X}^*$. However, in defining $\boldsymbol{p}$ we are allowed to choose arbitrary partitions $\Phi_\mathcal{U}$ only if $\mathcal{U}$ is not of the form $\mathcal{U} = f^{-1}(\mathcal{V})$, for $\mathcal{V} \in \mathrm{Cov}(Y)$. In the case when

$\mathcal{U} = (f^{-1}(V), V \in \mathcal{V})$, we define ϕ_U, for $U = f^{-1}(V)$, by putting $\phi_U = \psi_V f$, where $\Psi_{\mathcal{V}} = (\psi_V, V \in \mathcal{V})$ is the partition of unity used in defining $\boldsymbol{q}$. We now define a mapping $\boldsymbol{f} = (f, f_\mu): \boldsymbol{X} \to \boldsymbol{Y}$ as follows. If $\mu = \{\mathcal{V}_1, \dots, \mathcal{V}_n\}$, put $f(\mu) = \{f^{-1}(\mathcal{V}_1), \dots, f^{-1}(\mathcal{V}_n)\}$ and note that $f: M \to \Lambda$ is an increasing function. Then define a simpicial mapping $f_\mu: N_{f(\mu)} \to N_\mu$, by sending the vertex $\{f^{-1}(V_1), \dots, f^{-1}(V_n)\}$ to the vertex $\{V_1, \dots, V_n\}$. By definition, $f_\mu: X_{f(\mu)} \to Y_\mu$ is the corresponding piecewise linear mapping. It is readily seen that

$$f_\mu p_{\mu\mu'} = q_{\mu\mu'} f_{\mu'}, \ \mu \le \mu', \tag{9}$$

and thus, $\boldsymbol{f}: \boldsymbol{X} \to \boldsymbol{Y}$ is a mapping. Due to the special choice of the partitions of unity $\Phi_{\mathcal{U}}$, it is also easy to verify that

$$f_\mu p_{f(\mu)} = q_\mu f, \ \mu \in M. \tag{10}$$

We now extend $\boldsymbol{f}$ to a mapping $\boldsymbol{f}^* = (f^*, f^*_{\mu^*}): \boldsymbol{X}^* \to \boldsymbol{Y}^*$ as follows. If $\mu^* = (\mu, H) \in M^*$, we put $f^*(\mu^*) = (f(\mu), f_\mu^{-1}(H)) \in \Lambda^*$. Note that f^* is an increasing function. We define $f^*_{\mu^*}: X^*_{f^*(\mu^*)} \to Y^*_{\mu^*}$, by putting $f^*_{\mu^*} = f_\mu | f_\mu^{-1}(H)$. A straightforward verification shows that $\boldsymbol{f}^*$ is indeed a mapping. Moreover, (10) implies that

$$f^*_{\mu^*} p^*_{f^*(\mu^*)} = q^*_{\mu^*} f, \ \mu^* \in M^*, \tag{11}$$

which completes the proof that (p^*, q^*, f^*) is a polyhedral resolution of f. □

In the proof of Theorem 6.23, we need the following lemma.

LEMMA 6.25. *Let X be a topological space and let $f: X \to Q$ be a mapping into an* ANR *Q. Then there exists an* ANR *P, with density $d(P) \le d(X)$, and there exist maps $g: X \to P$, $h: P \to Q$ such that $f = hg$.*

Proof of Lemma 6.25. By the Kuratowski – Wojdisławski embedding theorem (see (Mardešić, Segal 1982), I. 3.1, Theorem 2), one can assume that $f(X)$ is contained in a normed vector space L and that $f(X)$ is closed in its convex hull $K \subseteq L$. It is easy to see that $d(K) = d(f(X)) \le d(X)$. Since Q is an ANR, the inclusion mapping $f(X) \to Q$ extends to a mapping $h: P \to Q$, where P is a sufficiently small open neighborhood of $f(X)$ in K. By Dugundji's extension theorem (see (Mardešić, Segal 1982), I. 3.1, Theorem 3), K is an ANR. Consequently, P is also an ANR. Moreover, $d(P) \le d(K) \le d(X)$. Finally, if one denotes by $g: X \to P$ the composition of $f: X \to f(X)$ with the inclusion $f(X) \to P$, then $f = hg$. □

Proof of Theorem 6.23. Let X be a space of density $d(X) = \kappa$. Let $\{p_\gamma: \gamma \in \Gamma\}$ be the set of all mappings $p_\gamma: X \to X_\gamma$ of X to ANR's, contained in the Tychonoff cube I^κ. Let $(\Lambda, \le)$ be the set of all finite subsets $\lambda = \{\gamma_1, \dots, \gamma_n\}$ of Γ, ordered by inclusion $\subseteq$. Put $X_\lambda = X_{\gamma_1} \times \dots \times X_{\gamma_n}$ and note that X_λ is also an ANR. For $\lambda \le \lambda' = \{\gamma_1, \dots, \gamma_n, \dots, \gamma_{n'}\}$, define $p_{\lambda\lambda'}: X_{\lambda'} \to X_\lambda$ to be the natural projection. Then $\boldsymbol{X} = (X_\lambda, p_{\lambda\lambda'}, \Lambda)$ is an inverse system of

ANR's. Define $p_\lambda: X \to X_\lambda$ to be the mapping $p_\lambda = p_{\gamma_1} \times \ldots \times p_{\gamma_n}$. Then $\boldsymbol{p} = (p_\lambda)$ is a mapping of systems.

Let us show that $\boldsymbol{p}$ has property (R1) for ANR's, hence, by Lemma 6.3, it also has property (R1) for polyhedra and by Theorem 6.7, it has property (B1). Let $f: X \to Q$ be a mapping into an arbitrary ANR Q. By Lemma 6.25, there exists an ANR P with $d(P) \leq d(X)$ and there exist mappings $g: X \to P$ and $h: P \to Q$ such that $f = hg$. It is well-known that the density and the weight of a metric space coincide ((Engelking 1977), Theorem 4.1.15). Therefore, the weight of P does not exceed κ. Consequently, one can assume that P is contained in the Tychonoff cube I^κ, and thus $g = p_\gamma$, for some $\gamma \in \Gamma$. Putting $\lambda = \{\gamma\}$, we see that $f = hp_\lambda$, which is a condition even stronger than (R1). In order to complete the proof, it now suffices to apply Lemma 6.24 to $\boldsymbol{p}$. It yields a resolution $\boldsymbol{p}^*: X \to \boldsymbol{X}^*$, whose terms are open subsets of ANR's, hence, they are also ANR's.

We now consider the case of a mapping $f: X \to Y$. Apply the above described construction to X and Y. It yields mappings $\boldsymbol{p}: X \to \boldsymbol{X}$ and $\boldsymbol{q} = (q_\mu): Y \to \boldsymbol{Y} = (Y_\mu, q_{\mu\mu'}, M)$ and ANR-resolutions $\boldsymbol{p}^*: X \to \boldsymbol{X}^*$ and $\boldsymbol{q}^*: Y \to \boldsymbol{Y}^*$. To define a mapping $\boldsymbol{f} = (f, f_\mu): \boldsymbol{X} \to \boldsymbol{Y}$, note that M consists of finite subsets $\mu = \{\delta_1, \ldots, \delta_n\}$ of a set Δ. For $\delta \in \Delta$, apply Lemma 6.25 to the mapping $q_\delta f: X \to Y_\delta$. It yields a factorization $q_\delta f = hg$ through an ANR P of density $d(P) \leq \kappa$. There is no loss of generality in assuming that P lies in I^κ. Therefore, $g = p_\gamma$, for some $\gamma \in \Gamma$. Putting $f(\delta) = \gamma$ and $f_\delta = h$, one obtains a function $f: \Delta \to \Gamma$ and mappings $f_\delta: X_{f(\delta)} \to Y_\delta$ such that

$$q_\delta f = f_\delta p_{f(\delta)}, \tag{12}$$

Now extend f to a function $f: M \to \Lambda$, by putting $f(\mu) = \{f(\delta_1), \ldots, f(\delta_n)\}$, for $\mu = \{\delta_1, \ldots, \delta_n\}$. Moreover, define $f_\mu: X_{f(\mu)} \to Y_\mu$ by putting $f_\mu = f_{\delta_1} \times \ldots \times f_{\delta_n}$. It is readily verified that $\boldsymbol{f}$ is indeed a mapping $\boldsymbol{f}: \boldsymbol{X} \to \boldsymbol{Y}$. Moreover, (12) implies that $\boldsymbol{qf} = \boldsymbol{fp}$. The extension of $\boldsymbol{f}$ to a mapping $\boldsymbol{f}^*: \boldsymbol{X}^* \to \boldsymbol{Y}^*$ now proceeds as in the case of Theorem 6.22. □

REMARK 6.26. Every 0-dimensional space X admits a resolution $\boldsymbol{p}: X \to \boldsymbol{X}$, where $\boldsymbol{X}$ consists of discrete spaces.

To verify this assertion one considers the set $\{p_\gamma: \gamma \in \Gamma\}$ of all mappings $p_\gamma: X \to X_\gamma$ of X to discrete spaces contained in I^κ, where $\kappa = d(X)$. Then one defines $\boldsymbol{X} = (X_\lambda, p_{\lambda\lambda'}, \Lambda)$ and $\boldsymbol{p} = (p_\lambda): X \to \boldsymbol{X}$ as in the proof of the first part of Theorem 6.23. Note that every X_λ, being a product of finitely many discrete spaces, is itself discrete. Since $\dim X = 0$, every normal covering $\mathcal{U}$ of X admits an open refinement $\mathcal{V}$, which consists of disjoint open sets. Let P be the discrete space, whose points are members of $\mathcal{V}$. Obviously, the quotient mapping $g: X \to P$ is continuous and the covering $\mathcal{W}$ of P, which consists of all the singletons has the property that $g^{-1}(\mathcal{W}) = \mathcal{V}$ refines $\mathcal{U}$. Since the weight $\mathrm{w}(P) = d(P) \leq d(X) = \kappa$, one can assume that P is embedded in I^κ and $g = p_\gamma$, for some $\gamma \in \Gamma$. Putting $\lambda = \{\gamma\}$, we see that

$p_\lambda = g$ and therefore, $p_\lambda^{-1}(\mathcal{W})$ refines $\mathcal{U}$, which establishes property (B1) for $\boldsymbol{p}$. Application of Lemma 6.24 to $\boldsymbol{p}$ now yields a resolution $\boldsymbol{p}^*: X \to \boldsymbol{X}^*$. The system $\boldsymbol{X}^*$ consists of discrete spaces because every subset of a discrete space is discrete.

REMARK 6.27. There exist compact Hausdorff spaces X with $\dim X = 1$, which do not admit resolutions consisting of 1 - dimensional polyhedra (1 - dimensional ANR's) (Pasynkov 1958), (Mardešić 1960), (Mardešić, Watanabe 1988).

REMARK 6.28. With every inverse system $\boldsymbol{X} = (X_\lambda, p_{\lambda\lambda'}, \Lambda)$ over a preordered set Λ, one can associate an inverse system $\boldsymbol{X}'$, indexed by an ordered set Λ' as follows. In each equivalence class of Λ with respect to the equivalence relation $\sim$ (see 1.1), one chooses a single element. These elements form the set $\Lambda' \subseteq \Lambda$. Clearly, the preordering $\leq'$ of Λ', inherited from Λ is antisymmetric, i.e., it is an ordering. The new system $\boldsymbol{X}'$ is obtained by restricting the system $\boldsymbol{X}$ to the subset Λ'. Furthermore, with every morphism of systems $\boldsymbol{p} = (p_\lambda): X \to \boldsymbol{X}$ one can associate a morphism $\boldsymbol{p}': X \to \boldsymbol{X}'$, by restricting $\boldsymbol{p}$ to Λ'. Similarly, if $\boldsymbol{f} = (f, f_\mu): \boldsymbol{X} \to \boldsymbol{Y}$ is a mapping, one can define a mapping $\boldsymbol{f}' = (f', f'_\mu): \boldsymbol{X}' \to \boldsymbol{Y}'$ as follows. If $\mu \in M'$, one takes for $f'(\mu)$ the only representative of the class of $f(\mu)$, which belongs to Λ'. Since $f(\mu) \sim f'(\mu)$, there is a well-defined homeomorphism $p_{f(\mu)f'(\mu)}: X_{f'(\mu)} \to X_{f(\mu)}$. One puts $f'_\mu = f_\mu p_{f(\mu)f'(\mu)}$. Note that, if $(\boldsymbol{p}, \boldsymbol{q}, \boldsymbol{f})$ satisfies $\boldsymbol{f}\boldsymbol{p} = \boldsymbol{q}f$, then $(\boldsymbol{p}', \boldsymbol{q}', \boldsymbol{f}')$ satisfies

$$\boldsymbol{f}'\boldsymbol{p}' = \boldsymbol{q}'f. \tag{13}$$

LEMMA 6.29. *If $\boldsymbol{p}: X \to \boldsymbol{X}$ is a resolution of X, then so is also $\boldsymbol{p}': X \to \boldsymbol{X}'$. If $(\boldsymbol{p}, \boldsymbol{q}, \boldsymbol{f})$ is a resolution of $f: X \to Y$, then so is $(\boldsymbol{p}', \boldsymbol{q}', \boldsymbol{f}')$.*

Proof. Let $f: X \to P$ be a mapping into a polyhedron P, let $\mathcal{V} \in \mathrm{Cov}(P)$, $\lambda \in \Lambda$, and let $g: X_\lambda \to P$ be a mapping such that $(gp_\lambda, f) \prec \mathcal{V}$. There is a (unique) $\lambda' \in \Lambda'$ such that $\lambda' \sim \lambda$. Clearly, the mapping $g' = gp_{\lambda\lambda'}: X_{\lambda'} \to P$ satisfies $(g'p_{\lambda'}, f) \prec \mathcal{V}$, which is property (R1), for $\boldsymbol{p}'$. Similarly, if a covering $\mathcal{V}'$ satisfies (R2), for $\boldsymbol{p}$, then it satisfies (R2) also for $\boldsymbol{p}'$. Indeed, if two mappings $f, f': X_{\lambda'} \to P$, $\lambda' \in \Lambda'$, have the property that $(fp_{\lambda'}, f'p_{\lambda'}) \prec \mathcal{V}'$, then there exists a $\lambda \geq \lambda'$ such that $(fp_{\lambda'\lambda}, f'p_{\lambda'\lambda}) \prec \mathcal{V}$. Let λ'' be the only element of Λ', which is equivalent to λ. Composing with $p_{\lambda\lambda''}$, we conclude that $(fp_{\lambda'\lambda''}, f'p_{\lambda'\lambda''}) \prec \mathcal{V}$, which establishes (R2), for $\boldsymbol{p}'$. This completes the proof of the first assertion. The second assertion is a consequence of the first assertion, for $\boldsymbol{p}'$ and $\boldsymbol{q}'$, and of (13). □

REMARK 6.30. With every inverse system $\boldsymbol{X} = (X_\lambda, p_{\lambda\lambda'}, \Lambda)$ over a directed ordered set Λ, one can associate an inverse system $\boldsymbol{X}^* = (X^*_\alpha, p^*_{\alpha\alpha'}, \Lambda^*)$, indexed by a cofinite set as follows. Λ^* is the set of all finite subsets $\alpha \subseteq \Lambda$, which have a terminal element, denoted by $\overline{\alpha}$. Note that $\overline{\alpha}$ is uniquely determined by α, because Λ is an ordered set. By definition, the ordering $\leq^*$ of Λ^* is the inclusion $\subseteq$. Clearly, this ordering is cofinite. To see that Λ^* is directed, consider two arbitrary elements $\alpha_1, \alpha_2 \in \Lambda^*$ and choose an element $\lambda \in \Lambda$, such that $\lambda \geq \overline{\alpha_1}, \overline{\alpha_2}$. Then $\alpha = \alpha_1 \cup \alpha_2 \cup \{\lambda\}$ belongs to Λ^* and $\alpha_1, \alpha_2 \leq^* \alpha$. Putting $X^*_\alpha = X_{\overline{\alpha}}$, $p^*_{\alpha\alpha'} = p_{\overline{\alpha}\,\overline{\alpha}'}$, one obtains the inverse system $\boldsymbol{X}^*$. Furthermore, with every morphism of systems $\boldsymbol{p} = (p_\lambda): X \to \boldsymbol{X}$ one can associate a morphism $\boldsymbol{p}^* = (p^*_\alpha): X \to \boldsymbol{X}^*$, by putting $p^*_\alpha = p_{\overline{\alpha}}$.

Similarly, if $\boldsymbol{f} = (f, f_\mu): \boldsymbol{X} \to \boldsymbol{Y}$ is a mapping and the index sets Λ and M are ordered, then one can define a mapping $\boldsymbol{f}^* = (f^*, f^*_\beta): \boldsymbol{X}^* \to \boldsymbol{Y}^*$ as follows. If $\beta \in M^*$, one puts $f^*(\beta) = f(\beta) \subseteq \Lambda$. Clearly, if $\overline{\beta}$ is the terminal object of β, then $f(\overline{\beta})$ is the terminal object of $f(\beta)$, which insures that $f^*(\beta) \in \Lambda^*$. Note that f^* is an increasing function. One defines $f^*_\beta: X^*_{f^*(\beta)} \to Y^*_\beta$ by putting $f^*_\beta = f_{\overline{\beta}}$. If $\boldsymbol{f}\boldsymbol{p} = \boldsymbol{q}f$, one readily verifies that

$$\boldsymbol{f}^*\boldsymbol{p}^* = \boldsymbol{q}^* f. \tag{14}$$

LEMMA 6.31. *If $\boldsymbol{p}: X \to \boldsymbol{X}$ is a resolution of X over a directed ordered index set Λ, then $\boldsymbol{p}^*: X \to \boldsymbol{X}^*$ is a resolution over the cofinite set Λ^*. If $(\boldsymbol{p}, \boldsymbol{q}, \boldsymbol{f})$ is a resolution of $f: X \to Y$ and the index sets Λ and M are ordered, then $(\boldsymbol{p}^*, \boldsymbol{q}^*, \boldsymbol{f}^*)$ is also a resolution of f.*

Proof. Let $f: X \to P$ be a mapping into a polyhedron P, let $\mathcal{V} \in \mathrm{Cov}(P)$, $\lambda \in \Lambda$, and let $g: X_\lambda \to P$ be a mapping such that $(gp_\lambda, f) \prec \mathcal{V}$. Note that $\alpha = \{\lambda\} \in \Lambda^*$, $X^*_\alpha = X_\lambda$ and $p^*_\alpha = p_\lambda$. Therefore, $g: X^*_\alpha \to P$ has the property required by (R1), for $\boldsymbol{p}^*$. Furthermore, if $\mathcal{V}' \in \mathrm{Cov}(P)$ satisfies property (R2) for $\boldsymbol{p}$, then it satisfies this property also for $\boldsymbol{p}^*$. Indeed, assume that two mappings $f, f': X^*_\alpha \to P$ satisfy $(fp^*_\alpha, f'p^*_\alpha) \prec \mathcal{V}'$. Since $X^*_\alpha = X_\lambda, p^*_\alpha = p_\lambda$, where $\lambda = \overline{\alpha}$, (R2) for $\boldsymbol{p}$ yields a $\lambda_1 \geq \lambda$ such that $(fp_{\lambda\lambda_1}, f'p_{\lambda\lambda_1}) \prec \mathcal{V}$. Consider the element $\alpha_1 = \alpha \cup \{\lambda_1\} \in \Lambda^*$. Clearly, $\alpha \leq^* \alpha_1$ and $\overline{\alpha_1} = \lambda_1$. Therefore, $p_{\lambda\lambda_1} = p^*_{\alpha\alpha_1}$, and we obtain the desired relation $(fp^*_{\alpha\alpha_1}, f'p^*_{\alpha\alpha_1}) \prec \mathcal{V}$. This completes the proof of the first assertion. The second assertion is a consequence of the first assertion for $\boldsymbol{p}^*$ and $\boldsymbol{q}^*$ and of (14). □

REMARK 6.32. Application of Lemmas 6.29 and 6.31 to Theorems 6.22 and 6.23 shows that both theorems can be strengthened by requiring cofiniteness of the resolutions. In the case of mappings, a well-known reindexing theorem (see e.g., (Mardešić, Segal 1982), I. 1.3, Theorem 3) enables one to achieve that in the resolution of f both systems $\boldsymbol{X}$ and $\boldsymbol{Y}$ are indexed by the same index set and the mapping $\boldsymbol{f}: \boldsymbol{X} \to \boldsymbol{Y}$ is a level mapping, i.e., the $(\boldsymbol{p}, \boldsymbol{q}, \boldsymbol{f})$ is a *level resolution.* This is an immediate consequence of the following construction and lemma.

Let $\boldsymbol{X}$ and $\boldsymbol{Y}$ be inverse systems indexed by cofinite sets Λ and M, respectively, and let $\boldsymbol{p}: X \to \boldsymbol{X}$, $\boldsymbol{q}: Y \to \boldsymbol{Y}$ and $\boldsymbol{f}: \boldsymbol{X} \to \boldsymbol{Y}$ be mappings such that

$$\boldsymbol{f}\boldsymbol{p} = \boldsymbol{q}f \tag{15}$$

Define new systems $\boldsymbol{X}' = (X'_\nu, p'_{\nu\nu'}, N)$ and $\boldsymbol{Y}' = (Y'_\nu, q'_{\nu\nu'}, N)$ and new mappings $\boldsymbol{p}' = (p'_\nu): X \to \boldsymbol{X}'$ and $\boldsymbol{q}' = (q'_\nu): Y \to \boldsymbol{Y}'$ as follows. Let N be the set of all pairs $\nu = (\lambda, \mu)$, where $f(\mu) \leq \lambda$. Put $\nu \leq \nu' = (\lambda', \mu')$, provided $\lambda \leq \lambda'$ and $\mu \leq \mu'$. Note that N is also cofinite. Put $X'_\nu = X_\lambda$, $p'_{\nu\nu'} = p_{\lambda\lambda'}$ and $Y'_\nu = Y_\mu$, $q'_{\nu\nu'} = q_{\mu\mu'}$. Moreover, put $p'_\nu = p_\lambda$ and $q'_\nu = q_\mu$. Also define a level mapping $\boldsymbol{f}' = (f'_\nu)$, by putting $f'_\nu = f_\mu p_{f(\mu)\lambda}$. It is readily seen that (15) implies $\boldsymbol{f}'\boldsymbol{p}' = \boldsymbol{q}'f$.

LEMMA 6.33. *If $(\boldsymbol{p}, \boldsymbol{q}, \boldsymbol{f})$ is a resolution of a mapping $f: X \to Y$, then $(\boldsymbol{p}', \boldsymbol{q}', \boldsymbol{f}')$ is a level resolution of f.*

Proof. We only need to show that $\boldsymbol{p}'$ and $\boldsymbol{q}'$ are resolutions. Since this is a straightforward verification, we will only show that $\boldsymbol{q}'$ has property (R2). Indeed, let $h, h': Y_\nu \to P$ be mappings to a polyhedron P and let $\mathcal{V}$ be an open covering of P. Since $q'_\nu = q_\mu$ and $Y'_\nu = Y_\mu$, one concludes that there exist a covering $\mathcal{V}'$ of P and a $\mu' \geq \mu$ such that $(hq_{\mu\mu'}, h'q_{\mu\mu'}) \prec \mathcal{V}$. Now choose $\lambda' \geq \lambda, f(\mu')$. Then $\nu' = (\lambda', \mu') \in N$ and $\nu \leq \nu'$. Moreover, $q'_{\nu\nu'} = q_{\mu\mu'}$ and thus, $(hq'_{\nu\nu'}, h'q'_{\nu\nu'}) \prec \mathcal{V}$. □

6.5 Resolutions of direct products and pairs

Let $\boldsymbol{p} = (p_\lambda): X \to \boldsymbol{X} = (X_\lambda, p_{\lambda\lambda'}, \Lambda)$ be a mapping. For any space Y, let $\boldsymbol{X} \times Y$ denote the inverse system $(X_\lambda \times Y, p_{\lambda\lambda'} \times 1, \Lambda)$ and let $\boldsymbol{p} \times 1: X \times Y \to \boldsymbol{X} \times Y$ denote the mapping of systems, given by the mappings of spaces $p_\lambda \times 1: X \times Y \to X_\lambda \times Y$. The main result of this subsection is the following theorem.

THEOREM 6.34. *If $\boldsymbol{p}: X \to \boldsymbol{X}$ is a resolution and Y is a compact Hausdorff space, then also $\boldsymbol{p} \times 1: X \times Y \to \boldsymbol{X} \times Y$ is a resolution.*

In the proof we will need a lemma on stacked coverings. Recall that a covering $\mathcal{W}$ of X and a collection $\{\mathcal{J}_W | W \in \mathcal{W}\}$ of coverings $\mathcal{J}_W$ of Y determine a covering $\mathcal{S}$ of $X \times Y$, which consists of all sets of the form $W \times J \subseteq X \times Y$, where $W \in \mathcal{W}$ and $J \in \mathcal{J}_W$. We call such a covering $\mathcal{S}$ a *stacked covering* of $X \times Y$. If $\mathcal{W}$ and all the coverings $\mathcal{J}_W$ are normal, then so is $\mathcal{S}$. Indeed, if $(\phi_W, W \in \mathcal{W})$ is a partition of unity on X, subordinated to $\mathcal{W}$ and $(\psi_J, J \in \mathcal{J}_W)$ is a partition of unity on Y, subordinated to $\mathcal{J}_W$, then the functions $\chi_{W\times J}: X \times Y \to I$, given by $\chi_{W\times J}(x,y) = \phi_W(x)\psi_J(y)$, form a partition of unity on $X \times Y$, subordinated to $\mathcal{S}$.

LEMMA 6.35. *Let X be a topological space and Y a compact Hausdorff space. Then every normal covering $\mathcal{V}$ of $X \times Y$ admits a refinement $\mathcal{S}$, which is a stacked covering, given by a normal covering $\mathcal{W}$ of X and by a collection $\{\mathcal{J}_W, W \in \mathcal{W}\}$ of finite open coverings $\mathcal{J}_W$ of Y.*

Proof. Let $x \in X$ be an arbitrary point. Using compactness of Y, we define a finite collection of open sets $W_1 \times J_1, \ldots, W_n \times J_n$ in $X \times Y$, whose restriction on $x \times Y$ covers $x \times Y$ and refines $\mathcal{V}|(x \times Y)$, and each of the sets $W_i \times J_i$ meets $x \times Y$. Putting $W = W_1 \cap \ldots \cap W_n$, $\mathcal{J}_W = \{J_1, \ldots, J_n\}$, we obtain a finite open covering $\{W \times J_1, \ldots, W \times J_n\}$ of $W \times Y$, which refines $\mathcal{V}|W \times Y$ and $x \times Y \subseteq W \times Y$. Repeating the construction for all $x \in X$, we obtain a stacked covering of $X \times Y$, which refines $\mathcal{V}$. If X is paracompact (in particular, metric), this completes the proof, because $\mathcal{W}$ must be normal.

In the general case, note that, $\mathcal{V}$ being a normal covering of $X \times Y$, it admits a metric space M, a mapping $F: X \times Y \to M$ and an open covering $\mathcal{V}'$ of M such that

$$F^{-1}(\mathcal{V}') \prec \mathcal{V}. \tag{1}$$

Consider the space M^Y of all mappings $\varphi: Y \to M$, endowed with the compact-open topology and recall that it is a metrizable space. Define a mapping $f: X \to M^Y$ by

$$(f(x))(y) = F(x, y), \ x \in X, \ y \in Y. \tag{2}$$

Let $e: M^Y \times Y \to M$ be the evaluation map,

$$e(\varphi, y) = \varphi(y), \ \varphi \in M^Y, \ y \in Y. \tag{3}$$

Since $F = e(f \times 1)$, we have

$$F^{-1}(\mathcal{V}') = (f \times 1)^{-1} e^{-1}(\mathcal{V}'). \tag{4}$$

Since $e^{-1}(\mathcal{V}')$ is an open covering of $M^Y \times Y$, the preceding argument yields a stacked covering $\mathcal{S}'$ of $M^Y \times Y$, such that

$$\mathcal{S}' \prec e^{-1}(\mathcal{V}'). \tag{5}$$

Assume that $\mathcal{S}'$ is given by an open covering $\mathcal{W}'$ of M^Y and by a collection $\{\mathcal{J}_{W'} | W' \in \mathcal{W}'\}$ of finite open coverings of Y. Then $\mathcal{W} = f^{-1}(\mathcal{W}')$ is a normal covering of X, because M^Y is a metrizable space. If $W = f^{-1}(W')$, $W' \in \mathcal{W}'$, is a member of $\mathcal{W}$, put $\mathcal{J}_W = \mathcal{J}_{W'}$. Let $\mathcal{S}$ be the stacked covering of $X \times Y$, given by $\mathcal{W}$ and by the finite coverings $\mathcal{J}_W$, $W \in \mathcal{W}$. By (5), every $W \times J \in \mathcal{S}$ admits a $V' \in \mathcal{V}'$ such that

$$\begin{aligned} &W \times J = f^{-1}(W') \times J = (f \times 1)^{-1}(W' \times J) \subseteq \\ &(f \times 1)^{-1} e^{-1}(V') = F^{-1}(V'). \end{aligned} \tag{6}$$

This shows that

$$\mathcal{S} \prec F^{-1}(\mathcal{V}') \prec \mathcal{V}. \ \square \tag{7}$$

Proof of Theorem 6.34. We will verify conditions (B1) and (B2). To verify (B1), consider a normal covering $\mathcal{V}$ of $X \times Y$. By Lemma 6.35, there exists a stacked refinement $\mathcal{S}'$ of $\mathcal{V}$, which is given by a normal covering $\mathcal{W}'$ of X and by a collection of finite open coverings $\mathcal{J}'_{W'}$, $W' \in \mathcal{W}'$, of Y. By (B1) for $\boldsymbol{p}$, there exist a $\lambda \in \Lambda$ and a normal covering $\mathcal{W}$ of X_λ, such that $p_\lambda^{-1}(\mathcal{W}) \prec \mathcal{W}'$. For each $W \in \mathcal{W}$, choose a $W' \in \mathcal{W}'$ such that $p_\lambda^{-1}(W) \subseteq W'$. Then put $\mathcal{J}_W = \mathcal{J}'_{W'}$. Clearly, $\mathcal{W}$ and the coverings $\mathcal{J}_W, W \in \mathcal{W}$, form a stacked covering $\mathcal{S}$ of $X_\lambda \times Y$ such that $(p_\lambda \times 1)^{-1}(\mathcal{S})$ refines $\mathcal{S}'$, hence, it also refines $\mathcal{V}$. Finally, $\mathcal{S}$ is a normal covering of $X_\lambda \times Y$, because $\mathcal{W}$ and all the coverings $\mathcal{J}_W$ are normal.

To verify (B2), consider a normal covering $\mathcal{U}$ of $X_\lambda \times Y$. By Lemma 6.35, this covering admits a stacked refinement $\mathcal{S}$, given by a normal covering $\mathcal{W}$ of X_λ and by a collection of finite open coverings $\mathcal{J}_W$ of Y, $W \in \mathcal{W}$. Since $\mathcal{W}$ is normal, (B2) for $\boldsymbol{p}$, yields a $\lambda' \geq \lambda$, such that

$$p_{\lambda\lambda'}(X_{\lambda'}) \subseteq \mathrm{St}(p_\lambda(X), \mathcal{W}). \tag{8}$$

Taking into account that

$$\mathrm{St}(p_\lambda(X), \mathcal{W}) \times Y = \mathrm{St}(p_\lambda(X) \times Y, \mathcal{S}), \tag{9}$$

(8) yields the desired relation

$$(p_{\lambda\lambda'} \times 1)(X_{\lambda'} \times Y) \subseteq \mathrm{St}(p_\lambda(X) \times Y, \mathcal{S}) \subseteq \mathrm{St}((p_\lambda \times 1)(X \times Y), \mathcal{U}). \;\square \tag{10}$$

EXAMPLE 6.36. In the previous theorem one cannot omit the compactness assumption for Y. Indeed, it is known that, for every $n > 0$, there exist normal locally compact spaces X, Y, whose covering dimensions $\dim X = \dim Y = 0$, but $X \times Y$ is a normal space with $\dim(X \times Y) = n > 0$ (Przymusiński 1979), (Tsuda 1982, 1985). However, every 0-dimensional space admits a resolution, which consists of 0-dimensional polyhedra, i.e., all of its terms are discrete spaces (see Remark 6.26). For such a resolution $\boldsymbol{p}: X \to \boldsymbol{X} = (X_\lambda, p_{\lambda\lambda'}, \Lambda)$ of X, the mapping $\boldsymbol{p} \times 1: X \times Y \to \boldsymbol{X} \times Y$ cannot be a resolution, because, $\dim(X_\lambda \times Y) = 0$ and Remark 6.15 would imply $\dim(X \times Y) = 0$.

By a *polyhedral pair* (P, P^0) we mean the geometric realization of a pair (K, K^0), which consists of a simplicial complex K and a subcomplex $K^0 \subseteq K$. Clearly, P^0 is always a closed subset of P. By an ANR-*pair* we mean a pair of spaces (X, X^0) such that both X and X^0 are ANR's and X^0 is a closed subset of X. A mapping of systems of pairs $\boldsymbol{p}: (X, X^0) \to (\boldsymbol{X}, \boldsymbol{X}^0)$ is a morphism of inv-Top_2. It is a *resolution of a pair* provided it satisfies conditions (R1) and (R2) for polyhedral pairs. These conditions differ from the corresponding conditions for single polyhedra in that all mappings and homotopies involved are mappings and homotopies of pairs of spaces. For more details see ((Mardešić, Segal 1982), I.6.5).

REMARK 6.37. A mapping $\boldsymbol{p}: (X, X^0) \to (\boldsymbol{X}, \boldsymbol{X}^0)$ is a resolution of pairs if and only if $\boldsymbol{p}: X \to \boldsymbol{X}$ is a resolution and the following condition holds.

(B2)* For every $\lambda \in \Lambda$ and every normal covering $\mathcal{V}$ of X_λ, there exists a $\lambda' \geq \lambda$ such that

$$p_{\lambda\lambda'}(X^0_{\lambda'}) \subseteq \mathrm{St}(p_\lambda(X^0), \mathcal{V}). \tag{11}$$

This is the analogue of Theorem 6.7, for pairs.

REMARK 6.38. The analogues of Theorems 6.22, 6.23 for pairs also hold. To prove the existence of ANR-resolutions of pairs one needs some non-trivial facts on ANR-pairs, established in (Mardešić 1984a).

REMARK 6.39. If $\boldsymbol{p}: (X, X^0) \to (\boldsymbol{X}, \boldsymbol{X}^0)$ is a polyhedral resolution and the restriction $\boldsymbol{p}^0: X^0 \to \boldsymbol{X}^0$ is a resolution, then X^0 is *normally embedded* (or $\mathcal{P}$-embedded) in X, i.e., every normal covering $\mathcal{U}^0 \in \mathrm{Cov}(X^0)$, admits a normal covering $\mathcal{U} \in \mathrm{Cov}(X)$, such that the restriction $\mathcal{U}|X^0$ refines $\mathcal{U}^0$. E.g., every closed subset X^0 of a collectionwise normal space is normally embedded in X. Pairs of spaces (X, X^0), where X^0 is normally embedded in X, are called *normal pairs*. In shape theory they have a role analogous to the role of cofibration pairs in homotopy theory. Conversely, if a subset X^0 is normally embedded in X, then there exists a polyhedral resolution $\boldsymbol{p}: (X, X^0) \to (\boldsymbol{X}, \boldsymbol{X}^0)$ such that the restriction $\boldsymbol{p}^0: X^0 \to \boldsymbol{X}^0$ is also a resolution. For more details see I.6.5 of (Mardešić, Segal 1982).

REMARK 6.40. Pointed resolutions $\boldsymbol{p}: (X, *) \to (\boldsymbol{X}, *)$ are defined similarly to resolutions for pairs. One only imposes the requirement that all mappings and homotopies are pointed. Another variation of the notion of resolution is the *resolution of a triad* $\boldsymbol{p}: (X, X', X'') \to (\boldsymbol{X}, \boldsymbol{X}', \boldsymbol{X}'')$ (for more details see (Mardešić 1984b)).

Bibliographic notes

The definition of resolution, using conditions (R1) and (R2), was introduced in (Mardešić 1981a, 1981b). In particular, condition (R2) there appeared for the first time. The same applies to the name *resolution*, which was suggested by the fact that, in the study of topological spaces, polyhedral resolutions play a role similar to the role of resolutions in homological algebra. Already in 1975, P. Bacon considered mappings of systems $\boldsymbol{p}: X \to \boldsymbol{X}$, closely related to resolutions, which he called *complements* (Bacon 1975). His conditions on $\boldsymbol{p}$ consisted of (B1) and a stronger form of (B2). Also in 1975, K. Morita considered inverse limits $\boldsymbol{p}: X \to \boldsymbol{X}$, which had an additional property (P). He called them *proper limits* (Morita 1975a). For a mapping $\boldsymbol{p}: X \to \boldsymbol{X}$, property (P) is defined as follows:

(P) For every normal covering $\mathcal{U}$ of X, for every $\lambda \in \Lambda$ and every normal covering $\mathcal{U}_\lambda$ of X_λ, there exists a $\mu \geq \lambda$ and a normal covering $\mathcal{V}_\mu$ of X_μ, which refines $p_{\lambda\mu}^{-1}(\mathcal{U}_\lambda)$ and is such that $p_\mu^{-1}(\mathcal{V}_\mu)$ refines $\mathcal{U}$. Moreover, the simplicial mapping, given by $V \mapsto p_\mu^{-1}(V)$, $V \in \mathcal{V}_\mu$, is an isomorphism of the nerves $N(\mathcal{V}_\mu) \to N(p_\mu^{-1}(\mathcal{V}_\mu))$.

It was proved later that condition (P) is equivalent to conditions (R1) and (R2) (Morita 1984). Condition (B2) was introduced by T. Watanabe (Watanabe 1987a). He was the first who proved Theorem 6.7. A slightly weaker version of this result appeared earlier in (Mardešić 1981b). Versions of Theorems

6.16, 6.20, 6.22 appeared in (Bacon 1975), (Morita 1975c), (Mardešić 1981b). That every paracompact space is the limit of an inverse system of polyhedra was first proved in (Alder 1974). Theorem 6.23 is taken from (Mardešić 1981a). Theorem 6.21 is taken from (Lončar 1987) (also see (Mardešić 1992)). Theorem 6.34 is taken from (Lisica, Mardešić 1984b). Statements in Remark 6.39 are taken from I.6.5 of (Mardešić, Segal 1982).

7. Strong expansions

In this section we define and study strong expansions of spaces. This is a technique for approximating spaces, which occupies an intermediate position between resolutions and homotopy expansions (used in ordinary shape theory). Strong expansions of spaces appear to be the correct way of approximating spaces, when one wants to develop strong shape theory. We will prove existence of polyhedral and ANR strong expansions by proving that resolutions are always strong expansions. The other important fact proved in this section is the invariance of strong expansions under coherent domination.

7.1 Strong expansions of spaces

Let $\boldsymbol{p} = (p_\lambda): X \to \boldsymbol{X} = (X_\lambda, p_{\lambda\lambda'}, \Lambda)$ be a mapping of systems. We will call it a *homotopy expansion* of X provided, for every polyhedron P, the following *conditions of Morita* (M1), (M2) are fulfilled.

(M1) *If $f: X \to P$ is a mapping, then there exists a $\lambda \in \Lambda$ and there exists a mapping $g: X_\lambda \to P$ such that the mappings gp_λ and f are homotopic,*

$$gp_\lambda \simeq f. \tag{1}$$

(M2) *If $\lambda \in \Lambda$ and $f_0, f_1: X_\lambda \to P$ are mappings such that $f_0 p_\lambda \simeq f_1 p_\lambda$, then there exist a $\lambda' \geq \lambda$ such that $f_0 p_{\lambda\lambda'} \simeq f_1 p_{\lambda\lambda'}$.*

REMARK 7.1. Application of the homotopy functor H: $\mathsf{Top} \to \mathrm{H}(\mathsf{Top})$ to a homotopy expansion $\boldsymbol{p}$ yields a $\mathrm{H}(\mathsf{Top})$ - expansion $[\boldsymbol{p}]: X \to [\boldsymbol{X}]$ in the sense of Morita (see (Mardešić, Segal 1982), I.4.1). Recall that the development of ordinary shape theory is based on polyhedral $\mathrm{H}(\mathsf{Top})$ - expansions.

In order to define strong expansions of X, we replace condition (M2) by the following stronger condition.

(S2) *If $\lambda \in \Lambda$, $f_0, f_1: X_\lambda \to P$ are mappings and $F: X \times I \to P$ is a homotopy, which connects $f_0 p_\lambda$ and $f_1 p_\lambda$, then there exist a $\lambda' \geq \lambda$ and a homotopy $H: X_{\lambda'} \times I \to P$, which connects $f_0 p_{\lambda\lambda'}$ and $f_1 p_{\lambda\lambda'}$. Moreover, the homotopies $H(p_{\lambda'} \times 1), F: X \times I \to P$ are connected by a homotopy $(X \times I) \times I \to P$, which is fixed on $X \times \partial I$, i.e.,*

$$H(p_{\lambda'} \times 1) \simeq F\,(\mathrm{rel}\,(X \times \partial I)). \tag{2}$$

A *strong expansion* of a space X is a mapping $\boldsymbol{p}: X \to \boldsymbol{X}$ such that, for any polyhedron P, conditions (S1) $\equiv$ (M1) and (S2) are fulfilled.

Note that the mappings $f_0, f_1: X_\lambda \to P$ define a mapping $f: X_\lambda \times \partial I \to P$, given by

$$f(y,0) = f_0(y),\ f(y,1) = f_1(y),\ y \in X_\lambda. \tag{3}$$

Therefore, the assumption on F becomes

$$F|X \times \partial I = f(p_\lambda \times 1), \tag{4}$$

while the first requirement on H becomes

$$H|X_{\lambda'} \times \partial I = f(p_{\lambda\lambda'} \times 1). \tag{5}$$

The following diagram illustrates the situation.

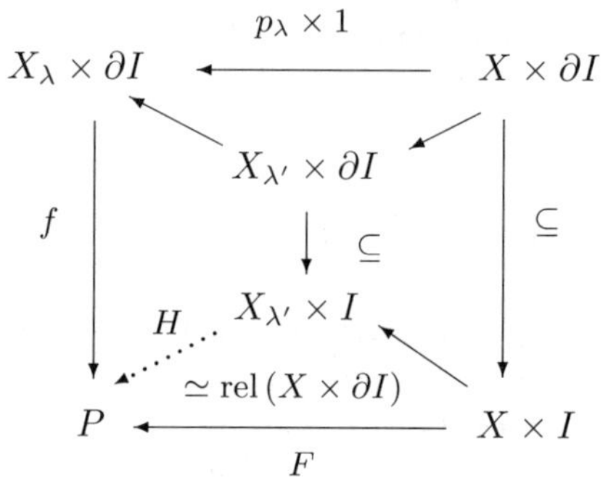

(6)

REMARK 7.2. Since (S2) implies (M2), every strong expansion is a homotopy expansion.

Conditions (S1) and (S2) make sense for any space P. We say that they hold for a class of spaces $\mathcal{C}$, if they hold, for every space P from $\mathcal{C}$.

LEMMA 7.3. *Condition* (Si), i=1,2, *for the class of polyhedra* Pol *is equivalent to the corresponding condition, for any of the following classes* $\mathcal{C}$: *polyhedra endowed with the metric topology* MPol, CW-*complexes* CW, *absolute neighborhood retracts for metric spaces* ANR, *spaces having the homotopy type of polyhedra* HPol.

A class of spaces $\mathcal{C}$ is *homotopy dominated* by a class $\mathcal{C}'$ provided every space $P \in \mathcal{C}$ is homotopy dominated by some space $P' \in \mathcal{C}'$, i.e., there exist mappings $\phi: P \to P'$ and $\psi: P' \to P$ such that $\psi\phi \simeq \mathrm{id}$. If also $\mathcal{C}'$ is homotopy dominated by $\mathcal{C}$, then $\mathcal{C}$ and $\mathcal{C}'$ are *homotopy equivalent* classes. It is well-known that Pol, MPol, CW, ANR and HPol are homotopy equivalent classes (see 6.1). It is now clear that Lemma 7.3 is a consequence of the following lemma.

LEMMA 7.4. *Let $\mathcal{C}$ be a class of spaces homotopy dominated by a class $\mathcal{C}'$. If condition* (Si), i=1,2, *holds for the class $\mathcal{C}'$, then it also holds for the class $\mathcal{C}$.*

Proof of Lemma 7.4. With each $P \in \mathcal{C}$ associate a $P' \in \mathcal{C}'$, mappings $\phi: P \to P'$, $\psi: P' \to P$, and a homotopy $\Phi: P \times I \to P$ such that

$$\Phi(y, 0) = y, \ \Phi(y, 1) = \psi\phi(y), \ y \in P. \tag{7}$$

First assume that $\boldsymbol{p}$ has property (S1) with respect to $\mathcal{C}'$. If $P \in \mathcal{C}$ and $f: X \to P$ is a mapping, then application of (S1) to $\phi f: X \to P'$ yields a $\lambda \in \Lambda$ and a mapping $g': X_\lambda \to P'$, such that $g' p_\lambda \simeq \phi f$. The mapping $g = \psi g': X_\lambda \to P$ has the desired property, because $g p_\lambda = \psi g' p_\lambda = \psi\phi f \simeq f$.

Now assume that $\boldsymbol{p}$ has property (S2) with respect to $\mathcal{C}'$. Consider $\lambda \in \Lambda$, maps $f_0, f_1: X_\lambda \to P$ and a homotopy $F: X \times I \to P$, such that

$$F(x, 0) = f_0 p_\lambda(x), \ F(x, 1) = f_1 p_\lambda(x), \ x \in X. \tag{8}$$

By (S2) with respect to $\mathcal{C}'$, applied to $\phi f_0, \phi f_1$ and ϕF, there exists a $\lambda' \geq \lambda$ and there exist homotopies $H': X_{\lambda'} \times I \to P', G': X \times I \times I \to P'$, such that

$$H'(z, 0) = \phi f_0 p_{\lambda\lambda'}(z), \ H'(z, 1) = \phi f_1 p_{\lambda\lambda'}(z), \ z \in X_{\lambda'}, \tag{9}$$

$$G'(x, t, 0) = H'(p_{\lambda'}(x), t), \ (x, t) \in X \times I, \tag{10}$$

$$G'(x, t, 1) = \phi F(x, t), \ (x, t) \in X \times I, \tag{11}$$

$$G'(x, t, s) = G'(x, t, 0), \ (x, t, s) \in X \times \partial I \times I. \tag{12}$$

Let us define $H: X_{\lambda'} \times I \to P$ by putting

$$H(z, t) = \begin{cases} \Phi(f_0 p_{\lambda\lambda'}(z), 3t), & 0 \leq t \leq 1/3, \\ \psi H'(z, 3t - 1), & 1/3 \leq t \leq 2/3, \\ \Phi(f_1 p_{\lambda\lambda'}(z), 3(1 - t)), & 2/3 \leq t \leq 1. \end{cases} \tag{13}$$

Using (7) and (9), it is readily seen that H is well defined and

$$H(z, 0) = f_0 p_{\lambda\lambda'}(z), \ H(z, 1) = f_1 p_{\lambda\lambda'}(z), \ z \in X_{\lambda'}. \tag{14}$$

We now define $G: X \times I \times I \to P$ by

$$G(x, t, s) = \begin{cases} \Phi(f_0 p_\lambda(x), 3t), & 0 \leq t \leq 1/3, \\ \psi G'(x, 3t - 1, s), & 1/3 \leq t \leq 2/3, \\ \Phi(f_1 p_\lambda(x), 3(1 - t)), & 2/3 \leq t \leq 1. \end{cases} \tag{15}$$

Using (7), (12), (10) and (9), one easily verifies that G is well defined. Furthermore, by (15), (10), (13) and (11),

$$G(x, t, 0) = H(p_{\lambda'}(x), t), \tag{16}$$

$$G(x,t,1) = \begin{cases} \Phi(f_0 p_\lambda(x), 3t), & 0 \le t \le 1/3, \\ \psi\phi F(x, 3t-1)), & 1/3 \le t \le 2/3, \\ \Phi(f_1 p_\lambda(x), 3(1-t)), & 2/3 \le t \le 1. \end{cases} \tag{17}$$

Moreover, by (15) and (7),

$$G(x,0,s) = f_0 p_\lambda(x),\ G(x,1,s) = f_1 p_\lambda(x),\ (x,s) \in X \times I. \tag{18}$$

Formulae (16) and (18) show that G is a homotopy rel$(X \times \partial I)$, which connects $H(p_\lambda \times I)$ with $G_1 : X \times I \to P$, where $G_1(x,t) = G(x,t,1)$. In order to complete the proof, it suffices to find a homotopy $K : X \times I \times I \to P(\mathrm{rel}(X \times \partial I))$, which connects G_1 with F. Such a homotopy is given by

$$K(x,t,s) = \begin{cases} \Phi(f_0 p_\lambda(x), 3t), & 0 \le t \le \frac{1-s}{3}, \\ \Phi(F(x, (3t+s-1)/(2s+1)), 1-s), & \frac{1-s}{3} \le t \le \frac{s+2}{3}, \\ \Phi(f_1 p_\lambda(x), 3(1-t)), & \frac{s+2}{3} \le t \le 1. \end{cases} \tag{19}$$

K is well defined, because of (8). Moreover, (19), (17) and (7) imply

$$K(x,t,0) = G_1(x,t),\ K(x,t,1) = F(x,t),\ (x,t) \in X \times I, \tag{20}$$

$$K(x,0,s) = f_0 p_\lambda(x),\ K(x,1,s) = f_1 p_\lambda(x),\ (x,s) \in X \times I, \tag{21}$$

which shows that K has the desired property. □

We close this subsection with a theorem needed in 8.

THEOREM 7.5. *If* $\boldsymbol{p}: X \to \boldsymbol{X}$ *is a strong expansion and* Y *is a compact metric space, then* $\boldsymbol{p} \times 1 : X \times Y \to \boldsymbol{X} \times Y$ *is also a strong expansion.*

Proof. We will verify conditions (S1) and (S2), for an arbitrary ANR P. To achieve this, we use the fact that the space P^Y of all mappings $\phi : Y \to P$, endowed with the compact-open topology, is an ANR ((Mardešić, Segal 1982), I.3.1, Theorem 4).

(S1). A mapping $f : X \times Y \to P$ determines the adjoint mapping $\tilde{f} : X \to P^Y$, given by

$$(\tilde{f}(x))(y) = f(x,y),\ x \in X, y \in Y. \tag{22}$$

Applying property (S1), for $\boldsymbol{p}$ to $\tilde{f}$, one obtains a $\lambda \in \Lambda$, a mapping $\tilde{g} : X_\lambda \to P^Y$ and a homotopy $\tilde{G} : X \times I \to P^Y$, such that

$$\tilde{G}(x,0) = \tilde{f}(x),\ \tilde{G}(x,1) = \tilde{g} p_\lambda(x),\ x \in X. \tag{23}$$

Let $g : X_\lambda \times Y \to P$ and $G : X \times Y \times I \to P$ be the adjoint mappings, given by

$$g(z,y) = (\tilde{g}(z))(y),\ (z,y) \in X_\lambda \times Y, \tag{24}$$

$$G(x,y,t) = (\tilde{G}(x,t))(y),\ (x,y,t) \in X \times Y \times I. \tag{25}$$

Then (25), (23), (22) and (24) yield

$$G(x,y,0) = f(x,y),\ G(x,y,1) = g(p_\lambda(x),y). \tag{26}$$

Consequently, the homotopy G connects f with $g(p_\lambda \times 1)$, which establishes (S1), for $\boldsymbol{p} \times 1$.

(S2). Consider a mapping $f: X_\lambda \times Y \times \partial I \to P$ and a homotopy $F: X \times Y \times I \to P$, which satisfy (8), which in this case assumes the form

$$F(x,y,t) = f(p_\lambda(x), y, t), \ (x,y,t) \in X \times Y \times \partial I. \tag{27}$$

Let $\tilde{f}: X_\lambda \times \partial I \to P^Y$ and $\tilde{F}: X \times I \to P^Y$ be the adjoint mappings, given by

$$(\tilde{f}(z,t))(y) = f(z,y,t), \ (z,y,t) \in X_\lambda \times Y \times \partial I, \tag{28}$$

$$(\tilde{F}(x,t))(y) = F(x,y,t), \ (x,y,t) \in X \times Y \times I. \tag{29}$$

Then, (28), (27) and (29) yield

$$(\tilde{f}(p_\lambda \times 1)(x,t))(y) = (\tilde{F}(x,t))(y), \ (x,y,t) \in X \times Y \times \partial I. \tag{30}$$

Applying (S2), for $\boldsymbol{p}$ to $\tilde{f}$ and $\tilde{F}$, one obtains a $\lambda' \geq \lambda$ and two homotopies $\tilde{H}: X_{\lambda'} \times I \to P^Y, \tilde{G}: X \times I \times I \to P^Y$, which satisfy

$$\tilde{H}(z,t) = \tilde{f}(p_{\lambda\lambda'}(z), t), \ (z,t) \in X_{\lambda'} \times \partial I, \tag{31}$$

$$\tilde{G}(x,t,0) = \tilde{H}(p_{\lambda'}(x), t), \ \tilde{G}(x,t,1) = \tilde{F}(x,t), \ (x,t) \in X \times I, \tag{32}$$

$$\tilde{G}(x,t,s) = \tilde{G}(x,t,1) = \tilde{F}(x,t), \ (x,t,s) \in X \times \partial I \times I. \tag{33}$$

Let $H: X_{\lambda'} \times Y \times I \to P$ and $G: X \times Y \times I \times I \to P$ be the adjoint mappings, given by

$$H(z,y,t) = (\tilde{H}(z,t))(y), \ (z,y,t) \in X_{\lambda'} \times Y \times I, \tag{34}$$

$$G(x,y,t,s) = (\tilde{G}(x,t,s))(y), \ (x,y,t,s) \in X \times Y \times I \times I. \tag{35}$$

Now (33), (30) and (28) yield

$$H|X_{\lambda'} \times Y \times \partial I = f(p_{\lambda\lambda'} \times 1 \times 1), \tag{36}$$

which corresponds to (5), for $\boldsymbol{p} \times 1$. Furthermore, (35), (32) and (34) yield

$$G(x,y,t,0) = H(p_{\lambda'}(x), y, t), \ (x,y,t) \in X \times Y \times I, \tag{37}$$

while (34), (31) and (29) yield

$$G(x,y,t,1) = F(x,y,t), \ (x,y,t) \in X \times Y \times I. \tag{38}$$

Consequently, G is a homotopy connecting $H(p_{\lambda'} \times 1 \times 1)$ with F. Finally, (35), (32) and (29) yield

$$G(x,y,t,s) = F(x,y,t), \ (x,y,t) \in X \times Y \times \partial I, \tag{39}$$

which shows that G is a homotopy rel$(X \times Y \times \partial I)$. □

7.2 Resolutions are strong expansions

The main purpose of the present subsection is to prove the following theorem.

THEOREM 7.6. *Every resolution is a strong expansion.*

Combining Theorem 7.6 with Theorems 6.22 and 6.23, one obtains the following theorem.

THEOREM 7.7. *Every topological space X admits a strong expansion, which consists of polyhedra* (ANR*′s*).

Since every strong expansion is a homotopy expansion, Theorem 7.6 implies the following result.

COROLLARY 7.8. *Every resolution is a homotopy expansion.*

In ((Mardešić and Segal 1982), I.6.1, Theorem 2) one finds a different proof of this corollary.

In order to prove Theorem 7.6, it suffices to verify conditions (S1) and (S2) in the case when P is an ANR (see Lemma 7.3). To achieve this, we need two lemmas.

LEMMA 7.9. *Let P be an* ANR *and let $\mathcal{U} \in$* Cov(P). *Then there exists a covering $\mathcal{V} \in$* Cov(P) *with the following property: whenever two mappings $g_0, g_1: Z \to P$ of a space Z into P are $\mathcal{V}$-near, then there exists a $\mathcal{U}$-homotopy $G: Z \times I \to P$, which connects g_0 and g_1 and is fixed on the subset $\{z \in Z | g_0(z) = g_1(z)\}$.*

For a proof of Lemma 7.9, we refer to ((Mardešić, Segal 1982), I.3.2, Theorem 6).

LEMMA 7.10. *Let $\boldsymbol{p}: X \to \boldsymbol{X}$ be a resolution, P an* ANR *and $\mathcal{U} \in$* Cov(P). *If $\lambda \in \Lambda$, $f_0, f_1: X_\lambda \to P$ are mappings and $F: X \times I \to P$ is a homotopy such that*

$$F(x, 0) = f_0 p_\lambda(x), \;\; F(x, 1) = f_1 p_\lambda(x), \;\; x \in X, \tag{1}$$

then there exist a $\lambda' \geq \lambda$ and a homotopy $H: X_{\lambda'} \times I \to P$ such that

$$H(y, 0) = f_0 p_{\lambda\lambda'}(y), \;\; H(y, 1) = f_1 p_{\lambda\lambda'}(y), \;\; y \in X_{\lambda'}, \tag{2}$$

$$(F, H(p_{\lambda'} \times 1)) \prec \mathcal{U}. \tag{3}$$

Note that Lemma 7.10 strengthens condition (M2), by adding the new requirement (3). We postpone its proof and proceed to prove the theorem.

Proof of Theorem 7.6. Let $\boldsymbol{p}: X \to \boldsymbol{X}$ be a resolution. By Lemma 7.10, it suffices to show that $\boldsymbol{p}$ has properties (Si), i=1,2, for ANR's. For an arbitrary ANR P, choose a covering $\mathcal{V} \in P$, by applying Lemma 7.9 to the trivial covering $\{P\}$. To verify (S1), consider any mapping $f: X \to P$. Applying

(R1) (for ANR's) to f and $\mathcal{V}$, one obtains a $\lambda \in \Lambda$ and a mapping $g: X_\lambda \to P$ such that $(gp_\lambda, f) \prec \mathcal{V}$ and therefore, by the choice of $\mathcal{V}$, $gp_\lambda \simeq f$.

To verify (S2), consider a $\lambda \in \Lambda$, mappings $f_0, f_1: X_\lambda \to P$ and a homotopy $F: X \times I \to P$, which satisfies (1). Applying Lemma 7.10 to $\boldsymbol{p}$, P and $\mathcal{V}$, one obtains a $\lambda' \geq \lambda$ and a homotopy $H: X_{\lambda'} \times I \to P$, which satisfies (2) and (3) (where $\mathcal{U}$ has been replaced by $\mathcal{V}$). Taking into account the properties of $\mathcal{V}$, one concludes that there exists a homotopy $G: X \times I \times I \to P$ such that

$$G(x,t,0) = F(x,t), G(x,t,1) = H(p_{\lambda'}(x),t), \ (x,t) \in X \times I. \tag{4}$$

Moreover, the homotopy G is fixed on the set

$$\{(x,t) \in X \times I | F(x,t) = H(p_{\lambda'}(x),t)\}. \tag{5}$$

To complete the proof, it suffices to show that G is a homotopy rel$(X \times \partial I)$. Consequently, it suffices to see that $X \times \partial I = X \times \{0,1\}$ is contained in the set (5). This is indeed the case, because (1) and (2) imply $F(x,0) = f_0 p_\lambda(x) = f_0 p_{\lambda\lambda'} p_{\lambda'}(x) = H(p_{\lambda'}(x),0)$ and similarly, $F(x,1) = H(p_{\lambda'}(x),1)$.

The proof of Lemma 7.10 is based on two further lemmas.

LEMMA 7.11. *Let $K: Y \times I \to P$ be a homotopy and let $\mathcal{V} \in \mathrm{Cov}(P)$. Then there exists a continuous function $\phi: Y \to I$ such that for every $y \in Y$, $\phi(y) > 0$, and whenever $|t - t'| \leq \phi(y)$, $t, t' \in I$, then there exists a $V \in \mathcal{V}$ such that $K(y,t), K(y,t') \in V$.*

Proof. The normal covering $K^{-1}(\mathcal{V})$ admits a stacked refinement $\mathcal{S}$, which is given by a normal covering $\mathcal{W}$ of Y and by a family of finite open coverings $\{\mathcal{J}_W | W \in \mathcal{W}\}$. For every $W \in \mathcal{W}$, let $\alpha(W) > 0$ be a Lebesgue number of the covering $\mathcal{J}_W$. Let $(\phi_W, W \in \mathcal{W})$ be a locally finite partition of unity, subordinated to $\mathcal{W}$, i.e., $\phi_W: Y \to I$ are continuous functions, whose sum equals 1, and $\phi_W(y) \neq 0$ implies $y \in W$. Put

$$\phi = \sup\{\alpha_W \phi_W | W \in \mathcal{W}\}. \tag{6}$$

By the local finiteness of the partition, every point $y \in Y$ admits a neighborhood U such that $\phi_W|U = 0$, except for a finite collection of indices $\{W_1, \dots, W_n\}$. Consequently,

$$\phi|U = \max\{\alpha_{W_1}\phi_{W_1}|U, \dots, \alpha_{W_n}\phi_{W_n}|U\}. \tag{7}$$

It follows that ϕ is continuous at the point y. Not all $\phi_W(y)$ can be 0, because this would contradict the fact that their sum is 1. Consequently, $\phi(y) > 0$. Now assume that $t, t' \in I$ and $|t - t'| \leq \phi(y)$. By (7), $y \in Y$ admits a $W \in \mathcal{W}$ such that

$$\phi(y) = \alpha_W \phi_W(y) \leq \alpha_W. \tag{8}$$

Clearly, $\phi(y) > 0$ implies $\phi_W(y) \neq 0$, and therefore, $y \in W$. Moreover, by (8), $|t - t'| \leq \alpha_W$. Since α_W is a Lebesgue number for the covering $\mathcal{J}_W$, one concludes that there exists a member $J \in \mathcal{J}_W$ such that $t, t' \in J$.

Consequently, $(y,t),(y,t') \in W \times J \in \mathcal{S}$. Taking into account that $\mathcal{S}$ refines $K^{-1}(\mathcal{V})$, one concludes that there exists a $V \in \mathcal{V}$ such that $K(y,t), K(y,t') \in V$. □

LEMMA 7.12. *Let P be an ANR and $\mathcal{V} \in \mathrm{Cov}(P)$. Let $K: Y \times I \to P$ be a homotopy and let $L, M: Y \times I \to P$ be $\mathcal{V}$-homotopies such that*

$$K_0 = L_1, \;\; K_1 = M_1. \tag{9}$$

Then there exists a homotopy $H: Y \times I \to P$ such that

$$H_0 = L_0, \;\; H_1 = M_0, \tag{10}$$

$$(H, K) \prec \mathrm{St}(\mathcal{V}). \tag{11}$$

Proof. Choose a function $\phi: Y \to I$, by applying Lemma 7.11 to K and $\mathcal{V}$. Without loss of generality one can assume that $\phi \leq \frac{1}{3}$ and therefore, $0 < \phi(y) < 1 - \phi(y) < 1$, for $y \in Y$. We then define $H: Y \times I \to P$ by putting

$$H(y,t) = \begin{cases} L(y, t/\phi(y)), & 0 \leq t \leq \phi(y), \\ K(y, (t - \phi(y))/(1 - 2\phi(y))), & \phi(y) \leq t \leq 1 - \phi(y), \\ M(y, (1-t)/\phi(y)), & 1 - \phi(y) \leq t \leq 1 \end{cases} \tag{12}$$

To see that H is well defined, use (9). Note that H satisfies (10). It thus remains to verify (11), i.e., to show that, for every $(y,t) \in Y \times I$, there exists a $V \in \mathcal{V}$ such that

$$H(y,t), K(y,t) \in \mathrm{St}(V, \mathcal{V}). \tag{13}$$

We distinguish three cases, depending on the position of t in I.

Case (a). $0 \leq t \leq \phi(y)$. In this case $H(y,t) = L(y,t')$, where $t' = t/\phi(y) \in I$. Moreover, by (9), $K(y,0) = L(y,1)$. Since L is a $\mathcal{V}$-homotopy, one concludes that there exists a set $V \in \mathcal{V}$ such that

$$H(y,t), K(y,0) \in V. \tag{14}$$

Moreover, by the properties of ϕ, we conclude that there exists a set $V_1 \in \mathcal{V}$ such that

$$K(y,0), K(y,t) \in V_1. \tag{15}$$

Clearly, (14) and (15) imply (13).

Case (b). $\phi(y) \leq t \leq 1 - \phi(y)$. In this case $H(y,t) = K(y,t')$, where $t' = (t - \phi(y))/(1 - 2\phi(y)) \in I$. Since $|1 - 2t| \leq 1 - 2\phi(y)$ and $t - t' = \phi(y)(1-2t)/(1-2\phi(y))$, we see that $|t-t'| \leq \phi(y)$. Therefore, by the properties of ϕ, there exists a set $V_1 \in \mathcal{V}$ such that $H(y,t), K(y,t) \in V_1$, and thus (13) holds.

Case (c). $1 - \phi(y) \leq t \leq 1$. $H(y,t) = M(y,t')$, where $t' = (1-t)/\phi(y) \in I$. Moreover, by (9), $K(y,1) = M(y,1)$. Since M is a $\mathcal{V}$-homotopy, one concludes that there exists a set $V \in \mathcal{V}$ such that

$$H(y,t), K(y,1) \in V. \tag{16}$$

However, in this case $1-t \leq \phi(y)$. Therefore, using the properties of ϕ, one obtains a set $V_1 \in \mathcal{V}$ such that

$$K(y,1), K(y,t) \in V_1. \tag{17}$$

Clearly, (16) and (17) imply (13). □

Proof of Lemma 7.10. Let $\mathcal{V}$ be a normal covering of P such that $\mathrm{St}^2(\mathcal{V}) = \mathrm{St}(\mathrm{St}(\mathcal{V}))$ refines $\mathcal{U}$. To prove the lemma, it suffices to exhibit a $\lambda' \geq \lambda$, a homotopy $K: X_{\lambda'} \times I \to P$, such that

$$(F, K(p_{\lambda'} \times 1)) \prec \mathcal{V}, \tag{18}$$

and two $\mathcal{V}$-homotopies $L, M: X_{\lambda'} \times I \to P$, such that (9) holds and also

$$L_0 = f_0 p_{\lambda\lambda'}, \quad M_0 = f_1 p_{\lambda\lambda'}. \tag{19}$$

Indeed, by applying Lemma 7.12, we will then obtain a homotopy $H: X_{\lambda'} \times I \to P$, which satisfies (10) and (11). Note that (10) and (19) imply (2) and (11) implies

$$(H(p_{\lambda'} \times 1), K(p_{\lambda'} \times 1)) \prec \mathrm{St}(\mathcal{V}) \tag{20}$$

However, (20) and (18) yield (3).

By Lemma 7.9, choose a covering $\mathcal{V}' \prec \mathcal{V}$ of P such that $\mathcal{V}'$-near mappings into P are $\mathcal{V}$-homotopic. Let $\mathcal{V}''$ be a star-refinement of $\mathcal{V}'$. By Theorem 6.34, $\boldsymbol{p} \times 1: X \times I \to \boldsymbol{X} \times I$ is also a resolution. Therefore, there exists a $\lambda'' \geq \lambda$ and a homotopy $G: X_{\lambda''} \times I \to P$ such that

$$(F, G(p_{\lambda''} \times 1)) \prec \mathcal{V}''. \tag{21}$$

Consider a stacked refinement $\mathcal{S}$ of the covering $G^{-1}(\mathcal{V}'')$, given by a normal covering $\mathcal{W}$ of $X_{\lambda''}$ and by a family of finite open coverings $\{\mathcal{J}_W | W \in \mathcal{W}\}$. We choose $\mathcal{W}$ so fine that

$$\mathcal{W} \prec (f_0 p_{\lambda\lambda''})^{-1}(\mathcal{V}''), \ \mathcal{W} \prec (f_1 p_{\lambda\lambda''})^{-1}(\mathcal{V}''). \tag{22}$$

Using property (B2), choose $\lambda' \geq \lambda''$ in such a way that

$$p_{\lambda''\lambda'}(X_{\lambda'}) \subseteq \mathrm{St}(p_{\lambda''}(X), \mathcal{W}). \tag{23}$$

Now define K by the formula

$$K = G(p_{\lambda''\lambda'} \times 1). \tag{24}$$

Since $K(p_{\lambda'} \times 1) = G(p_{\lambda''} \times 1)$, (21) applies and yields (18).

Let us now show that

$$(K_0, f_0 p_{\lambda\lambda'}) \prec \mathcal{V}', (K_1, f_1 p_{\lambda\lambda'}) \prec \mathcal{V}'. \tag{25}$$

By (23), for any point $y \in X_{\lambda'}$, there exist a point $x \in X$ and a set $W \in \mathcal{W}$ such that

$$p_{\lambda''\lambda'}(y), p_{\lambda''}(x) \in W. \tag{26}$$

Since $\mathcal{J}_W$ is a covering of I, there exists a $J \in \mathcal{J}_W$ such that $0 \in J$. Consequently, $(p_{\lambda''\lambda'}(y), 0), (p_{\lambda''}(x), 0) \in W \times J \in \mathcal{S}$. Then $\mathcal{S} \prec G^{-1}(\mathcal{V}'')$ implies the existence of a $V_1'' \in \mathcal{V}''$ such that

$$G(p_{\lambda''\lambda'}(y), 0), G(p_{\lambda''}(x), 0) \in V_1''. \tag{27}$$

Moreover, by (21), there exists a $V_2'' \in \mathcal{V}''$ such that

$$G(p_{\lambda''}(x), 0), F(x, 0) \in V_2''. \tag{28}$$

Furthermore, by (26) and (22), there is a $V_3'' \in \mathcal{V}''$ such that

$$f_0 p_{\lambda\lambda'}(y), f_0 p_{\lambda}(x) \in V_3''. \tag{29}$$

Since $\mathcal{V}'' \prec^* \mathcal{V}'$ and $K(y,0) = G(p_{\lambda''\lambda'}(y), 0)$, $F(x,0) = f_0 p_\lambda(x)$, (27), (28) and (29) yield a $V' \in \mathcal{V}'$ such that

$$K(y, 0), f_0 p_{\lambda\lambda'}(y) \in V', \tag{30}$$

which establishes the first relation in (25). The second one is established analogously.

Finally, since $\mathcal{V}'$-near mappings are $\mathcal{V}$-homotopic, (25) shows that there exist $\mathcal{V}$-homotopies $L, M: X_{\lambda'} \times I \to P$, which satisfy (9) and (19). □

REMARK 7.13. Lemma 7.9 remains valid if one replaces the assumption that P is an ANR by the assumption that P is a polyhedron. Indeed, if K is a triangulation of $P = |K|$ such that the covering $\mathcal{K}$, formed by the open stars $\mathrm{St}(v, K)$ of the vertices v of K, refines $\mathcal{U}$, then $\mathcal{V} = \mathcal{K}$ has the required properties. This is a consequence of the following fact (see (Cauty 1973), Theorem 2.6 or (Bacon 1975), Theorem 2.2). Let $g_0, g_1: Z \to P$ be two mappings of a space Z, which are $\mathcal{K}$-near. Then there exists a homotopy $H: Z \times I \to |K|$, which connects g_0 and g_1 and has two additional properties If for a point $z \in Z$, $g_0(z), g_1(z) \in \mathrm{St}(v, K)$, for some vertex v, then also $H(z \times I) \subseteq \mathrm{St}(v, K)$. Moreover, the homotopy H is fixed on the set $\{z \in Z | g_0(z) = g_1(z)\}$. It is now clear that Lemma 7.9 also holds for polyhedra P.

7.3 Invariance under coherent domination

Let $\boldsymbol{p}: X \to \boldsymbol{X}$ and $\boldsymbol{q}: X \to \boldsymbol{Y}$ be two mappings of the same space X. We say that $\boldsymbol{p}$ is *coherently dominated* by the mapping $\boldsymbol{q}$ provided there exist coherent mappings $\boldsymbol{f}: \boldsymbol{X} \to \boldsymbol{Y}$ and $\boldsymbol{g}: \boldsymbol{Y} \to \boldsymbol{X}$ such that

$$\boldsymbol{f} C(\boldsymbol{p}) \simeq C(\boldsymbol{q}), \tag{1}$$

$$\boldsymbol{g}\boldsymbol{f} \simeq C(\boldsymbol{1}). \tag{2}$$

The main purpose of this subsection is to prove the following theorem.

THEOREM 7.14. *If a mapping* $\boldsymbol{p}: X \to \boldsymbol{X}$ *is coherently dominated by a strong expansion* $\boldsymbol{q}: X \to \boldsymbol{Y}$, *then* $\mathbf{p}$ *itself is a strong expansion.*

We first prove two lemmas concerning the category H(pro-Top) (see 4.2).

LEMMA 7.15. *Let* $\boldsymbol{q}, \boldsymbol{q}': X \to \boldsymbol{Y} = (Y_\mu, q_{\mu\mu'}, M)$ *be two mappings which belong to the same class* $[\boldsymbol{q}] = [\boldsymbol{q}']$ *in* H(pro-Top). *If* $\boldsymbol{q}$ *is a strong expansion, then so is* $\boldsymbol{q}'$.

Proof. By assumption there exists a mapping $\boldsymbol{K} = (K_\mu): X \times I \to \boldsymbol{Y}$ such that, for every $\mu \in M$, K_μ connects q'_μ to q_μ, i.e.,

$$q'_\mu \simeq_{K_\mu} q_\mu. \tag{3}$$

Moreover,

$$q_{\mu\mu'} K_{\mu'} = K_\mu, \ \ \mu \le \mu'. \tag{4}$$

Now assume that $P \in$HPol and $\phi: X \to P$ is a mapping. By (S1) for $\boldsymbol{q}$, there exist a $\mu \in M$ and a mapping $\psi: Y_\mu \to P$ such that $\phi \simeq \psi q_\mu$. By (3), $q_\mu \simeq q'_\mu$ and thus, $\phi \simeq \psi q'_\mu$. However, this is the desired condition (S1) for $\boldsymbol{q}'$.

Now assume that $\mu \in M$, $\psi_0, \psi_1: Y_\mu \to P$ are mappings and $F': X \times I \to P$ is a homotopy such that

$$\psi_0 q'_\mu \simeq_{F'} \psi_1 q'_\mu. \tag{5}$$

Let $F: X \times I \to P$ be the homotopy obtained by juxtaposition of three homotopies according to the following formula.

$$F = \psi_0 K_\mu^- * F' * \psi_1 K_\mu, \tag{6}$$

where K^- denotes the opposite of the homotopy K, i.e., $K^-(x,t) = K(x, 1-t)$. The homotopy F is well defined and has the property that

$$\psi_0 q_\mu \simeq_F \psi_1 q_\mu. \tag{7}$$

Therefore, by condition (S2) for $\boldsymbol{q}$, there exist a $\mu' \ge \mu$ and a homotopy $H: Y_{\mu'} \times I \to P$ such that

$$\psi_0 q_{\mu\mu'} \simeq_H \psi_1 q_{\mu\mu'}. \tag{8}$$

Moreover,

$$F \simeq H(q_{\mu'} \times 1)\,(\mathrm{rel}(X \times \partial I)). \tag{9}$$

We shall prove that

$$F' \simeq H(q'_{\mu'} \times 1)\,(\mathrm{rel}(X \times \partial I)). \tag{10}$$

Clearly, (8) and (10) will establish the desired condition (S2) for $\boldsymbol{q}'$.

In order to prove (9), we define a homotopy $U: X \times I \times I \to P$ by putting

$$U(x,s,t) = H(K_{\mu'}(x,t), s). \tag{11}$$

Note that, by (3),

$$U(x,s,0)=H(q'_{\mu'}(x),s),\ U(x,s,1)=H(q_{\mu'}(x),s). \tag{12}$$

Moreover, by (8) and (4),

$$U(x,0,t)=\psi_0 q_{\mu\mu'}K_{\mu'}(x,t)=\psi_0 K_\mu(x,t), \tag{13}$$

$$U(x,1,t)=\psi_1 q_{\mu\mu'}K_{\mu'}(x,t)=\psi_1 K_\mu(x,t). \tag{14}$$

Let $V\colon X\times I\times I\to P$ be a homotopy, which realizes (9). Using U and V, we will now define a homotopy $W\colon X\times I\times I\to P$, which realizes (10). Divide the square $I\times I$ in two rectangles as shown on Fig. 7.1. Since V is a homotopy rel $(X\times\partial I)$ which connects F to $H(q_{\mu'}\times 1)$ and F is of the form (6), we can use V to fill up the lower rectangle as indicated on the figure. Then we use U^- to fill up the upper rectangle (observe the orientation of the upper rectangle on the figure).

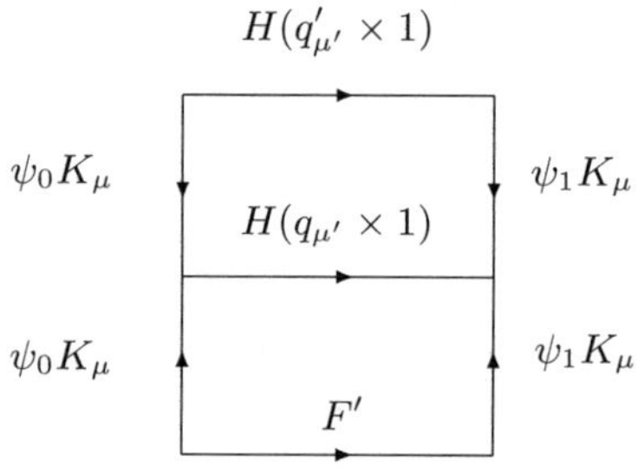

Fig. 7.1.

In this way we obtain a homotopy $W'\colon X\times I\times I\to P$ such that

$$W'|X\times I\times 0=F',\ W'|X\times I\times 1=H(q'_{\mu'}\times 1). \tag{15}$$

Moreover,

$$W'|X\times 0\times I=\psi_0(K_\mu * K_\mu^-),\ W'|X\times 1\times I=\psi_1(K_\mu * K_\mu^-). \tag{16}$$

Clearly, the restrictions of W' to the left and to the right sides of the two rectangles are homotopic to zero. Therefore, W' can be modified to a mapping $W\colon X\times D^2\to P$, where D^2 is a disc in such a way that, on the lower half S^- of the boundary ∂D^2, $W|X\times S^-$ coincides with F', while on the upper half S^+ of the boundary ∂D^2, $W|X\times S^+$ coincides with $H(q'_{\mu'}\times 1)$. Consequently, W can be viewed as the desired homotopy rel $(X\times\partial I)$. □

LEMMA 7.16. *Let* $\boldsymbol{p}\colon X\to\boldsymbol{X}$, $\boldsymbol{q}\colon X\to\boldsymbol{Y}$ *and* $\boldsymbol{f}\colon\boldsymbol{X}\to\boldsymbol{Y}$ *be mappings such that*

$$[\boldsymbol{f}][\boldsymbol{p}]=[\boldsymbol{q}] \tag{17}$$

in H(pro-Top). *Moreover, let* $g: \Lambda \to M$ *be an increasing function and let* $g_\lambda: Y_{g(\lambda)} \to X_\lambda$ *be mappings having the property that every* $\lambda \in \Lambda$ *admits a* $\lambda^* \geq \lambda, fg(\lambda)$ *such that*

$$p_{\lambda\lambda^*} \simeq g_\lambda f_{g(\lambda)} p_{fg(\lambda)\lambda^*}. \tag{18}$$

Then the assumption that $\boldsymbol{q}$ *is a strong expansion implies that also* $\mathbf{p}$ *is a strong expansion.*

Proof. It suffices to prove that the assertion holds, when (17) is replaced by the stronger assumption

$$\boldsymbol{fp} = \boldsymbol{q}. \tag{19}$$

Indeed, if $\boldsymbol{q}$ satisfies (17), then $\boldsymbol{q}' = \boldsymbol{fp}$ satisfies $[\boldsymbol{q}] = [\boldsymbol{q}']$ in H(pro-Top). Therefore, by Lemma 7.15, $\boldsymbol{q}'$ is also a strong expansion. However, $\boldsymbol{q}'$ satisfies the analogue of (19). Hence, the weaker version of Lemma 7.16 implies that $\boldsymbol{p}$ is a strong expansion.

We shall now prove the assertion of Lemma 7.16 assuming (19). For a mapping $\phi: X \to P \in$HPol, property (S1) for $\boldsymbol{q}$ yields a $\mu \in M$ and a mapping $\psi': Y_\mu \to P$ such that $\psi' q_\mu \simeq \phi$. However, by (19), $q_\mu = f_\mu p_{f(\mu)}$ and thus, $\lambda = f(\mu)$ and the mapping $\psi = \psi' f_\mu: X_\lambda \to P$ satisfy $\psi p_\lambda \simeq \phi$, which is the desired property (S1) for $\boldsymbol{p}$.

To establish (S2), let $\psi_0, \psi_1: X_\lambda \to P$, $\lambda \in \Lambda$, be mappings and let $F: X \times I \to P$ be a homotopy such that

$$\psi_0 p_\lambda \simeq_F \psi_1 p_\lambda. \tag{20}$$

Choose a $\lambda^* \geq \lambda, fg(\lambda)$ and a homotopy $K_\lambda: X_{\lambda^*} \times I \to P$, which realizes (18). Since $q_{g(\lambda)} = f_{g(\lambda)} p_{fg(\lambda)}$, one sees that $\psi_0 K_\lambda^-(p_{\lambda^*} \times 1)$ is a homotopy which connects $\psi_0 g_\lambda q_{g(\lambda)}$ to $\psi_0 p_\lambda$. Similarly, $\psi_1 K_\lambda(p_{\lambda^*} \times 1)$ is a homotopy which connects $\psi_1 p_\lambda$ to $\psi_1 g_\lambda q_{g(\lambda)}$. Therefore,

$$F' = \psi_0 K_\lambda^-(p_{\lambda^*} \times 1) * F * \psi_1 K_\lambda(p_{\lambda^*} \times 1). \tag{21}$$

is a well-defined homotopy $F': X \times I \to P$, which connects $\psi_0 g_\lambda q_{g(\lambda)}$ to $\psi_1 g_\lambda q_{g(\lambda)}$. Consequently, $\psi_0' = \psi_0 g_\lambda, \psi_1' = \psi_1 g_\lambda$ are mappings $Y_{g(\lambda)} \to P$ such that

$$\psi_0' q_{g(\lambda)} \simeq_{F'} \psi_1' q_{g(\lambda)}. \tag{22}$$

Using property (S2) for $\boldsymbol{q}$, we conclude that there exist an index $\mu' \geq g(\lambda)$ and a homotopy $H': Y_{\mu'} \times I \to P$ such that

$$\psi_0' q_{g(\lambda)\mu'} \simeq_{H'} \psi_1' q_{g(\lambda)\mu'}. \tag{23}$$

Moreover,

$$F' \simeq H'(q_{\mu'} \times 1)\,(\mathrm{rel}(X \times \partial I)). \tag{24}$$

Now choose a $\lambda' \geq \lambda^*, f(\mu')$. Note that $H'(f_{\mu'} p_{f(\mu')\lambda'} \times 1): X_{\lambda'} \times I \to P$ is a homotopy which connects the mapping $\psi_0' q_{g(\lambda)\mu'} f_{\mu'} p_{f(\mu')\lambda'} = \psi_0' f_{g(\lambda)} p_{fg(\lambda)\lambda'}$ to the mapping $\psi_1' f_{g(\lambda)} p_{fg(\lambda)\lambda'}$. Since K_λ realizes (18), we conclude that

$$H = \psi_0 K_\lambda(p_{\lambda^*\lambda'} \times 1) * H'(f_{\mu'} p_{f(\mu')\lambda'} \times 1) * \psi_1 K_\lambda^-(p_{\lambda^*\lambda'} \times 1) \tag{25}$$

is a well-defined homotopy $H: X_{\lambda'} \times I \to P$ such that

$$\psi_0 p_{\lambda\lambda'} \simeq_H \psi_1 p_{\lambda\lambda'}. \tag{26}$$

Hence, to complete the proof of Lemma 7.16, it suffices to prove that

$$F \simeq H(p_{\lambda'} \times 1)\,(\mathrm{rel}(X \times \partial I)). \tag{27}$$

Choose a homotopy U, which realizes (24). Clearly, it can be viewed as a mapping $U: X \times D^2 \to P$ such that $U|X \times S^- = F'$, while $U|X \times S^+ = H'(q_{\mu'} \times 1)$. By (21), $U|X \times S^-$ is the juxtaposition of three homotopies, defined on three consecutive arcs S_l^-, S_c^-, S_r^-. Now view the boundary ∂D^2 as divided in two arcs A^-, A^+. The arc $A^- = S_c^-$, while A^+ consists of the arcs S_l^-, S^+ and S_r^-, where S_l^- and S_r^- are taken with opposite orientations. Clearly, $U|X \times A^-$ can be viewed as F, while $U|X \times A^+$ can be viewed as the juxtaposition of homotopies which form $H(p_{\lambda'} \times 1)$ following (25). Consequently, U can be viewed as a homotopy realizing (27). □

REMARK 7.17. If $\boldsymbol{p}$, $\boldsymbol{q}$ and $\boldsymbol{f}$ are as in Lemma 7.16 and $\boldsymbol{g} = (g, g_\lambda): \boldsymbol{Y} \to \boldsymbol{X}$ is a mapping such that $[\boldsymbol{g}][\boldsymbol{f}] = [\mathbf{1}]$ in H(pro-Top), then the assumptions of Lemma 7.16 are fulfilled. Therefore, if $\boldsymbol{q}$ is a strong expansion, so is $\boldsymbol{p}$.

REMARK 7.18. If $\boldsymbol{p}$, $\boldsymbol{q}$ and $\boldsymbol{f}$ are as in Lemma 7.16, $\boldsymbol{f} = (f_\lambda)$ is a level homotopy equivalence and $g_\lambda: Y_\lambda \to X_\lambda$ are homotopy inverses of f_λ, $\lambda \in \Lambda$, then all the assumptions of Lemma 7.16 are fulfilled. Therefore, if $\boldsymbol{q}$ is a strong expansion, so is $\boldsymbol{p}$.

An important step in the proof of Theorem 7.14 is the following lemma on level homotopy equivalences.

LEMMA 7.19. *Let $\boldsymbol{p}: X \to \boldsymbol{X}$ be a strong expansion and let $\boldsymbol{f} = (f_\lambda): \boldsymbol{X} \to \boldsymbol{Y}$ be a level homotopy equivalence. Then $\boldsymbol{q} = \boldsymbol{f}\boldsymbol{p}: X \to \boldsymbol{Y}$ is also a strong expansion.*

Proof. If $\phi: X \to P \in$HPol is a mapping, there exist a $\lambda \in \Lambda$ and a mapping $\psi': X_\lambda \to P$ such that $\psi' p_\lambda \simeq \phi$. Since $1 \simeq g_\lambda f_\lambda$ and $q_\lambda = f_\lambda p_\lambda$, the mapping $\psi = \psi' g_\lambda: Y_\lambda \to P$ has the property that $\psi q_\lambda = \psi' g_\lambda f_\lambda p_\lambda \simeq \psi' p_\lambda \simeq \phi$, which establishes property (S1) for $\boldsymbol{q}$.

To verify property (S2) for $\boldsymbol{q}$, consider mappings $\psi_0, \psi_1: Y_\lambda \to P$ and a homotopy $F: X \times I \to P$ such that

$$\psi_0 q_\lambda \simeq_F \psi_1 q_\lambda. \tag{28}$$

Note that (28) implies

$$\psi_0' p_\lambda \simeq_F \psi_1' p_\lambda, \tag{29}$$

where $\psi_0' = \psi_0 f_\lambda$, $\psi_1' = \psi_1 f_\lambda$. Therefore, by assumption on $\boldsymbol{p}$, there exist a $\lambda' \geq \lambda$ and a homotopy $H': X_{\lambda'} \times I \to P$ such that

$$\psi'_0 p_{\lambda\lambda'} \simeq_{H'} \psi'_1 p_{\lambda\lambda'}. \tag{30}$$

Moreover,

$$F \simeq H'(p_{\lambda'} \times 1)\,(\mathrm{rel}(X \times \partial I)). \tag{31}$$

To continue the proof we apply Vogt's lemma, i.e., Lemma 4.7, to the homotopy equivalences f_λ, for every $\lambda \in \Lambda$. We thus obtain homotopies K_λ, L_λ such that

$$\mathrm{id} \simeq_{K_\lambda} g_\lambda f_\lambda,\ \mathrm{id} \simeq_{L_\lambda} f_\lambda g_\lambda, \tag{32}$$

$$L_\lambda(f_\lambda \times 1) \simeq f_\lambda K_\lambda\,(\mathrm{rel}(X_\lambda \times \partial I)). \tag{33}$$

Now note that the homotopy $H'(g_{\lambda'} \times 1)\colon Y_{\lambda'} \times I \to P$ connects $\psi_0 f_\lambda p_{\lambda\lambda'} g_{\lambda'} = \psi_0 q_{\lambda\lambda'} f_{\lambda'} g_{\lambda'}$ to $\psi_1 f_\lambda p_{\lambda\lambda'} g_{\lambda'} = \psi_1 q_{\lambda\lambda'} f_{\lambda'} g_{\lambda'}$. Therefore, formula

$$H = \psi_0 q_{\lambda\lambda'} L_{\lambda'} * H'(g_{\lambda'} \times 1) * \psi_1 q_{\lambda\lambda'} L^-_{\lambda'} \tag{34}$$

yields a well-defined homotopy $H\colon Y_{\lambda'} \times I \to P$, which connects $\psi_0 q_{\lambda\lambda'}$ to $\psi_1 q_{\lambda\lambda'}$. To complete the proof of Lemma 7.19, it remains to prove that

$$F \simeq H(q_{\lambda'} \times 1)\,(\mathrm{rel}(X \times \partial I)). \tag{35}$$

We first define a homotopy $U\colon X \times I \times I \to P$, by putting

$$U(x,s,t) = H'(K_{\lambda'}(p_{\lambda'}(x),t),s). \tag{36}$$

Note that

$$U(x,s,0) = H'(p_{\lambda'}(x),s), \tag{37}$$

$$U(x,s,1) = H'(g_{\lambda'} q_{\lambda'}(x),s), \tag{38}$$

$$U(x,0,t) = \psi_0 q_{\lambda\lambda'} f_{\lambda'} K_{\lambda'}(p_{\lambda'}(x),t), \tag{39}$$

$$U(x,1,t) = \psi_1 q_{\lambda\lambda'} f_{\lambda'} K_{\lambda'}(p_{\lambda'}(x),t). \tag{40}$$

In order to define a homotopy $W\colon X \times I \times I \to P$, which realizes (35), we first define a mapping $W'\colon X \times D \to P$, where D is the polygon, described by Fig. 7.2. It consists of four rectangles, denoted by D^+_l, D^+_c, D^+_r, and D^-. By definition, $W'|X \times D^+_c$ is given by the homotopy U, while $W'|X \times D^-$ is given by a homotopy V, which realizes (31). Note that (37) insures that the two definitions of W' on $X \times (D^+_c \cap D^-)$ coincide. We define $W'|D^+_l$ using the homotopy $\psi_0 q_{\lambda\lambda'} T_{\lambda'}(p_{\lambda'} \times 1)$, where $T_\lambda\colon X_\lambda \times I \times I \to Y_\lambda$ is a homotopy which realizes (33). More precisely,

$$T_\lambda(x,0,t) = L_\lambda(f_\lambda(x),t), \tag{41}$$

$$T_\lambda(x,1,t) = f_\lambda K_\lambda(x,t). \tag{42}$$

Moreover, $T_\lambda(x,s,0)$ and $T_\lambda(x,s,1)$ do not depend on s. Note that (39) and (42) insure that the two definitions of W' on $X \times (D^+_l \cap D^+_c)$ coincide.

Similarly, we define $W'|D^+_r$, using the homotopy $\psi_1 q_{\lambda\lambda'} T^-_{\lambda'}(p_{\lambda'} \times 1)$. Note that on each of the horizontal sides of the rectangles D^+_l and D^+_r and on the

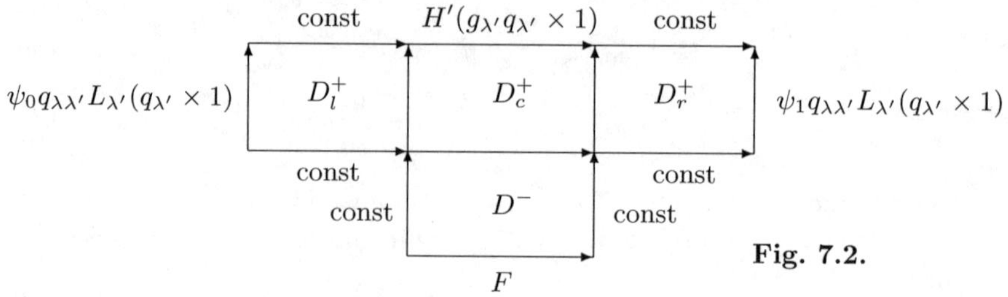

Fig. 7.2.

vertical sides of D_{-}, W' assumes constant values. Therefore, by collapsing each of these sides to a point, one obtains a mapping $W: X \times D^2 \to P$. Thereby, $W|S^-$ coincides with F, while $W|S^+$ coincides with the juxtaposition of the following three homotopies: $\psi_0 q_{\lambda\lambda'} L_{\lambda'}(q_{\lambda'} \times 1), H'(g_{\lambda'} q_{\lambda'} \times 1), \psi_1 q_{\lambda\lambda'} L^-_{\lambda'}(q_{\lambda'} \times 1)$. However, according to (34), this is just the homotopy $H(q_{\lambda'} \times 1)$. Hence, W can be viewed as a homotopy realizing (35). □

In the proof of Theorem 7.14 we will use the functors τ: inv-Top → inv-Top and $\boldsymbol{T}$: CH(pro-Top) → H(pro-Top) and the natural transformation ϕ between the identity functor on inv-Top and τ, all defined in 4.2. In particular, we will need the following facts.

(*i*) $\phi_{\boldsymbol{X}}: \boldsymbol{X} \to \tau(\boldsymbol{X}) = \boldsymbol{T}(\boldsymbol{X})$ is a level homotopy equivalence (see Theorem 4.30). If $\boldsymbol{X}$ is a single space X, then $\tau(X) = X$ and $\phi_X = \mathrm{id}$ (see Remark 4.28).

(*ii*) For every mapping $\boldsymbol{f}: \boldsymbol{X} \to \boldsymbol{Y}$, $[\tau(\boldsymbol{f})] = \boldsymbol{T}[C(\boldsymbol{f})]$ in H(pro-Top) (see Lemma 4.29).

Proof of Theorem 7.14. Let $\boldsymbol{p}: X \to \boldsymbol{X}$ be a mapping coherently dominated by a strong expansion $\boldsymbol{q}: X \to \boldsymbol{Y}$. Then there exist coherent mappings $\boldsymbol{f}: \boldsymbol{X} \to \boldsymbol{Y}$ and $\boldsymbol{g}: \boldsymbol{Y} \to \boldsymbol{X}$ such that (1) and (2) hold. Applying the functor $\boldsymbol{T}$ we conclude that

$$\boldsymbol{T}[\boldsymbol{f}]\boldsymbol{T}[C(\boldsymbol{p})] = \boldsymbol{T}[C(\boldsymbol{q})], \tag{43}$$

$$\boldsymbol{T}[\boldsymbol{g}]\boldsymbol{T}[\boldsymbol{f}] = [\boldsymbol{1}]. \tag{44}$$

Note that $\boldsymbol{T}[\boldsymbol{f}] = [\boldsymbol{T}(\boldsymbol{f})]$ is the class in H(pro-Top) which contains the mapping $\boldsymbol{T}(\boldsymbol{f}): \boldsymbol{X} \to \boldsymbol{Y}$ (see 4.2). Similarly, there is a mapping $\boldsymbol{T}(\boldsymbol{g}): \boldsymbol{Y} \to \boldsymbol{X}$ such that $\boldsymbol{T}[\boldsymbol{g}] = [\boldsymbol{T}(\boldsymbol{g})]$. Moreover, by (*ii*), $\boldsymbol{T}[C(\boldsymbol{p})] = [\tau(\boldsymbol{p})]$ and $\boldsymbol{T}[C(\boldsymbol{q})] = [\tau(\boldsymbol{q})]$. Consequently, (43) and (44) become

$$[\boldsymbol{T}(\boldsymbol{f})][\tau(\boldsymbol{p})] = [\tau(\boldsymbol{q})], \tag{45}$$

$$[\boldsymbol{T}(\boldsymbol{g})][\boldsymbol{T}(\boldsymbol{f})] = [\boldsymbol{1}]. \tag{46}$$

On the other hand, by the naturality of ϕ, the following diagram commutes.

$$\begin{array}{ccc} \boldsymbol{Y} & \xleftarrow{\ \boldsymbol{q}\ } & X \\ {\scriptstyle \phi_{\boldsymbol{Y}}}\downarrow & & \downarrow{\scriptstyle \phi_X = \mathrm{id}} \\ \tau(\boldsymbol{Y}) & \xleftarrow[\ \tau(\boldsymbol{q})\ ]{} & X \end{array} \tag{47}$$

In other words,

$$\phi_{\boldsymbol{Y}}\boldsymbol{q} = \tau(\boldsymbol{q}). \tag{48}$$

Since $\boldsymbol{q}$ is a strong expansion and $\phi_{\boldsymbol{Y}}$ is a level homotopy equivalence, Lemma 7.19 yields the conclusion that $\tau(\boldsymbol{q})$ is also a strong expansion. Applying Remark 7.17 to the mappings $\tau(\boldsymbol{p})\colon X \to \tau(\boldsymbol{X})$, $\tau(\boldsymbol{q})\colon X \to \tau(\boldsymbol{Y})$, $\boldsymbol{T}(\boldsymbol{f})\colon \tau(\boldsymbol{X}) \to \tau(\boldsymbol{Y})$ and $\boldsymbol{T}(\boldsymbol{g})\colon \tau(\boldsymbol{Y}) \to \tau(\boldsymbol{X})$ (taking into account (45) and (46)), one concludes that also $\tau(\boldsymbol{p})$ is a strong expansion. Now note that the analogue of (48) for $\boldsymbol{p}$ has the following form.

$$\phi_{\boldsymbol{X}}\boldsymbol{p} = \tau(\boldsymbol{p}). \tag{49}$$

Since $\phi_{\boldsymbol{X}}\colon \boldsymbol{X} \to \tau(\boldsymbol{X})$ is a level homotopy equivalence and $\tau(\boldsymbol{p})$ is a strong expansion, one can apply Remark 7.18 to $\boldsymbol{p}$, $\tau(\boldsymbol{p})$ and $\phi_{\boldsymbol{X}}$ and one obtains the desired conclusion that $\boldsymbol{p}$ is indeed a strong expansion. □

REMARK 7.20. A strong expansion of pairs $\boldsymbol{p}\colon (X, X^0) \to (\boldsymbol{X}, \boldsymbol{X}^0)$ is a mapping of systems of pairs, which satisfies conditions (S1), (S2) for polyhedral pairs. These conditions differ from the corresponding conditions for single polyhedra in that all mappings and homotopies involved are mappings and homotopies of pairs of spaces. All results established in this section, also hold for pairs of spaces. In particular, instead of using the class Pol_2 of polyhedral pairs (P, P^0) in (S1), (S2), one can use, equivalently, the classes ANR_2, MPol_2, CW_2 and HPol_2 of ANR-pairs, pairs of CW-complexes and pairs having the homotopy type of polyhedral pairs, respectively. For more details see (Mardešić 1991c).

Bibliographic notes

In the compact case, conditions (M1) and (M2) appeared in Lemmas 3 and 4 of (Mardešić, Segal 1971). In the general case they were introduced in (Morita 1975a). In (Mardešić, Segal 1982) these conditions were called (E1) and (E2). Strong expansions were introduced in (Günther 1989, 1992b) and in a more general situation (for mappings $\boldsymbol{p}\colon \boldsymbol{X} \to \boldsymbol{Y}$), they were introduced independently in (Dydak, Nowak 1991) (see 10.7). It was proved in (Mardešić 1991b) that resolutions are strong expansions. Theorem 7.14 is taken from (Mardešić 1998).

8. Strong shape

This section is devoted to the construction of the strong shape category SSh(Top) of topological spaces and the strong shape functor $\bar{S}$:H(Top)→ SSh(Top). Spaces X, Y are replaced by polyhedral coherent expansions $\boldsymbol{X}, \boldsymbol{Y}$ and the strong shape morphisms $F : X \to Y$ are given by homotopy classes $[\boldsymbol{f}]$ of coherent mappings $\boldsymbol{f} : \boldsymbol{X} \to \boldsymbol{Y}$. Existence of coherent expansions is a consequence of the fact that strong expansions (hence, also resolutions) are always coherent expansions. Conversely, coherent HPol-expansions are always strong expansions. It is also shown that mappings, which induce strong shape isomorphisms, can be characterized by rather simple conditions (SM1) and (SM2), which strengthen properties (M1) and (M2).

8.1 Coherent expansions of spaces

In order to introduce the strong shape category in a way analogous to the way used in (Mardešić, Segal 1982) in introducing the usual shape category, we need the notion of coherent expansion. However, it will be proved in this subsection that coherent HPol-expansions coincide with strong HPol-expansions (see Theorems 8.1 and 8.2).

A *coherent expansion* of a space X is a mapping $\boldsymbol{p}: X \to \boldsymbol{X}$, which has the following property:

(CH) *For every cofinite* HPol-*system* $\boldsymbol{Y}$ *and every morphism* $[\boldsymbol{f}]: X \to \boldsymbol{Y}$ *of* CH(pro-Top), *there exists a unique morphism* $[\boldsymbol{h}]: \boldsymbol{X} \to \boldsymbol{Y}$ *of* CH(pro-Top) *such that* $[\boldsymbol{h}]C(\boldsymbol{p}) = [\boldsymbol{f}]$.

Recall that $C(\boldsymbol{p}) = [C(\boldsymbol{p})]$ is a morphism of CH(pro-Top), induced by $\boldsymbol{p}$ (see (2.3. 33)). The main result of this subsection is the following theorem.

THEOREM 8.1. *Every strong expansion of a space is a coherent expansion of that space.*

In this subsection we will also establish the following converse of Theorem 8.1.

THEOREM 8.2. *Every cofinite coherent* HPol-*expansion of a space is also a strong expansion of that space.*

In the proof of Theorem 8.1 we will use the n-dimensional version, $n \geq 1$, of property (S2) from 7.1. For a mapping $\boldsymbol{p} = (p_\lambda): X \to \boldsymbol{X} = (X_\lambda, p_{\lambda\lambda'}, \Lambda)$ and a space P, this is the following statement.

$(S2)^n$ *If* $\lambda \in \Lambda$ *and* $f: X_\lambda \times \partial\Delta^n \to P$, $F: X \times \Delta^n \to P$ *are mappings, which satisfy*

$$F|X \times \partial\Delta^n = f(p_\lambda \times 1), \tag{1}$$

then there exist a $\lambda' \geq \lambda$ *and a mapping* $H: X_{\lambda'} \times \Delta^n \to P$, *such that*

$$H|X_{\lambda'} \times \partial\Delta^n = f(p_{\lambda\lambda'} \times 1), \tag{2}$$

$$H(p_{\lambda'} \times 1) \simeq F\,(\mathrm{rel}\,(X \times \partial\Delta^n)). \tag{3}$$

Condition $(S2)^n$ can be illustrated by the following diagram.

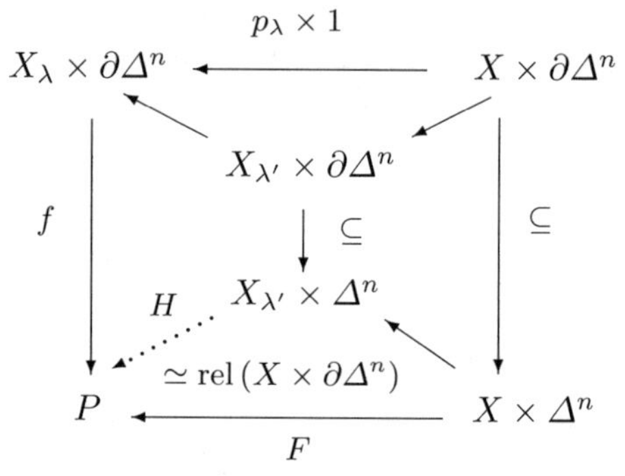

(4)

Note that property $(S2)^1$ coincides with (S2).

The following lemma plays an important role in the proof of Theorem 8.1.

LEMMA 8.3. *Every strong expansion* $\boldsymbol{p}: X \to \boldsymbol{X}$ *has property* $(S2)^n$, *for every* $n \geq 1$ *and every space* P *from* HPol.

Proof of Lemma 8.3. For a given $\lambda \in \Lambda$, let $f: X_\lambda \times \partial\Delta^n \to P$ and $F: X \times \Delta^n \to P$ satisfy (1). By Theorem 7.5, $\boldsymbol{p} \times 1: X \times \Delta^n \to \boldsymbol{X} \times \Delta^n$ is a strong expansion. Applying (S1) to this expansion, one obtains a $\lambda'' \geq \lambda$, a mapping $H'': X_{\lambda''} \times \Delta^n \to P$ and a homotopy $G: X \times \Delta^n \times I \to P$ such that

$$G(x,s,0) = H''(p_{\lambda''}(x), s),\ G(x,s,1) = F(x,s). \tag{5}$$

Note that $F'' = G|X \times \partial\Delta^n \times I$ is a homotopy which connects the mappings $f_0''(p_{\lambda''} \times 1)$ and $f_1''(p_{\lambda''} \times 1)$, where

$$f_0'' = H''|X_{\lambda''} \times \partial\Delta^n, \tag{6}$$

$$f_1'' = f(p_{\lambda\lambda''} \times 1)|X_{\lambda''} \times \partial\Delta^n. \tag{7}$$

Applying again Theorem 7.5, we conclude that $\boldsymbol{p}\times 1: X\times\partial\Delta^n \to \boldsymbol{X}\times\partial\Delta^n$ is also a strong expansion. By (S2) for this expansion, we obtain a $\lambda' \geq \lambda''$ and homotopies $K: X_{\lambda'} \times \partial\Delta^n \times I \to P, L: X \times \partial\Delta^n \times I \times I \to P$ such that

$$K(y,s,0) = H''(p_{\lambda''\lambda'}(y),s),\ K(y,s,1) = f(p_{\lambda\lambda'}(y),s),\ s \in \partial\Delta^n, \tag{8}$$

$$L(x,s,t,0) = K(p_{\lambda'}(x),s,t),\ L(x,s,t,1) = G(x,s,t),\ (s,t) \in \partial\Delta^n \times I, \tag{9}$$

$$L(x,s,t,u) = L(x,s,t,0) = K(p_{\lambda'}(x),s,t),\ (s,t,u) \in \partial\Delta^n \times \partial I \times I. \tag{10}$$

We now define a mapping $K': X_{\lambda'}\times J \to P$, where $J = (\partial\Delta^n\times I)\cup(\Delta^n\times 0) \subseteq \Delta^n \times I$, by putting

$$K'(y,s,t) = K(y,s,t),\ (s,t) \in \partial\Delta^n \times I, \tag{11}$$

$$K'(y,s,0) = H''(p_{\lambda''\lambda'}(y),s),\ s \in \Delta^n. \tag{12}$$

K' is well defined, because of (8).

Now note that J is homeomorphic to an n-ball. Therefore, there exists a retraction $r: \Delta^n \times I \to J$. Define $\tilde{K}: X_{\lambda'} \times \Delta^n \times I \to P$, by putting

$$\tilde{K}(x,s,t) = K'(y,r(s,t)). \tag{13}$$

Clearly, $\tilde{K}$ extends K'. Then define $H: X_{\lambda'} \times \Delta^n \to P$, by putting

$$H(x,s) = \tilde{K}(y,s,1). \tag{14}$$

Formulae (11) and (8) show that H satisfies (2).

Next, we consider the set

$$J' = (\Delta^n \times 0 \times I) \cup (\Delta^n \times I \times \partial I) \cup (\partial\Delta^n \times I \times I) \subseteq \Delta^n \times I \times I, \tag{15}$$

and define a mapping $L': X \times J' \to P$, by putting

$$L'(x,s,0,u) = H''(p_{\lambda''}(x),s),\ (s,u) \in \Delta^n \times I, \tag{16}$$

$$L'(x,s,t,0) = \tilde{K}(p_{\lambda'}(x),s,t),\ (s,t) \in \Delta^n \times I, \tag{17}$$

$$L'(x,s,t,1) = G(x,s,t),\ (s,t) \in \Delta^n \times I, \tag{18}$$

$$L'(x,s,t,u) = L(x,s,t,u),\ (s,t,u) \in \partial\Delta^n \times I \times I. \tag{19}$$

Using (5), (10), (8), it is easy to verify that L' is well defined.

Now note that J' consists of all the faces of $\Delta^n \times I \times I$, except the face $\Delta^n \times 1 \times I$, and is therefore, homeomorphic to an $(n+1)$-ball. Consequently, there exists a retraction $r': \Delta^n \times I \times I \to J'$. This enables us to define a mapping $\tilde{L}: X \times \Delta^n \times I \times I \to P$, by putting

$$\tilde{L}(x,s,t,u) = L'(x,r'(s,t,u)). \tag{20}$$

Clearly, $\tilde{L}$ extends L'.

Finally, we define a mapping $M: X \times \Delta^n \times I \to P$ by putting

$$M(x,s,u) = \tilde{L}(x,s,1,u) = L'(x,r'(s,1,u)). \tag{21}$$

Using (21), (17), (14), (18), (5), (19), (10) and (8), it is easy to see that

$$M(x,s,0) = H(p_{\lambda'}(x),s),\ M(x,s,1) = F(x,s),\ (x,s) \in X \times \Delta^n, \tag{22}$$

$$M(x,s,u) = f(p_\lambda(x),s),\ (x,s,u) \in X \times \partial\Delta^n \times I. \tag{23}$$

M is a homotopy rel$(X \times \partial\Delta^n)$, which connects $H(p_{\lambda'} \times 1)$ to F, because $f(p_\lambda(x),s)$ does not depend on u. This establishes (3). □

REMARK 8.4. If for given λ, f and F, the index λ' and the mapping H satisfy (2) and (3), then the mapping $H' = H(p_{\lambda'\lambda''} \times 1)\colon X_{\lambda''} \times \partial\Delta^n \to P$ satisfies the corresponding conditions, for any $\lambda'' \geq \lambda'$. Therefore, when using property $(\mathrm{S2})^n$, one can always assume that λ' is as large as desired.

In order to prove Theorem 8.1, it suffices to prove the following two lemmas.

LEMMA 8.5. *Every strong expansion $\boldsymbol{p}\colon X \to \boldsymbol{X}$ has the following property:*

(CH1) *For every cofinite* HPol - *system $\boldsymbol{Y}$ and every morphism $[\boldsymbol{f}]\colon X \to \boldsymbol{Y}$ of* CH(pro - Top), *there exists a morphism $[\boldsymbol{h}]\colon \boldsymbol{X} \to \boldsymbol{Y}$ of* CH(pro - Top) *such that $[\boldsymbol{f}] = [\boldsymbol{h}]C(\boldsymbol{p})$.*

LEMMA 8.6. *Every strong expansion $\boldsymbol{p}\colon X \to \boldsymbol{X}$ has the following property:*

(CH2) *If for a cofinite* HPol - *system $\boldsymbol{Y}$ two morphisms $[\boldsymbol{h}], [\boldsymbol{h}']\colon \boldsymbol{X} \to \boldsymbol{Y}$ of* CH(pro - Top) *satisfy $[\boldsymbol{h}]C(\boldsymbol{p}) = [\boldsymbol{h}']C(\boldsymbol{p})$, then $[\boldsymbol{h}] = [\boldsymbol{h}']$.*

Proof of Lemma 8.5. Let $\boldsymbol{X} = (X_\lambda, p_{\lambda\lambda'}, \Lambda), \boldsymbol{Y} = (q_\mu, q_{\mu\mu'}, M)$ and let $\boldsymbol{f} = (f_{\boldsymbol{\mu}})\colon X \to \boldsymbol{Y}$ be a coherent mapping. We will exhibit a function g and mappings $g_{\boldsymbol{\mu}}\colon X_{g(\boldsymbol{\mu})} \times \Delta^n \to Y_{\mu_0}$ having the following properties. To every non-degenerate multiindex $\boldsymbol{\mu} \in M_n$, the function g assigns an element $g(\boldsymbol{\mu}) \in \Lambda$ such that, for $0 \leq j \leq n$, $n > 0$,

$$g(\boldsymbol{\mu}) \geq g(d^j\boldsymbol{\mu}). \tag{24}$$

For $n > 0$,

$$g_{\boldsymbol{\mu}}(x, d_j t) = \begin{cases} q_{\mu_0\mu_1} g_{d^0\boldsymbol{\mu}}(p_{g(d^0\boldsymbol{\mu})g(\boldsymbol{\mu})}(x), t), & j = 0, \\ g_{d^j\boldsymbol{\mu}}(p_{g(d^j\boldsymbol{\mu})g(\boldsymbol{\mu})}(x), t). & 0 < j \leq n. \end{cases} \tag{25}$$

Moreover, we will exhibit a coherent homotopy $\boldsymbol{G} = (G_{\boldsymbol{\mu}})\colon X \times I \to \boldsymbol{Y}$, such that

$$G_{\boldsymbol{\mu}}(x,0,t) = f_{\boldsymbol{\mu}}(x,t),\ G_{\boldsymbol{\mu}}(x,1,t) = g_{\boldsymbol{\mu}}(p_{g(\boldsymbol{\mu})}(x),t). \tag{26}$$

Let us first show that this suffices to complete the proof of Lemma 8.5. Indeed, by Lemma 1.14, there exists a coherent mapping $\boldsymbol{h} = (h, h_{\boldsymbol{\mu}})\colon \boldsymbol{X} \to \boldsymbol{Y}$ such that

$$g(\boldsymbol{\mu}) \leq h(\mu_n), \tag{27}$$

$$h_{\boldsymbol{\mu}}(x,t) = g_{\boldsymbol{\mu}}(p_{g(\boldsymbol{\mu})h(\mu_n)}(x), t). \tag{28}$$

Therefore, the second equality in (26) implies

$$G_{\boldsymbol{\mu}}(x,1,t) = h_{\boldsymbol{\mu}}(p_{h(\mu_n)}(x), t). \tag{29}$$

Taking into account the first equality in (26), one concludes that $\boldsymbol{G}$ is a homotopy which connects $\boldsymbol{f}$ with the coherent mapping $X \to \boldsymbol{Y}$, given by the right side of (29). However, by Lemma 2.12, the latter is homotopic to $\boldsymbol{h}C(\boldsymbol{p})$. Consequently, $\boldsymbol{f} \simeq \boldsymbol{h}C(\boldsymbol{p})$.

We will now define $g(\boldsymbol{\mu})$, $g_{\boldsymbol{\mu}}$ and $G_{\boldsymbol{\mu}}$, for $\boldsymbol{\mu} \in M_n$, by induction on the length n of the sequence $\boldsymbol{\mu} = (\mu_0, \dots, \mu_n)$. If $n = 0$, condition (S1) = (M1) yields an index $g(\mu_0) \in \Lambda$, a mapping $g_{\mu_0}: X_{g(\mu_0)} \to Y_{\mu_0}$ and a homotopy $G_{\mu_0}: X \times I \to Y_{\mu_0}$, such that $G_{\mu_0}(x,0) = f_{\mu_0}(x)$ and $G_{\mu_0}(x,1) = g_{\mu_0} p_{g(\mu_0)}(x)$.

Assume that, for sequences $\boldsymbol{\mu}$ of length $< n$, $n \geq 1$, we have already defined $g(\boldsymbol{\mu})$, $g_{\boldsymbol{\mu}}$, $G_{\boldsymbol{\mu}}$, in accordance with (24)–(26). We will now define these data, for $\boldsymbol{\mu} = (\mu_0, \dots, \mu_n) \in M_n$.

Let $G': X \times ((0 \times \Delta^n) \cup (I \times \partial\Delta^n)) \to Y_{\mu_0}$ be the mapping, given by

$$G'(x,0,t) = f_{\boldsymbol{\mu}}(x,t), \ t \in \Delta^n, \tag{30}$$

$$G'(x,s,d_j t) = qG_{d^j\boldsymbol{\mu}}(x,s,t), \ 0 \leq j \leq n, \ (s,t) \in I \times \Delta^{n-1}. \tag{31}$$

For $s = 0$ and $t \in \Delta^{n-1}$, (31) and (26) yield $G'(x,0,d_j t) = qG_{d^j\boldsymbol{\mu}}(x,0,t) = qf_{d^j\boldsymbol{\mu}}(x,t) = f_{\boldsymbol{\mu}}(x,d_j t)$, and (30) gives the same value for $G'(x,0,d_j t)$. Therefore, G' is well defined. Also note that

$$G'(x,1,d_j t) = qg_{d^j\boldsymbol{\mu}}(p_{g(d^j\boldsymbol{\mu})}(x), t), \ 0 \leq j \leq n, \tag{32}$$

for $x \in X, \ t \in \Delta^{n-1}$.

Since $(0 \times \Delta^n) \cup (I \times \partial\Delta^n)$ is a retract of $I \times \Delta^n$, the mapping G' extends to a homotopy $G': X \times I \times \Delta^n \to Y_{\mu_0}$. Let $G'_1: X \times \Delta^n \to Y_{\mu_0}$ be the restriction of G' to the level $s = 1$, i.e., $G'_1(x,t) = G'(x,1,t)$. Choose an index $\lambda \in \Lambda$, such that $\lambda \geq g(d^j\boldsymbol{\mu})$, $0 \leq j \leq n$, and define a mapping $g': X_\lambda \times \partial\Delta^n \to Y_{\mu_0}$, by putting

$$g'(z,d_j t) = qg_{d^j\boldsymbol{\mu}}(p_{g(d^j\boldsymbol{\mu})\lambda}(z), t), \ 0 \leq j \leq n, \tag{33}$$

where $z \in X_\lambda, t \in \Delta^{n-1}$. By the induction hypothesis, the mappings $g_{\boldsymbol{\mu}'}$ satisfy the coherence relations (25), for multiindices $\boldsymbol{\mu}'$ of length $\leq n-1$. Using these relations, one readily concludes that g' is a well-defined mapping. Moreover, (33) and (32) yield

$$g'(p_\lambda(x), t) = G'_1(x,t), \ x \in X, \ t \in \partial\Delta^n. \tag{34}$$

This enables us to apply property $(S2)^n$ from Lemma 8.3, where for P, f and F we substitute $Y_{\mu_0}, g': X_\lambda \times \partial\Delta^n \to Y_{\mu_0}$ and $G'_1: X \times \Delta^n \to Y_{\mu_0}$. We obtain a $\lambda' \geq \lambda$ and a mapping $H: X_{\lambda'} \times \Delta^n \to Y_{\mu_0}$, which satisfy

$$H|X_{\lambda'} \times \partial\Delta^n = g'(p_{\lambda\lambda'} \times 1), \tag{35}$$

$$H(p_{\lambda'} \times 1) \simeq G_1' \,(\mathrm{rel}(X \times \partial\Delta^n)). \tag{36}$$

We now define $g(\boldsymbol{\mu}) \in \Lambda$ and $g_{\boldsymbol{\mu}}$, by putting

$$g(\boldsymbol{\mu}) = \lambda',\ g_{\boldsymbol{\mu}}(z,t) = H(z,t),\ z \in X_{g(\boldsymbol{\mu})},\ t \in \Delta^n. \tag{37}$$

Note that (35) implies

$$g_{\boldsymbol{\mu}}(z,t) = g'(p_{\lambda g(\boldsymbol{\mu})}(z),t),\ z \in X_{g(\boldsymbol{\mu})},\ t \in \partial\Delta^n. \tag{38}$$

This formula and (33) show that $g_{\boldsymbol{\mu}}$ satisfies the coherence conditions (25).

We now consider a homotopy $G'': X \times I \times \Delta^n \to Y_{\mu_0}$, which realizes condition (36). Then, by (37),

$$G''(x,0,t) = g_{\boldsymbol{\mu}}(p_{g(\boldsymbol{\mu})}(x),t),\ G''(x,1,t) = G_1'(x,t),\ x \in X,\ t \in \Delta^n, \tag{39}$$

$$G''(x,s,t) = G''(x,0,t),\ x \in X,\ s \in I,\ t \in \partial\Delta^n. \tag{40}$$

Next, we define a mapping $G: X \times [0,2] \times \Delta^n \to Y_{\mu_0}$, by putting

$$G(x,s,t) = \begin{cases} G'(x,s,t), & 0 \le s \le 1, \\ G''(x,2-s,t), & 1 \le s \le 2, \end{cases} \tag{41}$$

where $x \in X$, $t \in \Delta^n$. By (39), G is well defined. Moreover, by (30) and (39),

$$G(x,0,t) = f_{\boldsymbol{\mu}}(x,t),\ G(x,2,t) = g_{\boldsymbol{\mu}}(p_{g(\boldsymbol{\mu})}(x),t), \tag{42}$$

for $x \in X,\ t \in \Delta^n$.

We now need a mapping $\varphi^n: [0,2] \times \Delta^n \to [0,1]$, having the following properties.

$$\varphi^n(0,t) = 0,\ \varphi^n(2,t) = 1,\ t \in \Delta^n, \tag{43}$$

$$\varphi^n(s,t) = \begin{cases} s, & 0 \le s \le 1, & t \in \partial\Delta^n, \\ 1, & 1 \le s \le 2, & t \in \partial\Delta^n. \end{cases} \tag{44}$$

Moreover, for every $t \notin \partial\Delta^n$, $\varphi^n|[0,2] \times \{t\}: [0,2] \to [0,1]$ is a homeomorphism.

Such a mapping φ^n can be obtained by the formula

$$\varphi^n(s,t) = \begin{cases} s\tau(t), & 0 \le s \le 1, \\ \tau(t) + (s-1)(1-\tau(t)), & 1 \le s \le 2, \end{cases} \tag{45}$$

where $\tau(t) = 1 - \min\{t_0,\dots,t_n\}$. Note that $\tau(t) > 0$, for every $t \in \Delta^n$. Moreover, $\tau(t) = 1$ if and only if $t \in \partial\Delta^n$.

Now consider the mapping $\Phi: X \times [0,2] \times \Delta^n \to X \times I \times \Delta^n$, given by

$$\Phi(x,s,t) = (x, \varphi^n(s,t), t). \tag{46}$$

Note that for different points $(x,s,t) \ne (x',s',t')$ one has $\Phi(x,s,t) = \Phi(x',s',t')$ only when $x = x', t = t' \in \partial\Delta^n, s,s' \in [1,2]$. However, in this case, by (40) and (41), $G(x,s,t) = G(x',s',t')$. Therefore, G determines a unique mapping $G_{\boldsymbol{\mu}}: X \times I \times \Delta^n \to Y_{\mu_0}$, which satisfies

$$G_{\boldsymbol{\mu}}\Phi = G. \tag{47}$$

Using (47), (46), (43), (41), (30), and (39), one readily concludes that $G_{\boldsymbol{\mu}}$ satisfies (26). To see that $G_{\boldsymbol{\mu}}$ satisfies the coherence conditions, involving the face operators d_j, $0 \le j \le n$, note that, for $x \in X, s \in I, t \in \Delta^{n-1}$, (46), (44), (47), (41) and (31) yield

$$\begin{gathered} G_{\boldsymbol{\mu}}(x,s,d_jt) = G_{\boldsymbol{\mu}}\Phi(x,s,d_jt) = \\ G(x,s,d_jt) = G'(x,s,d_jt) = qG_{d^j\boldsymbol{\mu}}(x,s,t). \end{gathered} \tag{48}$$

By Lemma 1.13, there is a unique way to define $G_{\boldsymbol{\mu}}$, for $\boldsymbol{\mu}$ degenerate, and obtain a coherent mapping $\boldsymbol{G} = (G_{\boldsymbol{\mu}}): X \times I \to \boldsymbol{Y}$. Since we have already verified (26), for non-degenerate $\boldsymbol{\mu}$, the uniqueness part of the same lemma shows that (26) holds also for degenerate multiindices $\boldsymbol{\mu}$. □

Proof of Lemma 8.6. Let $\boldsymbol{h} = (h, h_{\boldsymbol{\mu}}), \boldsymbol{h}' = (h', h'_{\boldsymbol{\mu}}): \boldsymbol{X} \to \boldsymbol{Y}$ be coherent mappings. Consider the coherent mappings $f_{\boldsymbol{\mu}}, f'_{\boldsymbol{\mu}}: X \to Y_{\mu_0}$, given by

$$f_{\boldsymbol{\mu}} = h_{\boldsymbol{\mu}}(p_{h(\mu_n)}(x),t), \;\; f'_{\boldsymbol{\mu}} = h'_{\boldsymbol{\mu}}(p_{h'(\mu_n)}(x),t). \tag{49}$$

By Lemma 2.12, $\boldsymbol{f} \simeq \boldsymbol{h}C(\boldsymbol{p})$, $\boldsymbol{f}' \simeq \boldsymbol{h}'C(\boldsymbol{p})$ and thus, by assumption, $\boldsymbol{f} \simeq \boldsymbol{f}'$. Let $\boldsymbol{F} = (F_{\boldsymbol{\mu}}): X \times I \to \boldsymbol{Y}$ be a coherent homotopy which connects $\boldsymbol{f}$ to $\boldsymbol{f}'$, i.e.,

$$F_{\boldsymbol{\mu}}(x,0,t) = h_{\boldsymbol{\mu}}(p_{h(\mu_n)}(x),t), F_{\boldsymbol{\mu}}(x,1,t) = h'_{\boldsymbol{\mu}}(p_{h'(\mu_n)}(x),t). \tag{50}$$

In order to prove that $\boldsymbol{h} \simeq \boldsymbol{h}'$, it suffices to exhibit a function G and homotopies $G_{\boldsymbol{\mu}}: X_{G(\boldsymbol{\mu})} \times I \times \Delta^n \to Y_{\mu_0}$, having the following properties. To every non-degenerate multiindex $\boldsymbol{\mu} \in M_n$, the function G assigns an element $G(\boldsymbol{\mu}) \in \Lambda$ such that, for $0 \le j \le n$, $n > 0$,

$$G(\boldsymbol{\mu}) \ge G(d^j\boldsymbol{\mu}), \;\; G(\boldsymbol{\mu}) \ge h(\mu_n), h'(\mu_n). \tag{51}$$

For $n > 0$,

$$G_{\boldsymbol{\mu}}(x, d_jt') = \begin{cases} q_{\mu_0\mu_1}G_{d^0\boldsymbol{\mu}}(p_{G(d^0\boldsymbol{\mu})G(\boldsymbol{\mu})}(x),t'), & j=0, \\ G_{d^j\boldsymbol{\mu}}(p_{G(d^j\boldsymbol{\mu})G(\boldsymbol{\mu})}(x),t'). & 0 < j \le n. \end{cases} \tag{52}$$

Moreover,

$$G_{\boldsymbol{\mu}}(z,0,t) = h_{\boldsymbol{\mu}}(p_{h(\mu_n)G(\boldsymbol{\mu})}(z),t), \;\; G_{\boldsymbol{\mu}}(z,1,t) = h'_{\boldsymbol{\mu}}(p_{h'(\mu_n)G(\boldsymbol{\mu})}(z),t). \tag{53}$$

Indeed, Lemma 1.14, applied to $(G, G_{\boldsymbol{\mu}})$, yields a coherent homotopy $\boldsymbol{H} = (H, H_{\boldsymbol{\mu}}): \boldsymbol{X} \times I \to \boldsymbol{Y}$ such that

$$G(\boldsymbol{\mu}) \le H(\mu_n), \tag{54}$$

$$H_{\boldsymbol{\mu}}(x,s,t) = G_{\boldsymbol{\mu}}(p_{G(\boldsymbol{\mu})H(\mu_n)}(x),s,t). \tag{55}$$

Now (53) and (55) show that $\boldsymbol{H}$ connects $\boldsymbol{h}$ to $\boldsymbol{h}'$.

We will define $G(\boldsymbol{\mu})$ and $G_{\boldsymbol{\mu}}$ by induction on the length n of $\boldsymbol{\mu}$. In order to do this we need some additional mappings $Q_{\boldsymbol{\mu}}: X \times I \times I \times \Delta^n \to Y_{\mu_0}$, which satisfy the coherence conditions for the face operators and also satisfy the following conditions.

$$Q_{\boldsymbol{\mu}}(x,0,s,t) = F_{\boldsymbol{\mu}}(x,s,t),\ Q_{\boldsymbol{\mu}}(x,1,s,t) = G_{\boldsymbol{\mu}}(p_{G(\boldsymbol{\mu})}(x),s,t), \tag{56}$$

$$Q_{\boldsymbol{\mu}}(x,u,0,t) = h_{\boldsymbol{\mu}}(p_{h(\mu_n)}(x),s,t),\ Q_{\boldsymbol{\mu}}(x,u,1,t) = h'_{\boldsymbol{\mu}}(p_{h'(\mu_n)}(x),s,t). \tag{57}$$

Since the case $n = 0$ is also non-trivial, we begin our induction with the (empty) case $n = -1$. Note that for $n = 0$, $\partial\Delta^0 = \emptyset$ and one must disregard conditions and equations which involve this set.

For $n \geq 0$, choose a $\lambda \in \Lambda$ such that $\lambda \geq h(\boldsymbol{\mu}), g'(\boldsymbol{\mu}), G(d^j\boldsymbol{\mu})$, $0 \leq j \leq n$. Then define a mapping $G': X_\lambda \times \partial(I \times \Delta^n) \to Y_{\mu_0}$, by putting

$$G'(z,s,d_jt') = qG_{d^j\boldsymbol{\mu}}(p_{G(d^j\boldsymbol{\mu})\lambda}(z),s,t'), \tag{58}$$

for $z \in X_\lambda,\ s \in I,\ t' \in \Delta^{n-1}$,

$$G'(z,0,t) = h_{\boldsymbol{\mu}}(p_{g(\mu_n)\lambda}(z),t),\ G'(z,1,t) = h'_{\boldsymbol{\mu}}(p_{g'(\mu_n)\lambda}(z),t), \tag{59}$$

for $z \in X_\lambda,\ t \in \Delta^n$. The coherence relations for $G_{d^j\boldsymbol{\mu}}$, assumed by the induction hypothesis, insure that, by (58), $G'_{\boldsymbol{\mu}}$ is well defined on $X_\lambda \times I \times \partial\Delta^n$. Moreover, (53) for $d^j\boldsymbol{\mu}$ and the coherence conditions for $h_{\boldsymbol{\mu}}, h'\boldsymbol{\mu}$, prove the consistency of (58) and (59), so that G' is well defined on all of $X_\lambda \times \partial(I \times \Delta^n)$.

Next we define a homotopy $K: X \times I \times \partial(I \times \Delta^n) \to Y_{\mu_0}$, by putting

$$K(z,u,s,d_jt') = qQ_{d^j\boldsymbol{\mu}}(x,u,s,t'), \tag{60}$$

for $x \in X,\ (u,s) \in I \times I,\ t \in \Delta^{n-1}$,

$$K(x,u,0,t) = h_{\boldsymbol{\mu}}(p_{h(\mu_n)}(x),t),\ K(x,u,1,t) = h'_{\boldsymbol{\mu}}(p_{h'(\mu_n)}(x),t), \tag{61}$$

for $x \in X,\ t \in \Delta^n$. The coherence relations for $Q_{d^j\boldsymbol{\mu}}$, assumed by the induction hypothesis, (57) for $d^j\boldsymbol{\mu}$ and the coherence conditions for $h_{\boldsymbol{\mu}}, h'_{\boldsymbol{\mu}}$, insure that K is well defined on all of $X \times I \times \partial(I \times \Delta^n)$.

Putting $u = 0$ in (60), using the first equality in (56) for $d^j\boldsymbol{\mu}$ and the coherence conditions for $F_{\boldsymbol{\mu}}$, one concludes that

$$K(x,0,s,d_jt') = F_{\boldsymbol{\mu}}(x,s,d_jt'),\ x \in X,\ s \in I,\ t' \in \Delta^{n-1}. \tag{62}$$

Furthermore, (61) and (50) yield

$$K(x,0,0,t) = F_{\boldsymbol{\mu}}(x,0,t),\ K(x,0,1,t) = F_{\boldsymbol{\mu}}(x,1,t)\ x \in X,\ t \in \Delta^n, \tag{63}$$

so that $K_0: X \times \partial(I \times \Delta^n) \to Y_{\mu_0}$, defined by $K_0(x,s,t) = K(x,0,s,t)$, satisfies

$$K_0 = F_{\boldsymbol{\mu}}|X \times \partial(I \times \Delta^n). \tag{64}$$

Similarly, putting $u = 1$ in (60), using the second formula in (56) for $d^j\boldsymbol{\mu}$ and (58), one concludes that

$$K(x,1,s,d_jt') = G'(p_\lambda(x),s,d_jt'),\ x \in X,\ s \in I,\ t' \in \Delta^{n-1}. \tag{65}$$

Furthermore, (61) and (59) show that

$$K(x,1,0,t) = G'(p_\lambda(x),0,t),\ K(x,1,1,t) = G'(p_\lambda(x),1,t), \tag{66}$$

for $x \in X,\ t \in \Delta^n$, so that $K_1\colon X \times \partial(I \times \Delta^n) \to Y_{\mu_0}$, defined by $K_1(x,s,t) = K(x,1,s,t)$, satisfies

$$K_1 = G'(p_\lambda \times 1). \tag{67}$$

Next note that $F_{\boldsymbol{\mu}}$ can be viewed as defined on $X \times 0 \times I \times \Delta^n$. Moreover, $(0 \times I \times \Delta^n) \cup (I \times \partial(I \times \Delta^n))$ is a retract of $I \times I \times I \times \Delta^n$. Therefore, one can extend both mappings $F_{\boldsymbol{\mu}}$ and $K\colon X \times I \times \partial(I \times \Delta^n) \to Y_{\mu_0}$ to a mapping $K'\colon X \times I \times I \times \Delta^n \to Y_{\mu_0}$. By (67), the mapping $K_1'\colon X \times I \times \Delta^n \to Y_{\mu_0}$, defined by $K_1'(x,s,t) = K'(x,1,s,t)$, satisfies

$$K_1'|X \times \partial(I \times \Delta^n) = G'(p_\lambda \times 1). \tag{68}$$

The last equation enables us to apply to the strong expansion $\boldsymbol{p}\colon X \times I \to \boldsymbol{X} \times I$ property $(\mathrm{S2})^n$ from Lemma 8.3, where for P, f and F, we substitute Y_{μ_0}, G' and K_1', respectively. We obtain a $\lambda' \geq \lambda$ and a mapping $H\colon X_{\lambda'} \times I \times \Delta^n \to Y_{\mu_0}$, which satisfy

$$H|X_{\lambda'} \times I \times \partial\Delta^n = G'(p_{\lambda\lambda'} \times 1), \tag{69}$$

$$H(p_{\lambda'} \times 1) \simeq K_1'(\mathrm{rel}(X \times \partial\Delta^n). \tag{70}$$

We now put $G(\boldsymbol{\mu}) = \lambda'$ and $G_{\boldsymbol{\mu}} = H$. Formulae (69) and (59) show that $G_{\boldsymbol{\mu}}$ satisfies (53). Moreover, by (69), (67) and (58),

$$G_{\boldsymbol{\mu}}(z,s,d_jt') = G'(p_{\lambda\lambda'}(z),s,d_jt') = qG_{d^j\boldsymbol{\mu}}(p_{G(d^j\boldsymbol{\mu})G(\boldsymbol{\mu})}(z),s,t'), \tag{71}$$

which is the desired condition (52).

We define $Q_{\boldsymbol{\mu}}$ in a similar way. Let $K''\colon X \times I \times I \times \Delta^n \to Y_{\mu_0}$ be a homotopy, which realizes (70), i.e.,

$$K''(x,0,s,t) = G_{\boldsymbol{\mu}}(p_{G(\boldsymbol{\mu})}(x),s,t),\ K''(x,1,s,t) = K_1'(x,s,t), \tag{72}$$

for $x \in X,\ s \in I,\ t \in \Delta^n$,

$$K''(x,u,s,t) = K(x,0,s,t), \tag{73}$$

for $x \in X,\ (s,t) \in \partial(I \times \Delta^n)$. We define a homotopy $K^*\colon X \times [0,2] \times I \times \Delta^n \to Y_{\mu_0}$, by putting

$$K^*(x,u,s,t) = \begin{cases} K'(x,u,s,t), & 0 \leq u \leq 1, \\ K''(x,2-u,s,t), & 1 \leq u \leq 2. \end{cases} \tag{74}$$

By (72), K^* is well defined and

$$K^*(x,0,s,t) = F_\mu(x,s,t),\ K^*(x,2,s,t) = G_\mu(p_{G(\mu)}(x),s,t), \tag{75}$$

for $x \in X,\ s \in I,\ t \in \Delta^n$.

We now choose a homeomorphism $\eta^n: I \times \Delta^n \to \Delta^{n+1}$ and define a mapping $\psi^n: [0,2] \times I \times \Delta^n \to I$, by putting $\psi^n = \varphi^{n+1}(1 \times \eta^n)$, where $\varphi^{n+1}: [0,2] \times \Delta^{n+1} \to I$ is the mapping considered in the proof of Lemma 8.5. We then define a mapping $\Psi: X \times [0,2] \times I \times \Delta^n \to X \times I \times I \times \Delta^n$, by putting

$$\Psi(x,u,s,t) = (x, \varphi^n(u,s,t), s, t). \tag{76}$$

By properties of φ^{n+1}, it is easily seen that, for different points $(x,u,s,t) \neq (x',u',s',t')$, one has $\Psi(x,u,s,t) = \Psi(x',u',s',t')$, only when $x = x'$, $(s,t) = (s',t') \in \partial(I \times \Delta^n), u, u' \in [1,2]$. However, in this case, by (74) and (73), $K^*(x,u,s,t) = K^*(x',u',s',t')$. Therefore, K^* determines a unique mapping $Q_\mu: X \times I \times I \times \Delta^n \to Y_{\mu_0}$, which satisfies

$$Q_\mu = K^* \Psi. \tag{77}$$

Using (61), (66), (75) and (74), one readily verifies that Q_μ has the desired properties (56), (57). To see that the mappings G_μ satisfy the coherence conditions (52), note that, $(x,u,s,t) = \Psi(x,u,s,t)$, for $(s,t) \in \partial(I \times \Delta^n)$. Therefore, by (74) and (60),

$$\begin{aligned} Q_\mu(x,u,s,d_jt') &= K^*(x,u,s,d_jt') = \\ K'(x,u,s,d_jt') &= qQ_{d^j\mu}(x,u,s,t'). \ \square \end{aligned} \tag{78}$$

Proof of Theorem 8.2. Let $\boldsymbol{p}: X \to \boldsymbol{X}$ be a cofinite coherent expansion, where $\boldsymbol{X}$ consists of spaces from the class HPol. Choose a cofinite strong expansion $\boldsymbol{q}: X \to \boldsymbol{Y}$ such that $\boldsymbol{Y}$ also consists of spaces from HPol. Since $\boldsymbol{p}$ is a coherent expansion, there exists a coherent mapping $\boldsymbol{f}: \boldsymbol{X} \to \boldsymbol{Y}$ such that

$$\boldsymbol{f}C(\boldsymbol{p}) \simeq C(\boldsymbol{q}). \tag{79}$$

Now use the fact that $\boldsymbol{q}$ is also a coherent expansion, because it is a strong expansion (see Theorem 8.1). Since $\boldsymbol{X}$ is an HPol-system, we conclude that there exists a coherent mapping $\boldsymbol{g}: \boldsymbol{Y} \to \boldsymbol{X}$ such that

$$\boldsymbol{g}C(\boldsymbol{q}) \simeq C(\boldsymbol{p}), \tag{80}$$

and thus,

$$\boldsymbol{gf}C(\boldsymbol{p}) \simeq C(\boldsymbol{p}), \tag{81}$$

Now the uniqueness property of the coherent expansion $\boldsymbol{p}$ implies

$$\boldsymbol{gf} \simeq C(\boldsymbol{1}). \tag{82}$$

Formulae (81) and (82) show that $\boldsymbol{p}$ is coherently dominated by $\boldsymbol{q}$. Since $\boldsymbol{q}$ is a strong expansion, Theorem 7.14 yields the desired conclusion that also $\boldsymbol{p}$ is a strong expansion. □

Since resolutions are strong expansions (see Theorem 7.6), the following corollary is an immediate consequence of Theorem 8.1.

COROLLARY 8.7. *Every resolution of a space is a coherent expansion of that space.*

8.2 The strong shape category

We will now define the *strong shape category* SSh(Top). Its objects are all topological spaces. The morphisms $F: X \to Y$ are determined by triples $(\boldsymbol{p}, \boldsymbol{q}, [\boldsymbol{f}])$, i.e., by diagrams of the form

$$\begin{array}{ccc} \boldsymbol{X} & \xleftarrow{\;\boldsymbol{p}\;} & X \\ {\scriptstyle [\boldsymbol{f}]}\big\downarrow & & \\ \boldsymbol{Y} & \xleftarrow[\;\boldsymbol{q}\;]{} & Y, \end{array} \tag{1}$$

where $\boldsymbol{p}: X \to \boldsymbol{X}$ and $\boldsymbol{q}: Y \to \boldsymbol{Y}$ are *cofinite strong* HPol-*expansions* of X and Y, respectively and $[\boldsymbol{f}]: \boldsymbol{X} \to \boldsymbol{Y}$ is a morphism of CH(pro-Top).

Every topological space X admits such expansions, because, by Theorem 6.22 (Theorem 6.23) and Remarks 6.28, 6.30, X admits a cofinite polyhedral resolution (ANR-resolution) $\boldsymbol{p}: X \to \boldsymbol{X}$, and by Theorem 7.6, $\boldsymbol{p}$ is a strong expansion.

If $\boldsymbol{p}: X \to \boldsymbol{X}$ and $\boldsymbol{p}': X \to \boldsymbol{X}'$ are cofinite strong HPol-expansions of the same space X, then by Theorem 8.1, there exists a unique morphism $[\boldsymbol{i}]: \boldsymbol{X} \to \boldsymbol{X}'$ of CH(pro-Top) such that

$$[\boldsymbol{i}]C(\boldsymbol{p}) = C(\boldsymbol{p}'), \tag{2}$$

i.e., the following diagram commutes.

$$\begin{array}{ccccc} & & X & & \\ & {\scriptstyle C(\boldsymbol{p})}\swarrow & & \searrow{\scriptstyle C(\boldsymbol{p}')} & \\ \boldsymbol{X} & & \xrightarrow[\;[\boldsymbol{i}]\;]{} & & \boldsymbol{X}' \end{array} \tag{3}$$

Moreover, there exists a unique morphism $[\boldsymbol{i}^-]: \boldsymbol{X}' \to \boldsymbol{X}$ of CH(pro-Top) such that

$$[\boldsymbol{i}^-]C(\boldsymbol{p}') = C(\boldsymbol{p}). \tag{4}$$

Clearly, (2) and (4) imply

$$[\boldsymbol{i}^-][\boldsymbol{i}]C(\boldsymbol{p}) = C(\boldsymbol{p}), \;\; [\boldsymbol{i}][\boldsymbol{i}^-]C(\boldsymbol{p}') = C(\boldsymbol{p}'). \tag{5}$$

By uniqueness of such factorizations, one concludes that

$$[\boldsymbol{i}^-][\boldsymbol{i}] = [\mathbf{1}_{\boldsymbol{X}}], \;\; [\boldsymbol{i}][\boldsymbol{i}^-] = [\mathbf{1}_{\boldsymbol{X}'}], \tag{6}$$

which shows that $[\boldsymbol{i}]: \boldsymbol{X} \to \boldsymbol{X}'$ is an isomorphism of CH(pro-Top) and $[\boldsymbol{i}^-]: \boldsymbol{X}' \to \boldsymbol{X}$ is its inverse.

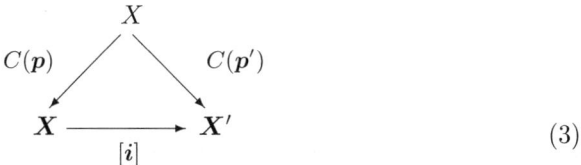

If we have a third expansion of the described type $\boldsymbol{p}'' : X \to \boldsymbol{X}''$, then we obtain unique isomorphisms $[\boldsymbol{i}'] : \boldsymbol{X}' \to \boldsymbol{X}''$ and $[\boldsymbol{i}''] : \boldsymbol{X} \to \boldsymbol{X}''$ such that

$$[\boldsymbol{i}']C(\boldsymbol{p}') = C(\boldsymbol{p}''), \; [\boldsymbol{i}'']C(\boldsymbol{p}) = C(\boldsymbol{p}''). \tag{7}$$

Using again uniqueness, one concludes that

$$[\boldsymbol{i}'][\boldsymbol{i}] = [\boldsymbol{i}'']. \tag{8}$$

Similarly, there is a unique isomorphism $[\boldsymbol{j}] : \boldsymbol{Y} \to \boldsymbol{Y}'$ of CH(pro-Top) such that

$$[\boldsymbol{j}]C(\boldsymbol{q}) = C(\boldsymbol{q}'). \tag{9}$$

We now define an equivalence relation $\sim$ between triples $(\boldsymbol{p}, \boldsymbol{q}, [\boldsymbol{f}])$, by putting $(\boldsymbol{p}, \boldsymbol{q}, [\boldsymbol{f}]) \sim (\boldsymbol{p}', \boldsymbol{q}', [\boldsymbol{f}'])$, provided

$$[\boldsymbol{f}'][\boldsymbol{i}] = [\boldsymbol{j}][\boldsymbol{f}], \tag{10}$$

i.e., the following diagram commutes.

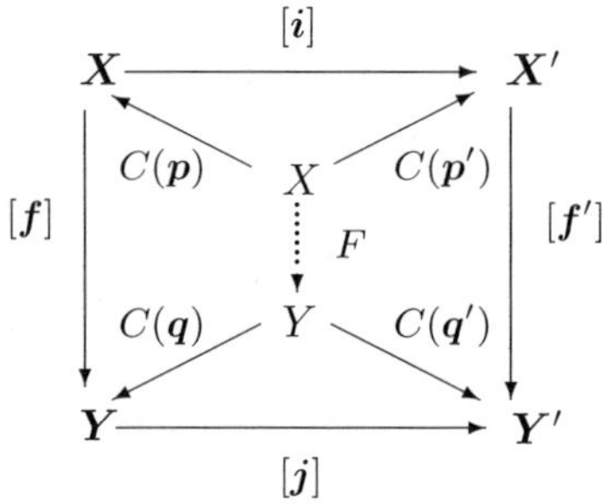

(11)

Using (6), (7) and the analogous formulae for $[\boldsymbol{j}], [\boldsymbol{j}^-], [\boldsymbol{j}']$ and $[\boldsymbol{j}'']$, it is readily seen that $\sim$ is indeed an equivalence relation.

We now define the *strong shape morphisms* $F : X \to Y$, i.e., the morphisms of SSh(Top), as equivalence classes of triples $(\boldsymbol{p}, \boldsymbol{q}, [\boldsymbol{f}])$, where $\boldsymbol{p} : X \to \boldsymbol{X}$ and $\boldsymbol{q} : Y \to \boldsymbol{Y}$ are cofinite strong HPol-expansions and $[\boldsymbol{f}] : \boldsymbol{X} \to \boldsymbol{Y}$ is a morphism of CH(pro-Top).

In order to define the *composition* of two strong shape morphisms $F : X \to Y$ and $G : Y \to Z$, assume that their representatives are triples $(\boldsymbol{p}, \boldsymbol{q}, [\boldsymbol{f}])$ and $(\boldsymbol{q}', \boldsymbol{r}, [\boldsymbol{g}'])$, respectively, where $\boldsymbol{p} : X \to \boldsymbol{X}, \boldsymbol{q} : Y \to \boldsymbol{Y}, \boldsymbol{q}' : Y' \to \boldsymbol{Y}', \boldsymbol{r} : Z \to \boldsymbol{Z}$ are cofinite strong HPol-expansions and $\boldsymbol{f} : \boldsymbol{X} \to \boldsymbol{Y}, \boldsymbol{g}' : \boldsymbol{Y}' \to \boldsymbol{Z}$ are coherent mappings. There is no loss of generality in assuming that the second triple is of the form $(\boldsymbol{q}, \boldsymbol{r}, [\boldsymbol{g}])$, i.e., that the same strong expansion $\boldsymbol{q}$ of Y appears in the triples representing both strong shape morphisms F and G. Indeed, if we put $[\boldsymbol{g}] = [\boldsymbol{g}'][\boldsymbol{j}] : \boldsymbol{X} \to \boldsymbol{Y}'$, where $[\boldsymbol{j}] : \boldsymbol{Y} \to \boldsymbol{Y}'$ denotes the unique morphism for which $[\boldsymbol{j}]C(\boldsymbol{q}) = C(\boldsymbol{q}')$, then $(\boldsymbol{q}', \boldsymbol{r}, [\boldsymbol{g}']) \sim (\boldsymbol{q}, \boldsymbol{r}, [\boldsymbol{g}])$, because $[\boldsymbol{g}'][\boldsymbol{j}] = [\boldsymbol{1}_Z][\boldsymbol{g}]$. By definition, $GF : X \to Z$ is the strong shape morphism whose representative is $(\boldsymbol{p}, \boldsymbol{r}, [\boldsymbol{g}][\boldsymbol{f}])$.

To see that GF is well defined, assume that $(\boldsymbol{p}', \boldsymbol{q}', [\boldsymbol{f}'])$ and $(\boldsymbol{q}', \boldsymbol{r}', [\boldsymbol{g}'])$ are other choices of representatives of F and G, respectively. If $[\boldsymbol{i}]: \boldsymbol{X} \to \boldsymbol{X}', [\boldsymbol{j}]: \boldsymbol{Y} \to \boldsymbol{Y}'$ and $[\boldsymbol{k}]: \boldsymbol{Z} \to \boldsymbol{Z}'$ are the unique isomorphisms defined by conditions of the form (2), then by (10), $[\boldsymbol{f}'][\boldsymbol{i}] = [\boldsymbol{j}][\boldsymbol{f}]$ and $[\boldsymbol{g}'][\boldsymbol{j}] = [\boldsymbol{k}][\boldsymbol{g}]$ and thus, $([\boldsymbol{g}'][\boldsymbol{f}'])[\boldsymbol{i}] = [\boldsymbol{k}]([\boldsymbol{g}][\boldsymbol{f}])$, which shows that $(\boldsymbol{p}, \boldsymbol{r}, [\boldsymbol{g}][\boldsymbol{f}]) \sim (\boldsymbol{p}, \boldsymbol{r}, [\boldsymbol{g}'][\boldsymbol{f}'])$.

It is easy to see that composition of strong shape morphisms is associative. Indeed, if we also have a morphism $H: Z \to W$, we can assume that H is represented by a triple $(\boldsymbol{r}, \boldsymbol{s}, [\boldsymbol{h}])$, where $\boldsymbol{s}: W \to \boldsymbol{W}$ is a cofinite strong HPol-expansion. Then $H(GF)$ is represented by the triple $(\boldsymbol{p}, \boldsymbol{s}, [\boldsymbol{h}]([\boldsymbol{g}][\boldsymbol{f}]))$, while $(HG)F$ is represented by the triple $(\boldsymbol{p}, \boldsymbol{s}, ([\boldsymbol{h}][\boldsymbol{g}])[\boldsymbol{f}])$. However, $[\boldsymbol{h}]([\boldsymbol{g}][\boldsymbol{f}]) = ([\boldsymbol{h}][\boldsymbol{g}])[\boldsymbol{f}]$.

By definition, the identity morphism $1_X: X \to X$ in SSh(Top) is the strong shape morphism, given by the triple $(\boldsymbol{p}, \boldsymbol{p}, [\mathbf{1}_{\boldsymbol{X}}])$. If $F: X \to Y$ is given by $(\boldsymbol{p}, \boldsymbol{q}, [\boldsymbol{f}])$, then the composition $F 1_X$ is given by $(\boldsymbol{p}, \boldsymbol{p}, [\boldsymbol{f}][\mathbf{1}_{\boldsymbol{X}}]) = (\boldsymbol{p}, \boldsymbol{p}, [\boldsymbol{f}])$, so $F 1_X = F$. Similarly, $1_Y F = F$. Consequently, SSh(Top) is indeed a category.

If two spaces X, Y are isomorphic objects of the strong shape category SSh(Top), we say that X and Y have the same *strong shape* and we write, $\mathrm{ssh}(X) = \mathrm{ssh}(Y)$. Similarly, if a strong shape morphism $F: X \to Y$ has a left inverse $G: Y \to X$, i.e., $GF = \mathrm{id}$, we say that the strong shape of Y *dominates* the strong shape of X and we write $\mathrm{ssh}(X) \le \mathrm{ssh}(Y)$.

REMARK 8.8. Every topological space X has the strong shape of a topologically complete space X'. Indeed, by Theorem 6.22, X admits a polyhedral resolution $\boldsymbol{p}: X \to \boldsymbol{X}$. Polyhedra are paracompact and thus, topologically complete spaces. Therefore, Theorem 6.5 yields a topologically complete space X' and a resolution $\boldsymbol{p}': X' \to \boldsymbol{X}$. Since in both resolutions $\boldsymbol{p}$ and $\boldsymbol{p}'$ the same polyhedral system $\boldsymbol{X}$ appears, the identity mapping on $\boldsymbol{X}$ determines an isomorphism of strong shape $X \to X'$.

We now proceed to define the strong shape functor $\bar{S}$: H(Top) $\to$ SSh(Top), which maps the homotopy category into the strong shape category. We first define a functor from the topological category Top to SSh(Top), also denoted by $\bar{S}$. This functor is fixed on objects, i.e., $\bar{S}(X) = X$, for every topological space X. If $f: X \to Y$ is a (continuous) mapping, we choose cofinite strong HPol-expansions $\boldsymbol{p}: X \to \boldsymbol{X}$, $\boldsymbol{q}: Y \to \boldsymbol{Y}$ and a morphism $[\boldsymbol{f}]: \boldsymbol{X} \to \boldsymbol{Y}$ of CH(pro-Top) such that

$$[\boldsymbol{f}]C(\boldsymbol{p}) = C(\boldsymbol{q})C(f). \tag{12}$$

By definition, $\bar{S}(f)$ is the morphism of SSh(Top), given by the triple $(\boldsymbol{p}, \boldsymbol{q}, [\boldsymbol{f}])$.

To see that $\bar{S}(f)$ is well defined, consider another choice of cofinite strong HPol-expansions $\boldsymbol{p}': X \to \boldsymbol{X}', \boldsymbol{q}': Y \to \boldsymbol{Y}'$ and let $[\boldsymbol{i}]: \boldsymbol{X} \to \boldsymbol{X}', [\boldsymbol{j}]: \boldsymbol{Y} \to \boldsymbol{Y}'$ be the above defined isomorphisms of CH(pro-Top). We must show that the triples $(\boldsymbol{p}, \boldsymbol{q}, [\boldsymbol{f}])$ and $(\boldsymbol{p}', \boldsymbol{q}', [\boldsymbol{f}'])$ are equivalent, where $[\boldsymbol{f}']: \boldsymbol{X}' \to \boldsymbol{Y}'$ is characterized by $[\boldsymbol{f}']C(\boldsymbol{p}') = C(\boldsymbol{q}')C(f)$. This amounts to showing that

$$[\boldsymbol{f}'][\boldsymbol{i}] = [\boldsymbol{j}][\boldsymbol{f}]. \tag{13}$$

Indeed, by (2) and (12), we have

$$[\boldsymbol{f}'][\boldsymbol{i}]C(\boldsymbol{p}) = [\boldsymbol{f}']C(\boldsymbol{p}') = C(\boldsymbol{q}')C(f) = [\boldsymbol{j}]C[\boldsymbol{q}] = [\boldsymbol{j}][\boldsymbol{f}]C(\boldsymbol{p}). \tag{14}$$

However, by the uniqueness of such factorizations, (14) implies (13).

If we also have a mapping $g: Y \to Z$, then $\bar{S}(g)$ is given by the triple $(\boldsymbol{q}, \boldsymbol{r}, [\boldsymbol{g}])$, where

$$[\boldsymbol{g}]C(\boldsymbol{q}) = C(\boldsymbol{r})C(g). \tag{15}$$

Therefore, by (12) and (15),

$$[\boldsymbol{g}][\boldsymbol{f}]C(\boldsymbol{p}) = ([\boldsymbol{g}]C(\boldsymbol{q}))C(f) = C(\boldsymbol{r})(C(g)C(f)) = C(\boldsymbol{r})C(gf), \tag{16}$$

and we see that $\bar{S}(gf)$ is given by the triple $(\boldsymbol{p}, \boldsymbol{r}, [\boldsymbol{g}][\boldsymbol{f}])$. However, $\bar{S}(g)\bar{S}(f)$ is given by the same triple, which shows that $\bar{S}(gf) = \bar{S}(g)\bar{S}(f)$. Moreover, $\boldsymbol{1_X}C(\boldsymbol{p}) = C(\boldsymbol{p}) = C(\boldsymbol{p})C(1_X)$, which shows that $\bar{S}(1_X)$ is given by the triple $(\boldsymbol{p}, \boldsymbol{p}, [\boldsymbol{1_X}])$, which determines the identity morphism 1_X of SSh(Top). Consequently, $\bar{S}(1_X) = 1_X$. Hence, we have proved that $\bar{S}$: Top $\to$ SSh(Top) is indeed a functor.

Let us show that $\bar{S}$ factors through the homotopy category H(Top) and thus, induces a functor $\bar{S}$: H(Top) $\to$ SSh(Top), called the *strong shape functor*. It suffices to show that, for homotopic mappings $f, f': X \to Y$, one has $\bar{S}(f) = \bar{S}(f')$. Indeed, let $, \bar{S}(f), \bar{S}(f')$ be given by triples $(\boldsymbol{p}, \boldsymbol{q}, [\boldsymbol{f}]), (\boldsymbol{p}, \boldsymbol{q}, [\boldsymbol{f}'])$, respectively. By Remark 2.15, the assumption $[f] = [f']$ implies $C(f) = C(f')$, and therefore, by (12),

$$[\boldsymbol{f}']C(\boldsymbol{p}) = C(\boldsymbol{q})C(f') = C(\boldsymbol{q})C(f) = [\boldsymbol{f}]C(\boldsymbol{p}), \tag{17}$$

which yields $[\boldsymbol{f}] = [\boldsymbol{f}']$, hence also $(\boldsymbol{p}, \boldsymbol{q}, [\boldsymbol{f}]) = (\boldsymbol{p}, \boldsymbol{q}, [\boldsymbol{f}]')$. This enables us to put $\bar{S}[f] = \bar{S}(f)$. It is now easy to see that

$$\bar{S}([g][f]) = \bar{S}[g]\bar{S}[f], \ \bar{S}[1_X] = 1_X, \tag{18}$$

which proves that $\bar{S}$: H(Top) $\to$ SSh(Top) is indeed a functor.

We will now define a functor $\bar{E}$: SSh(Top) $\to$ Sh(Top) from the strong shape category to the ordinary shape category, which we will call the *forgetful functor*. First recall that the objects of Sh(Top) are all topological spaces. A *shape morphism* $G: X \to Y$ is given by a triple $([\boldsymbol{p}], [\boldsymbol{q}], [\boldsymbol{g}])$, where $[\boldsymbol{p}]: X \to [\boldsymbol{X}]$ and $[\boldsymbol{q}]: Y \to [\boldsymbol{Y}]$ are HPol-expansions of X and Y respectively, and $[\boldsymbol{g}]: [\boldsymbol{X}] \to [\boldsymbol{Y}]$ is a morphism of pro-H(Top). More precisely (see (Mardešić, Segal 1982), I.2.3), a shape morphism $G: X \to Y$ is an equivalence class of triples $([\boldsymbol{p}], [\boldsymbol{q}], [\boldsymbol{g}])$, where the equivalence relation $([\boldsymbol{p}], [\boldsymbol{q}], [\boldsymbol{g}]) \sim ([\boldsymbol{p}'], [\boldsymbol{q}'], [\boldsymbol{g}'])$ holds, provided

$$[\boldsymbol{g}'][\boldsymbol{k}] = [\boldsymbol{l}][\boldsymbol{g}]. \tag{19}$$

Here $[\boldsymbol{k}]\colon [\boldsymbol{X}] \to [\boldsymbol{X}']$ is the unique isomorphism of pro-H(Top) such that

$$[\boldsymbol{k}][\boldsymbol{p}] = [\boldsymbol{p}]'. \tag{20}$$

$[\boldsymbol{l}]\colon [\boldsymbol{Y}] \to [\boldsymbol{Y}']$ is defined analogously.

We now proceed to define $\bar{E}$. For a space X, we put $\bar{E}(X) = X$. Now let $F\colon X \to Y$ be the strong shape morphism, given by a triple $(\boldsymbol{p}, \boldsymbol{q}, [\boldsymbol{f}])$. By Remark 7.2, $\boldsymbol{p}, \boldsymbol{q}$ are homotopy expansions. Therefore, application of the homotopy functor H transforms $\boldsymbol{p}$ and $\boldsymbol{q}$ to HPol-expansions $[\boldsymbol{p}]$ and $[\boldsymbol{q}]$ of X and Y. Furthermore, the forgetful functor E: CH(pro-Top) $\to$ pro-H(Top) transforms $[\boldsymbol{f}]$ to a morphism $E[\boldsymbol{f}]\colon [\boldsymbol{X}] \to [\boldsymbol{Y}]$ of pro-H(Top) (see 2.3). Consequently, the triple $([\boldsymbol{p}], [\boldsymbol{q}], E[\boldsymbol{f}])$ determines a shape morphism, which is, by definition, $\bar{E}(F)$.

In order to see that $\bar{E}(F)$ is well defined, we must show that the triple $([\boldsymbol{p}'], [\boldsymbol{q}'], E([\boldsymbol{f}']))$, assigned to a triple $(\boldsymbol{p}', \boldsymbol{q}', [\boldsymbol{f}']) \sim (\boldsymbol{p}, \boldsymbol{q}, [\boldsymbol{f}])$, is equivalent to $([\boldsymbol{p}], [\boldsymbol{q}], E[\boldsymbol{f}])$, i.e.,

$$E[\boldsymbol{f}'][\boldsymbol{k}] = [\boldsymbol{l}]E[\boldsymbol{f}]. \tag{21}$$

Indeed, E applied to (13) yields

$$E[\boldsymbol{f}']E[\boldsymbol{i}] = E[\boldsymbol{j}]E[\boldsymbol{f}]. \tag{22}$$

It thus suffices to show that

$$[\boldsymbol{k}] = E[\boldsymbol{i}], \ \ [\boldsymbol{l}] = E[\boldsymbol{j}]. \tag{23}$$

If we apply E to (2), we obtain

$$E[\boldsymbol{i}]EC(\boldsymbol{p}) = EC(\boldsymbol{p}'). \tag{24}$$

By, Remark 2.16, $EC(\boldsymbol{p}) = [\boldsymbol{p}], EC(\boldsymbol{p}') = [\boldsymbol{p}']$. Consequently, (24) becomes

$$E[\boldsymbol{i}][\boldsymbol{p}] = [\boldsymbol{p}'], \tag{25}$$

and comparison with (20) proves that indeed $[\boldsymbol{k}] = E[\boldsymbol{i}]$. The second equation in (23) is proved analogously.

THEOREM 8.9. *The composition of the strong shape functor* $\bar{S}$: H(Top) $\to$ SSh(Top) *with the forgetful functor* $\bar{E}$: SSh(Top) $\to$ Sh(Top) *yields the shape functor* S: H(Top) $\to$ Sh(Top), *i.e.,* $S = \bar{E}\bar{S}$.

Proof. We first recall the definition of the *shape functor* S (see (Mardešić, Segal 1982), I.2.3). It is fixed on objects. If $[f]\colon X \to Y$ is a homotopy class of mappings, then $S([f])\colon X \to Y$ is the shape morphism, given by the triple $([\boldsymbol{p}], [\boldsymbol{q}], [\boldsymbol{g}])$, where $[\boldsymbol{p}]\colon X \to [\boldsymbol{X}]$ and $[\boldsymbol{q}]\colon Y \to [\boldsymbol{Y}]$ are HPol-expansions of X and Y respectively, and $[\boldsymbol{g}]\colon [\boldsymbol{X}] \to [\boldsymbol{Y}]$ is the only morphism of pro-H(Top) such that

$$[\boldsymbol{g}][\boldsymbol{p}] = [\boldsymbol{q}][f]. \tag{26}$$

Let $f: X \to Y$ be a mapping representing the homotopy class $[f]: X \to Y$. Choose cofinite strong HPol-expansions $\boldsymbol{p}: X \to \boldsymbol{X}, \boldsymbol{q}: Y \to \boldsymbol{Y}$ and a morphism $[\boldsymbol{f}]: \boldsymbol{X} \to \boldsymbol{Y}$ of CH(pro-Top), such that (12) holds. Then $\bar{S}[f]$ is given by the triple $(\boldsymbol{p}, \boldsymbol{q}, [\boldsymbol{f}])$. Therefore, $\bar{E}\bar{S}[f]$ is given by the triple $([\boldsymbol{p}], [\boldsymbol{q}], E[\boldsymbol{f}])$. Applying E to (12), we obtain

$$E[\boldsymbol{f}][\boldsymbol{p}] = [\boldsymbol{q}][f]. \tag{27}$$

Comparing (27) to (26), we conclude that $[\boldsymbol{g}] = E[\boldsymbol{f}]$ and therefore, $S[f]$ is given by the same triple $([\boldsymbol{p}], [\boldsymbol{q}], E[\boldsymbol{f}])$, i.e., $\bar{E}\bar{S}[f] = S[f]$. □

REMARK 8.10. In the proof of Theorem 6.23, with every space X, we have associated a cofinite ANR-resolution $\boldsymbol{p}_X$ of X. If $[f]: X \to Y$ is a homotopy class of mappings, then Corollary 8.7 yields a unique morphism $[\boldsymbol{f}]: \boldsymbol{X} \to \boldsymbol{Y}$ of CH(pro-Top). Clearly, the assignment $[f] \mapsto [\boldsymbol{f}]$ is a functor H(Top) $\to$ CH(pro-Top), which can also be used to define the strong shape morphisms.

The following theorem states an important property of the strong shape functor $\bar{S}$.

THEOREM 8.11. *The strong shape functor* $\bar{S}$: H(Top) $\to$ SSh(Top) *admits a right adjoint functor* T: SSh(Top) $\to$ H(Top), *i.e., there exists a natural equivalence* η_{XY}: SSh(Top)$(X,Y) \to$ H(Top)$(X, T(Y))$.

Proof. To define the functor T, one associates with every space X a cofinite strong HPol-expansion $\boldsymbol{p}_X: X \to \boldsymbol{X}$. Then one defines $T(X)$ to be the cotelescope $T(\boldsymbol{X})$ of $\boldsymbol{X}$ (see 4.2). If $F: X \to Y$ is a strong shape morphism, there exists a unique morphism $[\boldsymbol{f}]: \boldsymbol{X} \to \boldsymbol{Y}$ of CH(pro-Top) such that F is given by the triple $(\boldsymbol{p}_X, \boldsymbol{p}_Y, [\boldsymbol{f}])$. In 4.2 we defined a functor T: CH(pro-Top) $\to$ H(Top). Therefore, $T[\boldsymbol{f}]: T(\boldsymbol{X}) \to T(\boldsymbol{Y})$ is a well-defined morphism of H(Top). We define $T(F): T(X) \to T(Y)$ by putting $T(F) = T[\boldsymbol{f}]$. It is readily seen that T is indeed a functor T: SSh(Top) $\to$ H(Top).

To define the natural equivalence η_{XY}, note that, for every strong shape morphism $F: X \to Y$, $[\boldsymbol{f}]C(\boldsymbol{p}_X): X \to \boldsymbol{Y}$ is a well-defined morphism of CH(pro-Top). Therefore, $R(F) = R([\boldsymbol{f}]C(\boldsymbol{p}_X)): X \to T(\boldsymbol{Y})$ is a well-defined homotopy class $R(F): X \to T(Y)$ (see Lemma 4.17). We now put $\eta_{XY}(F) = R(F)$.

To prove that η_{XY} is a bijection, consider an arbitrary homotopy class $[g]: X \to T(Y)$. By Corollary 4.19, there is a unique morphism $[\boldsymbol{h}]: X \to T(\boldsymbol{Y})$ of CH(pro-Top) such that $R[\boldsymbol{h}] = [g]$. Moreover, since $\boldsymbol{p}_X$ is a coherent expansion of X, there is a unique morphism $[\boldsymbol{f}]: \boldsymbol{X} \to \boldsymbol{Y}$ such that $[\boldsymbol{f}]C(\boldsymbol{p}_X) = [\boldsymbol{h}]$. Hence, the strong shape morphism $F: X \to Y$, given by the triple $(\boldsymbol{p}_X, \boldsymbol{p}_Y, [\boldsymbol{f}])$, is the only strong shape morphism such that $R(F) = R([\boldsymbol{f}]C(\boldsymbol{p}_X)) = R[\boldsymbol{h}] = [g]$. This shows that η_{XY} is a bijection.

To prove naturality of η_{XY} with respect to the first variable, consider a homotopy class $[f']: X' \to X$. The induced function SSh(Top)$(X,Y) \to$

$\mathrm{SSh}(\mathsf{Top})(X',Y)$ maps the strong shape morphism $F: X \to Y$, given by $(\boldsymbol{p}_X, \boldsymbol{p}_Y, [\boldsymbol{f}])$, to the strong shape morphism $F' = F\bar{S}[f']: X' \to Y$, given by the triple $(\boldsymbol{p}_{X'}, \boldsymbol{p}_Y, [\boldsymbol{f}][\boldsymbol{f}'])$, where $[\boldsymbol{f}']: \boldsymbol{X}' \to \boldsymbol{X}$ is the only morphism of CH(pro-Top), for which $[\boldsymbol{f}']C(\boldsymbol{p}_{X'}) = C(\boldsymbol{p}_X)C(f')$. Clearly, $\eta_{X'Y}$ maps F' to

$$R(F') = R([\boldsymbol{f}][\boldsymbol{f}']C(\boldsymbol{p}_{X'})) = R([\boldsymbol{f}]C(\boldsymbol{p}_X)C(f')). \tag{28}$$

Using naturality of $\eta_{\boldsymbol{XY}}$ in Corollary 4.19, we see that

$$R(F') = R([\boldsymbol{f}]C(\boldsymbol{p}_X))[f']. \tag{29}$$

On the other hand, the function $\mathrm{H}(\mathsf{Top})(X, T(Y)) \to \mathrm{H}(\mathsf{Top})(X', T(Y))$, induced by $[f']$, maps $R(F) = R([\boldsymbol{f}]C(\boldsymbol{p}_X))$ to

$$R(F)[f'] = R([\boldsymbol{f}]C(\boldsymbol{p}_X))[f']. \tag{30}$$

Therefore, $R(F\bar{S}[f']) = R(F)[f']$, which expresses naturality in the first variable.

To prove naturality in the second variable, consider a strong shape morphism $G: Y \to Y'$, given by a triple $(\boldsymbol{p}_Y, \boldsymbol{p}_{Y'}, [\boldsymbol{g}])$, where $[\boldsymbol{g}]: \boldsymbol{Y} \to \boldsymbol{Y}'$ is a morphism of CH(pro-Top). The induced function $\mathrm{SSh}(\mathsf{Top})(X,Y) \to \mathrm{SSh}(\mathsf{Top})(X,Y')$ maps the strong shape morphism $F: \boldsymbol{X} \to \boldsymbol{Y}$ to the strong shape morphism $F' = GF: \boldsymbol{X} \to \boldsymbol{Y}'$, given by the triple $(\boldsymbol{p}_X, \boldsymbol{p}_{Y'}, [\boldsymbol{g}][\boldsymbol{f}])$. Therefore,

$$R(GF) = R([\boldsymbol{g}][\boldsymbol{f}]C(\boldsymbol{p}_X)). \tag{31}$$

On the other hand, the function $\mathrm{H}(\mathsf{Top})(X, T(X)) \to \mathrm{H}(\mathsf{Top})(X, T(Y'))$, induced by G, maps $R(F)$ to $T(G)R(F) = T[\boldsymbol{g}]R([\boldsymbol{f}]C(\boldsymbol{p}_X))$. Using naturality of $\eta_{\boldsymbol{XY}}$ in Corollary 4.19, we see that

$$R([\boldsymbol{g}][\boldsymbol{f}]C(\boldsymbol{p}_X)) = T[\boldsymbol{g}]R([\boldsymbol{f}]C(\boldsymbol{p}_X)) \tag{32}$$

and thus, $R(GF) = T(G)R(F)$, which is the desired naturality condition. □

Using strong expansions for pairs of spaces, one also defines the *strong shape category of pairs* $\mathrm{SSh}(\mathsf{Top}_2)$ and the corresponding functors $\bar{S}$: $\mathrm{H}(\mathsf{Top}_2) \to \mathrm{SSh}(\mathsf{Top}_2)$ and $\bar{E}$: $\mathrm{SSh}(\mathsf{Top}_2) \to \mathrm{Sh}(\mathsf{Top}_2)$.

In full analogy with the definition of the strong shape category $\mathrm{SSh}(\mathsf{Top})$, for every $r \geq 0$, one defines the *strong shape category* $\mathrm{SSh}^{(r)}(\mathsf{Top})$ *of height* r. The objects of this category are topological spaces. Morphisms $F: X \to Y$ are determined by triples $(\boldsymbol{p}, \boldsymbol{q}, [\boldsymbol{f}])$, where $\boldsymbol{p}: X \to \boldsymbol{X}, \boldsymbol{q}: Y \to \boldsymbol{Y}$ are cofinite strong HPol-expansions of X and Y respectively, while $[\boldsymbol{f}]: \boldsymbol{X} \to \boldsymbol{Y}$ are morphisms of $\mathrm{CH}^{(r)}$(pro-Top) (instead of CH(pro-Top)).

One also defines the *forgetful functors* $E^{(r)}$: $\mathrm{SSh}(\mathsf{Top}) \to \mathrm{SSh}^{(r)}(\mathsf{Top})$ and $E^{(rr')}$: $\mathrm{SSh}^{(r')}(\mathsf{Top}) \to \mathrm{SSh}^{(r)}(\mathsf{Top})$, $r < r'$, using the functors $E^{(r)}$: CH(pro-Top) $\to \mathrm{CH}^{(r)}$(pro-Top) and $E^{(rr')}$: $\mathrm{CH}^{(r')}$(pro-Top) $\to \mathrm{CH}^{(r)}$(pro-Top), $r < r'$ (see 3.1). In particular, if $F: \boldsymbol{X} \to \boldsymbol{Y}$ is a morphism of $\mathrm{SSh}(\mathsf{Top})$, given by a triple $(\boldsymbol{p}, \boldsymbol{q}, [\boldsymbol{f}])$, where $[\boldsymbol{f}]: \boldsymbol{X} \to \boldsymbol{Y}$ is a morphism of CH(pro-Top), then $E^{(r)}(F): X \to Y$ is the morphism of $\mathrm{SSh}^{(r)}(\mathsf{Top})$, given by the triple $(\boldsymbol{p}, \boldsymbol{q}, E^{(r)}[\boldsymbol{f}])$.

Similar considerations yield the *strong shape functor of height* $r \geq 0$, denoted by $\bar{S}^{(r)}$: $\mathrm{H}(\mathsf{Top}) \to \mathrm{SSh}^{(r)}(\mathsf{Top})$. Note that

$$E^{(r)} \bar{S}^{(r)} = \bar{S}. \tag{33}$$

8.3 Strong shape equivalences

A mapping $f: X \to Y$ is called a *shape equivalence* provided the induced shape morphism $S[f]: X \to Y$ is an isomorphism of $\mathrm{Sh}(\mathsf{Top})$. Analogously, a mapping $f: X \to Y$ is called a *strong shape equivalence* provided the induced strong shape morphism $\bar{S}[f]: X \to Y$ is an isomorphism of $\mathrm{SSh}(\mathsf{Top})$. Shape equivalences are readily characterized by the following theorem.

THEOREM 8.12. *A mapping $f: X \to Y$ is a shape equivalence if and only if, for every* ANR *P, the following two conditions are satisfied:*

(SE1) *For every mapping $\phi: X \to P$, there exists a mapping $\psi: Y \to P$ such that $\phi \simeq \psi f$.*

(SE2) *If $\psi_0, \psi_1: Y \to P$ are mappings such that $\psi_0 f \simeq \psi_1 f$, then $\psi_0 \simeq \psi_1$.*

Proof. First assume that f is a shape equivalence and $\phi: X \to P$ is a mapping. Since P is an ANR, the shape morphism $S[\phi](S[f])^{-1}: Y \to P$ admits a mapping $\psi: Y \to P$ such that $S[\psi] = S[\phi](S[f])^{-1}$. Therefore, $S[\psi f] = S[\psi]S[f] = S[\phi]$, which implies $\psi f \simeq \phi$ and thus, (SE1) holds. Similarly, $[\psi_0 f] = [\psi_1 f]$ implies $S[\psi_0]S[f] = S[\psi_1]S[f]$. Therefore, $S[\psi_0] = S[\psi_1]$ and thus, $\psi_0 \simeq \psi_1$, which is (SE2).

To prove the converse, consider an HPol-homotopy expansion $\boldsymbol{q} = (q_\mu): Y \to \boldsymbol{Y} = (Y_\mu, q_{\mu\mu'}, M)$. Let us first show that the mappings $p_\mu = q_\mu f: X \to Y_\mu$, $\mu \in \mathrm{M}$, form an HPol-homotopy expansion $\boldsymbol{p}: X \to \boldsymbol{Y}$. Indeed, condition (M1) holds, because for any mapping $\phi: X \to P$, (SE1) yields a mapping $\psi: Y \to P$ such that $\phi \simeq \psi f$. Now (M1) for $\boldsymbol{q}$, yields an index $\mu \in \mathrm{M}$ and a mapping $h: Y_\mu \to P$ such that $hq_\mu \simeq \psi$ and thus, $hp_\mu \simeq \phi$. Now assume that $h_0, h_1: Y_\mu \to P$ are mappings such that $h_0 p_\mu \simeq h_1 p_\mu$. Then the mappings $\psi_0 = h_0 q_\mu, \psi_1 = h_1 q_\mu: Y \to P$ satisfy the condition $\psi_0 f \simeq \psi_1 f$. Therefore, property (SE2) implies $\psi_0 \simeq \psi_1$, i.e., $h_0 q_\mu \simeq h_1 q_\mu$. By property (M2) for $\boldsymbol{q}$, one obtains an index $\mu' \geq \mu$ such that $h_0 q_{\mu\mu'} \simeq h_1 q_{\mu\mu'}$, which is assertion (M2) for $\boldsymbol{p}$.

To show that $S[f]$ is a shape equivalence, it now suffices to consider the HPol-homotopy expansions $\boldsymbol{p}: X \to \boldsymbol{Y}$ and $\boldsymbol{q}: Y \to \boldsymbol{Y}$ and note that in pro-$\mathrm{H}(\mathsf{Top})$ one has $[\mathbf{1_Y}][\boldsymbol{p}] = [\boldsymbol{q}][\mathbf{f}]$. Therefore, $S[f]$ is determined by the triple $(\boldsymbol{p}, \boldsymbol{q}, [\mathbf{1_Y}])$. Clearly, $(\boldsymbol{q}, \boldsymbol{p}, [\mathbf{1_Y}])$ determines its inverse in $\mathrm{Sh}(\mathsf{Top})$, hence, $S[f]$ is an isomorphism of $\mathrm{Sh}(\mathsf{Top})$. □

Conditions, stated in the next theorem, characterize strong shape equivalences.

THEOREM 8.13. *A mapping $f: X \to Y$ is a strong shape equivalence if and only if, for every* ANR P, *in addition to* (SE1), *the following strengthening of condition* (SE2) *is fulfilled:*

(SSE2) *If $\psi_0, \psi_1: Y \to P$ are mappings and $F: X \times I \to P$ is a homotopy, which connects $\psi_0 f$ to $\psi_1 f$, then there exists a homotopy $G: Y \times I \to P$, which connects ψ_0 to ψ_1, and is such that*

$$G(f \times 1) \simeq F\,(\mathrm{rel}(X \times \partial I)). \tag{1}$$

Proof of the sufficiency. This part of the proof of Theorem 8.13 is similar to the proof of the corresponding part of Theorem 8.12. Assume that f has properties (SE1) and (SSE2), for every ANR P. Choose a strong ANR-expansion $\boldsymbol{q} = (q_\mu): Y \to \boldsymbol{Y} = (Y_\mu, q_{\mu\mu'}, M)$ and consider the mappings $p_\mu = q_\mu f: X \to Y_\mu$. Note that $\boldsymbol{p} = (p_\mu): X \to \boldsymbol{Y}$ is a mapping. Let us first show that $\boldsymbol{p}$ is a strong expansion of X.

Since (S1) coincides with (M1), this property is verified as in the case of Theorem 8.12. To establish property (S2) for $\boldsymbol{p}$, consider two mappings $\psi_0, \psi_1: Y_\mu \to P$ and a homotopy $F: X \times I \to P$, which connects $\psi_0 p_\mu = \psi_0 q_\mu f$ to $\psi_1 p_\mu = \psi_1 q_\mu f$. By property (SSE2), there exists a homotopy $G: Y \times I \to P$, which connects $\psi_0 q_\mu$ to $\psi_1 q_\mu$ and satisfies (1). Since $\boldsymbol{q}$ is a strong expansion of Y, there exist a $\mu' \geq \mu$ and a homotopy $H: Y_{\mu'} \times I \to P$, which connects $\psi_0 q_{\mu\mu'}$ to $\psi_1 q_{\mu\mu'}$ and is such that

$$H(q_{\mu'} \times 1) \simeq G\,(\mathrm{rel}(Y \times \partial I)). \tag{2}$$

Composing (2) with $(f \times 1)$, one obtains

$$H(p_{\mu'} \times 1) \simeq G(f \times 1)\,(\mathrm{rel}(X \times \partial I)). \tag{3}$$

Now (3) and (1) show that $H(p_{\mu'} \times 1) \simeq F\,(\mathrm{rel}(X \times \partial I))$, i.e., $\boldsymbol{p}$ too fulfills condition (S2).

It is now clear that the triple $(\boldsymbol{p}, \boldsymbol{q}, C[\mathbf{1_Y}])$ determines the strong shape morphism $\bar{S}[f]$ and the triple $(\boldsymbol{q}, \boldsymbol{p}, C[\mathbf{1_Y}])$ determines its inverse $(\bar{S}[f])^{-1}$. □

The proof of the *necessity* part of Theorem 8.13 is more involved and requires some preparation.

Recall that the *mapping cylinder* $M(f)$ of a mapping $f: X \to Y$ is obtained from the disjoint union $(X \times I) \sqcup Y$ by identifying points $(x, 1)$ and y, when $f(x) = y$. Let $\pi: (X \times I) \sqcup Y \to M(f)$ denote the quotient mapping. Then the mappings $i: X \to M(f)$ and $j: Y \to M(f)$, defined by $i(x) = \pi(x, 0)$ and $j(y) = \pi(y)$, respectively, are closed embeddings. Therefore, we often identify X and Y with $i(X)$ and $j(Y)$, respectively. The mapping $\hat{f}: M(f) \to Y$, given by $\hat{f}(\pi(x,t)) = f(x)$ and $\hat{f}(\pi(y)) = y$, has the property that $f = \hat{f} i$ and $\hat{f} j = \mathrm{id}$. Moreover, $j\hat{f} \simeq \mathrm{id}$, and thus, $\hat{f}$ is a homotopy equivalence. Finally, i is a cofibration (see e.g., (Dugundji 1966),

XVIII, 5.1). If $f_1: X_1 \to Y_1$ is another mapping and $p: X_1 \to X$, $q: Y_1 \to Y$ are mappings such that $fp = qf_1$, then one obtains an induced mapping $r: M(f_1) \to M(f)$, given by $r(\pi_1(x,t)) = \pi(p(x),t)$ and $r(\pi_1(y)) = \pi(q(y))$.

We will also need the *double mapping cylinder* $DM(f)$ of f. It is a special case of the *duplication* $D(Z,X)$ of a pair of spaces (Z,X), where X is closed in Z. To define $D(Z,X)$, one considers a copy (Z',X') of (Z,X) and the disjoint union $Z \sqcup Z'$. By definition, $D(Z,X)$ is obtained from $Z \sqcup Z'$ by identifying X with X'. Clearly, $D(X,X) = X$. There is a natural embedding $k: X \to D(Z,X)$. Moreover, there is a homeomorphism $s: D(Z,X) \to D(Z,X)$, which interchanges Z and Z', keeping X fixed. A mapping of pairs $r: (Z_2,X_2) \to (Z_1,X_1)$ induces a mapping $Dr: D(Z_2,X_2) \to D(Z_1,X_1)$, defined by $Dr|Z_2 = r$ and $Dr|Z_2' = r' = rs|Z_2'$. The double mapping cylinder $DM(f)$ is the duplication of the pair $(M(f),X)$, hence, it consists of $M(f)$ and its copy $M(f)'$. The mapping $k: X \to DM(f)$ is a closed embedding and $(DM(f),X)$ is a cofibered pair. There is a homeomorphism $s: DM(f) \to DM(f)$, which interchanges the copies $M(f)$ and $M(f)'$ and keeps X fixed. Mappings f, f_1, p, q as above induce a mapping of pairs $r: (M(f_1),X_1) \to (M(f),X)$, hence, they also induce the mapping $Dr: DM(f_1) \to DM(f)$, which on $M(f)$ and $M(f)'$ coincides with r and r' respectively. A mapping $h: M(f_1) \to X$ determines a mapping of pairs $h: (M(f_1),X_1) \to (X,X)$, hence, it induces a mapping $Dh: D(M(f_1),X_1) \to D(X,X)$, i.e., $Dh: DM(f_1) \to X$.

The double mapping cylinder $DM(f)$ can also be described as the subspace $DM^*(f)$ of $M(f) \times I$, given by the following formula.

$$DM^*(f) = (X \times I) \cup (M(f) \times \partial I). \tag{4}$$

There is an obvious homeomorphism $h: DM(f) \to DM^*(f)$, which takes X to $X \times \{0\}$. Note that $(M(f) \times I, DM^*(f))$ is a closed cofibered pair. To establish this fact, it suffices to consider the closed cofibered pair $(M(f),X)$ and apply the following lemma ((Strøm, 1968), Theorem 6).

LEMMA 8.14. *If (X,A) is a closed cofibered pair, then $(X \times I, A \times I \cup X \times \partial I)$ is also a cofibered pair.*

That every strong shape equivalence $f: X \to Y$ has properties (SE1) and (SSE2) is an immediate consequence of the following two theorems, which are of their own interest.

THEOREM 8.15. *If $f: X \to Y$ is a strong shape equivalence, then the natural inclusions $i: X \to M(f)$ and $k: X \to DM(f)$ are (ordinary) shape equivalences.*

THEOREM 8.16. *If $f: X \to Y$ is a mapping such that the natural inclusions $i: X \to M(f)$ and $k: X \to DM(f)$ are shape equivalences, then f has properties* (SE1) *and* (SSE2).

Proof of Theorem 8.16. Since $i: X \to M(f)$ is a shape equivalence and $\hat{f}$ is a homotopy equivalence, the equality $f = \hat{f}i$ implies that f is a shape equivalence. Hence, by Theorem 8.12, f has property (SE1). In order to show that f also has property (SSE2), first note that every mapping $g: X \to P$ extends to a mapping $\tilde{g}: M(f) \to P$ and every mapping $H: DM^*(f) \to P$ extends to a mapping $\tilde{H}: M(f) \times I \to P$. Indeed, by Theorem 8.12 applied to i, there exists a mapping $g': M(f) \to P$ such that $g'i \simeq g$. Since $(M(f), X)$ is a cofibered pair, the homotopy extension property yields a mapping $\tilde{g}$ such that $\tilde{g}|X = g$. Concerning the second assertion, note that the assumption that $k: X \to DM(f)$ is a shape equivalence implies that also the inclusion $l: DM^*(f) \to M(f) \times I$ is a shape equivalence. Indeed, for the above mentioned homeomorphism $h: DM(f) \to DM^*(f)$ and the mapping $u: X \to X \times I$, given by $u(x) = (x, 0)$, one has $(i \times 1)u = lhk$. Since the mappings $(i \times 1), u, h, k$ are shape equivalences, it follows that also l is a shape equivalence. However, $(M(f) \times I, DM^*(f))$ is a closed cofibered pair. Therefore, the argument used in proving the first assertion applies and shows that H admits an extension $\tilde{H}$.

Now assume that we have mappings $\psi_0, \psi_1: Y \to P$ and a homotopy $F: X \times I \to P$ which connects $\psi_0 f$ to $\psi_1 f$. We extend F to a mapping $H: DM^*(f) \to P$, by putting

$$H(z, 0) = \psi_0 \hat{f}(z), \;\; H(z, 1) = \psi_1 \hat{f}(z), \tag{5}$$

for $z \in M(f)$. Let $\tilde{H}: M(f) \times I \to P$ be an extension of H. Define a homotopy $G: Y \times I \to P$, by putting

$$G = \tilde{H}|Y \times I. \tag{6}$$

Note that $G(y, 0) = H(y, 0) = \psi_0(y)$ and $G(y, 1) = H(y, 1) = \psi_1(y)$, for $y \in Y$. Hence, G is a homotopy which connects ψ_0 with ψ_1. It remains to prove (1).

Consider the homotopy $K: X \times I \times I \to P$, given by

$$K = \tilde{H}(p \times 1), \tag{7}$$

where $p: X \times I \to M(f)$ now denotes the restriction of $\pi: (X \times I) \sqcup Y \to M(f)$ to $X \times I$. One readily verifies the following relations, which establish (1).

$$K(x, 0, s) = \tilde{H}(p(x, 0), s) = F(x, s), \tag{8}$$

$$K(x, 1, s) = \tilde{H}(p(x, 1), s) = G(f(x), s), \tag{9}$$

$$K(x, t, 0) = \tilde{H}(p(x, t), 0) = H(p(x, t), 0) = \psi_0 \hat{f} p(x, t) = \psi_0 f(x), \tag{10}$$

$$K(x, t, 1) = \tilde{H}(p(x, t), 1) = H(p(x, t), 1) = \psi_1 \hat{f} p(x, t) = \psi_1 f(x). \;\square \tag{11}$$

Proof of Theorem 8.15 (*Beginning of the proof*). The first assertion of Theorem 8.15 is easy to prove. By assumption, $f: X \to Y$ is a strong shape equivalence. Moreover, $\hat{f}: M(f) \to Y$ is a homotopy equivalence and $f = \hat{f}i$.

It follows that i too is a strong shape equivalence, hence, it is also a shape equivalence.

To prove the second assertion of Theorem 8.15, it suffices to find an HPol - resolution of $k: X \to DM(f)$, i.e., to find HPol - resolutions $\boldsymbol{p}: X \to \boldsymbol{X}$, $\boldsymbol{v}: DM(f) \to \boldsymbol{V}$ and a mapping $\boldsymbol{k}: \boldsymbol{X} \to \boldsymbol{V}$ such that $\boldsymbol{vk} = \boldsymbol{kp}$ and $\boldsymbol{k}$ induces an isomorphism of pro - H(Top).

Construction of the resolution for k. First apply Theorem 6.22, Remark 6.32 and Lemma 6.33 to the mapping f. One obtains cofinite polyhedral resolutions $\boldsymbol{p} = (p_\lambda): X \to \boldsymbol{X} = (X_\lambda, p_{\lambda\lambda'}, \Lambda)$, $\boldsymbol{q} = (q_\lambda): Y \to \boldsymbol{Y} = (Y_\lambda, q_{\lambda\lambda'}, \Lambda)$ and a level mapping $\boldsymbol{f} = (f_\lambda): \boldsymbol{X} \to \boldsymbol{Y}$ such that

$$f_\lambda p_\lambda = q_\lambda f, \;\; \lambda \in \Lambda. \tag{12}$$

We then consider the inverse system $\boldsymbol{M}(\boldsymbol{f})$, formed by the mapping cylinders $M(f_\lambda)$ and by the bonding mappings $r_{\lambda\lambda'}: M(f_{\lambda'}) \to M(f_\lambda)$, induced by the mappings $f_\lambda, f_{\lambda'}, p_{\lambda\lambda'}$ and $q_{\lambda\lambda'}$, since

$$f_\lambda p_{\lambda\lambda'} = q_{\lambda\lambda'} f_{\lambda'}. \tag{13}$$

We also consider the mapping $\boldsymbol{r}: M(f) \to \boldsymbol{M}(\boldsymbol{f})$, formed by the mappings $r_\lambda: M(f) \to M(f_\lambda)$, induced by f_λ, f, p_λ, and q_λ, due to (12). Moreover, we consider the level mapping $\boldsymbol{i}: \boldsymbol{X} \to \boldsymbol{M}(\boldsymbol{f})$, formed by the inclusions $i_\lambda: X_\lambda \to M(f_\lambda)$.

We will also consider the system $\boldsymbol{DM}(\boldsymbol{f})$, formed by the double mapping cylinders $DM(f_\lambda)$ and the mappings $Dr_{\lambda\lambda'}: DM(f_{\lambda'}) \to DM(f_\lambda)$. Moreover, we will consider the mapping $\boldsymbol{Dr}: DM(f) \to \boldsymbol{DM}(\boldsymbol{f})$, formed by the mappings $Dr_\lambda: DM(f) \to DM(f_\lambda)$ and the mapping $\boldsymbol{k}: \boldsymbol{X} \to \boldsymbol{DM}(\boldsymbol{f})$, formed by the inclusions $k_\lambda: X_\lambda \to DM(f_\lambda)$.

LEMMA 8.17. *$\boldsymbol{r}: M(f) \to \boldsymbol{M}(\boldsymbol{f})$ and $\boldsymbol{Dr}: DM(f) \to \boldsymbol{DM}(\boldsymbol{f})$ are cofinite* HPol *- resolutions of $M(f)$ and of $DM(f)$, respectively. Moreover,*

$$\boldsymbol{ri} = \boldsymbol{ip}, \tag{14}$$

$$(\boldsymbol{Dr})k = \boldsymbol{kp} \tag{15}$$

and thus, $\mathbf{p}$, $\boldsymbol{r}$, $\boldsymbol{i}$ and $\mathbf{p}$, $\boldsymbol{Dr}$, $\boldsymbol{k}$ form cofinite HPol *- resolutions of i and of k, respectively.*

In the proof of Lemma 8.17 we need the following fact (see (Dydak, Nowak 1991), Proposition 2.0).

LEMMA 8.18. *If $f: X \to Y$ is a mapping between two paracompact spaces, then $M(f)$ and $DM(f)$ are paracompact spaces.*

Proof. Let $\mathcal{U}$ be an open covering of $M(f)$. Since Y is paracompact, there exists a locally finite open covering $\mathcal{V}$ of Y, which refines $\mathcal{U}|Y$. For each $V \in \mathcal{V}$, choose a member $U_V \in \mathcal{U}$ such that $V \subseteq U_V$. Now associate with V the open set $U_V \cap (\hat{f})^{-1}(V) \subseteq M(f)$ and consider the family $\mathcal{U}' = (U_V \cap (\hat{f})^{-1}(V), V \in \mathcal{V})$. This family is locally finite, because the family $((\hat{f})^{-1}(V), V \in \mathcal{V})$ is locally finite. Moreover, $\mathcal{U}'$ refines $\mathcal{U}$ and covers Y.

On the other hand, the direct product of a paracompact space with a compact space is paracompact (see (Engelking 1977), Theorem 5.1.36), and a space which admits a locally finite closed cover formed by paracompact space, is itself paracompact (see (Engelking 1977), Theorem 5.1.34). Therefore, $M(f)\backslash Y \approx X \times [0,1)$ is paracompact. Hence, $M(f)\backslash Y$ can be covered by a locally finite refinement $\mathcal{U}''$ of $\mathcal{U}|(M(f)\backslash Y)$. Clearly, $\mathcal{U}' \cup \mathcal{U}''$ is a locally finite covering of $M(f)$, which refines $\mathcal{U}$. This establishes the first assertion of the lemma. The second assertion is an immediate consequence of the assertion concerning $M(f)$, because $DM(f)$ consists of two closed subsets, each homeomorphic to $M(f)$. □

Proof of Lemma 8.17. The validity of (14) and (15) readily follows from the construction. To prove that $\boldsymbol{r}: M(f) \to \boldsymbol{M(f)}$ is a resolution, it suffices to verify properties (B1) and (B2)$'$ (see Remark 6.13). Let $\mathcal{U}$ be a normal covering of $M(f)$. Choose a star-refinement $\mathcal{U}^*$ of $\mathcal{U}$. Since $\boldsymbol{p} \times 1: X \times I \to \boldsymbol{X} \times I$ and $\boldsymbol{q}: Y \to \boldsymbol{Y}$ are resolutions, there exists a $\lambda \in \Lambda$ and there exist open coverings $\mathcal{U}_\lambda$ of $X_\lambda \times I$ and $\mathcal{V}_\lambda$ of Y_λ such that $(p_\lambda \times 1)^{-1}(\mathcal{U}_\lambda)$ refines $\pi^{-1}(\mathcal{U}^*)|(X \times I)$ and $q_\lambda^{-1}(\mathcal{V}_\lambda)$ refines $\pi^{-1}(\mathcal{U}^*)|Y$, where $\pi: (X \times I) \sqcup Y \to M(f)$ denotes the natural quotient mapping. We will define an open covering $\mathcal{W}_\lambda$ of $M(f_\lambda)$ such that $r_\lambda^{-1}(\mathcal{W}_\lambda)$ refines $\mathcal{U}$.

For every $V_\lambda \in \mathcal{V}_\lambda$, define an open set $U(V_\lambda)$ of $X_\lambda \times I$ by putting

$$U(V_\lambda) = \bigcup_x U_\lambda(x), \tag{16}$$

where the union is taken over all $x \in f_\lambda^{-1}(V_\lambda)$ and $U_\lambda(x)$ is an open neighborhood of $(x,1)$ in $X_\lambda \times I$, defined as follows. Let U_λ be a member of $\mathcal{U}_\lambda$, which contains the point $(x,1)$. If $(x,1) \in \mathrm{Cl}\,(p_\lambda(X) \times \{1\})$, put $U_\lambda(x) = U_\lambda$. If $(x,1) \notin \mathrm{Cl}\,(p_\lambda(X) \times \{1\})$, put $U_\lambda(x) = U_\lambda \backslash \mathrm{Cl}\,(p_\lambda(X) \times \{1\})$. Note that $U(V_\lambda)$ contains $f_\lambda^{-1}(V_\lambda) \times \{1\}$.

Now consider the sets

$$W(V_\lambda) = j_\lambda(V_\lambda) \cup \pi_\lambda(U(V_\lambda) \cap (f_\lambda^{-1}(V_\lambda) \times I)) \subseteq M(f_\lambda), \tag{17}$$

where $j_\lambda: Y_\lambda \to M(f_\lambda)$ denotes the natural embedding and $\pi_\lambda: (X_\lambda \times I) \sqcup Y_\lambda \to M(f_\lambda)$ denotes the natural quotient mapping. By definition, $\mathcal{W}_\lambda$ consists of all sets $W(V_\lambda)$, where V_λ ranges over $\mathcal{V}_\lambda$ and of all sets $\pi_\lambda(U_\lambda \backslash (X_\lambda \times \{1\}))$, where U_λ ranges over $\mathcal{U}_\lambda$.

Note that $\mathcal{W}_\lambda$ covers $M(f_\lambda)$, because the sets $\pi_\lambda(U_\lambda \backslash (X_\lambda \times \{1\}))$ cover $M(f_\lambda) \backslash j_\lambda(Y_\lambda)$ and the sets $W(V_\lambda) \supseteq j_\lambda(V_\lambda)$ cover $j_\lambda(Y_\lambda)$. Let us prove that the members of $\mathcal{W}_\lambda$ are open sets of $M(f_\lambda)$. This is obvious for sets of the form $\pi_\lambda(U_\lambda \backslash (X_\lambda \times \{1\}))$. To prove that also the sets $W(V_\lambda)$ are open, it suffices to show that

$$\pi_\lambda^{-1}(W(V_\lambda)) = V_\lambda \cup (U(V_\lambda) \cap (f_\lambda^{-1}(V_\lambda) \times I)), \tag{18}$$

because the set on the right side of (18) is open in $Y_\lambda \sqcup (X_\lambda \times I)$.

That the right side of (18) is contained in the left side, is an immediate consequence of (17). To prove the opposite inclusion, consider a point $(x,t) \in X_\lambda \times I$, which belongs to the left side of (18). We will show that

$$(x,t) \in U(V_\lambda) \cap (f_\lambda^{-1}(V_\lambda) \times I). \tag{19}$$

First assume that $t < 1$. Then $\pi_\lambda(x,t)$ does not belong to $j_\lambda(Y_\lambda)$. Therefore, it must belong to $\pi_\lambda(U(V_\lambda) \cap (f_\lambda^{-1}(V_\lambda) \times I))$. Hence, there exists a point $(x',t') \in U(V_\lambda) \cap (f_\lambda^{-1}(V_\lambda) \times I)$ such that $\pi_\lambda(x,t) = \pi_\lambda(x',t')$. One cannot have $t' = 1$, because this would imply $\pi_\lambda(x',t') \in j_\lambda(Y_\lambda)$. Since $\pi_\lambda | X_\lambda \times [0,1)$ is an injection, one concludes that $(x,t) = (x',t')$ and thus, (19) holds. Now assume that $t = 1$. If $\pi_\lambda(x,1) = j_\lambda f_\lambda(x) \in j_\lambda(V_\lambda)$, then $(x,1) \in f_\lambda^{-1}(V_\lambda) \times \{1\} \subseteq U(V_\lambda)$, and (19) holds. The other possibility is that $\pi_\lambda(x,1) \in \pi_\lambda(U(V_\lambda) \cap (f_\lambda^{-1}(V_\lambda) \times I))$, i.e., $\pi_\lambda(x,1) = \pi_\lambda(x',t')$, for some point $(x',t') \in U(V_\lambda) \cap (f_\lambda^{-1}(V_\lambda) \times I)$. Since $\pi_\lambda(x,1) = j_\lambda f_\lambda(x) \in j_\lambda(Y_\lambda)$, one must have $t' = 1$ and thus, $\pi_\lambda(x',t') = f_\lambda(x')$. However, $(x',1) \in f_\lambda^{-1}(V_\lambda) \times I$ implies $f_\lambda(x') \in V_\lambda$, hence, also $f_\lambda(x) = f_\lambda(x') \in V_\lambda$. Consequently, $(x,1) \in f_\lambda^{-1}(V_\lambda) \times \{1\} \subseteq U(V_\lambda)$ and (19) holds again.

Now consider a point $y \in Y_\lambda$, which belongs to the left side of (18). We will show that

$$y \in V_\lambda, \tag{20}$$

hence, y belongs to the right side of (18). Indeed, if $\pi_\lambda(y) \in j_\lambda(V_\lambda)$, then $\pi_\lambda(y) = j_\lambda(y)$ implies (20). The other possibility is that $\pi_\lambda(y) = \pi_\lambda(x',t')$, where $(x',t') \in U(V_\lambda) \cap (f_\lambda^{-1}(V_\lambda) \times I) \subseteq f_\lambda^{-1}(V_\lambda) \times I$ and thus, $x' \in f_\lambda^{-1}(V_\lambda)$. Since $\pi_\lambda(y) \in j_\lambda(Y_\lambda)$, we must have $t' = 1$ and thus, $\pi_\lambda(x',t') = j_\lambda f_\lambda(x') \in j_\lambda(V_\lambda)$. Consequently, $j_\lambda(y) = \pi_\lambda(x',1) \in j_\lambda(V_\lambda)$, which shows that also in this case (20) holds.

Let us now prove that $r_\lambda^{-1}(\mathcal{W}_\lambda)$ refines the covering $\mathcal{U}$. The assertion is obvious for the members of $\mathcal{W}_\lambda$ of the form $\pi_\lambda(U_\lambda \backslash (X_\lambda \times \{1\}))$, because

$$\pi^{-1} r_\lambda^{-1}(\pi_\lambda(U_\lambda \backslash (X_\lambda \times \{1\})) = (p_\lambda \times 1)^{-1}(U_\lambda \backslash (X_\lambda \times \{1\})). \tag{21}$$

For a member of the form $W(V_\lambda)$, there exists a member $V_\lambda^* \in \mathcal{U}^*$ such that $q_\lambda^{-1}(V_\lambda) \subseteq \pi^{-1}(V_\lambda^*) \cap Y$. It now suffices to show that

$$\pi^{-1} r_\lambda^{-1}(W(V_\lambda)) \subseteq \mathrm{St}(\pi^{-1}(V_\lambda^*), \pi^{-1}(\mathcal{U}^*)), \tag{22}$$

Indeed, since $r_\lambda \pi = \pi_\lambda((p_\lambda \times 1) \sqcup q_\lambda)$, (18) yields

$$\pi^{-1} r_\lambda^{-1}(W(V_\lambda)) \subseteq (p_\lambda \times 1)^{-1}(U(V_\lambda) \cap (f_\lambda^{-1}(V_\lambda) \times I)) \cup q_\lambda^{-1}(V_\lambda). \tag{23}$$

To show that the first summand of the right side of (23) is also contained in $\mathrm{St}(\pi^{-1}(V_\lambda^*), \pi^{-1}(\mathcal{U}^*))$, it suffices to show that

$$(p_\lambda \times 1)^{-1}(U_\lambda(x)) \subseteq \mathrm{St}(\pi^{-1}(V_\lambda^*), \pi^{-1}(\mathcal{U}^*)), \tag{24}$$

for every $x \in f_\lambda^{-1}(V_\lambda)$.

First consider the case when $(x, 1) \in \mathrm{Cl}\,(p_\lambda(X) \times \{1\})$. Then $U_\lambda(x)$ is a member U_λ of $\mathcal{U}_\lambda$, which contains $(x, 1)$. Since $U_\lambda \cap (f_\lambda^{-1}(V_\lambda) \times \{1\})$ is an open neighborhood of $(x, 1)$, it intersects $p_\lambda(X) \times \{1\}$ and therefore, $(p_\lambda \times 1)^{-1}(U_\lambda \cap (f_\lambda^{-1}(V_\lambda) \times \{1\})$ is not empty. However, by construction, $(p_\lambda \times 1)^{-1}(U_\lambda)$ is contained in a member of $\pi^{-1}(\mathcal{U}^*)$ and $(p_\lambda \times 1)^{-1}(f_\lambda^{-1}(V_\lambda) \times \{1\}) = f^{-1}q_\lambda^{-1}(V_\lambda) \times \{1\} \subseteq \pi^{-1}(V_\lambda^*)$. Indeed, $q_\lambda^{-1}(V_\lambda) \subseteq \pi^{-1}(V_\lambda^*) \cap Y = j^{-1}(V_\lambda^*)$. Therefore, if $(x, 1) \in f^{-1}q_\lambda^{-1}(V_\lambda) \times \{1\}$, then $\pi(x, 1) = jf(x) \in jq_\lambda^{-1}(V_\lambda) \subseteq \pi^{-1}(V_\lambda^*)$, because $q_\lambda^{-1}(V_\lambda) \subseteq \pi^{-1}(V_\lambda^*) \cap Y$. We now conclude that $(p_\lambda \times 1)^{-1}(U_\lambda(x))$ is indeed contained in $\mathrm{St}(\pi^{-1}(V_\lambda^*), \pi^{-1}(\mathcal{U}^*))$. Now consider the case when $(x, 1) \notin \mathrm{Cl}\,(p_\lambda(X) \times \{1\})$. In this case, $U_\lambda(x) \cap (p_\lambda(X) \times \{1\}) = \emptyset$. Therefore, $(p_\lambda \times 1)^{-1}(U_\lambda(x)) = \emptyset$ and (24) holds again.

To conclude the verification of property (B1), note that the spaces X_λ, Y_λ are polyhedra, hence, paracompact spaces. Therefore, by Lemma 8.18, the mapping cylinder $M(f)$ is also paracompact. However, in a paracompact space every open covering is normal (see e.g., (Mardešić, Segal 1982), Appendix 1, 3.1, Corollary 1). Consequently, $\mathcal{W}_\lambda$ is a normal covering of $M(f_\lambda)$.

Property (B2)′ will be verified if we show that every open neighborhood W_λ of the set $\mathrm{Cl}\,(r_\lambda(M(f)))$ in $M(f_\lambda)$ admits a $\lambda' \geq \lambda$ such that

$$r_{\lambda\lambda'}(M(f_{\lambda'})) \subseteq W_\lambda. \tag{25}$$

Put

$$U_\lambda = \pi_\lambda^{-1}(W_\lambda) \cap (X_\lambda \times I), \tag{26}$$

$$V_\lambda = \pi_\lambda^{-1}(W_\lambda) \cap Y_\lambda \tag{27}$$

Note that

$$\pi_\lambda(p_\lambda \times 1)(X \times I) = r_\lambda \pi(X \times I) \subseteq r_\lambda(M(f)) \tag{28}$$

and therefore,

$$\mathrm{Cl}\,(p_\lambda \times 1)(X \times I) \subseteq U_\lambda. \tag{29}$$

Similarly, we conclude that

$$q_\lambda(Y) \subseteq V_\lambda. \tag{30}$$

Since $\boldsymbol{p} \times 1$ and $\boldsymbol{q}$ are resolutions and the spaces $X_\lambda \times I$ and Y_λ are normal, property (B2)′ holds for both $\boldsymbol{p} \times 1$ and $\boldsymbol{q}$ (see Remark 6.13). Therefore, we conclude that there exists a $\lambda' \geq \lambda$ such that

$$(p_{\lambda\lambda'} \times 1)(X_{\lambda'} \times I) \subseteq U_\lambda \subseteq \pi_\lambda^{-1}(W_\lambda), \tag{31}$$

$$q_{\lambda\lambda'}(Y_{\lambda'}) \subseteq V_\lambda \subseteq \pi_\lambda^{-1}(W_\lambda). \tag{32}$$

Since $r_{\lambda\lambda'}\pi_{\lambda'}(X_{\lambda'}\times I) = \pi_\lambda(p_{\lambda\lambda'}\times 1)(X_{\lambda'}\times I)$ and $r_{\lambda\lambda'}\pi_{\lambda'}(Y_{\lambda'}) = \pi_\lambda q_{\lambda\lambda'}$, one concludes that $r_{\lambda\lambda'}(M(f_{\lambda'})) \subseteq W_\lambda$. This completes the proof that $\boldsymbol{r}: M(f) \to \boldsymbol{M(f)}$ is a resolution.

To show that every space $M(f_\lambda)$ has the homotopy type of a polyhedron, i.e., belongs to the class HPol, it suffices to replace the mapping $f_\lambda: X_\lambda \to Y_\lambda$ by a simplicial approximation g_λ. Since $f_\lambda \simeq g_\lambda$, the pairs $(M(f_\lambda), X_\lambda)$ and $(M(g_\lambda), X_\lambda)$ have the same homotopy type rel(X_λ) (see e.g., (Dugundji 1966), Chapter XVIII, Theorem 4.3). Finally, it is known that for a simpicial mapping $g: P \to Q$ between polyhedra, $(M(g), P)$ is a polyhedral pair (see e.g., (Mardešić, Segal 1982), Appendix 1, Theorem 1.6). This completes the proof of the first assertion.

To prove the second assertion, consider a normal covering $\mathcal{U}$ of $DM(f)$. Let $\mathcal{U}^*$ be a star-refinement of $\mathcal{U}$. By the first assertion, there exists a $\lambda \in \Lambda$ and there exist open coverings $\mathcal{U}_\lambda$ of $M(f_\lambda)$ and $\mathcal{U}'_\lambda$ of $M(f_\lambda)'$ such that $r_\lambda^{-1}(\mathcal{U}_\lambda)$ and $(r'_\lambda)^{-1}(\mathcal{U}'_\lambda)$ refine $\mathcal{U}^*$. For every point $z \in DM(f_\lambda)$, we define an open neighborhood $V(z)$ as follows. If z belongs to $M(f_\lambda)\backslash X_\lambda$, we put $V(z) = U\backslash X_\lambda$, where U is any member of $\mathcal{U}_\lambda$, which contains z. Clearly, $(Dr_\lambda)^{-1}(V(z)) = r_\lambda^{-1}(V(z))$ is contained in a member of $\mathcal{U}^*$. For $z \in M(f_\lambda)'\backslash X_\lambda$, we define $V(z)$ analogously. Now assume that $z \in X_\lambda$. We distinguish two cases. In the first case, $z \notin \mathrm{Cl}\,(p_\lambda(X))$. For $V(z)$, choose any open neighborhhood of z in $DM(f_\lambda)$, which misses $\mathrm{Cl}\,(Dr_\lambda(DM(f)))$. Note that in this case $(Dr_\lambda)^{-1}(V(z)) = \emptyset$. Now assume that $z \in \mathrm{Cl}\,(p_\lambda(X))$. Choose a member $U(z)$ of $\mathcal{U}_\lambda$, which contains z and put $V(z) = U(z) \cup U'(z)$, where $U'(z) = s(U(z))$. Clearly, $V(z)$ is an open neighborhood of z in $DM(f_\lambda)$. Moreover, there exist members $U_1^*, U_2^* \in \mathcal{U}^*$ such that $r_\lambda^{-1}(U(z)) \subseteq U_1^*$ and $(r'_\lambda)^{-1}(U'(z)) \subseteq U_2^*$. Since $z \in \mathrm{Cl}\,(p_\lambda(X))$, both sets U_1^*, U_2^* intersect $p_\lambda^{-1}(U(z) \cap X_\lambda)$. Therefore, $U_1^* \cup U_2^*$ is contained in a memeber U of $\mathcal{U}$. However, $(Dr_\lambda)^{-1}(V(z)) = r_\lambda^{-1}(U(z)) \cup (r'_\lambda)^{-1}(U'(z)) \subseteq U_1^* \cup U_2^* \subseteq U$. This completes the verification of property (B1), for $\boldsymbol{Dr}$. Verification of property (B2)$'$ is easy and we omit it.

It remains to prove that every $DM(f_\lambda)$ has the homotopy type of a polyhedron. We already know that there exists a polyhedral pair (P_λ, X_λ) with the homotopy type of $(M(f_\lambda), X_\lambda)(\mathrm{rel}\, X_\lambda)$. Therefore, $D(M(f_\lambda), X_\lambda)$ has the homotopy type of the polyhedron $D(P_\lambda, X_\lambda)$ and thus, $DM(f_\lambda)$ belongs to the class HPol. □

Continuation of the proof of Theorem 8.15. We have already seen that $i: X \to M(f)$ is a strong shape equivalence. Since $\boldsymbol{p}: X \to \boldsymbol{X}$, $\boldsymbol{r}: M(f) \to \boldsymbol{M(f)}$ and $\boldsymbol{i}: \boldsymbol{X} \to \boldsymbol{M(f)}$ form a cofinite HPol-resolution of $i: X \to M(f)$, it follows that $\boldsymbol{i}: \boldsymbol{X} \to \boldsymbol{M(f)}$ induces an isomorphism of CH(pro-Top). Moreover, $\boldsymbol{p}: X \to \boldsymbol{X}$, $\boldsymbol{Dr}: DM(f) \to \boldsymbol{DM(f)}$ and $\boldsymbol{k}: \boldsymbol{X} \to \boldsymbol{DM(f)}$ form a cofinite HPol-resolution of $k: X \to M(f)$. Therefore, if we prove that $\boldsymbol{k}$ induces an isomorphism of pro-H(Top), it will follow that $k : X \to DM(f)$ is a shape equivalence and the proof of Theorem 8.15 will be completed. Hence, it suffices to establish the following Lemma.

LEMMA 8.19. *If* $\boldsymbol{i}\colon \boldsymbol{X} \to \boldsymbol{M}(\boldsymbol{f})$ *induces an isomorhism of* CH(pro-Top), *then* $\boldsymbol{k}\colon \boldsymbol{X} \to \boldsymbol{DM}(\boldsymbol{f})$ *induces an isomorphism of* pro-H(Top).

Lemma 8.19 is an easy consequence of the following lemma.

LEMMA 8.20. *If* $\boldsymbol{i}\colon \boldsymbol{X} \to \boldsymbol{M}(\boldsymbol{f})$ *induces an isomorhism of* CH(pro-Top), *then there exists a homotopy mapping* $\boldsymbol{h} = (h, h_\lambda)\colon \boldsymbol{M}(\boldsymbol{f}) \to \boldsymbol{X}$ *such that* $h \geq \mathrm{id}$,

$$h_\lambda i_{h(\lambda)} = p_{\lambda h(\lambda)}, \tag{33}$$

$$i_\lambda h_\lambda \simeq r_{\lambda h(\lambda)}\,(\mathrm{rel}(X_{h(\lambda)})). \tag{34}$$

Proof of Lemma 8.19. By Lemma 8.20, there exists a homotopy mapping $\boldsymbol{h}\colon \boldsymbol{M}(\boldsymbol{f}) \to \boldsymbol{X}$, such that $h \geq \mathrm{id}$ and relations (33) and (34) hold. Note that $h_\lambda\colon M(f_{h(\lambda)}) \to X_\lambda$ induces a mapping $Dh_\lambda\colon DM(f_{h(\lambda)}) \to X_\lambda$. Let us show that h and the mappings Dh_λ form a homotopy mapping $\boldsymbol{Dh}\colon \boldsymbol{DM}(\boldsymbol{f}) \to \boldsymbol{X}$, whose class $[\boldsymbol{Dh}]$ is the inverse of $[\boldsymbol{k}]$ in pro-H(Top).

Since $\boldsymbol{h}$ is a homotopy mapping, there exists a homotopy $H_{\lambda\lambda'}\colon M(f_{h(\lambda')}) \times I \to X_\lambda$, $\lambda \leq \lambda'$, which connects $p_{\lambda\lambda'}h_{\lambda'}$ and $h_\lambda r_{h(\lambda)h(\lambda')}$. Consider the duplication $DH_{\lambda\lambda'}\colon D(M(f_{h(\lambda')}) \times I, X \times I) \to X_\lambda$. Note that $D(M(f_{h(\lambda')}) \times I, X \times I)$ can be identified with $DM(f_{h(\lambda')}) \times I$. Therefore, $DH_{\lambda\lambda'}$ can be viewed as a homotopy $DH_{\lambda\lambda'}\colon DM(f_{h(\lambda')}) \times I \to X_\lambda$. Clearly, this homotopy connects the mappings $p_{\lambda\lambda'}Dh_{\lambda'}$ and $(Dh_\lambda)(Dr_{h(\lambda)h(\lambda')})$. This proves that $\boldsymbol{Dh} = (h, Dh_\lambda)$ is indeed a homotopy mapping.

It is clear that (33) implies

$$Dh_\lambda k_{h(\lambda)} = p_{\lambda h(\lambda)}. \tag{35}$$

Furthermore, for every $\lambda \in \Lambda$, (34) yields a homotopy H_λ, which is a mapping of pairs $H_\lambda\colon (M(f_{h(\lambda)}) \times I, X_{h(\lambda)}) \times I) \to (M(f_\lambda), X_\lambda)$. Consequently, its duplication is a mapping $DH_\lambda\colon D(M(f_{h(\lambda)}) \times I, X_{h(\lambda)}) \times I) \to D(M(f_\lambda), X_\lambda) = DM(f_\lambda)$. After identifications as above, this duplication can be viewed as a homotopy $DH_\lambda\colon DM(f_{h(\lambda)}) \times I \to DM(f_\lambda)$. Clearly, DH_λ connects the mappings $Dr_{\lambda h(\lambda)}$ and $k_\lambda Dh_\lambda$, i.e., one has

$$Dr_{\lambda h(\lambda)} \simeq k_\lambda Dh_\lambda. \tag{36}$$

Formulae (35) and (36) show that in pro-H(Top) the class $[\boldsymbol{Dh}]$ is indeed the inverse of the class $[\boldsymbol{k}]$. □

To complete the proof of Theorem 8.15, it remains to prove Lemma 8.20. Since the category CH(pro-Top) is difficult to work with, we pass to the category H(pro-Top), which is much simpler. This is achieved by using the functors τ: inv-Top→ inv-Top and $\boldsymbol{T}$: CH(pro-Top) → H(pro-Top), defined in 4.2. We need the following lemma.

LEMMA 8.21. *The level mapping* $\tau(\boldsymbol{i})\colon \boldsymbol{T}(\boldsymbol{X}) \to \boldsymbol{T}(\boldsymbol{M}(\boldsymbol{f}))$, *induced by* $\boldsymbol{i}\colon \boldsymbol{X} \to \boldsymbol{M}(\boldsymbol{f})$ *consists of mappings* $\tau_\lambda\colon T(\boldsymbol{X}_\lambda) \to T(\boldsymbol{M}(\boldsymbol{f})_\lambda)$, *which are closed embeddings and cofibrations. If* $\boldsymbol{i}$ *induces an isomorphism of* CH(pro-Top), *then* $\tau(\boldsymbol{i})$ *is an isomorphism of* H(pro-Top).

In the proof of lemma 8.21, we will use the following characterization of cofibered pairs (see (Strøm 1966), Theorem 2).

LEMMA 8.22. *A closed pair of spaces (Z, X) is a cofibered pair if and only if it has the following property. There exists a neighborhood U of X in Z and a mapping $\gamma: Z \to I$ such that $X = \gamma^{-1}(0)$ and $\gamma|(Z\backslash U) = 1$. Moreover, there exists a deformation $\Gamma: U \times I \to Z(\mathrm{rel}\, X)$ such that $\Gamma(z, 1) \in X$, for all $z \in Z$.*

Proof of Lemma 8.21. Recall that $T(\boldsymbol{X}_\lambda)$ consists of collections $\omega = (\omega_{\boldsymbol{\nu}})$, $\boldsymbol{\nu} \leq \lambda$, of mappings $\omega_{\boldsymbol{\nu}}: \Delta^n \to X_{\lambda_0}$, which satisfy the coherence conditions (4.2.2) and (4.2.3). Also recall that (see 4.2.54) and (4.2.55), $\tau_\lambda(\omega) = (\zeta) = (\zeta_{\boldsymbol{\nu}})$, where $\zeta_{\boldsymbol{\nu}} = i_{\lambda_0}\omega_{\boldsymbol{\nu}}$, i.e., $\zeta_{\boldsymbol{\nu}}$ is just $\omega_{\boldsymbol{\nu}}$, viewed as a mapping of Δ^n into $M(f_{\lambda_0})$. Therefore, τ_λ is an injection. If $\zeta \in T(\boldsymbol{M}(\boldsymbol{f})_\lambda)\backslash T(\boldsymbol{X}_\lambda)$, then there exists a $\boldsymbol{\nu} \leq \lambda$ such that $\zeta_{\boldsymbol{\nu}}$ maps some point $t_0 \in \Delta^n$ to a point $\zeta_{\boldsymbol{\nu}}(t_0) \notin X_{\lambda_0}$. Clearly, all $\zeta' \in T(\boldsymbol{M}(\boldsymbol{f})_\lambda)$, which have the property that $\zeta'_{\boldsymbol{\nu}}(t_0) \notin X_{\lambda_0}$, form an open neihghborhood of ζ, which misses $T(\boldsymbol{X}_\lambda)$. This shows that $T(\boldsymbol{X}_\lambda)$ is a closed subset of $T(\boldsymbol{M}(\boldsymbol{f})_\lambda)$.

To prove that τ_λ is a cofibration, it suffices to verify the conditions of Lemma 8.22. We first need an open neighborhood U of $T(\boldsymbol{X}_\lambda)$ in $T(\boldsymbol{M}(\boldsymbol{f})_\lambda)$ and a mapping $\gamma: T(\boldsymbol{M}(\boldsymbol{f})_\lambda) \to I$. For $\nu \in \Lambda$, denote by U_ν the open set $M(f_\nu)\backslash Y_\nu$ of $M(f_\nu)$. For $\boldsymbol{\nu} \in \Lambda_n$, let $U_{\boldsymbol{\nu}}$ denote the set of all $\omega \in M(f_{\nu_0})^{\Delta^n}$ such that $\omega(\Delta^n) \subseteq U_{\nu_0}$. Clearly, $U_{\boldsymbol{\nu}}$ is an open subset of $M(f_{\nu_0})^{\Delta^n}$. Finally, let U be the set of all $\zeta = (\zeta_{\boldsymbol{\nu}}) \in T(\boldsymbol{M}(\boldsymbol{f})_\lambda)$, which have the property that $\zeta_{\boldsymbol{\nu}} \in U_{\boldsymbol{\nu}}$, for all non-degenerate $\boldsymbol{\nu} \leq \lambda$. Since there are only finitely many such multiindices $\boldsymbol{\nu}$, it follows that U is an open subset of $T(\boldsymbol{M}(\boldsymbol{f})_\lambda)$. Clearly, $T(\boldsymbol{X}_\lambda) \subseteq U$.

In order to define γ, with every $\nu \in \Lambda$ associate the mapping $\rho_\nu: M(f_\nu) \to I$, induced by the second projection $X_\nu \times I \to I$ and by the constant mapping $Y_\nu \to \{1\}$. Then, for every $\boldsymbol{\nu} \leq \lambda$ of length n, let $\gamma_{\boldsymbol{\nu}}: T(\boldsymbol{M}(\boldsymbol{f})_\lambda) \to I$ be the mapping given by $\gamma_{\boldsymbol{\nu}}(\zeta) = \max(\rho_{\nu_0}\zeta_{\boldsymbol{\nu}})$. Clearly, $\gamma_{\boldsymbol{\nu}}$ is the composition of three continuous mappings, which insures its continuity. The first one is given by the projection $\zeta \mapsto \zeta_{\boldsymbol{\nu}}$. The second one is given by $\zeta_{\boldsymbol{\nu}} \mapsto \rho_{\nu_0}\zeta_{\boldsymbol{\nu}}$. Its continuity is a well-known property of the compact-open topology (see e.g., Theorem 2.1 of Chapter XII of (Dugundji, 1966)). The third mapping is the mapping $\max: I^{\Delta^n} \to I$. Finally, $\gamma = \max\{\gamma_{\boldsymbol{\nu}} \mid \boldsymbol{\nu} \text{ non-degenerate}\}$. Since there are only finitely many non-degenerate $\boldsymbol{\nu} \leq \lambda$, γ is a well-defined continuous mapping. It is readily seen that

$$T(\boldsymbol{X}_\lambda) = \gamma^{-1}(0), \;\; \gamma|T(\boldsymbol{M}(\boldsymbol{f})_\lambda)\backslash U = 1. \tag{37}$$

In particular, note that $\gamma(\zeta) = 0$, implies $\rho_{\nu_0}\zeta_{\boldsymbol{\nu}} = 0$ and thus, $\zeta_{\boldsymbol{\nu}}(\Delta^n) \subseteq X_{\nu_0}$, for all non-degenerate $\boldsymbol{\nu} \leq \lambda$. However, the same relation also holds for degenerate $\boldsymbol{\nu}$, because of the coherence conditions (4.2.3). Consequently, $\zeta = (\zeta_{\boldsymbol{\nu}})$ belongs to $T(\boldsymbol{X}_\lambda)$.

In order to define the deformation $\Gamma: U \times I \to T(\boldsymbol{M}(\boldsymbol{f})_\lambda)$, note that π_ν maps $X_\nu \times [0,1)$ onto U_ν homeomorphically and one can identify these two sets. Let $\Gamma_\nu: X_\nu \times [0,1) \times I \to X_\nu \times [0,1)$ be the mapping defined by

$$\Gamma_\nu((x,s),u) = (x, s(1-u)). \tag{38}$$

Since $r_{\nu\nu'}|U_{\nu'}: U_{\nu'} \to U_\nu$ identifies with $p_{\nu\nu'} \times 1: X_{\nu'} \times [0,1) \to X_\nu \times [0,1)$, one has $r_{\nu\nu'}\Gamma_{\nu'}((x,s),u) = r_{\nu\nu'}(x, s(1-u)) = (p_{\nu\nu'}(x), s(1-u))$. On the other hand, $\Gamma_\nu(r_{\nu\nu'}(x,s),u) = \Gamma_\nu((p_{\nu\nu'}(x),s),u) = (p_{\nu\nu'}(x), s(1-u))$. Consequently,

$$r_{\nu\nu'}\Gamma_{\nu'} = \Gamma_\nu(r_{\nu\nu'} \times 1). \tag{39}$$

For $\zeta = (\zeta_{\boldsymbol{\nu}}) \in U$ and $u \in I$, we now put $\Gamma(\zeta,u) = \eta = (\eta_{\boldsymbol{\nu}})$, where $\eta_{\boldsymbol{\nu}}: \Delta^n \to U_{\nu_0}$ is given by

$$\eta_{\boldsymbol{\nu}}(t) = \Gamma_{\nu_0}(\zeta_{\boldsymbol{\nu}}(t), u). \tag{40}$$

To show that η belongs to $T(\boldsymbol{M}(\boldsymbol{f})_\lambda)$, we must verify the coherence conditions.

If $0 < j \le n$, we have

$$\eta_{\boldsymbol{\nu}}(d_j t') = \Gamma_{\nu_0}(\zeta_{\boldsymbol{\nu}}(d_j t'), u) = \Gamma_{\nu_0}(\zeta_{d^j\boldsymbol{\nu}}(t'), u) = \eta_{d^j\boldsymbol{\nu}}(t'). \tag{41}$$

Now assume that $j = 0$. Then

$$\eta_{\boldsymbol{\nu}}(d_0 t') = \Gamma_{\nu_0}(\zeta_{\boldsymbol{\nu}}(d_0 t'), u) = \Gamma_{\nu_0}(r_{\nu_0\nu_1}\zeta_{d^0\boldsymbol{\nu}}(t'), u). \tag{42}$$

Using (39), we conclude that

$$\eta_{\boldsymbol{\nu}}(d_0 t') = r_{\nu_0\nu_1}\Gamma_{\nu_1}(\zeta_{d^0\boldsymbol{\nu}}(t'), u). \tag{43}$$

However, by (40), the right side of (43) coincides with $r_{\nu_0\nu_1}\eta_{d^0\boldsymbol{\nu}}(t')$ and we obtain the desired condition $\eta_{\boldsymbol{\nu}}(d_0 t') = r_{\nu_0\nu_1}\eta_{d^0\boldsymbol{\nu}}(t')$.

If $u = 0$, one has $\eta_{\boldsymbol{\nu}}(t) = \zeta_{\boldsymbol{\nu}}(t)$ and thus, $\Gamma(\zeta, 0) = \zeta$, i.e., $\Gamma(\,.\,,0)$ is the identity mapping. If $u = 1$, $\Gamma_{\nu_0}((x,s),1) = (x,0) \in X_{\nu_0}$ and thus, $\Gamma(\,.\,,1)$ is a mapping into $T(\boldsymbol{X}_\lambda)$. Finally, if ζ belongs to $T(\boldsymbol{X}_\lambda)$, then $\zeta_{\boldsymbol{\nu}}(t) \in X_{\nu_0} \times \{0\}$. However, by (38), $\Gamma_{\nu_0}((x,0),u) = (x,0)$ and therefore, $\eta_{\boldsymbol{\nu}}(t) = \zeta_{\boldsymbol{\nu}}(t)$, for all $u \in I$, which shows that $\Gamma(\zeta,u)$ does not depend on u, i.e., Γ is a homotopy rel $(T(\boldsymbol{X}_\lambda))$.

Recall that $[C(\boldsymbol{i})]$ is an isomorphism of CH(pro-Top). Since $\boldsymbol{T}$: CH(pro-Top) $\to$ H(pro-Top) is a functor (Theorem 4.25), $\boldsymbol{T}[C(\boldsymbol{i})]$ is an isomorphism of H(pro-Top). However, by Lemma 4.29, $[\tau(\boldsymbol{i})] = [\boldsymbol{T}C(\boldsymbol{i})] = \boldsymbol{T}[C(\boldsymbol{i})]$. □

To prove Lemma 8.20, we also need a lemma concerning the category H(pro-Top).

LEMMA 8.23. *Let $(\boldsymbol{Z}, \boldsymbol{X})$ be a cofinite system of closed cofibered pairs and let the inclusion mapping $\boldsymbol{i}: \boldsymbol{X} \to \boldsymbol{Z}$ induce an isomorphism of* H(pro-Top). *Then, there exists a homotopy mapping $\boldsymbol{g} = (g, g_\lambda): \boldsymbol{Z} \to \boldsymbol{X}$ such that $g \geq \mathrm{id}$,*

$$g_\lambda i_{g(\lambda)} = p_{\lambda g(\lambda)}, \tag{44}$$

$$r_{\lambda g(\lambda)} \simeq i_\lambda g_\lambda \,(\mathrm{rel}(X_{g(\lambda)})). \tag{45}$$

We first prove an auxilliary lemma.

LEMMA 8.24. *Let $(\boldsymbol{Z}, \boldsymbol{X}) = ((Z_\lambda, X_\lambda), r_{\lambda\lambda'}, \Lambda)$ be a cofinite system of closed cofibered pairs and let $\boldsymbol{i}: \boldsymbol{X} \to \boldsymbol{Z}$ be the level mapping formed by inclusions $i_\lambda: X_\lambda \to Z_\lambda$. If $[\boldsymbol{i}]: \boldsymbol{X} \to \boldsymbol{Z}$ is an isomorphism of* H(pro-Top), *then there exists a homotopy mapping $\boldsymbol{k} = (k, k_\lambda): \boldsymbol{Z} \to \boldsymbol{X}$ such that $k \geq \mathrm{id}$ and*

$$k_\lambda i_{k(\lambda)} = p_{\lambda k(\lambda)}, \tag{46}$$

where $p_{\lambda\lambda'}: X_{\lambda'} \to X_\lambda$ is the restriction of $r_{\lambda\lambda'}$ to $X_{\lambda'}$. Moreover, there exist homotopies $H_\lambda: Z_{k(\lambda)} \times I \to Z_\lambda$, $\lambda \in \Lambda$, such that

$$H_\lambda(z, 0) = r_{\lambda k(\lambda)}(z), \tag{47}$$

$$H_\lambda(z, 1) = i_\lambda k_\lambda(z), \tag{48}$$

$$r_{\lambda\lambda'} H_{\lambda'} | X_{k(\lambda')} \times I = H_\lambda(r_{k(\lambda)k(\lambda')} \times 1) | X_{k(\lambda')} \times I, \ \lambda \leq \lambda'. \tag{49}$$

Proof. By assumption, there exists a mapping $\boldsymbol{h} = (h, h_\lambda): \boldsymbol{Z} \to \boldsymbol{X}$ such that

$$[\boldsymbol{1}] = [\boldsymbol{i}][\boldsymbol{h}], \tag{50}$$

$$[\boldsymbol{h}][\boldsymbol{i}] = [\boldsymbol{1}], \tag{51}$$

i.e., there are mappings $\boldsymbol{H}' = (H', H'_\lambda): \boldsymbol{Z} \times I \to \boldsymbol{Z}$, $\boldsymbol{H}'' = (H'', H''_\lambda): \boldsymbol{X} \times I \to \boldsymbol{X}$, which realize (50) and (51), respectively. Note that, $H' \geq h, \mathrm{id}$ and for every $z \in Z_{H'(\lambda)}$, one has

$$H'_\lambda(z, 0) = r_{\lambda H'(\lambda)}(z). \tag{52}$$

$$H'_\lambda(z, 1) = h_\lambda r_{h(\lambda) H'(\lambda)}(z), \tag{53}$$

Moreover, for $\lambda \leq \lambda'$,

$$H'_\lambda(r_{H'(\lambda) H'(\lambda')} \times 1) = r_{\lambda\lambda'} H'_{\lambda'}. \tag{54}$$

Similarly, $H'' \geq h, \mathrm{id}$ and for every $x \in X_{H''(\lambda)}$, one has

$$H''_\lambda(x, 0) = h_\lambda p_{h(\lambda) H''(\lambda)}(x), \tag{55}$$

$$H''_\lambda(x, 1) = p_{\lambda H''(\lambda)}, \tag{56}$$

where $p_{\lambda\lambda'} = r_{\lambda\lambda'} | X_{\lambda'}$. Moreover,

$$H''_\lambda(p_{H''(\lambda)H''(\lambda')} \times 1) = p_{\lambda\lambda'} H''_{\lambda'}. \tag{57}$$

Since there exists an increasing function $H \geq H', H''$, there is no loss of generality in assuming that $H' = H'' = H$.

We now extend H''_λ to a mapping $H''_\lambda \colon (X_{H(\lambda)} \times I) \cup (Z_{H(\lambda)} \times 0) \to X_\lambda$, by putting

$$H''_\lambda(z, 0) = h_\lambda r_{h(\lambda)H(\lambda)}(z), \;\; z \in Z_{H(\lambda)}. \tag{58}$$

By (55), H''_λ is well defined. Since $(Z_{H(\lambda)}, X_{H(\lambda)})$ is a cofibered pair, H''_λ extends further to a mapping $H''_\lambda \colon Z_{H(\lambda)} \times I \to X_\lambda$.

Now put $k = H$ and define mappings $k_\lambda \colon Z_{k(\lambda)} \to X_\lambda$ by putting

$$k_\lambda(z) = H''_\lambda(z, 1). \tag{59}$$

Note that (58) implies $k_\lambda \simeq h_\lambda r_{h(\lambda)k(\lambda)}$. The function k and the mappings k_λ form a homotopy mapping $\boldsymbol{k} \colon Z_\lambda \to X_\lambda$ (which need not be a mapping). Indeed, for $\lambda \leq \lambda'$, one has

$$k_\lambda r_{k(\lambda)k(\lambda')} \simeq h_\lambda r_{h(\lambda)k(\lambda')} = p_{\lambda\lambda'} h_{\lambda'} r_{h(\lambda')k(\lambda')}. \tag{60}$$

On the other hand,

$$p_{\lambda\lambda'} k_{\lambda'} \simeq p_{\lambda\lambda'} h_{\lambda'} r_{h(\lambda')k(\lambda')}. \tag{61}$$

The mapping k_λ satisfies (46), because of (56).

Now define the homotopy $H_\lambda \colon Z_{k(\lambda)} \times I \to Z_\lambda$ as the juxtaposition of the homotopies H'_λ and $i_\lambda H''_\lambda$,

$$H_\lambda = H'_\lambda * i_\lambda H''_\lambda. \tag{62}$$

By (53) and (58), $H'_\lambda(z, 1) = i_\lambda H''_\lambda(z, 0)$, which makes H_λ well-defined. Clearly, (52) implies (47) and (59) implies (48). Moreover, (54) and (57) imply (49). □

Proof of Lemma 8.23. By Lemma 8.24, there exists a homotopy mapping $\boldsymbol{k} = (k, k_\lambda) \colon \boldsymbol{Z} \to \boldsymbol{X}$, $k \geq \mathrm{id}$, and there exist homotopies $H_\lambda \colon Z_{k(\lambda)} \times I \to Z_\lambda$, $\lambda \in \Lambda$, such that (46)–(49) hold. Put $g = k^2$ and define $g_\lambda \colon Z_{g(\lambda)} \to X_\lambda$ by

$$g_\lambda = p_{\lambda k(\lambda)} k_{k(\lambda)}. \tag{63}$$

Since $\boldsymbol{k}$ is a homotopy mapping, so is also $\boldsymbol{g} = (g, g_\lambda)$. Note that (46) implies (44). To establish (45), we first define a mapping $K \colon (Z_{g(\lambda)} \times I \times 0) \cup (Z_{g(\lambda)} \times \partial I \times I) \to Z_\lambda$, by putting

$$K(z, t, 0) = r_{\lambda k(\lambda)} H_{k(\lambda)}(z, t), \tag{64}$$

$$K(z, 0, s) = r_{\lambda g(\lambda)}(z), \tag{65}$$

$$K(z, 1, s) = H_\lambda(i_{k(\lambda)} k_{k(\lambda)}(z), 1 - s). \tag{66}$$

K is well defined. Indeed, by (47), (64) yields $K(z, 0, 0) = r_{\lambda g(\lambda)}(z)$, which agrees with (65). By (48), (64) also yields $K(z, 1, 0) = r_{\lambda k(\lambda)} i_{k(\lambda)} k_{k(\lambda)}(z)$. However, by (47), (66) yields the same value for $K(z, 1, 0)$.

We now extend K also to $X_{g(\lambda)} \times I \times I$ by putting

$$K(x,t,s) = r_{\lambda k(\lambda)} H_{k(\lambda)}(x,(1-s)t). \tag{67}$$

Using (46), it is readily seen that (64) and (67) yield the same value $K(x,t,0) = r_{\lambda k(\lambda)} H_{k(\lambda)}(x,t)$. Similarly, both formulae (65) and (67) yield $K(x,0,s) = r_{\lambda g(\lambda)}(x)$. Finally, for $x \in X_{g(\lambda)}$, by (46),

$$k_{k(\lambda)}(x) = p_{k(\lambda)g(\lambda)}(x) \tag{68}$$

and therefore, (66) yields

$$K(x,1,s) = H_\lambda(p_{k(\lambda)g(\lambda)}(x), 1-s). \tag{69}$$

However, (67) assigns to $K(x,1,s)$ the same value, because of (49), for $\lambda' = k(\lambda)$.

By assumption, $(Z_{g(\lambda)}, X_{g(\lambda)})$ is a closed cofibered pair. Lemma 8.14, applied to this pair, enables us to use the homotopy extension property and further extend the mapping K to all of $Z_{g(\lambda)} \times I \times I$. Finally, define a homotopy $G: Z_{g(\lambda)} \times I \to Z_\lambda$ by putting

$$G(z,t) = K(z,t,1), \tag{70}$$

Now (65) implies

$$G(z,0) = r_{\lambda g(\lambda)}(z), \tag{71}$$

Similarly, (66) and (47) yield

$$G(z,1) = r_{\lambda k(\lambda)} i_{k(\lambda)} k_{k(\lambda)}(z) = i_\lambda p_{\lambda k(\lambda)} k_{k(\lambda)}(z) = i_\lambda g_\lambda(z). \tag{72}$$

Finally, (67) and (47) show that

$$G(x,t) = r_{\lambda g(\lambda)}(x), \;\; x \in X_{g(\lambda)}. \tag{73}$$

Formulae (71), (72) and (73) establish (45). □

Proof of Lemma 8.20. Since $\boldsymbol{i}: \boldsymbol{X} \to \boldsymbol{M}(\boldsymbol{f})$ induces an isomorphism of CH(pro-Top), Lemma 8.22 shows that $\tau(\boldsymbol{i})$ is an isomorphism of H(pro-Top). Moreover, $(\boldsymbol{T}(\boldsymbol{M}(\boldsymbol{f})), \boldsymbol{T}(\boldsymbol{X}))$ is an inverse system of closed cofibered pairs (we view the embeddings $\tau_\lambda : T(\boldsymbol{X}_\lambda) \to T(\boldsymbol{M}(\boldsymbol{f})_\lambda)$ as inclusions). Therefore, Lemma 8.23 applies and yields a homotopy mapping $\boldsymbol{g} = (g, g_\lambda): \boldsymbol{T}(\boldsymbol{M}(\boldsymbol{f})) \to \boldsymbol{T}(\boldsymbol{X})$ such that $g \geq \mathrm{id}$,

$$g_\lambda \tau_{g(\lambda)} = u_{\lambda g(\lambda)}, \tag{74}$$

$$v_{\lambda g(\lambda)} \simeq \tau_\lambda g_\lambda \,(\mathrm{rel}(T(\boldsymbol{X}_{g(\lambda)}))), \tag{75}$$

where $u_{\lambda\lambda'}$ and $v_{\lambda\lambda'}$ denote the bonding mappings of $\boldsymbol{T}(\boldsymbol{X})$ and $\boldsymbol{T}(\boldsymbol{M}(\boldsymbol{f}))$, respectively. Now recall the level homotopy equivalence $\phi_{\boldsymbol{X}} = (\phi_\lambda): \boldsymbol{X} \to \boldsymbol{T}(\boldsymbol{X})$ and the homotopy inverses $\psi_\lambda: T(\boldsymbol{X}_\lambda) \to X_\lambda$ of ϕ_λ, defined in 4.2. According to (4.2.69), one has

$$\psi_\lambda \phi_\lambda = \mathrm{id}. \tag{76}$$

Moreover, if we put $\phi_{\boldsymbol{M}(\boldsymbol{f})} = (\phi'_\lambda)$ and $\psi_{\boldsymbol{M}(\boldsymbol{f})} = (\psi'_\lambda)$, then

$$\tau_\lambda \phi_\lambda = \phi'_\lambda i_\lambda, \tag{77}$$

$$i_\lambda \psi_\lambda = \psi'_\lambda \tau_\lambda. \tag{78}$$

The first of these equalities follows from (4.2.67) and the second one follows from Remark 4.33. Also recall that the mappings ψ_λ form a level homotopy mapping $\psi_{\boldsymbol{X}}: \boldsymbol{T}(\boldsymbol{X}) \to \boldsymbol{X}$ (see Remark 4.32). Therefore, we obtain a homotopy mapping $\boldsymbol{h} = (h, h_\lambda): \boldsymbol{M}(\boldsymbol{f}) \to \boldsymbol{X}$, by putting $h = g$ and

$$h_\lambda = \psi_\lambda g_\lambda \phi'_{g(\lambda)}. \tag{79}$$

Now (77), (74) and (76) yield

$$\begin{aligned} h_\lambda i_{g(\lambda)} = \psi_\lambda g_\lambda \phi'_{g(\lambda)} i_{g(\lambda)} = \psi_\lambda g_\lambda \tau_{g(\lambda)} \phi_{g(\lambda)} = \\ \psi_\lambda u_{\lambda g(\lambda)} \phi_{g(\lambda)} = \psi_\lambda \phi_\lambda p_{\lambda g(\lambda)} = p_{\lambda g(\lambda)}. \end{aligned} \tag{80}$$

Similarly, (78) implies

$$i_\lambda h_\lambda = i_\lambda \psi_\lambda g_\lambda \phi'_{g(\lambda)} = \psi'_\lambda \tau_\lambda g_\lambda \phi'_{g(\lambda)}, \tag{81}$$

By (75), there exists a homotopy $G_\lambda: T(\boldsymbol{M}(\boldsymbol{f})_{g(\lambda)}) \times I \to T(\boldsymbol{M}(\boldsymbol{f})_\lambda)$, fixed on $T(\boldsymbol{X}_{g(\lambda)})$), which connects $v_{\lambda g(\lambda)}$ to $\tau_\lambda g_\lambda$. Therefore, $\psi'_\lambda G_\lambda(\phi'_{g(\lambda)} \times 1): M(\boldsymbol{f})_\lambda) \times I \to M(f_{g(\lambda)})$ is a homotopy rel$(X_{g(\lambda)})$, which connects $\psi'_\lambda v_{\lambda g(\lambda)} \phi'_{g(\lambda)}$ and $\psi'_\lambda \tau_\lambda g_\lambda \phi'_{g(\lambda)}$. However, by (76), one has

$$\psi'_\lambda v_{\lambda g(\lambda)} \phi'_{g(\lambda)} = \psi'_\lambda \phi'_\lambda r_{\lambda g(\lambda)} = r_{\lambda g(\lambda)}. \tag{82}$$

Now (81) and (82) establish (34). □

Bibliographic notes

Coherent expansions of spaces, the strong shape category and the strong shape functor were defined in (Lisica, Mardešić 1983, 1984a, 1984b). The introduction of strong expansions, led to Theorem 8.1 (Mardešić 1991a, 1991c) and thus, split the rather lengthy original proof of Corollary 8.7 (Lisica, Mardešić 1984b) in two parts, i.e., Theorem 7.6 and Theorem 8.1. The proof of Theorem 8.2 is taken from (Mardešić 1998). B. Günther stated (without proof) that coherent expansions are always strong expansions (see (Günther 1991a), Remark on p.149). He probably meant polyhedral coherent expansions. Theorem 8.9 is from (Lisica, Mardešić 1984b). Theorem 8.11 is from (Günther 1992a). However, Günther states that the result is implicitely contained already in Ch. I, Theorem 6 of (Lisica, Mardešić 1984b). Theorem 8.13 was first proved by B. Günther (see (Günther 1989), Korollar 1.12), who used his description of strong shape. The proof of Theorem 8.16 is taken from (Dydak, Nowak 1991), Theorem 1.3. The starting point of that paper is the definition of strong shape equivalences of mappings between spaces as mappings having properties (SE1) and (SSE2). Their next step consists in generalizing the notion of strong shape equivalence to mappings between inverse systems, $\boldsymbol{f}\colon \boldsymbol{X} \to \boldsymbol{Y}$. The definition of this notion is given in 10.7 under the name of strong expansion. Only in the sequel of their paper Dydak and Nowak defined the strong shape category, using fibrant expansions of spaces. However, their construction of the latter contains a technical error. The first use of resolutions in constructing the strong shape category for topological spaces is due to F.W. Cathey and J. Segal. Their construction can be described just as the one given in the present book, the only difference being that, instead of the category CH(pro-Top), they use the Edwards – Hastings category Ho(pro-Top). However, since these categories are isomorphic, the Cathey – Segal strong shape category (Cathey, Segal 1983) is isomorphic to the category SSh(Top).

9. Strong shape of metric compacta

This section is devoted to the strong shape theory of metric compacta. Specialization of the theory to this important case brings considerable simplification. The reason for this is that metric compacta admit polyhedral resolutions, which are sequences, and the coherent category of sequences is rather simple (see 3). In the first subsection it is shown that the strong shape category of metric compacta SSh(CM) has an elementary description, first introduced by J.B. Quigley. The second subsection is devoted to the complement theorem, which relates strong shape to proper homotopy.

9.1 The Quigley strong shape category

Let SSh(CM) denote the restriction of the category SSh(Top) to metric compacta and let $\mathrm{SSh}^{(1)}(\mathsf{CM})$ denote the corresponding restriction of $\mathrm{SSh}^{(1)}(\mathsf{Top})$.

THEOREM 9.1. $E^{(1)}$: SSh (CM) $\to \mathrm{SSh}^{(1)}(\mathsf{CM})$ *is an isomorphism of categories.*

Proof. Every metric compactum X is the limit of an inverse sequence of compact polyhedra. Therefore, for metric compacta X, Y there exist polyhedral resolutions $\boldsymbol{p}: X \to \boldsymbol{X}$, $\boldsymbol{q}: Y \to \boldsymbol{Y}$, such that $\boldsymbol{X}$ and $\boldsymbol{Y}$ are inverse sequences. By definition, every morphism $F: X \to Y$ of SSh(CM) is given by a triple $(\boldsymbol{p}, \boldsymbol{q}, [\boldsymbol{f}])$, where $[\boldsymbol{f}]: \boldsymbol{X} \to \boldsymbol{Y}$ is a morphism of CH(tow-Top). By definition, $E^{(1)}(F)$ is given by the triple $(\boldsymbol{p}, \boldsymbol{q}, E^{(1)}[\boldsymbol{f}])$. However, by Theorem 3.7, $E^{(1)}$: CH(tow-Top) $\to \mathrm{CH}^{(1)}$(tow-Top) is an isomorphism of categories. Therefore, $E^{(1)}$ induces a bijection between morphisms $F: X \to Y$ of SSh(CM) and $\mathrm{SSh}^{(1)}(\mathsf{CM})$. □

The above theorem shows that, in the case of metric compacta, strong shape is a relatively simple notion. An even simpler description of the strong shape category, restricted to closed subsets of the Hilbert cube Q, was given in 1973 by J.B. Quigley (Quigley 1973). In this subsection we denote the *Quigley strong shape category* by QSh. We will show that this category is equivalent to SSh(CM).

The objects of QSh are closed subsets of the Hilbert cube $Q = I^{\omega}$. To describe morphisms, consider the *rays* $R_{\tau} = [\tau, \infty) \subseteq \mathbb{R}$, for real numbers

$\tau \geq 0$. An *approaching mapping* $\phi: X \to Y$ between closed subsets $X, Y \subseteq Q$ is a (continuous) mapping $\phi: Q \times R_0 \to Q$, which satisfies the following condition:

(Q) *For every neighborhood* V *of* Y *in* Q*, there exist a neighborhood* U *of* X *in* Q *and a real number* $\tau \geq 0$*, such that*

$$\phi(U \times R_\tau) \subseteq V. \tag{1}$$

The composition $\chi = \psi\phi: X \to Z$ of approaching mappings $\phi: X \to Y, \psi: Y \to Z$ is defined by

$$\chi(u, t) = \psi(\phi(u, t), t), \tag{2}$$

where $u \in Q$, $t \in R_0$. To verify that χ is an approaching mapping, consider a neighborhood W of Z in Q. Choose a neighborhood V of Y and $\tau' \geq 0$ so that

$$\psi(V \times R_{\tau'}) \subseteq W. \tag{3}$$

Then choose a neighborhood U of X and $\tau \geq 0$ so that (1) holds. Clearly, for $\tau'' = \max\{\tau, \tau'\}$, one has $\chi(U \times R_{\tau''}) \subseteq \psi(V \times R_{\tau''}) \subseteq W$. Note that composition of approaching mappings is associative. We also consider a particular approaching mapping $\iota: X \to X$, given by $\iota(u, t) = u$, $u \in Q \times R_0$. Note that, for every approaching mapping $\phi : X \to Y$, one has $\phi\iota = \phi$ and $\iota\phi = \phi$.

An *approaching homotopy* $\Phi: X \times I \to Y$ is a mapping $\Phi: Q \times I \times R_0 \to Q$, which satisfies the analogous condition:

(QH) *For every neighborhood* V *of* Y *in* Q*, there exist a neighborhood* U *of* X *in* Q *and a real number* $\tau \in R_0$ *such that*

$$\Phi(U \times I \times R_\tau) \subseteq V. \tag{4}$$

Two approaching mappings $\phi: X \to Y, \phi': X \to Y$ are *homotopic*, $\phi \simeq \phi'$, provided there exists an approaching homotopy Φ, which connects them, i.e.,

$$\Phi(u, 0, t) = \phi(u, t),\ \Phi(u, 1, t) = \phi'(u, t), \tag{5}$$

for $u \in Q$, $t \in R_0$.

Clearly, $\simeq$ is an equivalence relation, which decomposes the set of approaching mappings $\phi: X \to Y$ into equivalence classes $[\phi]$. Moreover, $\phi \simeq \phi'$ implies $\psi\phi \simeq \psi\phi'$, because the mapping $H: Q \times I \times R_0 \to Q$, given by $H(u, s, t) = \psi(\Phi(u, s, t), t))$, is an approaching homotopy $X \times I \times R_0 \to Z$, which connects $\psi\phi$ to $\psi\phi'$. Indeed, for any neighborhood W of Z in Q, there exist a neighborhood V of Y in Q and a $\tau' \geq 0$ such that (3) holds. Choosing U and $\tau \geq \tau'$ so that (4) holds, one concludes that $H(U \times I \times R_\tau) \subseteq W$. Similarly, one concludes that $\psi \simeq \psi'$ implies $\psi\phi \simeq \psi'\phi$. Therefore, one can define the composition of homotopy classes by composing their representatives, $[\psi][\phi] = [\psi\phi]$. It is now clear that composition of homotopy classes is associative. Moreover, the homotopy class $[\iota]$ has the property that $[\phi][\iota] = [\phi]$ and $[\iota][\phi] = [\phi]$. We have thus, established the following theorem.

THEOREM 9.2. *For closed subsets of the Hilbert cube, homotopy classes of approaching mappings form a category, called the Quigley strong shape category* QSh.

The main result of this subsection is the following theorem.

THEOREM 9.3. *The strong shape category* SSh(CM) *is equivalent to the Quigley category* QSh.

In view of Theorem 9.1, Theorem 9.3 is an immediate consequence of the following assertion.

THEOREM 9.4. *The strong shape category* $\mathrm{SSh}^{(1)}(\mathsf{CM})$ *of height* 1 *is equivalent to the Quigley category* QSh.

Denote by $\mathrm{SSh}^{(1)}(\mathsf{Q})$ the full subcategory of $\mathrm{SSh}^{(1)}(\mathsf{CM})$, whose objects are closed subsets of the Hilbert cube. Since every compact metric space X admits an embedding in Q, and homeomorphisms induce isomorphisms of strong shape, it is clear that the inclusion $\mathrm{SSh}^{(1)}(\mathsf{Q}) \to \mathrm{SSh}^{(1)}(\mathsf{CM})$ is an equivalence of categories. Therefore, Theorem 9.4 is an immediate consequence of the following theorem.

THEOREM 9.5. *The categories* QSh *and* $\mathrm{SSh}^{(1)}(\mathsf{Q})$ *are isomorphic.*

To prove Theorem 9.5, we need an isomorphism T: $\mathsf{QSh} \to \mathrm{SSh}^{(1)}(\mathsf{Q})$. On objects, we put $T(X) = X$. In order to define T on morphisms, we associate with every closed subset $X \subseteq Q$ a decreasing sequence of closed neighborhoods X_m of X, such that $\cap_m X_m = X$ and each X_m is of the form $P \times (\prod_{i>n} I)$, where $P \subseteq I^n$ is a compact polyhedron. Clearly, X_m has the homotopy type of P. Let $p_m: X \to X_m$ and $p_{mm'}: X_{m'} \to X_m$, $m \leq m'$, be inclusion mappings. Then $\boldsymbol{X} = (X_m, p_{mm'}, \mathbb{N})$ is an inverse sequence and $\boldsymbol{p} = (p_m): X \to \boldsymbol{X}$ is its limit.

We will say that a coherent mapping $\boldsymbol{f} = (f, f_m, f_{mm'}): \boldsymbol{X} \to \boldsymbol{Y}$ of height 1 is *associated* with an approaching mapping $\phi: X \to Y$, provided $f: \mathbb{N} \to \mathbb{N}$ is an increasing function such that

$$\phi(X_{f(k)} \times [f(k), \infty)) \subseteq Y_k, \; k \in \mathbb{N}. \tag{6}$$

Moreover, the mappings $f_m: X_{f(m)} \to Y_m$, $f_{mm'}: X_{f(m')} \times I \to Y_m$, $m \leq m'$, satisfy the following conditions:

$$f_m(x) = \phi(x, f(m)). \tag{7}$$

$$f_{mm'}(x,t) = \phi(x, (1-t)f(m) + tf(m')). \tag{8}$$

Note that the coherence conditions imply:

$$f_{mm'}(x,0) = f_m(x), \; f_{mm'}(x,1) = f_{m'}(x). \tag{9}$$

$$f_{mm}(x,t) = f_m(x). \tag{10}$$

Every approaching mapping ϕ admits an associated coherent mapping of height 1. Indeed, by property (Q), there exists a function $f: \mathbb{N} \to \mathbb{N}$ satisfying (6). There is no loss of generality in assuming that f increases and is as large as desired. One defines the mappings f_m and $f_{mm'}$ using (7) and (8). In view of Lemma 3.3, it remains to show that

$$(f_{mm'}|(X_{f(m'')} \times I)) * f_{m'm''} \simeq f_{mm''}(\text{rel}\,\{0,1\}),\ m \le m' \le m''. \tag{11}$$

Indeed, for a given $x \in X_{f(m'')}$, the homotopy on the left side of (11) determines a path, which is the composition of the paths $\phi|\{x\} \times [f(m), f(m')]$ and $\phi|\{x\} \times [f(m'), f(m'')]$. However, this path is homotopic rel $\{0,1\}$ to the path $\phi|\{x\} \times [f(m), f(m'')]$, determined by the right side of (11).

LEMMA 9.6. *The homotopy class $[\boldsymbol{f}]$ of a coherent mapping $\boldsymbol{f}$ of height 1, associated with ϕ, does not depend of the choice of the function f.*

Proof. Let $\boldsymbol{f}' = (f', f'_m, f'_{mm'})$ be another coherent mapping of height 1, associated with ϕ. We need to show that $\boldsymbol{f} \simeq \boldsymbol{f}'$. Clearly, it suffices to consider the case when $f \le f'$. To define a 1-coherent homotopy $\boldsymbol{F} = (f', F_m, F_{mm'})$, which connects $\boldsymbol{f}$ to $\boldsymbol{f}'$, we first define homotopies $F_m: X_{f'(m)} \times I \to Y_m$, by putting

$$F_m(x,s) = \phi(x, (1-s)f(m) + sf'(m)). \tag{12}$$

Note that

$$F_m(x,0) = f_m(x),\ F_m(x,1) = f'_m(x). \tag{13}$$

In order to define homotopies $F_{mm'}: X_{f'(m')} \times I \times I \to Y_m$, $m \le m'$, we first define a mapping $w_{mm'}: I \times I \to [f(m), \infty)$, by putting

$$w_{mm'}(s,t) = (1-s)((1-t)f(m)+tf'(m))+s((1-t)f'(m)+tf'(m')). \tag{14}$$

It is readily seen that

$$w_{mm'}(s,0) = (1-s)f(m) + sf'(m), \tag{15}$$
$$w_{mm'}(s,1) = (1-s)f(m') + sf'(m'), \tag{16}$$
$$w_{mm'}(0,t) = (1-t)f(m) + tf(m'), \tag{17}$$
$$w_{mm'}(1,t) = (1-t)f'(m) + tf'(m'), \tag{18}$$
$$w_{mm}(s,t) = (1-s)f(m) + sf'(m). \tag{19}$$

Now let

$$F_{mm'}(x,s,t) = \phi(x, w_{mm'}(s,t)). \tag{20}$$

Formulae (15)–(19) and (8) readily yield

$$F_{mm'}(x,s,0) = F_m(x,s), \tag{21}$$
$$F_{mm'}(x,s,1) = F_{m'}(x,s), \tag{22}$$
$$F_{mm'}(x,0,t) = f_{mm'}(x,t), \tag{23}$$
$$F_{mm'}(x,1,t) = f'_{mm'}(x,t). \tag{24}$$
$$F_{mm}(x,s,t) = F_m(x,s,t). \tag{25}$$

Now Lemma 3.4 implies that $\boldsymbol{F} = (f', F_m, F_{mm'})$ is indeed a coherent homotopy of height 1, which connects $\boldsymbol{f}$ and $\boldsymbol{f}'$. □

LEMMA 9.7. *Let $\phi, \phi': X \to Y$ be two approaching mappings and let $\boldsymbol{f}, \boldsymbol{f}': \boldsymbol{X} \to \boldsymbol{Y}$ be associated coherent mappings of height 1. If $\phi \simeq \phi'$, then $\boldsymbol{f} \simeq \boldsymbol{f}'$. Consequently, $T([\phi]) = [\boldsymbol{f}]$ yields a well-defined function $T: \mathsf{QSh}(X,Y) \to \mathrm{CH}^{(1)}(\text{tow-}\mathsf{Top})(\boldsymbol{X},\boldsymbol{Y})$.*

Proof. Let $\Phi: X \times I \to Y$ be an approaching homotopy, which connects ϕ to ϕ'. Choose an increasing function $F: \mathbb{N} \to \mathbb{N}$ such that

$$\Phi(X_{F(m)} \times I \times [F(m), \infty)) \subseteq Y_m, \ m \in \mathbb{N}. \tag{26}$$

By (6),

$$\phi(X_{F(m)} \times [F(m), \infty)) \subseteq Y_m, \ \phi'(X_{F(m)} \times [F(m), \infty)) \subseteq Y_m. \tag{27}$$

By Lemma 9.6, one can assume that $\boldsymbol{f} = (F, f_m, f_{mm'})$, $\boldsymbol{f}' = (F, f'_m, f'_{mm'})$. We now define $F_m: X_{F(m)} \times I \to Y_m$ and $F_{mm'}: X_{F(m')} \times I \times I \to Y_m$, by putting

$$F_m(x,s) = \Phi(x, s, F(m)), \tag{28}$$

$$F_{mm'}(x,s,t) = \Phi(x, s, (1-t)F(m) + tF(m')). \tag{29}$$

It is now easy to verify that $(F, F_m, F_{mm'})$ is a coherent homotopy of height 1, which connects $\boldsymbol{f}$ to $\boldsymbol{f}'$. $\square$

Assigning to $[\phi] \in \mathsf{QSh}(X,Y)$ the strong shape morphism of height 1, given by the triple $(\boldsymbol{p}, \boldsymbol{q}, T[\phi])$, one obtains a function T_{XY}: $\mathsf{QSh}(X,Y) \to \mathrm{SSh}^{(1)}(\mathsf{Q})\,(X,Y)$. The next lemma shows that the functions T_{XY} define a functor T: $\mathsf{QSh} \to \mathrm{SSh}^{(1)}(\mathsf{Q})$, which preserves objects, i.e., $T(X) = X$.

LEMMA 9.8. *If $\phi: X \to Y$ and $\psi: Y \to Z$ are approaching mappings, then $T([\psi][\phi]) = T[\psi]T[\phi]$. Moreover, $T[\iota] = \mathrm{id}$.*

In the proof of Lemma 9.8, we will use the following lemma.

LEMMA 9.9. *Let $A: X \times I \to Y$ and $B: Y \times I \to Z$ be homotopies connecting mappings $a, a': X \to Y$ and $b, b': Y \to Z$, respectively. Let $C, C': X \times I \to Z$ be homotopies given by*

$$C(x,t) = B(A(x,t), t), \tag{30}$$

$$C'(x,t) = \begin{cases} b(A(x,2t)), & 0 \le t \le \frac{1}{2}, \\ B(a'(x), 2t-1), & \frac{1}{2} \le t \le 1. \end{cases} \tag{31}$$

Then there exists a homotopy $G: X \times I \times I \to Z$ such that

$$G(x,s,0) = b(a(x)), \tag{32}$$

$$G(x,s,1) = b'(a'(x)), \tag{33}$$

$$G(x,0,t) = C(x,t), \tag{34}$$

$$G(x,1,t) = C'(x,t). \tag{35}$$

Moreover, if A and B are constant homotopies, i.e., $A(x,t) = a(x) = a'(x)$ and $B(y,t) = b(y) = b'(y)$, for all $t \in I$, then G is also a constant homotopy, given by $G(x,s,t) = b(a(x)) = b'(a'(x))$, for all $(s,t) \in I \times I$.

Proof. We first define two homotopies $E: X\times I\times I\to Y$ and $F: Y\times I\times I\to Y$, by putting

$$E(x,s,t)=\begin{cases} A(x,\frac{2t}{2-s}), & s\le 2(1-t),\\ a'(x), & 2(1-t)\le s,\end{cases} \tag{36}$$

$$F(y,s,t)=\begin{cases} b(y), & 2t\le s,\\ B(y,\frac{2t-s}{2-s}), & s\le 2t.\end{cases} \tag{37}$$

Note that

$$E(x,s,0)=a(x),\ E(x,s,1)=a'(x), \tag{38}$$

$$E(x,0,t)=A(x,t). \tag{39}$$

$$E(x,1,t)=\begin{cases} A(x,2t), & 0\le t\le \frac12,\\ a'(x), & \frac12\le t\le 1.\end{cases} \tag{40}$$

$$F(y,s,0)=b(y),\ F(y,s,1)=b'(y), \tag{41}$$

$$F(y,0,t)=B(y,t). \tag{42}$$

$$F(y,1,t)=\begin{cases} b(y), & 0\le t\le \frac12,\\ B(y,2t-1), & \frac12\le t\le 1.\end{cases} \tag{43}$$

A homotopy G with desired properties is now given by

$$G(x,s,t)=F(E(x,s,t),s,t).\ \square. \tag{44}$$

Proof of Lemma 9.8. Let $\phi: X\to Y$ and $\psi: Y\to Z$ be approaching mappings and let $\chi=\psi\phi: X\to Z$ be their composition, defined by (2). Let $\boldsymbol{f}=(f,f_m,f_{mm'})$ and $\boldsymbol{g}=(g,g_m,g_{mm'})$ be coherent mappings of height 1, associated with ϕ and ψ, respectively. Then $h=fg:\mathbb{N}\to\mathbb{N}$ is an increasing function and $h\ge f,g$. Since $\chi(u,t)=\psi(\phi(u,t),t)$, it follows that

$$\chi(X_{h(m)}\times[h(m),\infty))\subseteq Z_m. \tag{45}$$

Putting

$$h_m(x)=\psi(\phi(x,h(m)),h(m)), \tag{46}$$

$$h_{mm'}(x,t)=\psi(\phi(x,(1-t)h(m)+th(m')),(1-t)h(m)+th(m')), \tag{47}$$

one obtains a coherent mapping $\boldsymbol{h}=(h,h_m,h_{mm'})$ of height 1, associated with χ. Since $T[\phi]$ is represented by $\boldsymbol{f}$ and $T[\psi]$ is represented by $\boldsymbol{g}$, it follows that $T[\psi]T[\phi]$ is represented by the composition $\boldsymbol{h}'=(h',h'_m,h'_{mm'})=\boldsymbol{gf}$. According to (1.3.1), $h'=fg=h$,

$$h'_m(x)=\psi(\phi(x,h(m)),g(m)), \tag{48}$$

$$h'_{mm'}(x,t)=$$

$$\begin{cases} \psi(\phi(x,(1-2t)h(m)+2th(m')),g(m)), & 0\le t\le\frac12,\\ \psi(\phi(x,h(m')),2(1-t)g(m)+(2t-1)g(m')), & \frac12\le t\le 1.\end{cases} \tag{49}$$

To complete the proof, we need to show that $\boldsymbol{h} \simeq \boldsymbol{h}'$. We will do this by considering a third coherent mapping $\boldsymbol{h}'' = (h'', h''_m, h''_{mm'})$ of height 1 and by showing that $\boldsymbol{h} \simeq \boldsymbol{h}'' \simeq \boldsymbol{h}'$. We put $h'' = h$,

$$h''_m(x) = h'_m(x), \tag{50}$$

$$h''_{mm'}(x,t) = \psi(\phi(x,(1-t)h(m) + th(m')), (1-t)g(m) + tg(m')). \tag{51}$$

In order to prove that $\boldsymbol{h} \simeq \boldsymbol{h}''$, we define homotopies $H_m: X_{h(m)} \times I \to Z_m$ and $H_{mm'}: X_{h(m')} \times I \times I \to Z_m$, by putting

$$H_m(x,t) = \psi(\phi(x,h(m)), (1-t)h(m) + tg(m)), \tag{52}$$

$$H_{mm'}(x,s,t) = \psi(\phi(x,(1-t)h(m) + th(m')), u_{mm'}(s,t)), \tag{53}$$

where $u_{mm'}: I \times I \to [g(m), \infty)$ is the mapping given by

$$u_{mm'}(s,t) = (1-s)((1-t)h(m) + th(m')) + s((1-t)g(m) + tg(m')). \tag{54}$$

It is readily seen that

$$H_{mm'}(x,s,0) = H_m(x,s), \tag{55}$$

$$H_{mm'}(x,s,1) = H_{m'}(x,s), \tag{56}$$

$$H_{mm'}(x,0,t) = h_{mm'}(x,t), \tag{57}$$

$$H_{mm'}(x,1,t) = h''_{mm'}(x), \tag{58}$$

$$H_{mm}(x,s,t) = H_m(x,s), \tag{59}$$

which shows that $\boldsymbol{H} = (h, H_m, H_{mm'})$ is a coherent homotopy of height 1. Moreover,

$$H_m(x,0) = h_m(x), \tag{60}$$

$$H_m(x,1) = h''_m(x,t), \tag{61}$$

which shows that $\boldsymbol{H}$ connects $\boldsymbol{h}$ with $\boldsymbol{h}''$ and thus, $\boldsymbol{h} \simeq \boldsymbol{h}''$.

In order to find a coherent homotopy $\boldsymbol{G} = (G, G_m, G_{mm'})$ of height 1, which connects $\boldsymbol{h}''$ to $\boldsymbol{h}'$, put $G = h$, and

$$G_m(x,t) = h''_m(x) = h'_m(x). \tag{62}$$

Moreover, consider mappings $A: X \times I \to Y$ and $B: Y \times I \to Z$, defined by

$$A(x,t) = \phi(x,(1-t)h(m) + th(m')), \tag{63}$$

$$B(y,t) = \psi(y,(1-t)g(m) + tg(m')). \tag{64}$$

Note that

$$a(x) = A(x,0) = \phi(x,h(m)),\ \ a'(x) = A(x,1) = \phi(x,h(m')), \tag{65}$$

$$b(y) = B(y,0) = \psi(y,g(m)),\ \ b'(y) = B(y,1) = \psi(y,g(m')). \tag{66}$$

Therefore, (30) and (51) imply

$$C(x,t) = h''_{mm'}(x,t). \tag{67}$$

while (31) and (49) imply

$$C'(x,t) = h'_{mm'}(x,t). \tag{68}$$

Consequently, Lemma 9.9 yields a homotopy $G_{mm'}: X_{h(m')} \times I \times I \to Z_m$ such that

$$G_{mm'}(x,s,0) = \psi(\phi(x,h(m)),g(m)) = G_m(x,s), \tag{69}$$

$$G_{mm'}(x,s,1) = \psi(\phi(x,h(m')),g(m')) = G_{m'}(x,s), \tag{70}$$

$$G_{mm'}(x,0,t) = h''_{mm'}(x,t), \tag{71}$$

$$G_{mm'}(x,1,t) = h'_{mm'}(x,t). \tag{72}$$

Finally, if $m = m'$, then A and B are constant homotopies and thus, G_{mm} is also a constant homotopy, given by

$$G_{mm}(x,s,t) = \psi(\phi(x,h(m)),g(m)) = G_m(x,s). \tag{73}$$

Formulae (69)–(73) prove that $\boldsymbol{G}$ is a coherent homotopy of height 1, which connects $\boldsymbol{h}''$ to $\boldsymbol{h}'$.

A coherent mapping $\boldsymbol{i} = (i, i_m, i_{mm'})$, associated with the approaching mapping $\iota: X \to X$, is given by $i(m) = m$, $i_m(x) = x$, $i_{mm'}(x,t) = x$. Therefore, $T[\iota] = \mathrm{id}$. □

LEMMA 9.10. *If $\boldsymbol{f} = (f, f_m, f_{mm'}): \boldsymbol{X} \to \boldsymbol{Y}$ is a coherent mapping of height 1, then there exists an approaching mapping $\phi: X \to Y$ such that $T[\phi] = [\boldsymbol{f}]$.*

Proof. Without loss of generality we can assume that $f: \mathbb{N} \to \mathbb{N}$ is a strictly increasing function. To simplify formulae which follow, we introduce some notation (see Fig. 9.1).

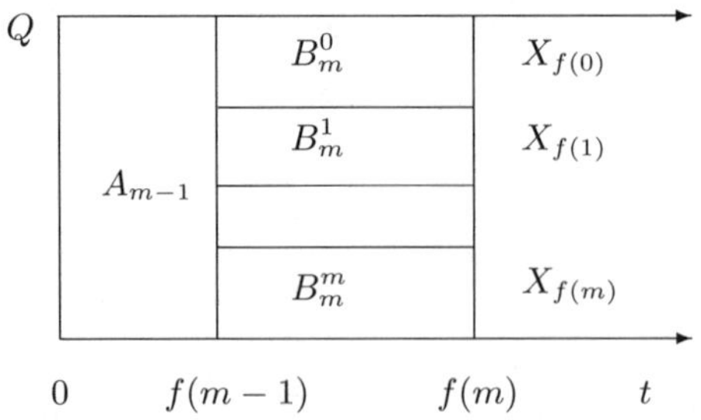

Fig. 9.1.

$$A_m = Q \times [0, f(m)],\ m \geq 1, \tag{74}$$

$$B_m^k = X_{f(k)} \times [f(m-1), f(m)],\ 0 \leq k \leq m. \tag{75}$$

Here $f(0) = 0, X_0 = Y_0 = Q$. Note that

$$B_m^m \subseteq B_m^{m-1} \subseteq \ldots \subseteq B_m^0 = Q \times [f(m-1), f(m)], \tag{76}$$

$$A_m = A_{m-1} \cup B_m^0. \tag{77}$$

We will first construct mappings $\phi_m \colon A_m \to Q$, $m \geq 1$, such that

$$\phi_m | A_{m-1} = \phi_{m-1},\ m \geq 2, \tag{78}$$

$$\phi_m(B_m^k) \subseteq Y_k,\ 0 \leq k \leq m, \tag{79}$$

$$\phi_m(x, (1-t)f(m-1) + tf(m)) = f_{m-1,m}(x,t),\ (x,t) \in X_{f(m)} \times I. \tag{80}$$

Here $f_{01}(x,s) = f_1(x)$, for $(x,s) \in X_{f(1)} \times I$. The desired approaching mapping $\phi \colon Q \times [0, \infty) \to Q$ is then given by

$$\phi(x,t) = \phi_m(x,t),\ (x,t) \in A_m. \tag{81}$$

Clearly, (78) implies that ϕ is well defined. To see that ϕ is indeed an approaching mapping $\phi \colon X \to Y$, it suffices to verify (6). Consider any point $(x,t) \in X_{f(k)} \times [f(k), \infty)$. Since f is strictly increasing, there is a largest integer $m \geq 1$ such that $f(m-1) \leq t$. Clearly, $t < f(m)$, $k \leq m-1$ and $(x,t) \in B_m^k \subseteq A_m$. Consequently, (81) and (79) imply $\phi(x,t) = \phi_m(x,t) \in Y_k$, which is the desired condition.

Now note that (80), (81) and (9) imply

$$\phi(x, f(m)) = \phi_m(x, f(m)) = f_{m-1,m}(x, 1) = f_m(x),\ x \in X_{f(m)}. \tag{82}$$

Therefore, one obtains a coherent mapping $\boldsymbol{f}' = (f', f'_m, f'_{mm'})$ of height 1, associated with ϕ, by putting $f' = f, f'_m = f_m$ and

$$f'_{mm'}(x,t) = \phi(x, (1-t)f(m) + tf(m')),\ (x,t) \in X_{f(m')} \times I. \tag{83}$$

Note that $f'_{mm}(x,t) = \phi(x, f(m)) = f_m(x) = f_{mm}(x,t)$. Moreover, comparing (83) with (8), one concludes that

$$f'_{m-1,m}(x,t) = f_{m-1,m}(x,t). \tag{84}$$

However, for $m < m'$,

$$f_{mm'} \simeq f_{m,m+1} * \ldots * f_{m'-1,m'} (\mathrm{rel}\, \{0,1\}), \tag{85}$$

$$f'_{mm'} \simeq f'_{m,m+1} * \ldots * f'_{m'-1,m'} (\mathrm{rel}\, \{0,1\}). \tag{86}$$

This and (84) show that

$$f_{mm'} \simeq f_{mm'} (\mathrm{rel}\, \{0,1\}). \tag{87}$$

Consequently, $[\boldsymbol{f}] = [\boldsymbol{f}']$.

To complete the proof of Lemma 9.10, we will now construct the mappings ϕ_m, by induction on m. We begin the induction by defining $\phi_1 \colon A_1 \to Q$, putting $\phi_1(x,t) = f_1(x)$. Note that, for $(x,t) \in B_1^1 = X_{f(1)} \times [0, f(1)]$, $f_1(x) \in Y_1$ and thus, (79) is fulfilled. Clearly, (80) is also satisfied. Now assume that we have already defined mappings $\phi_1, \dots, \phi_{m-1}$, $m \geq 2$, which satisfy all our requirements. In order to define $\phi_m \colon A_m \to Q$, we will define (by induction on i) mappings $\phi_m^i \colon A_{m-1} \cup B_m^{m-i} \to Q$, $0 \leq i \leq m$, which have the following properties:

$$\phi_m^i | A_{m-1} = \phi_{m-1},\ 0 \leq i \leq m, \tag{88}$$

$$\phi_m^0(x, (1-t)f(m-1) + tf(m)) = f_{m-1,m}(x,t),\ (x,t) \in X_{f(m)} \times I, \tag{89}$$

$$\phi_m^i | B_m^{m-i+1} = \phi_m^{i-1} | B_m^{m-i+1},\ 0 \leq i \leq m, \tag{90}$$

$$\phi_m^i(B_m^{m-i}) \subseteq Y_{m-i},\ 0 \leq i \leq m. \tag{91}$$

Once the mappings ϕ_m^i, $0 \leq i \leq m$, are construcetd, we define $\phi_m \colon A_m \to Q$, by putting $\phi_m = \phi_m^m$. Property (78) follows from (88). If $(x,t) \in B_m^k$, then (90) and (91) imply

$$\phi_m(x,t) = \phi_m^m(x,t) = \ldots = \phi_m^{m-k}(x,t) \in Y_k, \tag{92}$$

so that (79) holds. Finally, formula (80) is a consequence of (90) and (89), because

$$\begin{aligned} &\phi_m^m(x, (1-t)f(m-1) + tf(m)) = \ldots = \\ &\phi_m^0(x, (1-t)f(m-1) + tf(m)) = f_{m-1,m}(x,t), \end{aligned} \tag{93}$$

for $(x,t) \in X_{f(m)} \times I$.

Let us now define the mappings ϕ_m^i, beginning with ϕ_m^0. We define $\phi_m^0 | A_{m-1}$, using (88), and we define $\phi_m^0 | B_m^m$, using (89). To see that ϕ_m^0 is well defined, notice that $A_{m-1} \cap B_m^m = X_{f(m)} \times \{f(m-1)\}$. Then notice that, for $x \in X_{f(m)}$, (88) and (80) yield $\phi_m^0(x, f(m-1)) = \phi_{m-1}(x, f(m-1)) = f_{m-2,m-1}(x,1) = f_{m-1}(x)$. However, the same value is obtained, using (89), because $f_{m-1,m}(x,0) = f_{m-1}(x)$. Now assume that we have already defined the mappings ϕ_m^j, for $0 \leq j < i \leq m$. We first define ϕ_m^i on $A_{m-1} \cup B_{m-i+1}^m$, using (88) on the first summand and (90) on the second summand. By (88) (for $i-1$), $\phi_m^{i-1} | A_{m-1} = \phi_{m-1}$, which implies that ϕ_m^i is well defined on $A_{m-1} \cup B_{m-i+1}^m$. Moreover, by (91), $\phi_m^{i-1}(B_{m-i+1}^m) \subseteq Y_{m-i+1} \subseteq Y_{m-i}$. We now extend ϕ_m^{i-1} further to a mapping $\phi_m^{i-1} : A_{m-1} \cup B_{m-i}^m \to Q$ such that $\phi_m^i(B_m^{m-i}) \subseteq Y_{m-i}$, i.e., (91) holds. This can be achieved using the homotopy extension theorem. Indeed, $B_{m-i}^m = X_{f(m-i)} \times [f(m-1), f(m)]$ and ϕ_m^i is already defined on $B_{m-i+1}^m = X_{f(m-i)} \times [f(m-1), f(m)]$ and also on $X_{f(m-i)} \times \{f(m-1)\}$, because the latter set belongs to $B_{m-1}^{m-i} \subseteq B_{m-1}^0 \subseteq A_{m-1}$. Moreover, by (91), $\phi_m^i(B_{m-i+1}^m) = \phi_m^{i-1}(B_{m-i+1}^m) \subseteq Y_{m-i+1} \subseteq Y_{m-i}$. □

To complete the proof of Theorem 9.5, we still need the next lemma.

LEMMA 9.11. *Let $\phi, \phi' : X \to Y$ be two approaching mappings. If $T[\phi] = T[\phi']$, then these approaching mappings are homotopic, i.e., $[\phi] = [\phi']$.*

Proof. Let $\boldsymbol{f} = (f, f_m, f_{mm'}) : \boldsymbol{X} \to \boldsymbol{Y}$ and $\boldsymbol{f}' = (f', f'_m, f'_{mm'}) : \boldsymbol{X} \to \boldsymbol{Y}$ be coherent mappings of height 1, associated with ϕ and ϕ', respectively. By assumption, there exists a coherent homotopy $\boldsymbol{F} = (F, F_m, F_{mm'}) : \boldsymbol{X} \times I \to \boldsymbol{Y}$ of height 1, which connects $\boldsymbol{f}$ to $\boldsymbol{f}'$. Therefore, $F \geq f, f'$,

$$F_m(x,0) = f_m(x),\ F_m(x,1) = f'_m(x),\ x \in X_{F(m)}, \tag{94}$$

$$F_{mm'}(x,0,t) = f_{mm'}(x,t),\ F_{mm'}(x,1,t) = f'_{mm'}(x,t),\ (x,t) \in X_{F(m')}. \tag{95}$$

There is no loss of generality in assuming that F is strictly increasing. Similarly to the arguments used in the proof of Lemma 9.10, we consider sets

$$A_m = Q \times I \times [0, F(m)],\ m \geq 1, \tag{96}$$

$$B_m^k = X_{F(k)} \times I \times [f(m-1), f(m)],\ 0 \leq k \leq m, \tag{97}$$

$(F(0) = 0)$ and define mappings $\Phi_m : A_m \to Q$, $m \in \mathbb{N}$, such that

$$\Phi_m | A_{m-1} = \Phi_{m-1},\ m \geq 2, \tag{98}$$

$$\Phi_m(B_m^k) \subseteq Y_k,\ 0 \leq k \leq m, \tag{99}$$

$$\Phi_m(x, s, (1-t)F(m-1) + tF(m)) = F_{m-1,m}(x,s,t), \tag{100}$$

for $(x,s,t) \in X_{F(m)} \times I \times I$; here $F_{01}(x,s,t) = F_1(x,s)$, for $(x,s) \in X_{F(1)} \times I$. Moreover, we require that

$$\Phi_m(x,0,t) = \phi(x,t),\ (x,t) \in Q \times [0,\infty), \tag{101}$$

$$\Phi_m(x,1,t) = \phi'(x,t),\ (x,t) \in Q \times [0,\infty). \tag{102}$$

We then define a mapping $\Phi : Q \times I \times [0,\infty) \to Q$, by putting

$$\Phi(x,s,t) = \Phi_m(x,s,t),\ (x,s,t) \in A_m. \tag{103}$$

One easily verifies that Φ is a well-defined approaching homotopy, which connects ϕ and ϕ', because

$$\Phi(x,0,t) = \phi(x,t),\ \Phi(x,1,t) = \phi'(x,t),\ (x,t) \in Q \times [0,\infty). \tag{104}$$

Consequently, $\phi \simeq \phi'$.

The mappings Φ_m are defined, by induction on m. For $(x,s,t) \in A_1$, put $\Phi_1(x,s,t) = F_1(x,s) \in Y_1 \subseteq Q$. Clearly, (99) and (100) are satisfied. Now assume that we have already defined mappings $\Phi_1, \ldots, \Phi_{m-1}$, which satisfy our requirements. In order to define $\Phi_m : A_m \to Q$, one defines, by induction on i, using the homotopy extension theorem, mappings $\Phi_m^i : A_{m-1} \cup B_m^{m-i} \to Q$, $0 \leq i \leq m$, which have the following properties:

$$\Phi_m^i | A_{m-1} = \Phi_{m-1},\ 0 \leq i \leq m, \tag{105}$$

$$\begin{gathered}\Phi_m^0(x, s, (1-t)F(m-1) + tF(m)) = F_{m-1,m}(x,s,t),\\ (x,s,t) \in X_{F(m)} \times I \times I,\end{gathered} \tag{106}$$

$$\Phi_m^i | B_m^{m-i+1} = \Phi_m^{i-1} | B_m^{m-i+1},\ 0 \leq i \leq m, \tag{107}$$

$$\Phi_m^i(B_m^{m-i}) \subseteq Y_{m-i},\ 0 \leq i \leq m. \tag{108}$$

$$\Phi_m^i(x,0,t) = \phi(x,t),\ \Phi_m^i(x,1,t) = \phi'(x,t),\ \text{on } A_m \cup B_m^{m-i}. \tag{109}$$

Then $\Phi_m = \Phi_m^m : Q \times I \times I \to Q$ has all the desired properties. □

9.2 Complement theorems

A mapping $f: X \to Y$ between topological spaces is called *proper* (or *compact*) provided $f^{-1}(B) \subseteq X$ is compact, whenever $B \subseteq Y$ is compact. Clearly, composition of proper mappings and the identity mapping are proper mappings. Consequently, topological spaces and proper mappings form a category, denoted by P(Top). Two proper mappings $f, f': X \to Y$ are *properly homotopic*, $f \simeq_P f'$, provided there exists a homotopy $F: X \times I \to Y$, which connects f and f', and F is a proper mapping. Clearly, $\simeq_P$ is an equivalence relation. We denote the equivalence class of f by $[f]_P$. If $[f]_P = [f']_P: X \to Y$ and $[g]_P = [g']_P: Y \to Z$, then $[gf]_P = [g'f']_P$. Therefore, one can define composition of proper homotopy classes, by composing representatives, $[f]_P[g]_P = [gf]_P$. The *proper homotopy category* thus obtained, is denoted by PH(Top).

REMARK 9.12. If Y is a Hausdorff space and every compact set $B \subseteq Y$ admits a compact set $A \subseteq X$ such that

$$f(X \backslash A) \subseteq Y \backslash B, \tag{1}$$

then $f: X \to Y$ is a proper mapping. Indeed, B is closed in Y and therefore, $f^{-1}(B) \subseteq A$ is closed in X, hence, also in A. Consequently, $f^{-1}(B)$ is compact. Conversely, if f is proper, for every compact set $B \subseteq Y$, the set $A = f^{-1}(B)$ satisfies (1).

REMARK 9.13. It is easy to see that, for metric spaces a mapping $f: X \to Y$ is proper if and only if it is *perfect*, i.e., it is a closed mapping with compact fibers $f^{-1}(y)$, $y \in Y$. For sufficiency of this condition (even for Y arbitrary and X Hausdorff) see Theorem 3.7.2 of (Engelking 1977). To prove necessity, consider a proper mapping f and a closed set $A \subseteq X$. To verify that $f(A)$ is closed, one needs to show that, for every sequence of points $a_n \in A$ with $\lim f(a_n) = y \in Y$, the point y belongs to $f(A)$. Since $B = \{y, f(a_1), f(a_2), \ldots\}$ is a compact set, the sets $f^{-1}(B)$ and $A \cap f^{-1}(B)$ are also compact. However, (a_n) is a sequence from $A \cap f^{-1}(B)$ and thus, there exists a subsequence (a_{n_k}), which converges towards some point $a \in A$. Consequently, the sequence $(f(a_{n_k}))$ converges towards $f(a) = y$, hence, $y \in f(A)$.

In order to state the complement theorems, we need the notion of a Z-*set*, first introduced by R.D. Anderson in his study of infinite-dimensional topology (Anderson 1967). In a metric space M, a Z-set of M is a closed set $X \subseteq M$, which has the property that, for every $\varepsilon > 0$ and every mapping $g: Q \to M$ of the Hilbert cube Q, there exists a mapping $f: Q \to M$ such that the distance $d(f, g) < \varepsilon$ and $f(Q) \subseteq M \backslash X$. In particular, a closed subset X of the Hilbert cube Q is a Z-set of Q provided, for every $\varepsilon > 0$, there exists a mapping $f: Q \to Q \backslash X$, whose distance from the identity mapping id: $Q \to Q$ is $< \varepsilon$. Clearly, every compact set contained in the end-slice

$\{x = (x_n) \in Q: x_1 = 1\}$ of Q is a Z-set of Q. Moreover, compacta contained in the pseudointerior $s = \prod_i (0,1)$ of Q or in the pseudoboundary $Q \backslash s$ of Q are Z-sets of Q. A very important result in the theory of Z-sets is the homeomorphism extension theorem. It asserts that every homeomorphism $h: X \to X'$ between two Z- sets of Q can be extended to a homeomorphism $Q \to Q$. Clearly, complements $Q \backslash X$ of Z-sets X of Q are locally compact ANR's. They are even AR's, because they are contractible. This is obvious if X is contained in the end-slice. The general case follows, by applying the homeomorphism extension theorem. For more information on Z-sets see (Chapman 1976) and (Mill 1989).

LEMMA 9.14. *If X is a Z-set of the Hilbert cube Q, then there exists a homotopy $H_X: Q \times I \to Q$ such that*

$$H_X(x, 0) = x, \ x \in Q, \tag{2}$$

$$H_X(x, s) \in Q \backslash X, \ x \in Q, \ s > 0. \tag{3}$$

Proof. In the special case, when X belongs to the end-slice of Q, the assertion is obvious. The general case is reduced to the special case, by applying the homeomorphism extension theorem. □

The main purpose of this subsection is to prove the following *complement theorem.*

THEOREM 9.15. *There exists an isomorphism T between the proper homotopy category of complements $Q \backslash X$ of Z-sets $X \subseteq Q$ and the strong shape category of Z-sets $X \subseteq Q$. For every Z-set X, $T(Q \backslash X) = X$.*

Due to Theorem 9.3, it suffices to establish an isomorphism T between the proper homotopy category of complements of Z-sets of Q and the Quigley category, restricted to Z-sets of Q. By definition, $T(Q \backslash X) = X$. In order to define T on morphisms, we associate with every Z-set $X \subseteq Q$ a mapping $\alpha_X: Q \to I = [0,1]$ such that $(\alpha_X)^{-1}(0) = X$. We then define a mapping $j_X: Q \backslash X \to Q \times [0, \infty)$, by putting

$$j_X(x) = (x, 1/\alpha_X(x)), \ x \in Q \backslash X. \tag{4}$$

Also put

$$K_X = j_X(Q \backslash X) \subseteq (Q \backslash X) \times [0, \infty). \tag{5}$$

$j_X: Q \backslash X \to K_X$ is a homeomorphism with inverse $j_X^{-1} = \pi | K_X: K_X \to Q \backslash X$, where $\pi: (Q \backslash X) \times [0, \infty) \to Q \backslash X$ denotes the first projection. Observe that K_X is a closed subset of $Q \times [0, \infty)$. Therefore, $(A \times [0, \infty)) \cap K_X$ is compact, whenever $A \subseteq Q$ is compact.

LEMMA 9.16. *For $x \in Q \backslash X$, let*

$$r_X(x,t) = \begin{cases} (j_X H_X)(x, u_X(x,t)), & 0 \le t \le 1/\alpha_X(x), \\ j_X(x), & 1/\alpha_X(x) \le t, \end{cases} \tag{6}$$

where

$$u_X(x,t) = \frac{1 - t\alpha_X(x)}{1+t}, \ x \in Q\backslash X, \ 0 \leq t \leq 1/\alpha_X(x). \tag{7}$$

For $x \in X$, let

$$r_X(x,t) = j_X H_X(x, 1/(1+t)). \tag{8}$$

Then $r_X: Q \times [0,\infty) \to K_X$ is a well-defined retraction.

Proof. Note that $u_X(x,t) = 0$, for $t = 1/\alpha_X(x)$. Furthermore, when $x \in Q\backslash X$ tends to X, $\alpha_X(x)$ tends to 0 and therefore, $u_X(x,t)$ tends to $1/(1+t)$. □

Proof of Theorem 9.15. Let $X, Y \subseteq Q$ be Z-sets and let $f: Q\backslash X \to Q\backslash Y$ be a proper mapping. We associate with f a mapping $\phi: Q \times [0,\infty) \to Q$, given by

$$\phi = i_Y f j_X^{-1} r_X, \tag{9}$$

where $i_Y: Q\backslash Y \to Q$ denotes inclusion. First note that ϕ is an approaching mapping $\phi: X \to Y$ (in the sense of 9.1). Indeed, if V is an open neighborhood of Y in Q, then $Q\backslash Y$ is a compact set. Therefore, $f^{-1}(Q\backslash V)$ is a compact subset of $Q\backslash X$. Consequently, there exists an $\alpha > 0$ such that

$$j_X(f^{-1}(Q\backslash V)) \subseteq Q \times [0, \alpha]. \tag{10}$$

Note that, for $x \in X$, $1/\alpha_X(H_X(x, 1/(1+t))$ tends to 0, when t tends to ∞. Consequently, by (4) and (8), the second coordinate of $r_X(x,t)$ exceeds α, for t sufficiently large. Clearly, this also holds for all x from a sufficiently small neighborhood U of X. Therefore, there exists a $t \geq 0$ so large that

$$r_X(U \times [t,\infty)) \subseteq Q \times (\alpha, \infty). \tag{11}$$

Comparing (10) and (11), one concludes that

$$r_X(U \times [t,\infty)) \cap j_X(f^{-1}(Q\backslash V)) = \emptyset. \tag{12}$$

Since j_X^{-1} is injective, it follows that the sets $(j_X^{-1} r_X)(U \times [t,\infty))$ and $f^{-1}(Q\backslash V)$ are disjoint and thus,

$$\phi(U \times [t,\infty)) \subseteq V, \tag{13}$$

as desired. We now put $T(f) = \phi$.

If $f': Q\backslash X \to Q\backslash Y$ is another proper mapping and $F: (Q\backslash X) \times I \to Q\backslash Y$ is a proper homotopy, which connects f and f', then the approaching mappings $\phi = T(f)$ and $\phi' = T(f')$ are homotopic. Indeed, let

$$\Phi(x,s,t) = i_Y F(j_X^{-1} r_X(x,t), s), \ (x,s,t) \in Q \times I \times [0,\infty). \tag{14}$$

An argument as the one above shows that Φ is an approaching homotopy. Moreover, $\Phi(x,0,t) = (i_Y f j_X^{-1} r_X)(x,t) = \phi(x,t)$ and $\Phi(x,1,t) = \phi'(x,t)$, which shows that Φ connects ϕ and ϕ'. This enables us to define T on proper homotopy classes $[f]$, by putting $T[f] = [T(f)]$.

The function T of proper homotopy classes is injective, i.e., $T(f) \simeq T(f')$ implies $f \simeq_p f'$. Indeed, assume that $\Phi: Q \times I \times [0, \infty) \to Q$ is an approaching homotopy, which connects $\phi = T(f)$ and $\phi' = T(f')$. Consider the mapping $F: K_X \times I \to Q$, defined by

$$F(x, t, s) = H_Y(\Phi(x, s, t), s(1-s)/(1+t)). \tag{15}$$

Note that, for $s \neq 0, 1$, (3) implies $F(x, t, s) \in Q \backslash Y$, while

$$F(x, t, 0) = \Phi(x, 0, t) = \phi(x, t) = f j_X^{-1} r_X(x, t) \in Q \backslash Y. \tag{16}$$

$$F(x, t, 1) = \Phi(x, 1, t) = \phi'(x, t) = f' j_X^{-1} r_X(x, t) \in Q \backslash Y. \tag{17}$$

Therefore, F maps $K_X \times I$ to $Q \backslash Y$. This mapping is proper. Indeed, if $B \subseteq Q \backslash Y$ is compact, then $V = Q \backslash B$ is an open neighborhood of Y. For $(y, s) \in Y \times I$ and t sufficiently large, $H_Y(y, s(1-s)/(1+t))$ is arbitrarily close to $H_Y(y, 0) = y$. Clearly, the same holds for y in a sufficiently small neighborhood W of Y. Since Φ is an approaching homotopy, there exists an open neighborhood U of X such that $\Phi(x, t, s) \in W$, for $x \in U$, $s \in I$ and t large enough. Hence, there exists an $\alpha > 0$, such that

$$F((U \times [\alpha, \infty) \times I) \cap (K_X \times I)) \subseteq V. \tag{18}$$

Therefore, the set $F^{-1}(B)$ must be contained in the union of the compact sets $\overline{U} \times [0, \alpha] \times I$ and $((Q \backslash U) \times [0, \infty)) \cap (K_X \times I)$. Finally, we define a proper homotopy $F': (Q \backslash U) \times I \to Q \backslash Y$, by putting

$$F' = F(j_X \times 1). \tag{19}$$

(16) and (17) show that F' connects f and f', because $F'(x, 0) = F(j_X(x), 0) = f j_X^{-1} r_X(j_X(x)) = f(x)$ and $F'(x, 1) = F(j_X(x), 1) = f' j_X^{-1} r_X(j_X(x)) = f'(x)$.

To prove that the function T of proper homotopy classes is surjective, consider an approaching mapping $\phi: X \to Y$. Define a mapping $\Phi': Q \times I \times [0, \infty) \to Q$, by putting

$$\Phi'(x, s, t) = H_Y(\phi(x, t), s/(1+t)). \tag{20}$$

It is readily seen that Φ' is an approaching homotopy $X \times I \to Y$. Indeed, for a point $y \in Y$ and t sufficiently large, $H_Y(y, s/(1+t))$ is arbitrarily close to $H_Y(y, 0) = y$. Therefore, for a neighborhood V of Y, there exist an $\alpha > 0$ and a neighborhood W of Y such that $H_Y(w, s/(1+t)) \in V$, for $t \geq \alpha$ and $w \in W$. Since $\phi: X \to Y$ is an approaching mapping, there exist a neighborhood U of X and an $\alpha' \geq \alpha$, such that $\phi(u, t) \in W$, for $u \in U$ and $t \geq \alpha'$. Consequently,

$$\Phi'(U \times I \times [\alpha', \infty)) \subseteq V, \tag{21}$$

as required. Note that the mapping $\phi': Q \times [0, \infty) \to Q$, defined by

$$\phi'(x, t) = \Phi'(x, 1, t) = H_Y(\phi(x, t), 1/(1+t)), \tag{22}$$

is an approaching mapping $\phi'\colon X \to Y$, which is homotopic to ϕ, because

$$\Phi'(x,0,t) = H_Y(\phi(x,t),0) = \phi(x,t). \tag{23}$$

Consequently, to prove surjectivity, it suffices to exhibit a proper mapping $f\colon Q\backslash X \to Q\backslash Y$ such that $T(f)$ is homotopic to ϕ'.

First notice that $\phi'|K_X\colon K_X \to Q\backslash Y$ is a proper mapping. Indeed, if $B \subseteq Q\backslash Y$ is a compact set, then $V = Q\backslash B$ is an open neighborhood of Y. Since ϕ' is an approaching mapping, there exist an open neighborhood U of X and an $\alpha > 0$, such that

$$\phi'(U \times (\alpha,\infty)) \subseteq V. \tag{24}$$

Consequently, $(\phi'|K_X)^{-1}(B)$ is contained in the union of the compact sets $(\overline{U} \times [0,\alpha]) \cap K_X$ and $((Q\backslash U) \times [0,\infty)) \cap K_X$.

We now define $f\colon Q\backslash X \to Q\backslash Y$, by putting

$$f(u) = \phi'(j_X(u)),\ \ u \in Q\backslash X. \tag{25}$$

Since $j_X\colon Q\backslash X \to K_X$ is a homeomorphism, it follows that f is also a proper mapping. The corresponding approaching mapping $\phi'' = T(f)$ is given by

$$\phi'' = i_Y f j_X^{-1} r_X = \phi' j_X j_X^{-1} r_X = \phi' r_X. \tag{26}$$

It remains to show that $\phi' \simeq \phi''$.

To achieve this, it suffices to show that there exists a mapping $G\colon Q \times I \times [0,\infty) \to Q \times [0,\infty)$ such that

$$G(x,0,t) = (x,t), G(x,1,t) = r_X(x,t),\ \ (x,t) \in Q \times I, \tag{27}$$

$$G(x,s,t) = (x,t),\ \ (x,t) \in K_X,\ \ s \in I. \tag{28}$$

Moreover, for any $\beta > 0$, there exist a neighborhood U of X and an $\alpha > 0$ such that

$$G(X \times I \times [\alpha,\infty)) \subseteq U \times [\beta,\infty). \tag{29}$$

Indeed, if one has such a mapping G, then $\phi' G\colon Q \times I \times [0,\infty) \to Q$ is an approaching homotopy, which connects ϕ' to $\phi'' = \phi' r_X$.

A mapping G, having the desired properties, is defined explicitly as follows. If $x \in Q\backslash X$, and $0 \leq t \leq 1/\alpha_X(x)$, let $G(x,s,t)$ assume the value

$$(H_X(x, s u_X(x,t)), t - s\,(t - 1/\alpha_X(H_X(x, u_X(x,t))))). \tag{30}$$

If $x \in Q\backslash X$ and if $1/\alpha_X(x) \leq t$, let $G(x,s,t)$ assume the value

$$(x, (1-s)t + s/\alpha_X(x)). \tag{31}$$

Finally, if $x \in X$, let $G(x,s,t)$ be given by

$$(H_X(x, s/(1+t)), t - s\,(t - 1/\alpha_X(H_X(x, 1/(1+t))))). \tag{32}$$

To complete the proof of the theorem, it remains to show that, for proper mappings $f: Q\backslash X \to Q\backslash Y$ and $g: Q\backslash Y \to Q\backslash Z$, one has $T(gf) \simeq T(g)T(f)$. Moreover, for the identity mapping $\mathrm{id}: Q\backslash X \to Q\backslash X$, the approaching mapping $T(\mathrm{id})$ is homotopic to the identity approaching mapping.

By definition, we see that

$$T(g)T(f)(x,t) = i_Z g(j_Y)^{-1} r_Y(i_Y f(j_X)^{-1} r_X(x,t), t). \tag{33}$$

On the other hand,

$$T(gf)(x,t) = i_Z(gf)(j_X)^{-1} r_X(x,t). \tag{34}$$

Consequently, we obtain a homotopy $H: Q \times [0,\infty) \times I \to Q$, which connects $T(g)T(f)$ to $T(gf)$, by putting

$$\begin{aligned} &H(x,t,s) = i_Z g(j_Y)^{-1} r_Y \\ &(i_Y f(j_X)^{-1} r_X(x,t), (1-s)t + s(\pi'' j_Y f(j_X)^{-1} r_X(x,t))), \end{aligned} \tag{35}$$

where $\pi'': Q \times [0,\infty) \to [0,\infty)$ is the second projection.

Finally, note that $\phi = T(\mathrm{id})$ is given by $\phi(x,t) = i_X j_X^{-1} r_X(x,t)$, while the approaching identity mapping ι is given by $\iota(x,t) = x$. Therefore the mapping $\pi G: Q \times I \times [0,\infty) \to Q$ is an approaching homotopy which connects ι to $i_X j_X^{-1} r_X = \phi$. □

REMARK 9.17. The above proof also applies to absolute retracts M and closed subsets $X \subseteq M$, which satisfy the assertions of Lemma 9.14.

Two proper mappings $f, g: M \to N$ are said to be *weakly properly homotopic* provided every compact set $B \subseteq N$ admits a compact set $A \subseteq M$ and a homotopy $H: M \times I \to N$, which connects f and g and is such that

$$H((M\backslash A) \times I) \subseteq N\backslash B. \tag{36}$$

It is readily seen that weak proper homotopy is an equivalence relation. Clearly, proper homotopy implies weak proper homotopy, but not conversely. Locally compact spaces and weak proper homotopy classes of proper mappings form the *weak proper homotopy category.* T.A. Chapman established the following analogue of Theorem 9.15, for ordinary shape (Chapman 1972).

THEOREM 9.18. *There exists an isomorphism T between the weak proper homotopy category of complements $Q\backslash X$ of Z-sets $X \subseteq Q$ and the usual shape category of Z-sets $X \subseteq Q$. On objects $T(Q\backslash X) = X$.*

Comparison of Theorems 9.15 and 9.18 shows a distinct advantage of strong shape over usual shape. Theorem 9.18 preceded Theorem 9.15. In fact, Edwards and Hastings engaged in the study of strong shape to obtain an analogue of Chapman's theorem with weak proper homotopy replaced by proper homotopy.

In general, there are many more morphisms of strong shape than morphisms of usual shape. E.g., there are uncountably many different strong shape morphisms of a point into a solenoid. However, there is only one shape morphisms between these spaces. Nevertheless, the two classifications of metric compacta coincide as shown by the following result.

THEOREM 9.19. *Two metric compacta have the same strong shape if and only if they have the same shape.*

A proof of Theorem 9.19 follows from Theorems 9.15, 9.18 and the following result from (Edwards, Hastings 1976b).

THEOREM 9.20. *Every weak proper homotopy equivalence $f: X \to Y$ between σ - compact locally compact Hausdorff spaces is weakly properly homotopic to a proper homotopy equivalence $g: X \to Y$.*

Proof of Theorem 9.19. If $\mathrm{ssh}(X) = \mathrm{ssh}(Y)$, then there exists a strong shape equivalence $F: X \to Y$. Since $E: \mathrm{SSh}(\mathsf{Top}) \to \mathrm{Sh}(\mathsf{Top})$ is a functor, one concludes that $E(F): X \to Y$ is a shape equivalence and thus, $\mathrm{sh}(X) = \mathrm{sh}(Y)$. Conversely, if $\mathrm{sh}(X) = \mathrm{sh}(Y)$, there exists a shape equivalence $F: X \to Y$. By Theorem 9.18, there exists a weak proper homotopy equivalence $f: Q\backslash X \to Q\backslash Y$. However, Theorem 9.20 implies the existence of a proper homotopy equivalence $f': Q\backslash X \to Q\backslash Y$. Now Theorem 9.15 yields a strong shape equivalence $F: X \to Y$, which establishes the desired relation $\mathrm{ssh}(X) = \mathrm{ssh}(Y)$. □

REMARK 9.21. The following problem, raised in (Chapman, Siebenmann 1976) is still open. Is every weak proper homotopy equivalence between Q - manifolds a proper homotopy equivalence?

Since homeomorphisms are obviously, proper homotopy equivalences, Theorem 9.19 is also an immediate consequence of the following *Chapman's complement theorem* (Chapman 1972, 1976).

THEOREM 9.22. *Two Z - sets X, Y in the Hilbert cube Q have the same shape if and only if their complements $Q\backslash X$, $Q\backslash Y$ are homeomorphic.*

In the study of strong shape of metric compacta a very useful tool is the *contractible telescope*. Let $\boldsymbol{X} = (X_n, p_{nn'}, \mathbb{N})$ be an inverse sequence of metric compacta with $X_1 = *$. For each n let Z_n be the mapping cylinder of $p_{nn+1}: X_{n+1} \to X_n$,

$$Z_n = X_n \cup (X_{n+1} \times I)/\sim, \tag{37}$$

where $\sim$ identifies the points $(x, 1) \in X_{n+1} \times I$ with $p_{nn+1}(x) \in X_n$.

The contractible telescope $\mathrm{CTel}\, \boldsymbol{X}$ is obtained from the disjoint union $\cup_n Z_n$, by identifying the points $[x, 0] \in Z_{n-1}$ with $[x] \in Z_n$, where $x \in X_n$ (see Fig. 9.2).

The contractible telescope admits a metric compactification X^*,

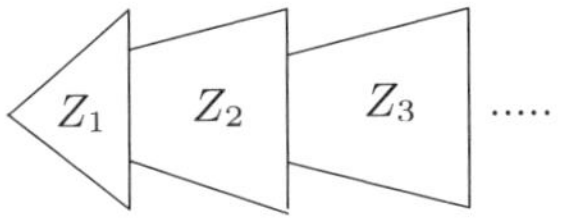

Fig. 9.2. Contractible telescope

$$X^* = X \cup \mathrm{CTel}\,\boldsymbol{X}, \tag{38}$$

where $X = \lim \boldsymbol{X}$. It is obvious how to define neighborhoods around points $z \in \mathrm{CTel}\,\boldsymbol{X}$. For points $z \in X$, a basis of neighborhoods is given by sets of the form $(p_n)^{-1}(U_n) \cup V(U_n)$, where U_n is an open subset of X_n, $n \in \mathbb{N}$, and $V(U_n)$ is the part of $\mathrm{CTel}\,\boldsymbol{X}$, which corresponds to the set

$$\cup_{n \geq m} (p_{nm})^{-1}(U_n) \times I. \tag{39}$$

If $\boldsymbol{X}$ is an inverse sequence of compact ANR's, then X^* is a compact AR (Krasinkiewicz 1976). Moreover, it is readily seen that X satisfies the assertions of Lemma 9.14. Therefore, by Remark 9.17, there is a bijection between strong shape morphisms of metric compacta $X \to Y$ and proper homotopy classes of mappings between the contractible telescopes $\mathrm{CTel}\,\boldsymbol{X} \to \mathrm{CTel}\,\boldsymbol{Y}$, where $\boldsymbol{X}$ and $\boldsymbol{Y}$ are ANR - sequences with $\lim \boldsymbol{X} = X$, $\lim \boldsymbol{Y} = Y$.

Bibliographic notes

The definition of the Quigley category, in particular Theorem 9.2, is taken from (Quigley 1973). The same category was discovered independently a little later in (Kodama, Ono 1979). Theorem 9.3 and its proof are from (Lisica, Mardešić 1985a). Lemma 9.14 is from (Chapman 1972). The complement theorem (Theorem 9.15) was first obtained by Edwards and Hastings (see (3.7.20) of (Edwards, Hastings 1976)). They defined the strong shape category of metric compacta using morphisms of Ho(pro-Top) between ANR-sequences, associated with these compacta. The proof presented in the present book is taken from (Kodama, Ono 1979). Theorems 9.18 and 9.22 are due to Chapman (Chapman 1972). Complement theorems triggered much further research in shape theory. In particular, there exist complement theorems in finite-dimensional ambient manifolds, complement theorems in fibered shape, complement theorems involving the uniform structure of the complements and the stable structure of the complements. For a survey of results see (Sher 1981, 1987). Also see (Mrozik 1984, 1985, 1986, 1991).

10. Selected results on strong shape

This section is a survey of selected results from strong shape theory. For proofs we refer to the appropriate literature.

10.1 Normal pairs of spaces

Recall that a pair of topological spaces (X, A) is *normal* provided A is normally embedded in X, i.e., every normal covering $\mathcal{V} \in \mathrm{Cov}(A)$ admits a normal covering $\mathcal{U} \in \mathrm{Cov}(X)$, such that the restriction $\mathcal{U}|A$ refines $\mathcal{V}$. A neighborhood V of a subset $A \subseteq X$ is called *normal* provided there exists a mapping $\phi: X \to I$ such that $\phi|A = 1$ and $\phi|(X \backslash V) = 0$. The following characterization of normal pairs was obtained in (Günther 1991c).

THEOREM 10.1. *A pair of spaces (X, A) is normal if and only if every mapping $f: A \to P$ to an* ANR *P and every open covering $\mathcal{U}$ of P admit a normal neighborhood V of A in X and a mapping $g: V \to P$ such that the mappings f and $g|A$ are $\mathcal{U}$-near.*

The following result is also from (Günther 1991c) (in the compact metric case the result is already in (Dydak, Segal 1981)).

THEOREM 10.2. *Let (X, A) be a normal pair of spaces. The inclusion mapping $i: A \to X$ is a strong shape equivalence if and only if the inclusion mapping $j: (A, A) \to (X, A)$ is an equivalence of ordinary shape.*

For a pair of spaces (X, A), consider the pair

$$(X', A') = (X \times \{1\} \cup A \times I, A \times \{0\}) \tag{1}$$

and the first projection $p: X' \to X$. Note that p is a homotopy equivalence and $p|A': A' \to A$ is a homeomorphism. In general (even for normal pairs), $p: (X', A') \to (X, A)$ fails to be a homotopy equivalence of pairs. However, in strong shape the following result holds (Günther 1992a).

THEOREM 10.3. *If (X, A) is a normal pair of spaces, $\bar{S}(p): (X', A') \to (X, A)$ is a strong shape equivalence of pairs.*

The proof is non-trivial. It uses Theorem 8.11 (the version for pairs) as well as Lemmas 4.7 and 4.8. Note that A' is a closed subset of X' and the inclusion $A' \to X'$ is a cofibration. Consequently, Theorem 10.3 implies the following corollary, which clarifies the role of normal pairs in strong shape theory.

COROLLARY 10.4. *Every normal pair of spaces (X, A) has the strong shape of a closed cofibration pair (Y, B).*

Clearly, if $F\colon (X, A) \to (Y, B)$ is a strong shape equivalence of pairs, then the restrictions $F\colon X \to Y$ and $F\colon A \to B$ are also strong shape equivalences. In (Günther 1992a) it is proved that, for normal pairs, also the converse holds.

THEOREM 10.5. *Let (X, A) and (Y, B) be normal pairs of spaces and let $F\colon (X, A) \to (Y, B)$ be a strong shape morphism of pairs. If the restrictions $F\colon X \to Y$ and $F\colon A \to B$ are strong shape equivalences, then $F\colon (X, A) \to (Y, B)$ is also a strong shape equivalence.*

The proof uses Theorem 8.11, Corollary 10.4 and the following assertion from homotopy theory.

LEMMA 10.6. *Let $f\colon (X, A) \to (Y, B)$ be a mapping of pairs such that the restrictions $f\colon X \to Y$, $f\colon A \to B$ are homotopy equivalences and let (Z, C) be a cofibration pair. Then f induces a bijection between the set of mappings $(Z, C) \to (X, A)$ and the set of mappings $(Z, C) \to (Y, B)$.*

Theorem 10.5 has several important consequences.

COROLLARY 10.7. *A mapping of pointed spaces $f\colon (X, *) \to (Y, *)$ is a pointed strong shape equivalence if and only if $f\colon X \to Y$ is a strong shape equivalence.*

COROLLARY 10.8. *Let (X, A) be a normal pair and let the subspace A have trivial shape. Then the quotient mapping $q\colon X \to X/A$ is a strong shape equivalence.*

Note that a space has the shape of a point if and only if it has the strong shape of a point (Koyama, Mardešić, Watanabe 1988).

10.2 Normal triads of spaces

By a *triad of spaces* (X, A', A'') we mean a topological space X and two subsets $A', A'' \subseteq X$ such that $X = A' \cup A''$. A triad is called *normal* provided A' and A'' are closed subsets of X and $A' \cap A''$ is normally embedded in X. In this case the pairs (X, A'), (X, A'') must be normal (Günther 1991c).

Strengthening a result from (Nowak, Spież 1981), Günther established the following theorem (Günther 1991c).

THEOREM 10.9. *Let (X, A', A'') be a normal triad. Then the shape dimension* $\mathrm{Sd}\, X$ *satisfies the following inequality.*

$$\mathrm{Sd}\, X \leq \max\{\mathrm{Sd}\, A', \mathrm{Sd}\, A'', 1 + \mathrm{Sd}\,(A' \cap A'')\}. \tag{1}$$

Two strong shape morphisms $F' \colon A' \to Y$, $F'' \colon A'' \to Y$ are said to be *compatible* provided their restrictions to $A' \cap A''$ coincide, i.e.,

$$F'|A' \cap A'' = F''|A' \cap A''. \tag{2}$$

Here $F'|A' \cap A''$ denotes the composition $F'\bar{S}(i)$, where $i \colon A' \cap A'' \to A'$ is the inclusion mapping. The following *pasting theorem* is a property specific for strong shape.

THEOREM 10.10. *Let (X, A', A'') be a normal triad and let $F' \colon A' \to Y$ and $F'' \colon A'' \to Y$ be compatible strong shape morphisms. Then there exists a strong shape morphism $F \colon X \to Y$ such that*

$$F|A' = F', \ F|A'' = F''. \tag{3}$$

Note that the pasting theorem does not assert uniqueness of F. Indeed, in general uniqueness of F fails, as seen by the following simple example. Let $X = Y = S^1$, let $A', A'' \subseteq X$ be two semicircles covering X and let F, G be the strong shape morphisms induces by mappings $f, g \colon X \to Y$, which belong to different homotopy classes. Then $F \neq G$, because the induced shape morphisms are different. Nevertheless, $F|A' = G|A'$ and $F|A'' = G|A''$, because all four morphisms are induced by inessential mappings.

A slightly weaker form of Theorem 10.10 was obtained in (Lisica, Mardešić 1985b). The above version is from (Günther 1991c), where a different proof was given.

EXAMPLE 10.11. In ordinary shape the analogue of the pasting theorem fails. We will show this by an example from (Lisica, Mardešić 1985b). Let $A' = P^2$ be the real projective plane and let $B = P^1 \subseteq P^2$ be the real projective line. Let A'' be a 2-cell D^2 with boundary S^1. Identify P^1 in P^2 with S^1 in D^2 to obtain $X = A' \cup A''$. Clearly, $A' \cap A'' = B$. Consider the inverse sequence $\boldsymbol{Y} = (Y_m, q_{mm'}, \mathbb{N})$, where $Y_m = S^2$ and $q_{mm+1} = q \colon S^2 \to S^2$ is a mapping of degree 3. Put $Y = \lim \boldsymbol{Y}$ and let $q_m \colon Y \to Y_m$ be the natural projections. Let $g' \colon P^2 \to S^2$ be a mapping belonging to the only non-trivial homotopy class of $[P^2, S^2] \approx H^2(P^2, \mathbb{Z}) \approx \mathbb{Z}/2$. Note that $qg' \simeq g'$. Therefore, the mappings $g'_m = g' \colon A' \to Y_m$ determine a shape morphism $F' \colon A' \to Y$. On the other hand, let $F'' \colon A'' \to Y$ be the shape morphism given by the constant mappings $g''_m = g'' \colon A'' \to Y_m$. Note that the shape morphisms F', F'' are compatible, because $\pi_1(S^2) = 0$ implies that both mappings $g'_m|B$, $g''_m|B$ are inessential. Now assume that $F \colon X \to Y$ is a shape morphism such that $F|A' = F'$, $F|A'' = F''$. Since $A'' = D^2$, one concludes that X has the homotopy type of $X/A'' \approx P^2/P_1 \approx S^2$. Let $\phi \colon X \to S^2$ be a homotopy equivalence with a homotopy inverse $\psi \colon S^2 \to X$.

Clearly, $F = FS[\psi]S[\phi]$. However, any shape morphism $H: S^2 \to Y$ is given by a sequence of maps $h_m: S^2 \to Y_m = S^2$ such that $h_m \simeq qh_{m+1}$. If h_m has degree d_m, then $3d_{m+1} = d_m$, which implies that d_m is divisible by arbitrarily high powers of 3, hence, $d_m = 0$. Consequently, $h_m \simeq 0$. This shows that H is a trivial shape morphism. In particular, $FS[\psi]$ is trivial and therefore, also F and $F' = F|A'$ must be trivial. However, this is a contradiction, because $S[q_1]F' = S[g_1'] = S[g']$ and $[g']$ is a non-trivial homotopy class. □

REMARK 10.12. There are easy examples which show that the pasting theorem fails also in the homotopy category of metric continua (Lisica, Mardešić 1985b).

10.3 Strong shape using the Vietoris system

With every topological space X one can associate a particular inverse system in Top, called the *Vietoris system* $\boldsymbol{V}(X) = (X_\lambda, p_{\lambda\lambda'}, \Lambda)$. It is indexed by the set $\Lambda = \mathrm{Cov}(X)$ of all normal coverings $\mathcal{U}$ of X. If $\lambda = \mathcal{U}$, $\lambda' = \mathcal{U}'$, one puts $\lambda \le \lambda'$, provided $\mathcal{U}'$ refines $\mathcal{U}$. For $\lambda = \mathcal{U}$, X_λ is the geometric realization $|K_\lambda|$ of a simplicial complex K_λ, defined as follows. The vertices of K_λ are all points $x \in X$. Vertices $x_0, \dots, x_n$ span a simplex of K_λ provided there exists a set $U \in \mathcal{U}$ such that $x_0, \dots, x_n \in U$. Clearly, if $\lambda \le \lambda'$, then $K_{\lambda'}$ is a subcomplex of K_λ. The bonding mapping $p_{\lambda\lambda'}: X_{\lambda'} \to X_\lambda$ is induced by the inclusion $K_{\lambda'} \subseteq K_\lambda$. It is easy to verify that $(\Lambda, \le)$ is a directed set and $\boldsymbol{V}(X)$ is an inverse system. Furthermore, with every mapping $f: X \to Y$ one can associate a mapping of systems $\boldsymbol{V}(f): \boldsymbol{V}(X) \to \boldsymbol{V}(Y) = (Y_\mu, q_{\mu\mu'}, M)$. It is given by an increasing function $f: M = \mathrm{Cov}(Y) \to \mathrm{Cov}(X) = \Lambda$ and by mappings $f_\mu: X_{f(\mu)} = |K_{f(\mu)}| \to |L_\mu| = Y_\mu$, defined as follows. If $\mu = \mathcal{V}$, then $f(\mu) = f^{-1}(\mathcal{V}) \in \mathrm{Cov}(X)$ and f_μ is induced by the simplicial mapping $f_\mu: K_{f(\mu)} \to L_\mu$, which sends the vertex $x \in K_{f(\mu)}$ to the vertex $f(x) \in L_\mu$. It is easy to verify that the Vietoris system, thus defined, is a functor $\boldsymbol{V}$: Top $\to$ inv-Top.

Application of the homology functor $H_n(\,.\,, G)$ to $\boldsymbol{V}(X)$ yields an abelian pro-group, whose limit is the *Vietoris homology group* of X, first defined for metric compacta in (Vietoris 1927). The same procedure, applied to the *Čech system* $\boldsymbol{C}(X)$ (see e.g., (Mardešić, Segal 1982), I.4.2) yields the *Čech homology groups* (Čech 1932). The construction of the Čech system is in a sense dual to the construction of the Vietoris system (Dowker 1952). However, in contrast to $\boldsymbol{C}(X)$, which is only a system in the homotopy category H(Top), the Vietoris system $\boldsymbol{V}(X)$ has the advantage of being a system in Top. Consequently, $\boldsymbol{V}(X)$ is an object of CH(pro-Top), while $\boldsymbol{C}(X)$ is not. For arbitrary spaces, the Vietoris and the Čech homology groups are naturally isomorphic, because the homotopy functor H transforms $\boldsymbol{V}(X)$ into a system $H\boldsymbol{V}(X)$, which is isomorphic to $\boldsymbol{C}(X)$ in pro-H(Top) (Dowker 1952). Unfortunately, the Vietoris system $\boldsymbol{V}(X)$ does not admit projections $p_\lambda: X \to X_\lambda$ such that

$p_{\lambda\lambda'}p_{\lambda'} = p_\lambda$, for $\lambda \le \lambda'$. This was one of the reasons for introducing resolutions and strong expansions.

The Vietoris system was considered in (Porter 1973) in order to define "Čech homotopy". It was used in (Edwards, Hastings 1976a) to study strong shape of metric compacta. Subsequently, Hastings defined strong shape morphisms $X \to Y$ of arbitrary spaces as morphisms $\boldsymbol{V}(X) \to \boldsymbol{V}(Y)$ from the category Ho (pro-Top) (Hastings 1977). The question whether this definition is equivalent to the one given in the present book was answered affirmatively by B. Günther, who proved the following theorem (Günther 1992c).

THEOREM 10.13. *For every space X the Vietoris system $\boldsymbol{V}(X)$ is naturally isomorphic in* CH(pro-Top) *to a strong* ANR*-expansion of X.*

The proof is non-trivial and consists in constructing an ANR-system $\boldsymbol{V}'(X)$, a level mapping $\boldsymbol{i}: \boldsymbol{V}(X) \to \boldsymbol{V}'(X)$ and a mapping $\boldsymbol{q}: X \to \boldsymbol{V}'(X)$ such that $\boldsymbol{i}$ is a level homotopy equivalence and $\boldsymbol{q}$ is a strong expansion of X. Clearly, $\boldsymbol{i}$ induces an isomorphism of $\boldsymbol{V}(X)$ and $\boldsymbol{V}'(X)$ in CH(pro-Top).

10.4 The Bauer – Günther description of strong shape

Rigidifying Mardešić's description of the ordinary shape category (Mardešić 1973), F.W. Bauer defined a strong shape category of topological spaces, using the techniques of 2-categories (Bauer 1976). In a subsequent paper, he sketched a more general approach, based on ∞-categories (Bauer 1978). His ideas were restated in simplicial terms and provided with all the needed details by B. Günther (Günther 1989, 1992b). In this subsection, we reproduce the main notions of Günther's description of strong shape.

The starting point is a *simplicial class* $\mathfrak{Top}$, which generalizes the category Top. Simplicial classes differ from simplicial sets only in that their n-simplices are not required to form a set. This generalization is needed, because simplicial classes are designed to generalize categories. The 0-simplices of $\mathfrak{Top}$ are all topological spaces X_0. Its 1-simplices are all mappings $\rho_{10}: X_0 \to X_1$ and its 2-simplices consist of mappings $\rho_{10}: X_0 \to X_1$, $\rho_{21}: X_1 \to X_2$, $\rho_{20}: X_0 \to X_2$ and of homotopies $\rho_{20}: X_0 \times I \to X_2$, which connect $\rho_{21}\rho_{10}: X_0 \to X_2$ to $\rho_{20}: X_0 \to X_2$. Generally, an n-simplex consists of a sequence of spaces $(X_0, X_1, \dots, X_n)$ and of a sequence of mappings $\rho_{mk}: X_k \times I^{m-k-1} \to X_m$, $0 \le k < m \le n$, which, for $0 \le l < k < m \le n$, make the following diagrams commutative.

$$
\begin{array}{ccc}
X_l \times I^{k-l-1} \times I^{m-k-1} & \xrightarrow{\rho_{kl} \times \mathrm{id}} & X_k \times I^{m-k-1} \\
\Big\downarrow {\scriptstyle \mathrm{id} \times \chi} & & \Big\downarrow {\scriptstyle \rho_{mk}} \\
X_l \times I^{m-l-1} & \xrightarrow[\rho_{ml}]{} & X_m
\end{array}
\qquad (1)
$$

where

$$\chi(s_1, \dots, s_{k-l-1}, t_1, \dots, t_{m-k-1}) = (s_1, \dots, s_{k-l-1}, 0, t_1, \dots, t_{m-k-1}). \quad (2)$$

The boundary operator $d_i \colon \mathfrak{Top}_n \to \mathfrak{Top}_{n-1}$ maps the n-simplex, consisting of the mappings $\rho_{mk} \colon X_k \times I^{m-k-1} \to X_m$, $0 \leq k < m \leq n$, to the $(n-1)$-simplex, consisting of the mappings $r_{mk} \colon Y_k \times I^{m-k-1} \to Y_m$, $0 \leq k < m \leq n-1$, where $Y_k = X_k$, for $k < i$, and $Y_k = X_{k+1}$, for $k \geq i$. Moreover,

$$r_{mk}(x, s_1, \dots, s_{m-k-1}) =$$
$$\begin{cases} \rho_{mk}(x, s_1, \dots, s_{m-k-1}), & m < i, \\ \rho_{m+1,k}(x, s_1, \dots, s_{i-k-1}, 1, s_{i-k}, \dots, s_{m-k-1}), & k < i \leq m, \\ \rho_{m+1,k+1}(x, s_1, \dots, s_{m-k-1}), & i \leq k. \end{cases} \quad (3)$$

The degeneracy operator $s_i \colon \mathfrak{Top}_n \to \mathfrak{Top}_{n+1}$ maps the n-simplex (ρ_{mk}), $0 \leq k < m \leq n$, to the $(n+1)$-simplex (r_{mk}), $0 \leq k < m \leq n+1$, where $Y_k = X_k$, for $k \leq i$, and $Y_k = X_{k-1}$, for $k > i$. Moreover,

$$r_{mk}(x, s_1, ..., s_{m-k-1}) =$$
$$\begin{cases} \rho_{m-1,k-1}(x, s_1, ..., s_{m-k-1}), & i < k < m, \\ \rho_{m-1,k}(x, s_2, ..., s_{m-k-1}) & i = k < m-1, \\ x, & i = k = m-1, \\ \rho_{m-1,k}(x, s_1, ..., s_{i-k-1}, s_{i-k}s_{i-k+1}, s_{i-k+2}, ..., s_{m-k-1}), & k < i < m-1, \\ \rho_{m-1,k}(x, s_1, ..., s_{m-k-2}), & k < i = m-1, \\ \rho_{mk}(x, s_1, ..., s_{m-k-1}), & k < m \leq i. \end{cases} \quad (4)$$

REMARK 10.14. $\mathfrak{Top}$ coincides with the homotopy coherent nerve of the simplicially enriched category Top_s, introduced in (Vogt 1973) and studied in (Cordier 1982).

With every topological space X, one associates a simplicial class $\mathfrak{Top}(X)$. Its n-simplices are $(n+1)$-simplices of $\mathfrak{Top}$, where in $(X_0, \dots, X_{n+1})$ the first term $X_0 = X$. The boundary and degeneracy operators d_i, s_i in $\mathfrak{Top}(X)$ coincide with the boundary and degeneracy operators d_{i+1}, s_{i+1} of $\mathfrak{Top}$. Moreover, the boundary operator d_0 of $\mathfrak{Top}$ induces a simplicial mapping

$p_X\colon \mathfrak{Top}(X) \to \mathfrak{Top}$. To every mapping $f\colon X \to Y$ one assigns a simplicial mapping $\mathfrak{Top}(f)\colon \mathfrak{Top}(Y) \to \mathfrak{Top}(X)$, defined by $\mathfrak{Top}(f)(\rho_{mk}) = (r_{mk})$, where

$$r_{mk} = \begin{cases} \rho_{mk}(f \times \mathrm{id}), & k = 0, \\ \rho_{mk}, & k > 0. \end{cases} \tag{5}$$

Let $\mathfrak{ANR}$ denote the full simplicial subclass of $\mathfrak{Top}$, spanned by all ANR's. Let $\mathfrak{A}(X)$ be the simpicial class $\mathfrak{A}(X) = (p_X)^{-1}(\mathfrak{ANR})$. Note that the following diagram commutes.

$$\begin{array}{ccc} \mathfrak{A}(X) & \xrightarrow{\subseteq} & \mathfrak{Top}(X) \\ {\scriptstyle p_X}\downarrow & & \downarrow{\scriptstyle p_X} \\ \mathfrak{ANR} & \xrightarrow[\subseteq]{} & \mathfrak{Top} \end{array} \tag{6}$$

By definition, a *strong shape function* $\phi\colon X \to Y$ is a simplicial mapping $\phi\colon \mathfrak{A}(Y) \to \mathfrak{A}(X)$ such that the following diagram commutes.

$$\begin{array}{ccccc} \mathfrak{A}(Y) & & \xrightarrow{\phi} & & \mathfrak{A}(X) \\ & {\scriptstyle p_Y}\searrow & & \swarrow{\scriptstyle p_X} & \\ & & \mathfrak{ANR} & & \end{array} \tag{7}$$

Note that the vertices of $\mathfrak{A}(Y)$ are mappings $f\colon Y \to P$, where P is an ANR. The simplicial mapping p_Y maps the vertex f into the target space P. The 1-simplices of $\mathfrak{A}(Y)$ consist of Y, ANR's P_1, P_2, mappings $f_1\colon Y \to P_1$, $f_2\colon Y \to P_2$, $r\colon P_1 \to P_2$ and a homotopy $H\colon Y \times I \to P_2$, which connects rf_1 to f_2. A strong shape function $\phi\colon X \to Y$ maps $f\colon Y \to P$ to $\phi(f)\colon X \to P$ and it maps the described 1-simplex to the 1-simplex, which consists of X, P_1, P_2, of mappings $\phi(f_1)\colon X \to P_1$, $\phi(f_2)\colon X \to P_2$, $r\colon P_1 \to P_2$ and of a homotopy $\phi(H)\colon X \times I \to P_2$, which connects $r\phi(f_1)$ to $r(f_2)$.

A homotopy connecting two strong shape functions $\phi, \psi\colon X \to Y$ is a strong shape function $\Phi\colon X \times I \to Y$, which satisfies the appropriate boundary conditions. By definition, the strong shape morphisms $F\colon X \to Y$ are homotopy classes of strong shape functions $\phi\colon X \to Y$.

In (Günther 1991a) it is proved that the strong shape category, thus defined, is isomorphic to the category SSh(Top), defined in 8.2 of this book. A special feature of the Bauer – Günther approach to strong shape is the fact that their construction does not involve different inverse systems approximating the spaces. In fact, in their theory every space is approximated by a universal (generalized) inverse system $\mathfrak{A}(X)$.

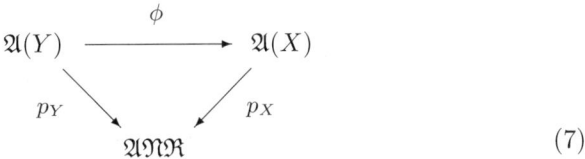

10.5 Strong shape of compacta via multi-valued maps

In 1992 J.M.R. Sanjurjo gave an alternative description of the shape theory of metric compacta, based on the notion of a multi-net (Sanjurjo 1992). By a *multi-valued map* $F: X \to Y$ is meant an upper semi-continuous multi-valued function, which assigns to points $x \in X$ non-empty closed subsets $F(x) \subseteq Y$. A *multi-net* $\boldsymbol{F}$ from X to Y is a sequence of multi-valued mappings $F_n: X \to Y$, $n = 1, 2, \ldots$, having the property that, for every $\varepsilon > 0$ and every sufficiently large n, F_n and F_{n+1} are ε-multi-homotopic. The latter condition means that there exists an ε-small multi-valued homotopy $H: X \times I \to Y$, which connects F to G, i.e., $\mathrm{diam}(H(x,t)) < \varepsilon$, for every $(x,t) \in X \times I$ and $H(x,0) = F(x)$, $H(x,1) = G(x)$. Two multi-nets $\boldsymbol{F}: X \to Y$ and $\boldsymbol{G}: X \to Y$ are considered homotopic provided, for every $\varepsilon > 0$, F_n and G_n are ε-multi-homotopic, for n sufficiently large. If $\boldsymbol{F} = (F_n): X \to Y$ and $\boldsymbol{G} = (G_k): Y \to Z$ are multi-nets, $(G_n F_n)$ need not be a multi-net from X to Z. However, for a suitable choice of subsequences (F_{k_n}) of (F_n), $(G_n F_{k_n})$ is a multi-net from X to Z, whose homotopy class depends only on the homotopy classes of the multi-nets $\boldsymbol{F}$ and $\boldsymbol{G}$. In this way one obtains a category, which is isomorphic to the ordinary shape category $\mathrm{Sh}(\mathsf{CM})$. Sanjurjo's theory was generalized to paracompact spaces in (Morón, Ruiz del Portal 1994) and to arbitrary topological spaces in (Čerin 1993).

In 1995 A. Giraldo and J.M.R. Sanjurjo gave a similar description of the strong shape theory of metric compacta. Multi-nets are replaced by *fine multi-valued mappings*. These are multi-valued mappings $F: X \times R_0 \to Y$, $R_0 = [0, \infty)$, having the property that, for every $\varepsilon > 0$, the diameter $\mathrm{diam}(F(x,t) < \varepsilon$, for every $x \in X$ and every sufficiently large $t \in R_0$. Two fine multi-valued mappings $F, G: X \times R_0 \to Y$ are homotopic provided there exists a fine multi-valued mapping $H: X \times I \times R_0 \to Y$ such that $H(x,0,t) = F(x,t)$, $H(x,1,t) = G(x,t)$, for every $(x,t) \in X \times R_0$. In order to define composition of homotopy classes $[F]: X \times R_0 \to Y$ and $[G]: Y \times R_0 \to Z$, Giraldo and Sanjurjo introduced *stretching mappings*. These are increasing mappings $\alpha: R_0 \to R_0$, which admit null-sequences (ε_n), (η_n) such that $\mathrm{diam}(G(K \times \{t\})) < \varepsilon_n$, for every $K \subseteq Y$ with $\mathrm{diam}(K) < \eta_n$ and every $t \in [n, n+1]$. One also requires that $\mathrm{diam}(F(x,t)) < \eta_n$, for every $x \in X$ and $t > \alpha(n)$. Stretching mappings always exist. Moreover, if $F: X \times R_0 \to Y$ and $G: Y \times R_0 \to Z$ are fine multi-valued mappings, then $H(x,t) = G(F(x, \alpha(t)), t)$ defines a fine multi-valued mapping $H: X \times R_0 \to Z$, whose homotopy class $[H]$ depends only on the homotopy classes $[F]$ and $[G]$. In this way one obtains a category, which is isomorphic to the strong shape category $\mathrm{SSh}(\mathsf{CM})$ (Giraldo, Sanjurjo 1995).

Let us mention that Ju.T. Lisica already in 1983 established a relationship between strong shape morphisms of metric compacta and upper semi-continuous multi-valued mappings (Lisica 1983).

10.6 Strong shape using approximate systems

A classical result (Freudenthal 1937) asserts that every metric compactum X with $\dim X = n$ is the limit of an inverse system of compact polyhedra of dimension $\leq n$. Surprisingly, this result does not generalize to Hausdorff compact spaces (Pasynkov 1958), (Mardešić 1960). This and other reasons led S. Mardešić and L.R. Rubin to introduce a more flexible type of systems, called approximate inverse systems (Mardešić, Rubin 1989). The usual requirement for inverse systems, that the mappings $p_{\lambda\lambda'}p_{\lambda'\lambda''}$ and $p_{\lambda\lambda''}$ coincide, for $\lambda \leq \lambda' \leq \lambda''$, is replaced by a weaker condition, which requires that these two mappings are arbitrarily near, when λ' is large enough. It was proved in (Mardešić, Rubin 1989) that a compact Hausdorff space X has $\dim X \leq n$ if and only if it is the limit of an approximate inverse system consisting of compact polyhedra X_λ of dimension $\leq n$. These authors considered only approximate systems of metric compacta. A general theory of approximate systems and approximate resolutions for arbitrary spaces was developed in (Mardešić, Watanabe 1989). A technically simpler generalization was given in (Charalambous 1991) and subsequently in (Mardešić 1993a). T. Watanabe proved that a topological space X has $\dim X \leq n$ if and only if it admits an approximate resolution consisting of polyhedra X_λ of dimension $\dim X_\lambda \leq n$ (Watanabe 1991a).

An *approximate system* $\boldsymbol{X} = (X_\lambda, p_{\lambda\lambda'}, \Lambda)$ consists of the same data as an inverse system. However, beside the usual requirement that $p_{\lambda\lambda} = \mathrm{id}$, one now requires that the following condition holds.

(A) For every $\lambda \in \Lambda$ and every $\mathcal{U} \in \mathrm{Cov}(X_\lambda)$, there exists a $\lambda' \geq \lambda$ such that, for $\lambda_2 \geq \lambda_1 \geq \lambda'$, the mappings $p_{\lambda\lambda_1}p_{\lambda_1\lambda_2}$ and $p_{\lambda\lambda_2}$ are $\mathcal{U}$-near.

An *approximate mapping* $X \to \boldsymbol{X}$ of a space X to an approximate system is a collection of mappings $f_\lambda: X \to X_\lambda$, $\lambda \in \Lambda$, such that the process of adding to $\boldsymbol{X}$ the space $X = X_\infty$ and the mappings f_λ yields an approximate system, indexed by the directed set $\Lambda \cup \{\infty\}$, where $\infty \geq \lambda$, for all $\lambda \in \Lambda$. An *approximate resolution* of a space X consists of an approximate system $\boldsymbol{X}$ and an approximate mapping $\boldsymbol{f}: X \to \boldsymbol{X}$ such that, for any polyhedron P and any open covering $\mathcal{V}$ of P, the following conditions hold (they modify conditions (R1) and (R2) of a resolution).

(AR1) For every mapping $f: X \to P$, there exists a $\lambda \in \Lambda$ such that, for $\lambda' \geq \lambda$, there exist mappings $h_{\lambda'}: X_{\lambda'} \to P$ having the property that $h_{\lambda'}p_{\lambda'}$ and f are $\mathcal{V}$-near mappings.

(AR2) There exists an open covering $\mathcal{V}'$ of P, such that, whenever, for a $\lambda \in \Lambda$ and for two mappings $h_\lambda, h'_\lambda: X_\lambda \to P$, the mappings $h_\lambda p_\lambda$, $h'_\lambda p_\lambda$ are $\mathcal{V}'$-near, then there exists a $\lambda' \geq \lambda$ such that the mappings $h_\lambda p_{\lambda\lambda''}, h'_\lambda p_{\lambda\lambda''}$ are $\mathcal{V}$-near, for $\lambda'' \geq \lambda'$.

Since approximate resolutions with desired properties often exist, when usual resolutions do not, it is natural to raise the question whether in strong shape theory one can also use approximate resolutions. A positive answer was provided by B. Günther, who showed that an approximate polyhedral resolution $\boldsymbol{p}: X \to \boldsymbol{X}$ of a space X completely determines the strong shape of X (Günther 1994b).

More precisely, given an approximate system $\boldsymbol{X} = (X_\lambda, p_{\lambda\lambda'}, \Lambda)$, Günther first describes a procedure of reindexing and of adding homotopies of all orders, which converts $\boldsymbol{X}$ into a *coherent system* $\boldsymbol{X}^* = (X_\lambda, p_{\lambda_0 \ldots \lambda_n}, \Lambda_*)$. The latter notion can be defined as a simplicial mapping $\Lambda_* \to \mathfrak{Top}$, where Λ_* is the simplicial set of finite sequences in Λ (defined in 1.2) and $\mathfrak{Top}$ is the simplicial class defined in 10.4. Similarly, an approximate mapping $\boldsymbol{p}: X \to \boldsymbol{X}$ is converted into a coherent mapping $\boldsymbol{p}^*: X \to \boldsymbol{X}^*$. If $\boldsymbol{p}$ is an approximate resolution, then $\boldsymbol{p}^*: X \to \boldsymbol{X}^*$ satisfies conditions which correspond to conditions (S1), (S2) of a strong expansion. Günther now proceeds by applying to $\boldsymbol{p}^*$ a process of rigidification, using function spaces as in 4.2. As a result he obtains a genuine strong expansion $\hat{\boldsymbol{p}}: X \to \hat{\boldsymbol{X}}$. Since $\boldsymbol{X}$ was a polyhedral system, $\hat{\boldsymbol{X}}$ turns out to be an HPol-system. Finally, there is a bijection between SSh(Top)(X, Y) and H(pro-Top)$(\hat{\boldsymbol{X}}, \hat{\boldsymbol{Y}})$.

10.7 Strong shape and localization

In 1981 A. Calder and H.M. Hastings proved that the strong shape category of metric compacta SSh(CM) can be obtained by localizing the category CM of metric compacta and continuous mappings at strong shape equivalences. They also showed that localization of CM at ordinary shape equivalences yields a category, which is not equivalent to Sh(CM) (Calder, Hastings 1981). Since the homotopy category H(CM) can be obtained from CM by localization at homotopy equivalences (see Theorem 4.35), it follows that SSh(CM) can also be obtained by localizing H(CM) at strong shape equivalences.

F.W. Cathey showed that the same category is obtained by localizing H(CM) at *strong shape deformation retractions* (abbreviated as SSDR - maps) (Cathey 1981). The latter were defined by J. Dydak and J. Segal as inclusions $i: X \to Y$ of metric compacta, which have the following property. For Y embedded in the Hilbert cube Q and arbitrary neighborhoods V of Y and U of X in Q, there exists a homotopy $H: Y \times I \to V$ (rel X), such that $H_0 = \mathrm{id}$ and $H_1(Y) \subseteq U$ (see Theorem 6.2 of (Dydak, Segal 1978b)).

In the case of closed inclusions $i: X \to Y$ of arbitrary topological spaces, SSDR - maps can be defined as mappings which have the property that every mapping $f: X \to P$ to an ANR P has an extension to all of Y and any two such extensions are homotopic rel (X). It turns out that SSDR - maps coincide with closed inclusions, which are strong shape equivalences and have the homotopy extension property for all $P \in$ ANR. B. Günther considered the restriction of SSh(Top) to the class of spaces, which have the strong shape of compact Hausdorff spaces, and he showed that this category can be obtained from the homotopy category of these spaces by localization at SSDR - mappings (Günther 1991b).

J. Dydak and S. Nowak have defined a class of morphisms of inverse systems $\boldsymbol{f}: \boldsymbol{X} \to \boldsymbol{Y}$ of pro-Top, which they named *strong shape equivalences* (Dydak, Nowak 1991). The defining conditions (pro-SE1) and (pro-SSE2) generalize conditions (SE1) and (SSE2) (see Theorem 8.13).

(pro-SE1) For every mapping $\phi: X_\lambda \to P$, $\lambda \in \Lambda$, $P \in \mathsf{ANR}$, there exist a $\mu \in M$ and a mapping $\psi: Y_\mu \to P$ such that $\psi f_\mu p_{f(\mu)\lambda'} \simeq \phi p_{\lambda\lambda'}$, for some $\lambda' \geq \lambda, f(\mu)$.

(pro-SSE 2) If $\psi_0, \psi_1: Y_\mu \to P$ are mappings and $F: X_\lambda \times I \to P$ is a homotopy, $\lambda \geq f(\mu)$, which connects $\psi_0 f_\mu p_{f(\mu)\lambda}$ to $\psi_1 f_\mu p_{f(\mu)\lambda}$, then there exists a homotopy $G: Y_{\mu'} \times I \to P$, $\mu' \geq \mu$, which connects $\psi_0 q_{\mu\mu'}$ to $\psi_1 q_{\mu\mu'}$, and there exists a $\lambda' \geq \lambda, f(\mu')$, such that $G(f_{\mu'} p_{f(\mu')\lambda'} \times 1) \simeq F(p_{f(\mu)\lambda'} \times 1)\,(\mathrm{rel}\,(X_{\lambda'} \times \partial I))$.

Strong shape equivalences in the above sense are also called *strong expansions*, because the assertion that a mapping $\boldsymbol{p}: X \to \boldsymbol{X}$ of a space X is a strong shape equivalence is nothing else but the assertion that $\boldsymbol{p}$ is a strong expansion in the sense of 7.

Recently, A.V. Prasolov defined the *strong shape category* SSh(pro-Top) *of systems* by localizing pro-Top (or equivalently, H(pro-Top)) at strong shape equivalences in the above sense. He then proved that the restriction (full subcategory) of the category SSh(pro-Top) to spaces is just the strong shape category of spaces SSh(Top), as defined in the present book (Prasolov to appear). The category SSh(pro-Top) was defined already in 1991 by Dydak and Nowak.

10.8 Stable strong shape

The stable homotopy category Stab-Ho of E.H. Spanier and J.H.C. Whitehead has as objects all compactly generated spaces, while the set of morphisms Stab-Ho$(X, Y) = \{X, Y\}$ is the colimit of the direct sequence

$$[X, Y] \xrightarrow{\Sigma} [\Sigma X, \Sigma Y] \xrightarrow{\Sigma} [\Sigma^2 X, \Sigma^2 Y] \xrightarrow{\Sigma} \dots; \tag{1}$$

here Σ denotes the suspension operator. The classical Whitehead – Spanier duality theorem establishes an isomorphism C between the stable homotopy category Stab-Ho(Pol_n) of compact polyhedra $X \subseteq S^n$ and the stable homotopy category Stab-Ho(CPol_n) of their complements $S^n \backslash X$. Hereby, $C(X) = S^n \backslash X$ and $C\{i\} = \{j\}$, where $i: X \to Y$ and $j: S^n \backslash Y \to S^n \backslash X$ are inclusions (Spanier, Whitehead 1955).

The Spanier – Whitehead theorem admits a generalization to compacta $X \subseteq S^n$ involving shape theory. The first results were obtained already in 1959, when E.L. Lima introduced a stable shape category (Lima 1959). Lima made no attempts to develop the shape categories Sh(CM) or SSh(CM) and his paper was "discovered" by the shape-theorists with considerable delay.

Subsequently, the stable shape category was studied in (Puppe, Dold 1980), (Henn 1981) and (Nowak 1987).

The shape-theoretic duality theorem establishes an isomorphism C between the *stable shape category* Stab - Sh(CM_n) of compacta $X \subseteq S^n$ and the stable weak homotopy category Stab - Ho_w(CCM_n) of their complements $S^n \backslash X$. In order to define the morphisms $X \to Y$ of Stab - Sh(CM_n)(X, Y), one considers the direct sequence

$$\mathrm{Sh}(X,Y) \xrightarrow{\Sigma} \mathrm{Sh}(\Sigma X, \Sigma Y) \xrightarrow{\Sigma} \mathrm{Sh}(\Sigma^2 X, \Sigma^2 Y) \xrightarrow{\Sigma} \dots . \tag{2}$$

By definition, the desired morphisms are elements of the colimit of this sequence. The set of morphisms Stab - Ho_w(CCM_n)(U, V) is the colimit

$$[U,V]_w \xrightarrow{\Sigma} [\Sigma U, \Sigma V]_w \xrightarrow{\Sigma} [\Sigma^2 U, \Sigma^2 V]_w \xrightarrow{\Sigma} \dots , \tag{3}$$

where $[U, V]_w$ denotes the set of weak homotopy classes $f: U \to V$, i.e., classes with respect to the equivalence relation of weak homotopy $\simeq_w$. Recall that $g \simeq_w g': U \to V$ means that $gf \simeq g'f$, for every mapping $f: P \to U$, where P is a compact polyhedron.

Q. Haxhibeqiri and S. Nowak have obtained a Spanier – Whitehead duality theorem for the *stable strong shape category* Stab - SSh(CM_n) of pointed compacta $X \subseteq S^n$ and the stable homotopy category Stab - Ho(CCM_n) of their complements (Haxhibeqiri, Nowak 1987, 1989). One defines the set of morphisms in the first of these categories as in the case of the category Stab - Sh(CM_n), the only difference being that one uses pointed compacta and the sets $\mathrm{Sh}(\Sigma^k X, \Sigma^k Y)$ are replaced by the sets $\mathrm{SSh}(\Sigma^k X, \Sigma^k Y)$. This theorem is another example of a situation, where strong shape shows an advantage over ordinary shape, because stable weak homotopy gets replaced by stable homotopy. In (Bauer 1991) this result is derived from a very general version of the Alexander duality theorem (Bauer 1984). A generalization of the theorem to arbitrary subsets of S^n has been obtained in (Bauer 1995b), using compactly generated strong shape. Also consider (Bauer 1997).

III. HIGHER DERIVED LIMITS

11. The derived functors of lim

In contrast to Čech cohomology groups, which have a long history of successful applications in algebraic topology, Čech homology groups did not prove very useful. This is because the Čech cohomology groups satisfy the exactness axiom, while the Čech homology groups do not. The reason for this difference in behavior lies in the fact that colim (direct limit) is an exact functor, while lim (inverse limit) is only left exact. Over a long period of time considerable efforts were devoted to the development of exact homology groups, called Steenrod or strong homology groups. These groups have successfully replaced the Čech homology groups (see Chapter IV). Proper understanding of strong homology requires understanding of the derived functors $\lim^n$ of the functor lim. Therefore, all of Chapter III is devoted to these functors.

In homological algebra additive left exact functors, defined on abelian categories with enough injective objects, give rise to a sequence of right derived functors. The category of inverse systems of modules indexed by a fixed preordered set Λ and the limit functor lim fulfill these conditions. Therefore, there exists a sequence of right derived functors of lim, which measure the discrepancy from exactness of the functor lim. These are the higher derived limits $\lim^n$. Instead of applying the results of the general theory, in this section we develop these results in the particular case of lim. This should enable also readers not familiar with homological algebra to understand the definition and the basic properties of the functors $\lim^n$. We also give the axiomatic characterization of these functors as well as explicit formulae which define them. In the last subsection we show that, in the case of inverse sequences $\boldsymbol{X}$, the derived limits $\lim^n \boldsymbol{X}$ vanish, for $n \geq 2$.

11.1 Inverse systems of modules

Let Mod_R, or just Mod, denote the category of modules over a commutative ring R with unit $1 \neq 0$. In this subsection we will consider the category Mod^Λ, whose objects are inverse systems of R-modules $\boldsymbol{X} = (X_\lambda, p_{\lambda\lambda'}, \Lambda)$, indexed by a fixed preordered set Λ, and the morphisms are level morphisms $\boldsymbol{f} = (f_\lambda): \boldsymbol{X} \to \boldsymbol{Y} = (Y_\lambda, q_{\lambda\lambda'}, \Lambda)$ (see 1.1). In the special case $R = \mathbb{Z}$, Mod_R is the category Ab of abelian groups and one obtains the category Ab^Λ. The main purpose of this subsection is to establish the following theorem.

THEOREM 11.1. *The category* Mod^{Λ} *is an abelian category.*

Recall that a category $\mathcal{A}$ is *abelian* if it has the following properties.

(*i*) It has a zero object.

(*ii*) Every morphism has a kernel and a cokernel. Every monomorphism is a kernel and every epimorphism is a cokernel.

(*iii*) It has finite products and finite coproducts.

The lemmas which follow will show that the category Mod^{Λ} does have properties (*i*)–(*iii*).

First recall that a morphism $f: X \to Y$ in a category $\mathcal{A}$ is an epimorphism provided, for arbitrary morphisms $v, v': Y \to V$, $vf = v'f$ implies $v = v'$. Analogously, $f: X \to Y$ is a monomorphism provided, for arbitrary morphisms $u, u': U \to X$, $fu = fu'$ implies $u = u'$.

LEMMA 11.2. *A morphism* $\boldsymbol{f} = (f_\lambda): \boldsymbol{X} \to \boldsymbol{Y}$ *in* Mod^{Λ} *is an epimorphism if and only if all* $f_\lambda: X_\lambda \to Y_\lambda$, $\lambda \in \Lambda$, *are epimorphisms in the category* Mod, *i.e., are surjections. A morphism* $\boldsymbol{f}$ *is a monomorphism if and only if all* $f_\lambda: X_\lambda \to Y_\lambda$ *are monomorphisms in* Mod, *i.e., are injections. A morphism* $\boldsymbol{f}$ *is an isomorphism if and only if all* $f_\lambda: X_\lambda \to Y_\lambda$ *are isomorphisms in* Mod, *i.e., are bijections.*

Proof. Let $\boldsymbol{f}$ be an epimorphism. To show that, for an arbitrary $\mu \in \Lambda$, f_μ is an epimorphism, assume that $v, v': Y_\mu \to V$ are homomorphisms such that $vf_\mu = v'f_\mu$. Put $V_\lambda = V$ and $v_{\lambda\lambda'} = \mathrm{id}$, if $\mu \le \lambda$, and put $V_\lambda = 0$ and $v_{\lambda\lambda'} = 0$, otherwise. Then, $\boldsymbol{V} = (V_\lambda, v_{\lambda\lambda'}, \Lambda)$ is an inverse system. Now define a morphism $\boldsymbol{v} = (v_\lambda): \boldsymbol{Y} \to \boldsymbol{V}$, by putting $v_\lambda = vq_{\mu\lambda}$, if $\mu \le \lambda$, and $v_\lambda = 0$, otherwise. Clearly, $\boldsymbol{v} = (v_\lambda): \boldsymbol{Y} \to \boldsymbol{V}$ is a morphism. Define $\boldsymbol{v}'$ analogously, using v' instead of v. Note that $\boldsymbol{vf} = \boldsymbol{v'f}$, because $vq_{\mu\lambda}f_\lambda = vf_\mu p_{\mu\lambda} = v'q_{\mu\lambda}f_\lambda$. Consequently, $\boldsymbol{v} = \boldsymbol{v}'$ and thus, $v = v_\mu = v'_\mu = v'$. The converse implication is obvious.

Now assume that $\boldsymbol{f}$ is a monomorphism. For an arbitrary $\mu \in \Lambda$, let $u, u': U \to X_\mu$ be homomorphisms such that $f_\mu u = f_\mu u'$. Put $U_\lambda = U$ and $u_{\lambda\lambda'} = \mathrm{id}$, if $\lambda \le \mu$, and put $U_\lambda = 0$ and $u_{\lambda\lambda'} = 0$, otherwise. Then, $\boldsymbol{U} = (U_\lambda, u_{\lambda\lambda'}, \Lambda)$ is an inverse system. Now define a morphism $\boldsymbol{u} = (u_\lambda): \boldsymbol{U} \to \boldsymbol{X}$, by putting $u_\lambda = p_{\lambda\mu}u$, if $\lambda \le \mu$, and $u_\lambda = 0$, otherwise. Clearly, $\boldsymbol{u} = (u_\lambda): \boldsymbol{U} \to \boldsymbol{X}$ is a morphism. Define $\boldsymbol{u}'$ analogously, using u' instead of u. Note that $\boldsymbol{fu} = \boldsymbol{fu}'$, because $f_\lambda p_{\lambda\mu}u = p_{\lambda\mu}f_\mu u = f_\lambda p_{\lambda\mu}u'$. Consequently, $\boldsymbol{u} = \boldsymbol{u}'$ and thus, $u = u_\mu = u'_\mu = u'$. The converse implication is obvious. □

Recall that a *zero object* in a category $\mathcal{A}$ is an object 0, which is both *initial* and *terminal*, i.e., for every object X there is a unique morphism $0 \to X$ and a unique morphism $X \to 0$. Zero objects are unique up to isomorphism. A *zero morphisms* is a morphism 0 which factors through some zero object. The *kernel* $\mathrm{Ker}(f)$ of a morphism $f: X \to Y$ is a morphism $u: U \to X$ with the property that $fu = 0$ and whenever $u': U' \to X$ is a morphism such that $fu' = 0$, then there exists a unique morphism $h: U' \to U$ such that $u' = uh$. When they exist, kernels are unique up to isomorphism, i.e., if both u and

u' are kernels, then h is an isomorphism. Every kernel is a monomorphism. Indeed, if $k, k': U' \to U$ are morphisms such that $uk = uk' = u'$, then $fu' = 0$ and therefore, by uniqueness of the factorization, $k = k'$. Dually, the *cokernel* $\mathrm{Coker}(f)$ of a morphism $f: X \to Y$ is a morphism $v: Y \to V$ with the property that $vf = 0$ and whenever $v': Y \to V'$ is a morphism such that $v'f = 0$, then there exists a unique morphism $k: V \to V'$ such that $v' = kv$. When they exist, cokernels are unique up to isomorphism. Every cokernel is an epimorphism.

With every morphism f one associates two more morphisms, the *coimage* $\mathrm{Coim}(f)$ of f, which is defined by

$$\mathrm{Coim}(f) = \mathrm{Coker}(\mathrm{Ker} f), \tag{1}$$

and the *image* $\mathrm{Im}(f)$ of f, which is defined by

$$\mathrm{Im}(f) = \mathrm{Ker}(\mathrm{Coker} f). \tag{2}$$

We will now consider the described notions in the category Mod^{Λ}. A well-defined object of this category is the *zero system* $\mathbf{0}$. This is the inverse system which consists of copies of the module 0. For any two systems $\boldsymbol{X}, \boldsymbol{Y}$, the *zero morphism* $\mathbf{0}: \boldsymbol{X} \to \boldsymbol{Y}$ is defined. It consists of the homomorphisms $0: X_\lambda \to Y_\lambda$, $\lambda \in \Lambda$.

For a morphism $\boldsymbol{f} = (f_\lambda): \boldsymbol{X} \to \boldsymbol{Y}$, we define its *canonical kernel* $\mathrm{Ker}(\boldsymbol{f})$ as the inverse system $\boldsymbol{U} = (U_\lambda, u_{\lambda\lambda'}, \Lambda)$, where $U_\lambda = f_\lambda^{-1}(0)$ and $u_{\lambda\lambda'}: U_{\lambda'} \to U_\lambda$ is the homomorphism induced by $p_{\lambda\lambda'}: X_{\lambda'} \to X_\lambda$. With $\boldsymbol{U}$ we associate a canonical morphism $\boldsymbol{u} = (u_\lambda): \boldsymbol{U} \to \boldsymbol{X}$, where $u_\lambda: U_\lambda \to X_\lambda$ are inclusions.

The *canonical cokernel* $\mathrm{Coker}(\boldsymbol{f})$ of $\boldsymbol{f}$, sometimes denoted by $\boldsymbol{Y}/\boldsymbol{f}(\boldsymbol{X})$, is the inverse system $\boldsymbol{V} = (V_\lambda, v_{\lambda\lambda'}, \Lambda)$, where $V_\lambda = Y_\lambda / f_\lambda(X_\lambda)$ and $v_{\lambda\lambda'}: V_{\lambda'} \to V_\lambda$ is the homomorphism induced by $q_{\lambda\lambda'}: Y_{\lambda'} \to Y_\lambda$. With $\boldsymbol{V}$ we associate a canonical morphism $\boldsymbol{v} = (v_\lambda): \boldsymbol{Y} \to \boldsymbol{V}$, where $v_\lambda: Y_\lambda \to V_\lambda$ are the quotient homomorphisms.

We also define the *canonical image* $\mathrm{Im}(\boldsymbol{f})$ of $\boldsymbol{f}$, sometimes denoted by $\boldsymbol{f}(\boldsymbol{X})$ or just $\boldsymbol{f}\boldsymbol{X}$, as the inverse system $\boldsymbol{S} = (S_\lambda, s_{\lambda\lambda'}, \Lambda)$, where $S_\lambda = f_\lambda(X_\lambda)$ and $s_{\lambda\lambda'}$ is the homomorphism obtained from $q_{\lambda\lambda'}$ by appropriate restrictions. With $\boldsymbol{S}$ we associate a canonical morphism $\boldsymbol{s} = (s_\lambda): \boldsymbol{S} \to \boldsymbol{Y}$, where $s_\lambda: f_\lambda(X_\lambda) \to Y_\lambda$ are inclusions. Analogously, *canonical coimage* $\mathrm{Coim}(\boldsymbol{f})$ of $\boldsymbol{f}$ is the system $\boldsymbol{T} = (T_\lambda, t_{\lambda\lambda'}, \Lambda)$, where $T_\lambda = X_\lambda / f_\lambda^{-1}(0)$ and $t_{\lambda\lambda'}$ are homomorphisms induced by $p_{\lambda\lambda'}$. With $\boldsymbol{T}$ we associate a canonical morphism $\boldsymbol{t} = (t_\lambda): \boldsymbol{X} \to \boldsymbol{T}$, where $t_\lambda: X_\lambda \to X_\lambda / f_\lambda^{-1}(0)$ are quotient homomorphisms.

LEMMA 11.3. *In the category* Mod^{Λ} *the zero system* $\mathbf{0}$ *is the only zero object. Between any two objects* $\boldsymbol{X}$, $\boldsymbol{Y}$, $\mathbf{0}: \boldsymbol{X} \to \boldsymbol{Y}$ *is the only zero morphism. For every morphism* $\boldsymbol{f}: \boldsymbol{X} \to \boldsymbol{Y}$, *the canonical kernel* $\boldsymbol{u}$, *cokernel* $\boldsymbol{v}$, *image* $\boldsymbol{s}$ *and coimage* $\boldsymbol{t}$ *are a kernel, a cokernel, an image and a coimage of* $\boldsymbol{f}$, *respectively. Moreover, there exists a unique isomorphism* $\boldsymbol{w}: \mathrm{Coim}\boldsymbol{f} \to \mathrm{Im}\boldsymbol{f}$ *such that* $\boldsymbol{f} = \boldsymbol{s}\boldsymbol{w}\boldsymbol{t}$. *Every monomorphism is a kernel and every epimorphism is a cokernel.*

Proof. All the statements are easily verified. E.g., if $\boldsymbol{f} = (f_\lambda)$, then every $f_\lambda: X_\lambda \to Y_\lambda$ induces an isomorphism $w_\lambda: X_\lambda / f_\lambda^{-1}(0) \to f(X_\lambda)$ such that

$$f_\lambda = s_\lambda w_\lambda t_\lambda. \tag{3}$$

Note that w_λ maps the class $[x_\lambda]$ of an element $x_\lambda \in X_\lambda$ to $f_\lambda(x_\lambda)$. Therefore, for $\lambda \leq \lambda'$ and $x_{\lambda'} \in X_{\lambda'}$, one has $t_{\lambda\lambda'} w_{\lambda'}[x_{\lambda'}] = q_{\lambda\lambda'} f_{\lambda'}(x_{\lambda'})$ and $w_\lambda s_{\lambda\lambda'}[x_{\lambda'}] = w_\lambda[p_{\lambda\lambda'}(x_{\lambda'})] = f_\lambda p_{\lambda\lambda'}(x_{\lambda'})$, which shows that the mappings $w_{\lambda\lambda'}$ form a morphism $\boldsymbol{w}: \mathrm{Coim}\boldsymbol{f} \to \mathrm{Im}\boldsymbol{f}$ such that $\boldsymbol{f} = \boldsymbol{swt}$. This morphism is unique, because, by Lemma 11.2, $\boldsymbol{s}$ is a monomorphism and $\boldsymbol{t}$ is an epimorphism. If $\boldsymbol{f}$ is a monomorphism, then $\mathrm{Ker}(\boldsymbol{f}) = \boldsymbol{0}$ and thus, $\mathrm{Coim}(\boldsymbol{f}) = \boldsymbol{X}$ and $\boldsymbol{t} = \mathrm{id}$. Consequently, $\boldsymbol{f} = \boldsymbol{sw}$. Since $\boldsymbol{s}: \mathrm{Im}(\boldsymbol{f}) \to \boldsymbol{Y}$ is the kernel of $\boldsymbol{v}: \boldsymbol{Y} \to \mathrm{Coker}(\boldsymbol{f})$ and $\boldsymbol{w}$ is an isomorphism, it follows that $\boldsymbol{f}$ is also a kernel of $\boldsymbol{v}$. If $\boldsymbol{f}$ is an epimorphism, then $\mathrm{Im}(\boldsymbol{f}) = \boldsymbol{Y}$ and $\boldsymbol{s} = \mathrm{id}$. Consequently, $\boldsymbol{f} = \boldsymbol{wt}$. Since $\boldsymbol{t}: \boldsymbol{X} \to \mathrm{Coim}(\boldsymbol{f})$ is the cokernel of $\boldsymbol{u}: \mathrm{Ker}(\boldsymbol{f}) \to \boldsymbol{X}$ and $\boldsymbol{w}$ is an isomorphism, it follows that $\boldsymbol{f}$ is also a cokernel $\boldsymbol{u}$. □

LEMMA 11.4. *The category* Mod^Λ *has arbitrary products and coproducts.*

Proof. Let $\mathcal{X} = (\boldsymbol{X}^\alpha, \alpha \in A)$ be a family of objects $\boldsymbol{X}^\alpha = (X_\lambda^\alpha, p_{\lambda\lambda'}^\alpha, \Lambda)$ from Mod^Λ. If one puts $X_\lambda = \prod_\alpha X_\lambda^\alpha$, $p_{\lambda\lambda'} = \prod_\alpha p_{\lambda\lambda'}^\alpha$, then $\boldsymbol{X} = (X_\lambda, p_{\lambda\lambda'}, \Lambda)$ is also an object of Mod^Λ. If for every $\alpha \in A$ one defines $p_\lambda^\alpha: X_\lambda \to X_\lambda^\alpha$ as the natural projection, then $\boldsymbol{p}^\alpha = (p_\lambda^\alpha): \boldsymbol{X} \to \boldsymbol{X}^\alpha$ becomes a morphism of Mod^Λ, called the *canonical projection.* It is easy to verify the universal property of products, i.e., for any collection of morphisms $\boldsymbol{f}^\alpha: \boldsymbol{Y} \to \boldsymbol{X}^\alpha$, there is a unique morphism $\boldsymbol{f}: \boldsymbol{Y} \to \boldsymbol{X}$ such that $\boldsymbol{p}^\alpha \boldsymbol{f} = \boldsymbol{f}^\alpha$, $\alpha \in A$.

Similarly, if one puts $X_\lambda = \bigoplus_\alpha X_\lambda^\alpha$, $p_{\lambda\lambda'} = \bigoplus_\alpha p_{\lambda\lambda'}^\alpha$, then $\boldsymbol{X} = (X_\lambda, p_{\lambda\lambda'}, \Lambda)$ is an object of Mod^Λ. If for every $\alpha \in A$ one defines $i_\lambda^\alpha: X_\lambda^\alpha \to X_\lambda$ as the natural injection, then $\boldsymbol{i}^\alpha = (i_\lambda^\alpha): \boldsymbol{X}^\alpha \to \boldsymbol{X}$ becomes a morphism of Mod^Λ, called the *canonical injection.* It is easy to verify the universal property of coproducts, which is dual to the one of products. □

Proof of Theorem 11.1. Lemmas 11.3 and 11.4 contain assertions, which include conditions (*i*)–(*iii*) from the definition of an abelian category. □

A sequence $\ldots \to X' \xrightarrow{f'} X \xrightarrow{f} X'' \to \ldots$ in an abelian category is said to be *exact* at X provided the image of f' is the kernel of f. A sequence is exact if it is exact at every one of its terms. A *short exact sequence* is an exact sequence of the form $0 \to X' \xrightarrow{f'} X \xrightarrow{f} X'' \to 0$. Exactness of such a sequence is equivalent to the assertion that f' is a monomorphism, f is an epimorphism and f' is a kernel of f (see e.g., (Mitchell, 1965), p. 19).

LEMMA 11.5. *A sequence in* Mod^Λ

$$\boldsymbol{X}' \xrightarrow{\boldsymbol{f}'} \boldsymbol{X} \xrightarrow{\boldsymbol{f}} \boldsymbol{X}'' \tag{4}$$

is exact at $\boldsymbol{X}$ *if and only if, for every* $\lambda \in \Lambda$*, the sequence of modules*

$$X'_\lambda \xrightarrow{f'_\lambda} X_\lambda \xrightarrow{f_\lambda} X''_\lambda \tag{5}$$

is exact at X_λ*, i.e.,* $f'_\lambda(X'_\lambda) = f_\lambda^{-1}(0)$.

Proof. The assertion is an immediate consequence of the description of canonical kernels and canonical images of morphisms in Mod^Λ. □

If $\boldsymbol{f} = (f_\lambda), \boldsymbol{f}' = (f'_\lambda): \boldsymbol{X} \to \boldsymbol{Y}$ are morphisms of Mod^Λ, one defines their *sum* $\boldsymbol{f} + \boldsymbol{f}': \boldsymbol{X} \to \boldsymbol{Y}$ as the morphism consisting of the homomorphisms $f_\lambda + f'_\lambda: X_\lambda \to Y_\lambda$, $\lambda \in \Lambda$.

LEMMA 11.6. *The set* $\mathrm{Hom}(\boldsymbol{X}, \boldsymbol{Y})$ *of all morphisms* $\boldsymbol{f}: \boldsymbol{X} \to \boldsymbol{Y}$*, endowed with addition* $+$*, is an abelian group and the following distribution laws hold:*

$$\boldsymbol{g}(\boldsymbol{f} + \boldsymbol{f}') = \boldsymbol{g}\boldsymbol{f} + \boldsymbol{g}\boldsymbol{f}', \tag{6}$$

$$(\boldsymbol{g} + \boldsymbol{g}')\boldsymbol{f} = \boldsymbol{g}\boldsymbol{f} + \boldsymbol{g}'\boldsymbol{f}. \tag{7}$$

Proof. The assertion is obtained by straightforward verification. □

LEMMA 11.7. *Let* $\boldsymbol{X} = \boldsymbol{X}' \times \boldsymbol{X}'' = \boldsymbol{X}' \oplus \boldsymbol{X}''$ *and let* $\boldsymbol{p}': \boldsymbol{X} \to \boldsymbol{X}'$, $\boldsymbol{p}'': \boldsymbol{X} \to \boldsymbol{X}''$ *and* $\boldsymbol{i}': \boldsymbol{X}' \to \boldsymbol{X}$, $\boldsymbol{i}'': \boldsymbol{X}'' \to \boldsymbol{X}$ *be the canonical projections and injections, respectively. Then*

$$\boldsymbol{p}'\boldsymbol{i}' = \mathbf{1}_{\boldsymbol{X}'},\ \boldsymbol{p}''\boldsymbol{i}'' = \mathbf{1}_{\boldsymbol{X}''},\ \boldsymbol{p}'\boldsymbol{i}'' = \mathbf{0},\ \boldsymbol{p}''\boldsymbol{i}' = \mathbf{0}, \tag{8}$$

$$\boldsymbol{i}'\boldsymbol{p}' + \boldsymbol{i}''\boldsymbol{p}'' = \mathbf{1}_{\boldsymbol{X}}. \tag{9}$$

Proof. Formulae (8) and (9) are immediate consequences of the analogous formulae, for $X_\lambda = X'_\lambda \times X''_\lambda = X'_\lambda \oplus X''_\lambda$, the projections $p'_\lambda: X_\lambda \to X'_\lambda$, $p''_\lambda: X \to X''_\lambda$ and the injections $i'_\lambda: X'_\lambda \to X_\lambda$, $i''_\lambda: X''_\lambda \to X_\lambda$. □

In general, we refer to morphisms $\boldsymbol{i}', \boldsymbol{i}'', \boldsymbol{p}', \boldsymbol{p}''$ which satisfy formulae (8) and (9) as to a *presentation of the direct sum* $\boldsymbol{X} = \boldsymbol{X}' \oplus \boldsymbol{X}''$.

REMARK 11.8. Properties established by Lemmas 11.6 and 11.7 make Mod^Λ an *additive category.* Note that abelian categories admit unique additive structures (see e.g., Theorems 1.2.7 and 1.6.4 in Vol 2 of (Borceux 1994)). In additive categories finite products and coproducts coincide (see e.g., Proposition 1.2.4 in Vol 2 of (Borceux 1994)). Instead of $X \times Y = X \prod Y$ and $X \coprod Y$ we will use the notation $X \oplus Y$.

REMARK 11.9. In $\mathrm{Hom}\,(\boldsymbol{X}, \boldsymbol{Y})$ one can also define multiplication with scalars $r \in R$. If $\boldsymbol{f} = (f_\lambda) \in \mathrm{Hom}(\boldsymbol{X}, \boldsymbol{Y})$ is a morphism, one puts $r\boldsymbol{f} = (rf_\lambda)$. Since R was assumed commutative, $\mathrm{Hom}(\boldsymbol{X}, \boldsymbol{Y})$ is an R-module.

If $\boldsymbol{i}'$ and $\boldsymbol{p}''$ are as in Lemma 11.7, then

$$\mathbf{0} \to \boldsymbol{X}' \xrightarrow{\boldsymbol{i}'} \boldsymbol{X}' \oplus \boldsymbol{X}'' \xrightarrow{\boldsymbol{p}''} \boldsymbol{X}'' \to \mathbf{0} \tag{10}$$

is a short exact sequence. A short exact sequence

$$\mathbf{0} \to \boldsymbol{X}' \xrightarrow{\boldsymbol{f}'} \boldsymbol{X} \xrightarrow{\boldsymbol{f}} \boldsymbol{X}'' \to \mathbf{0} \tag{11}$$

splits if it is isomorphic to the sequence (10), i.e., if there exists an isomorphism $\boldsymbol{h}: \boldsymbol{X} \to \boldsymbol{X}' \oplus \boldsymbol{X}''$ such that $\boldsymbol{h}\boldsymbol{f}' = \boldsymbol{i}'$ and $\boldsymbol{p}''\boldsymbol{h} = \boldsymbol{f}$.

LEMMA 11.10. *For an exact sequence* (11) *the following conditions are equivalent.*

(*i*) *The sequence splits.*

(*ii*) $\boldsymbol{f}'$ *has a left inverse* $\boldsymbol{g}': \boldsymbol{X} \to \boldsymbol{X}'$, $\boldsymbol{g}'\boldsymbol{f}' = \mathrm{id}$.

(*iii*) $\boldsymbol{f}$ *has a right inverse* $\boldsymbol{g}: \boldsymbol{X}'' \to \boldsymbol{X}$, $\boldsymbol{f}\boldsymbol{g} = \mathrm{id}$.

Proof. That (*i*) implies (*ii*) and (*iii*) follows immediately from Lemma 11.7. To prove that (*ii*) implies (*i*), consider the morphism $\boldsymbol{h}: \boldsymbol{X} \to \boldsymbol{X}' \oplus \boldsymbol{X}''$, determined by conditions

$$\boldsymbol{p}'\boldsymbol{h} = \boldsymbol{g}', \ \boldsymbol{p}''\boldsymbol{h} = \boldsymbol{f}. \tag{12}$$

Then

$$\boldsymbol{h}\boldsymbol{f}' = \boldsymbol{i}', \tag{13}$$

because, by (13), $\boldsymbol{h}\boldsymbol{f}' = \boldsymbol{i}'\boldsymbol{p}'\boldsymbol{h}\boldsymbol{f}' + \boldsymbol{i}''\boldsymbol{p}''\boldsymbol{h}\boldsymbol{f}' = \boldsymbol{i}'\boldsymbol{g}'\boldsymbol{f}' + \boldsymbol{i}''\boldsymbol{f}\boldsymbol{f}' = \boldsymbol{i}'$. It remains to show that $\boldsymbol{h}$ is an isomorphism, i.e., that all $h_\lambda: X_\lambda \to X'_\lambda \oplus X''_\lambda$ are isomorphisms of modules. This readily follows by a diagram chasing argument, applied to the following commutative diagram with exact rows (special case of the "five lemma").

$$\begin{array}{ccccccccc}
0 & \longrightarrow & X'_\lambda & \xrightarrow{f'_\lambda} & X_\lambda & \xrightarrow{f_\lambda} & X''_\lambda & \longrightarrow & 0 \\
 & & {\scriptstyle =}\downarrow & & \downarrow{\scriptstyle h_\lambda} & & \downarrow{\scriptstyle =} & & \\
0 & \longrightarrow & X'_\lambda & \xrightarrow[i'_\lambda]{} & X'_\lambda \oplus X''_\lambda & \xrightarrow[p''_\lambda]{} & X''_\lambda & \longrightarrow & 0\,.
\end{array} \tag{14}$$

The proof that (*iii*) implies (*i*) is dual. If $\boldsymbol{g}: \boldsymbol{X}'' \to \boldsymbol{X}$ is a morphism such that $\boldsymbol{f}\boldsymbol{g} = \mathrm{id}$, then there exists a morphism $\boldsymbol{k}: \boldsymbol{X}' \oplus \boldsymbol{X}'' \to \boldsymbol{X}$ such that $\boldsymbol{k}\boldsymbol{i}' = \boldsymbol{f}'$ and $\boldsymbol{f}\boldsymbol{k} = \boldsymbol{p}''$. A diagram chase shows that $\boldsymbol{k}$ is an isomorphism. □

11.2 Projective and injective systems

In this subsection in an inverse system of R-modules $\boldsymbol{X} = (X_\lambda, p_{\lambda\lambda'}, \Lambda)$ we view the modules $X_\lambda, X_{\lambda'}, \lambda \neq \lambda'$, as disjoint. Therefore, every element $x \in \cup_\lambda X_\lambda$ admits a unique index $\lambda \in \Lambda$ such that $x \in X_\lambda$. We denote this index by $\lambda(x)$. A set $E \subseteq \cup_\lambda X_\lambda$ is a *set of generators* for $\boldsymbol{X}$ provided, for every $x \in \cup_\lambda X_\lambda, x \neq 0$, there exist elements $e_i \in E$, $i = 1, \dots, n$, such that $\lambda(x) \leq \lambda(e_i)$ and there exist elements $r_i \in R, i = 1, \dots, n$, satisfying

$$x = \sum_{i=1}^{n} r_i p_{\lambda(x)\lambda(e_i)}(e_i). \tag{1}$$

$\boldsymbol{X}$ is *finitely generated* if it admits a finite set of generators E. A set of generators is a *basis* if the coefficients r_i in (1) are uniquely determined by x. An inverse system is called *free* if it admits a basis. A free system is of *finite type* if it admits a finite basis. $\boldsymbol{X}$ is *free cyclic*, provided it has a basis consisting of a single element e. In this case, for $\lambda \leq \mu = \lambda(e)$, $r \mapsto r\, p_{\lambda\mu}$ defines an isomorphism $R \to X_\lambda$. For any other λ, $X_\lambda = 0$. Consequently,

$$X_\lambda = \begin{cases} R, & \lambda \leq \mu, \\ 0, & \lambda \nleq \mu, \end{cases} \tag{2}$$

$$p_{\lambda\lambda'} = \begin{cases} \text{id}, & \lambda' \leq \mu, \\ 0, & \lambda' \nleq \mu. \end{cases} \tag{3}$$

Similarly, if $\boldsymbol{X}$ is a free inverse system with a basis E, then X_λ is a free R-module with a basis E_λ, which consists of all $p_{\lambda\lambda(e)}(e)$, where $\lambda \leq \lambda(e)$, $e \in E$. Every free system is a direct sum of free cyclic systems. A system $\boldsymbol{X}$ is *finitely presented* if it is the quotient $\boldsymbol{X} = \boldsymbol{F}/\boldsymbol{G}$ of a free system of finite type $\boldsymbol{F}$ and of a finitely generated subsystem $\boldsymbol{G}$.

LEMMA 11.11. *Let $\boldsymbol{X} = (X_\lambda, p_{\lambda\lambda'}, \Lambda)$ be a free inverse system with a basis E and let $\boldsymbol{Y} = (Y_\lambda, q_{\lambda\lambda'}, \Lambda)$ be an arbitrary inverse system. Then, for any choice of values $\lambda(e) \in \Lambda$ and $y(e) \in Y_{\lambda(e)}$, $e \in E$, there exists a unique morphism $\boldsymbol{f} = (f_\lambda) \colon \boldsymbol{X} \to \boldsymbol{Y}$ such that $f_{\lambda(e)}(e) = y(e)$.*

Proof. First assume that $\boldsymbol{X}$ is free cyclic with a basis consisting of a unique element e. For $\lambda \leq \lambda(e)$, define $f_\lambda \colon X_\lambda \to Y_\lambda$, by putting $f_\lambda(rp_{\lambda\lambda(e)}(e)) = rq_{\lambda\lambda(e)}(y(e))$, for $r \in R$. If $\lambda \nleq \lambda(e)$, $X_\lambda = 0$ and therefore, put $f_\lambda = 0$. Using (2) and (3), one readily verifies that the homomorphisms f_λ form a morphism $\boldsymbol{f}$ having the desired property. The uniqueness of $\boldsymbol{f}$ is obvious. The general case follows, using the universal property of direct sums and the fact that $\boldsymbol{X}$ is a direct sum of free cyclic systems. □

LEMMA 11.12. *Every inverse system $\boldsymbol{Y} = (Y_\lambda, q_{\lambda\lambda'}, \Lambda)$ admits an epimorphism $\boldsymbol{f} = (f_\lambda) \colon \boldsymbol{F} \to \boldsymbol{Y}$, where $\boldsymbol{F} = (F_\lambda, p_{\lambda\lambda'}, \Lambda)$ is a free inverse system. If $\boldsymbol{Y}$ is finitely generated, one can achieve that $\boldsymbol{F}$ has a finite basis.*

Proof. For each $\mu \in \Lambda$, choose a free R-module F^μ with a basis E^μ and an epimorphism $r^\mu\colon F^\mu \to Y_\mu$. For $\lambda, \mu \in \Lambda$, define an R-module F^μ_λ, by putting

$$F^\mu_\lambda = \begin{cases} F^\mu, & \lambda \le \mu, \\ 0, & \lambda \not\le \mu. \end{cases} \tag{4}$$

For $\lambda \le \lambda'$ and $\mu \in \Lambda$, define $p^\mu_{\lambda\lambda'}\colon F^\mu_{\lambda'} \to F^\mu_\lambda$, by putting

$$p^\mu_{\lambda\lambda'} = \begin{cases} \mathrm{id}, & \lambda' \le \mu, \\ 0, & \lambda' \not\le \mu. \end{cases} \tag{5}$$

It is readily seen that $\boldsymbol{F}^\mu = (F^\mu_\lambda, p^\mu_{\lambda\lambda'}, \Lambda)$ is a free inverse system having E^μ for a basis. Clearly, the direct sum $\boldsymbol{F} = \bigoplus_\mu \boldsymbol{F}^\mu$ is a also a free system $\boldsymbol{F} = (F_\lambda, p_{\lambda\lambda'}, \Lambda)$, having $E = \cup_\mu E^\mu$ for a basis. Note that

$$F_\lambda = \bigoplus_\mu F^\mu_\lambda \approx \bigoplus_{\mu \ge \lambda} F^\mu, \tag{6}$$

$$p_{\lambda\lambda'} = \bigoplus_\mu p^\mu_{\lambda\lambda'}. \tag{7}$$

The canonical injections $\boldsymbol{i}^\mu\colon \boldsymbol{F}^\mu \to \boldsymbol{F}$ consist of natural injections $i^\mu_\lambda\colon F^\mu_\lambda \to F_\lambda$.

Next we define homomorphisms $f^\mu_\lambda\colon F^\mu_\lambda \to Y_\lambda$, by putting

$$f^\mu_\lambda = \begin{cases} q_{\lambda\mu} r^\mu, & \lambda \le \mu, \\ 0, & \lambda \not\le \mu. \end{cases} \tag{8}$$

It is readily seen that the homomorphisms f^μ_λ, $\lambda \in \Lambda$, form a morphism $\boldsymbol{f}^\mu\colon \boldsymbol{F}^\mu \to \boldsymbol{Y}$. Since $\boldsymbol{F}$ is a coproduct, one obtains a morphism $\boldsymbol{f} = (f_\lambda)\colon \boldsymbol{F} \to \boldsymbol{Y}$ such that

$$\boldsymbol{f}\boldsymbol{i}^\mu = \boldsymbol{f}^\mu, \ \mu \in \Lambda. \tag{9}$$

Notice that (9) implies

$$f_\lambda i^\mu_\lambda = f^\mu_\lambda, \ \lambda, \mu \in \Lambda. \tag{10}$$

Since $f^\lambda_\lambda = r^\lambda\colon F^\lambda \to Y_\lambda$ is an epimorphism, (10) for $\lambda = \mu$ shows that each f_λ is an epimorphism. Hence, Lemma 11.2 implies that $\boldsymbol{f}$ is an epimorphism.

If G is a finite set of generators of $\boldsymbol{Y}$, G is contained in a finite union $Y_{\mu_1} \cup Y_{\mu_k}$ and $G \cap Y_{\mu_i}$ is a finite set of generators for Y_{μ_i}, for every $i \in \{1, \dots, k\}$. Therefore, one can assume that every module F^{μ_i} has a finite basis. Consequently, $\boldsymbol{F} = \cup_{1 \le i \le k} \boldsymbol{F}^{\mu_i}$ is a free system having a finite basis. Finally, the morphisms $\boldsymbol{f}^{\mu_i}$ determine the desired epimorphism $\boldsymbol{f}\colon \boldsymbol{F} \to \boldsymbol{Y}$. □

Projective and injective objects are defined in any category $\mathcal{C}$. In particular, an object P is *projective* if for any epimorphism $h\colon A \to B$ and any morphism $k\colon P \to B$ there exists a morphism $g\colon P \to A$ such that $hg = k$.

Putting $k = \mathrm{id}$, we see that every epimorphism $h: A \to P$ has a right inverse $g: P \to A$, $hg = \mathrm{id}$. Conversely, if for an object P, every epimorphism $h: A \to P$ admits a right inverse, then P is projective.

In the category Mod_R of R-modules, free modules are projective. In the category Ab of abelian groups, i.e., for $R = \mathbb{Z}$, also the converse holds. This is because subgroups of a free abelian group are again free abelian groups. In general, the converse fails. E.g., if $R = \mathbb{Z}/6$, then $\mathbb{Z}/2$ and $\mathbb{Z}/3$ are projective R-modules, whose direct sum $\mathbb{Z}/2 \oplus \mathbb{Z}/3 \approx \mathbb{Z}/6$. Nevertheles, these modules are not free.

LEMMA 11.13. *Every free inverse system is projective.*

Proof. Let $\boldsymbol{F}$ be a free system with a basis E and assume that $\boldsymbol{h} = (h_\lambda): \boldsymbol{A} \to \boldsymbol{B}$ is an epimorphism and $\boldsymbol{k} = (k_\lambda): \boldsymbol{F} \to \boldsymbol{B}$ is a morphism. Since each h_λ is surjective, for each $e \in E$, there exists a point $a(e) \in A_{\lambda(e)}$ such that $h_{\lambda(e)}(a(e)) = k_{\lambda(e)}(e)$. By Lemma 11.11, there exists a unique $\boldsymbol{g} = (g_\lambda): \boldsymbol{F} \to \boldsymbol{A}$ such that $g_{\lambda(e)}(e) = a(e)$, for each $e \in E$. Clearly, $\boldsymbol{hg} = \boldsymbol{k}$. □

The next theorem is an immediate consequence of Lemmas 11.12 and 11.13.

THEOREM 11.14. *The category* Mod^Λ *has enough projective objects, i.e., every inverse system* $\boldsymbol{X} \in \mathsf{Mod}^\Lambda$ *admits a projective system* $\boldsymbol{P}$ *and an epimorphism* $\boldsymbol{f}: \boldsymbol{P} \to \boldsymbol{X}$.

LEMMA 11.15. *For an inverse system* $\boldsymbol{P}$ *the following two properties are equivalent.*

(*i*) $\boldsymbol{P}$ *is projective.*

(*ii*) $\boldsymbol{P}$ *is a direct summand of a free system* $\boldsymbol{F}$.

Proof. By Lemma 11.12, there exist a free system $\boldsymbol{F}$ and an epimorphism $\boldsymbol{f}: \boldsymbol{F} \to \boldsymbol{P}$. Since $\boldsymbol{P}$ is projective, $\boldsymbol{f}$ admits a right inverse $\boldsymbol{g}: \boldsymbol{P} \to \boldsymbol{F}$, $\boldsymbol{fg} = \mathrm{id}$. Apply Lemma 11.10 to the exact sequence $\boldsymbol{0} \to \mathrm{Ker}(\boldsymbol{f}) \to \boldsymbol{F} \xrightarrow{\boldsymbol{f}} \boldsymbol{P} \to \boldsymbol{0}$. It follows that $\boldsymbol{P}$ is a direct summand of $\boldsymbol{F}$. Conversely, let $\boldsymbol{F} = \boldsymbol{P} \oplus \boldsymbol{X}$, where $\boldsymbol{F}$ is a free system, and let $\boldsymbol{i}: \boldsymbol{P} \to \boldsymbol{F}$ and $\boldsymbol{p}: \boldsymbol{F} \to \boldsymbol{P}$ be the canonical injection and the canonical projection, respectively. Then $\boldsymbol{pi} = \mathrm{id}$. However, by Lemma 11.13, $\boldsymbol{F}$ is projective and it immediately follows that $\boldsymbol{P}$ is also a projective system. □

REMARK 11.16. If $\boldsymbol{P} = (P_\lambda, p_{\lambda\lambda'}, \Lambda)$ is a projective system, then every P_λ is a projective R-module.

Indeed, for $\mu \in \Lambda$, let $k: P_\mu \to B$ be a homomorphism and let $h: A \to B$ be an epimorphism of modules. We need a homomorphism $g: P_\mu \to A$ such that $hg = k$. It suffices to consider inverse systems $\boldsymbol{A} = (A_\lambda, a_{\lambda\lambda'}, \Lambda)$, $\boldsymbol{B} = (B_\lambda, b_{\lambda\lambda'}, \Lambda)$, an epimorphism $\boldsymbol{h} = (h_\lambda): \boldsymbol{A} \to \boldsymbol{B}$ and a morphism $\boldsymbol{k} = (k_\lambda): \boldsymbol{P} \to \boldsymbol{B}$ such that $A_\mu = A$, $B_\mu = B$, $h_\mu = h$ and $k_\mu = k$. Indeed, in that case, one concludes that there exists a morphism $\boldsymbol{g} = (g_\lambda): \boldsymbol{P} \to \boldsymbol{A}$

such that $\boldsymbol{hg} = \boldsymbol{k}$ and thus, $g = g_\mu$ has the desired property. One defines the required objects, by putting $A_\lambda = A$, $B_\lambda = B$, $h_\lambda = h$, $k_\lambda = kp_{\mu\lambda}$, if $\mu \leq \lambda$, and $A_\lambda = B_\lambda = h_\lambda = k_\lambda = 0$, if $\mu \not\leq \lambda$, and by putting $a_{\lambda\lambda'} = b_{\lambda\lambda'} = \mathrm{id}$, if $\mu \leq \lambda$, and $a_{\lambda\lambda'} = b_{\lambda\lambda'} = 0$, if $\mu \not\leq \lambda$.

The following is an example of an inverse system which consists of projective R-modules, but is not a projective object of Mod^Λ.

EXAMPLE 11.17. For a directed set Λ let $\boldsymbol{\Delta}(\Lambda)$, shorter $\boldsymbol{\Delta}$, be the *diagonal inverse system* $\boldsymbol{\Delta} = (\Delta_\lambda, u_{\lambda\lambda'}, \Lambda)$, i.e., $\Delta_\lambda = R$ and $u_{\lambda\lambda'} = \mathrm{id}$, for all $\lambda \leq \lambda'$. If Λ has no terminal element (e.g., if $\Lambda = \mathbb{N}$), then $\boldsymbol{\Delta}$ is not a projective object of Mod^Λ.

To verify the assertion, consider the system $\boldsymbol{P} = (P_\lambda, i_{\lambda\lambda'}, \Lambda)$, where

$$P_\lambda = \bigoplus_{\mu \geq \lambda} R, \tag{11}$$

and all the bonding homomorphisms $i_{\lambda\lambda'}$ are natural inclusions. For $\mu \geq \lambda$, let $\langle\mu\rangle \in P_\lambda$ denote the generator $1 \in R$ of the summand corresponding to the index μ. Let $f_\lambda\colon P_\lambda \to \Delta_\lambda = R$ be the homomorphism, given by $f_\lambda\langle\mu\rangle = 1$, for all $\mu \geq \lambda$. Clearly, $f_\lambda u_{\lambda\lambda'} = i_{\lambda\lambda'} f_{\lambda'}$, for $\lambda \leq \lambda'$, and therefore, $\boldsymbol{f} = (f_\lambda)\colon \boldsymbol{P} \to \boldsymbol{\Delta}$ is a morphism. Since all f_λ are epimorphisms, $\boldsymbol{f}$ is also an epimorphism. The assumption that $\boldsymbol{\Delta}$ is projective would imply the existence of a right inverse $\boldsymbol{g}\colon \boldsymbol{\Delta} \to \boldsymbol{P}$ of $\boldsymbol{f}$, $\boldsymbol{fg} = \mathrm{id}$. However, this is impossible, because every morphism $\boldsymbol{g} = (g_\lambda)\colon \boldsymbol{\Delta} \to \boldsymbol{P}$ equals $\boldsymbol{0}$.

Indeed, for an arbitrary $\lambda \in \Lambda$, one has $g_\lambda(1) \in P_\lambda$. By (1), there exists a finite sequence of distinct indices $\mu_1, \ldots, \mu_k \geq \lambda$ and there exists a sequence of elements $r_1, \ldots, r_k \in R$, such that

$$g_\lambda(1) = r_1\langle\mu_1\rangle + \ldots + r_k\langle\mu_k\rangle. \tag{12}$$

Since Λ has no terminal element, there exists a $\lambda' \in \Lambda$, such that $\lambda' > \lambda, \mu_1, \ldots, \mu_k$ (see Remark 1.3). Now, $g_\lambda(1) = i_{\lambda\lambda'} g_{\lambda'}(1) = g_{\lambda'}(1)$ and $g_{\lambda'}(1)$ is a linear combination of elements of the form $\langle\mu\rangle$, where $\mu \geq \lambda'$. Therefore, the coefficients $r_1, \ldots, r_k$, which belong to the indices $\mu_1, \ldots, \mu_k < \lambda'$ all vanish and thus, $g_\lambda(1) = 0$, i.e., $g_\lambda = 0$.

An object I in a category $\mathcal{C}$ is *injective* if for any monomorphism $h\colon B \to A$ and any morphism $k\colon B \to I$, there exists a morphism $g\colon A \to I$ such that $gh = k$. Putting $k = \mathrm{id}$, we see that every monomorphism $h\colon I \to A$ has a left inverse $g\colon A \to I$, $gh = \mathrm{id}$. Conversely, if for an object I, every monomorphism $h\colon I \to A$ admits a left inverse, then I is injective.

In the category Ab injective objects coincide with *divisible* abelian groups, i.e., abelian groups I such that, for every $n \in \mathbb{Z}$, $n \neq 0$, every element $x \in I$ admits an element $y \in I$ such that $x = ny$ (see e.g., (Hilton, Stammbach 1971), Ch. I, Theorem 7.1). In particular, the group of rationals $\mathbb{Q}$ is injective, while the group of integers $\mathbb{Z}$ is not. Every abelian group X embeds as a subgroup in a divisible (hence injective) abelian group. Indeed, there exists a free abelian group F and an epimorphism $f: F \to X$. If K is the kernel of f, then X is isomorphic to F/K. However, F is isomorphic to a direct sum $\oplus \mathbb{Z}$ of copies of $\mathbb{Z}$. Since $\oplus \mathbb{Z} \subseteq \oplus \mathbb{Q}$, we see that X is isomorphic to a subgroup of $I = (\oplus \mathbb{Q})/K$. Clearly, the group $\oplus \mathbb{Q}$ is divisible and so must be its homomorphic image I.

In general, every R-module X embeds in an injective R-module X (see e.g., (Hilton, Stammbach 1971), Ch. I, Proposition 8.3). We will now establish the analoguous assertion for inverse systems of R-modules and thus obtain the analogue of Theorem 11.14.

THEOREM 11.18. *The category* Mod^{Λ} *has enough injective objects, i.e., every inverse system* $\boldsymbol{X} \in \mathsf{Mod}^{\Lambda}$ *admits an injective system* $\boldsymbol{I}$ *and a monomorphism* $\boldsymbol{f}: \boldsymbol{X} \to \boldsymbol{I}$.

Proof. Consider any system $\boldsymbol{X} = (X_\lambda, p_{\lambda\lambda'}, \Lambda)$. For each $\mu \in \Lambda$ choose an injective R-module I^{μ} and a monomorphism $s^{\mu}: X_{\mu} \to I^{\mu}$. For $\lambda, \mu \in \Lambda$, define an R-module I^{μ}_{λ} by putting

$$I^{\mu}_{\lambda} = \begin{cases} I^{\mu}, & \mu \leq \lambda, \\ 0, & \mu \not\leq \lambda. \end{cases} \tag{13}$$

Moreover, for $\lambda \leq \lambda'$ and $\mu \in \Lambda$, define $q^{\mu}_{\lambda\lambda'}: I^{\mu}_{\lambda'} \to I^{\mu}_{\lambda}$ by putting

$$q^{\mu}_{\lambda\lambda'} = \begin{cases} \mathrm{id}, & \mu \leq \lambda, \\ 0, & \mu \not\leq \lambda. \end{cases} \tag{14}$$

It is readily seen that $\boldsymbol{I}^{\mu} = (I^{\mu}_{\lambda}, q^{\mu}_{\lambda\lambda'}, \Lambda)$ is an object of Mod^{Λ}. Now consider the direct product $\boldsymbol{I} = \prod_{\mu} \boldsymbol{I}^{\mu}$. Note that $\boldsymbol{I} = (I_\lambda, q_{\lambda\lambda'}, \Lambda)$, where

$$I_\lambda = \prod_{\mu} I^{\mu}_{\lambda} \approx \prod_{\mu \leq \lambda} I^{\mu}, \tag{15}$$

$$q_{\lambda\lambda'} = \prod_{\mu} q^{\mu}_{\lambda\lambda'}. \tag{16}$$

The canonical morphisms $\boldsymbol{p}^{\mu} = (\pi^{\mu}_{\lambda}): \boldsymbol{I} \to \boldsymbol{I}^{\mu}$, $\mu \in \Lambda$, consist of natural projections $\pi^{\mu}_{\lambda}: I_\lambda \to I^{\mu}_{\lambda}$. Next define homomorphisms $f^{\mu}_{\lambda}: X_\lambda \to I^{\mu}_{\lambda}$, by putting

$$f^{\mu}_{\lambda} = \begin{cases} s^{\mu} p_{\mu\lambda}, & \mu \leq \lambda, \\ 0, & \mu \not\leq \lambda. \end{cases} \tag{17}$$

It is readily seen that the homomorphisms f^{μ}_{λ}, $\lambda \in \Lambda$, form a morphism $\boldsymbol{f}^{\mu}: \boldsymbol{X} \to \boldsymbol{I}^{\mu}$. Since $\boldsymbol{I}$ is a product, one obtains a morphism $\boldsymbol{f} = (f_\lambda): \boldsymbol{X} \to \boldsymbol{I}$ such that

$$\boldsymbol{p}^{\mu} \boldsymbol{f} = \boldsymbol{f}^{\mu}, \ \mu \in \Lambda. \tag{18}$$

In order to conclude that $\boldsymbol{f}$ is a monomorphism, note that

$$\pi^{\mu}_{\lambda} f_\lambda = f^{\mu}_{\lambda}, \ \lambda, \mu \in \Lambda. \tag{19}$$

Since $f_\lambda^\lambda = s^\lambda$ is a monomorphism, (19) for $\lambda = \mu$, shows that also f_λ is a monomorphism, for all $\lambda \in \Lambda$. Hence, by Lemma 11.2, $\boldsymbol{f}$ is a monomorphism.

To complete the proof it remains to show that $\boldsymbol{I}$ is an injective system. Since a product of injective objects is always injective, it suffices to see that $\boldsymbol{I}^\mu$ is injective. Consider objects $\boldsymbol{A} = (A_\lambda, a_{\lambda\lambda'}, \Lambda), \boldsymbol{B} = (B_\lambda, b_{\lambda\lambda'}, \Lambda)$ and morphisms $\boldsymbol{h} = (h_\lambda)\colon \boldsymbol{B} \to \boldsymbol{A}, \boldsymbol{k} = (k_\lambda)\colon \boldsymbol{B} \to \boldsymbol{I}^\mu$ and assume that $\boldsymbol{h}$ is a monomorphism, i.e., all h_λ are injective. We must produce a morphism $\boldsymbol{g} = (g_\lambda)\colon \boldsymbol{A} \to \boldsymbol{I}^\mu$ such that $\boldsymbol{gh} = \boldsymbol{k}$. Since I^μ is injective, there exists a homomorphism $u_\mu\colon A_\mu \to I^\mu$ such that

$$u_\mu h_\mu = k_\mu. \tag{20}$$

We then define $g_\lambda\colon A_\lambda \to I_\lambda^\mu$ by putting

$$g_\lambda = \begin{cases} u_\mu a_{\mu\lambda}, & \mu \le \lambda, \\ 0, & \mu \not\le \lambda. \end{cases} \tag{21}$$

For $\lambda \le \lambda'$, one obtains

$$q_{\lambda\lambda'}^\mu g_{\lambda'} = g_\lambda a_{\lambda\lambda'}. \tag{22}$$

Indeed, for $\mu \le \lambda$, (22) assumes the form $u_\mu a_{\mu\lambda'} = u_\mu a_{\mu\lambda} a_{\lambda\lambda'}$. For the remaining μ, both sides of (22) equal 0. Now (22) shows that $\boldsymbol{g} = (g_\lambda)\colon \boldsymbol{A} \to \boldsymbol{I}^\mu$ is a morphism of Mod^Λ.

In order to prove that $\boldsymbol{gh} = \boldsymbol{k}$, i.e.,

$$g_\lambda h_\lambda = k_\lambda, \tag{23}$$

one can assume that $\mu \le \lambda$, for in the remaining cases both sides equal 0. Using (21),(24), (20),(25),(14) and the relations

$$a_{\mu\lambda} h_\lambda = h_\mu b_{\mu\lambda}, \tag{24}$$

$$k_\mu b_{\mu\lambda} = q_{\mu\lambda}^\mu k_\lambda, \tag{25}$$

one obtains, indeed,

$$g_\lambda h_\lambda = u_\mu a_{\mu\lambda} h_\lambda = u_\mu h_\mu b_{\mu\lambda} = k_\mu b_{\mu\lambda} = q_{\mu\lambda}^\mu k_\lambda = k_\lambda. \ \square \tag{26}$$

REMARK 11.19. If $\boldsymbol{I} = (I_\lambda, q_{\lambda\lambda'}, \Lambda)$ is an injective system, then every I_μ, $\mu \in \Lambda$, is an injective R-module. A proof of this assertion is obtained by dualizing the proof of Remark 11.16.

LEMMA 11.20. *If in a short exact sequence* (11) $\boldsymbol{X}'$ *is injective, or* $\boldsymbol{X}''$ *is projective, then the sequence splits.*

Proof. If $\boldsymbol{X}'$ is injective, the monomorphism $f'\colon \boldsymbol{X}' \to \boldsymbol{X}$ has a left inverse and if $\boldsymbol{X}''$ is projective, the epimorphism $f\colon \boldsymbol{X} \to \boldsymbol{X}''$ has a right inverse. Hence, in both cases Lemma 11.10 shows that the sequence splits. □

A *projective resolution* of an object $\boldsymbol{X}$ of Mod^Λ is an exact sequence

$$\boldsymbol{0} \leftarrow \boldsymbol{X} \xleftarrow{e} \boldsymbol{P}_0 \xleftarrow{d_1} \boldsymbol{P}_1 \xleftarrow{d_2} \boldsymbol{P}_2 \leftarrow \dots, \tag{27}$$

where every $\boldsymbol{P}_n$, $n \geq 0$, is a projective system from Mod^Λ. Consequently, the resolution (27) consists of the epimorphism $\boldsymbol{e} : \boldsymbol{P}_0 \to \boldsymbol{X}$ and of the sequence

$$\mathcal{P} = (\boldsymbol{0} \leftarrow \boldsymbol{P}_0 \xleftarrow{d_1} \boldsymbol{P}_1 \xleftarrow{d_2} \boldsymbol{P}_2 \leftarrow \dots). \tag{28}$$

For (27) we will sometimes use the notation $(\mathcal{P}, \boldsymbol{e})$.

Dually, an *injective resolution* of $\boldsymbol{X}$ is an exact sequence

$$\boldsymbol{0} \to \boldsymbol{X} \xrightarrow{e} \boldsymbol{I}^0 \xrightarrow{d^1} \boldsymbol{I}^1 \xrightarrow{d^2} \boldsymbol{I}^2 \to \dots, \tag{29}$$

where every $\boldsymbol{I}^n$, $n \geq 0$, is an injective system from Mod^Λ. Consequently, the resolution (29) consists of the monomorphism $\boldsymbol{e}\colon \boldsymbol{P}_0 \to \boldsymbol{X}$ and of the sequence

$$\mathcal{I} = (\boldsymbol{0} \to \boldsymbol{I}^0 \xrightarrow{d^1} \boldsymbol{I}^1 \xrightarrow{d^2} \boldsymbol{I}^2 \to \dots). \tag{30}$$

For (29) we will sometimes use the notation $(\mathcal{I}, \boldsymbol{e})$.

THEOREM 11.21. *In the category* Mod^Λ *every object* $\boldsymbol{X}$ *admits projective and injective resolutions.*

Proof. To prove the first assertion, note that Theorem 11.14 yields a projective system $\boldsymbol{P}_0$ and an epimorphism $\boldsymbol{e}\colon \boldsymbol{P}_0 \to \boldsymbol{X}$. Taking for $\boldsymbol{u}_1\colon \boldsymbol{R}_0 \to \boldsymbol{P}_0$ the kernel of $\boldsymbol{e}$, one obtains an exact sequence

$$\boldsymbol{0} \leftarrow \boldsymbol{X} \xleftarrow{e} \boldsymbol{P}_0 \xleftarrow{u_1} \boldsymbol{R}_0 \leftarrow \boldsymbol{0}, \tag{31}$$

with $\boldsymbol{P}_0$ projective, called a *projective presentation* of $\boldsymbol{X}$. In the same way one can produce, by induction, projective presentations

$$\boldsymbol{0} \leftarrow \boldsymbol{R}_{i-1} \xleftarrow{e_i} \boldsymbol{P}_i \xleftarrow{u_{i+1}} \boldsymbol{R}_i \leftarrow \boldsymbol{0},\ i = 1, 2, \dots. \tag{32}$$

Putting $\boldsymbol{d}_i = \boldsymbol{u}_i \boldsymbol{e}_i$, one obtains a projective resolution (27). The dual argument, which uses Theorem 11.18 and *injective presentations,* yields an injective resolution (29). □

11.3 lim and its right derived functors

Let $\boldsymbol{X} = (X_\lambda, p_{\lambda\lambda'}, \Lambda)$ be an inverse system of R-modules. An *inverse limit* of $\boldsymbol{X}$ consists of an R-module X, denoted by $X = \lim \boldsymbol{X}$, and of homomorphisms $p_\lambda : X \to X_\lambda$ (called *projections*) such that $p_{\lambda\lambda'} p_{\lambda'} = p_{\lambda'}$, for $\lambda \leq \lambda'$. Moreover, if $p'_\lambda : X' \to X_\lambda$ are homomorphisms satisfying $p_{\lambda\lambda'} p'_{\lambda'} = p'_\lambda$, then there is a unique homomorphism $u: X' \to X$ such that $p_\lambda u = p'_\lambda$, for all $\lambda \in \Lambda$. Inverse limits of systems $\boldsymbol{X}$ from Mod^Λ exist and are unique up to natural isomorphism. The canonical construction defines X as the submodule of $\prod_\lambda X_\lambda$, which consists of all the points $x = (x_\lambda)$, which satisfy $x_\lambda = p_{\lambda\lambda'}(x_{\lambda'})$, for $\lambda \leq \lambda'$, and $p_\lambda(x) = x_\lambda$. Here $x_\lambda = \pi_\lambda(x)$, where $\pi_\lambda : \prod_\lambda X_\lambda \to X_\lambda$ is the canonical projection. A morphism $\boldsymbol{f} = (f_\lambda) : \boldsymbol{X} \to \boldsymbol{Y} = (Y_\lambda, q_{\lambda\lambda'}, \Lambda)$ of Mod^Λ induces a homomorphism $f = \lim \boldsymbol{f} : X \to Y = \lim \boldsymbol{Y}$. It is the only homomorphism f, which satisfies $q_\lambda f = f_\lambda p_\lambda$, for all $\lambda \in \Lambda$. Clearly, if $x = (x_\lambda) \in X$, then $f(x) = (f_\lambda(x_\lambda)) \in Y$. It is readily seen that $\lim \boldsymbol{1}_{\boldsymbol{X}} = 1_X$ and $\lim \boldsymbol{g}\boldsymbol{f} = \lim \boldsymbol{g} \lim \boldsymbol{f}$, which shows that lim is a functor $\lim : \mathsf{Mod}^\Lambda \to \mathsf{Mod}$.

LEMMA 11.22. *The functor* $\lim : \mathsf{Mod}^\Lambda \to \mathsf{Mod}$ *is additive and left exact.*

Proof. The assertion that lim is an *additive functor* means that, for any two morphisms $\boldsymbol{f}, \boldsymbol{f}' : \boldsymbol{X} \to \boldsymbol{Y}$, one has $\lim(\boldsymbol{f} + \boldsymbol{f}') = \lim \boldsymbol{f} + \lim \boldsymbol{f}'$. This is an immediate consequence of the equality

$$q_\lambda (f + f') = q_\lambda f + q_\lambda f' = f_\lambda p_\lambda + f'_\lambda p_\lambda = (f_\lambda + f'_\lambda) p_\lambda, \tag{1}$$

where $f = \lim \boldsymbol{f}$, $f' = \lim \boldsymbol{f}'$.

The assertion that lim is *left exact* means that exactness of the sequence

$$\boldsymbol{0} \to \boldsymbol{X}' \xrightarrow{\boldsymbol{f}'} \boldsymbol{X} \xrightarrow{\boldsymbol{f}} \boldsymbol{X}'' \tag{2}$$

implies exactness of the sequence

$$0 \to \lim \boldsymbol{X}' \xrightarrow{\lim \boldsymbol{f}'} \lim \boldsymbol{X} \xrightarrow{\lim \boldsymbol{f}} \lim \boldsymbol{X}''. \tag{3}$$

In order to prove this assertion note that (2) implies exactness of the sequence

$$0 \to X'_\lambda \xrightarrow{f'_\lambda} X_\lambda \xrightarrow{f_\lambda} X''_\lambda. \tag{4}$$

If $x' = (x'_\lambda) \in X' = \lim \boldsymbol{X}'$ and $f'(x') = (f'_\lambda(x'_\lambda)) = 0$, then by (4), $x'_\lambda = 0$, for all $\lambda \in \Lambda$. Hence, $x' = 0$ and the exactness of (3) at X' is established. Now assume that $x = (x_\lambda) \in X = \lim \boldsymbol{X}$ has the property that $f(x) = (f_\lambda(x_\lambda)) = 0$. Then by (4), there exist elements $x'_\lambda \in X'_\lambda$, $\lambda \in \Lambda$, such that $f'_\lambda(x'_\lambda) = x_\lambda$. Note that $x' = (x'_\lambda) \in X'$, i.e.,

$$p'_{\lambda\lambda'}(x'_{\lambda'}) = x'_\lambda, \ \lambda \leq \lambda'. \tag{5}$$

Indeed,

$$f'_\lambda p'_{\lambda\lambda'}(x'_{\lambda'}) = p_{\lambda\lambda'} f'_{\lambda'}(x'_{\lambda'}) = p_{\lambda\lambda'}(x_{\lambda'}) = x_\lambda = f'_\lambda(x'_\lambda). \tag{6}$$

However, f'_λ is an injection, and therefore, (6) implies (5). Since $f'(x') = x$, we have proved that $\mathrm{Ker}(f) \subseteq \mathrm{Im}(f')$. The opposite inclusion holds because $\boldsymbol{f}\boldsymbol{f}' = \boldsymbol{0}$ implies $ff' = 0$. $\square$

In general, the functor lim is not *exact*, i.e., from the exactness of

$$\boldsymbol{0} \to \boldsymbol{X}' \xrightarrow{\boldsymbol{f}'} \boldsymbol{X} \xrightarrow{\boldsymbol{f}} \boldsymbol{X}'' \to \boldsymbol{0} \tag{7}$$

it does not follow the exactness of

$$0 \to \lim \boldsymbol{X}' \xrightarrow{\lim \boldsymbol{f}'} \lim \boldsymbol{X} \xrightarrow{\lim \boldsymbol{f}} \lim \boldsymbol{X}'' \to 0. \tag{8}$$

This is shown by the following example of inverse systems of abelian groups.

EXAMPLE 11.23. Let $\Lambda = \mathbb{N}$ be the set of natural numbers and let

$$\boldsymbol{X}' = \boldsymbol{X} = (\mathbb{Z} \xleftarrow{2} \mathbb{Z} \xleftarrow{2} \mathbb{Z} \leftarrow \ldots),\ \boldsymbol{X}'' = (\mathbb{Z}/3 \xleftarrow{2} \mathbb{Z}/3 \xleftarrow{2} \mathbb{Z}/3 \leftarrow \ldots), \tag{9}$$

where $2\colon \mathbb{Z} \to \mathbb{Z}$ denotes multiplication by 2 and $2\colon \mathbb{Z}/3 \to \mathbb{Z}/3$ denotes the induced homomorphism. Let $\boldsymbol{f}'$ consist of homomorphisms $3\colon \mathbb{Z} \to \mathbb{Z}$ and let $\boldsymbol{f}$ consist of quotient homomorphisms $\mathbb{Z} \to \mathbb{Z}/3$. Then (7) is exact. However, $f = \lim \boldsymbol{f}$ is not a surjection. Indeed, $X = \lim \boldsymbol{X} = 0$, while $X'' = \lim \boldsymbol{X}'' \approx \mathbb{Z}/3$, because all the bonding homomorphisms $p''_{\lambda\lambda'}$ are isomorphisms.

LEMMA 11.24. *The functor lim preserves finite products.*

Proof. Let $\boldsymbol{X} = \boldsymbol{X}' \oplus \boldsymbol{X}''$ and let $\boldsymbol{i}'\colon \boldsymbol{X}' \to \boldsymbol{X}$, $\boldsymbol{i}''\colon \boldsymbol{X}'' \to \boldsymbol{X}$, $\boldsymbol{p}'\colon \boldsymbol{X} \to \boldsymbol{X}'$ and $\boldsymbol{p}''\colon \boldsymbol{X} \to \boldsymbol{X}''$ be canonical injections and projections. By Lemma 11.7, their limits X, i', i'', p' and p'' satisfy

$$p'i' = 1_{X'},\ p''i'' = 1_{X''},\ p'i'' = 0,\ p''i' = 0, \tag{10}$$

$$i'p' + i''p'' = 1_X. \tag{11}$$

However, these relations imply that X is the direct sum $X' \times X'' = X' \oplus X''$ and i', i'', p' and p'' are the canonical injections and projections. $\square$

REMARK 11.25. If in an exact sequence (7) $\boldsymbol{X}'$ is injective or $\boldsymbol{X}''$ is projective, then the sequence (8) is exact. Indeed, by Lemma 11.22, the sequence (7) splits and the assertion follows from Lemma 11.24.

EXAMPLE 11.26. If R is a field, R-modules are vector spaces. In this case, lim is an exact functor. Indeed, it is easy to show that every inverse system of vector spaces has a basis (use Zorn's lemma), i.e., it is a free system. Therefore, by Lemma 11.13, it is also projective. Therefore, Remark 11.25 applies.

Whenever $\mathcal{A}$ is an abelian category with enough injective objects and $S: \mathcal{A} \to \mathsf{Mod}$ is an additive left exact functor, then a standard procedure of homological algebra yields a sequence of additive functors S^n, $n \geq 0$, called the *right derived functors* of S. These functors measure discrepancy from exactness of S and S can be identified with S^0 (see e.g., (Bucur, Deleanu 1968), Ch.7 or (Hilton, Stammbach 1971), Ch.IV). Since Mod^Λ is an abelian category with enough injective objects and lim: $\mathsf{Mod}^\Lambda \to \mathsf{Mod}$ is an additive left exact functor, the procedure applies and yields a sequence of additive functors, called the *right derived functors* of the functor lim or the *higher limits.* They are denoted by $\lim^n : \mathsf{Mod}^\Lambda \to \mathsf{Mod}$, $n = 0, 1, 2, \ldots$. These functors measure discrepancy from exactness of the functor $\lim = \lim^0$. We shall now describe the procedure in the case of lim.

To every inverse system $\boldsymbol{X}$ assign a (fixed) injective resolution $(\mathcal{I}, \boldsymbol{e}) = (\mathcal{I}(\boldsymbol{X}), \boldsymbol{e}(\boldsymbol{X}))$ of $\boldsymbol{X}$,

$$\mathbf{0} \to \boldsymbol{X} \xrightarrow{\boldsymbol{e}} \boldsymbol{I}^0 \xrightarrow{\boldsymbol{d}^1} \boldsymbol{I}^1 \xrightarrow{\boldsymbol{d}^2} \boldsymbol{I}^2 \to \ldots, \tag{12}$$

This is possible due to Theorem 11.21. Application of the functor lim to (12) yields a sequence of modules

$$0 \to X \xrightarrow{e} I^0 \xrightarrow{d^1} I^1 \xrightarrow{d^2} I^2 \to \ldots, \tag{13}$$

where $X = \lim \boldsymbol{X}$, $I^n = \lim \boldsymbol{I}^n$, $e = \lim \boldsymbol{e}$, $d^n = \lim \boldsymbol{d}^n$. Note that

$$I = I_{\boldsymbol{X}} = (0 \to I^0 \xrightarrow{d^1} I^1 \xrightarrow{d^2} I^2 \to \ldots) \tag{14}$$

is a *cochain complex*, i.e., a sequence of modules I^i and homomorphisms $d^i: I^{i-1} \to I^i$ such that $d^{i+1}d^i = 0$, for $i = 1, 2, \ldots$. By definition, the n-th *derived limit* $\lim^n \boldsymbol{X}$ of $\boldsymbol{X}$ is the n-th cohomology module of $I_{\boldsymbol{X}}$, i.e.,

$$\lim^n \boldsymbol{X} = H^n(I) = \operatorname{Ker}(d^{n+1})/\operatorname{Im}(d^n), \quad n = 0, 1, \ldots, \tag{15}$$

where $d^0 = 0$.

Since lim is a left exact functor and the sequence

$$\mathbf{0} \to \boldsymbol{X} \xrightarrow{\boldsymbol{e}} \boldsymbol{I}^0 \xrightarrow{\boldsymbol{d}^1} \boldsymbol{I}^1 \tag{16}$$

is exact, it follows that the induced sequence of modules

$$0 \to X \xrightarrow{e} I^0 \xrightarrow{d^1} I^1 \tag{17}$$

is also exact. Therefore, $e: X \to I^0$ is a monomorphism and $e(X) = \operatorname{Ker}(d^1)$. On the other hand, it follows from (15) that $\operatorname{Ker}(d^1) = \lim^0 \boldsymbol{X}$. Consequently,

$$e: \lim \boldsymbol{X} \to \lim^0 \boldsymbol{X} \tag{18}$$

is an isomorphism, which identifies $\lim \boldsymbol{X}$ with $\lim^0 \boldsymbol{X}$. Hence, $\lim^0 \boldsymbol{X}$ does not depend on the choice of the resolution $(\mathcal{I}(\boldsymbol{X}), \boldsymbol{e}(\boldsymbol{X}))$.

In order to show that, also for $n \geq 1$, $\lim^n \boldsymbol{X}$ does not depend on this choice and to define homomorphisms $\lim^n \boldsymbol{h}: \lim^n \boldsymbol{X} \to \lim^n \boldsymbol{Y}$, induced by morphisms $\boldsymbol{h}: \boldsymbol{X} \to \boldsymbol{Y}$, we need two lemmas.

LEMMA 11.27. *Let $\boldsymbol{h}: \boldsymbol{X} \to \boldsymbol{Y}$ be a morphism* Mod^{Λ} *and let $(\mathcal{I}_X, \boldsymbol{e}_X)$ and $(\mathcal{I}_Y, \boldsymbol{e}_Y)$ be injective resolutions of $\boldsymbol{X}$ and $\boldsymbol{Y}$, respectively. Then there exist morphisms $\boldsymbol{h}^n: \boldsymbol{I}_X^n \to \boldsymbol{I}_Y^n$, $n \geq 0$, which make the following diagram commutative*

$$\begin{array}{ccccccccc}
0 & \longrightarrow & \boldsymbol{X} & \xrightarrow{\boldsymbol{e}_X} & \boldsymbol{I}_X^0 & \xrightarrow{\boldsymbol{d}_X^1} & \boldsymbol{I}_X^1 & \longrightarrow & \cdots \\
 & & \downarrow{\scriptstyle \boldsymbol{h}} & & \downarrow{\scriptstyle \boldsymbol{h}^0} & & \downarrow{\scriptstyle \boldsymbol{h}^1} & & \\
0 & \longrightarrow & \boldsymbol{Y} & \xrightarrow[\boldsymbol{e}_Y]{} & \boldsymbol{I}_Y^0 & \xrightarrow[\boldsymbol{d}_Y^1]{} & \boldsymbol{I}_Y^1 & \longrightarrow & \cdots .
\end{array} \tag{19}$$

We will refer to $(\boldsymbol{h}^n)$ as to an *injective resolution* of the morphism $\boldsymbol{h}$.

LEMMA 11.28. *Let $\boldsymbol{h}'^n: \boldsymbol{I}_X^n \to \boldsymbol{I}_Y^n$, $n \geq 0$, be another sequence of morphisms of* Mod^{Λ} *which make diagram* (19) *commutative. Then there exist morphisms $\boldsymbol{D}^n: \boldsymbol{I}_X^n \to \boldsymbol{I}_Y^{n-1}$, $n \geq 1$, such that*

$$\boldsymbol{D}^1 \boldsymbol{d}_X^1 = \boldsymbol{h}^0 - \boldsymbol{h}'^0, \tag{20}$$

$$\boldsymbol{d}_Y^n \boldsymbol{D}^n + \boldsymbol{D}^{n+1} \boldsymbol{d}_X^{n+1} = \boldsymbol{h}^n - \boldsymbol{h}'^n,\ n \geq 1. \tag{21}$$

To prove Lemmas 11.27 and 11.28, we need another lemma.

LEMMA 11.29. *Let $\boldsymbol{X}' \xrightarrow{\boldsymbol{f}'} \boldsymbol{X} \xrightarrow{\boldsymbol{f}} \boldsymbol{X}''$ be an exact sequence and let $\boldsymbol{g}: \boldsymbol{X} \to \boldsymbol{I}$ be a morphism to an injective system $\boldsymbol{I}$. If $\boldsymbol{g}\boldsymbol{f}' = \boldsymbol{0}$, then there exists a morphism $\boldsymbol{g}'': \boldsymbol{X}'' \to \boldsymbol{I}$ such that $\boldsymbol{g}''\boldsymbol{f} = \boldsymbol{g}$.*

Proof. Let $\boldsymbol{f} = \boldsymbol{swt}$ and $\boldsymbol{f}' = \boldsymbol{s}'\boldsymbol{w}'\boldsymbol{t}'$ be decompositions of $\boldsymbol{f}$ and $\boldsymbol{f}'$, as described in Lemma 11.3. By exactness at $\boldsymbol{X}$, $\boldsymbol{t}: \boldsymbol{X} \to \mathrm{Coim}(\boldsymbol{f})$, which consists of quotient homomorphisms $t_\lambda: X_\lambda \to X_\lambda / f_\lambda^{-1}(0) = X_\lambda / f'_\lambda(X'_\lambda)$ is the cokernel of $\boldsymbol{t}': \mathrm{Im}(\boldsymbol{f}) \to \mathrm{Im}(\boldsymbol{f}')$, which consists of inclusions $f'_\lambda(X'_\lambda) \to X_\lambda$. By assumption $\boldsymbol{g}\boldsymbol{s}'\boldsymbol{w}'\boldsymbol{t}' = \boldsymbol{0}$, and since $\boldsymbol{w}'\boldsymbol{t}'$ is an epimorphism, also $\boldsymbol{g}\boldsymbol{s}' = \boldsymbol{0}$. Consequently, there exists a morphism $\boldsymbol{k}: \boldsymbol{X}'' \to \boldsymbol{I}$ such that $\boldsymbol{kt} = \boldsymbol{g}$. Now note that $\boldsymbol{sw}$ is a monomorphism and $\boldsymbol{I}$ is injective. Therefore, there exists a morphism $\boldsymbol{g}'': \boldsymbol{X}'' \to \boldsymbol{I}$ such that $\boldsymbol{k} = \boldsymbol{g}''\boldsymbol{sw}$ and thus, also $\boldsymbol{g} = \boldsymbol{g}''\boldsymbol{f}$. □

Proof of Lemma 11.27. Since $\boldsymbol{e}_X$ is a monomorphism and $\boldsymbol{I}_Y^0$ is injective, there exists a morphism $\boldsymbol{h}^0: \boldsymbol{I}_X^0 \to \boldsymbol{I}_Y^0$ such that $\boldsymbol{e}_Y\boldsymbol{h} = \boldsymbol{h}^0\boldsymbol{e}_X$. To obtain $\boldsymbol{h}^1$, one first notes that the morphism $\boldsymbol{d}_Y^1\boldsymbol{h}^0: \boldsymbol{I}_X^0 \to \boldsymbol{I}_Y^1$ has the property that $(\boldsymbol{d}_Y^1\boldsymbol{h}^0)\boldsymbol{e}_X = (\boldsymbol{d}_Y^1\boldsymbol{e}_Y)\boldsymbol{h} = \boldsymbol{0}$. Therefore, Lemma 11.29 applies and yields a morphism $\boldsymbol{h}^1: \boldsymbol{I}_X^1 \to \boldsymbol{I}_Y^1$, which has the property that $\boldsymbol{h}^1\boldsymbol{d}_X^1 = \boldsymbol{d}_Y^1\boldsymbol{h}^0$. One proceeds by induction. □

Proof of Lemma 11.28. The morphisms $\boldsymbol{D}^n$, $n \geq 1$, are also constructed by induction on n, using Lemma 11.29. To define $\boldsymbol{D}^1$ it suffices to note that $(\boldsymbol{h}^0 - \boldsymbol{h}'^0)\boldsymbol{e}_X = \boldsymbol{e}_Y(\boldsymbol{h} - \boldsymbol{h}) = \boldsymbol{0}$. To define $\boldsymbol{D}^{n+1}$, we consider the morphism

$\boldsymbol{h}^n - \boldsymbol{h}'^n - \boldsymbol{d}_Y{}^n\boldsymbol{D}^n\colon \boldsymbol{I}_X{}^n \to \boldsymbol{I}_Y{}^n$. We need to show that its composition with $\boldsymbol{d}_X^n$ equals $\mathbf{0}$. Indeed, since $(\boldsymbol{h}^n - \boldsymbol{h}'^n)\boldsymbol{d}_X^n = \boldsymbol{d}_Y^n(\boldsymbol{h}^{n-1} - \boldsymbol{h}'^{n-1})$ and (21) holds for $n-1$, this composition equals $\boldsymbol{d}'^n(\boldsymbol{h}^{n-1} - \boldsymbol{h}'^{n-1}) - \boldsymbol{D}^n\boldsymbol{d}_X^n = \boldsymbol{d}_Y^n\boldsymbol{d}_Y^{n-1}\boldsymbol{D}^{n-1} = \mathbf{0}$. □

To define $\lim^n \boldsymbol{h}$, apply Lemma 11.27 to the resolutions $(\mathcal{I}(\boldsymbol{X}), \boldsymbol{e}(\boldsymbol{X}))$ and $(\mathcal{I}(\boldsymbol{Y}), \boldsymbol{e}(\boldsymbol{Y}))$ associated with $\boldsymbol{X}$ and $\boldsymbol{Y}$ and to the morphism $\boldsymbol{h}\colon \boldsymbol{X} \to \boldsymbol{Y}$. Application of the functor lim to (19) yields a commutative diagram of modules.

$$\begin{array}{ccccccccc} 0 & \longrightarrow & X & \xrightarrow{e_X} & I_X^0 & \xrightarrow{d_X^1} & I_X^1 & \longrightarrow & \cdots \\ & & \downarrow{\scriptstyle h} & & \downarrow{\scriptstyle h^0} & & \downarrow{\scriptstyle h^1} & & \\ 0 & \longrightarrow & Y & \xrightarrow[e_Y]{} & I_Y^0 & \xrightarrow[d_Y^1]{} & I_Y^1 & \longrightarrow & \cdots . \end{array} \tag{22}$$

The homomorphisms h^n form a *cochain mapping* $h^{\#}\colon I(\boldsymbol{X}) \to I(\boldsymbol{Y})$, i.e., $d_Y^n h^n = h^{n+1} d_X^n$, for $n \geq 1$. Therefore, $h^{\#}$ induces homomorphisms $H^n(h^{\#})$ between the corresponding cohomology modules $H^n(I(\boldsymbol{X}))$ and $H^n(I(\boldsymbol{Y}))$. By definition, these are the homomorphisms $\lim^n \boldsymbol{h}\colon \lim^n \boldsymbol{X} \to \lim^n \boldsymbol{Y}$. Diagram (22) also shows that $e(\boldsymbol{X}) = \boldsymbol{e}_X$ and $e(\boldsymbol{Y}) = \boldsymbol{e}_Y$ are monomorphisms, which identify $h = \lim \boldsymbol{h}$ with $\lim^0 \boldsymbol{h}$.

To show that $\lim^n \boldsymbol{h}$ does not depend on the choice of the morphisms $\boldsymbol{h}^n$, we use Lemma 11.28. Application of the functor lim yields homomorphisms $D^n\colon I_X^n \to I_Y^{n-1}$, $n \geq 1$, which, because of (20) and (21), satisfy

$$D^1 d_X^1 = h^0 - h'^0, \tag{23}$$

$$d_Y^n D^n + D^{n+1} d_X^{n+1} = h^n - h'^n, \; n \geq 1. \tag{24}$$

These relations show that the homomorphisms D^n form a *homotopy of cochain mappings* $D^{\#}$ connecting the cochain mappings $h^{\#}, h'^{\#}\colon I(\boldsymbol{X}) \to I(\boldsymbol{Y})$. Hence, $h^{\#}$ and $h'^{\#}$ induce the same homomorphisms of cohomology modules, i.e., $\lim^n \boldsymbol{h}\colon \lim^n \boldsymbol{X} \to \lim^n \boldsymbol{Y}$ does not depend on the choice of the morphisms $\boldsymbol{h}^n$, but depends only on $\boldsymbol{h}\colon \boldsymbol{X} \to \boldsymbol{Y}$ (and on the choice of the fixed resolutions $(\mathcal{I}(\boldsymbol{X}), \boldsymbol{e}(\boldsymbol{X}))$, $(\mathcal{I}(\boldsymbol{Y}), \boldsymbol{e}(\boldsymbol{Y}))$).

LEMMA 11.30. $\lim^n\colon \mathsf{Mod}^\Lambda \to \mathsf{Mod}$, $n \geq 0$, *is an additive functor.*

Proof. For the identity morphism $\mathbf{1}_X$, one has

$$\lim{}^n \mathbf{1}_X = \mathrm{id}, \tag{25}$$

because the morphisms $\boldsymbol{h}^n = \mathrm{id}$ make the corresponding diagram (19) commutative. For morphisms $\boldsymbol{h}\colon \boldsymbol{X} \to \boldsymbol{Y}$, $\boldsymbol{k}\colon \boldsymbol{Y} \to \boldsymbol{Z}$, one has

$$\lim{}^n (\boldsymbol{kh}) = \lim{}^n \boldsymbol{k} \lim{}^n \boldsymbol{h}, \tag{26}$$

because diagrams (19) for $\boldsymbol{h}$ and $\boldsymbol{k}$, put together, yield diagram (19) for $\boldsymbol{kh}$.

The functor $\lim^n$ is additive, i.e., for morphisms $\boldsymbol{h}, \boldsymbol{h}' \colon \boldsymbol{X} \to \boldsymbol{Y}$ of Mod^Λ,

$$\lim^n (\boldsymbol{h} + \boldsymbol{h}') = \lim^n \boldsymbol{h} + \lim^n \boldsymbol{h}'. \tag{27}$$

Indeed, summing up diagram (19) with the corresponding diagram for $\boldsymbol{h}'$, one obtains diagram (19), for $\boldsymbol{h} + \boldsymbol{h}'$. Since lim is an additive functor, one concludes that the cochain mapping $(h + h')^\# = h^\# + h'^\#$, which in turn induces (27). A similar argument shows that $\lim^n (r\boldsymbol{h}) = r\lim^n (\boldsymbol{h})$, for $r \in R$. □

Now assume that $\lim'^n$ is the functor obtained using a different choice of injective resolutions, say $(\mathcal{I}'(\boldsymbol{X}), \boldsymbol{e}'(\boldsymbol{X}))$. Then the functors $\lim^n$ and $\lim'^n$ are naturally isomorphic. Indeed, using Lemma 11.27, one can associate with the identity morphism $\boldsymbol{h} = \boldsymbol{k} = \boldsymbol{1}_{\boldsymbol{X}}$ two commutative diagrams of type (19), which contain morphisms $\boldsymbol{h}^n \colon \boldsymbol{I}^n_X \to \boldsymbol{I'}_X{}^n$ and $\boldsymbol{k}^n \colon \boldsymbol{I'}_X{}^n \to \boldsymbol{I}^n_X$, respectively. Application of the functor lim yields two cochain mappings $h^\# \colon I(\boldsymbol{X}) \to I'(\boldsymbol{X})$ and $k^\# \colon I'(\boldsymbol{X}) \to I(\boldsymbol{X})$ such that $k^\# h^\#$ and $h^\# k^\#$ are homotopic to the identity cochain mappings. Consequently, $h^\#$ induces isomorphisms of cohomology modules $h^{n*} = \lim^n \boldsymbol{h} \colon \lim^n \boldsymbol{X} \to \lim'^n \boldsymbol{X}$. Moreover, $h^{0*} e_X = e'_X$. The naturality of the isomorphisms is also easily verified using again Lemmas 11.27 and 11.28.

LEMMA 11.31. *If* $\boldsymbol{X}$ *is an injective object of* Mod^Λ, *then* $\lim^n \boldsymbol{X} = 0$, *for* $n \geq 1$.

Proof. Since $\boldsymbol{X}$ is injective,

$$\boldsymbol{0} \to \boldsymbol{X} \xrightarrow{\text{id}} \boldsymbol{X} \to \boldsymbol{0} \tag{28}$$

is an injective resolution of $\boldsymbol{X}$ with $\boldsymbol{I}^n_X = \boldsymbol{0}$, for $n \geq 1$. Using this resolution in the construction of $\lim^n \boldsymbol{X}$, the assertion becomes trivial. □

The next theorem states the main properties of the functors $\lim^n$. It is a special case of a general theorem on properties of the right derived functors of an additive left exact functor from an abelian category with enough injective objects to another abelian category (see e.g., (Bucur, Deleanu 1968), Theorem 7.11).

THEOREM 11.32. *Let*

$$E = (\boldsymbol{0} \to \boldsymbol{X}' \xrightarrow{\boldsymbol{f}'} \boldsymbol{X} \xrightarrow{\boldsymbol{f}} \boldsymbol{X}'' \to \boldsymbol{0}) \tag{29}$$

be a short exact sequence in Mod^Λ. *Then there exist homomorphisms* $\theta^n_E : \lim^n \boldsymbol{X}'' \to \lim^{n+1} \boldsymbol{X}'$ *such that the following sequence of modules is exact*

$$\begin{gathered} \boldsymbol{0} \to \lim \boldsymbol{X}' \xrightarrow{\lim \boldsymbol{f}'} \lim \boldsymbol{X} \xrightarrow{\lim \boldsymbol{f}} \lim \boldsymbol{X}'' \xrightarrow{\theta^0_E} \lim^1 \boldsymbol{X}' \to \dots \\ \dots \to \lim^n \boldsymbol{X}' \xrightarrow{\lim^n \boldsymbol{f}'} \lim^n \boldsymbol{X} \xrightarrow{\lim^n \boldsymbol{f}} \lim^n \boldsymbol{X}'' \xrightarrow{\theta^n_E} \lim^{n+1} \boldsymbol{X}' \to \dots . \end{gathered} \tag{30}$$

Moreover, if

$$F = (\mathbf{0} \to \boldsymbol{Y}' \xrightarrow{g'} \boldsymbol{Y} \xrightarrow{g} \boldsymbol{Y}'' \to \mathbf{0}) \tag{31}$$

is another short exact sequence and $h: E \to F$ *is a morphism of short exact sequences, i.e.,* $h = (\boldsymbol{h}', \boldsymbol{h}, \boldsymbol{h}'')$ *is a triple of morphisms* $\boldsymbol{h}': \boldsymbol{X}' \to \boldsymbol{Y}'$, $\boldsymbol{h}: \boldsymbol{X} \to \boldsymbol{Y}$, $\boldsymbol{h}'': \boldsymbol{X}'' \to \boldsymbol{Y}''$ *such that the diagram*

$$\begin{array}{ccccccccc} \mathbf{0} & \longrightarrow & \boldsymbol{X}' & \xrightarrow{\boldsymbol{f}'} & \boldsymbol{X} & \xrightarrow{\boldsymbol{f}} & \boldsymbol{X}'' & \longrightarrow & \mathbf{0} \\ & & \downarrow{\scriptstyle \boldsymbol{h}'} & & \downarrow{\scriptstyle \boldsymbol{h}} & & \downarrow{\scriptstyle \boldsymbol{h}''} & & \\ \mathbf{0} & \longrightarrow & \boldsymbol{Y}' & \xrightarrow[\boldsymbol{g}']{} & \boldsymbol{Y} & \xrightarrow[\boldsymbol{g}]{} & \boldsymbol{Y}'' & \longrightarrow & \mathbf{0} \end{array} \tag{32}$$

commutes, then for every $n \geq 0$ *also commutes the diagram*

$$\begin{array}{ccc} \lim^n \boldsymbol{X}'' & \xrightarrow{\theta_E^n} & \lim^{n+1} \boldsymbol{X}' \\ \downarrow{\scriptstyle \lim^n \boldsymbol{h}''} & & \downarrow{\scriptstyle \lim^{n+1} \boldsymbol{h}'} \\ \lim^n \boldsymbol{Y}'' & \xrightarrow[\theta_F^n]{} & \lim^{n+1} \boldsymbol{Y}' . \end{array} \tag{33}$$

To prove Theorem 11.32 we need two lemmas.

LEMMA 11.33. *Let* E *be an exact sequence* (29) *and let* $(\mathcal{I}', \mathbf{e}')$ *and* $(\mathcal{I}'', \mathbf{e}'')$ *be injective resolutions of* $\boldsymbol{X}'$ *and* $\boldsymbol{X}''$*, respectively. Then there exists an injective resolution* $(\mathcal{I}, \mathbf{e})$ *of* $\boldsymbol{X}$ *and there exist morphisms* $\boldsymbol{f}'^n: \boldsymbol{I}'^n \to \boldsymbol{I}^n$ *and* $\boldsymbol{f}^n: \boldsymbol{I}^n \to \boldsymbol{I}''^n$ *of* Mod^Λ *such that the following diagram commutes and has exact rows.*

$$\begin{array}{ccccccccc} & & \mathbf{0} & & \mathbf{0} & & \mathbf{0} & & \\ & & \downarrow & & \downarrow & & \downarrow & & \\ \mathbf{0} & \longrightarrow & \boldsymbol{X}' & \xrightarrow{\boldsymbol{f}'} & \boldsymbol{X} & \xrightarrow{\boldsymbol{f}} & \boldsymbol{X}'' & \longrightarrow & \mathbf{0} \\ & & \downarrow{\scriptstyle \mathbf{e}'} & & \downarrow{\scriptstyle \mathbf{e}} & & \downarrow{\scriptstyle \mathbf{e}''} & & \\ \mathbf{0} & \longrightarrow & \boldsymbol{I}'^0 & \xrightarrow{\boldsymbol{f}'^0} & \boldsymbol{I}^0 & \xrightarrow{\boldsymbol{f}^0} & \boldsymbol{I}''^0 & \longrightarrow & \mathbf{0} \\ & & \downarrow{\scriptstyle \boldsymbol{d}'^1} & & \downarrow{\scriptstyle \boldsymbol{d}^1} & & \downarrow{\scriptstyle \boldsymbol{d}''^1} & & \\ \mathbf{0} & \longrightarrow & \boldsymbol{I}'^1 & \xrightarrow{\boldsymbol{f}'^1} & \boldsymbol{I}^1 & \xrightarrow{\boldsymbol{f}^1} & \boldsymbol{I}''^1 & \longrightarrow & \mathbf{0} \\ & & \downarrow & & \downarrow & & \downarrow & & \\ & & \vdots & & \vdots & & \vdots & & \end{array} \tag{34}$$

Proof. First note that if one has a commutative diagram (34) with exact rows, then the sequence $\boldsymbol{0} \to \boldsymbol{I}'^n \xrightarrow{\boldsymbol{f}'^n} \boldsymbol{I}^n \xrightarrow{\boldsymbol{f}^n} \boldsymbol{I}''^n \to \boldsymbol{0}$ splits, because $\boldsymbol{I}'^n$ is injective (see Lemma 11.22). Therefore, we put $\boldsymbol{I}^n = \boldsymbol{I}'^n \oplus \boldsymbol{I}''^n$ and we take for $\boldsymbol{f}'^n$ the canonical injection $\boldsymbol{i}'^n : \boldsymbol{I}'^n \to \boldsymbol{I}^n$ and for $\boldsymbol{f}^n$ the canonical projection $\boldsymbol{p}''^n : \boldsymbol{I}^n \to \boldsymbol{I}''^n$. Note that injectivity of $\boldsymbol{I}'^n$ and $\boldsymbol{I}''^n$ implies injectivity of $\boldsymbol{I}$, because a product of injective objects is injective. It thus remains to define $\boldsymbol{e}$ and $\boldsymbol{d}^n$ in such a way that diagram (34) commutes and its middle column is exact. We will refer to such a diagram as a *standard injective resolution* of the sequence E.

We begin the construction by defining $\boldsymbol{e} : \boldsymbol{X} \to \boldsymbol{I}^0 = \boldsymbol{I}'^0 \oplus \boldsymbol{I}''^0$. First choose a morphism $\tilde{\boldsymbol{e}} : \boldsymbol{X} \to \boldsymbol{I}'^0$ such that $\tilde{\boldsymbol{e}}\boldsymbol{f}' = \boldsymbol{e}'$. Since $\boldsymbol{I}'^0$ is injective and $\boldsymbol{f}'$ is a monomorphism, $\tilde{\boldsymbol{e}}$ exists. Let $\boldsymbol{e} : \boldsymbol{X} \to \boldsymbol{I}^0$ be the only morphism, for which $\boldsymbol{p}'^0 \boldsymbol{e} = \tilde{\boldsymbol{e}}$ and $\boldsymbol{p}''^0 \boldsymbol{e} = \boldsymbol{e}''\boldsymbol{f}$, where $\boldsymbol{p}'^0 : \boldsymbol{I} \to \boldsymbol{I}'^0$ is the canonical projection. Since $\boldsymbol{f}^0 = \boldsymbol{p}''^0$, the top right square in diagram (34) commutes. To see that also the top left square commutes, recall that $\boldsymbol{1} = \boldsymbol{i}'^0\boldsymbol{p}'^0 + \boldsymbol{i}''^0\boldsymbol{p}''^0$, where $\boldsymbol{i}''^0 : \boldsymbol{I}''^0 \to \boldsymbol{I}^0$ is the canonical injection. Therefore, $\boldsymbol{e}\boldsymbol{f}' = \boldsymbol{i}'^0\boldsymbol{p}'^0\boldsymbol{e}\boldsymbol{f}' + \boldsymbol{i}''^0\boldsymbol{p}''^0\boldsymbol{e}\boldsymbol{f}' = \boldsymbol{i}'^0\tilde{\boldsymbol{e}}\boldsymbol{f}' + \boldsymbol{i}''^0\boldsymbol{e}''\boldsymbol{f}\boldsymbol{f}' = \boldsymbol{i}'^0\boldsymbol{e}'$. To see that $\boldsymbol{e}$ is a monomorphism, one can use a diagram chasing argument, applied to the diagrams obtained by restricting (34) to levels λ. Indeed, assume that $x \in X_\lambda$ is such that $e_\lambda(x) = 0$. Then $e''_\lambda f_\lambda(x) = f^0_\lambda e_\lambda(x) = 0$. Since e''_λ is a monomorphism, it follows that $f_\lambda(x) = 0$. Therefore, there exists an element $x' \in X'_\lambda$ such that $f'_\lambda(x') = x_\lambda$. However, $f_\lambda'^0 e'_\lambda(x') = e_\lambda f'_\lambda(x') = e_\lambda(x) = 0$. Since $f_\lambda'^0$ and e'_λ are monomorphisms, it follows that $x' = 0$ and thus, also $x = 0$.

To define $\boldsymbol{d}^1$, we apply Lemma 11.3 and decompose $\boldsymbol{d}'^1$ into an epimorphism $\overline{\boldsymbol{d}}'^1 : \boldsymbol{I}'^0 \to \boldsymbol{d}'^1(\boldsymbol{I}'^0) = \mathrm{Im}(\boldsymbol{d}'^1)$ and a monomorphism $\tilde{\boldsymbol{d}}'^1 : \mathrm{Im}(\boldsymbol{d}'^1) \to \boldsymbol{I}'^1$, $\boldsymbol{d}'^1 = \tilde{\boldsymbol{d}}'^1\overline{\boldsymbol{d}}'^1$. We also consider the analogous decomposition $\boldsymbol{d}''^1 = \tilde{\boldsymbol{d}}''^1\overline{\boldsymbol{d}}''^1$ and the cokernel $\overline{\boldsymbol{d}}^1 : \boldsymbol{I}^0 \to \boldsymbol{I}^0/e(\boldsymbol{X})$. Now consider the following diagram (solid arrows only):

$$
\begin{array}{ccccccccc}
 & & \boldsymbol{0} & & \boldsymbol{0} & & \boldsymbol{0} & & \\
 & & \downarrow & & \downarrow & & \downarrow & & \\
\boldsymbol{0} & \longrightarrow & \boldsymbol{X}' & \xrightarrow{\ \boldsymbol{f}'\ } & \boldsymbol{X} & \xrightarrow{\ \boldsymbol{f}\ } & \boldsymbol{X}'' & \longrightarrow & \boldsymbol{0} \\
 & & \Big\downarrow{\scriptstyle \boldsymbol{e}'} & & \Big\downarrow{\scriptstyle \boldsymbol{e}} & & \Big\downarrow{\scriptstyle \boldsymbol{e}''} & & \\
\boldsymbol{0} & \longrightarrow & \boldsymbol{I}'^0 & \xrightarrow{\ \boldsymbol{f}'^0\ } & \boldsymbol{I}^0 & \xrightarrow{\ \boldsymbol{f}^0\ } & \boldsymbol{I}''^0 & \longrightarrow & \boldsymbol{0} \\
 & & \Big\downarrow{\scriptstyle \overline{\boldsymbol{d}}'^1} & & \Big\downarrow{\scriptstyle \overline{\boldsymbol{d}}^1} & & \Big\downarrow{\scriptstyle \overline{\boldsymbol{d}}''^1} & & \\
\boldsymbol{0} & \longrightarrow & \boldsymbol{d}'^1(\boldsymbol{I}'^0) & \overset{\overline{\boldsymbol{f}}'^0}{\dashrightarrow} & \boldsymbol{I}^0/\boldsymbol{e}(\boldsymbol{X}) & \overset{\overline{\boldsymbol{f}}^0}{\dashrightarrow} & \boldsymbol{d}''^1(\boldsymbol{I}''^0) & \longrightarrow & \boldsymbol{0}\,. \\
 & & \downarrow & & \downarrow & & \downarrow & & \\
 & & \boldsymbol{0} & & \boldsymbol{0} & & \boldsymbol{0} & &
\end{array}
\tag{35}
$$

The first two rows are exact and so are all three columns. Therefore, there exist unique morphisms $\overline{\boldsymbol{f}}\,'^0 : \boldsymbol{d}'^1(\boldsymbol{I}\,'^0) \to \boldsymbol{I}\,^0/e(\boldsymbol{X})$ and $\overline{\boldsymbol{f}}\,^0 : \boldsymbol{I}\,^0/e(\boldsymbol{X}) \to \boldsymbol{d}''^1(\boldsymbol{I}\,''^0)$, which preserve commutativity of the diagram. Moreover, the newly obtained last row is also exact.

We now consider the following diagram (solid arrows only):

$$\begin{array}{ccccccccc}
 & & \mathbf{0} & & \mathbf{0} & & \mathbf{0} & & \\
 & & \downarrow & & \downarrow & & \downarrow & & \\
\mathbf{0} & \longrightarrow & \boldsymbol{d}'^1(\boldsymbol{I}\,'^0) & \xrightarrow{\overline{\boldsymbol{f}}\,'^0} & \boldsymbol{I}\,^0/e(\boldsymbol{X}) & \xrightarrow{\overline{\boldsymbol{f}}\,^0} & \boldsymbol{d}''^1(\boldsymbol{I}\,''^0) & \longrightarrow & \mathbf{0} \\
 & & \tilde{\boldsymbol{d}}'^1 \big\downarrow & & \vdots\, \tilde{\boldsymbol{d}}^1 & & \big\downarrow \tilde{\boldsymbol{d}}''^1 & & \\
\mathbf{0} & \longrightarrow & \boldsymbol{I}\,'^1 & \xrightarrow{\boldsymbol{f}\,'^1} & \boldsymbol{I}\,^1 & \xrightarrow{\boldsymbol{f}\,^1} & \boldsymbol{I}\,''^1 & \longrightarrow & \mathbf{0}\,.
\end{array} \tag{36}$$

We are now in the same situation as when we constructed $\boldsymbol{e}$. Therefore, the same argument yields a monomorphism $\tilde{\boldsymbol{d}}^1$ which makes the diagram commutative. We then put together diagrams (35) and (36). Putting $\boldsymbol{d}^1 = \tilde{\boldsymbol{d}}^1\overline{\boldsymbol{d}}^1$, we obtain the first three rows of diagram (34). Repeating the procedure, we obtain, by induction, the whole diagram (34). □

LEMMA 11.34. *Let* $(\boldsymbol{h}', \boldsymbol{h}, \boldsymbol{h}''): E \to F$ *be a morphism of exact sequences* (32). *Let* $((\mathcal{I}_{X'}, \boldsymbol{e}_{X'}), (\mathcal{I}_X, \boldsymbol{e}_X), (\mathcal{I}_{X''}, \boldsymbol{e}_{X''}), (\boldsymbol{f}'^{\,n}), (\boldsymbol{f}^{\,n}))$ *be a standard injective resolution of* E *and let* $((\mathcal{I}_{Y'}, \boldsymbol{e}_{Y'}), (\mathcal{I}_Y, \boldsymbol{e}_Y), (\mathcal{I}_{Y''}, \boldsymbol{e}_{Y''}), (\boldsymbol{g}'^n), (\boldsymbol{g}^n))$ *be a standard injective resolution of* F. *Furthermore, let the morphisms* $\boldsymbol{h}'^n : \boldsymbol{I}\,'^n \to \boldsymbol{J}\,'^n$ *and* $\boldsymbol{h}''^n : \boldsymbol{I}\,''^n \to \boldsymbol{J}\,''^n$ *form an injective resolution of* $\boldsymbol{h}'$ *and* $\boldsymbol{h}''$, *respectively. Then there are morphisms* $\boldsymbol{h}^n : \boldsymbol{I}\,^n \to \boldsymbol{J}\,^n$ *such that* $(\boldsymbol{h}'^n, \boldsymbol{h}^n, \boldsymbol{h}''^n)$ *is a morphism between exact sequences* $E_n = (\mathbf{0} \to \boldsymbol{I}\,'^n \to \boldsymbol{I}\,^n \to \boldsymbol{I}\,''^n \to \mathbf{0})$ *and* $F_n = (\mathbf{0} \to \boldsymbol{J}\,'^n \to \boldsymbol{J}\,^n \to \boldsymbol{J}\,''^n \to \mathbf{0})$ *and* $(\boldsymbol{h}^n)$ *is an injective resolution of* $\boldsymbol{h}$.

Proof. Recall that $\boldsymbol{i}'^n, \boldsymbol{i}''^n$ and $\boldsymbol{p}'^n, \boldsymbol{p}''^n$ denote the canonical injections and projections for $\boldsymbol{I}^n = \boldsymbol{I}\,'^n \oplus \boldsymbol{I}\,''^n$ and let $\boldsymbol{j}'^n, \boldsymbol{j}''^n$ and $\boldsymbol{q}'^n, \boldsymbol{q}''^n$ denote the canonical injections and projections for $\boldsymbol{J}^n = \boldsymbol{J}\,'^n \oplus \boldsymbol{J}\,''^n$. The assertion that $(\boldsymbol{h}'^n, \boldsymbol{h}^n, \boldsymbol{h}''^n)$ is a morphism between exact sequences E_n and F_n, $n \geq 0$, means that

$$\boldsymbol{h}^n\boldsymbol{i}'^n = \boldsymbol{j}'^n\boldsymbol{h}'^n, \tag{37}$$

$$\boldsymbol{q}''^n\boldsymbol{h}^n = \boldsymbol{h}''^n\boldsymbol{p}''^n. \tag{38}$$

On the other hand, the assertion that $(\boldsymbol{h}^n)$ is an injective resolution of $\boldsymbol{h}$ means that

$$\boldsymbol{e}_Y\boldsymbol{h} = \boldsymbol{h}^0\boldsymbol{e}_X, \tag{39}$$

$$\boldsymbol{d}_Y^{n+1}\boldsymbol{h}^n = \boldsymbol{h}^{n+1}\boldsymbol{d}_X^{n+1}. \tag{40}$$

To define morphisms $\boldsymbol{h}^n$ which satisfy (37)–(40), we will define certain morphisms $\boldsymbol{k}^n : \boldsymbol{I}\,''^n \to \boldsymbol{J}\,'^n$, $n \geq 0$. Then, we will put

$$\boldsymbol{h}^n = \boldsymbol{j}'^n\boldsymbol{k}^n\boldsymbol{p}''^n + \boldsymbol{j}'^n\boldsymbol{h}'^n\boldsymbol{p}'^n + \boldsymbol{j}''^n\boldsymbol{h}''^n\boldsymbol{p}''^n, \quad n \geq 0. \tag{41}$$

Using formulae (11.1.8)–(11.1.9), we readily see that, regardless of the definition of $\boldsymbol{k}^n$, (41) implies (37) and (38).

In analyzing conditions (39) and (40), it is convenient to introduce the following abbreviations.

$$\boldsymbol{u}^0 = \boldsymbol{p}'^0 \boldsymbol{e}_X \colon \boldsymbol{X} \to \boldsymbol{I}'^0, \tag{42}$$

$$\boldsymbol{u}^n = \boldsymbol{p}'^n \boldsymbol{d}_X^n \boldsymbol{i}''^{n-1} \colon \boldsymbol{I}''^{n-1} \to \boldsymbol{I}'^n,\ n \geq 1. \tag{43}$$

In an analogous way one also defines morphisms $\boldsymbol{v}^0 \colon \boldsymbol{Y} \to \boldsymbol{J}'^0$, $\boldsymbol{v}^n \colon \boldsymbol{J}''^{n-1} \to \boldsymbol{J}'^n$, $n \geq 1$. Using only formulae (11.1.8)–(11.1.9) and the commutativity and exactness of the diagram which represents the injective resolution of E, it is easy to verify the following relations:

$$\boldsymbol{e}_X = \boldsymbol{i}'^0 \boldsymbol{u}^0 + \boldsymbol{i}''^0 \boldsymbol{e}_{X''} \boldsymbol{f}, \tag{44}$$

$$\boldsymbol{e}_{X'} = \boldsymbol{u}^0 \boldsymbol{f}', \tag{45}$$

$$\boldsymbol{d}_{X'}^1 \boldsymbol{u}^0 + \boldsymbol{u}^1 \boldsymbol{e}_{X''} \boldsymbol{f} = \boldsymbol{0}, \tag{46}$$

$$\boldsymbol{d}_X^{n+1} = \boldsymbol{i}''^{n+1} \boldsymbol{d}_{X''}^{n+1} \boldsymbol{p}''^n + \boldsymbol{i}'^{n+1} \boldsymbol{u}^{n+1} \boldsymbol{p}''^n + \boldsymbol{i}'^{n+1} \boldsymbol{d}_{X'}^{n+1} \boldsymbol{p}'^n, \tag{47}$$

$$\boldsymbol{u}^{n+1} \boldsymbol{d}_{X''}^n + \boldsymbol{d}_{X'}^n \boldsymbol{u}^n = \boldsymbol{0}. \tag{48}$$

Note that in verifying (46), it suffices to show that the left side, composed with $\boldsymbol{i}'^1$, equals $\boldsymbol{0}$, because $\boldsymbol{i}'^1$ is a monomorphism. However, this follows using the fact that $\boldsymbol{d}_X^1 \boldsymbol{e}_X = 0$ and $\boldsymbol{d}_{X''}^1 \boldsymbol{e}_{X''} = 0$. In verifying (47), it suffices to show that the right side, composed with $\boldsymbol{p}'^{n+1}$ and $\boldsymbol{p}''^{n+1}$, yields $\boldsymbol{p}'^{n+1} \boldsymbol{d}_X^{n+1}$ and $\boldsymbol{p}''^{n+1} \boldsymbol{d}_X^{n+1}$, respectively. In verifying (48), it suffices to show that the left side, composed with $\boldsymbol{i}'^{n+1}$, equals $\boldsymbol{0}$, because $\boldsymbol{i}'^{n+1}$ is a monomorphism. However, this follows, using the fact that $\boldsymbol{d}_X^{n+1} \boldsymbol{d}_X^n = \boldsymbol{0}$ and $\boldsymbol{d}_{X''}^{n+1} \boldsymbol{d}_{X''}^n = \boldsymbol{0}$. Analogous formulae hold for $\boldsymbol{g}', \boldsymbol{g}$, $\boldsymbol{e}_{Y'}, \boldsymbol{e}_Y, \boldsymbol{e}_{Y''}$, $\boldsymbol{j}'^n, \boldsymbol{j}^n, \boldsymbol{j}''^n, \boldsymbol{q}'^n, \boldsymbol{q}^n, \boldsymbol{q}''^n, \boldsymbol{d}_{Y'}^n, \boldsymbol{d}_Y^n, \boldsymbol{d}_{Y'}^n$ and $\boldsymbol{v}^n$.

Using (41) and (44), we see that the right side of (39) is given by

$$\begin{aligned} \boldsymbol{h}^0 \boldsymbol{e}_X = (\boldsymbol{j}'^0 \boldsymbol{k}^0 \boldsymbol{p}''^0 + \boldsymbol{j}'^0 \boldsymbol{h}'^0 \boldsymbol{p}'^0 + \boldsymbol{j}''^0 \boldsymbol{h}''^0 \boldsymbol{p}''^0)(\boldsymbol{i}'^0 \boldsymbol{u}^0 + \boldsymbol{i}''^0 \boldsymbol{e}_{X''} \boldsymbol{f}) = \\ \boldsymbol{j}'^0 \boldsymbol{h}'^0 \boldsymbol{u}^0 + \boldsymbol{j}'^0 \boldsymbol{k}^0 \boldsymbol{e}_{X''} \boldsymbol{f} + \boldsymbol{j}''^0 \boldsymbol{h}''^0 \boldsymbol{e}_{X''} \boldsymbol{f}. \end{aligned} \tag{49}$$

Similarly, (41) and the analogue of (44) show that the left side of (39) equals

$$\boldsymbol{e}_Y \boldsymbol{h} = \boldsymbol{j}'^0 \boldsymbol{v}^0 \boldsymbol{h} + \boldsymbol{j}''^0 \boldsymbol{e}_{Y''} \boldsymbol{g} \boldsymbol{h} = \boldsymbol{j}'^0 \boldsymbol{v}^0 \boldsymbol{h} + \boldsymbol{j}''^0 \boldsymbol{h}''^0 \boldsymbol{e}_{X''} \boldsymbol{f}. \tag{50}$$

Comparing (49) with (50), we conclude that (39) is equivalent to

$$\boldsymbol{h}'^0 \boldsymbol{u}^0 + \boldsymbol{k}^0 \boldsymbol{e}_{X''} \boldsymbol{f} = \boldsymbol{v}^0 \boldsymbol{h}. \tag{51}$$

Furthermore, using (41) and the analogue of (47), it is readily seen that

$$\begin{aligned} \boldsymbol{d}_Y^{n+1} \boldsymbol{h}^n = \boldsymbol{j}'^{n+1} (\boldsymbol{d}_{Y'}^{n+1} \boldsymbol{k}^n + \boldsymbol{v}^{n+1} \boldsymbol{h}''^n) \boldsymbol{p}''^n + \\ \boldsymbol{j}'^{n+1} \boldsymbol{h}'^{n+1} \boldsymbol{d}_{X'}^{n+1} \boldsymbol{p}'^n + \boldsymbol{j}''^{n+1} \boldsymbol{h}''^{n+1} \boldsymbol{d}_{X''}^{n+1} \boldsymbol{p}''^n. \end{aligned} \tag{52}$$

Similarly,

$$\begin{aligned} \boldsymbol{h}^{n+1}\boldsymbol{d}_X^{n+1} &= \boldsymbol{j}'^{n+1}(\boldsymbol{k}^{n+1}\boldsymbol{d}_{X''}^{n+1} + \boldsymbol{h}'^{n+1}\boldsymbol{u}^{n+1})\boldsymbol{p}''^n + \\ &\boldsymbol{j}'^{n+1}\boldsymbol{h}'^{n+1}\boldsymbol{d}_{X'}^{n+1}\boldsymbol{p}'^n + \boldsymbol{j}''^{n+1}\boldsymbol{h}''^{n+1}\boldsymbol{d}_{X''}^{n+1}\boldsymbol{p}''^n. \end{aligned} \tag{53}$$

Therefore, (40) is equivalent to

$$\boldsymbol{d}_{Y'}^{n+1}\boldsymbol{k}^n + \boldsymbol{v}^{n+1}\boldsymbol{h}''^n = \boldsymbol{k}^{n+1}\boldsymbol{d}_{X''}^{n+1} + \boldsymbol{h}'^{n+1}\boldsymbol{u}^{n+1}. \tag{54}$$

Consequently, the morphisms $\boldsymbol{h}^n$, $n \geq 0$, given by (41), will satisfy (37)–(40) if and only if the morphisms $\boldsymbol{k}^n$, $n \geq 0$, satisfy conditions (51) and (54). We will construct such morphisms $\boldsymbol{k}^n$ by induction on n, using Lemma 11.29.

To define $\boldsymbol{k}^0$, we apply the lemma to the exact sequence

$$\mathbf{0} \to \boldsymbol{X} \xrightarrow{\boldsymbol{f}'} \boldsymbol{X} \xrightarrow{e_{X''}\boldsymbol{f}} \boldsymbol{I}''^0 \to \mathbf{0} \tag{55}$$

and to the morphism $\boldsymbol{v}^0\boldsymbol{h} - \boldsymbol{h}'^0\boldsymbol{u}^0 \colon \boldsymbol{X} \to \boldsymbol{J}'^0$. To conclude that there exists a morphism $\boldsymbol{k}^0 \colon \boldsymbol{I}''^0 \to \boldsymbol{J}'^0$, which satisfies (51), we only need to verify that

$$(\boldsymbol{v}^0\boldsymbol{h} - \boldsymbol{h}'^0\boldsymbol{u}^0)\boldsymbol{f}' = \mathbf{0}. \tag{56}$$

Indeed, by (45) and its analogue, we see that $\boldsymbol{h}'^0\boldsymbol{u}^0\boldsymbol{f}' = \boldsymbol{h}'^0 e_{X'}$ and $\boldsymbol{v}^0\boldsymbol{h}\boldsymbol{f}' = \boldsymbol{v}^0\boldsymbol{g}'\boldsymbol{h}' = e_{Y'}\boldsymbol{h}'$. However, $\boldsymbol{h}'^0 e_{X'} = e_{Y'}\boldsymbol{h}'$.

To define $\boldsymbol{k}^1$, we apply Lemma 11.29 to the exact sequence

$$\mathbf{0} \to \boldsymbol{X} \xrightarrow{e_{X''}\boldsymbol{f}} \boldsymbol{I}''^0 \xrightarrow{\boldsymbol{d}_{X''}^1} \boldsymbol{I}''^1 \to \mathbf{0} \tag{57}$$

and to the morphism $\boldsymbol{d}_{Y'}^1\boldsymbol{k}^0 + \boldsymbol{v}^1\boldsymbol{h}''^0 - \boldsymbol{h}'^1\boldsymbol{u}^1$. To conclude that there exists a morphism $\boldsymbol{k}^1 \colon \boldsymbol{I}''^1 \to \boldsymbol{J}'^1$, which satisfies (54), for $n = 0$, we only need to verify that

$$(\boldsymbol{d}_{Y'}^1\boldsymbol{k}^0 + \boldsymbol{v}^1\boldsymbol{h}''^0 - \boldsymbol{h}'^1\boldsymbol{u}^1)e_{X''}\boldsymbol{f} = \mathbf{0}. \tag{58}$$

Indeed, by (51) and the analogue of (46), $\boldsymbol{d}_{Y'}^1\boldsymbol{k}^0 e_{X''}\boldsymbol{f} = -\boldsymbol{v}^1 e_{Y''}\boldsymbol{g}\boldsymbol{h} - \boldsymbol{d}_{Y'}^1\boldsymbol{h}'^0\boldsymbol{u}^0$. On the other hand, since $\boldsymbol{h}''^0 e_{X''} = e_{Y''}\boldsymbol{h}''$, (46) implies $(\boldsymbol{v}^1\boldsymbol{h}''^0 - \boldsymbol{h}'^1\boldsymbol{u}^1)e_{X''}\boldsymbol{f} = \boldsymbol{v}^1 e_{Y''}\boldsymbol{h}''\boldsymbol{f} + \boldsymbol{h}'^1\boldsymbol{d}_{X'}^1\boldsymbol{u}^0$. The two expressions add up to zero, because $\boldsymbol{g}\boldsymbol{h} = \boldsymbol{h}''\boldsymbol{f}$ and $\boldsymbol{d}_{Y'}^1\boldsymbol{h}'^0 = \boldsymbol{h}'^1\boldsymbol{d}_{X'}^1$.

Finally, we define $\boldsymbol{k}^{n+1}$, for $n \geq 1$, by applying Lemma 11.29 to the exact sequence

$$\mathbf{0} \to \boldsymbol{I}''^{n-1} \xrightarrow{\boldsymbol{d}_{X''}^n} \boldsymbol{I}''^n \xrightarrow{\boldsymbol{d}_{X''}^{n+1}} \boldsymbol{I}''^{n+1} \to \mathbf{0} \tag{59}$$

and to the morphism $\boldsymbol{d}_{Y'}^{n+1}\boldsymbol{k}^n + \boldsymbol{v}^{n+1}\boldsymbol{h}''^n - \boldsymbol{h}'^{n+1}\boldsymbol{u}^{n+1}$. We only need to verify that

$$(\boldsymbol{d}_{Y'}^{n+1}\boldsymbol{k}^n + \boldsymbol{v}^{n+1}\boldsymbol{h}''^n - \boldsymbol{h}'^{n+1}\boldsymbol{u}^{n+1})\boldsymbol{d}_{X''}^n = \mathbf{0}. \tag{60}$$

By the induction hypothesis, (54) yields

$$\begin{aligned} \boldsymbol{d}_{Y'}^{n+1}\boldsymbol{k}^n\boldsymbol{d}_{X''}^n &= \boldsymbol{d}_{Y'}^{n+1}(\boldsymbol{d}_{Y'}^n\boldsymbol{k}^{n-1} + \boldsymbol{v}^n\boldsymbol{h}''^{n-1} - \boldsymbol{h}'^n\boldsymbol{u}^n) = \\ &\boldsymbol{d}_{Y'}^{n+1}\boldsymbol{v}^n\boldsymbol{h}''^{n-1} - \boldsymbol{h}'^{n+1}\boldsymbol{d}_{X'}^{n+1}\boldsymbol{h}'^n\boldsymbol{u}^n. \end{aligned} \tag{61}$$

On the other hand,

$$(\boldsymbol{v}^{n+1}\boldsymbol{h}''^{n} - \boldsymbol{h}'^{n+1}\boldsymbol{u}^{n+1})\boldsymbol{d}^{n}_{X''} = \boldsymbol{v}^{n+1}\boldsymbol{d}^{n}_{Y''}\boldsymbol{h}''^{n-1} - \boldsymbol{h}'^{n+1}\boldsymbol{d}^{n+1}_{\boldsymbol{X}'}\boldsymbol{h}'^{n}\boldsymbol{u}^{n}. \qquad (62)$$

Adding up (61) and (62) and taking into account (48) and its analogue, we obtain (60). □

Proof of Theorem 11.32. By Lemma 11.33, there exists an injective resolution of E. Application of the functor lim to this resolution yields cochain mappings $f'^{\#}: I(\boldsymbol{X}') \to I(\boldsymbol{X})$ and $f^{\#}: I(\boldsymbol{X}) \to I(\boldsymbol{X}'')$ (see (22)). By Remark 11.25, the sequences $0 \to I'^n \to I^n \to I''^n \to 0$, $n \geq 0$, are exact. Therefore, the sequence of cochain complexes

$$0 \to I(\boldsymbol{X}') \xrightarrow{f'^{\#}} I(\boldsymbol{X}) \xrightarrow{f^{\#}} I(\boldsymbol{X}'') \to 0 \qquad (63)$$

is also exact. Identifying lim with $\lim^0$, one sees that (30) is just the long exact sequence of cohomology modules associated with (63) (see e.g., Theorem 2.1 of Chapter IV in (Hilton, Stammbach 1971)).

To prove the second assertion of Theorem 11.32, we apply Lemma 11.34 to $(\boldsymbol{h}', \boldsymbol{h}, \boldsymbol{h}'')$. Application of the functor lim to the obtained diagram yields a morphism (h', h, h'') between exact sequences of cochain complexes,

$$\begin{array}{ccccccccc}
0 & \longrightarrow & I(\boldsymbol{X}') & \xrightarrow{f'^{\#}} & I(\boldsymbol{X}) & \xrightarrow{f^{\#}} & I(\boldsymbol{X}'') & \longrightarrow & 0 \\
 & & \big\downarrow{\scriptstyle h'^{\#}} & & \big\downarrow{\scriptstyle h^{\#}} & & \big\downarrow{\scriptstyle h''^{\#}} & & \\
0 & \longrightarrow & I(\boldsymbol{Y}') & \xrightarrow[g'^{\#}]{} & I(\boldsymbol{Y}) & \xrightarrow[g^{\#}]{} & I(\boldsymbol{Y}'') & \longrightarrow & 0.
\end{array} \qquad (64)$$

However, this morphism induces a morphism between the corresponding long exact sequences of cohomology modules (see e.g., Theorem 2.1 of Chapter IV in (Hilton, Stammbach 1971)).

$$\begin{array}{ccccccccccc}
\cdots & \to & \lim^n \boldsymbol{X}' & \xrightarrow{\lim^n \boldsymbol{f}'} & \lim^n \boldsymbol{X} & \xrightarrow{\lim^n \boldsymbol{f}} & \lim^n \boldsymbol{X}'' & \xrightarrow{\theta^n_E} & \lim^{n+1} \boldsymbol{X}' & \to & \cdots \\
 & & \big\downarrow{\scriptstyle \lim^n \boldsymbol{h}'} & & \big\downarrow{\scriptstyle \lim^n \boldsymbol{h}} & & \big\downarrow{\scriptstyle \lim^n \boldsymbol{h}''} & & \big\downarrow{\scriptstyle \lim^{n+1} \boldsymbol{h}'} & & \\
\cdots & \to & \lim^n \boldsymbol{Y}' & \xrightarrow[\lim^n \boldsymbol{g}']{} & \lim^n \boldsymbol{Y} & \xrightarrow[\lim^n \boldsymbol{g}]{} & \lim^n \boldsymbol{Y}'' & \xrightarrow[\theta^n_F]{} & \lim^{n+1} \boldsymbol{Y}' & \to & \cdots
\end{array} \qquad (65)$$

The commutativity of the first two rectangles in (65) is a consequence of the fact that $\lim^n$ is a functor. Hence, the only new information is the commutativity of the third rectangle, i.e., the commutativity of (33). □

11.4 Axiomatic characterization of the functors $\lim^n$

In order to characterize axiomatically the functors $\lim^n$, one needs the notion of an (*upper*) *connected sequence of functors* (also known as a *covariant cohomology functor*) from an abelian category $\mathcal{A}$ to Mod. This is a sequence S of additive functors $S^n: \mathcal{A} \to \mathsf{Mod}$, $n = 0, 1, \ldots$, together with a sequence of homomorphisms $\theta_E^n: S^n(X'') \to S^{n+1}(X')$, associated with every short exact sequence

$$E = (0 \to X' \xrightarrow{f'} X \xrightarrow{f} X'' \to 0) \tag{1}$$

in $\mathcal{A}$, called the *connecting homomorphisms.* These homomorphisms make the following sequence exact.

$$\begin{array}{c} 0 \to S^0(X') \xrightarrow{S^0(f')} S^0(X) \xrightarrow{S^0(f)} S^0(X'') \xrightarrow{\theta_E^0} S^1(X') \to \ldots \\ \ldots \to S^n(X') \xrightarrow{S^n(f')} S^n(X) \xrightarrow{S^n(f)} S^n(X'') \xrightarrow{\theta_E^n} S^{n+1}(X') \to \ldots \end{array} \tag{2}$$

Moreover, if $h = (h', h, h''): E \to F$ is a morphism of short exact sequences, then, for every $n \geq 0$, the following diagram commutes.

$$\begin{array}{ccc} S^n(X'') & \xrightarrow{\theta_E^n} & S^{n+1}(X') \\ {\scriptstyle S^n(h'')}\downarrow & & \downarrow{\scriptstyle S^{n+1}(h')} \\ S^n(Y'') & \xrightarrow[\theta_F^n]{} & S^{n+1}(Y') . \end{array} \tag{3}$$

REMARK 11.35. Theorem 11.32 shows that the functors $\lim^n$, together with the connecting homomorphisms θ_E^n, $n = 0, 1, \ldots$, form a connected sequence of functors from Mod^Λ to Mod.

Let functors $T^n: \mathcal{A} \to \mathsf{Mod}$, $n = 0, 1, \ldots$, together with homomorphisms $\omega_E^n: T^n(X'') \to T^{n+1}(X')$ form another connected sequence of functors from $\mathcal{A}$ to Mod. A *morphism of connected sequences of functors* from $S = (S^n, \theta_E^n)$ to $T = (T^n, \omega_E^n)$ consists of a sequence of natural transformations $\phi^n: S^n \to T^n$ such that, for every exact sequence (1), the following diagram commutes.

$$\begin{array}{ccc} S^n(X'') & \xrightarrow{\theta_E^n} & S^{n+1}(X') \\ {\scriptstyle \phi_{X''}^n}\downarrow & & \downarrow{\scriptstyle \phi_{X'}^{n+1}} \\ T^n(X'') & \xrightarrow[\omega_E^n]{} & T^{n+1}(X') . \end{array} \tag{4}$$

If all ϕ^n are natural equivalences, i.e., $\phi_X^n: S^n(X) \to T^n(X)$ is an isomorphism, for every object X from $\mathcal{A}$, then one has an *isomorphism* of connected sequences of functors. A connected sequence of functors $S = (S^n, \theta_E^n)$

is said to be *universal* provided, for any other connected sequence of functors $T = (T^n, \omega_E^n)$ and any natural transformation $\phi: S^0 \to T^0$, there exists a unique morphism (ϕ^n) from S to T such that $\phi^0 = \phi$.

REMARK 11.36. If both connected sequences of functors S and T are universal and $\phi: S^0 \to T^0$ is a natural equivalence, then there is a unique isomorphism (ϕ^n) between the two sequences, such that $\phi^0 = \phi$. In other words, universal connected sequences of functors S are completely determined by the functor S^0.

The following theorem from homological algebra gives a criterion for the universality of a connected sequence of functors.

THEOREM 11.37. *Let $S = (S^n, \theta_E^n)$ be a connected sequence of functors from an abelian category $\mathcal{A}$ to* Mod. *If $\mathcal{A}$ has enough injective objects and $S^n(I) = 0$, $n \geq 1$, for every injective object I of $\mathcal{A}$, then S is universal.*

Proof. Let $T = (T^n, \omega_E^n)$ be another connected sequence of functors and let $\phi^0: S^0 \to T^0$ be a natural transformation. We shall construct unique natural transformations $\phi^n: S^n \to T^n$, which make (4) commutative. In order to define the homomorphisms $\phi_X^1: S^1(X) \to T^1(X)$, for systems X, consider an injective presentation of X,

$$F = (0 \to X \to I \xrightarrow{f} I/X \to 0) \tag{5}$$

and consider the following diagram (solid arrows only).

$$\begin{array}{ccccccc} S^0(I) & \xrightarrow{S^0(f)} & S^0(I/X) & \xrightarrow{\theta_F^0} & S^1(X) & \longrightarrow & 0 \\ \phi_I^0 \downarrow & & \downarrow \phi_{I/X}^0 & & \vdots\, \phi_X^1 & & \\ T^0(I) & \xrightarrow[T^0(f)]{} & T^0(I/X) & \xrightarrow[\omega_F^0]{} & T^1(X) . & & \end{array} \tag{6}$$

Since $S^1(I) = 0$, (2) shows that the rows in the diagram are exact. The square commutes by naturality of ϕ^0. A diagram chasing argument shows that there exists a unique homomorphism $\phi_X^1: S^1(X) \to T^1(X)$ (dotted arrow), which completes (6) to a commutative diagram.

The next step consists in proving that ϕ_X^1 does not depend on the choice of the injective presentation (5). Indeed, assume that another injective presentation F' of X yields $\phi_X'^1$. Using the assumption that I' is injective, one can embed the two sequences F and F' in a commutative diagram

$$\begin{array}{ccccccccc} 0 & \longrightarrow & X & \longrightarrow & I & \xrightarrow{f} & I/X & \longrightarrow & 0 \\ & & 1_X \downarrow & & h \downarrow & & \downarrow k & & \\ 0 & \longrightarrow & X & \longrightarrow & I' & \xrightarrow[f']{} & I'/X & \longrightarrow & 0. \end{array} \tag{7}$$

Note that naturality of ϕ^0 yields the commutative diagram

$$\begin{array}{ccc} S^0(I/X) & \xrightarrow{S^0(k)} & S^0(I'/X) \\ {\scriptstyle \phi^0_{I/X}}\downarrow & & \downarrow{\scriptstyle \phi^0_{I'/X}} \\ T^0(I/X) & \xrightarrow[T^0(k)]{} & T^0(I'/X)\ . \end{array} \tag{8}$$

Also note that (3), applied to (7) for S and T, yields

$$\theta^0_F = \theta^0_{F'}S^0(k),\ \omega^0_F = \omega^0_{F'}T^0(k). \tag{9}$$

Moreover, by (6) and its analogue for F', one has

$$\phi^1_X\theta^0_F = \omega^0_F\phi^0_{I/X},\ \phi'^1_X\theta^0_{F'} = \omega^0_{F'}\phi^0_{I'/X}. \tag{10}$$

From (10), (9) and (8), it follows that $\phi^1_X\theta^0_F = \phi'^1_X\theta^0_F$. However, this equality implies the desired conclusion, $\phi^1_X = \phi'^1_X$, because, by the exactness of the first row in (6), θ^0_F is an epimorphism.

A similar argument is used to prove naturality of ϕ^1, i.e., to show that, for any $g: X \to X'$, one has $\phi^1_{X'}S^1(g) = T^1(g)\phi^1_X$. Since θ^0_F is an epimorphism, it suffices to show that

$$\phi^1_{X'}S^1(g)\theta^0_F = T^1(g)\phi^1_X\theta^0_F. \tag{11}$$

In order to prove (11), one chooses for X' an injective presentation G and one embeds F and G in a commutative diagram,

$$\begin{array}{ccccccccc} 0 & \to & X & \to & I & \xrightarrow{f} & I/X & \to & 0 \\ & & {\scriptstyle g}\downarrow & & {\scriptstyle h}\downarrow & & \downarrow{\scriptstyle k} & & \\ 0 & \to & X' & \to & I' & \xrightarrow[f']{} & I'/X' & \to & 0. \end{array} \tag{12}$$

Naturality of ϕ^0 yields the commutative diagram

$$\begin{array}{ccc} S^0(I/X) & \xrightarrow{S^0(k)} & S^0(I'/X') \\ {\scriptstyle \phi^0_{I/X}}\downarrow & & \downarrow{\scriptstyle \phi^0_{I'/X'}} \\ T^0(I/X) & \xrightarrow[T^0(k)]{} & T^0(I'/X')\ . \end{array} \tag{13}$$

To obtain (11), one first applies (3), for S and T, to (12). This yields the following equalities.

$$S^1(g)\theta^0_F = \theta^0_G S^0(k),\ T^1(g)\omega^0_F = \omega^0_G T^0(k). \tag{14}$$

Moreover, (6) and its analogue for G yield

$$\phi^1_X\theta^0_F = \omega^0_F\phi^0_{I/X},\ \phi^1_{X'}\theta^0_G = \omega^0_G\phi^0_{I'/X'}. \tag{15}$$

Now (11) follows from (14), (15) and (13).

As far as the natural transformation ϕ^1 is concerned, the final step of the proof consists in showing that exactness of the sequence (1) implies commutativity of diagram (4), for $n = 0$. To achieve this, one chooses an injective presentation H of X' and embeds it, together with the sequence (1), in a commutative diagram

$$\begin{array}{ccccccccc} 0 & \longrightarrow & X' & \xrightarrow{f'} & X & \longrightarrow & X'' & \longrightarrow & 0 \\ & & \downarrow{\scriptstyle 1_{X'}} & & \downarrow{\scriptstyle h} & & \downarrow{\scriptstyle k} & & \\ 0 & \longrightarrow & X' & \longrightarrow & I' & \xrightarrow[g]{} & I'/X' & \longrightarrow & 0. \end{array} \tag{16}$$

Naturality of ϕ^0 yields the commutative diagram

$$\begin{array}{ccc} S^0(X'') & \xrightarrow{S^0(k)} & S^0(I'/X') \\ \downarrow{\scriptstyle \phi^0_{X''}} & & \downarrow{\scriptstyle \phi^0_{I'/X'}} \\ T^0(X'') & \xrightarrow[T^0(k)]{} & T^0(I'/X'). \end{array} \tag{17}$$

By definition of $\phi^1_{X'}$ one has

$$\phi^1_{X'}\theta^0_H = \omega^0_H\phi^0_{I'/X'}. \tag{18}$$

Moreover, (3) applied to diagram (16) for S and T yields

$$\theta^0_E = \theta^0_H S^0(k),\ \omega^0_E = \omega^0_H T^0(k). \tag{19}$$

Now (19), (18) and (17) yield the desired relation $\phi^1_{X'}\theta^0_E = \omega^0_E\phi^0_{X''}$.

Construction of the natural transformations ϕ^n, for $n \geq 2$, proceeds by induction. □

Theorem 11.32 and Lemma 11.31 enable us to apply Theorem 11.37 and obtain the following result.

THEOREM 11.38. *The functors* $\lim^n\colon \mathsf{Mod}^\Lambda \to \mathsf{Mod}$, *together with the connecting homomorphisms* θ^n_E, $n = 0, 1, \ldots$, *defined in* 11.3, *form a universal connected sequence of functors.*

Since universal sequences of functors are unique up to natural isomorphism (see Remark 11.36), the above results give an *axiomatic characterization* of the derived limits $\lim^n$.

11.5 Explicit formulae for $\lim^n$

With every inverse system of R-modules $\boldsymbol{X} = (X_\lambda, p_{\lambda\lambda'}, \Lambda)$ we associate a cochain complex

$$K(\boldsymbol{X}) = (0 \to K^0(\boldsymbol{X}) \xrightarrow{\delta^1} K^1(\boldsymbol{X}) \xrightarrow{\delta^2} K^2(\boldsymbol{X}) \to \ldots) \tag{1}$$

as follows. Let

$$K^n = \prod_{\boldsymbol{\lambda} \in \Lambda_n} X_{\lambda_0} = \prod_{\lambda_0 \le \ldots \le \lambda_n} X_{\lambda_0}, \tag{2}$$

i.e., let $K^n = K^n(\boldsymbol{X})$ consist of all functions $c: \Lambda_n \to \cup_\lambda X_\lambda$ such that $c(\boldsymbol{\lambda}) \in X_{\lambda_0}$. Recall that Λ_n is the set of all multiindices $\boldsymbol{\lambda} = (\lambda_0, \ldots, \lambda_n)$ from Λ of length n (see 1.2). For the values $c(\boldsymbol{\lambda})$, we sometimes use the notation $c_{\boldsymbol{\lambda}}$ or $c_{\lambda_0 \ldots \lambda_n}$. Being a product of R-modules, K^n is an R-module. The homomorphisms $\delta^n: K^{n-1} \to K^n$, $n \ge 1$, called the *Nöbeling – Roos operators*, are given by the formula

$$(\delta^n c)(\boldsymbol{\lambda}) = p_{\lambda_0\lambda_1} c(d^0 \boldsymbol{\lambda}) + \sum_{j=1}^{n} (-1)^j c(d^j \boldsymbol{\lambda}), \; c \in K^{n-1}, \; \boldsymbol{\lambda} \in \Lambda_n. \tag{3}$$

Recall that $d^j \boldsymbol{\lambda}$ is the sequence obtained from $\boldsymbol{\lambda}$ by omitting λ_j (see 1.2). Supressing the indices of p one can give (3) the shorter form,

$$(\delta^n c)(\boldsymbol{\lambda}) = \sum_{j=0}^{n} (-1)^j p c(d^j \boldsymbol{\lambda}) \; ; \tag{4}$$

The missing indices can always be recovered by considering the domain and the codomain of the homomorphisms p. In particular, for $j > 0$, $d^j \boldsymbol{\lambda}$ begins with λ_0. Therefore, $c(d^j \boldsymbol{\lambda}) \in X_{\lambda_0}$ and both, the domain and the codomain of p equal λ_0. Hence, for $j > 0$, p stands for $p_{\lambda_0\lambda_0} = \mathrm{id}$.

LEMMA 11.39. *$K(\boldsymbol{X})$ is a cochain complex, i.e., $\delta^{n+1}\delta^n = 0$, $n \ge 1$.*

Proof. For $c \in K^{n-1}$ and $\boldsymbol{\lambda} \in \Lambda_n$, one has

$$(\delta\delta c)(\boldsymbol{\lambda}) = \sum_{j=0}^{n} (-1)^j p(\delta c)(d^j \boldsymbol{\lambda}) = \sum_{j=0}^{n} \sum_{k=0}^{n-1} (-1)^{j+k} p c(d^k d^j \boldsymbol{\lambda}). \tag{5}$$

Decompose the set $A = \{(j,k): 0 \le j \le n, \; 0 \le k \le n-1\}$, over which the summation is performed, in two disjoint subsets, $B = \{(j,k): 0 \le j \le n, \; 0 \le k \le j-1\}$ and $C = \{(j,k): 0 \le j \le n-1, \; j \le k \le n-1\}$. It is readily seen that the function $\phi: B \to C$, given by $\phi(j,k) = (k, j-1)$, is a bijection. Therefore, to show that the double sum in (5) equals 0, it suffices to show that

$$(-1)^{j+k} c(d^k d^j \boldsymbol{\lambda}) + (-1)^{k+j-1} c(d^{j-1} d^k \boldsymbol{\lambda}) = 0, \; (j,k) \in B. \tag{6}$$

However, (6) holds, because $d^k d^j = d^{j-1} d^k$, $0 \le k < j \le n$ (see 1.2.19). □

For every $n = 0, 1, \ldots$, we now consider the n-th cohomology module of the complex $K(\boldsymbol{X})$ and we put

$$S^n(\boldsymbol{X}) = H^n(K(\boldsymbol{X})). \tag{7}$$

We also define homomorphisms $S^n(\boldsymbol{f})\colon S^n(\boldsymbol{X}) \to S^n(\boldsymbol{X}'')$, induced by a morphism $\boldsymbol{f}\colon \boldsymbol{X} \to \boldsymbol{X}''$. If $\boldsymbol{f} = (f_\lambda)$, where $f_\lambda\colon X_\lambda \to X''_\lambda$, we first define a mapping of cochain complexes $\boldsymbol{f}^\# = (f^n)\colon K(\boldsymbol{X}) \to K(\boldsymbol{X}'')$. If $c \in K^n(\boldsymbol{X})$, then $f^n c \in K^n(\boldsymbol{X}'')$ is given by

$$(f^n c)(\boldsymbol{\lambda}) = f_{\lambda_0}(c(\boldsymbol{\lambda})), \ \boldsymbol{\lambda} \in \Lambda_n. \tag{8}$$

The next lemma shows that $\boldsymbol{f}^\#$ is indeed a mapping of cochain complexes.

LEMMA 11.40. *$f^n \delta^n = \delta^n f^{n-1}$, for $n \ge 1$.*

Proof. If $c \in K^{n-1}(\boldsymbol{X})$, $\boldsymbol{\lambda} \in \Lambda_n$, then

$$(f^n \delta^n c)(\boldsymbol{\lambda}) = f_{\lambda_0}(\delta^n c(\boldsymbol{\lambda})) = f_{\lambda_0} p_{\lambda_0 \lambda_1}(c(d^0 \boldsymbol{\lambda})) + \sum_{j=1}^{n} (-1)^j f_{\lambda_0}(c(d^j \boldsymbol{\lambda})) \ , \tag{9}$$

$$\begin{aligned} (\delta^n f^{n-1} c)(\boldsymbol{\lambda}) &= p''_{\lambda_0 \lambda_1}((f^{n-1} c)(d^0 \boldsymbol{\lambda})) + \sum_{j=1}^{n} (-1)^j (f^{n-1} c)(d^j \boldsymbol{\lambda}) \\ &= p''_{\lambda_0 \lambda_1} f_{\lambda_1}(c(d^0 \boldsymbol{\lambda})) + \sum_{j=1}^{n} (-1)^j f_{\lambda_0}(c(d^j \boldsymbol{\lambda})). \end{aligned} \tag{10}$$

The two expressions are equal because $f_{\lambda_0} p_{\lambda_0 \lambda_1} = p''_{\lambda_0 \lambda_1} f_{\lambda_1}$. □

The mapping of cochain complexes $\boldsymbol{f}^\# = (f^n)\colon K(\boldsymbol{X}) \to K(\boldsymbol{X}'')$ induces a homomorphism of cohomology modules $\boldsymbol{f}^*\colon H^n(K(\boldsymbol{X})) \to H^n(K(\boldsymbol{X}''))$. If $[z]$ denotes the cohomology class of a cocyle z of $K^n(\boldsymbol{X})$, then $\boldsymbol{f}^*[z] = [\boldsymbol{f}^\#(z)]$. This homomorphism is, by definition, the induced homomorphism $S^n(\boldsymbol{f})\colon S^n(\boldsymbol{X}) \to S^n(\boldsymbol{X}'')$.

LEMMA 11.41. *For every $n = 0, 1, \ldots$, $S^n = H^n K\colon \mathsf{Mod}^\Lambda \to \mathsf{Mod}$ is an additive functor.*

Proof. Consider morphisms $\boldsymbol{f}'\colon \boldsymbol{X}' \to \boldsymbol{X}$, $\boldsymbol{f}\colon \boldsymbol{X} \to \boldsymbol{X}''$ and the identity morphism $\boldsymbol{1_X}\colon \boldsymbol{X} \to \boldsymbol{X}$. To prove that

$$(\boldsymbol{f}\boldsymbol{f}')^* = \boldsymbol{f}^* \boldsymbol{f}'^*, \ \boldsymbol{1}^*_{\boldsymbol{X}} = \mathrm{id}, \tag{11}$$

it suffices to prove the corresponding relations for the induced mappings of cochain complexes, i.e.,

$$(\boldsymbol{f}\boldsymbol{f}')^\# = \boldsymbol{f}^\# \boldsymbol{f}'^\#, \ \boldsymbol{1}^\#_{\boldsymbol{X}} = \mathrm{id}. \tag{12}$$

Putting $\boldsymbol{f}\boldsymbol{f}' = \boldsymbol{f}''$, we see that

$$\begin{aligned}(f''^n c)(\boldsymbol{\lambda}) &= f''_{\lambda_0}(c(\boldsymbol{\lambda})) = f_{\lambda_0} f'_{\lambda_0}(c(\boldsymbol{\lambda})) \\ &= f_{\lambda_0}((f'^n c)(\boldsymbol{\lambda})) = (f^n f'^n c)(\boldsymbol{\lambda}),\end{aligned} \tag{13}$$

which establishes the first equality in (12). The identity morphism $\mathbf{1}_{\boldsymbol{X}} \colon \boldsymbol{X} \to \boldsymbol{X}$ is given by the identity homomorphisms $1_\lambda \colon X_\lambda \to X_\lambda$ and therefore, $\mathbf{1}^\#_{\boldsymbol{X}} \colon K(\boldsymbol{X}) \to K(\boldsymbol{X})$ is given by homomorphisms $1^n \colon K^n(\boldsymbol{X}) \to K^n(\boldsymbol{X})$, where $(1^n c)(\boldsymbol{\lambda}) = c(\boldsymbol{\lambda})$. This establishes the second equality in (12).

In order to prove additivity, i.e., to show that $(\boldsymbol{f} + \boldsymbol{g})^* = \boldsymbol{f}^* + \boldsymbol{g}^*$, it suffices to see that $(\boldsymbol{f} + \boldsymbol{g})^\# = \boldsymbol{f}^\# + \boldsymbol{g}^\#$. Putting $\boldsymbol{f} + \boldsymbol{g} = \boldsymbol{h} = (h^n)$, we see that $(h^n c)(\boldsymbol{\lambda}) = (f_{\lambda_0} + g_{\lambda_0})(c(\boldsymbol{\lambda})) = (f^n c)(\boldsymbol{\lambda}) + (g^n c)(\boldsymbol{\lambda})$. □

We will now associate with every $n \geq 0$ and every short exact sequence of inverse systems

$$E = (\mathbf{0} \to \boldsymbol{X}' \xrightarrow{\boldsymbol{f}'} \boldsymbol{X} \xrightarrow{\boldsymbol{f}} \boldsymbol{X}'' \to \mathbf{0}) \tag{14}$$

a connecting homomorphism $\omega^n_E \colon S^n(\boldsymbol{X}'') \to S^{n+1}(\boldsymbol{X}')$. To do this we need the following lemma.

LEMMA 11.42. *If the sequence of inverse systems* (14) *is exact, then the induced sequence of cochain complexes*

$$K(E) = (0 \to K(\boldsymbol{X}') \xrightarrow{\boldsymbol{f}'^\#} K(\boldsymbol{X}) \xrightarrow{\boldsymbol{f}^\#} K(\boldsymbol{X}'') \to 0) \tag{15}$$

is also exact.

Proof. It suffices to prove that the sequence of modules

$$K^n(E) = (0 \to K^n(\boldsymbol{X}') \xrightarrow{f'^n} K^n(\boldsymbol{X}) \xrightarrow{f^n} K^n(\boldsymbol{X}'') \to 0) \tag{16}$$

is exact, for every $n \geq 0$. By (14), for every $\lambda_0 \in \Lambda$, the following sequence of modules is exact

$$0 \to X'_{\lambda_0} \xrightarrow{f'_{\lambda_0}} X_{\lambda_0} \xrightarrow{f_{\lambda_0}} X''_{\lambda_0} \to 0. \tag{17}$$

Since f^n can be interpreted as

$$f^n = \prod_{\lambda_0 \leq \ldots \leq \lambda_n} f_{\lambda_0} \colon \prod_{\lambda_0 \leq \ldots \leq \lambda_n} X_{\lambda_0} \to \prod_{\lambda_0 \leq \ldots \leq \lambda_n} X''_{\lambda_0} \tag{18}$$

and f'^n has an analogous interpretation, it is clear that the exactness of (17) implies the desired exactness of (16). □

It is well known in homological algebra that a short exact sequence of cochain complexes generates a long exact sequence of cohomology modules (see e.g., (Hilton, Stammbach 1971), Ch. IV, Theorem 2.1). In particular, the sequence (15) yields the desired connecting homomorphisms $\omega^n_E \colon S^n(\boldsymbol{X}) = H^n K(\boldsymbol{X}'') \to H^{n+1} K(\boldsymbol{X}') = S^{n+1}(\boldsymbol{X}')$.

LEMMA 11.43. *The functors $S^n = H^nK\colon \mathsf{Mod}^\Lambda \to \mathsf{Mod}$ and the homomorphisms $\omega_E^n\colon S^n(\boldsymbol{X}'') \to S^{n+1}(\boldsymbol{X}')$ form a connected sequence of functors.*

Proof. We have already proved that every short exact sequence (14) generates a long exact sequence of the form (11.4.2). It remains to show that, for a commutative diagram of the form (11.3.32) and with exact rows, one obtains a commutative diagram of the form (11.4.3). First note that (12) yields a commutative diagram of cochain complexes:

$$\begin{array}{ccccccccc} 0 & \longrightarrow & K(\boldsymbol{X}') & \xrightarrow{\boldsymbol{f}'^{\#}} & K(\boldsymbol{X}) & \xrightarrow{\boldsymbol{f}^{\#}} & K(\boldsymbol{X}'') & \longrightarrow & 0 \\ & & \downarrow{\scriptstyle \boldsymbol{h}'^{\#}} & & \downarrow{\scriptstyle \boldsymbol{h}^{\#}} & & \downarrow{\scriptstyle \boldsymbol{h}''^{\#}} & & \\ 0 & \longrightarrow & K(\boldsymbol{Y}') & \xrightarrow[\boldsymbol{g}'^{\#}]{} & K(\boldsymbol{Y}) & \xrightarrow[\boldsymbol{g}^{\#}]{} & K(\boldsymbol{Y}'') & \longrightarrow & 0\,. \end{array} \tag{19}$$

By Lemma 11.42, the rows of this diagram are exact. However, it is well known in homological algebra (see e.g., (Hilton, Stammbach 1971), Chapter IV, Theorem 2.1) that the connecting homomorphisms ω_E^n, ω_F^n of the cohomology modules then make the following diagram commutative.

$$\begin{array}{ccc} S^n(\boldsymbol{X}'') & \xrightarrow{\omega_E^n} & S^{n+1}(\boldsymbol{X}') \\ {\scriptstyle S^n(\boldsymbol{h}'')}\downarrow & & \downarrow{\scriptstyle S^{n+1}(\boldsymbol{h}')} \\ S^n(\boldsymbol{Y}'') & \xrightarrow[\omega_F^n]{} & S^{n+1}(\boldsymbol{Y}')\,. \end{array} \qquad \square \tag{20}$$

LEMMA 11.44. *If $\boldsymbol{X}$ is an injective object of Mod^Λ, then $S^n(\boldsymbol{X}) = 0$, for all $n \geq 1$.*

Proof. Associate with $\boldsymbol{X}$ the injective system $\boldsymbol{I}$ and the monomorphism $\boldsymbol{f}\colon \boldsymbol{X} \to \boldsymbol{I}$ constructed in the proof of Theorem 11.18. In order to prove that $S^n(\boldsymbol{X}) = 0$, $n \geq 1$, it suffices to show that $S^n(\boldsymbol{I}) = 0$. Indeed, since $\boldsymbol{X}$ is injective, there exists a morphism $\boldsymbol{g}\colon \boldsymbol{I} \to \boldsymbol{X}$ such that $\boldsymbol{gf} = \mathbf{1}$. Clearly, $S^n(\boldsymbol{I}) = 0$ implies $1_{S^n(X)} = S^n(\boldsymbol{g})S^n(\boldsymbol{f}) = 0$ and thus, $S^n(\boldsymbol{X}) = 0$.

By construction, $\boldsymbol{I}$ is the direct product

$$\boldsymbol{I} = \prod_{\mu \in \Lambda} \boldsymbol{I}^\mu \tag{21}$$

of a collection of systems $\boldsymbol{I}^\mu$, $\mu \in \Lambda$. This implies that the cochain complex $K(\boldsymbol{I})$ is a direct product of cochain complexes $K(\boldsymbol{I}^\mu)$. Consequently, to show that $S^n(\boldsymbol{I}) = H^nK(\boldsymbol{I}) = 0$, $n \geq 1$, it suffices to show that

$$H^nK(\boldsymbol{I}^\mu) = 0, \ n \geq 1. \tag{22}$$

We now take into account the fact that all X_λ, $\lambda \in \Lambda$, are injective modules (see Remark 11.19). Therefore, in the construction of $\boldsymbol{I}^\mu = (I^\mu_\lambda, q^\mu_{\lambda\lambda'}, \Lambda)$, one can take $s^\mu = \mathrm{id}$ and obtain

$$I^\mu_\lambda = \begin{cases} X_\mu, & \mu \le \lambda, \\ 0, & \mu \not\le \lambda. \end{cases} \tag{23}$$

$$q^\mu_{\lambda\lambda'} = \begin{cases} \mathrm{id}, & \mu \le \lambda, \\ 0, & \mu \not\le \lambda. \end{cases} \tag{24}$$

To prove (22), consider any n-cocycle $z \in K^n(\boldsymbol{I}^\mu)$ and consider the $(n-1)$-cochain $c \in K^{n-1}(\boldsymbol{I}^\mu)$, given by

$$c(\lambda_0, \dots, \lambda_{n-1}) = \begin{cases} z(\mu, \lambda_0, \dots, \lambda_{n-1}), & \mu \le \lambda_0, \\ 0, & \mu \not\le \lambda_0. \end{cases} \tag{25}$$

Then, for $\boldsymbol{\lambda} = (\lambda_0, \dots, \lambda_n)$ and $\mu \le \lambda_0$, one has

$$\begin{aligned} (\delta^n c)(\boldsymbol{\lambda}) &= q^\mu_{\lambda_0\lambda_1}(c(d^0\boldsymbol{\lambda})) + \textstyle\sum_{j=1}^n (-1)^j c(d^j\boldsymbol{\lambda}) \\ &= z(\mu, d^0\boldsymbol{\lambda}) + \textstyle\sum_{j=1}^n (-1)^j z(\mu, d^j\boldsymbol{\lambda}). \end{aligned} \tag{26}$$

On the other hand, $\delta^{n+1} z = 0$ shows that

$$(\delta^{n+1} z)(\mu, \boldsymbol{\lambda}) = z(\boldsymbol{\lambda}) - \sum_{j=0}^n (-1)^j z(\mu, d^j\boldsymbol{\lambda}) = 0. \tag{27}$$

Comparing (26) with (27), one concludes that, for $\mu \le \lambda_0$,

$$(\delta^n c)(\boldsymbol{\lambda}) = z(\boldsymbol{\lambda}). \tag{28}$$

If $\mu \not\le \lambda_0$, (28) holds also, because both sides equal 0, since $I^\mu_{\lambda_0} = 0$. □

Lemmas 11.43 and 11.44 enable us to apply Theorem 11.37 and obtain the following result.

THEOREM 11.45. *The functors $S^n = H^n K \colon \mathsf{Mod}^\Lambda \to \mathsf{Mod}$ and the homomorphisms $\omega^n_E \colon S^n(\boldsymbol{X}'') \to S^{n+1}(\boldsymbol{X}')$ form a universal connected sequence of functors.*

We will now exhibit a natural equivalence ϕ between the functors lim and $S^0 = H^0 K$. Let $\phi_{\boldsymbol{X}} \colon \lim \boldsymbol{X} \to K^0(\boldsymbol{X})$ be the homomorphism, which maps $x = (x_{\lambda_0}) \in \lim \boldsymbol{X}$ to $c \in K^0(\boldsymbol{X})$, where $c(\lambda_0) = x_{\lambda_0}$. Note that, for $\lambda_0 \le \lambda_1$, (3) yields $(\delta^1 c)(\lambda_0, \lambda_1) = p_{\lambda_0\lambda_1}(c(\lambda_1)) - c(\lambda_0) = p_{\lambda_0\lambda_1}(x_{\lambda_1}) - x_{\lambda_0} = 0$. Consequently, $\phi_{\boldsymbol{X}}$ maps $\lim \boldsymbol{X}$ to $\mathrm{Ker}(\delta^1) = H^0 K(\boldsymbol{X})$. Clearly, $\phi_{\boldsymbol{X}} \colon \lim \boldsymbol{X} \to H^0 K(\boldsymbol{X})$ is a bijection and thus, an isomorphism. Naturality of ϕ is easily verified. Indeed, if $\boldsymbol{f} \colon \boldsymbol{X} \to \boldsymbol{X}''$ is a morphism of Mod^Λ, then the homomorphism $f^0 \colon K^0(\boldsymbol{X}) \to K^0(\boldsymbol{X}'')$ maps $c = \phi_{\boldsymbol{X}}(x)$ to $c'' \in K^0(\boldsymbol{X}'')$, where $c''(\lambda_0) = f_{\lambda_0}(c(\lambda_0)) = f_{\lambda_0}(x_{\lambda_0})$. On the other hand, $\lim \boldsymbol{f}$ maps x to $(f_{\lambda_0}(x_{\lambda_0}))$, which in turn is mapped by $\phi_{\boldsymbol{X}''}$ also to c''. We have thus established the following lemma.

LEMMA 11.46. *The homomorphisms $\phi_{\boldsymbol{X}}$ define a natural equivalence of functors $\phi\colon \lim \to H^0K$.*

If we now take into account Theorems 11.38 and 11.45, as well as Remark 11.36, we obtain the following corollary.

COROLLARY 11.47. *ϕ induces a natural isomorphism between connected sequences of functors $(\lim^n, \theta_E^n)$ and (H^nK, ω_E^n). In particular, it induces a natural isomorphism*

$$\lim^n \boldsymbol{X} \approx H^nK(\boldsymbol{X}),\ n \geq 0. \tag{29}$$

The corollary shows that the constructions described in this subsection give an explicit description of the functors $\lim^n$.

11.6 $\lim^n$ for sequences

In this subsection we will consider the functors $\lim^n$, for the special case of inverse sequences, i.e., for the case $\Lambda = \mathbb{N}$. We will show that, for an arbitrary inverse sequence $\boldsymbol{X} = (X_n, p_{nn'}, \mathbb{N})$ of R-modules, $\lim^n \boldsymbol{X} = 0$, for all $n \geq 2$, while $\lim^1 \boldsymbol{X}$ admits a relatively simple description (see e.g., (Mardešić, Segal 1982)), temporarily denoted by $\underline{\lim}^1$. We begin by recalling this description, using the cochain complex $K = K(\boldsymbol{X})$, introduced in 11.5,

$$K = (0 \to K^0 \xrightarrow{\delta^1} K^1 \xrightarrow{\delta^2} K^2 \ldots). \tag{1}$$

Consider the submodule $M^0 = M^0(\boldsymbol{X}) \subseteq K^0 = \prod_{n\in\mathbb{N}} X_n$, which consists of all $x = (x_n) \in K^0$, $x_n \in X_n$ such that there exists an element $u \in K^0$ having the property that

$$x_n = p_{n\,n+1}(u_{n+1}) - u_n = (\delta^1 u)_{n,n+1}. \tag{2}$$

Put $\underline{\lim}^1 \boldsymbol{X} = K^0/M^0$. A morphism $\boldsymbol{f} = (f_n)\colon \boldsymbol{X}' \to \boldsymbol{X}$ induces a homomorphism $\underline{\lim}^1 \boldsymbol{f}\colon \underline{\lim}^1 \boldsymbol{X}' \to \underline{\lim}^1 \boldsymbol{X}$, defined as follows. If $x' = (x'_n) \in K^0(\boldsymbol{X}')$, $x'_n \in X'_n$, then $x_n = f_n(x'_n) \in X_n$ determines an element $x = (x_n) \in K^0(\boldsymbol{X})$ and $x' \mapsto x$ is a homomorphism $K^0(\boldsymbol{X}') \to K^0(\boldsymbol{X})$. Moreover, if $x' \in M^0(\boldsymbol{X}')$, i.e., if there exists an element $u' = (u'_n) \in K^0(\boldsymbol{X}')$ such that $x'_n = p'_{n\,n+1}(u'_{n+1}) - u'_n$, then $f_n(x'_n) = p_{n\,n+1} f_{n+1}(u'_{n+1}) - f_n(u'_n)$, i.e., (x_n) satisfies (2), for $u_n = f_n(u'_n)$. This shows that the homomorphism $x' \mapsto x$ maps $M^0(\boldsymbol{X}')$ to $M^0(\boldsymbol{X})$ and thus, induces a homomorphism $\underline{\lim}^1 \boldsymbol{X}' \to \underline{\lim}^1 \boldsymbol{X}$. This is the desired homomorphism $\underline{\lim}^1 \boldsymbol{f}$.

For every inverse sequence $\boldsymbol{X}$, we will now define a natural homomorphism $\psi_{\boldsymbol{X}}\colon \lim^1 \boldsymbol{X} \to \underline{\lim}^1 \boldsymbol{X}$ as follows. First recall that, by 11.5, $K^1(\boldsymbol{X}) =$

$\prod_{n\le n'} X_n$ and $\lim^1 \boldsymbol{X} = \operatorname{Ker}\delta^2/\operatorname{Im}\delta^1$. If $z \in \operatorname{Ker}\delta^2$, $z = (z_{nn'})$, $z_{nn'} \in X_n$, put $\psi(z) = x = (x_n)$, where

$$x_n = z_{n,n+1}. \tag{3}$$

Since $z_{n,n+1} \in X_n$, (3) defines a homomorphism $\psi\colon \operatorname{Ker}\delta^2 \to K^0$. If $z \in \operatorname{Im}\delta^1$, then there exists a $u = (u_n) \in K^0$ such that $z = \delta^1 u$. Therefore, $z_{n,n+1} = (\delta^1 u)_{n,n+1}$, which shows that $\psi(z) \in M^0$. Mapping the class of z to the class of $\psi(z)$, we obtain the desired homomorphism $\psi_{\boldsymbol{X}}\colon \lim^1 \boldsymbol{X} \to \underline{\lim}^1 \boldsymbol{X}$.

LEMMA 11.48. *The homomorphisms $\psi_{\boldsymbol{X}}\colon \lim^1 \boldsymbol{X} \to \underline{\lim}^1 \boldsymbol{X}$ define a natural transformation between the functors $\lim^1$ and $\underline{\lim}^1$.*

Proof. We must show that, for any morphism $\boldsymbol{f}\colon \boldsymbol{X}' \to \boldsymbol{X}$, the following diagram commutes.

$$\begin{array}{ccc} \lim^1 \boldsymbol{X}' & \xrightarrow{\lim^1 \boldsymbol{f}} & \lim^1 \boldsymbol{X} \\ \Big\downarrow{\scriptstyle \psi_{\boldsymbol{X}'}} & & \Big\downarrow{\scriptstyle \psi_{\boldsymbol{X}}} \\ \underline{\lim}^1 \boldsymbol{X}' & \xrightarrow[\underline{\lim}^1 \boldsymbol{f}]{} & \underline{\lim}^1 \boldsymbol{X}. \end{array} \tag{4}$$

Indeed, let $z' = (z'_{nn'}) \in K^1(\boldsymbol{X}')$, $z'_{nn'} \in X'_n$, and let $\delta^2 z' = 0$. Then $\psi_{\boldsymbol{X}'}$ maps the class of z' to the class of $x' = (x'_n)$, where $x'_n = z'_{n,n+1} \in X'_n$. Furthermore, $\underline{\lim}^1 \boldsymbol{f}$ maps that class to the class of $x = (x_n)$, where $x_n = f_n(x'_n) = f_n(z'_{n,n+1})$. On the other hand, $\lim^1 \boldsymbol{f}$ maps the class of z' to the class of $z = (z_{nn'})$, where $z_{nn'} = f_n(z'_{nn'})$. However, $\psi_{\boldsymbol{X}}$ maps that class also to the class of $(f_n(z'_{n,n+1}))$. □

LEMMA 11.49. *For every inverse sequence $\boldsymbol{X}$, $\psi_{\boldsymbol{X}}\colon \lim^1 \boldsymbol{X} \to \underline{\lim}^1 \boldsymbol{X}$ is an isomorphism of modules.*

Proof. In order to show that $\psi_{\boldsymbol{X}}\colon \lim^1 \boldsymbol{X} \to \underline{\lim}^1 \boldsymbol{X}$ is an epimorphism, it suffices to show that $\psi\colon \operatorname{Ker}\delta^2 \to K^0$ is surjective. To prove the latter assertion, consider any $x = (x_n) \in K^0$, $x_n \in X_n$. We shall construct an element $z = (z_{nn'}) \in \operatorname{Ker}\delta^2$ such that $\psi(z) = x$. The construction of $z_{nn'} \in X_n$ is by induction on $k = n' - n \ge 0$. We begin the induction, by putting

$$z_{nn} = 0,\ z_{n,n+1} = x_n,\ n \in \mathbb{N}. \tag{5}$$

If $z_{n,n+k} \in X_n$ is already defined, we define $z_{n,n+k+1} \in X_n$, by putting

$$z_{n,n+k+1} = p_{n,n+k}(x_{n+k}) + z_{n,n+k}. \tag{6}$$

Clearly, $z = (z_{nn'}) \in K^1$. We will now prove that $\delta^2 z = 0$, i.e.,

$$(\delta^2 z)_{n,n+k,n+l} = 0,\ 0 \le k \le l. \tag{7}$$

The proof proceeds by induction on l, while n and k have arbitrary fixed values. If $l = k$, $(\delta^2 z)_{n,n+k,n+l} = p_{n\,n+k}(z_{n+k,n+k}) = 0$, because, by the first equality in (5), $z_{n+k,n+k} = 0$. For $l = k+1$, the second equality in (5) shows that the assertion coincides with (6). Now assume that the assertion holds for a given $l \geq 1$. Using the fact that $\delta^3\delta^2 = 0$, we will prove it for $l+1$. Indeed, we have

$$\begin{aligned} 0 &= (\delta^3\delta^2 z)_{n,n+k,n+l,n+l+1} \\ &= p_{n\,n+k}((\delta^2 z)_{n+k,n+l,n+l+1}) - (\delta^2 z)_{n,n+l,n+l+1} \\ &+ (\delta^2 z)_{n,n+k,n+l+1} - (\delta^2 z)_{n,n+k,n+l}. \end{aligned} \tag{8}$$

Of the four summands in (8), the first two vanish by the first step in the induction. The last term vanishes by the preceding step in the induction. Consequently, (8) reduces to the desired relation

$$(\delta^2 z)_{n,n+k,n+l+1} = 0. \tag{9}$$

Finally, by (3) and (5), $\psi(z) = (z_{n,n+1}) = (x_n) = x$.

In order to prove that $\psi_{\mathbf{X}}$ is also a monomorphism, consider any $z = (z_{nn'}) \in K^1$, $\delta^2 z = 0$, such that $\psi(z) \in M^0$, i.e., such that there exists $u = (u_n) \in K^0$, satisfying

$$(\delta^1 u)_{n,n+1} = z_{n,n+1}. \tag{10}$$

The assertion will be proved, if we show that $\delta^1 u = z$, i.e.,

$$(\delta^1 u)_{n,n+k} = z_{n,n+k}\ , k \geq 0. \tag{11}$$

For $k = 0$, (11) holds because both sides vanish. Indeed, $(\delta^1 u)_{nn} = u_n - u_n = 0$, while $0 = (\delta^2 z)_{nnn} = z_{nn} - z_{nn} + z_{nn} = z_{nn}$. For $k = 1$, (11) reduces to (10). We proceed by induction on k, using the fact that $\delta^2\delta^1 = 0$ and therefore,

$$\begin{aligned} 0 &= (\delta^2\delta^1 u)_{n,n+k,n+k+1} \\ &= p_{n\,n+k}((\delta^1 u)_{n+k,n+k+1}) - (\delta^1 u)_{n,n+k+1} + (\delta^1 u)_{n,n+k}. \end{aligned} \tag{12}$$

By (10), the first of the three summands coincides with $p_{n\,n+k}(z_{n+k,n+k+1})$ and, by the induction hypothesis, the last summand coincides with $z_{n,n+k}$. Consequently, (12) becomes

$$(\delta^1 u)_{n,n+k+1} = p_{n\,n+k}(z_{n+k,n+k+1}) + z_{n,n+k}. \tag{13}$$

On the other hand, $\delta^2 z = 0$ implies

$$0 = (\delta^2 z)_{n,n+k,n+k+1} = p_{n\,n+k}(z_{n+k,n+k+1}) - z_{n,n+k+1} + z_{n,n+k}. \tag{14}$$

Now (13) and (14) yield the desired relation

$$(\delta^1 u)_{n,n+k+1} = z_{n,n+k+1}. \;\square \tag{15}$$

Lemmas 11.48 and 11.49 yield the following theorem.

THEOREM 11.50. *The homomorphisms* $\psi_{\boldsymbol{X}}: \lim^1 \boldsymbol{X} \to \underline{\lim}^1 \boldsymbol{X}$ *define a natural equivalence* ψ *between the functors* $\lim^1$ *and* $\underline{\lim}^1$.

COROLLARY 11.51. *For every short exact sequence of towers*

$$E = (\mathbf{0} \to \boldsymbol{X}' \xrightarrow{\boldsymbol{f}'} \boldsymbol{X} \xrightarrow{\boldsymbol{f}} \boldsymbol{X}'' \to \mathbf{0}), \tag{16}$$

the following long sequence of modules is exact:

$$\begin{aligned} &\mathbf{0} \to \lim \boldsymbol{X}' \xrightarrow{\lim \boldsymbol{f}'} \lim \boldsymbol{X} \xrightarrow{\lim \boldsymbol{f}} \lim \boldsymbol{X}'' \xrightarrow{\theta_E^0} \lim^1 \boldsymbol{X}' \\ &\xrightarrow{\lim^1 \boldsymbol{f}'} \lim^1 \boldsymbol{X} \xrightarrow{\lim^1 \boldsymbol{f}} \lim^1 \boldsymbol{X}'' \to \mathbf{0}. \end{aligned} \tag{17}$$

Proof. It suffices to show that exactness of $\boldsymbol{X} \xrightarrow{\boldsymbol{f}} \boldsymbol{X}'' \to 0$ implies exactness of

$$\lim^1 \boldsymbol{X} \xrightarrow{\lim^1 \boldsymbol{f}} \lim^1 \boldsymbol{X}'' \to 0. \tag{18}$$

In view of Theorem 11.50, this assertion is equivalent to the analogous assertion for $\underline{\lim}^1$, which is true. Indeed, every $f_n: X_n \to X_n''$ is surjective, hence, also $\prod f_n: K^0(\boldsymbol{X}) = \prod X_n \to \prod X_n'' = K^0(\boldsymbol{X}'')$ is surjective. Since $\underline{\lim}^1 \boldsymbol{f}$ is the induced homomorphism $K^0(\boldsymbol{X})/M^0(\boldsymbol{X}) \to K^0(\boldsymbol{X}'')/M^0(\boldsymbol{X}'')$, it is also surjective.

THEOREM 11.52. *For every inverse sequence* $\boldsymbol{X}$ *and* $n \geq 2$, $\lim^n \boldsymbol{X} = 0$.

Proof. By Corollary 11.51, the functors $T^0 = \lim, T^1 = \lim^1$, $T^n = 0$, for $n \geq 2$, and the connecting homomorphisms $\omega_E^0: \lim \boldsymbol{X}'' \to \lim^1 \boldsymbol{X}'$ and $\omega_E^m = 0$, for $m \geq 1$, form a universal connected sequence of functors $\mathsf{Mod}^{\mathbb{N}} \to \mathsf{Mod}$ (see Theorem 11.37). On the other hand, by Theorem 11.38, the functors $S^n = \lim^n$ and the connecting homomorphisms θ_E^n, $n \geq 0$, also form a universal connecting sequence. Since $S^0 = T^0$ and $S^1 = T^1$, it follows, by uniqueness (see Remark 11.35), that the two sequences are isomorphic. In particular, $S^n \approx T^n$, for all $n \geq 0$. In particular, $\lim^n(\boldsymbol{X}) = 0$. However, $T^n = 0$, for $n \geq 2$, and thus, also $\lim^n \boldsymbol{X} = 0$, for $n \geq 2$. □

Bibliographic notes

Subsections 11.1 and 11.2 are based on (Nöbeling 1961). Construction and characterization of higher derived limits as described in 11.3 and 11.4 follow the general theory, as described, e.g., in (Bucur, Deleanu 1968). Explicit formulae for $\lim^n$ were discovered independently and simultaneously by G. Nöbeling and J.-E. Roos and 11.5 is based on their papers (Nöbeling 1961), (Roos 1961) (also see (Deheuvels 1960, 1962)). The derived functors of lim, for $n \geq 2$, were first studied in (Yeh 1959). The alternative description of $\lim^1$, for inverse sequences, is attributed to S. Eilenberg (see (Jensen 1972), p. 13).

12. $\lim^n$ and the extension functors Ext^n

In this section we first define and analyze in detail the extension products $\operatorname{Ext}^n(\boldsymbol{A}, \boldsymbol{X})$ of two inverse systems of modules. We then show that $\lim^n \boldsymbol{X}$ coincides with $\operatorname{Ext}^n(\boldsymbol{\Delta}(\Lambda), \boldsymbol{X})$, where $\boldsymbol{\Delta}(\Lambda)$ is the diagonal inverse system. The advantage of this description of $\lim^n \boldsymbol{X}$ over the description given in 11 lies in the fact that $\lim^n \boldsymbol{X}$ can be determined using, the same projective resolution of $\boldsymbol{\Delta}(\Lambda)$, for all $\boldsymbol{X}$.

12.1 The bifunctors Ext^n

For any fixed inverse system $\boldsymbol{A} \in \mathsf{Mod}^\Lambda$, $\operatorname{Hom}(\boldsymbol{A}, .): \mathsf{Mod}^\Lambda \to \mathsf{Mod}$ is a covariant functor. To a system $\boldsymbol{X}$, it assigns the R-module $\operatorname{Hom}(\boldsymbol{A}, \boldsymbol{X})$ consisting of all morphisms $\boldsymbol{h}: \boldsymbol{A} \to \boldsymbol{X}$. To a morphism $\boldsymbol{f}: \boldsymbol{X} \to \boldsymbol{X}''$, it assigns the homomorphism $\operatorname{Hom}(\boldsymbol{A}, \boldsymbol{f}): \operatorname{Hom}(\boldsymbol{A}, \boldsymbol{X}) \to \operatorname{Hom}(\boldsymbol{A}, \boldsymbol{X}'')$, which maps $\boldsymbol{h} \in \operatorname{Hom}(\boldsymbol{A}, \boldsymbol{X})$ to $\boldsymbol{f}\boldsymbol{h} \in \operatorname{Hom}(\boldsymbol{A}, \boldsymbol{X}'')$.

LEMMA 12.1. *The functor* $\operatorname{Hom}(\boldsymbol{A}, .)$ *is additive and left exact. Moreover, if* $\boldsymbol{A}$ *is a projective object of* Mod^Λ, *then* $\operatorname{Hom}(\boldsymbol{A}, .)$ *is an exact functor.*

Proof. The additivity of $\operatorname{Hom}(\boldsymbol{A}, .)$ follows from Lemma 11.6. To prove left exactness, consider an exact sequence of inverse systems

$$\boldsymbol{0} \to \boldsymbol{X}' \xrightarrow{\boldsymbol{f}'} \boldsymbol{X} \xrightarrow{\boldsymbol{f}} \boldsymbol{X}''. \tag{1}$$

We must prove that the induced sequence

$$\boldsymbol{0} \to \operatorname{Hom}(\boldsymbol{A}, \boldsymbol{X}') \xrightarrow{\boldsymbol{f}'_*} \operatorname{Hom}(\boldsymbol{A}, \boldsymbol{X}) \xrightarrow{\boldsymbol{f}_*} \operatorname{Hom}(\boldsymbol{A}, \boldsymbol{X}'') \tag{2}$$

is exact; here $\boldsymbol{f}'_* = \operatorname{Hom}(\boldsymbol{A}, \boldsymbol{f}')$, $\boldsymbol{f}_* = \operatorname{Hom}(\boldsymbol{A}, \boldsymbol{f})$. By Lemma 11.5, the sequence of modules

$$0 \to X'_\lambda \xrightarrow{f'_\lambda} X_\lambda \xrightarrow{f_\lambda} X''_\lambda \tag{3}$$

is exact, for every $\lambda \in \Lambda$. It is well known that, for any module A, the functor $\operatorname{Hom}(A, .): \mathsf{Mod} \to \mathsf{Mod}$ is left exact. Therefore, (3) yields exactness of

$$0 \to \operatorname{Hom}(A_\lambda, X'_\lambda) \to \operatorname{Hom}(A_\lambda, X_\lambda) \to \operatorname{Hom}(A_\lambda, X''_\lambda). \tag{4}$$

Now let $\boldsymbol{h} = (h_\lambda) \in \mathrm{Hom}\,(\boldsymbol{A}, \boldsymbol{X})$ belong to the kernel of $\mathrm{Hom}\,(\boldsymbol{A}, \boldsymbol{f})$, i.e., $\boldsymbol{fh} = 0$. In order to show that $\boldsymbol{h}$ belongs to the image of $\mathrm{Hom}\,(\boldsymbol{A}, \boldsymbol{f}\,')$, we must find a morphism $\boldsymbol{h}' = (h'_\lambda) \in \mathrm{Hom}\,(\boldsymbol{A}, \boldsymbol{X}\,')$, such that $\boldsymbol{f}\,'\boldsymbol{h}' = \boldsymbol{h}$, i.e., $f'_\lambda h'_\lambda = h_\lambda$. By assumption, $f_\lambda h_\lambda = 0$ and therefore, by exactness of (4), there exist unique homomorphisms $h'_\lambda \colon A_\lambda \to X'_\lambda$ such that

$$f'_\lambda h'_\lambda = h_\lambda. \tag{5}$$

It remains to verify that the homomorphisms h'_λ form a morphism $\boldsymbol{h}' \colon \boldsymbol{A} \to \boldsymbol{X}\,'$, i.e., that they satisfy the equalities

$$p'_{\lambda\lambda'} h'_{\lambda'} = h'_\lambda a_{\lambda\lambda'},\ \lambda \leq \lambda', \tag{6}$$

where $p'_{\lambda\lambda'}$ and $a_{\lambda\lambda'}$ denote the bonding homomorphisms of $\boldsymbol{X}\,'$ and $\boldsymbol{A}$, respectively. Using (5), for λ' and λ, as well as the fact that $\boldsymbol{f}\,'$ and $\boldsymbol{h}$ are morphisms, one readily obtains

$$f'_\lambda p'_{\lambda\lambda'} h'_{\lambda'} = p_{\lambda\lambda'} f'_{\lambda'} h'_{\lambda'} = p_{\lambda\lambda'} h_{\lambda'} = h_\lambda a_{\lambda\lambda'} = f'_\lambda h'_\lambda a_{\lambda\lambda'}. \tag{7}$$

Since f'_λ is a monomorphism, (7) implies the desired formula (6). The remaining parts of the proof of the left exactness of $\mathrm{Hom}\,(\boldsymbol{A}, \,.\,)$ are straightforward and we omit them.

Now assume that $\boldsymbol{A}$ is projective and $\boldsymbol{f} \colon \boldsymbol{X} \to \boldsymbol{X}\,''$ is an epimorphism. We must show that $\mathrm{Hom}\,(\boldsymbol{A}, \boldsymbol{f}) \colon \mathrm{Hom}\,(\boldsymbol{A}, \boldsymbol{X}) \to \mathrm{Hom}\,(\boldsymbol{A}, \boldsymbol{X}\,'')$ is also an epimorphism, i.e., that every $\boldsymbol{h}'' \in \mathrm{Hom}\,(\boldsymbol{A}, \boldsymbol{X}\,'')$ admits a $\boldsymbol{h} \in \mathrm{Hom}\,(\boldsymbol{A}, \boldsymbol{X})$ such that $\boldsymbol{fh} = \boldsymbol{h}''$. However, this is just the property used in the definition of a projective object. This completes the proof of Lemma 12.1. □

As pointed out in 11.3, the construction of right derived functors, applies to additive left exact functors in abelian categories with enough injective objects. Therefore, it can be applied to the functor $\mathrm{Hom}\,(\boldsymbol{A}, \,.\,) \colon \mathsf{Mod}^\Lambda \to \mathsf{Mod}$, for any fixed inverse system $\boldsymbol{A} \in \mathsf{Mod}^\Lambda$. It yields a sequence of additive functors, denoted by $\mathrm{Ext}^n(\boldsymbol{A}, \,.\,) \colon \mathsf{Mod}^\Lambda \to \mathsf{Mod}$, $n \geq 0$. Moreover, for exact sequences

$$E = (\boldsymbol{0} \to \boldsymbol{X}\,' \xrightarrow{\boldsymbol{f}\,'} \boldsymbol{X} \xrightarrow{\boldsymbol{f}} \boldsymbol{X}\,'' \to \boldsymbol{0}), \tag{8}$$

it yields connecting homomorphisms $\theta^n_E \colon \mathrm{Ext}^n(\boldsymbol{A}, \boldsymbol{X}\,'') \to \mathrm{Ext}^{n+1}(\boldsymbol{A}, \boldsymbol{X}\,')$. We now briefly repeat that construction, omitting details already given for $\lim^n$ in 11.3.

For $\boldsymbol{X} \in \mathsf{Mod}^\Lambda$, one considers an injective resolution $(\mathcal{I}_X, \boldsymbol{e}_X)$, where

$$\mathcal{I}_X = (\boldsymbol{0} \to \boldsymbol{I}^0_X \xrightarrow{d^1_X} \boldsymbol{I}^1_X \xrightarrow{d^2_X} \boldsymbol{I}^2_X \to \ldots) \tag{9}$$

and $\boldsymbol{e}_X \colon \boldsymbol{X} \to \boldsymbol{I}^0_X$ is a monomorphism. Then one considers the induced cochain complex

$$\mathrm{Hom}\,(\boldsymbol{A}, \mathcal{I}_X) = (0 \to \mathrm{Hom}\,(\boldsymbol{A}, \boldsymbol{I}^0_X) \to \mathrm{Hom}\,(\boldsymbol{A}, \boldsymbol{I}^1_X) \to \ldots). \tag{10}$$

By definition, $\operatorname{Ext}^n(\boldsymbol{A}, \boldsymbol{X})$ is the n-th cohomology module of this complex, i.e.,

$$\operatorname{Ext}^n(\boldsymbol{A}, \boldsymbol{X}) = H^n(\operatorname{Hom}(\boldsymbol{A}, \mathcal{I}_X)). \tag{11}$$

Occasionally Ext^1 is abbreviated to Ext.

In order to define $\operatorname{Ext}^n(\boldsymbol{A}, \boldsymbol{f})\colon \operatorname{Ext}^n(\boldsymbol{A}, \boldsymbol{X}) \to \operatorname{Ext}^n(\boldsymbol{A}, \boldsymbol{X}'')$, induced by a morphism $\boldsymbol{f}\colon \boldsymbol{X} \to \boldsymbol{X}''$, one considers an injective resolution of the morphism $\boldsymbol{f}$, i.e., a commutative diagram (see Lemma 11.27)

$$\begin{array}{ccccccc} \boldsymbol{0} \longrightarrow & \boldsymbol{X} & \xrightarrow{\boldsymbol{e}_X} & \boldsymbol{I}_X^0 & \xrightarrow{\boldsymbol{d}_X^1} & \boldsymbol{I}_X^1 & \longrightarrow \cdots \\ & \boldsymbol{f}\big\downarrow & & \big\downarrow \boldsymbol{f}^0 & & \big\downarrow \boldsymbol{f}^1 & \\ \boldsymbol{0} \longrightarrow & \boldsymbol{X}'' & \xrightarrow[\boldsymbol{e}_{X''}]{} & \boldsymbol{I}_{X''}^0 & \xrightarrow[\boldsymbol{d}_{X''}^1]{} & \boldsymbol{I}_{X''}^1 & \longrightarrow \cdots \end{array} \tag{12}$$

where the rows are injective resolutions of $\boldsymbol{X}$ and $\boldsymbol{X}''$, respectively. Application of the functor $\operatorname{Hom}(\boldsymbol{A},\,.\,)$ to (12) yields a mapping of cochain complexes $\operatorname{Hom}(\boldsymbol{A}, \boldsymbol{f}) : \operatorname{Hom}(\boldsymbol{A}, \mathcal{I}_X) \to \operatorname{Hom}(\boldsymbol{A}, \mathcal{I}_{X''})$,

$$\begin{array}{ccccccc} 0 \longrightarrow & \operatorname{Hom}(\boldsymbol{A}, \boldsymbol{I}_X^0) & \longrightarrow & \operatorname{Hom}(\boldsymbol{A}, \boldsymbol{I}_X^1) & \longrightarrow & \operatorname{Hom}(\boldsymbol{A}, \boldsymbol{I}_X^2) & \longrightarrow \ldots \\ & f^0 \big\downarrow & & f^1 \big\downarrow & & f^2 \big\downarrow & \\ 0 \longrightarrow & \operatorname{Hom}(\boldsymbol{A}, \boldsymbol{I}_{X''}^0) & \longrightarrow & \operatorname{Hom}(\boldsymbol{A}, \boldsymbol{I}_{X''}^1) & \longrightarrow & \operatorname{Hom}(\boldsymbol{A}, \boldsymbol{I}_{X''}^2) & \longrightarrow \ldots. \end{array} \tag{13}$$

By definition, $\operatorname{Ext}^n(\boldsymbol{A}, \boldsymbol{f})$ is the induced homomorphism of the corresponding cohomology modules.

For a short exact sequence (8), consider a standard injective resolution of E (see 11.33). It is given by a commutative diagram of type (11.3.34). Its columns and rows are exact and all the modules $\boldsymbol{I}_{X'}^n, \boldsymbol{I}_X^n$ and $\boldsymbol{I}_{X''}^n$ are injective. Application of the functor $\operatorname{Hom}(\boldsymbol{A},\,.\,)$ to the exact sequence

$$\boldsymbol{0} \to \boldsymbol{I}_{X'}^n \xrightarrow{\boldsymbol{i}'^n} \boldsymbol{I}_X^n \xrightarrow{\boldsymbol{p}''^n} \boldsymbol{I}_{X''}^n \to \boldsymbol{0}, \; n \geq 0, \tag{14}$$

yields an exact sequence of modules

$$0 \to \operatorname{Hom}(\boldsymbol{A}, \boldsymbol{I}_{X'}^n) \to \operatorname{Hom}(\boldsymbol{A}, \boldsymbol{I}_X^n) \to \operatorname{Hom}(\boldsymbol{A}, \boldsymbol{I}_{X''}^n) \to 0. \tag{15}$$

Indeed, in view of Lemma 12.1, it suffices to show that $\operatorname{Hom}(\boldsymbol{A}, \boldsymbol{p}''^n)$ is an epimorphism. However, $\boldsymbol{p}''^n \boldsymbol{i}''^n = \mathrm{id}$ implies $\operatorname{Hom}(\boldsymbol{A}, \boldsymbol{p}''^n)\operatorname{Hom}(\boldsymbol{A}, \boldsymbol{i}''^n) = \mathrm{id}$ and thus, $\operatorname{Hom}(\boldsymbol{A}, \boldsymbol{p}''^n)$ is indeed an epimorphism. Extactness of (15) shows that

$$0 \to \operatorname{Hom}(\boldsymbol{A}, \mathcal{I}_{X'}) \to \operatorname{Hom}(\boldsymbol{A}, \mathcal{I}_X) \to \operatorname{Hom}(\boldsymbol{A}, \mathcal{I}_{X''}) \to 0 \tag{16}$$

is a short exact sequence of cochain complexes. Consequently, it induces a long exact sequence of cohomology modules. Its connecting homomorphisms are, by definition, the connecting homomorphisms

$$\theta_E^n \colon \operatorname{Ext}^n(\boldsymbol{A}, \boldsymbol{X}'') \to \operatorname{Ext}^{n+1}(\boldsymbol{A}, \boldsymbol{X}'). \tag{17}$$

The next theorem states properties of the functors $\operatorname{Ext}^n(\boldsymbol{A}, .)$ and the homomorphisms θ_E^n.

THEOREM 12.2.

(i) $\operatorname{Ext}^n(\boldsymbol{A}, .)$, $n \geq 0$, *is an additive functor from* Mod^Λ *to* Mod.

(ii) The homomorphisms $\operatorname{Hom}(\boldsymbol{A}, \boldsymbol{e}_X) \colon \operatorname{Hom}(\boldsymbol{A}, \boldsymbol{X}) \to \operatorname{Hom}(\boldsymbol{A}, \boldsymbol{I}_X^0)$ *induce a natural equivalence of functors* $\operatorname{Hom}(\boldsymbol{A}, .) \to \operatorname{Ext}^0(\boldsymbol{A}, .)$.

(iii) For every short exact sequence (8), *the following long sequence of modules is exact,*

$$\begin{gathered} 0 \to \operatorname{Hom}(\boldsymbol{A}, \boldsymbol{X}') \to \operatorname{Hom}(\boldsymbol{A}, \boldsymbol{X}) \to \operatorname{Hom}(\boldsymbol{A}, \boldsymbol{X}'') \overset{\theta_E^0}{\to} \operatorname{Ext}^1(\boldsymbol{A}, \boldsymbol{X}') \to \dots \\ \operatorname{Ext}^n(\boldsymbol{A}, \boldsymbol{X}') \to \operatorname{Ext}^n(\boldsymbol{A}, \boldsymbol{X}) \to \operatorname{Ext}^n(\boldsymbol{A}, \boldsymbol{X}'') \overset{\theta_E^n}{\to} \operatorname{Ext}^{n+1}(\boldsymbol{A}, \boldsymbol{X}') \to \dots . \end{gathered} \tag{18}$$

(iv) A morphism of exact sequences (11.3.32) *induces a commutative diagram,*

$$\begin{array}{ccccccccc} \to & \operatorname{Ext}^n(\boldsymbol{A}, \boldsymbol{X}') & \to & \operatorname{Ext}^n(\boldsymbol{A}, \boldsymbol{X}) & \to & \operatorname{Ext}^n(\boldsymbol{A}, \boldsymbol{X}'') & \overset{\theta_E^n}{\to} & \operatorname{Ext}^{n+1}(\boldsymbol{A}, \boldsymbol{X}') & \to \\ & \downarrow & & \downarrow & & \downarrow & & \downarrow & \\ \to & \operatorname{Ext}^n(\boldsymbol{A}, \boldsymbol{Y}') & \to & \operatorname{Ext}^n(\boldsymbol{A}, \boldsymbol{Y}) & \to & \operatorname{Ext}^n(\boldsymbol{A}, \boldsymbol{Y}'') & \underset{\theta_F^n}{\to} & \operatorname{Ext}^{n+1}(\boldsymbol{A}, \boldsymbol{Y}') & \to \end{array} \tag{19}$$

(v) If $\boldsymbol{X}$ *is injective, then*

$$\operatorname{Ext}^n(\boldsymbol{A}, \boldsymbol{X}) = 0, \ n \geq 1. \tag{20}$$

The sequence (18) is called the *first exact sequence* for Ext^n.

REMARK 12.3. The only new information contained in (19) is the assertion that the last rectangle is commutative. The commutativity of the two remaining rectangles is a consequence of the fact that $\operatorname{Ext}^n(\boldsymbol{A}, .)$ is a functor. In view of Theorem 11.37, Theorem 12.2 implies that the functors $\operatorname{Ext}^n(\boldsymbol{A}, .)$ together with the connecting homomorphisms θ_E^n form a *universal connected sequence of functors.*

We will now consider the contravariant functor $\mathrm{Hom}(\,.\,,\boldsymbol{X})\colon \mathsf{Mod}^{\Lambda} \to \mathsf{Mod}$, where $\boldsymbol{X} \in \mathsf{Mod}^{\Lambda}$ is a fixed inverse system. It assigns to every $\boldsymbol{A} \in \mathsf{Mod}^{\Lambda}$ the module $\mathrm{Hom}(\boldsymbol{A},\boldsymbol{X})$. Moreover, to a morphism $\boldsymbol{g}'\colon \boldsymbol{A}' \to \boldsymbol{A}$, it assigns the homomorphism $\mathrm{Hom}(\boldsymbol{g}',\boldsymbol{X})\colon \mathrm{Hom}(\boldsymbol{A},\boldsymbol{X}) \to \mathrm{Hom}(\boldsymbol{A}',\boldsymbol{X})$, which maps $\boldsymbol{h} \in \mathrm{Hom}(\boldsymbol{A},\boldsymbol{X})$ to $\boldsymbol{h}\boldsymbol{g}' \in \mathrm{Hom}(\boldsymbol{A}',\boldsymbol{X})$.

LEMMA 12.4. *The functor* $\mathrm{Hom}(\,.\,,\boldsymbol{X})$ *is additive and left exact. Moreover, if* $\boldsymbol{X}$ *is injective, then* $\mathrm{Hom}(\,.\,,\boldsymbol{X})$ *is exact.*

Proof. Since $\mathrm{Hom}(\,.\,,\boldsymbol{X})$ is a contravariant functor, left exactness means that exactness of a sequence

$$\boldsymbol{A}' \xrightarrow{\boldsymbol{g}'} \boldsymbol{A} \xrightarrow{\boldsymbol{g}} \boldsymbol{A}'' \to \boldsymbol{0} \tag{21}$$

implies exactness of the induced sequence

$$0 \to \mathrm{Hom}(\boldsymbol{A}'',\boldsymbol{X}) \xrightarrow{\boldsymbol{g}} \mathrm{Hom}(\boldsymbol{A},\boldsymbol{X}) \xrightarrow{\boldsymbol{g}'} \mathrm{Hom}(\boldsymbol{A}',\boldsymbol{X}). \tag{22}$$

The argument is dual to the one used in the proof of Lemma 12.1. By Lemma 11.5, the sequence

$$A'_\lambda \xrightarrow{g'_\lambda} A_\lambda \xrightarrow{g_\lambda} A''_\lambda \to 0, \tag{23}$$

is exact, for every $\lambda \in \Lambda$. It is well known that $\mathrm{Hom}(\,.\,,X_\lambda)$ is a left exact contravariant functor and therefore, the sequence

$$0 \to \mathrm{Hom}(A''_\lambda, X_\lambda) \to \mathrm{Hom}(A_\lambda, X_\lambda) \to \mathrm{Hom}(A'_\lambda, X_\lambda) \tag{24}$$

is exact. If $\boldsymbol{h}$ is in the kernel of $\mathrm{Hom}(\boldsymbol{g}',\boldsymbol{X})$, then h_λ is in the kernel of $\mathrm{Hom}(g'_\lambda, X_\lambda)$. Hence, there exists a unique $h''_\lambda \in \mathrm{Hom}(A''_\lambda, X_\lambda)$ such that $h''_\lambda g_\lambda = h_\lambda$. Now

$$p_{\lambda\lambda'} h''_{\lambda'} g_{\lambda'} = p_{\lambda\lambda'} h_{\lambda'} = h_\lambda a_{\lambda\lambda'} = h''_\lambda g_\lambda a_{\lambda\lambda'} = h''_\lambda a''_{\lambda\lambda'} g_{\lambda'}. \tag{25}$$

However, (24) shows that $g_{\lambda'}$ is an epimorphism and can be cancelled from (25). The relation which one obtains in this way shows that $\boldsymbol{h}'' = (h''_\lambda)$ is indeed a morphism of Mod^{Λ}. Clearly, $\boldsymbol{h}''\boldsymbol{g} = \boldsymbol{h}$. The rest of the proof of the left exactness is straightforward. The last assertion is an immediate consequence of the definition of injective objects. □

With every morphism $\boldsymbol{g}'\colon \boldsymbol{A}' \to \boldsymbol{A}$ one can associate an induced homomorphism $\mathrm{Ext}^n(\boldsymbol{g}',\boldsymbol{X})\colon \mathrm{Ext}^n(\boldsymbol{A},\boldsymbol{X}) \to \mathrm{Ext}^n(\boldsymbol{A}',\boldsymbol{X})$. To define it, one applies the functors $\mathrm{Hom}(\boldsymbol{A},\,.\,)$ and $\mathrm{Hom}(\boldsymbol{A}',\,.\,)$ to (9). This yields the following commutative diagram,

$$\begin{array}{ccccccc}
0 \to & \mathrm{Hom}(\boldsymbol{A},\boldsymbol{I}^0_X) & \to & \mathrm{Hom}(\boldsymbol{A},\boldsymbol{I}^1_X) & \to & \mathrm{Hom}(\boldsymbol{A},\boldsymbol{I}^2_X) & \to \dots \\
& \downarrow g'^0 & & \downarrow g'^1 & & \downarrow g'^2 & \\
0 \to & \mathrm{Hom}(\boldsymbol{A}',\boldsymbol{I}^0_X) & \to & \mathrm{Hom}(\boldsymbol{A}',\boldsymbol{I}^1_X) & \to & \mathrm{Hom}(\boldsymbol{A}',\boldsymbol{I}^2_X) & \to \dots .
\end{array} \tag{26}$$

(26) is a mapping of cochain complexes $\mathrm{Hom}\,(\boldsymbol{A},\mathcal{I}_X) \to \mathrm{Hom}\,(\boldsymbol{A}',\mathcal{I}_X)$ and thus, it induces homomorphisms of the corresponding cohomology modules. By definition, these are the homomorphisms $\mathrm{Ext}^n(\boldsymbol{g}',\boldsymbol{X})$.

THEOREM 12.5.

(i) For an exact sequence (8), every morphism $\boldsymbol{g}'\colon \boldsymbol{A}' \to \boldsymbol{A}$ induces a morphism of long exact sequences (18). Hence, the following diagram commutes.

$$\begin{array}{ccccccccc}
\to & \mathrm{Ext}^n(\boldsymbol{A},\boldsymbol{X}') & \to & \mathrm{Ext}^n(\boldsymbol{A},\boldsymbol{X}) & \to & \mathrm{Ext}^n(\boldsymbol{A},\boldsymbol{X}'') & \xrightarrow{\theta^n_E} & \mathrm{Ext}^{n+1}(\boldsymbol{A},\boldsymbol{X}') & \to \\
 & \downarrow & & \downarrow & & \downarrow & & \downarrow & \\
\to & \mathrm{Ext}^n(\boldsymbol{A}',\boldsymbol{X}') & \to & \mathrm{Ext}^n(\boldsymbol{A}',\boldsymbol{X}) & \to & \mathrm{Ext}^n(\boldsymbol{A}',\boldsymbol{X}'') & \xrightarrow[\theta^n_E]{} & \mathrm{Ext}^{n+1}(\boldsymbol{A},\boldsymbol{X}') & \to
\end{array} \tag{27}$$

(ii) $\mathrm{Ext}^n\,(\,.\,,\,.\,)$, $n \geq 0$, is a bifunctor, i.e., $\mathrm{Ext}^n\,(\boldsymbol{A},\,.\,)$ is a covariant functor, $\mathrm{Ext}^n\,(\,.\,,\boldsymbol{X})$ is a contravariant functor and for arbitrary morphisms $\boldsymbol{f}'\colon \boldsymbol{X}' \to \boldsymbol{X}$, $\boldsymbol{g}'\colon \boldsymbol{A}' \to \boldsymbol{A}$, the following diagram commutes.

$$\begin{array}{ccc}
\mathrm{Ext}^n(\boldsymbol{A},\boldsymbol{X}') & \longrightarrow & \mathrm{Ext}^n(\boldsymbol{A},\boldsymbol{X}) \\
\downarrow & & \downarrow \\
\mathrm{Ext}^n(\boldsymbol{A}',\boldsymbol{X}') & \longrightarrow & \mathrm{Ext}^n(\boldsymbol{A}',\boldsymbol{X})
\end{array} \tag{28}$$

(iii) For $\boldsymbol{A}$ projective and $\boldsymbol{X}$ arbitrary,

$$\mathrm{Ext}^n(\boldsymbol{A},\boldsymbol{X}) = 0,\ n \geq 1. \tag{29}$$

Proof. In order to prove (i), consider the short exact sequence of cochain complexes (16) and its analogue for $\boldsymbol{A}'$. The morphism $\boldsymbol{g}'$ induces mappings of cochain complexes which make the following diagram commutative.

$$\begin{array}{ccccccccc}
0 & \to & \mathrm{Hom}(\boldsymbol{A},\mathcal{I}_{X'}) & \longrightarrow & \mathrm{Hom}(\boldsymbol{A},\mathcal{I}_X) & \longrightarrow & \mathrm{Hom}(\boldsymbol{A},\mathcal{I}_{X''}) & \to & 0 \\
 & & \downarrow & & \downarrow & & \downarrow & & \\
0 & \to & \mathrm{Hom}(\boldsymbol{A}',\mathcal{I}_{X'}) & \longrightarrow & \mathrm{Hom}(\boldsymbol{A}',\mathcal{I}_X) & \longrightarrow & \mathrm{Hom}(\boldsymbol{A}',\mathcal{I}_{X''}) & \to & 0\,.
\end{array} \tag{30}$$

Indeed, commutativity of the left rectangle of (30) is an immediate consequence of the commutativity of the following diagrams of modules.

$$\begin{array}{ccc}
\mathrm{Hom}\,(\boldsymbol{A},\boldsymbol{I}^n_{X'}) & \longrightarrow & \mathrm{Hom}\,(\boldsymbol{A},\boldsymbol{I}^n_X) \\
\downarrow & & \downarrow \\
\mathrm{Hom}\,(\boldsymbol{A}',\boldsymbol{I}^n_{X'}) & \longrightarrow & \mathrm{Hom}\,(\boldsymbol{A}',\boldsymbol{I}^n_X).
\end{array} \tag{31}$$

A similar argument proves commutativity of the right rectangle. Finally, diagram (30) implies commutativity of the induced diagram of cohomology modules, which is just diagram (27).

In order to prove (*ii*), notice that every monomorphism $\boldsymbol{f}': \boldsymbol{X}' \to \boldsymbol{X}$ embeds in an exact sequence E. Therefore, the first rectangle of (27) shows that (28) is commutative for monomorphisms. Similarly, every epimorphism embeds in an exact sequence and therefore, the second rectangle of (27) shows that (28) is commutative also for epimorphisms. Finally, since every morphism is the composition of an epimorphism and a monomorphism, the commutativity of (28) holds in general.

In order to prove (*iii*), it suffices to show that the sequence (10) is exact at places $n \geq 1$, because $\mathrm{Ext}^n(\boldsymbol{A}, \boldsymbol{X})$ is the corresponding cohomology module. Note that the resolution $(\mathcal{I}_X, \boldsymbol{e}_X)$ yields short exact sequences

$$\begin{array}{l} \mathbf{0} \to \boldsymbol{X} \xrightarrow{\boldsymbol{e}_X} \boldsymbol{I}_X^0 \xrightarrow{\boldsymbol{d}_X^1} \boldsymbol{d}_X^1(\boldsymbol{I}_X^0) \to \mathbf{0}, \\ \mathbf{0} \to \boldsymbol{d}_X^1(\boldsymbol{I}_X^0) \xrightarrow{\boldsymbol{i}_X} \boldsymbol{I}_X^1 \xrightarrow{\boldsymbol{d}_X^2} \boldsymbol{d}_X^2(\boldsymbol{I}_X^1) \to \mathbf{0}, \\ \mathbf{0} \to \boldsymbol{d}_X^2(\boldsymbol{I}_X^1) \xrightarrow{\boldsymbol{i}_X} \boldsymbol{I}_X^2 \xrightarrow{\boldsymbol{d}_X^3} \boldsymbol{d}_X^3(\boldsymbol{I}_X^2) \to \mathbf{0}, \\ \dots\dots\dots\dots\dots\dots\dots\dots\dots\dots\dots\dots \end{array} \tag{32}$$

where $\boldsymbol{i}_X$ is given by the corresponding inclusions. Since, by Lemma 12.1, $\mathrm{Hom}(\boldsymbol{A}, \,.\,)$ is an exact functor, (32) yields exact sequences

$$\begin{array}{l} 0 \to \mathrm{Hom}(\boldsymbol{A}, \boldsymbol{X}) \longrightarrow \mathrm{Hom}(\boldsymbol{A}, \boldsymbol{I}_X^0) \longrightarrow \mathrm{Hom}(\boldsymbol{A}, \boldsymbol{d}^1(\boldsymbol{I}_X^0)) \to 0, \\ 0 \to \mathrm{Hom}(\boldsymbol{A}, \boldsymbol{d}^1(\boldsymbol{I}_X^0)) \longrightarrow \mathrm{Hom}(\boldsymbol{A}, \boldsymbol{I}_X^1) \longrightarrow \mathrm{Hom}(\boldsymbol{A}, \boldsymbol{d}^2(\boldsymbol{I}_X^1)) \to 0, \\ 0 \to \mathrm{Hom}(\boldsymbol{A}, \boldsymbol{d}^2(\boldsymbol{I}_X^1)) \longrightarrow \mathrm{Hom}(\boldsymbol{A}, \boldsymbol{I}_X^2) \longrightarrow \mathrm{Hom}(\boldsymbol{A}, \boldsymbol{d}^3(\boldsymbol{I}_X^2)) \to 0, \\ \dots\dots\dots\dots\dots\dots\dots\dots\dots\dots\dots\dots\dots\dots\dots\dots\dots\dots \end{array} \tag{33}$$

However, exactness of the sequences (33) implies exactness of the long sequence (10) at places $n \geq 1$. □

REMARK 12.6. The assertions (*i*) of Theorem 11.1 and (*iv*) of Theorem 12.2 show that the first exact sequence for Ext^n (18) is natural with respect to morphisms in the first variable and to morphisms of short exact sequences in the second variable.

We will now develop a *second exact sequence for* Ext^n. Consider a short exact sequence,

$$G = (\mathbf{0} \to \boldsymbol{A}' \xrightarrow{\boldsymbol{g}'} \boldsymbol{A} \xrightarrow{\boldsymbol{g}} \boldsymbol{A}'' \to \mathbf{0}). \tag{34}$$

Since all $\boldsymbol{I}_X^n$ in (10) are injective objects, Lemma 12.4 shows that

$$0 \to \mathrm{Hom}(\boldsymbol{A}'', \boldsymbol{I}_X^n) \xrightarrow{g^*} \mathrm{Hom}(\boldsymbol{A}, \boldsymbol{I}_X^n) \xrightarrow{g'^*} \mathrm{Hom}(\boldsymbol{A}', \boldsymbol{I}_X^n) \to 0 \tag{35}$$

is exact. Therefore, one obtains an exact sequence of cochain complexes,

$$0 \to \mathrm{Hom}(\boldsymbol{A}'', \mathcal{I}_X) \to \mathrm{Hom}(\boldsymbol{A}, \mathcal{I}_X) \to \mathrm{Hom}(\boldsymbol{A}', \mathcal{I}_X) \to 0. \tag{36}$$

The short exact sequence (36) yields a long exact sequence of cohomology modules. By (11), these modules are just the extension products of $\boldsymbol{A}'', \boldsymbol{A}$ and $\boldsymbol{A}'$ with $\boldsymbol{X}$. The connecting homomorphisms of this sequence are, by definition, the *connecting homomorphisms* $\omega_G^n\colon \mathrm{Ext}^n(\boldsymbol{A}', \boldsymbol{X}) \to \mathrm{Ext}^{n+1}(\boldsymbol{A}'', \boldsymbol{X})$.

THEOREM 12.7.

(i) $\mathrm{Ext}^n(\,.\,, \boldsymbol{X})$, $n \geq 0$, *is an additive contravariant functor from* Mod^Λ *to* Mod.

(ii) For every short exact sequence (34), *the following long sequence of modules is exact*

$$\begin{array}{l} 0 \to \mathrm{Hom}(\boldsymbol{A}'', \boldsymbol{X}) \to \mathrm{Hom}(\boldsymbol{A}, \boldsymbol{X}) \to \mathrm{Hom}(\boldsymbol{A}', \boldsymbol{X}) \xrightarrow{\omega_G^0} \mathrm{Ext}^1(\boldsymbol{A}'', \boldsymbol{X}) \to \dots \\ \to \mathrm{Ext}^n(\boldsymbol{A}'', \boldsymbol{X}) \to \mathrm{Ext}^n(\boldsymbol{A}, \boldsymbol{X}) \to \mathrm{Ext}^n(\boldsymbol{A}', \boldsymbol{X}) \xrightarrow{\omega_G^n} \mathrm{Ext}^{n+1}(\boldsymbol{A}'', \boldsymbol{X}) \to \end{array} \tag{37}$$

(iii) A morphism of exact sequences $k = (\boldsymbol{k}', \boldsymbol{k}, \boldsymbol{k}'')\colon G \to H$,

$$\begin{array}{ccccccccc} 0 & \longrightarrow & \boldsymbol{A}' & \xrightarrow{g'} & \boldsymbol{A} & \xrightarrow{g} & \boldsymbol{A}'' & \longrightarrow & 0 \\ & & \downarrow{\scriptstyle \boldsymbol{k}'} & & \downarrow{\scriptstyle \boldsymbol{k}} & & \downarrow{\scriptstyle \boldsymbol{k}''} & & \\ 0 & \longrightarrow & \boldsymbol{B}' & \xrightarrow[h']{} & \boldsymbol{B} & \xrightarrow[h]{} & \boldsymbol{B}'' & \longrightarrow & 0 \end{array} \tag{38}$$

induces a commutative diagram

$$\begin{array}{ccccccc} \to \mathrm{Ext}^n(\boldsymbol{B}'', \boldsymbol{X}) & \to & \mathrm{Ext}^n(\boldsymbol{B}, \boldsymbol{X}) & \to & \mathrm{Ext}^n(\boldsymbol{B}', \boldsymbol{X}) & \xrightarrow{\omega_G^n} & \mathrm{Ext}^{n+1}(\boldsymbol{B}'', \boldsymbol{X}) \to \\ \downarrow & & \downarrow & & \downarrow & & \downarrow \\ \to \mathrm{Ext}^n(\boldsymbol{A}'', \boldsymbol{X}) & \to & \mathrm{Ext}^n(\boldsymbol{A}, \boldsymbol{X}) & \to & \mathrm{Ext}^n(\boldsymbol{A}', \boldsymbol{X}) & \xrightarrow[\omega_H^n]{} & \mathrm{Ext}^{n+1}(\boldsymbol{A}'', \boldsymbol{X}) \to \end{array} \tag{39}$$

(iv) For a fixed exact sequence (34) *and a morphism* $\boldsymbol{f}\colon \boldsymbol{X} \to \boldsymbol{Y}$, *the following diagram is commutative,*

$$\begin{array}{ccccccc} \to \mathrm{Ext}^n(\boldsymbol{A}'', \boldsymbol{X}) & \to & \mathrm{Ext}^n(\boldsymbol{A}, \boldsymbol{X}) & \to & \mathrm{Ext}^n(\boldsymbol{A}', \boldsymbol{X}) & \xrightarrow{\omega_G^n} & \mathrm{Ext}^{n+1}(\boldsymbol{A}'', \boldsymbol{X}) \to \\ \downarrow & & \downarrow & & \downarrow & & \downarrow \\ \to \mathrm{Ext}^n(\boldsymbol{A}'', \boldsymbol{Y}) & \to & \mathrm{Ext}^n(\boldsymbol{A}, \boldsymbol{Y}) & \to & \mathrm{Ext}^n(\boldsymbol{A}', \boldsymbol{Y}) & \xrightarrow[\omega_G^n]{} & \mathrm{Ext}^{n+1}(\boldsymbol{A}'', \boldsymbol{Y}) \to \end{array} \tag{40}$$

REMARK 12.8. The only new information contained in (37) and (38) is the assertion that the last rectangles in these diagrams are commutative. The commutativity of the remaining rectangles in (37) is a consequence of (*i*) and in (38) is a consequence of (28).

Proof of Theorem 12.7. The arguments given above have already established (i) and (ii). In order to prove (iii), note that diagram (38) induces a commutative diagram of cochain complexes,

$$\begin{array}{ccccccccc}
0 & \to & \mathrm{Hom}(\boldsymbol{B}'',\mathcal{I}_X) & \to & \mathrm{Hom}(\boldsymbol{B},\mathcal{I}_X) & \to & \mathrm{Hom}(\boldsymbol{B}',\mathcal{I}(\boldsymbol{X})) & \to & 0 \\
 & & \downarrow & & \downarrow & & \downarrow & & \\
0 & \to & \mathrm{Hom}(\boldsymbol{A}'',\mathcal{I}_X) & \to & \mathrm{Hom}(\boldsymbol{A},\mathcal{I}_X) & \to & \mathrm{Hom}(\boldsymbol{A}',\mathcal{I}_X) & \to & 0.
\end{array} \tag{41}$$

By Lemma 12.4, the rows in diagram (41) are exact. Consequently, the diagram yields a morphism of the induced long exact sequences of cohomology modules, which is just the desired diagram (39).

(iv) is a consequence of the commutativity of the following diagram of cochain complexes,

$$\begin{array}{ccccccccc}
0 & \to & \mathrm{Hom}(\boldsymbol{A}'',\mathcal{I}_X) & \to & \mathrm{Hom}(\boldsymbol{A},\mathcal{I}_X) & \to & \mathrm{Hom}(\boldsymbol{A}',\mathcal{I}(\boldsymbol{X})) & \to & 0 \\
 & & \downarrow & & \downarrow & & \downarrow & & \\
0 & \to & \mathrm{Hom}(\boldsymbol{A}'',\mathcal{I}(\boldsymbol{Y})) & \to & \mathrm{Hom}(\boldsymbol{A},\mathcal{I}(\boldsymbol{Y})) & \to & \mathrm{Hom}(\boldsymbol{A}',\mathcal{I}(\boldsymbol{Y})) & \to & 0. \quad \square
\end{array} \tag{42}$$

REMARK 12.9. Theorem 12.7 shows that the second exact sequence (37) is natural with respect to morphisms of short exact sequences in the first variable and to morphisms in the second variable. This and Remark 12.6 show that the long exact sequences of Ext^{n} are natural in all possible ways.

Dualizing the definitions given in 11.4, one obtains the notion of an (*upper*) *connected sequence of contravariant functors* (also called a *contravariant cohomology functor*) and of a *universal connected sequence of contravariant functors.* Dualizing the reasoning used in 11.4, one can prove the dual of Theorem 11.37, which assumes the following form.

THEOREM 12.10. *Let $T = (T^n, \omega_E^n)$ be a connected sequence of contravariant functors from Mod^{Λ} to Mod. If T has the property that $T^n(\boldsymbol{P}) = 0$, $n \geq 1$, for every projective system $\boldsymbol{P}$, then T is universal.*

It is now clear that Theorems 12.5 and 12.7 yield the following result.

THEOREM 12.11. *The functors* $\mathrm{Ext}^n(\,.\,,\boldsymbol{X})$, *together with the connecting morphisms ω_G^n, $n = 0, 1, \ldots$, form a universal connected sequence of contravariant functors, which begins with* $\mathrm{Ext}^0(\,.\,,\boldsymbol{X}) = \mathrm{Hom}(\,.\,,\boldsymbol{X})$.

Dualizing the construction of right derived functors, described in 11.3, to contravariant left exact functors, one obtains a universal connected sequence of contravariant functors (S^n, θ_E^n), beginning with the functor $\mathrm{Hom}(\,.\,, \boldsymbol{X})$. Consequently, by the dual of Remark 11.36, this sequence is isomorphic to the sequence $(\mathrm{Ext}^n\,(\,.\,, \boldsymbol{X}), \omega_G^n)$, $n = 0, 1, \ldots$. Consequently, one has an alternative way to obtain the latter sequence. In other words, the following theorem holds.

THEOREM 12.12. *Let* $(\mathcal{P}_A, \boldsymbol{e}_A)$ *be a projective resolution of* $\boldsymbol{A} \in \mathsf{Mod}^\Lambda$, *where*

$$\mathcal{P}_A = (\mathbf{0} \leftarrow \boldsymbol{P}_0 \xleftarrow{d_1} \boldsymbol{P}_1 \xleftarrow{d_2} \boldsymbol{P}_2 \leftarrow \ldots). \tag{43}$$

Let $\mathrm{Hom}(\mathcal{P}_A, \boldsymbol{X}) = (0 \to \mathrm{Hom}(\boldsymbol{P}_0, \boldsymbol{X}) \to (\boldsymbol{P}_1, \boldsymbol{X}) \to \ldots)$ *be the induced cochain complex. The cohomology modules of this cochain complex along with the connecting homomorphisms form a universal connected sequence of contravariant functors which begins with* $\mathrm{Hom}(\,.\,, \boldsymbol{X})$. *Consequently, this sequence is isomorphic to the sequence* $(\mathrm{Ext}^n(\,.\,, \boldsymbol{X}), \omega_E^n)$. *In particular, there are natural isomorphisms*

$$\mathrm{Ext}^n(\boldsymbol{A}, \boldsymbol{X}) \approx H^n(\mathrm{Hom}(\mathcal{P}_A, \boldsymbol{X})). \tag{44}$$

12.2 Expressing $\lim^n$ in terms of Ext^n

In this subsection we will show that the functors $\lim^n$ can be identified with the functors $\mathrm{Ext}^n(\boldsymbol{\Delta}, \,.\,)$, where $\boldsymbol{\Delta} = \boldsymbol{\Delta}(\Lambda)$ is the diagonal inverse system from Mod^Λ, described in Example 11.17. Recall that $\boldsymbol{\Delta}(\Lambda)$ is indexed by Λ, all the terms equal the ground ring R and all the bonding homomorphisms are identities.

Consider the functor $\mathrm{Hom}(\boldsymbol{\Delta}, \,.\,)\colon \mathsf{Mod}^\Lambda \to \mathsf{Mod}$. If $\boldsymbol{X} = (X_\lambda, p_{\lambda\lambda'}, \Lambda) \in \mathsf{Mod}^\Lambda$ and $\boldsymbol{h} = (h_\lambda) \in \mathrm{Hom}(\boldsymbol{\Delta}, \boldsymbol{X})$ is a morphism, then the collection $(h_\lambda(1))$ is an element of $\lim \boldsymbol{X}$, because, $h_\lambda(1) = p_{\lambda\lambda'} h_{\lambda'}(1)$, for $\lambda \le \lambda'$. Consequently,

$$\phi_{\boldsymbol{X}}(\boldsymbol{h}) = (h_\lambda(1)) \tag{1}$$

determines a homomorphism $\phi_{\boldsymbol{X}}\colon \mathrm{Hom}\,(\boldsymbol{\Delta}, \boldsymbol{X}) \to \lim \boldsymbol{X}$.

LEMMA 12.13. *The homomorphisms* $\phi_{\boldsymbol{X}}$, $\boldsymbol{X} \in \mathsf{Mod}^\Lambda$, *define a natural equivalence of functors* $\phi\colon \mathrm{Hom}\,(\boldsymbol{\Delta}, \,.\,) \to \lim$.

Proof. Let $\boldsymbol{f}\colon \boldsymbol{X} \to \boldsymbol{X}''$ be a morphism given by homomorphisms $f_\lambda\colon X_\lambda \to X''_\lambda$. Then the diagram

$$\begin{array}{ccc} \mathrm{Hom}\,(\boldsymbol{\Delta}, \boldsymbol{X}) & \xrightarrow{\mathrm{Hom}\,(\boldsymbol{\Delta}, \boldsymbol{f})} & \mathrm{Hom}\,(\boldsymbol{\Delta}, \boldsymbol{X}'') \\ \phi_{\boldsymbol{X}} \downarrow & & \downarrow \phi_{\boldsymbol{X}''} \\ \lim \boldsymbol{X} & \xrightarrow[\lim \boldsymbol{f}]{} & \lim \boldsymbol{X}'' \end{array} \tag{2}$$

commutes, which makes ϕ a natural transformation. Indeed, if $\boldsymbol{h}: \boldsymbol{\Delta} \to \boldsymbol{X}$ is a morphism, given by homomorphisms $h_\lambda: R \to X_\lambda$, then

$$(\lim \boldsymbol{f})(\phi_{\boldsymbol{X}}(\boldsymbol{h})) = (\lim \boldsymbol{f})(h_\lambda(1)) = (f_\lambda h_\lambda(1)). \tag{3}$$

On the other hand,

$$\phi_{\boldsymbol{X}''}((\mathrm{Hom}(\boldsymbol{\Delta}, \boldsymbol{f}))(\boldsymbol{h})) = \phi_{\boldsymbol{X}''}(\boldsymbol{f}\boldsymbol{h}) = (f_\lambda h_\lambda(1)). \tag{4}$$

Finally, every $\phi_{\boldsymbol{X}}$ is an isomorphism. Indeed, it has an inverse $\phi_{\boldsymbol{X}}^{-1}$, which maps $x = (x_\lambda) \in \lim \boldsymbol{X}$ to $\boldsymbol{h} = (h_\lambda): \boldsymbol{\Delta} \to \boldsymbol{X}$, where the homomorphism $h_\lambda: R \to X_\lambda$ is completely determined by putting $h_\lambda(1) = x_\lambda$. □

Remark 12.3 and Theorem 11.38 show that the functors $\mathrm{Ext}^n\,(\boldsymbol{\Delta}, .)$ and $\lim^n$, together with the corresponding connecting morphisms, form universal connected sequences of functors $\mathsf{Mod}^\Lambda \to \mathsf{Mod}$, beginning with $\mathrm{Ext}^0\,(\boldsymbol{\Delta}, .) \approx \mathrm{Hom}\,(\boldsymbol{\Delta}, .)$ and $\lim^0 \approx \lim$, respectively. Since $\phi: \mathrm{Hom}\,(\boldsymbol{\Delta}, .) \to \lim$ is a natural equivalence, Remark 11.36 applies and yields the following theorem.

THEOREM 12.14. *The natural equivalence ϕ extends in a unique way to an isomorphism between the connected sequences of functors* $\mathrm{Ext}^n\,(\boldsymbol{\Delta}, .)$ *and* $\lim^n$. *In particular, for any integer $n \geq 0$, the functors* $\mathrm{Ext}^n(\boldsymbol{\Delta}, .)$ *and* $\lim^n$ *are naturally equivalent,*

$$\mathrm{Ext}^n(\boldsymbol{\Delta}, \boldsymbol{X}) \approx \lim{}^n \boldsymbol{X}. \tag{5}$$

As a first application of Theorem 12.14 we will now prove a result, needed in 17.4.

COROLLARY 12.15. *For any collection $(\boldsymbol{X}^\alpha)$, $\alpha \in A$, of inverse systems from* Mod^Λ, *there exists an isomorphism*

$$u: \prod_\alpha (\lim{}^n \boldsymbol{X}^\alpha) \to \lim{}^n(\prod_\alpha \boldsymbol{X}^\alpha). \tag{6}$$

Moreover, for any collection of morphisms $\boldsymbol{f}^\alpha: \boldsymbol{X}' \to \boldsymbol{X}^\alpha$, $\alpha \in A$, the induced homomorphisms make the following diagram commutative.

$$\begin{array}{ccc} \lim^n \boldsymbol{X}' & & \\ \downarrow & \searrow & \\ \prod_\alpha (\lim^n \boldsymbol{X}^\alpha) & \xrightarrow[u]{} & \lim^n(\prod_\alpha \boldsymbol{X}^\alpha)\,. \end{array} \tag{7}$$

Proof. Choose an arbitrary projective resolution of $\boldsymbol{\Delta} = \boldsymbol{\Delta}(\Lambda)$,

$$\mathbf{0} \leftarrow \boldsymbol{\Delta} \xleftarrow{e} \boldsymbol{P}_0 \xleftarrow{d_1} \boldsymbol{P}_1 \xleftarrow{d_2} \boldsymbol{P}_2 \leftarrow \dots . \tag{8}$$

Then $\lim^n(\prod_\alpha \boldsymbol{X}^\alpha)$ is the n-th cohomology module of the cochain complex

$$0 \to \operatorname{Hom}(\boldsymbol{P}_0, \prod_\alpha \boldsymbol{X}^\alpha) \to \dots \to \operatorname{Hom}(\boldsymbol{P}_i, \prod_\alpha \boldsymbol{X}^\alpha) \to \dots . \tag{9}$$

A collection of morphisms $\boldsymbol{h}^\alpha \colon \boldsymbol{P}_i \to \boldsymbol{X}^\alpha$, $\alpha \in A$, determines a morphism $\boldsymbol{h}_i \colon \boldsymbol{P}_i \to \prod_\alpha \boldsymbol{X}^\alpha$ and vice versa. In this way one obtains a sequence of isomorphisms $u_i \colon \prod_\alpha \operatorname{Hom}(\boldsymbol{P}_i, \boldsymbol{X}^\alpha) \to \operatorname{Hom}(\boldsymbol{P}_i, \prod_\alpha \boldsymbol{X}^\alpha)$, which define an isomorphism $\boldsymbol{u}$ of the cochain complex

$$0 \to \prod_\alpha \operatorname{Hom}(\boldsymbol{P}_0, \boldsymbol{X}^\alpha) \to \dots \to \prod_\alpha \operatorname{Hom}(\boldsymbol{P}_i, \boldsymbol{X}^\alpha) \to \dots . \tag{10}$$

to the cochain complex (9). Clearly, this isomorphism induces isomorphisms of the corresponding cohomology modules. However, the n-th cohomology module of (10) is the product of the n-th cohomology modules of the complexes

$$0 \to \operatorname{Hom}(\boldsymbol{P}_0, \boldsymbol{X}^\alpha) \to \dots \to \operatorname{Hom}(\boldsymbol{P}_i, \boldsymbol{X}^\alpha) \to \dots , \tag{11}$$

i.e., it coincides with $\prod_\alpha(\lim^n \boldsymbol{X}^\alpha)$. The commutativity of (7) is a consequence of the commutativity of the diagrams

$$\begin{array}{ccc} \operatorname{Hom}(\boldsymbol{P}_i, \boldsymbol{X}') & & \\ \downarrow & \searrow & \\ \prod_\alpha(\operatorname{Hom}(\boldsymbol{P}_i, \boldsymbol{X}^\alpha) & \xrightarrow[u_i]{} & \operatorname{Hom}(\boldsymbol{P}_i, \prod_\alpha \boldsymbol{X}^\alpha) . \end{array} \quad \square \tag{12}$$

According to Theorem 12.14, one can determine $\lim^n \boldsymbol{X}$ using any projective resolution of $\boldsymbol{\Delta}$. We will now derive the explicit formulae for $\lim^n \boldsymbol{X}$ (already given in 11.5) by using a particularly simple projective resolution (8) of $\boldsymbol{\Delta}$, here called the *standard projective resolution.*

We define $\boldsymbol{P}_n = (P_{n\lambda}, i_{n\lambda\lambda'}, \Lambda)$ as follows. Let

$$P_n = \bigoplus_{\lambda \in \Lambda_n} R, \tag{13}$$

i.e., P_n is the direct sum of a collection of copies of the ground ring R, indexed by all multiindices $\boldsymbol{\lambda} = (\lambda_0, \dots, \lambda_n)$, $\lambda_0 \le \dots \le \lambda_n$, of length n. We denote by $\langle \boldsymbol{\lambda} \rangle = \langle \lambda_0, \dots, \lambda_n \rangle$ the element of P_n, which assumes the value $1 \in R$ at $\boldsymbol{\lambda} = (\lambda_0, \dots, \lambda_n)$ and has values 0, for all other multiindices. Consequently, P_n is a free R-module with a basis consisting of all elements of the form $\langle \boldsymbol{\lambda} \rangle$. We define $P_{n\lambda}$, $\lambda \in \Lambda$, as the direct summand of P_n, generated by all $\langle \boldsymbol{\lambda} \rangle = \langle \lambda_0, \dots, \lambda_n \rangle$, where $\lambda \le \lambda_0$, i.e., by putting

$$P_{n\lambda} = \bigoplus_{\lambda \leq \boldsymbol{\lambda}} R\,. \tag{14}$$

For $\lambda \leq \lambda'$, we define $i_{n\lambda\lambda'}: P_{n\lambda'} \to P_{n\lambda}$ as the natural inclusion. Clearly, $\boldsymbol{P}_n = (P_{n\lambda}, i_{n\lambda\lambda'}, \Lambda)$ is an inverse system of R-modules over Λ.

LEMMA 12.16. *For every $n \geq 0$, $\boldsymbol{P}_n$ is a projective object of* Mod^Λ.

Proof. Let $\boldsymbol{A} = (A_\lambda, a_{\lambda\lambda'}, \Lambda)$ and $\boldsymbol{B} = (B_\lambda, b_{\lambda\lambda'}, \Lambda)$ be inverse systems and let $\boldsymbol{h} = (h_\lambda): \boldsymbol{A} \to \boldsymbol{B}$ and $\boldsymbol{k} = (k_\lambda): \boldsymbol{P}_n \to \boldsymbol{B}$ be morphisms. Assuming that $\boldsymbol{h}$ is an epimorphism, we must prove the existence of a morphism $\boldsymbol{g} = (g_\lambda): \boldsymbol{P}_n \to \boldsymbol{A}$, such that $\boldsymbol{hg} = \boldsymbol{k}$.

In order to define the homomorphisms $g_\lambda: P_{n\lambda} \to A_\lambda$, for $\lambda \in \Lambda$, it suffices to define $g_\lambda\langle\boldsymbol{\lambda}\rangle$, for all multiindices $\boldsymbol{\lambda}$ in Λ, for which $\lambda \leq \boldsymbol{\lambda}$. Note that $\langle\boldsymbol{\lambda}\rangle \in P_{n\lambda_0}$ and therefore, $k_{\lambda_0}\langle\boldsymbol{\lambda}\rangle$ is a well-defined element of B_{λ_0}. Since h_{λ_0} is surjective, there exists an element $a_{\boldsymbol{\lambda}} \in A_{\lambda_0}$, such that $h_{\lambda_0}(a_{\boldsymbol{\lambda}}) = k_{\lambda_0}\langle\boldsymbol{\lambda}\rangle$. Now put

$$g_{\lambda_0}\langle\boldsymbol{\lambda}\rangle = a_{\boldsymbol{\lambda}}. \tag{15}$$

Note that $h_{\lambda_0} g_{\lambda_0}\langle\boldsymbol{\lambda}\rangle = k_{\lambda_0}\langle\boldsymbol{\lambda}\rangle$. For other $\lambda \leq \lambda_0$, put

$$g_\lambda\langle\boldsymbol{\lambda}\rangle = a_{\lambda\lambda_0} g_{\lambda_0}\langle\boldsymbol{\lambda}\rangle. \tag{16}$$

If $\lambda \leq \lambda'$ and $\lambda' \leq \lambda_0$, then (16) implies

$$a_{\lambda\lambda'} g_{\lambda'}\langle\boldsymbol{\lambda}\rangle = a_{\lambda\lambda'} a_{\lambda'\lambda_0} g_{\lambda_0}\langle\boldsymbol{\lambda}\rangle = g_\lambda\langle\boldsymbol{\lambda}\rangle, \tag{17}$$

which shows that the homomorphisms g_λ define a morphism $\boldsymbol{g}: \boldsymbol{P}_n \to \boldsymbol{A}$. Finally,

$$\begin{aligned} h_\lambda g_\lambda\langle\boldsymbol{\lambda}\rangle &= h_\lambda a_{\lambda\lambda_0} g_{\lambda_0}\langle\boldsymbol{\lambda}\rangle = b_{\lambda\lambda_0} h_{\lambda_0} g_{\lambda_0}\langle\boldsymbol{\lambda}\rangle \\ &= b_{\lambda\lambda_0} k_{\lambda_0}\langle\boldsymbol{\lambda}\rangle = k_\lambda\langle\boldsymbol{\lambda}\rangle, \end{aligned} \tag{18}$$

which yields the desired relation

$$h_\lambda g_\lambda = k_\lambda. \quad \square \tag{19}$$

We now define morphisms $\boldsymbol{e} = (e_\lambda): \boldsymbol{P}_0 \to \boldsymbol{\Delta}$ and $\boldsymbol{d}_n = (d_{n\lambda}): \boldsymbol{P}_n \to \boldsymbol{P}_{n-1}$ as follows. Let $e: P_0 \to R$ be given by

$$e\langle\lambda_0\rangle = 1, \tag{20}$$

and let $e_\lambda = e|P_{0\lambda}$. Let $d_n: P_n \to P_{n-1},\ n \geq 1$, be given by

$$d_n\langle\boldsymbol{\lambda}\rangle = \sum_{j=0}^{n} (-1)^j \langle d^j \boldsymbol{\lambda}\rangle, \tag{21}$$

and let $d_{n\lambda} = d_n|P_{n\lambda}$. Clearly, the homomorphisms e_λ define a morphism $\boldsymbol{e}: \boldsymbol{P}_0 \to \boldsymbol{\Delta}$ and the homomorphisms $d_{n\lambda}$ define morphisms $\boldsymbol{d}_n: \boldsymbol{P}_n \to \boldsymbol{P}_{n-1}$.

LEMMA 12.17. *The systems* $\boldsymbol{P}_n$ *and the morphisms* $\boldsymbol{e}$ *and* $\boldsymbol{d}_n$ *form a projective resolution* $(\mathcal{P}, \boldsymbol{e})$ *of* $\boldsymbol{\Delta} = \boldsymbol{\Delta}(\Lambda)$, *called the standard projective resolution of* $\boldsymbol{\Delta}(\Lambda)$.

Proof. It suffices to prove that, for every $\lambda \in \Lambda$ the following sequence of modules is exact.

$$0 \leftarrow R \overset{e_\lambda}{\leftarrow} P_{0\lambda} \overset{d_{1\lambda}}{\leftarrow} P_{1\lambda} \leftarrow \dots . \tag{22}$$

Formula (21) is the usual expression for the boundary of a simplex. Therefore, $d_n d_{n+1}\langle\lambda_0, \dots, \lambda_{n+1}\rangle = 0$, for $n \geq 1$, and thus, $\mathrm{Im}(d_{n+1\lambda}) \subseteq \mathrm{Ker}(d_{n\lambda})$. Moreover, $\mathrm{Im}(d_{1\lambda}) \subseteq \mathrm{Ker}(e_\lambda)$, because $ed_1\langle\lambda_0, \lambda_1\rangle = e\langle\lambda_1\rangle - e\langle\lambda_0\rangle = 1 - 1 = 0$. In order to prove the converse inclusions, one considers an operation, which corresponds to the formation of a cone over a chain. More precisely, if $\boldsymbol{\lambda} = (\lambda_0, \dots, \lambda_n)$ is a multiindex in Λ and $\mu \in \Lambda$ has the property that $\lambda_n \leq \mu$, then $\langle\boldsymbol{\lambda}, \mu\rangle = \langle\lambda_0, \dots, \lambda_n, \mu\rangle$ is a well-defined element of P_{n+1}. More generally, every element $x \in P_n$ is a finite sum of the form

$$x = \sum_{i \in I} r_i \langle\boldsymbol{\lambda}^i\rangle, \tag{23}$$

where all $\boldsymbol{\lambda}^i$ belong to Λ_n, $i \in I$, and all $r_i \in R$. If $\mu \in \Lambda$ satisfies $\lambda_n^i \leq \mu$, for all $i \in I$, then an element $\langle x, \mu\rangle$ of P_{n+1} is well defined, by putting

$$\langle x, \mu\rangle = \sum_{i \in I} r_i \langle\boldsymbol{\lambda}^i, \mu\rangle. \tag{24}$$

It is readily verified that (21), (23) and (24) yield

$$d_{n+1}\langle x, \mu\rangle = \langle d_n x, \mu\rangle + (-1)^{n+1} x, \; n \geq 1. \tag{25}$$

Now assume that $x \in \mathrm{Ker}(d_{n\lambda})$, i.e., $d_n x = 0$. Since Λ is directed and I is finite, one can find an index $\mu \in \Lambda$, such that $\lambda_n^i \leq \mu$, for all $i \in I$. Then (25) applies and yields

$$x = (-1)^{n+1} d_{n+1\lambda}\langle x, \mu\rangle \in \mathrm{Im}(d_{n+1\lambda}), \tag{26}$$

as desired. If $n = 0$, then (21), (23) and (24) yield

$$d_1\langle x, \mu\rangle = \left(\sum_{i \in I} r_i\right) \langle\mu\rangle - x. \tag{27}$$

However, if $x \in \mathrm{Ker}(e_\lambda)$, i.e., if $e(x) = 0$, then $\sum_{i \in I} r_i = 0$ and thus, (27) yields $x = -d_1\langle x, \mu\rangle \in \mathrm{Im}(d_{1\lambda})$. Finally, $e_\lambda\colon P_0 \to R$ is a surjection. □

Application of Theorem 12.12 now yields the following lemma.

LEMMA 12.18. *Let $(\mathcal{P}, \boldsymbol{e})$ be the standard projective resolution of $\boldsymbol{\Delta}(\Lambda)$. Furthermore, for every inverse system $\boldsymbol{X} \in \mathsf{Mod}^{\Lambda}$, let $L(\boldsymbol{X})$ be the cochain complex, which consists of the R-modules*

$$L^n(\boldsymbol{X}) = \mathrm{Hom}(\boldsymbol{P}_n, \boldsymbol{X}) \tag{28}$$

and of the coboundary operators

$$d^n = \mathrm{Hom}\,(\boldsymbol{d}_n, \boldsymbol{X})\colon L^{n-1}(\boldsymbol{X}) \to L^n(\boldsymbol{X}). \tag{29}$$

Then $\mathrm{Ext}^n(\boldsymbol{\Delta}(\Lambda), \boldsymbol{X})$ *is naturally isomorphic to the n-th cohomology module of $L(\boldsymbol{X})$, i.e.,*

$$\mathrm{Ext}^n(\boldsymbol{\Delta}(\Lambda), \boldsymbol{X}) \approx H^n L(\boldsymbol{X}). \tag{30}$$

Our next goal is to see that the cochain complex $L(\boldsymbol{X})$ is isomorphic to the cochain complex $K(\boldsymbol{X})$, defined in 11.5. Taking into account Theorem 12.14, this gives an alternative proof of Corollary 11.47.

We first define homomorphisms $\phi^n_{\mathbf{X}}\colon L^n(\boldsymbol{X}) \to K^n(\boldsymbol{X})$. If $\boldsymbol{f} \in L^n(\boldsymbol{X}) = \mathrm{Hom}\,(\boldsymbol{P}_n, \boldsymbol{X})$ is given by homomorphisms $f_\lambda\colon P_{n\lambda} \to X_\lambda$, then $\phi^n_{\mathbf{X}}(\boldsymbol{f}) \in K^n$ is the n-cochain defined by

$$(\phi^n_{\mathbf{X}}(\boldsymbol{f}))(\boldsymbol{\lambda}) = f_{\lambda_0}\langle\boldsymbol{\lambda}\rangle \in X_{\lambda_0}. \tag{31}$$

LEMMA 12.19. *The homomorphisms $\phi^n_{\mathbf{X}}\colon L^n(\boldsymbol{X}) \to K^n(\boldsymbol{X})$, $n \geq 0$, form an isomorphism of cochain complexes $\phi_{\mathbf{X}}\colon L(\boldsymbol{X}) \to K(\boldsymbol{X})$.*

Proof. Let us first verify the commutativity of the diagram

$$\begin{array}{ccc}
L^{n-1}(\boldsymbol{X}) & \xrightarrow{\;d^n\;} & L^n(\boldsymbol{X}) \\
\phi^{n-1}_{\mathbf{X}} \Big\downarrow & & \Big\downarrow \phi^n_{\mathbf{X}} \\
K^{n-1}(\boldsymbol{X}) & \xrightarrow[\;\delta^n\;]{} & K^n(\boldsymbol{X})\,.
\end{array} \tag{32}$$

If $\boldsymbol{f} = (f_\lambda)\colon \boldsymbol{P}_{n-1} \to \boldsymbol{X}$, then

$$(\phi^{n-1}_{\mathbf{X}}(\boldsymbol{f}))(d^j\boldsymbol{\lambda}) = f_{\lambda_0}\langle d^j\boldsymbol{\lambda}\rangle,\ j > 0, \tag{33}$$

$$(\phi^{n-1}_{\mathbf{X}}(\boldsymbol{f}))(d^0\boldsymbol{\lambda}) = f_{\lambda_1}\langle d^0\boldsymbol{\lambda}\rangle,\ j = 0. \tag{34}$$

Consequently,

$$((\delta^n\phi^{n-1}_{\mathbf{X}})(\boldsymbol{f}))(\boldsymbol{\lambda}) = p_{\lambda_0\lambda_1} f_{\lambda_1}\langle d^0\boldsymbol{\lambda}\rangle + \sum_{j=1}^{n}(-1)^j f_{\lambda_0}\langle d^j\boldsymbol{\lambda}\rangle. \tag{35}$$

On the other hand, by (29), $\phi^n_{\mathbf{X}}(d^n(\boldsymbol{f})) = \phi^n_{\mathbf{X}}(\boldsymbol{f}\boldsymbol{d}_n)$. Therefore,

$$((\phi^n_{\mathbf{X}} d^n)(\boldsymbol{f}))(\boldsymbol{\lambda}) = f_{\lambda_0}\langle d^0\boldsymbol{\lambda}\rangle + \sum_{j=1}^{n}(-1)^j f_{\lambda_0}\langle d^j\boldsymbol{\lambda}\rangle \ . \tag{36}$$

The right sides of (35) and (36) coincide, because $p_{\lambda_0\lambda_1} f_{\lambda_1} = f_{\lambda_0}|P_{n-1\lambda_1}$.

In order to show that $\phi^n_{\mathbf{X}}$ is a monomorphism, assume that, for a given $\boldsymbol{f}$, one has $\phi^n_{\mathbf{X}}(\boldsymbol{f}) = 0$, i.e.,

$$f_{\lambda_0}\langle\boldsymbol{\lambda}\rangle = 0, \tag{37}$$

for every multiindex $\boldsymbol{\lambda} = (\lambda_0, \ldots, \lambda_n)$. One must show that $f_\lambda = 0$, for every $\lambda \in \Lambda$. It suffices to show that

$$f_\lambda\langle\boldsymbol{\lambda}\rangle = 0, \tag{38}$$

whenever $\lambda \le \lambda_0$, because for such $\boldsymbol{\lambda}$, the elements $\langle\boldsymbol{\lambda}\rangle$ form a basis of $P_{n\lambda}$. However, for $\lambda \le \lambda_0$, one has $f_\lambda = p_{\lambda\lambda_0} f_{\lambda_0}$, and (38) follows from (37).

It remains to show that $\phi^n_{\mathbf{X}}$ is an epimorphism. Consider an arbitrary element $c \in K^n(\boldsymbol{X})$. By putting

$$f_\lambda\langle\boldsymbol{\lambda}\rangle = p_{\lambda\lambda_0}(c(\boldsymbol{\lambda})) \in X_\lambda, \tag{39}$$

one obtains a well-defined homomorphism $f_\lambda : P_{n\lambda} \to X_\lambda$. The homomorphisms f_λ determine a morphism $\boldsymbol{f} : \boldsymbol{P}_n \to \boldsymbol{X}$, because

$$f_\lambda = p_{\lambda\lambda'} f_{\lambda'}, \ \lambda \le \lambda'. \tag{40}$$

This is easily verified, because both sides of (40) yield the same values at the generators $\langle\boldsymbol{\lambda}\rangle$, $\lambda' \le \lambda_0$, of $P_{n\lambda'}$. Finally, for $\lambda \le \lambda_0$, one has

$$(\phi^n_{\mathbf{X}}(\boldsymbol{f}))(\boldsymbol{\lambda}) = f_{\lambda_0}\langle\boldsymbol{\lambda}\rangle = x(\boldsymbol{\lambda}), \tag{41}$$

which shows that, indeed, $c = \phi^n_{\mathbf{X}}(\boldsymbol{f}) \in \mathrm{Im}(\phi^n_{\mathbf{X}})$. □

Bibliographic notes

The bifunctor Ext^n is a standard object of study in homological algebra. In our exposition in subsection 12.1, we primarily followed (Hilton, Stammbach 1971). Theorem 12.14 was the starting point of Mitchell's work on higher limits (Mitchell 1972). The standard projective resolution of $\boldsymbol{\Delta}(\Lambda)$ also appears in Mitchell's paper and was patterned after (Osofsky 1968a, 1968b, 1971).

13. The vanishing theorems

In general, the computation of higher derived limits $\lim^n \boldsymbol{X}$ of an inverse system of modules is very difficult. Therefore, most applications of these functors depend on the information whether $\lim^n \boldsymbol{X}$ vanishes or not. Consequently, it is very important to have conditions, which imply $\lim^n \boldsymbol{X} = 0$, as well as conditions, which imply $\lim^n \boldsymbol{X} \neq 0$. This section is devoted to both cases.

13.1 Homological dimension

For an inverse system of R-modules $\boldsymbol{A} \in \mathsf{Mod}^\Lambda$ one defines its *homological dimension* $\mathrm{hd}(\boldsymbol{A}) \geq 0$ by putting $\mathrm{hd}(\boldsymbol{A}) \leq n$, provided $\boldsymbol{A}$ admits a projective resolution of length n,

$$\mathbf{0} \leftarrow \boldsymbol{A} \xleftarrow{e} \boldsymbol{P}_0 \xleftarrow{d_1} \boldsymbol{P}_1 \leftarrow \ldots \leftarrow \boldsymbol{P}_{n-1} \xleftarrow{d_n} \boldsymbol{P}_n \leftarrow \mathbf{0}. \tag{1}$$

If there is no such n, one puts $\mathrm{hd}(\boldsymbol{A}) = \infty$. Since $\mathbf{0}$ is projective, a projective resolution (1) of length n can be extended to a projective resolution of length $m > n$ by inserting at the right end of (1) additional zero systems. This means that $\mathrm{hd}(\boldsymbol{A}) \leq n$ implies $\mathrm{hd}(\boldsymbol{A}) \leq m$, for $n \leq m$.

Clearly, if $\boldsymbol{A}$ is projective, then $\mathbf{0} \leftarrow \boldsymbol{A} \xleftarrow{\mathrm{id}} \boldsymbol{A} \leftarrow \mathbf{0}$ is a projective resolution of length 0 and, therefore, $\mathrm{hd}(\boldsymbol{A}) = 0$. Conversely, if $\mathrm{hd}(\boldsymbol{A}) = 0$, then $\boldsymbol{A}$ admits a projective resolution $\mathbf{0} \leftarrow \boldsymbol{A} \xleftarrow{e} \boldsymbol{P}_0 \leftarrow \mathbf{0}$ and $\boldsymbol{A} \approx \boldsymbol{P}_0$ is projective. We saw in Example 11.17 that the diagonal system $\boldsymbol{\Delta}(\Lambda)$, for $\Lambda = \mathbb{Z}$, is not projective and therefore, $\mathrm{hd}(\boldsymbol{\Delta}(\Lambda)) \geq 1$.

REMARK 13.1. The above definition applies to any abelian category with enough projective objects. In particular, it applies to the category Mod_R of R-modules. Note that, in the case of R-modules over a principal ideal domain R (e.g., for abelian groups), $\mathrm{hd}(A) \leq 1$, for every R-module A, because then projective modules coincide with free modules and every submodule of a free module is free.

If $\boldsymbol{A}$ has dimension $\mathrm{hd}(\boldsymbol{A}) \leq n$ and $\boldsymbol{X} \in \mathsf{Mod}^\Lambda$, one can determine the modules $\mathrm{Ext}^m(\boldsymbol{A}, \boldsymbol{X})$ using a projective resolution (1) of length $\leq n$. Since in this case $\boldsymbol{P}_{n+1} = \mathbf{0}$, the group of $(n+1)$-cochains $\mathrm{Hom}\,(\boldsymbol{P}_{n+1}, \boldsymbol{X})$ vanishes and one concludes, by Theorem 12.12, that

$$\operatorname{Ext}^{n+1}(\boldsymbol{A}, \boldsymbol{X}) = 0, \tag{2}$$

for every $\boldsymbol{X} \in \mathsf{Mod}^{\Lambda}$. In this subsection we will prove that also the converse implication holds. More precisely, we will prove the following theorem.

THEOREM 13.2. *For a system $\boldsymbol{A} \in \mathsf{Mod}^{\Lambda}$ the following assertions are equivalent:*

(i) $\operatorname{hd}(\boldsymbol{A}) \leq n$.

(ii) $\operatorname{Ext}^{n+1}(\boldsymbol{A}, \boldsymbol{X}) = 0$, *for every* $\boldsymbol{X} \in \mathsf{Mod}^{\Lambda}$.

(iii) An epimorphism $\boldsymbol{f}: \boldsymbol{X} \to \boldsymbol{X}''$ *induces an epimorphism*

$$\operatorname{Ext}^{n}(\boldsymbol{A}, \boldsymbol{f}): \operatorname{Ext}^{n}(\boldsymbol{A}, \boldsymbol{X}) \to \operatorname{Ext}^{n}(\boldsymbol{A}, \boldsymbol{X}''). \tag{3}$$

(iv) If

$$\boldsymbol{0} \leftarrow \boldsymbol{A} \xleftarrow{e} \boldsymbol{P}_0 \xleftarrow{d_1} \boldsymbol{P}_1 \leftarrow \ldots \leftarrow \boldsymbol{P}_{n-1} \xleftarrow{i} \boldsymbol{Q} \leftarrow \boldsymbol{0} \tag{4}$$

is an exact sequence and $\boldsymbol{P}_0, \ldots, \boldsymbol{P}_{n-1}$ *are projective objects, then also* $\boldsymbol{Q}$ *is projective.*

Proof. We already saw that *(i)* ⇒ *(ii)*. To show that *(ii)* ⇒ *(iii)*, complete $\boldsymbol{X} \xrightarrow{f} \boldsymbol{X}'' \to \boldsymbol{0}$ to an exact sequence $E = (\boldsymbol{0} \to \boldsymbol{X}' \to \boldsymbol{X} \xrightarrow{f} \boldsymbol{X}'' \to \boldsymbol{0})$. Then, by (12.1.18), $\operatorname{Ext}^{n+1}(\boldsymbol{A}, \boldsymbol{X}') = 0$ implies that $\operatorname{Ext}^{n}(\boldsymbol{A}, \boldsymbol{f})$ is an epimorphism.

In order to show that *(iv)* ⇒ *(i)*, take a projective resolution $(\mathcal{P}_A, \boldsymbol{e})$ of $\boldsymbol{A}$, where $\mathcal{P}_A = (\boldsymbol{0} \leftarrow \boldsymbol{P}_0 \leftarrow \boldsymbol{P}_1 \leftarrow \ldots)$. Then put $\boldsymbol{Q} = \operatorname{Im}(\boldsymbol{d}_n)$ and take for $\boldsymbol{i}$ the inclusion $\boldsymbol{i}: \boldsymbol{Q} \to \boldsymbol{P}_{n-1}$. One obtains an exact sequence (4), where all $\boldsymbol{P}_i$, $i = 0, \ldots, n-1$, are projective. Therefore, *(iv)* implies that also $\boldsymbol{Q}$ is projective, which shows that (4) is a projective resolution of $\boldsymbol{A}$ of length n. Hence, $\operatorname{hd}(\boldsymbol{A}) \leq n$.

It remains to prove the only non-trivial implication *(iii)* ⇒ *(iv)*. Consider an exact sequence (4), where $\boldsymbol{P}_0, \ldots, \boldsymbol{P}_{n-1}$ are projective objects. We must show that also $\boldsymbol{Q}$ is projective. By definition, this is the same as showing that, whenever $\boldsymbol{f}: \boldsymbol{X} \to \boldsymbol{X}''$ is an epimorphism, then also $\operatorname{Hom}(\boldsymbol{Q}, \boldsymbol{f}): \operatorname{Hom}(\boldsymbol{Q}, \boldsymbol{X}) \to \operatorname{Hom}(\boldsymbol{Q}, \boldsymbol{X}'')$ is an epimorhism. By assumption, $\boldsymbol{P}_{n-1}$ is projective and therefore, $\operatorname{Hom}(\boldsymbol{P}_{n-1}, \boldsymbol{f}): \operatorname{Hom}(\boldsymbol{P}_{n-1}, \boldsymbol{X}) \to \operatorname{Hom}(\boldsymbol{P}_{n-1}, \boldsymbol{X}'')$ is an epimorhism. We also know, by *(iii)*, that $\operatorname{Ext}^{n}(\boldsymbol{A}, \boldsymbol{f})$: $\operatorname{Ext}^{n}(\boldsymbol{A}, \boldsymbol{X}) \to \operatorname{Ext}^{n}(\boldsymbol{A}, \boldsymbol{X}'')$ is an epimorphism. In order to show that these two facts imply that $\operatorname{Hom}(\boldsymbol{Q}, \boldsymbol{f})$ is an epimorhism, we put $\boldsymbol{Q}_{i+1} = \operatorname{Ker}(\boldsymbol{d}_i) = \operatorname{Im}(\boldsymbol{d}_{i+1})$ and consider the obvious monomorphism $\boldsymbol{Q}_{i+1} \to \boldsymbol{P}_i$ and the obvious epimorphism $\boldsymbol{P}_i \to \boldsymbol{Q}_i$. We thus obtain a sequence of projective presentations

$$\begin{gathered} G_0 = (\boldsymbol{0} \leftarrow \boldsymbol{Q}_0 \leftarrow \boldsymbol{P}_0 \leftarrow \boldsymbol{Q}_1 \leftarrow \boldsymbol{0}), \\ \cdots\cdots\cdots\cdots\cdots\cdots \\ G_i = (\boldsymbol{0} \leftarrow \boldsymbol{Q}_i \leftarrow \boldsymbol{P}_i \leftarrow \boldsymbol{Q}_{i+1} \leftarrow \boldsymbol{0}), \\ \cdots\cdots\cdots\cdots\cdots\cdots \\ G_{n-1} = (\boldsymbol{0} \leftarrow \boldsymbol{Q}_{n-1} \leftarrow \boldsymbol{P}_{n-1} \leftarrow \boldsymbol{Q}_n \leftarrow \boldsymbol{0}), \end{gathered} \tag{5}$$

where $\boldsymbol{Q}_0 = \boldsymbol{A}, \boldsymbol{Q}_n = \boldsymbol{Q}$. Since $\boldsymbol{P}_i$ is projective, for $0 \le i \le n-1$, by Theorem 12.5(*iii*), $\operatorname{Ext}^m(\boldsymbol{P}_i, \boldsymbol{X}) = 0$, for all $m \ge 1$. Therefore, the second exact sequence for Ext^n (see Theorem 12.7(*ii*)), induced by G_i, implies that $\omega_{G_i}\colon \operatorname{Ext}^{n-i-1}(\boldsymbol{Q}_{i+1}, \boldsymbol{X}) \to \operatorname{Ext}^{n-i}(\boldsymbol{Q}_i, \boldsymbol{X})$ is an isomorphism, for $0 \le i \le n-2$. The analogous assertion for $\boldsymbol{X}''$ also holds. Furthermore, the following diagram commutes

$$\begin{array}{ccc} \operatorname{Ext}^{n-i-1}(\boldsymbol{Q}_{i+1}, \boldsymbol{X}) & \xrightarrow{\omega_{G_i}} & \operatorname{Ext}^{n-i}(\boldsymbol{Q}_i, \boldsymbol{X}) \\ {\scriptstyle f_{i+1}}\downarrow & & \downarrow{\scriptstyle f_i} \\ \operatorname{Ext}^{n-i-1}(\boldsymbol{Q}_{i+1}, \boldsymbol{X}'') & \xrightarrow[\omega_{G_i}]{} & \operatorname{Ext}^{n-i}(\boldsymbol{Q}_i, \boldsymbol{X}''), \end{array} \tag{6}$$

where $f_i = \operatorname{Ext}^{n-i}(\boldsymbol{Q}_i, \boldsymbol{f})$, $0 \le i \le n$, because it is part of diagram (12.1.40), for G_i. Therefore, the assumption that f_i is an epimorphism implies that also f_{i+1} is an epimorphism, for $i \le n-2$. Since, by (*iii*), $f_0 = \operatorname{Ext}^n(\boldsymbol{A}, \boldsymbol{f})$ is an epimorphism, we conclude, by induction, that also $f_{n-1} = \operatorname{Ext}^1(\boldsymbol{Q}_{n-1}, \boldsymbol{f})$ is an epimorphism. We now consider the following commutative diagram, which is part of diagram (12.1.40) for G_{n-1},

$$\begin{array}{ccccccc} \operatorname{Hom}(\boldsymbol{P}_{n-1}, \boldsymbol{X}) & \longrightarrow & \operatorname{Hom}(\boldsymbol{Q}, \boldsymbol{X}) & \longrightarrow & \operatorname{Ext}^1(\boldsymbol{Q}_{n-1}, \boldsymbol{X}) & \longrightarrow & 0 \\ \downarrow & & \downarrow & & \downarrow{\scriptstyle f_{n-1}} & & \\ \operatorname{Hom}(\boldsymbol{P}_{n-1}, \boldsymbol{X}'') & \longrightarrow & \operatorname{Hom}(\boldsymbol{Q}, \boldsymbol{X}'') & \longrightarrow & \operatorname{Ext}^1(\boldsymbol{Q}_{n-1}, \boldsymbol{X}'') & \longrightarrow & 0. \end{array} \tag{7}$$

Since the rows are exact and the first and the third vertical arrows are epimorphisms, the same is true of the second one, i.e., $\operatorname{Hom}(\boldsymbol{Q}, \boldsymbol{f})$ is indeed an epimorphism. □

We now define the *homological dimension* $\operatorname{hd}_R(\Lambda)$ (or just $\operatorname{hd}(\Lambda)$) of a directed set Λ, by putting

$$\operatorname{hd}(\Lambda) = \operatorname{hd}(\boldsymbol{\Delta}(\Lambda)). \tag{8}$$

THEOREM 13.3. *For a directed set Λ, the following conditions are equivalent.*

(*i*) $\operatorname{hd}(\Lambda) \le n$.

(*ii*) $\lim^{n+1} \boldsymbol{X} = 0$, *for every inverse system* $\boldsymbol{X} \in \mathsf{Mod}^\Lambda$.

(*iii*) *For every projective resolution*

$$\boldsymbol{0} \leftarrow \boldsymbol{\Delta}(\Lambda) \xleftarrow{e} \boldsymbol{P}_0 \xleftarrow{d_1} \boldsymbol{P}_1 \leftarrow \ldots \leftarrow \boldsymbol{P}_{n-1} \xleftarrow{d_n} \boldsymbol{P}_n \leftarrow \cdots \tag{9}$$

of $\boldsymbol{\Delta}(\Lambda)$, $\boldsymbol{d}_n \boldsymbol{P}_n$ is a projective object of Mod^Λ.

(*iv*) *There exists a projective resolution* (9) *of $\boldsymbol{\Delta}(\Lambda)$ such that $\boldsymbol{d}_n \boldsymbol{P}_n$ is projective.*

Proof. By Theorem 13.2, $\mathrm{hd}(\Lambda) \leq n$, i.e., $\mathrm{hd}(\boldsymbol{\Delta}(\Lambda)) \leq n$, if and only if $\mathrm{Ext}^{n+1}(\boldsymbol{\Delta}(\Lambda), \boldsymbol{X}) = 0$, for all $\boldsymbol{X} \in \mathsf{Mod}^{\Lambda}$. Since, by Theorem 12.14, $\lim^{n+1} \boldsymbol{X} \approx \mathrm{Ext}^{n+1}(\boldsymbol{\Delta}(\Lambda), \boldsymbol{X})$, (i) and (ii) are equivalent. To show that (i) implies (iii), notice that (9) yields an exact sequence

$$\mathbf{0} \leftarrow \boldsymbol{\Delta} \leftarrow \boldsymbol{P}_0 \leftarrow \boldsymbol{P}_1 \leftarrow \ldots \leftarrow \boldsymbol{P}_{n-1} \leftarrow \boldsymbol{d}_n \boldsymbol{P}_n \leftarrow \mathbf{0}, \tag{10}$$

where $\boldsymbol{P}_0, \ldots, \boldsymbol{P}_{n-1}$ are projective. Then, Theorem 13.2, applied to $\boldsymbol{A} = \boldsymbol{\Delta}$, shows that (i) implies projectivity of $\boldsymbol{d}_n \boldsymbol{P}_n$. It is obvious that (iii) implies (iv). Finally, if (iv) holds and (9) is a projective resolution such that $\boldsymbol{d}_n \boldsymbol{P}_n$ is projective, then (10) is a projective resolution of length n, and therefore, $\mathrm{hd}(\Lambda) \leq n$, i.e., (i) holds. □

REMARK 13.4. If $\lim^n \boldsymbol{X} = 0$, for every inverse system $\boldsymbol{X}$ from Mod^{Λ}, then also $\lim^m \boldsymbol{X} = 0$, for every $m > n$. Indeed, by the implication $(ii) \Rightarrow (i)$ of Theorem 13.3, one concludes that $\mathrm{hd}(\Lambda) \leq n-1$ and thus, also $\mathrm{hd}(\Lambda) \leq m-1$. Now the converse implication $(i) \Rightarrow (ii)$ shows that $\lim^m \boldsymbol{X} = 0$.

THEOREM 13.5. *Let $\boldsymbol{A}$ and $\boldsymbol{X}$ be inverse systems of R-modules, let*

$$\mathbf{0} \leftarrow \boldsymbol{A} \xleftarrow{e} \boldsymbol{P}_0 \xleftarrow{d_1} \boldsymbol{P}_1 \leftarrow \ldots \leftarrow \boldsymbol{P}_n \xleftarrow{d_{n+1}} \boldsymbol{P}_{n+1} \leftarrow \ldots \tag{11}$$

be a projective resolution of $\boldsymbol{A}$ and let $\boldsymbol{i}_n \colon \boldsymbol{d}_{n+1}\boldsymbol{P}_{n+1} \to \boldsymbol{P}_n$ be the natural inclusion morphism. Then $\mathrm{Ext}^{n+1}(\boldsymbol{A}, \boldsymbol{X}) = 0$ if and only if

$$\mathrm{Hom}(\boldsymbol{i}_n, \boldsymbol{X}) \colon \mathrm{Hom}(\boldsymbol{P}_n, \boldsymbol{X}) \to \mathrm{Hom}(\boldsymbol{d}_{n+1}\boldsymbol{P}_{n+1}, \boldsymbol{X}) \tag{12}$$

is an epimorphism.

Proof. Consider the induced exact sequences

$$\begin{aligned} G_0 &= (\mathbf{0} \leftarrow \boldsymbol{A} \leftarrow \boldsymbol{P}_0 \leftarrow \boldsymbol{d}_1 \boldsymbol{P}_1 \leftarrow \mathbf{0}), \\ &\ldots\ldots\ldots\ldots\ldots\ldots\ldots\ldots\ldots\ldots \\ G_i &= (\mathbf{0} \leftarrow \boldsymbol{d}_i \boldsymbol{P}_i \leftarrow \boldsymbol{P}_i \leftarrow \boldsymbol{d}_{i+1} \boldsymbol{P}_{i+1} \leftarrow \mathbf{0}), \\ &\ldots\ldots\ldots\ldots\ldots\ldots\ldots\ldots\ldots\ldots \\ G_n &= (\mathbf{0} \leftarrow \boldsymbol{d}_n \boldsymbol{P}_n \leftarrow \boldsymbol{P}_n \leftarrow \boldsymbol{d}_{n+1} \boldsymbol{P}_{n+1} \leftarrow \mathbf{0}). \end{aligned} \tag{13}$$

Since $\mathrm{Ext}^m(\boldsymbol{P}_i, \boldsymbol{X}) = 0$, for every $i \geq 0$ and $m \geq 1$, the second exact sequence (12.1.37), applied to (G_i), shows that the connecting homomorphism $\omega_{G_i}^{n-i} \colon \mathrm{Ext}^{n-i}(\boldsymbol{d}_{i+1}\boldsymbol{P}_{i+1}, \boldsymbol{X}) \to \mathrm{Ext}^{n-i+1}(\boldsymbol{d}_i \boldsymbol{P}_i, \boldsymbol{X})$ is an isomorphism, for $0 \leq i \leq n-1$. Consequently, $\omega_{G_0}^n \cdots \omega_{G_{n-1}}^1 \colon \mathrm{Ext}^1(\boldsymbol{d}_n \boldsymbol{P}_n, \boldsymbol{X}) \to \mathrm{Ext}^{n+1}(\boldsymbol{A}, \boldsymbol{X})$ is also an isomorphism. Therefore,

$$\mathrm{Ext}^{n+1}(\boldsymbol{A}, \boldsymbol{X}) = 0 \iff \mathrm{Ext}^1(\boldsymbol{d}_n \boldsymbol{P}_n, \boldsymbol{X}) = 0. \tag{14}$$

Moreover,(12.1.37) for G_n, yields the exact sequence

$$\mathrm{Hom}(\boldsymbol{P}_n, \boldsymbol{X}) \to \mathrm{Hom}(\boldsymbol{d}_{n+1}\boldsymbol{P}_{n+1}, \boldsymbol{X}) \to \mathrm{Ext}^1(\boldsymbol{d}_n \boldsymbol{P}_n, \boldsymbol{X}) \to 0 \tag{15}$$

and thus, $\mathrm{Ext}^1(\boldsymbol{d}_n\boldsymbol{P}_n, \boldsymbol{X}) = 0$ if and only if the morphism $\mathrm{Hom}(\boldsymbol{P}_n, \boldsymbol{X}) \to \mathrm{Hom}(\boldsymbol{d}_{n+1}\boldsymbol{P}_{n+1}, \boldsymbol{X})$ is an epimorphism. □

Putting $\boldsymbol{A} = \boldsymbol{\Delta}$ and taking into account Theorem 12.14, the above theorem yields the following corollary.

COROLLARY 13.6. *Let $\boldsymbol{X}$ be an inverse system of R-modules. Let*

$$\mathbf{0} \leftarrow \boldsymbol{\Delta} \xleftarrow{e} \boldsymbol{P}_0 \xleftarrow{d_1} \boldsymbol{P}_1 \leftarrow \ldots \leftarrow \boldsymbol{P}_n \xleftarrow{d_{n+1}} \boldsymbol{P}_{n+1} \leftarrow \ldots \tag{16}$$

be a projective resolution of $\boldsymbol{\Delta}$ and let $\boldsymbol{i}_n : \boldsymbol{d}_{n+1}\boldsymbol{P}_{n+1} \to \boldsymbol{P}_n$ be the inclusion morphism. Then, $\lim^{n+1} \boldsymbol{X} = 0$ if and only if the induced homomorphism

$$\mathrm{Hom}(\boldsymbol{i}_n, \boldsymbol{X}) : \mathrm{Hom}(\boldsymbol{P}_n, \boldsymbol{X}) \to \mathrm{Hom}(\boldsymbol{d}_{n+1}\boldsymbol{P}_{n+1}, \boldsymbol{X}) \tag{17}$$

is an epimorphism.

THEOREM 13.7. *Let* (16) *be a projective resolution of $\boldsymbol{\Delta}(\Lambda)$. Then the following conditions are equivalent.*

(*i*) $\mathrm{hd}(\Lambda) \le n$.

(*ii*) $\lim^{n+1} \boldsymbol{d}_{n+1}\boldsymbol{P}_{n+1} = 0$.

(*iii*) $\boldsymbol{d}_{n+1}\boldsymbol{P}_{n+1}$ *is a retract of* $\boldsymbol{P}_n$.

Proof. (*i*) ⇒ (*ii*) is a consequence of Theorem 13.3. If (*ii*) holds, then Corollary 13.6 implies that

$$\mathrm{Hom}(\boldsymbol{P}_n, \boldsymbol{d}_{n+1}\boldsymbol{P}_{n+1}) \to \mathrm{Hom}(\boldsymbol{d}_{n+1}\boldsymbol{P}_{n+1}, \boldsymbol{d}_{n+1}\boldsymbol{P}_{n+1}) \tag{18}$$

is an epimorphism. Especially, for the identity morphism on $\boldsymbol{d}_{n+1}\boldsymbol{P}_{n+1}$, there is a morphism $\boldsymbol{r}_n : \boldsymbol{P}_n \to \boldsymbol{d}_{n+1}\boldsymbol{P}_{n+1}$, such that $\mathrm{Hom}(\boldsymbol{i}_n, \boldsymbol{d}_{n+1}\boldsymbol{P}_{n+1})(\boldsymbol{r}_n) = \boldsymbol{r}_n\boldsymbol{i}_n = \mathrm{id}$. Consequently, $\boldsymbol{d}_{n+1}\boldsymbol{P}_{n+1}$ is a retract of $\boldsymbol{P}_n$, i.e., (*iii*) holds.

Now assume (*iii*) and consider an arbitrary system $\boldsymbol{X}$. By Corollary 13.6 and Theorem 13.3, in order to prove (*i*), it suffices to show that (17) is an epimorphism. Let $\boldsymbol{r}_n : \boldsymbol{P}_n \to \boldsymbol{d}_{n+1}\boldsymbol{P}_{n+1}$ be a retraction, i.e., a morphism such that $\boldsymbol{r}_n\boldsymbol{i}_n = \mathrm{id}$. For a morphism $\boldsymbol{f} \in \mathrm{Hom}(\boldsymbol{d}_{n+1}\boldsymbol{P}_{n+1}, \boldsymbol{X})$, put $\boldsymbol{g} = \boldsymbol{f}\boldsymbol{r}_n \in \mathrm{Hom}(\boldsymbol{P}_n, \boldsymbol{X})$. Clearly, $\mathrm{Hom}(\boldsymbol{i}_n, \boldsymbol{X})(\boldsymbol{g}) = \boldsymbol{g}\boldsymbol{i}_n = \boldsymbol{f}\boldsymbol{r}_n\boldsymbol{i}_n = \boldsymbol{f}$, which shows that (17) is indeed an epimorphism. □

COROLLARY 13.8. *If* $\mathrm{hd}(\Lambda) = n$, *then* $\lim^n \boldsymbol{P}_n \ne 0$.

Proof. By the implication (*i*) ⇒ (*ii*) of Theorem 13.7, we see that $\mathrm{hd}(\Lambda) = n$ implies $\lim^{n+1} \boldsymbol{d}_{n+1}\boldsymbol{P}_{n+1} = 0$. On the other hand, one cannot have $\lim^n \boldsymbol{d}_n\boldsymbol{P}_n = 0$, because Theorem 13.7 would imply that $\mathrm{hd}(\Lambda) \le n-1$, which is not the case. Consequently, $\lim^n \boldsymbol{d}_n\boldsymbol{P}_n \ne 0$. Furthermore, exactness of (16) implies exactness of the sequence

$$\mathbf{0} \to \boldsymbol{d}_{n+1}\boldsymbol{P}_{n+1} \to \boldsymbol{P}_n \to \boldsymbol{d}_n\boldsymbol{P}_n \to \mathbf{0}. \tag{19}$$

Therefore, also the sequence

$$\lim^n \boldsymbol{P}_n \to \lim^n \boldsymbol{d}_n\boldsymbol{P}_n \to \lim^{n+1} \boldsymbol{d}_{n+1}\boldsymbol{P}_{n+1} \tag{20}$$

is exact. Since $\lim^n \boldsymbol{d}_n\boldsymbol{P}_n \ne 0$ and $\lim^{n+1} \boldsymbol{d}_{n+1}\boldsymbol{P}_{n+1} = 0$, (20) shows that $\lim^n \boldsymbol{P}_n$ cannot vanish. □

13.2 Goblot's vanishing theorem

In this subsection we prove a theorem, which gives sufficient conditions for the vanishing of $\lim^m \boldsymbol{X}$ in terms of the *cofinality* $\mathrm{cof}(\Lambda)$ of the index set Λ of $\boldsymbol{X}$. Recall that $\mathrm{cof}(\Lambda)$ is the smallest cardinal κ such that there exists a subset $M \subseteq \Lambda$ of cardinality κ, which is *cofinal* in Λ, i.e., for every $\lambda \in \Lambda$ there is a $\mu \in M$ such that $\lambda \leq \mu$.

THEOREM 13.9. *If Λ is a directed set of cofinality $\mathrm{cof}(\Lambda) \leq \aleph_n$, where $n \geq 0$ is an integer, then the homological dimension $\mathrm{hd}(\Lambda) \leq n+1$ and thus, for every inverse system $\boldsymbol{X}$ of R-modules indexed by Λ and every $m \geq n+2$,*

$$\lim^m \boldsymbol{X} = 0. \tag{1}$$

In the proof of Theorem 13.9 we need the following lemma.

LEMMA 13.10. *If Λ is directed and $\mathrm{cof}(\Lambda) = \aleph_n$, $n \geq 0$, then there exists an increasing family of directed subsets $\Lambda^\alpha \subseteq \Lambda$, indexed by the set of ordinals $\Omega_n = \{\alpha | \alpha < \omega_n\}$, such that the cofinality $\mathrm{cof}(\Lambda^\alpha) < \aleph_n$, for all $\alpha \in \Omega_n$. Moreover,*

$$\Lambda^\beta = \cup_{\alpha<\beta} \Lambda^\alpha, \tag{2}$$

for every limit ordinal $\beta < \omega_n$, and

$$\Lambda = \cup_{\alpha<\omega_n} \Lambda^\alpha. \tag{3}$$

The condition $\mathrm{cof}(\Lambda^\alpha) < \aleph_0$ means that $\mathrm{cof}(\Lambda^\alpha)$ is a finite cardinal. Since Λ^α is directed, this is equivalent to Λ^α having a terminal element.

Proof of Lemma 13.10. The proof is by induction on $n \geq 0$. We first consider the case $n = 0$. By assumption, there exists a function $\phi \colon \Omega_0 \to \Lambda$ such hat $\phi(\Omega_0)$ is cofinal in Λ. Since Ω_0 is cofinite, one can apply Lemma 1.2, and obtain an increasing function $\psi \colon \Omega_0 \to \Lambda$ such that $\psi \geq \phi$. For $\alpha < \omega_0$, let Λ^α denote the set of all elements $\lambda \in \Lambda$ such that $\lambda \leq \psi(\alpha)$. Clearly, $\alpha \leq \alpha'$ implies $\Lambda^\alpha \subseteq \Lambda^{\alpha'}$ and the sets Λ^α, $\alpha < \omega_0$, have property (3). Moreover, $\psi(\alpha)$ is a terminal element of Λ^α. Condition (2) does not apply.

Now assume that $n > 0$ and that the assertion holds for $m < n$. We will first construct a function h, defined on all infinite subsets M of Λ, having as values directed subsets of Λ. We require that h has the following three properties. The cardinal $\mathrm{card}(h(M)) = \mathrm{card}(M)$, $M \subseteq M'$ implies $h(M) \subseteq h(M')$ and, for every linearly ordered family of infinite subsets (M_ν), $h(\cup_\nu M_\nu) = \cup_\nu h(M_\nu)$. In order to construct h, one associates with every nonempty finite subset $\Gamma \subseteq \Lambda$ an element $\Gamma^* \in \Lambda$ such that $\gamma \leq \Gamma^*$, for every $\gamma \in \Gamma$. If Γ consists of a single element γ, one requires that $\Gamma^* = \gamma$. Then, for an infinite subset $M \subseteq \Lambda$, one puts $g(M) = \{\Gamma^* | \Gamma \subseteq M,\ \mathrm{card}(\Gamma) < \aleph_0\}$. Note that $M \subseteq g(M)$ and thus, $\mathrm{card}(M) \leq \mathrm{card}(g(M))$. On the other hand, since M is infinite, $\mathrm{card}(g(M)) \leq \mathrm{card}(M)$ and thus, $\mathrm{card}(g(M)) = \mathrm{card}(M)$.

Furthermore, $M \subseteq M'$ implies $g(M) \subseteq g(M')$. Finally, for any family of infinite subsets (M_ν), $\cup_\nu g(M_\nu) \subseteq g(\cup_\nu M_\nu)$. If the family is linearly ordered, then also the opposite inclusion holds, because every finite set $\Gamma \subseteq \cup_\nu M_\nu$ is already contained in some M_ν. We now define h by putting

$$h(M) = M \cup g(M) \cup g(g(M)) \cup \dots . \tag{4}$$

It is easy to see that h has all the desired properties.

Since $\operatorname{cof}(\Lambda) = \aleph_n$, there exists an injective mapping $\phi\colon \Omega_n \to \Lambda$ such that $\phi(\Omega_n)$ is cofinal in Λ. For every α, $\omega_0 \le \alpha < \omega_n$, let Λ^α be the set of all elements $\lambda \in \Lambda$ such that $\lambda \le \mu$, for some element $\mu \in h(\phi(\,.\,,\alpha))$, where $(\,.\,,\alpha)$ denotes the set of all ordinals $\xi < \alpha$. For $0 \le \alpha < \omega_0$, let $\Lambda^\alpha = \Lambda^{\omega_0}$. Clearly, for $\omega_0 \le \alpha < \omega_n$, $h(\phi(\,.\,,\alpha))$ is a directed subset of Λ of cardinality $\operatorname{card}(h(\phi(\,.\,,\alpha))) = \operatorname{card}(\phi(\,.\,,\alpha)) = \operatorname{card}((\,.\,,\alpha)) < \aleph_n$. Consequently, since the set $h(\phi(\,.\,,\alpha))$ is cofinal in Λ^α, the latter is a directed set and $\operatorname{cof}(\Lambda^\alpha) \le \operatorname{card}(h(\phi(\,.\,,\alpha))) < \aleph_n$. Moreover, $\alpha \le \alpha'$ implies $\phi(\,.\,,\alpha) \subseteq \phi(\,.\,,\alpha')$ and the monotonicity of h implies $\Lambda^\alpha \subseteq \Lambda^{\alpha'}$. If β is a limit ordinal, $\omega_0 < \beta < \omega_n$, then

$$\phi(\,.\,,\beta) = \bigcup_{\omega_0 \le \alpha < \beta} \phi(\,.\,,\alpha) \tag{5}$$

and since the sets $\phi(\,.\,,\alpha)$, $\omega_0 \le \alpha < \beta$, form a linearly ordered family of infinite sets, one concludes that

$$h(\phi(\,.\,,\beta)) = \bigcup_{\omega_0 \le \alpha < \beta} h(\phi(\,.\,,\alpha)) \tag{6}$$

and thus, (2) holds. Finally, (3) is an easy consequence of the fact that $\phi(\,.\,,\omega_n)$ is cofinal in Λ. □

Proof of Theorem 13.9. The proof is by induction on n. We begin the induction with $n = -1$. In this case the assumption is interpeted as Λ being finite. Since Λ is directed, it has a maximum $\lambda_0 \in \Lambda$. Clearly, the element $1 \in R$ in the corresponding term of $\boldsymbol{\Delta}(\Lambda)$ forms a basis of $\boldsymbol{\Delta}(\Lambda)$. Consequently, the system $\boldsymbol{\Delta}(\Lambda)$ is free, hence, also projective. Therefore, $\mathbf{0} \leftarrow \boldsymbol{\Delta}(\Lambda) \xleftarrow{\text{id}} \boldsymbol{\Delta}(\Lambda) \leftarrow \mathbf{0}$ is a projective resolution of $\boldsymbol{\Delta}(\Lambda)$ of length 0, which shows that $\operatorname{hd}(\Lambda) = 0$.

Let us now assume that the asertion holds for $n-1$ and $n \ge 0$. We will prove that it also holds for n. Recall that $\lim^m \boldsymbol{X}$ is the m-th cohomology module of the cochain complex $K(\boldsymbol{X})$, described in 11.5. Also recall that an m-cochain $x \in K^m(\boldsymbol{X})$ is a function, defined on all (increasing) multiindices $\boldsymbol{\lambda} = (\lambda_0, \dots, \lambda_m)$ from Λ, which takes values in X_{λ_0}, denoted by $x(\boldsymbol{\lambda})$. We must show that, for $m \ge n+2$, every m-cocyle $z \in K^m(\boldsymbol{X})$ is the coboundary of an $(m-1)$-cochain $x \in K^{m-1}(\boldsymbol{X})$.

Choose directed subsets $\Lambda^\alpha \subseteq \Lambda$, $0 \le \alpha < \omega_n$, as in Lemma 13.10, and let $\boldsymbol{X}^\alpha$ denote the restriction of the system $\boldsymbol{X}$ to Λ^α. For a cochain $x \in K^m(\boldsymbol{X})$, let $x|\Lambda^\alpha \in K^m(\boldsymbol{X}^\alpha)$ denote the restriction of x to the multiindices of Λ^α, i.e., let $(x|\Lambda^\alpha)(\boldsymbol{\lambda}) = x(\boldsymbol{\lambda})$, for multiindices $\boldsymbol{\lambda}$ from Λ^α.

To prove the theorem, it suffices to construct cochains $x^\alpha \in K^{m-1}(\boldsymbol{X}^\alpha)$, $\alpha < \omega_n$, having the following two properties.

$$\delta x^\alpha = z|\Lambda^\alpha, \tag{7}$$

$$x^{\alpha'}|\Lambda^\alpha = x^\alpha, \tag{8}$$

whenever $\alpha < \alpha' < \omega_n$.

Indeed, if such cochains x^α are given, one defines the desired cochains $x \in K^{m-1}(\boldsymbol{X})$, by putting

$$x(\boldsymbol{\lambda}) = x^\alpha(\boldsymbol{\lambda}),\ \boldsymbol{\lambda} \in \Lambda_{m-1}, \tag{9}$$

where $\alpha < \omega_n$ is an ordinal for which $\boldsymbol{\lambda} \in \Lambda^\alpha$. By (8), $x(\boldsymbol{\lambda})$ does not depend on α and thus, $x \in K^{m-1}(\boldsymbol{X})$ is well defined. Moreover, $\delta x = z$. Indeed, for an arbitrary multiindex $\boldsymbol{\lambda} = (\lambda_0, \dots, \lambda_m) \in \Lambda_m$, choose an $\alpha < \omega_n$ so large that $\lambda_0, \dots, \lambda_m \in \Lambda^\alpha$, i.e., $\boldsymbol{\lambda} \in \Lambda^\alpha$. By (9), $x(d^j\boldsymbol{\lambda}) = x^\alpha(d^j\boldsymbol{\lambda})$, for $0 \le j \le m$. Therefore, by the definition of δ (see (11.5.4)) and by formula (7), one concludes that $(\delta x)(\boldsymbol{\lambda}) = (\delta x^\alpha)(\boldsymbol{\lambda}) = z(\boldsymbol{\lambda})$.

In order to define the $(m-1)$-cochains $x^\alpha \in K^{m-1}(\boldsymbol{X}^\alpha)$, $\alpha < \omega_n$, we first define $(m-1)$-cochains $y^\alpha \in K^{m-1}(\boldsymbol{X}^\alpha)$ as follows. Since $z \in K^m(\boldsymbol{X})$ is an m-cocyle, its restriction $z|\Lambda^\alpha$ is an m-cocyle in $K^m(\boldsymbol{X}^\alpha)$. However, $\mathrm{cof}(\Lambda^\alpha) \le \aleph_{n-1}$ and $m \ge n+1$. Therefore, by the induction hypothesis, there exists a cochain $y^\alpha \in K^{m-1}(\boldsymbol{X}^\alpha)$ such that

$$\delta y^\alpha = z|\Lambda^\alpha. \tag{10}$$

We will now define the desired $(m-1)$-cochains x^α, by transfinite induction on $\alpha < \omega_n$. For $\alpha = 0$, we put $x^0 = y^0$. Now consider $0 < \beta < \omega_n$ and assume that we have already defined the $(m-1)$-cochains x^α, for all $0 \le \alpha < \beta$, in such a way that they satisfy (7) and (8).

If β is of the form $\beta = \alpha + 1$, then x^α is already defined and

$$\delta(y^\beta|\Lambda^\alpha - x^\alpha) = \delta(y^\beta)|\Lambda^\alpha - \delta x^\alpha = (z|\Lambda^\beta)|\Lambda^\alpha - z|\Lambda^\alpha = 0. \tag{11}$$

Consequently, $y^\beta|\Lambda^\alpha - x^\alpha$ is an $(m-1)$-cocycle of $K(\boldsymbol{X}^\alpha)$. Since $\mathrm{cof}(\Lambda^\alpha) \le \aleph_{n-1}$ and $m - 1 \ge n + 1$, the induction hypothesis implies $\lim^{m-1}(\boldsymbol{X}^\alpha) = 0$. Therefore, there exists a cochain $u^\alpha \in K^{m-2}(\boldsymbol{X}^\alpha)$ such that

$$\delta u^\alpha = y^\beta|\Lambda^\alpha - x^\alpha. \tag{12}$$

We now extend u^α to a cochain $u^\beta \in K^{m-2}(\boldsymbol{X}^\beta)$, by putting

$$u^\beta(\lambda_0, ..., \lambda_{m-2}) = \begin{cases} u^\alpha(\lambda_0, ..., \lambda_{m-2}), & \lambda_0, \dots, \lambda_{m-2} \in \Lambda^\alpha, \\ 0, & \text{otherwise.} \end{cases} \tag{13}$$

Finally, we put

$$x^\beta = y^\beta - \delta u^\beta. \tag{14}$$

Clearly, $x^\beta \in K^{m-1}(\boldsymbol{X}^\beta)$. Moreover, by (14) and (10),

$$\delta x^\beta = \delta y^\beta = z|\Lambda^\beta, \tag{15}$$

as required by (7). To prove (8), it suffices to prove it in the case of the pair $\alpha, \alpha' = \beta$. However, by (13), $u^\beta|\Lambda^\alpha = u^\alpha$ and therefore, (12) yields

$$x^\beta|\Lambda^\alpha = y^\beta|\Lambda^\alpha - \delta(u^\beta|\Lambda^\alpha) = y^\beta|\Lambda^\alpha - \delta u^\alpha = x^\alpha. \tag{16}$$

If β is a limit ordinal, we define x^β by putting

$$x^\beta(\lambda_0, ..., \lambda_{m-1}) = x^\alpha(\lambda_0, ..., \lambda_{m-1}),\ (\lambda_0, \dots, \lambda_{m-1}) \in \Lambda_{m-1}, \tag{17}$$

where one chooses $\alpha < \omega_n$ so large that $\lambda_0, \dots, \lambda_{m-1} \in \Lambda^\alpha$. An argument, similar to the one used in the case of the cochain x, shows that x^β is well defined. Moreover, (7) and (8) are satisfied. □

13.3 Systems with non-vanishing $\lim^n$

The main result of this subsection is the following theorem (generalized in Theorem 14.14 to arbitrary directed sets Λ).

THEOREM 13.11. *If Λ is a linearly ordered set with* $\mathrm{cof}(\Lambda) \geq \aleph_n$, *then* $\mathrm{hd}(\Lambda) \geq n+1$ *and thus, there exists a system $\boldsymbol{X}$ from* Mod^Λ *such that* $\lim^{n+1}\boldsymbol{X} \neq 0$.

The following corollary yields explicit examples of non-vanishing limits of arbitrarily high order n.

COROLLARY 13.12. *Let Λ be a linearly ordered set with* $\mathrm{cof}(\Lambda) = \aleph_n$ *(e.g., the set of ordinals $\{\alpha|\alpha < \omega_n\}$) and let $\boldsymbol{P}_n$ be the n-th term of the standard projective resolution of* $\boldsymbol{\Delta}(\Lambda)$ (see 12.2). *Then*

$$\lim^{n+1}\boldsymbol{P}_{n+1} \neq 0. \tag{1}$$

Proof. By Theorem 13.9, $\mathrm{hd}(\Lambda) \leq n+1$. However, by Theorem 13.11, $\mathrm{hd}(\Lambda) \geq n+1$ and thus, $\mathrm{hd}(\Lambda) = n+1$. Consequently, Corollary 13.8 implies (1). □

Theorem 13.11 is an easy consequence of the following lemma.

LEMMA 13.13. *Let Λ be a linearly ordered set. If* $\mathrm{cof}(\Lambda) \geq \aleph_{n+1}$ *and* $\mathrm{hd}(\Lambda) \leq k$, $k \geq 1$, *then there exists a subset $M \subseteq \Lambda$ such that* $\mathrm{cof}(M) \geq \aleph_n$ *and* $\mathrm{hd}(M) \leq k-1$.

Proof of Theorem 13.11. The proof is by induction on n. If $n = 0$, then Λ is a directed set with no terminal element. Therefore, by Example 11.17, $\boldsymbol{\Delta}(\Lambda)$ is not projective. Hence, $\mathrm{hd}(\Lambda) \geq 1$. Now assume that the assertion of the theorem is true, for a given $n \geq 0$. If it were false for $n+1$, one could find a linearly ordered set Λ with $\mathrm{cof}(\Lambda) \geq \aleph_{n+1}$ and $\mathrm{hd}(\Lambda) \leq n+1$. Then, Lemma

13.13 would yield a subset $M \subseteq \Lambda$ with $\mathrm{cof}(M) \geq \aleph_n$ and $\mathrm{hd}(M) \leq n$. Since M is also linearly ordered, this contradicts the induction hypothesis. □

We will prove Lemma 13.13 using another lemma (Lemma 13.14). To state it, we associate with every directed set Λ and everyone of its directed subsets $M \subseteq \Lambda$ a sequence of inverse systems $\boldsymbol{Q}_k(M)$, $k = 0, 1, \ldots$, indexed by Λ, and a sequence of morphisms $\boldsymbol{d}_k: \boldsymbol{Q}_k(M) \to \boldsymbol{Q}_{k-1}(M)$. Recall that the k-th term of the standard projective resolution of $\boldsymbol{\Delta}(\Lambda)$, here denoted by $\boldsymbol{P}_k(\Lambda)$, is a system consisting of R-modules

$$P_{k\lambda}(\Lambda) = \bigoplus_{\lambda \leq \boldsymbol{\lambda} \in \Lambda_k} R, \ \lambda \in \Lambda, \tag{2}$$

and of natural inclusions $i_{k\lambda\lambda'}$ (see 12.2). By definition, $\boldsymbol{Q}_k(M)$ is a subsystem of $\boldsymbol{P}_k(\Lambda)$ consisting of R-modules

$$Q_{k\lambda}(M) = \bigoplus_{\lambda \leq \boldsymbol{\mu} \in M_k} R, \ \lambda \in \Lambda, \tag{3}$$

and of natural inclusions. The morphisms $\boldsymbol{d}_k = (d_{k\lambda}): \boldsymbol{Q}_k(M) \to \boldsymbol{Q}_{k-1}(M)$ are defined as natural restrictions of the morphisms $\boldsymbol{d}_k = (d_{k\lambda}): \boldsymbol{P}_k(\Lambda) \to \boldsymbol{P}_{k-1}(\Lambda)$.

LEMMA 13.14. *If Λ is linearly ordered,* $\mathrm{cof}(\Lambda) \geq \aleph_{n+1}$ *and* $\mathrm{hd}(\Lambda) \leq k$, $k \geq 1$, *then there exists a subset $M \subseteq \Lambda$, having the following properties:*

(*i*) $\mathrm{cof}(M) \geq \aleph_n$,

(*ii*) $\mathrm{card}(M) = \aleph_n$,

(*iii*) $\boldsymbol{d}_k\boldsymbol{Q}_k(M)$ *is a retract of* $\boldsymbol{d}_k\boldsymbol{P}_k(\Lambda)$.

Proof of Lemma 13.13. Let Λ be a linearly ordered set with $\mathrm{cof}(\Lambda) \geq \aleph_{n+1}$ and $\mathrm{hd}(\Lambda) \leq k$, $k \geq 1$. By Lemma 13.14, there exists a subset $M \subseteq \Lambda$ with properties (*i*)–(*iii*). To complete the proof it remains to show that $\mathrm{hd}(M) \leq k-1$. In view of Theorem 13.7, it suffices to show that $\boldsymbol{d}_k\boldsymbol{P}_k(M)$ is a retract of $\boldsymbol{P}_{k-1}(M)$, where $\boldsymbol{P}_k(M)$ denotes the k-th term of the standard projective resolution of $\boldsymbol{\Delta}$(M). Actually, it suffices to construct a retraction $\boldsymbol{s}: \boldsymbol{Q}_{k-1}(M) \to \boldsymbol{d}_k\boldsymbol{Q}_k(M)$. Indeed, $\boldsymbol{P}_k(M)$ consists of modules

$$P_{k\mu}(M) = \bigoplus_{\mu \leq \boldsymbol{\mu} \in M_k} R, \ \mu \in M, \tag{4}$$

and of natural inclusions and the restriction of $\boldsymbol{s}$ to M is a retraction $\boldsymbol{P}_{k-1}(M) \to \boldsymbol{d}_k\boldsymbol{P}_k(M)$, as desired.

First note that the set M is not cofinal in Λ, because $\mathrm{card}(M) = \aleph_n$, but $\mathrm{cof}(\Lambda) \geq \aleph_{n+1}$. Since Λ is linearly ordered, one concludes that there exists an element $\nu \in \Lambda \backslash M$ such that $\mu < \nu$, for every $\mu \in M$. Consequently, for every increasing sequence $\mu_0 \leq \ldots \leq \mu_{k-1}$ in M, $\langle \mu_0, \ldots, \mu_{k-1}, \nu \rangle$ is a well-defined element of $P_k(\Lambda)$. Consider the inverse system $\boldsymbol{R}_k(M) \subseteq \boldsymbol{P}_k(\Lambda)$, which consists of modules $R_{k\lambda}(M)$ and of inclusion homomorphisms. Here $R_{k\lambda}(M) \subseteq P_{k\lambda}(\Lambda)$ is generated by all elements of the form $\langle\mu_0, \ldots, \mu_{k-1}, \nu\rangle$, where $(\mu_0, \ldots, \mu_{k-1}) \in M_{k-1}$ and $\nu \in \Lambda$.

Let us first show that

$$\boldsymbol{d}_k\boldsymbol{Q}_k(M) \subseteq \boldsymbol{d}_k\boldsymbol{R}_k(M). \tag{5}$$

Indeed, $Q_{k\lambda}(M)$ is generated by elements of the form $x = \langle\mu_0, \ldots, \mu_k\rangle$, where $(\mu_0, \ldots, \mu_k) \in M_k$ and $\lambda \le \mu_0$. Therefore, by (12.2.24) and (12.2.25), $\langle x, \nu\rangle = \langle\mu_0, \ldots, \mu_k, \nu\rangle \in P_{k+1,\lambda}(\Lambda)$ and

$$d_{k+1}\langle x, \nu\rangle = \langle d_k x, \nu\rangle + (-1)^{k+1}x. \tag{6}$$

Consequently, application of d_k yields

$$0 = d_k\langle d_k x, \nu\rangle + (-1)^{k+1}d_k x. \tag{7}$$

Now notice that $\langle d_k x, \nu\rangle \in R_{k\lambda}(M)$, and thus $d_k\langle d_k x, \nu\rangle \in d_k R_{k\lambda}(M)$. Therefore, (7) yields the desired conclusion that $d_{k\lambda}x \in d_{k\lambda}R_{k\lambda}(M)$.

In order to complete the proof, we consider the following commutative diagram of systems and morphisms given by inclusions (solid arrows).

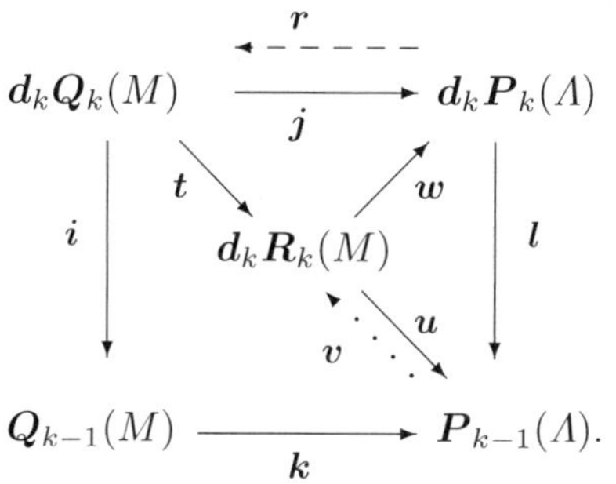

(8)

Next we define a morphism $\boldsymbol{v}\colon \boldsymbol{P}_{k-1}(\Lambda) \to \boldsymbol{d}_k\boldsymbol{R}_k(M)$, for which

$$\boldsymbol{vu} = \mathrm{id}. \tag{9}$$

It suffices to define v_λ on the generators $\langle\lambda_0, \ldots, \lambda_{k-1}\rangle$ of $P_{k-1\lambda}(\Lambda)$. If all λ_i belong to M, put

$$v_\lambda\langle\lambda_0, \ldots, \lambda_{k-1}\rangle = (-1)^k d_k\langle\lambda_0, \ldots, \lambda_{k-1}, \nu\rangle, \tag{10}$$

which belongs to $d_{k\lambda}R_{k\lambda}(M)$. If at least one $\lambda_i \notin M$, put $v_\lambda\langle\lambda_0, \ldots, \lambda_{k-1}\rangle = 0$. To prove (9), it suffices to verify that on the generators $d_k\langle\mu_0, \ldots, \mu_{k-1}, \nu\rangle$ of $d_{k\lambda}R_{k\lambda}(M)$, v_λ is the identity. First note that

$$d_k\langle\mu_0, \ldots, \mu_{k-1}, \nu\rangle = \langle d_{k-1}\langle\mu_0, \ldots, \mu_{k-1}\rangle, \nu\rangle + (-1)^k\langle\mu_0, \ldots, \mu_{k-1}\rangle. \tag{11}$$

Also note that $\langle d_{k-1}\langle \mu_0, \ldots, \mu_{k-1}\rangle, \nu\rangle$ is a linear combination of elements of the form $\langle \mu_0, \ldots, \widehat{\mu_i}, \ldots, \mu_{k-1}, \nu\rangle$ and v_λ maps each of these elements to 0, because $\nu \notin M$. Consequently, by (10), the application of v_λ to (11) yields

$$v_\lambda(d_k\langle \mu_0, \ldots, \mu_{k-1}, \nu\rangle) = d_k\langle \mu_0, \ldots, \mu_{k-1}, \nu\rangle. \tag{12}$$

By *(iii)* from Lemma 13.14, there is a morphism $\boldsymbol{r}\colon \boldsymbol{d}_k\boldsymbol{P}_k(\Lambda) \to \boldsymbol{d}_k\boldsymbol{P}_k(M)$ such that $\boldsymbol{rj} = \mathrm{id}$. It is now easy to see that $\boldsymbol{s} = \boldsymbol{rwvk}\colon \boldsymbol{Q}_{k-1}(M) \to \boldsymbol{d}_k\boldsymbol{Q}_k(M)$ is a retraction. Indeed,

$$\boldsymbol{si} = \boldsymbol{rwvki} = \boldsymbol{rwvlj} = \boldsymbol{rwvlwt} = \boldsymbol{rwvut} = \boldsymbol{rwt} = \boldsymbol{rj} = \mathrm{id}. \ \square \tag{13}$$

Proof of Lemma 13.14. By assumption, $\mathrm{hd}(\Lambda) \leq k$, $k \geq 1$. Therefore, by Theorem 13.3, $\boldsymbol{d}_k\boldsymbol{P}_k(\Lambda)$ is projective. Since $\boldsymbol{d}_k\colon \boldsymbol{P}_k(\Lambda) \to \boldsymbol{d}_k\boldsymbol{P}_k(\Lambda)$ is an epimorphism, one concludes that there exists a morphism $\boldsymbol{u} = (u_\lambda)\colon \boldsymbol{d}_k\boldsymbol{P}_k(\Lambda) \to \boldsymbol{P}_k(\Lambda)$ such that

$$\boldsymbol{d}_k\boldsymbol{u} = \mathrm{id}. \tag{14}$$

In order to construct the set M, we will first define, for every ordinal $\alpha < \omega_n$, a set $M_\alpha \subseteq \Lambda$ and a point $\mu_\alpha \in M_\alpha$, having the following properties:

$$M_\alpha \subseteq M_{\alpha'}, \ \text{for } \alpha \leq \alpha'. \tag{15}$$

$$\mu < \mu_{\alpha+1}, \ \text{for every } \mu \in M_\alpha, \tag{16}$$

$$\mathrm{card}(M_\alpha) < \aleph_n, \tag{17}$$

$$u_\lambda d_{k\lambda} Q_{k\lambda}(M_\alpha) \subseteq Q_{k\lambda}(M_{\alpha+1}), \ \lambda \in \Lambda. \tag{18}$$

The sets M_α, $\alpha < \omega_n$, are constructed by transfinite induction. One takes for M_0 any non-empty subset of Λ with $\mathrm{card}(M_0) < \aleph_n$, and one takes for μ_0 an arbitrary point of M_0. If we have already defined the sets M_α and the points μ_α, for $\alpha < \alpha' < \omega_n$, and if $\alpha' = \alpha+1$, one considers the set Γ_α, formed by all the generators $\langle \mu_0, \ldots, \mu_k\rangle$ of $Q_{k\lambda}(M_\alpha)$. Clearly, $\mathrm{card}(M_\alpha) < \aleph_n$ implies $\mathrm{card}(\Gamma_\alpha) < \aleph_n$. For each of these generators, $u_\lambda d_k\langle \mu_0, \ldots, \mu_k\rangle \in P_{k\lambda}(\Lambda)$ can be expressed as a finite linear combination of generators $\langle \lambda_0, \ldots, \lambda_k\rangle$ of $P_{k\lambda}(\Lambda)$. Let H_α be the set consisting of all elements λ_i, which appear in the expressions $u_\lambda d_k\langle \mu_0, \ldots, \mu_k\rangle$, when $\langle \mu_0, \ldots, \mu_k\rangle$ ranges through the set Γ_α. Clearly, $\mathrm{card}(\mathrm{H}_\alpha) = \mathrm{card}(\Gamma_\alpha) < \aleph_n$ and

$$u_\lambda d_{k\lambda}(Q_{k\lambda}(M_\alpha)) \subseteq Q_{k\lambda}(\mathrm{H}_\alpha). \tag{19}$$

Since $\mathrm{card}(M_\alpha \cup \mathrm{H}_\alpha) < \aleph_n$ and $\mathrm{cof}(\Lambda) \geq \aleph_{n+1}$, the set $M_\alpha \cup \mathrm{H}_\alpha$ is not cofinal in Λ. Taking into account the assumption that Λ is linearly ordered, one concludes that there exists a $\mu_{\alpha+1} \in \Lambda$, such that $\mu < \mu_{\alpha+1}$, for every $\mu \in M_\alpha \cup \mathrm{H}_\alpha$. We now put $M_{\alpha'} = M_{\alpha+1} = M_\alpha \cup \mathrm{H}_\alpha \cup \{\mu_{\alpha+1}\}$. Clearly, conditions (16)–(18) are satisfied. Moreover, $M_\alpha \subseteq M_{\alpha+1}$. If α' is a limit ordinal, we put

$$M_{\alpha'} = \cup_{\alpha<\alpha'} M_\alpha \tag{20}$$

and we choose for $\mu_{\alpha'}$ any point from $M_{\alpha'}$. Finally, note that $\mathrm{card}(M_{\alpha'}) < \aleph_n$ is an immediate consequence of $\mathrm{card}(M_{\alpha'}) < \aleph_n$ and of (17), for $\alpha < \alpha'$.

We now define the set M by putting

$$M = \cup_{\alpha<\omega_n} M_\alpha. \tag{21}$$

Let us first show that M has property (i). For $n = 0$, this is obvious, because (16) implies that M is infinite. Now assume that $n > 0$ and M contains a cofinal subset N of cardinality $\mathrm{card}(N) < \aleph_n$. Then there exists a set A of ordinals $\alpha < \omega_n$, such that $\mathrm{card}(A) < \aleph_n$ and

$$N \subseteq \cup_{\alpha\in A} M_\alpha \tag{22}$$

Since $\mathrm{card}(A) < \aleph_n$, one can find an ordinal $\alpha_0 < \omega_n$ such that $\alpha \leq \alpha_0$, and thus, $M_\alpha \subseteq M_{\alpha_0}$, for all $\alpha \in A$. Therefore, (22) implies

$$N \subseteq M_{\alpha_0}. \tag{23}$$

Now, using (16), one concludes that $\nu < \mu_{\alpha_0+1}$, for every $\nu \in N$. However, since N is cofinal in M, this implies that also $\mu < \mu_{\alpha_0+1}$, for every $\mu \in M$, which is, clearly, a contradiction, because $\mu_{\alpha_0+1} \in M$. Since $\mathrm{card}(\{\alpha | \alpha < \omega_n\}) = \aleph_n$, and $\mathrm{card}(M_\alpha) < \aleph_n$ for all α, one concludes that $\mathrm{card}(M) \leq \aleph_n$. Condition (ii) now follows from (i).

In order to verify condition (iii), let us first show that

$$\boldsymbol{u}\boldsymbol{d}_k\boldsymbol{Q}_k(M) \subseteq \boldsymbol{Q}_k(M). \tag{24}$$

Indeed, (18) implies

$$\cup_{\alpha<\omega_n} u_\lambda d_{k\lambda} Q_{k\lambda}(M_\alpha) \subseteq \cup_{\alpha<\omega_n} Q_{k\lambda}(M_{\alpha+1}), \tag{25}$$

while (15) and (21) imply

$$M = \cup_{\alpha<\omega_n} M_{\alpha+1}. \tag{26}$$

Therefore, the left side of (25) equals $u_\lambda d_{k\lambda} Q_{k\lambda}(M)$ and the right side equals $Q_{k\lambda}(M)$.

To verify (iii), consider the commutative diagram (solid arrows)

$$\begin{array}{ccc}
\boldsymbol{d}_k\boldsymbol{Q}_k(M) & \underset{\boldsymbol{v}}{\overset{\boldsymbol{d}_k}{\leftarrow\!\dashrightarrow}} & \boldsymbol{Q}_k(M) \\
\downarrow \boldsymbol{j} & & \boldsymbol{p} \uparrow \; \downarrow \boldsymbol{i} \\
\boldsymbol{d}_k\boldsymbol{P}_k(\Lambda) & \underset{\boldsymbol{d}_k}{\overset{\boldsymbol{u}}{\dashrightarrow\!\leftarrow}} & \boldsymbol{P}_k(\Lambda)\ ,
\end{array} \tag{27}$$

where $\boldsymbol{i}$ and $\boldsymbol{j}$ are morphisms given by inclusions and thus, are monomorphisms. Because of (24), there exists a unique morphism $\boldsymbol{v}: \boldsymbol{d}_k\boldsymbol{Q}_k(M) \to \boldsymbol{Q}_k(M)$, such that

$$\boldsymbol{iv} = \boldsymbol{uj}. \tag{28}$$

Now notice that $Q_{k\lambda}(M)$ is a direct summand of $P_{k\lambda}(\Lambda)$. Therefore, natural projections define an epimorphism $\boldsymbol{p}: \boldsymbol{P}_k(\Lambda) \to \boldsymbol{Q}_k(M)$ such that

$$\boldsymbol{pi} = \mathrm{id}. \tag{29}$$

By (27), (28) and (14),

$$\boldsymbol{j}\boldsymbol{d}_k\boldsymbol{v} = \boldsymbol{d}_k\boldsymbol{iv} = \boldsymbol{d}_k\boldsymbol{uj} = \boldsymbol{j}. \tag{30}$$

Since $\boldsymbol{j}$ is a monomorphism, (30) implies

$$\boldsymbol{d}_k\boldsymbol{v} = \mathrm{id}. \tag{31}$$

Now (28), (29) and (31) yield

$$(\boldsymbol{d}_k\boldsymbol{pu})\boldsymbol{j} = \boldsymbol{d}_k\boldsymbol{piv} = \boldsymbol{d}_k\boldsymbol{v} = \mathrm{id}, \tag{32}$$

which shows that $\boldsymbol{d}_k\boldsymbol{pu}: \boldsymbol{d}_k\boldsymbol{P}_k(\Lambda) \to \boldsymbol{d}_k\boldsymbol{Q}_k(M)$ is a retraction, as desired. □

Bibliographic notes

The systematic study of homological (projective) dimension hd(A) of R-modules A was initiated in the book (Cartan, Eilenberg 1956). In particular, one finds there the analogue of Theorem 13.2 for R-modules (see Ch. VI, Prop. 2.1). Theorem 13.2 itself appears as Theorem 7.20 in the book (Bucur, Deleanu 1968). Theorem 13.3 is an immediate consequence of Theorem 13.2 and the fact that $\mathrm{Ext}^{n+1}(\boldsymbol{\Delta}(\Lambda), \boldsymbol{X}) \approx \lim^{n+1} \boldsymbol{X}$ (Theorem 12.14). Theorems 13.5 and 13.7 appeared in (Mardešić 1996a) as Lemmas 5 and 4. Theorem 13.9 was first proved in (Goblot 1970). Our proof is different, but uses Lemma 13.10, which is taken from Goblot's paper. Theorem 13.11 is due to B. Mitchell. In the comprehensive paper (Mitchell 1972), he developed a theory of modules over rings with several objects. It generalizes usual rings and modules, which can be viewed as small additive categories with one object and additive functors from such categories to the category Ab, respectively. This generalization includes inverse systems of modules indexed by Λ, by viewing Λ as a ring with more objects. This new point of view made it possible to adapt the proofs of already known result, concerning the homological dimension of modules and rings, to obtain new results concerning the homological dimension of directed sets. In particular, Mitchell obtained Theorem 13.11, by adapting the proof of a theorem of Barbara Osofsky (Osofsky 1968a). The proof of Theorem 13.11, given in the present book, is patterned after Mitchell's proof (hence, also Osofsky's proof). Another result from (Osofsky

1968b) enabled Mitchell to give a new proof of Goblot's teorem (Mitchell 1972). Further relevant information on homological dimension of modules and rings and on the vanishing of higher limits can be found in (Roos 1961), (Jensen 1972, 1977), (Osofsky 1971, 1974) and (Gruson, Jensen 1981). It is difficult to find in the literature sufficiently simple, explicitly described, inverse systems of modules $\boldsymbol{X}$ with $\lim^n \boldsymbol{X} \neq 0$, $n \geq 2$. An example for $n = 2$ is given in (Kuz'minov 1967). The example for n arbitrary, given by Corollary 13.12 is from (Mardešić 1996a).

14. The cofinality theorem

In this section we associate with every increasing function $\phi: M \to \Lambda$ and inverse system $\boldsymbol{Y}$ indexed by Λ, an inverse system $\phi^*(\boldsymbol{Y})$ indexed by M and a morphism $\boldsymbol{\Phi}: \boldsymbol{Y} \to \phi^*(\boldsymbol{Y})$. The main goal is to prove that, for cofinal functions, $\boldsymbol{\Phi}$ induces an isomorphism of the derived limits $\lim^n \boldsymbol{Y} \to \lim^n \phi^*(\boldsymbol{Y})$. If ϕ is an injection, the proof is rather elementary (see Remark 14.18). In the general case, the proof is more involved and uses tensor products $\boldsymbol{A} \otimes \boldsymbol{B}$ of a direct and an inverse system of modules. This notion, as well as its basic properties, are developed in the first subsection. The general case of the cofinality theorem is needed in order to generalize Theorem 13.11 to index sets Λ, which are not linearly ordered.

14.1 Colimits and tensor products

A *direct system* of R-modules, indexed by a fixed directed set Λ, $\boldsymbol{A} = (A^\lambda, a^{\lambda\lambda'}, \Lambda)$ consists of R-modules A^λ, $\lambda \in \Lambda$, and of homomorphisms $a^{\lambda\lambda'}: A^\lambda \to A^{\lambda'}$, $\lambda \leq \lambda'$, such that $a^{\lambda\lambda} = \mathrm{id}$ and $a^{\lambda'\lambda''} a^{\lambda\lambda'} = a^{\lambda\lambda''}$, for $\lambda \leq \lambda' \leq \lambda''$. A (level) morphism $\boldsymbol{f}: \boldsymbol{A} \to \boldsymbol{B} = (B^\lambda, b^{\lambda\lambda'}, \Lambda)$ consist of homomorphisms $f^\lambda: A^\lambda \to B^\lambda$ such that $f^{\lambda'} a^{\lambda\lambda'} = b^{\lambda\lambda'} f^\lambda$. Composition $\boldsymbol{h} = \boldsymbol{g}\boldsymbol{f}: \boldsymbol{A} \to \boldsymbol{C}$ of the morphisms $\boldsymbol{f} = (f^\lambda): \boldsymbol{A} \to \boldsymbol{B}$ and $\boldsymbol{g} = (g^\lambda): \boldsymbol{B} \to \boldsymbol{C} = (C^\lambda, c^{\lambda\lambda'}, \Lambda)$ is given by the compositions $h^\lambda = g^\lambda f^\lambda$. The identity morphism $\boldsymbol{1}^{\boldsymbol{A}}: \boldsymbol{A} \to \boldsymbol{A}$ is given by the identity homomorphisms $A^\lambda \to A^\lambda$. In this way one obtains a category Mod_Λ.

In subsections 11.1 and 11.2 (except in Example 11.17), we did not use the assumption that Λ is directed. Therefore, all results proved in those subsections for inverse systems also hold for direct systems. In particular, a sequence of direct systems

$$\boldsymbol{A}' \xrightarrow{\boldsymbol{f}'} \boldsymbol{A} \xrightarrow{\boldsymbol{f}} \boldsymbol{A}'' \tag{1}$$

is exact at $\boldsymbol{A}$ if and only if the sequence of modules

$$A'^\lambda \xrightarrow{f'^\lambda} A^\lambda \xrightarrow{f^\lambda} A''^\lambda \tag{2}$$

is exact at A^λ, for every $\lambda \in \Lambda$.

The *direct limit* or *colimit* of a direct system $\boldsymbol{A}$ is the quotient module

$$A = \operatorname{colim} \boldsymbol{A} = (\bigoplus_{\lambda \in \Lambda} A^{\lambda})/K, \tag{3}$$

where K is the submodule of $\oplus A^{\lambda}$, generated by all elements of the form

$$a^{\lambda\lambda'}(u^{\lambda}) - u^{\lambda},\ u^{\lambda} \in A^{\lambda}. \tag{4}$$

One also considers the *canonical homomorphisms* $a^{\lambda}: A^{\lambda} \to A$, which are defined by composing the natural injections $A^{\lambda} \to \oplus A^{\lambda}$ with the quotient homomorphism $q: \oplus A^{\lambda} \to A$. Clearly, $a^{\lambda\lambda'} a^{\lambda'} = a^{\lambda}$, for $\lambda \leq \lambda'$. The direct limit is characterized by the following universal property. If $f^{\lambda}: A^{\lambda} \to C$ are homomorphisms such that $f^{\lambda'} a^{\lambda\lambda'} = f^{\lambda}$, for $\lambda \leq \lambda'$, then there exists a unique morphism $f: A \to C$ such that $fa^{\lambda} = f^{\lambda}$, for all $\lambda \in \Lambda$. For a morphism $\boldsymbol{f} = (f^{\lambda}): \boldsymbol{A} \to \boldsymbol{B}$, the homomorphism $\oplus f^{\lambda}: \oplus A^{\lambda} \to \oplus B^{\lambda}$ induces a homomorphism $f = \operatorname{colim} \boldsymbol{f}: A \to B$, where $A = \operatorname{colim} \boldsymbol{A}$, $B = \operatorname{colim} \boldsymbol{B}$ such that $fa^{\lambda} = b^{\lambda} f^{\lambda}$, for every $\lambda \in \Lambda$.

LEMMA 14.1. *If* $u^{\lambda_i} \in A^{\lambda_i}$, $i = 1, \dots, n$, *then the element* $\sum_i u^{\lambda_i} \in \bigoplus A^{\lambda}$ *belongs to the submodule* K *if an only if there exists a* $\lambda' \geq \lambda_1, \dots, \lambda_n$ *such that*

$$\sum_i a^{\lambda_i \lambda'}(u^{\lambda_i}) = 0. \tag{5}$$

In particular, if $u^{\lambda_i} \in A^{\lambda_i}$, $i = 1, 2$, *then* $a^{\lambda_1}(u^{\lambda_1}) = a^{\lambda_2}(u^{\lambda_2})$ *if and only if there exists a* $\lambda' \geq \lambda_1, \lambda_2$ *such that* $a^{\lambda_1\lambda'}(u^{\lambda_1}) = a^{\lambda_2\lambda'}(u^{\lambda_2})$.

Proof. Let K' denote the set of all elements $\sum_i u^{\lambda_i} \in \bigoplus A^{\lambda}$, which have the property that there exists a $\lambda' \geq \lambda_1, \dots, \lambda_n$ satisfying (5). Using directedness of Λ, it is readily seen that K' is a submodule of $\bigoplus A^{\lambda}$. Clearly, every generator (4) of K belongs to K' and thus $K \subseteq K'$. Conversely, if $\sum_i u^{\lambda_i}$ belongs to K', then there is a $\lambda' \geq \lambda_i$ such that (5) holds and thus,

$$\sum_i u^{\lambda_i} = \sum_i (u^{\lambda_i} - a^{\lambda_i \lambda'}(u^{\lambda_i})), \tag{6}$$

which shows that $\sum_i u^{\lambda_i} \in K$, i.e., $K' \subseteq K$. Hence, $K' = K$. Moreover, $a^{\lambda_1}(u^{\lambda_1}) = a^{\lambda_2}(u^{\lambda_2})$ implies $u^{\lambda_1} - u^{\lambda_2} \in K = K'$. Hence, there exists a $\lambda' \geq \lambda_1, \lambda_2$ such that $a^{\lambda_1\lambda'}(u^{\lambda_1}) - a^{\lambda_2\lambda'}(u^{\lambda_2}) = 0$. The opposite inclusion is obvious. □

In contrast to inverse limits, for direct limits (indexed by directed sets) the following lemma holds.

LEMMA 14.2. colim: $\mathsf{Mod}_{\Lambda} \to \mathsf{Mod}$ *is an exact functor, i.e., exactness of* (1) *implies exactness of*

$$\operatorname{colim} \boldsymbol{A}' \xrightarrow{f'} \operatorname{colim} \boldsymbol{A} \xrightarrow{f} \operatorname{colim} \boldsymbol{A}'', \tag{7}$$

where $f' = \operatorname{colim} \boldsymbol{f}', f = \operatorname{colim} \boldsymbol{f}$.

Proof. Let $u \in A = \operatorname{colim} \boldsymbol{A}$ be such that $f(u) = 0$. There exist a $\lambda \in \Lambda$ and an $u^\lambda \in A^\lambda$ such that $a^\lambda(u^\lambda) = u$. Therefore, $a''^\lambda f^\lambda(u^\lambda) = fa^\lambda(u^\lambda) = f(u) = 0$. By Lemma 14.1, there exists a $\lambda' \geq \lambda$ such that $a''^{\lambda\lambda'} f^\lambda(u^\lambda) = 0$. Hence, also $f^{\lambda'} a^{\lambda\lambda'}(u^\lambda) = 0$. By (2), one concludes that there exists a $u'^{\lambda'} \in A'^{\lambda'}$ such that $f'^{\lambda'}(u'^{\lambda'}) = a^{\lambda\lambda'}(u^\lambda)$. Since $a^{\lambda'} a^{\lambda\lambda'} = a^\lambda$, one concludes that

$$f'(a'^{\lambda'}(u'^{\lambda'})) = a^{\lambda'} f'^{\lambda'}(u'^{\lambda'}) = a^{\lambda'} a^{\lambda\lambda'}(u^\lambda) = a^\lambda(u^\lambda) = u. \square \tag{8}$$

The usual definition of the tensor product $A \otimes_R B$, or just $A \otimes B$, of two R-modules generalizes to the following notion of *tensor product* $\boldsymbol{A} \otimes \boldsymbol{B}$ of a direct system of R-modules $\boldsymbol{A} = (A^\lambda, a^{\lambda\lambda'}, \Lambda)$ and an inverse system of R-modules $\boldsymbol{B} = (B_\lambda, b_{\lambda\lambda'}, \Lambda)$. By definition, $\boldsymbol{A} \otimes \boldsymbol{B}$ is the R-module

$$\boldsymbol{A} \otimes \boldsymbol{B} = (\bigoplus_{\lambda \in \Lambda} (A^\lambda \otimes B_\lambda))/L, \tag{9}$$

where L is the submodule generated by all elements of the form

$$a^{\lambda\lambda'}(u^\lambda) \otimes v_{\lambda'} - u^\lambda \otimes b_{\lambda\lambda'}(v_{\lambda'}),\ u^\lambda \in A^\lambda,\ v_{\lambda'} \in B_{\lambda'},\ \lambda \leq \lambda'. \tag{10}$$

A morphism $\boldsymbol{f} = (f^\lambda)\colon \boldsymbol{A} \to \boldsymbol{A}'$ induces a homomorphism $\boldsymbol{f} \otimes \boldsymbol{1}\colon \boldsymbol{A} \otimes \boldsymbol{B} \to \boldsymbol{A}' \otimes \boldsymbol{B}$, determined by $\oplus(f^\lambda \otimes 1)\colon \oplus(A^\lambda \otimes B_\lambda) \to \oplus(A'^\lambda \otimes B_\lambda)$. Note that $\oplus(f^\lambda \otimes 1)$ maps (10) to

$$\begin{aligned} &f^{\lambda'} a^{\lambda\lambda'}(u^\lambda) \otimes v_{\lambda'} - f^\lambda(u^\lambda) \otimes b_{\lambda\lambda'}(v_{\lambda'}) = \\ &a'^{\lambda\lambda'}(f^\lambda(u^\lambda)) \otimes v_{\lambda'} - f^\lambda(u^\lambda) \otimes b_{\lambda\lambda'}(v_{\lambda'}), \end{aligned} \tag{11}$$

which belongs to the corresponding submodule L'. Clearly, $\otimes \boldsymbol{B}$ is a covariant functor $\mathsf{Mod}_\Lambda \to \mathsf{Mod}$. Analogously, one defines the functor $\boldsymbol{A}\otimes : \mathsf{Mod}^\Lambda \to \mathsf{Mod}$. Actually, $\otimes$ is a bifunctor $\mathsf{Mod}_\Lambda \times \mathsf{Mod}^\Lambda \to \mathsf{Mod}$.

EXAMPLE 14.3. For every direct system $\boldsymbol{A}$, there is a natural isomorphism $\Phi\colon \boldsymbol{A} \otimes \boldsymbol{\Delta} \to \operatorname{colim} \boldsymbol{A}$. Indeed, there are natural isomorphisms $\phi^\lambda\colon A^\lambda \otimes R \to A^\lambda$ such that $\phi^\lambda(u \otimes 1) = u$, for $u \in A^\lambda$. Then $\phi = \oplus_\lambda \phi^\lambda$ is also a natural isomorphism. ϕ induces a natural isomorphism $\Phi\colon \boldsymbol{A} \otimes \boldsymbol{\Delta} \to \operatorname{colim}\boldsymbol{A}$, because $\phi(L) = K$. The latter equality is a consequence of the equality $\oplus\phi^\lambda(a^{\lambda\lambda'}(u) \otimes 1 - u \otimes 1) = a^{\lambda\lambda'}(u) - u = 0$

REMARK 14.4. In analogy with modules, we call an inverse system $\boldsymbol{B}$ from Mod^Λ *flat* provided, for every exact sequence of direct systems from Mod_Λ,

$$\boldsymbol{0} \to \boldsymbol{A}' \to \boldsymbol{A} \to \boldsymbol{A}'' \to \boldsymbol{0}, \tag{12}$$

the sequence of modules

$$0 \to \boldsymbol{A}' \otimes \boldsymbol{B} \to \boldsymbol{A} \otimes \boldsymbol{B} \to \boldsymbol{A}'' \otimes \boldsymbol{B} \to 0 \tag{13}$$

is exact. By Example 14.3 and Lemma 14.2, the diagonal system $\boldsymbol{\Delta} = \boldsymbol{\Delta}(\Lambda)$ is flat.

The main result of this subsection is the following lemma.

LEMMA 14.5. *Let*

$$\mathbf{0} \leftarrow \boldsymbol{\Delta}(\Lambda) \overset{e}{\longleftarrow} \boldsymbol{P}_0 \overset{d_1}{\longleftarrow} \boldsymbol{P}_1 \overset{d_2}{\longleftarrow} \boldsymbol{P}_2 \leftarrow \dots \tag{14}$$

be a projective resolution of $\boldsymbol{\Delta}(\Lambda)$ *and let* $\boldsymbol{A}$ *be a direct system from* Mod_Λ. *Then*

$$0 \leftarrow \boldsymbol{A} \otimes \boldsymbol{\Delta}(\Lambda) \overset{1\otimes e}{\longleftarrow} \boldsymbol{A} \otimes \boldsymbol{P}_0 \overset{1\otimes d_1}{\longleftarrow} \boldsymbol{A} \otimes \boldsymbol{P}_1 \overset{1\otimes d_2}{\longleftarrow} \boldsymbol{A} \otimes \boldsymbol{P}_2 \leftarrow \dots \tag{15}$$

is an exact sequence of modules.

Note that it suffices to prove the assertion in the special case when (14) is the standard projective resolution of $\boldsymbol{\Delta}(\Lambda)$ (see 12.2). Indeed, any two projective resolutions of $\boldsymbol{\Delta}(\Lambda)$ are homotopy equivalent (augmented) chain complexes. Therefore, the functor $\boldsymbol{A}\otimes$ maps them into (augmented) chain complexes, which are also homotopy equivalent. Hence, if one of them is acyclic, so must be the other one.

We will first show that, in the case when (14) is the standard projective resolution of $\boldsymbol{\Delta}(\Lambda)$, then the sequence (15) is isomorphic to a sequence

$$(\boldsymbol{L}, \eta) = (0 \leftarrow \operatorname{colim} \boldsymbol{A} \overset{\eta}{\longleftarrow} L_0 \overset{\partial}{\longleftarrow} L_1 \overset{\partial}{\longleftarrow} L_2 \leftarrow \dots), \tag{16}$$

which has a rather simple explicit description.

$$L_n = \bigoplus_{\boldsymbol{\lambda} \in \Lambda_n} A^{\lambda_0}, \; n \geq 0. \tag{17}$$

$\boldsymbol{\lambda} = (\lambda_0, \dots, \lambda_n) \in \Lambda_n$. For $n < 0$, $L_n = 0$. If $i_{\boldsymbol{\lambda}} \colon A^{\lambda_0} \to L_n$ denotes the natural injection of the summand A^{λ_0}, which belongs to the index $\boldsymbol{\lambda}$, then the boundary operator $\partial \colon L_n \to L_{n-1}$, $n \geq 1$, is given by

$$\partial(i_{\boldsymbol{\lambda}}(y)) = i_{d^0 \boldsymbol{\lambda}}(a^{\lambda_0 \lambda_1}(y)) + \sum_{j=1}^{n} (-1)^j i_{d^j \boldsymbol{\lambda}}(y), \; y \in A^{\lambda_0}. \tag{18}$$

The homomorphism η is given by

$$\eta(i_{\lambda_0}(y)) = [y], \tag{19}$$

where $[y]$ denotes the class of $y \in A^{\lambda_0}$ in $\operatorname{colim} \boldsymbol{A}$. A straightforward computation shows that $\partial^2 = 0$ and $\eta\partial = 0$, so that (16) is indeed an (augmented) chain complex.

LEMMA 14.6. *If* (14) *is the standard projective resolution of* $\boldsymbol{\Delta}(\Lambda)$, *then the (augmented) chain complex* (15) *is isomorphic to the (augmented) chain complex* $(\boldsymbol{L}, \eta)$ *of* (16).

Proof. We need isomorphisms $\Phi_n\colon \boldsymbol{A}\otimes \boldsymbol{P}^n \to L_n$, $n \geq 0$, and an isomorphism $\Phi\colon \boldsymbol{A}\otimes\boldsymbol{\Delta} \to \operatorname{colim}\boldsymbol{A}$, such that the following diagrams commute.

$$\begin{array}{ccccccc}
\boldsymbol{A}\otimes\boldsymbol{\Delta} & \xleftarrow{1\otimes \boldsymbol{e}} & \boldsymbol{A}\otimes\boldsymbol{P}_0 & \qquad & \boldsymbol{A}\otimes\boldsymbol{P}_{n-1} & \xleftarrow{1\otimes \boldsymbol{d}} & \boldsymbol{A}\otimes\boldsymbol{P}_n \\
\Phi\downarrow & & \downarrow\Phi_0 & & \downarrow\Phi_{n-1} & & \downarrow\Phi_n \\
\operatorname{colim}\boldsymbol{A} & \xleftarrow[\eta]{} & L_0 & & L_{n-1} & \xleftarrow[\partial]{} & L_n\,.
\end{array} \tag{20}$$

Let $\varphi_n\colon \oplus_\lambda(A^\lambda\otimes P_{n\lambda}) \to L_n$, $n\geq 0$, be the homomorphism, which maps elements of the form $x\otimes\langle\boldsymbol{\lambda}\rangle \in A^\lambda\otimes P_{n\lambda}$ to $i_{\boldsymbol{\lambda}} a^{\lambda\lambda_0}(x)\in L_n$, where $x\in A^\lambda$ and $\langle\boldsymbol{\lambda}\rangle = \langle\lambda_0,\ldots,\lambda_n\rangle$ is a generator of $P_{n\lambda}$ (see 12.2). The homomorphism φ_n induces a homomorphism $\Phi_n\colon \boldsymbol{A}\otimes\boldsymbol{P}_n \to L_n$. Indeed, if $x\in A^\lambda$ and $\langle\boldsymbol{\lambda}\rangle\in P_{n\lambda'}$, and thus $\lambda'\leq\lambda_0$, then, for $\lambda\leq\lambda'$,

$$\varphi_n(x\otimes i_{\lambda\lambda'}\langle\boldsymbol{\lambda}\rangle) = i_{\boldsymbol{\lambda}} a^{\lambda\lambda_0}(x), \tag{21}$$

because $\lambda\leq\lambda_0$ and $i_{\lambda\lambda'}\langle\boldsymbol{\lambda}\rangle = \langle\boldsymbol{\lambda}\rangle$. However, one also has

$$\varphi_n(a^{\lambda\lambda'}(x)\otimes\langle\boldsymbol{\lambda}\rangle) = i_{\boldsymbol{\lambda}}(a^{\lambda'\lambda_0}a^{\lambda\lambda'}(x)) = i_{\boldsymbol{\lambda}} a^{\lambda\lambda_0}(x). \tag{22}$$

To show that Φ_n is an isomorphism, we consider the homomorphism $\psi_n\colon L_n \to \oplus_\lambda(A^\lambda\otimes P_{n\lambda})$, given by

$$\psi_n(i_{\boldsymbol{\lambda}}(y)) = y\otimes\langle\boldsymbol{\lambda}\rangle \in A^{\lambda_0}\otimes P_{n\lambda_0} \subseteq \oplus_\lambda(A^\lambda\otimes P_{n\lambda}), \tag{23}$$

where $y\in A^{\lambda_0}$. The desired homomorphism Ψ_n is obtained by composing ψ_n with the quotient homomorphism $\oplus_\lambda(A^\lambda\otimes P_{n\lambda}) \to \boldsymbol{A}\otimes\boldsymbol{P}_n$.

The composition $\Psi_n\Phi_n = \mathrm{id}$, because, for $x\in A^\lambda$,

$$\begin{gathered}\Psi_n\Phi_n[x\otimes\langle\boldsymbol{\lambda}\rangle] = \Psi_n(i_{\boldsymbol{\lambda}} a^{\lambda\lambda_0}(x)) = \\ [a^{\lambda\lambda_0}(x)\otimes\langle\boldsymbol{\lambda}\rangle] = [x\otimes\langle\boldsymbol{\lambda}\rangle].\end{gathered} \tag{24}$$

One also has $\Phi_n\Psi_n = \mathrm{id}$. Indeed, for $y\in A^{\lambda_0}$,

$$\begin{gathered}\Phi_n\Psi_n(i_{\boldsymbol{\lambda}}(y)) = \Phi_n[y\otimes\langle\boldsymbol{\lambda}\rangle] = \\ \varphi_n(y\otimes\langle\boldsymbol{\lambda}\rangle) = i_{\boldsymbol{\lambda}}(y).\end{gathered} \tag{25}$$

In order to prove commutativity of the second diagram in (20), note that

$$\begin{gathered}\partial\Phi_n[x\otimes\langle\boldsymbol{\lambda}\rangle] = \partial(i_{\boldsymbol{\lambda}} a^{\lambda\lambda_0}(x)) = \\ i_{d^0\langle\boldsymbol{\lambda}\rangle}(a^{\lambda\lambda_1}(x)) + \textstyle\sum_{j=1}^n (-1)^j i_{d^j\boldsymbol{\lambda}}(a^{\lambda\lambda_0}(x)).\end{gathered} \tag{26}$$

On the other hand,

$$\Phi_{n-1}(1\otimes\boldsymbol{d})[x\otimes\langle\boldsymbol{\lambda}\rangle] = \sum_{j=0}^n (-1)^j\Phi_{n-1}[x\otimes d^j\boldsymbol{\lambda}], \tag{27}$$

which equals the right side of (26). The isomorphism $\Phi\colon \boldsymbol{A} \otimes \boldsymbol{\Delta} \to \operatorname{colim}\boldsymbol{A}$ was already defined in Example 14.3. It is readily seen that it makes the first diagram in (20) commutative. □

LEMMA 14.7. *The sequence $(\boldsymbol{L}, \eta)$ is exact.*

Proof. We will represent the augmented chain complex $(\boldsymbol{L}, \eta)$ as the colimit of a direct system $((\boldsymbol{L}, \eta)^\mu, s^{\mu\mu'}, \Lambda)$ of augmented chain complexes $(\boldsymbol{L}, \eta)^\mu$. We will then show that each $(\boldsymbol{L}, \eta)^\mu$ is an exact sequence. By Lemma 14.2, this will imply that also $(\boldsymbol{L}, \eta)$ is exact.

To define $\boldsymbol{L}^\mu = (L_n^\mu, \partial)$, for $\mu \in \Lambda$, put

$$L_n^\mu = \bigoplus_{\mu \geq \boldsymbol{\lambda} \in \Lambda_n} A^{\lambda_0}, \tag{28}$$

for $n \geq 0$ and put $L_n^\mu = 0$, for $n < 0$. Take for $\partial\colon L_n^\mu \to L_{n-1}^\mu$ the restriction of $\partial\colon L_n \to L_{n-1}$ to L_n^μ. Define $\eta^\mu\colon L_0^\mu \to A^\mu$, by putting $\eta^\mu(i_{\lambda_0}(x)) = a^{\lambda_0\mu}(x)$, for $x \in A^{\lambda_0}$. For $\mu \leq \mu'$, define the chain mapping $s^{\mu\mu'}\colon \boldsymbol{L}^\mu \to \boldsymbol{L}^{\mu'}$, using the natural inclusions $L_n^\mu \to L_n^{\mu'}$. It is extended to augmented complexes by the homomorphism $a^{\mu\mu'}\colon A^\mu \to A^{\mu'}$. Moreover, the natural inclusions $L_n^\mu \to L_n$ define a chain mapping $s^\mu\colon \boldsymbol{L}^\mu \to \boldsymbol{L}$, which extends to the augmented complexes by the natural homomorphism $A^\mu \to \operatorname{colim}\boldsymbol{A}$. It is straightforward to verify that in this way $(\boldsymbol{L}, \eta)$ is represented as the colimit of the described direct system of augmented chain complexes $(\boldsymbol{L}, \eta)^\mu$.

To prove exactness of $(\boldsymbol{L}, \eta)^\mu$, it suffices to show that this chain complex is contractible, i.e., there exist homomorphisms $\varepsilon^\mu\colon A^\mu \to L_0^\mu$ and $c_n^\mu\colon L_n^\mu \to L_{n+1}^\mu$, $n \geq 0$, such that

$$\eta^\mu \varepsilon^\mu = \mathrm{id}, \;\; \varepsilon^\mu \eta^\mu - \partial c_0^\mu = \mathrm{id}. \tag{29}$$

$$c_{n-1}^\mu \partial - \partial c_n^\mu = (-1)^n \mathrm{id}. \tag{30}$$

Such homomorphisms are obtained by putting

$$\varepsilon^\mu(x) = i_\mu(x), \;\; x \in A^\mu, \tag{31}$$

$$c_n^\mu(i_{\boldsymbol{\lambda}}(x)) = i_{\boldsymbol{\lambda}\mu}(x), \;\; \boldsymbol{\lambda} \leq \mu. \tag{32}$$

To verify (30), first note that

$$\begin{aligned} &\partial c_n^\mu(i_{\boldsymbol{\lambda}}(x)) = i_{d^0(\boldsymbol{\lambda}\mu)}\, a^{\lambda_0\lambda_1}(x) + \\ &\textstyle\sum_{j=1}^n (-1)^j i_{d^j(\boldsymbol{\lambda}\mu)}(x) + (-1)^{n+1} i_{d^{n+1}(\boldsymbol{\lambda}\mu)}(x). \end{aligned} \tag{33}$$

Then take into account the fact that $d^j(\boldsymbol{\lambda}\mu) = (d^j\boldsymbol{\lambda})\mu$, for $0 \leq j \leq n$, while $d^{n+1}(\boldsymbol{\lambda}\mu) = \boldsymbol{\lambda}$. Therefore, the right side of (33) assumes the value

$$c_{n-1}^\mu \partial(i_{\boldsymbol{\lambda}}(x)) + (-1)^{n+1} i_{\boldsymbol{\lambda}}(x). \tag{34}$$

Formulae (29) are also easily verified. □

REMARK 14.8. There is another proof of Lemma 14.5. First note that the results of 11.2 also hold for direct systems. Hence, the category Mod_Λ has enough injective and enough projective objects. For a fixed inverse system $\boldsymbol{B}$, the (covariant) functor $\otimes\boldsymbol{B}\colon \mathsf{Mod}_\Lambda \to \mathsf{Mod}$ is an additive right exact functor. Therefore, one can define its left derived functors $\mathrm{Tor}^n(\,.\,,\boldsymbol{B})$ using projective resolutions of direct systems $\boldsymbol{A}$, say, $\boldsymbol{0} \leftarrow \boldsymbol{A} \leftarrow \boldsymbol{Q}_0 \leftarrow \boldsymbol{Q}_1 \leftarrow \dots$. By definition, $\mathrm{Tor}^n(\,.\,,\boldsymbol{B})$ equals the n-th homology module of the chain complex $0 \leftarrow \boldsymbol{Q}_0 \otimes \boldsymbol{B} \leftarrow \boldsymbol{Q}_1 \otimes \boldsymbol{B} \leftarrow \dots$. Moreover, morphisms $\boldsymbol{A} \to \boldsymbol{A}''$ induce morphisms $\mathrm{Tor}^n(\boldsymbol{A}, \boldsymbol{B}) \to \mathrm{Tor}^n(\boldsymbol{A}'', \boldsymbol{B})$ and one also obtains connecting homomorphisms $\theta^n_E\colon \mathrm{Tor}^n(\boldsymbol{A}'', \boldsymbol{B}) \to \mathrm{Tor}^{n-1}(\boldsymbol{A}', \boldsymbol{B})$, associated with exact sequences $E = (\boldsymbol{0} \to \boldsymbol{A}' \to \boldsymbol{A} \to \boldsymbol{A}'' \to \boldsymbol{0})$. Finally, a projective system $\boldsymbol{A}$ admits a projective resolution $\boldsymbol{0} \leftarrow \boldsymbol{A} \leftarrow \boldsymbol{A} \leftarrow \boldsymbol{0}$, which implies $\mathrm{Tor}^n(\boldsymbol{A}, \boldsymbol{B}) = 0$, for $n \geq 1$. Consequently, $\mathrm{Tor}^n(\,.\,,\boldsymbol{B})$ is a universal connected sequence of functors. Now put $\boldsymbol{B} = \boldsymbol{\Delta}$ and note that $\mathrm{Tor}^0(\,.\,,\boldsymbol{\Delta}) = \otimes\boldsymbol{\Delta}$ is naturally equivalent to the functor colim (see Example 14.3). Since colim is an exact functor (Lemma 14.2), it is clear that the sequence of functors $T^0 = \mathrm{colim}$, $T^n = 0$, $n \geq 1$, with trivial connecting homomorphisms $\omega^n_E = 0$ also defines a universal connected sequence of functors. Therefore, by the uniqueness theorem (see Remark 11.36), one concludes that $\mathrm{Tor}^n(\boldsymbol{A}, \boldsymbol{\Delta}) = 0$, for $n \geq 1$. In analogy with Ext^n (see 12.1), there is another way of obtaining the modules $\mathrm{Tor}^n(\boldsymbol{A}, \boldsymbol{\Delta})$. One considers the functor $\boldsymbol{A}\otimes$ and any projective resolution $\boldsymbol{0} \leftarrow \boldsymbol{\Delta} \leftarrow \boldsymbol{P}_0 \leftarrow \boldsymbol{P}_1 \leftarrow \dots$ of $\boldsymbol{\Delta}$. Then $0 \leftarrow \boldsymbol{A} \otimes \boldsymbol{P}_0 \leftarrow \boldsymbol{A} \otimes \boldsymbol{P}_1 \leftarrow \dots$ is a chain complex and $\mathrm{Tor}^n(\boldsymbol{A}, \boldsymbol{\Delta})$ is its n-th homology module. Since $\mathrm{Tor}^n(\boldsymbol{A}, \boldsymbol{\Delta}) = 0$, for $n \geq 1$, one concludes that the sequence (15) is indeed exact at places $n \geq 1$.

14.2 The cofinality theorem for $\lim^n$

Let Λ and M be directed sets and let $\phi\colon M \to \Lambda$ be an increasing function. Then, with every inverse system of R-modules $\boldsymbol{Y} = (Y_\lambda, q_{\lambda\lambda'}, \Lambda)$, indexed by Λ, one associates an inverse system of R-modules $\phi^*(\boldsymbol{Y}) = \boldsymbol{Y}^*$, indexed by M. By definition, $\boldsymbol{Y}^* = (Y^*_\mu, q^*_{\mu\mu'}, M)$, where $Y^*_\mu = Y_{\phi(\mu)}$, $q^*_{\mu\mu'} = q_{\phi(\mu)\phi(\mu')}$. Moreover, every morphism in $\mathrm{inv}\text{-}\mathsf{Mod}$ $\boldsymbol{g} = (g_\lambda)\colon \boldsymbol{Y} \to \boldsymbol{Y}' = (Y'_\lambda, q'_{\lambda\lambda'}, \Lambda)$ induces a morphism $\phi^*(\boldsymbol{g}) = \boldsymbol{g}^* = (g^*_\mu)\colon \phi^*(\boldsymbol{Y}) \to \phi^*(\boldsymbol{Y}')$, where $g^*_\mu = g_{\phi(\mu)}$. Clearly, ϕ^* is a functor $\mathsf{Mod}^\Lambda \to \mathsf{Mod}^M$.

A function $\phi\colon M \to \Lambda$ is said to be *cofinal*, provided $\phi(M)$ is cofinal in Λ, i.e., every $\lambda \in \Lambda$ admits a $\mu \in M$ such that $\lambda \leq \phi(\mu)$. The main purpose of this section is to prove the following *cofinality theorem.*

THEOREM 14.9. *Every increasing cofinal function $\phi\colon M \to \Lambda$ induces an isomorphism of derived limits,*

$$\lim{}^n \boldsymbol{Y} \approx \lim{}^n \phi^*(\boldsymbol{Y}),\ n \geq 0. \tag{1}$$

For $n = 0$, the proof is elementary and we give it in a separate lemma, before embarking on the rather non-trivial general case.

LEMMA 14.10. *An increasing cofinal function $\phi: M \to \Lambda$ induces an isomorphism of inverse limits,*

$$\lim \boldsymbol{Y} \approx \lim \phi^*(\boldsymbol{Y}). \tag{2}$$

Proof. First note that ϕ induces a morphism $\boldsymbol{\Phi} = (\phi, \phi_\mu): \boldsymbol{Y} \to \phi^*(\boldsymbol{Y})$ in inv - Mod, where $\phi_\mu = \mathrm{id}: Y_{\phi(\mu)} \to Y^*_\mu = Y_{\phi(\mu)}$. The morphism $\boldsymbol{\Phi}$ induces a homomorphism $f: Y \to Y^*$ between the limits $Y = \lim \boldsymbol{Y}$ and $Y^* = \lim \phi^*(\boldsymbol{Y})$. By definition, f assigns to $y = (y_\lambda) \in Y$ the element $f(y) = x = (x_\mu) \in Y^*$, where $x_\mu = y_{\phi(\mu)} \in Y^*_\mu$. If $\mu \le \mu'$, one has $q^*_{\mu\mu'}(x_{\mu'}) = q_{\phi(\mu)\phi(\mu')}(y_{\phi(\mu')}) = y_{\phi(\mu)} = x_\mu$, and thus, $x \in Y^*$.

To show that f is a monomorphism, consider any $y = (y_\lambda) \in Y$ such that $f(y) = x = 0$, i.e., $x_\mu = 0$, for all $\mu \in M$. Then $y_{\phi(\mu)} = x_\mu = 0$, for all μ. Since ϕ is cofinal, every $\lambda \in \Lambda$ admits a $\mu \in M$ such that $\lambda \le \phi(\mu)$. Consequently, $y_\lambda = q_{\lambda\phi(\mu)}(y_{\phi(\mu)}) = 0$, for all $\lambda \in \Lambda$, i.e., $y = 0$.

To show that f is also an epimorphism, consider any $x = (x_\mu) \in Y^*$. For every $\lambda \in \Lambda$, choose a $\mu \in M$ such that $\lambda \le \phi(\mu)$, and put $y_\lambda = q_{\lambda\phi(\mu)}(x_\mu)$. Notice that the choice of a $\mu' \ge \mu$ yields the same value for y_λ. Indeed, $x_\mu = q^*_{\mu\mu'}(x_{\mu'}) = q_{\phi(\mu)\phi(\mu')}(x_{\mu'})$ and therefore, $q_{\lambda\phi(\mu)}(x_\mu) = q_{\lambda\phi(\mu')}(x_{\mu'})$. By directedness, the same conclusion also holds for an arbitrary μ', satisfying $\lambda \le \phi(\mu')$. Next one verifies that $y = (y_\lambda)$ is a point of Y. Indeed, if $\lambda \le \lambda'$, one can choose a μ' such that $\phi(\mu') \ge \lambda' \ge \lambda$. Then $q_{\lambda\lambda'}(y_{\lambda'}) = q_{\lambda\lambda'}q_{\lambda'\phi(\mu')}(x_{\mu'}) = q_{\lambda\phi(\mu')}(x_{\mu'}) = y_\lambda$. Finally, $f(y) = x$, because $f_{\phi(\mu)}(y_{\phi(\mu)}) = y_{\phi(\mu)} = x_\mu$. □

In order to prove Theorem 14.9, for n arbitrary, we shall exhibit a functor $\psi: \mathsf{Mod}^M \to \mathsf{Mod}^\Lambda$, which is left adjoint to the functor $\phi^*: \mathsf{Mod}^\Lambda \to \mathsf{Mod}^M$. This means that there exist natural isomorphisms of modules

$$\eta_{\boldsymbol{XY}}: \mathrm{Hom}(\psi(\boldsymbol{X}), \boldsymbol{Y}) \to \mathrm{Hom}(\boldsymbol{X}, \phi^*(\boldsymbol{Y})). \tag{3}$$

Consequently, if $\boldsymbol{f} = (f_\mu) : \boldsymbol{X}' \to \boldsymbol{X}$ is a morphism of Mod^M, then the following diagram commutes,

$$\begin{array}{ccc} \mathrm{Hom}(\psi(\boldsymbol{X}), \boldsymbol{Y}) & \xrightarrow{\eta_{\boldsymbol{XY}}} & \mathrm{Hom}(\boldsymbol{X}, \phi^*(\boldsymbol{Y})) \\ \downarrow & & \downarrow \\ \mathrm{Hom}(\psi(\boldsymbol{X}'), \boldsymbol{Y}) & \xrightarrow[\eta_{\boldsymbol{X}'\boldsymbol{Y}}]{} & \mathrm{Hom}(\boldsymbol{X}', \phi^*(\boldsymbol{Y})) \; , \end{array} \tag{4}$$

where the vertical arrows are $\mathrm{Hom}(\psi(\boldsymbol{f}), 1)$ and $\mathrm{Hom}(\boldsymbol{f}, 1)$, respectively. Moreover, if $\boldsymbol{f} = (f_\mu) : \boldsymbol{Y} \to \boldsymbol{Y}'$ is a morphism of Mod^Λ, then the following diagram commutes,

$$\begin{array}{ccc} \mathrm{Hom}(\psi(\boldsymbol{X}),\boldsymbol{Y}) & \xrightarrow{\eta_{\boldsymbol{XY}}} & \mathrm{Hom}(\boldsymbol{X},\phi^*(\boldsymbol{Y})) \\ \downarrow & & \downarrow \\ \mathrm{Hom}(\psi(\boldsymbol{X}),\boldsymbol{Y}') & \xrightarrow[\eta_{\boldsymbol{XY}'}]{} & \mathrm{Hom}(\boldsymbol{X},\phi^*(\boldsymbol{Y}')) \end{array}, \tag{5}$$

where the vertical arrows are $\mathrm{Hom}(1,\boldsymbol{f})$ and $\mathrm{Hom}(1,\phi^*(\boldsymbol{f}))$, respectively.

In order to define ψ, consider the direct systems $\boldsymbol{\Gamma}_\lambda = (\Gamma_\lambda^\mu, \gamma_\lambda^{\mu\mu'}, M)$, $\lambda \in \Lambda$, where

$$\Gamma_\lambda^\mu = \begin{cases} R, & \lambda \le \phi(\mu), \\ 0, & \lambda \not\le \phi(\mu), \end{cases} \tag{6}$$

and $\gamma_\lambda^{\mu\mu'}\colon \Gamma_\lambda^\mu \to \Gamma_\lambda^{\mu'}$, $\mu \le \mu'$, are natural inclusions, i.e.,

$$\gamma_\lambda^{\mu\mu'} = \begin{cases} \mathrm{id}, & \lambda \le \phi(\mu), \\ 0, & \lambda \not\le \phi(\mu). \end{cases} \tag{7}$$

If $\lambda \le \lambda'$, let $\boldsymbol{\gamma}_{\lambda\lambda'}\colon \boldsymbol{\Gamma}_{\lambda'} \to \boldsymbol{\Gamma}_\lambda$ be the morphism, defined by the natural inclusions $\gamma_{\lambda\lambda'}^\mu\colon \Gamma_{\lambda'}^\mu \to \Gamma_\lambda^\mu$. One obtains in this way an inverse system $\boldsymbol{\Gamma} = (\boldsymbol{\Gamma}_\lambda, \boldsymbol{\gamma}_{\lambda\lambda'}, \Lambda)$ of direct systems $\boldsymbol{\Gamma}_\lambda$.

For every $\lambda \in \Lambda$ and $\boldsymbol{X} = (X_\mu, p_{\mu\mu'}, M)$ from Mod^M, $\boldsymbol{\Gamma}_\lambda \otimes \boldsymbol{X}$ is the R-module

$$\boldsymbol{\Gamma}_\lambda \otimes \boldsymbol{X} = \Big(\bigoplus_{\lambda \le \phi(\mu)} (R \otimes X_\mu) \Big) / L, \tag{8}$$

where the submodule L has been defined in (14.1.10). We define $\psi(\boldsymbol{X})$ as the inverse system

$$\psi(\boldsymbol{X}) = (\boldsymbol{\Gamma}_\lambda \otimes \boldsymbol{X}, \boldsymbol{\gamma}_{\lambda\lambda'} \otimes 1, \Lambda). \tag{9}$$

Every morphism $\boldsymbol{f} = (f_\mu)\colon \boldsymbol{X}' \to \boldsymbol{X}$ induces a morphism $\psi(\boldsymbol{f})\colon \psi(\boldsymbol{X}') \to \psi(\boldsymbol{X})$, which consists of homomorphisms $1 \otimes \boldsymbol{f}\colon \boldsymbol{\Gamma}_\lambda \otimes \boldsymbol{X}' \to \boldsymbol{\Gamma}_\lambda \otimes \boldsymbol{X}$. It is readily seen that ψ is indeed a functor.

LEMMA 14.11. *The functor* $\psi\colon \mathsf{Mod}^M \to \mathsf{Mod}^\Lambda$ *is left adjoint to the functor* $\phi^*\colon \mathsf{Mod}^\Lambda \to \mathsf{Mod}^M$.

Proof. For a morphism $\boldsymbol{g}\colon \psi(\boldsymbol{X}) \to \boldsymbol{Y}$, which consists of homomorphisms $g_\lambda\colon \boldsymbol{\Gamma}_\lambda \otimes \boldsymbol{X} \to Y_\lambda$, $\lambda \in \Lambda$, we define the morphism $\eta_{\boldsymbol{XY}}(\boldsymbol{g}) = \boldsymbol{h} = (h_\mu)\colon \boldsymbol{X} \to \phi^*(\boldsymbol{Y})$ as follows. For $\mu \in M$, one has $\Gamma_{\phi(\mu)}^\mu = R$, and therefore, every $x \in X_\mu$ determines elements $1 \otimes x \in \Gamma_{\phi(\mu)}^\mu \otimes X_\mu$ and $[1 \otimes x] \in \boldsymbol{\Gamma}_{\phi(\mu)} \otimes \boldsymbol{X}$, where $[1 \otimes x]$ is the class of $1 \otimes x$. Put

$$h_\mu(x) = g_{\phi(\mu)}[1 \otimes x] \in Y_{\phi(\mu)} = Y_\mu^*. \tag{10}$$

To see that $\boldsymbol{h}$ is a morphism, we must show that $h_\mu p_{\mu\mu'} = q_{\phi(\mu)\phi(\mu')} h_{\mu'}$, for $\mu \le \mu'$. Indeed, for $x' \in X_{\mu'}$, we have

$$h_\mu p_{\mu\mu'}(x') = g_{\phi(\mu)}[1 \otimes p_{\mu\mu'}(x')], \tag{11}$$

$$\begin{aligned} q_{\phi(\mu)\phi(\mu')}h_{\mu'}(x') &= q_{\phi(\mu)\phi(\mu')}g_{\phi(\mu')}[1\otimes x'] \\ &= g_{\phi(\mu)}(\gamma_{\phi(\mu)\phi(\mu')}\otimes 1)[1\otimes x']. \end{aligned} \tag{12}$$

However, by the definition of L, $(\gamma_{\phi(\mu)\phi(\mu')}\otimes 1)[1\otimes x'] = [1\otimes x'] = [1\otimes p_{\mu\mu'}(x')]$. Therefore, the left sides of (11) and (12) coincide. An inspection of formula (10) immediately shows that $\eta_{\boldsymbol{XY}}(\boldsymbol{g}+\boldsymbol{g}') = \eta_{\boldsymbol{XY}}(\boldsymbol{g}) + \eta_{\boldsymbol{XY}}(\boldsymbol{g}')$ and $\eta_{\boldsymbol{XY}}(r\boldsymbol{g}) = r\eta_{\boldsymbol{XY}}(\boldsymbol{g})$, which means that $\eta_{\boldsymbol{XY}}$ is a homomorphism of modules.

The inverse $\varepsilon_{\boldsymbol{XY}}$ of $\eta_{\boldsymbol{XY}}$ associates with a morphism $\boldsymbol{h} = (h_\mu)\colon \boldsymbol{X}\to \phi^*(\boldsymbol{Y})$ a morphism $\varepsilon_{\boldsymbol{XY}}(\boldsymbol{h}) = \boldsymbol{g} = (g_\lambda)\colon \psi(\boldsymbol{X})\to \boldsymbol{Y}$, defined as follows. For a given $\lambda\in\Lambda$ and $\mu\in M$, let $g^\mu_\lambda\colon \Gamma^\mu_\lambda\otimes X_\mu\to Y_\lambda$ be the homomorphism, determined by

$$g^\mu_\lambda(r\otimes x) = \begin{cases} rq_{\lambda\phi(\mu)}h_\mu(x), & \lambda\le\phi(\mu),\\ 0, & \lambda\not\le\phi(\mu). \end{cases} \tag{13}$$

Clearly, the homomorphisms g^μ_λ define a homomorphism $\oplus_\mu g^\mu_\lambda\colon \oplus_\mu(\Gamma^\mu_\lambda\otimes X_\mu)\to Y_\lambda$, which maps

$$\gamma^{\mu\mu'}_\lambda(r)\otimes x' - r\otimes p_{\mu\mu'}(x') \tag{14}$$

to

$$g^{\mu'}_\lambda(\gamma^{\mu\mu'}_\lambda(r)\otimes x') - g^\mu_\lambda(r\otimes p_{\mu\mu'}(x')). \tag{15}$$

If $\lambda\le\phi(\mu)$, (15) equals $rq_{\lambda\phi(\mu')}h_{\mu'}(x') - rq_{\lambda\phi(\mu)}h_\mu p_{\mu\mu'}(x') = 0$. If $\lambda\not\le\phi(\mu)$, (15) also vanishes, because $r\in\Gamma^\mu_\lambda = 0$. Therefore, $\oplus g^\mu_\lambda$ induces a homomorphism $g_\lambda\colon \boldsymbol{\Gamma}_\lambda\otimes\boldsymbol{X}\to Y_\lambda$. If $\lambda\le\lambda'$, then $g_\lambda(\boldsymbol{\gamma}_{\lambda\lambda'}\otimes 1) = q_{\lambda\lambda'}g_{\lambda'}$. Indeed, consider a generator $[r\otimes x]\in\boldsymbol{\Gamma}_{\lambda'}\otimes\boldsymbol{X}$, where $r\in R$, $x\in X_\mu$ and $\lambda'\le\phi(\mu)$. Then $g_{\lambda'}[r\otimes x] = g^\mu_{\lambda'}(r\otimes x) = rq_{\lambda'\phi(\mu)}h_\mu(x)$ and thus, $q_{\lambda\lambda'}g_{\lambda'}[r\otimes x] = rq_{\lambda\phi(\mu)}h_\mu(x)$. On the other hand, since also $\lambda\le\phi(\mu)$, one has $(\gamma_{\lambda\lambda'}\otimes 1)[r\otimes x] = [\gamma^\mu_{\lambda\lambda'}(r)\otimes x] = [r\otimes x]$ and thus also, $g_\lambda(\gamma_{\lambda\lambda'}\otimes 1)[r\otimes x] = g_\lambda[r\otimes x] = g^\mu_\lambda(r\otimes x) = rq_{\lambda\phi(\mu)}h_\mu(x)$.

We will now show that the two compositions coincide with the identities. If $\boldsymbol{h} = \eta_{\boldsymbol{XY}}(\boldsymbol{g})$ and $\boldsymbol{g}' = \varepsilon_{\boldsymbol{XY}}(\boldsymbol{h})$, then, for $\lambda\le\phi(\mu)$, one has $g'_\lambda[r\otimes x] = rq_{\lambda\phi(\mu)}h_\mu(x) = q_{\lambda\phi(\mu)}g_{\phi(\mu)}[r\otimes x] = g_\lambda(\gamma_{\lambda\phi(\mu)}\otimes 1)[r\otimes x] = g_\lambda[r\otimes x], r\in R, x\in X_\mu$. Hence, $g'_\lambda = g_\lambda$ and thus, $\boldsymbol{g}' = \boldsymbol{g}$. Similarly, if $\boldsymbol{g} = \varepsilon_{\boldsymbol{XY}}(\boldsymbol{h})$ and $\boldsymbol{h}' = \eta_{\boldsymbol{XY}}(\boldsymbol{g})$, then, $h'_\mu(x) = g_{\phi(\mu)}[1\otimes x] = h_\mu(x)$, so that $h'_\mu = h_\mu$ and thus, $\boldsymbol{h}' = \boldsymbol{h}$.

To verify commutativity of (4), note that $\mathrm{Hom}(\psi(\boldsymbol{f}),1)$ maps $\boldsymbol{g} = (g_\lambda)\in\mathrm{Hom}(\psi(\boldsymbol{X}),\boldsymbol{Y})$ to $\boldsymbol{g}' = (g'_\lambda) = \boldsymbol{g}\psi(\boldsymbol{f})\in\mathrm{Hom}(\psi(\boldsymbol{X}'),\boldsymbol{Y})$, where $g'_\lambda = g_\lambda(1\otimes\boldsymbol{f})\colon\boldsymbol{\Gamma}_\lambda\otimes\boldsymbol{X}'\to Y_\lambda$. Furthermore, $\eta_{\boldsymbol{X}'\boldsymbol{Y}}$ maps $\boldsymbol{g}'$ to $\boldsymbol{h}' = (h'_\mu)\in\mathrm{Hom}(\boldsymbol{X}',\phi^*(\boldsymbol{Y}))$, where $h'_\mu(x') = g'_{\phi(\mu)}[1\otimes x']$, for $x'\in X'_\mu$. Consequently,

$$h'_\mu(x') = g_{\phi(\mu)}(1\otimes\boldsymbol{f})[1\otimes x'] = g_{\phi(\mu)}[1\otimes f_\mu(x')]. \tag{16}$$

On the other hand, $\eta_{\boldsymbol{XY}}$ maps $\boldsymbol{g}$ to $\boldsymbol{h} = (h_\mu)\in\mathrm{Hom}(\boldsymbol{X},\phi^*(\boldsymbol{Y}))$, given by (10), and $\mathrm{Hom}(\boldsymbol{f},1)$ maps $\boldsymbol{h}$ to $\boldsymbol{h}'' = (h''_\mu) = \boldsymbol{hf}$, where

$$h''_\mu(x') = h_\mu f_\mu(x') = g_{\phi(\mu)}[1\otimes f_\mu(x')]. \tag{17}$$

Now comparison of (16) with (17) shows that $h'_\mu = h''_\mu$ and thus, $\boldsymbol{h}' = \boldsymbol{h}''$, which establishes the assertion.

To verify commutativity of (5), note that $\mathrm{Hom}(1, \boldsymbol{f})$ maps $\boldsymbol{g} = (g_\lambda) \in \mathrm{Hom}(\psi(\boldsymbol{X}), \boldsymbol{Y})$ to $\boldsymbol{g}' = (g'_\lambda) = \boldsymbol{f}\boldsymbol{g} \in \mathrm{Hom}(\psi(\boldsymbol{X}), \boldsymbol{Y}')$, where $g'_\lambda = f_\lambda g_\lambda$. Therefore, $\eta_{\boldsymbol{X}\boldsymbol{Y}'}$ maps $\boldsymbol{g}'$ to $\boldsymbol{h}' = (h'_\mu) \in \mathrm{Hom}(\boldsymbol{X}, \phi^*(\boldsymbol{Y}'))$, where $h'_\mu(x) = f_{\phi(\mu)} g_{\phi(\mu)}[1 \otimes x]$, for $x \in X_\mu$. On the other hand, $\eta_{\boldsymbol{X}\boldsymbol{Y}}$ maps $\boldsymbol{g}$ to $\boldsymbol{h} = (h_\mu) \in \mathrm{Hom}(\boldsymbol{X}, \phi^*(\boldsymbol{Y}))$, given by (10). Therefore, $\mathrm{Hom}(1, \phi^*(\boldsymbol{f}))$ maps $\boldsymbol{h}$ to $\boldsymbol{h}'' = (h''_\mu)$, where, $h''_\mu(x) = (\phi^*(\boldsymbol{f}))_\mu h_\mu(x) = f_{\phi(\mu)} g_{\phi(\mu)}[1 \otimes x]$. □

The following lemma is an easy consequence of Lemma 14.11.

LEMMA 14.12. *If $\boldsymbol{X}$ is projective in Mod^M, then $\psi(\boldsymbol{X})$ is projective in Mod^Λ.*

Proof. Let $\boldsymbol{h}: \boldsymbol{A} \to \boldsymbol{B}$ be an epimorphism in Mod^Λ. We must show that every morphism $\boldsymbol{f}: \psi(\boldsymbol{X}) \to \boldsymbol{B}$ admits a morphism $\boldsymbol{g}: \psi(\boldsymbol{X}) \to \boldsymbol{A}$ such that $\boldsymbol{h}\boldsymbol{g} = \boldsymbol{f}$. First note that $\phi^*(\boldsymbol{h}) = \boldsymbol{h}^*: \phi^*(\boldsymbol{A}) \to \phi^*(\boldsymbol{B})$ is also an epimorphism. Indeed, for every $\mu \in M$, $h^*_\mu: A^*_\mu \to B^*_\mu$ equals $h_{\phi(\mu)}: A_{\phi(\mu)} \to B_{\phi(\mu)}$, which is an epimorphism, by Lemma 11.2. Now the same theorem shows that $\phi^*(\boldsymbol{h})$ is also an epimorphism. Since $\boldsymbol{X}$ was assumed projective, there exists a morphism $\boldsymbol{k}: \boldsymbol{X} \to \phi^*(\boldsymbol{A})$, such that $\phi^*(\boldsymbol{h})\boldsymbol{k} = \eta_{\boldsymbol{X}\boldsymbol{B}}(\boldsymbol{f})$. Applying Lemma 14.11, we conclude that there exists a morphism $\boldsymbol{g}: \psi(\boldsymbol{X}) \to \boldsymbol{A}$, such that $\eta_{\boldsymbol{X}\boldsymbol{A}}(\boldsymbol{g}) = \boldsymbol{k}$. Using the commutativity of (5) (for $\boldsymbol{Y} = \boldsymbol{A}, \boldsymbol{Y}' = \boldsymbol{B}, \boldsymbol{f} = \boldsymbol{h}$), we see that $\eta_{\boldsymbol{X}\boldsymbol{B}}(\boldsymbol{h}\boldsymbol{g}) = \phi^*(\boldsymbol{h})\eta_{\boldsymbol{X}\boldsymbol{A}}(\boldsymbol{g})$ and thus, $\eta_{\boldsymbol{X}\boldsymbol{B}}(\boldsymbol{h}\boldsymbol{g}) = \phi^*(\boldsymbol{h})\boldsymbol{k} = \eta_{\boldsymbol{X}\boldsymbol{B}}(\boldsymbol{f})$. Since $\eta_{\boldsymbol{X}\boldsymbol{B}}$ is a bijection, we conclude that $\boldsymbol{h}\boldsymbol{g} = \boldsymbol{f}$, as desired. □

LEMMA 14.13. *If ϕ is cofinal, the systems $\psi(\boldsymbol{\Delta}(\mathrm{M}))$ and $\boldsymbol{\Delta}(\Lambda)$ are isomorphic in Mod^Λ.*

Proof. We must exhibit morphisms $\boldsymbol{u} = (u_\lambda): \psi(\boldsymbol{\Delta}(\mathrm{M})) \to \boldsymbol{\Delta}(\Lambda)$ and $\boldsymbol{v} = (v_\lambda): \boldsymbol{\Delta}(\Lambda) \to \psi(\boldsymbol{\Delta}(\mathrm{M}))$ such that, for every $\lambda \in \Lambda$,

$$v_\lambda u_\lambda = \mathrm{id}, \;\; u_\lambda v_\lambda = \mathrm{id}. \tag{18}$$

In order to define $u_\lambda: \boldsymbol{\Gamma}_\lambda \otimes \boldsymbol{\Delta}(\mathrm{M}) \to R$, note that

$$\boldsymbol{\Gamma}_\lambda \otimes \boldsymbol{\Delta}(\mathrm{M}) = (\oplus_\mu (\Gamma^\mu_\lambda \otimes R))/L, \tag{19}$$

where the generators of L are given by (14.1.10). Consider the homomorphisms $u^\mu_\lambda: \Gamma^\mu_\lambda \otimes R \to R$, determined by $u^\mu_\lambda(r \otimes s) = rs$. Recall that $\Gamma^\mu_\lambda = R$, if $\lambda \le \phi(\mu)$, and $\Gamma^\mu_\lambda = 0$, if $\lambda \not\le \phi(\mu)$. It is readily seen that $\oplus_\mu u^\mu_\lambda: \oplus_\mu(\Gamma^\mu_\lambda \otimes R) \to R$ induces a homomorphism $u_\lambda: \boldsymbol{\Gamma}_\lambda \otimes \boldsymbol{\Delta}(\mathrm{M}) \to R$ such that

$$u_\lambda[r \otimes s] = rs, \tag{20}$$

and the homomorphisms u_λ determine a level morphism $\boldsymbol{u}: \psi(\boldsymbol{\Delta}(\mathrm{M})) \to \boldsymbol{\Delta}(\Lambda)$.

To describe $\boldsymbol{v}$, we first define homomorphisms $v^\mu_\lambda: R \to \Gamma^\mu_\lambda \otimes R = R \otimes R$, for $\lambda \le \phi(\mu)$, by putting $v^\mu_\lambda(r) = 1 \otimes r$. Note that cofinality of ϕ implies that every

$\lambda \in \Lambda$ admits such $\mu \in M$. Clearly, for $\mu \le \mu'$, one has $(\gamma_\lambda^{\mu\mu'} \otimes 1)v_\lambda^\mu = v_\lambda^{\mu'}$. Therefore, the class $[v_\lambda^\mu(r)] = [1 \otimes r] \in \boldsymbol{\Gamma}_\lambda \otimes \boldsymbol{\Delta}(\mathrm{M})$ does not depend on μ, provided μ is sufficiently large, and v_λ is well defined by putting

$$v_\lambda(r) = [1 \otimes r]. \tag{21}$$

It is readily seen that $v_\lambda = \gamma_{\lambda\lambda'} v_{\lambda'}$ and thus, $\boldsymbol{v} = (v_\lambda)\colon \boldsymbol{\Delta}(\Lambda) \to \psi(\boldsymbol{\Delta}(\mathrm{M}))$ is a level morphism. The equalities (18) are now easily verified. □

Proof of Theorem 14.9. Consider projective resolutions for $\boldsymbol{\Delta}(\Lambda)$ and $\boldsymbol{\Delta}(\mathrm{M})$, respectively,

$$\boldsymbol{0} \leftarrow \boldsymbol{\Delta}(\Lambda) \xleftarrow{\boldsymbol{e}} \boldsymbol{P}_0 \xleftarrow{\boldsymbol{d}_1} \boldsymbol{P}_1 \leftarrow \dots, \tag{22}$$

$$\boldsymbol{0} \leftarrow \boldsymbol{\Delta}(\mathrm{M}) \xleftarrow{\boldsymbol{e}} \boldsymbol{R}_0 \xleftarrow{\boldsymbol{d}_1} \boldsymbol{R}_1 \leftarrow \dots . \tag{23}$$

Application of the functor ψ to (23), yields a chain complex of inverse systems from Mod^Λ.

$$\boldsymbol{0} \leftarrow \psi(\boldsymbol{\Delta}(\mathrm{M})) \xleftarrow{\psi(\boldsymbol{e})} \psi(\boldsymbol{R}_0) \xleftarrow{\psi(\boldsymbol{d}_1)} \psi(\boldsymbol{R}_1) \leftarrow \dots . \tag{24}$$

On the level $\lambda \in \Lambda$, (24) reduces to the chain complex of modules

$$0 \leftarrow \boldsymbol{\Gamma}_\lambda \otimes \boldsymbol{\Delta}(\mathrm{M}) \xleftarrow{1 \otimes \boldsymbol{e}} \boldsymbol{\Gamma}_\lambda \otimes \boldsymbol{R}_0 \xleftarrow{1 \otimes \boldsymbol{d}_1} \boldsymbol{\Gamma}_\lambda \otimes \boldsymbol{R}_1 \leftarrow \dots . \tag{25}$$

By Lemma 14.5, applied to $\boldsymbol{\Delta}(\mathrm{M})$, the sequence (25) is exact. This implies that also (24) is exact. However, by Lemma 14.12, $\psi(\boldsymbol{R}_n)$ is projective, for every $n \ge 0$. Consequently, (24) is a projective resolution of $\psi(\boldsymbol{\Delta}(\mathrm{M}))$. By Lemma 14.13, there exists an isomorphism $\boldsymbol{u}\colon \psi(\boldsymbol{\Delta}(\mathrm{M})) \to \boldsymbol{\Delta}(\Lambda)$. Since (22) is a projective resolution of $\boldsymbol{\Delta}(\Lambda)$, $\boldsymbol{u}$ induces a homotopy equivalence of chain complexes of inverse systems,

$$\begin{array}{ccccccccc}
\boldsymbol{0} & \leftarrow & \psi(\boldsymbol{\Delta}(\mathrm{M})) & \leftarrow & \psi(\boldsymbol{R}_0) & \leftarrow & \psi(\boldsymbol{R}_1) & \leftarrow & \dots \\
 & & \downarrow{\scriptstyle \boldsymbol{u}} & & \downarrow{\scriptstyle \boldsymbol{u}_0} & & \downarrow{\scriptstyle \boldsymbol{u}_1} & & \\
\boldsymbol{0} & \leftarrow & \boldsymbol{\Delta}(\Lambda) & \leftarrow & \boldsymbol{P}_0 & \leftarrow & \boldsymbol{P}_1 & \leftarrow & \dots
\end{array} \tag{26}$$

Application of the functor $\mathrm{Hom}(\,.\,, \boldsymbol{Y})$ to (26) yields a homotopy equivalence of cochain complexes of modules,

$$\begin{array}{ccccccc}
0 & \rightarrow & \mathrm{Hom}(\psi(\boldsymbol{R}_0), \boldsymbol{Y}) & \rightarrow & \mathrm{Hom}(\psi(\boldsymbol{R}_1), \boldsymbol{Y}) & \rightarrow & \dots \\
 & & \uparrow & & \uparrow & & \\
0 & \rightarrow & \mathrm{Hom}(\boldsymbol{P}_0, \boldsymbol{Y}) & \rightarrow & \mathrm{Hom}(\boldsymbol{P}_1, \boldsymbol{Y}) & \rightarrow & \dots
\end{array} \tag{27}$$

Applying Lemma 14.11, we can replace the first row in (27) by the isomorphic cochain complex

$$0 \to \mathrm{Hom}(\boldsymbol{R}_0, \phi^*(\boldsymbol{Y})) \to \mathrm{Hom}(\boldsymbol{R}_1, \phi^*(\boldsymbol{Y})) \to \dots . \tag{28}$$

We thus obtain a homotopy equivalence of cochain complexes

$$\begin{array}{ccccccc} 0 & \longrightarrow & \mathrm{Hom}(\boldsymbol{R}_0, \phi^*(Y)) & \longrightarrow & \mathrm{Hom}(\boldsymbol{R}_1, \phi^*(\boldsymbol{Y})) & \longrightarrow & \cdots \\ & & {\scriptstyle w_0}\uparrow & & \uparrow{\scriptstyle w_1} & & \\ 0 & \longrightarrow & \mathrm{Hom}(\boldsymbol{P}_0, \boldsymbol{Y}) & \longrightarrow & \mathrm{Hom}(\boldsymbol{P}_1, \boldsymbol{Y}) & \longrightarrow & \cdots \end{array} \tag{29}$$

The homotopy equivalence $\boldsymbol{w} = (w_n)$ induces isomorphisms of the n-th cohomology modules of the rows of (29). Since (22) and (23) are projective resolutions of $\boldsymbol{\Delta}(\Lambda)$ and $\boldsymbol{\Delta}(\mathrm{M})$, respectively, these modules are naturally isomorphic to $\mathrm{Ext}^n(\boldsymbol{\Delta}(\Lambda), \boldsymbol{Y})$ and $\mathrm{Ext}^n(\boldsymbol{\Delta}(M), \phi^*(\boldsymbol{Y}))$, respectively (see Theorem 12.12). However, by Theorem 12.14, the latter modules are naturally isomorphic to $\lim^n \boldsymbol{Y}$ and $\lim^n \phi^*(\boldsymbol{Y})$, respectively. □

One of the applications of Theorem 14.9 is an extension of Theorem 13.11 to the case of directed sets Λ, which need not be linearly ordered.

THEOREM 14.14. *If Λ is a directed set with* $\mathrm{cof}(\Lambda) = \aleph_n$, *then* $\mathrm{hd}(\Lambda) \geq n+1$.

In the proof we need the following lemma.

LEMMA 14.15. *Let Λ be a directed set of cofinality* $\mathrm{cof}(\Lambda) = \aleph_n$ *and let Ω_n be the set of ordinals $\Omega_n = \{\alpha | \alpha < \omega_n\}$. Then there exists a cofinal increasing function $\phi: \Lambda \to \Omega_n$.*

Proof. By assumption on Λ, there exists a set $\Lambda' \subseteq \Lambda$, which is cofinal in Λ and has cardinality $\mathrm{card}(\Lambda') = \aleph_n$. Since $\mathrm{card}(\Omega_n) = \aleph_n$, there exists a bijection $\psi: \Omega_n \to \Lambda'$. By the cofinality of Λ', for every $\lambda \in \Lambda$, the set $\{\alpha < \omega_n | \lambda \leq \psi(\alpha)\} \neq \emptyset$. Let $\phi(\lambda)$ be its minimal element. Then $\phi: \Lambda \to \Omega_n$ is a well-defined function. Note that

$$\lambda \leq \psi\phi(\lambda), \text{ for every } \lambda \in \Lambda. \tag{30}$$

Moreover, if $\lambda \leq \lambda'$, then $\lambda' \leq \psi\phi(\lambda')$ implies $\lambda \leq \psi(\phi(\lambda'))$. Therefore, one concludes that $\phi(\lambda) \leq \phi(\lambda')$, which proves that ϕ is increasing. We claim that ϕ is also cofinal. Indeed, if this were not the case, one could find a $\beta < \omega_n$ such that $\phi(\lambda) < \beta$, for every $\lambda \in \Lambda$. However, (30) would then imply that $\lambda \leq \psi\phi(\lambda) \in \psi([0, \beta))$, showing that $\psi([0, \beta))$ is a cofinal subset of Λ of cardinality $\mathrm{card}(\psi([0, \beta))) \leq \mathrm{card}([0, \beta)) < \aleph_n$. However, this contradicts the assumption. □

Proof of Theorem 14.14. In view of Theorem 13.3, we must produce an inverse system $\boldsymbol{X}$, indexed by Λ, such that $\lim^{n+1} \boldsymbol{X} \neq 0$. Since Ω_n is linearly ordered and $\mathrm{cof}(\Omega_n) = \aleph_n$, Corollary 13.12 yields a system $\boldsymbol{Y}$, indexed by Ω_n, such that $\lim^{n+1} \boldsymbol{Y} \neq 0$. By Lemma 14.15, there exists an increasing cofinal function $\phi\colon \Lambda \to \Omega_n$. Therefore, by Theorem 14.9, $\boldsymbol{X} = \phi^*(\boldsymbol{Y})$ is a system, indexed by Λ, such that $\lim^{n+1}\boldsymbol{X} = \lim^{n+1}\phi^*(\boldsymbol{Y}) \approx \lim^{n+1}\boldsymbol{Y} \neq 0$. □

Theorem 14.14, together with Goblot's theorem (Theorem 13.9), yields the following result (Mitchell 1973).

THEOREM 14.16. *A directed set Λ of cofinality* $\mathrm{cof}(\Lambda) = \aleph_n$ *has homological dimension* $\mathrm{hd}(\Lambda) = n+1$, $n \geq 0$.

Now Corollary 13.8 yields the following result.

COROLLARY 14.17. *Let Λ be a directed set of cofinality* $\mathrm{cof}(\Lambda) = \aleph_n$. *If*

$$\boldsymbol{0} \leftarrow \boldsymbol{\Delta}(\Lambda) \overset{\boldsymbol{e}}{\leftarrow} \boldsymbol{P}_0 \overset{d_1}{\longleftarrow} \boldsymbol{P}_1 \leftarrow \ldots \leftarrow \boldsymbol{P}_n \overset{d_{n+1}}{\longleftarrow} \boldsymbol{P}_{n+1} \leftarrow \ldots \tag{31}$$

is a projective resolution of the diagonal system $\boldsymbol{\Delta}(\Lambda)$, then

$$\lim{}^{n+1}\boldsymbol{P}^{n+1} \neq 0. \tag{32}$$

REMARK 14.18. There is a much easier proof (Nöbeling 1961) of the cofinality theorem (Theorem 14.9) in the special case, when Λ is a cofinite directed set and $M \subseteq \Lambda$ is a cofinal subset.

Proof. Let

$$\boldsymbol{0} \to \boldsymbol{Y} \overset{\boldsymbol{e}}{\longrightarrow} \boldsymbol{I}^0 \overset{\boldsymbol{d}^1}{\longrightarrow} \boldsymbol{I}^1 \overset{\boldsymbol{d}^2}{\longrightarrow} \boldsymbol{I}^2 \to \ldots . \tag{33}$$

be an injective resolution. Consider the commutative diagram

$$\begin{array}{ccccccccc}
\boldsymbol{0} & \longrightarrow & \boldsymbol{Y} & \overset{\boldsymbol{e}}{\longrightarrow} & \boldsymbol{I}^0 & \overset{\boldsymbol{d}^1}{\longrightarrow} & \boldsymbol{I}^1 & \longrightarrow & \cdots \\
 & & {\scriptstyle \boldsymbol{i}}\downarrow & & {\scriptstyle \boldsymbol{i}^0}\downarrow & & {\scriptstyle \boldsymbol{i}^1}\downarrow & & \\
\boldsymbol{0} & \longrightarrow & \boldsymbol{Y}|M & \underset{\boldsymbol{e}'}{\longrightarrow} & \boldsymbol{I}^0|M & \underset{\boldsymbol{d}'^1}{\longrightarrow} & \boldsymbol{I}^1|M & \longrightarrow & \cdots,
\end{array} \tag{34}$$

where systems in the second row are obtained from those in the first row by restriction to M; vertical arrows are the level morphisms induced by the inclusion $i\colon M \to \Lambda$. Exactness of the second row is an immediate consequence of Lemma 11.5. In the lemma which follows we will prove that all systems $\boldsymbol{I}^n|M$ are injective and therefore, the second row is an injective resolution of $\boldsymbol{Y}|M$. Application of the functor lim yields a commutative diagram of modules

$$\begin{array}{ccccccccc}
0 & \longrightarrow & \lim \boldsymbol{I}^0 & \overset{\lim \boldsymbol{d}^0}{\longrightarrow} & \lim \boldsymbol{I}^1 & \overset{\lim \boldsymbol{d}^1}{\longrightarrow} & \lim \boldsymbol{I}^2 & \longrightarrow & \ldots \\
 & & {\scriptstyle \lim \boldsymbol{i}^0}\downarrow & & {\scriptstyle \lim \boldsymbol{i}^1}\downarrow & & \downarrow{\scriptstyle \lim \boldsymbol{i}^2} & & \\
0 & \longrightarrow & \lim \boldsymbol{I}^0|M & \underset{\lim \boldsymbol{d}^0}{\longrightarrow} & \lim \boldsymbol{I}^1|M & \underset{\lim \boldsymbol{d}^1}{\longrightarrow} & \lim \boldsymbol{I}^2|M & \longrightarrow & \ldots .
\end{array} \tag{35}$$

The rows of this diagram can be viewed as cochain complexes. By definition, their n-th cohomology modules equal $\lim^n \boldsymbol{Y}$ and $\lim^n \boldsymbol{Y}|M$, respectively (see 11.3). However, by Lemma 14.10, the vertical arrows in diagram (35) are isomorphisms. Hence, they define an isomorphism of cochain complexes and thus, induce an isomorphism of the cohomology modules, i.e., an isomorphism $\lim^n \boldsymbol{Y} \to \lim^n \boldsymbol{Y}|M$. □

LEMMA 14.19. *Let $\boldsymbol{I}$ be an injective system in Mod^{Λ}, where Λ is a cofinite directed set. If $M \subseteq \Lambda$ is a cofinal subset, the restriction $\boldsymbol{I}|M$ is an injective system in Mod^M.*

Proof. We first define an increasing retraction $\rho: \Lambda \to M$, such that $\lambda \leq \rho(\lambda)$, for every $\lambda \in \Lambda$. By definition, $\rho|M$ is the identity. One extends ρ to all of Λ by induction on the number of predecessors of $\lambda \in \Lambda$. This is possible because λ has finitely many predecessors and M is directed (see the proof of Lemma 1.2).

Consider a morphism $\boldsymbol{k}: \boldsymbol{B} \to \boldsymbol{I}|M$ and a monomorphism $\boldsymbol{h}: \boldsymbol{B} \to \boldsymbol{A}$ of Mod^M. We must exhibit a morphism $\boldsymbol{g}: \boldsymbol{A} \to \boldsymbol{I}|M$ of Mod^M such that $\boldsymbol{gh} = \boldsymbol{k}$. If $\boldsymbol{A} = (A_\mu, a_{\mu\mu'}, M)$, we associate with $\boldsymbol{A}$ a system $\boldsymbol{A}' = (A'_\lambda, a'_{\lambda\lambda'}, \Lambda)$, indexed by Λ, by putting $A'_\lambda = A_{\rho(\lambda)}, a'_{\lambda\lambda'} = a_{\rho(\lambda)\rho(\lambda')}$. Similarly, we associate with $\boldsymbol{B} = (B_\mu, b_{\mu\mu'}, M)$ a system $\boldsymbol{B}' = (B'_\lambda, b'_{\lambda\lambda'}, \Lambda)$, where $B'_\lambda = B_{\rho(\lambda)}, b'_{\lambda\lambda'} = b_{\rho(\lambda)\rho(\lambda')}$. Moreover, we associate with $\boldsymbol{h} = (h_\mu)$ a morphism $\boldsymbol{h}' = (h'_\lambda): \boldsymbol{B}' \to \boldsymbol{A}'$, where $h'_\lambda = h_{\rho(\lambda)}$. If $\boldsymbol{I} = (I_\lambda, r_{\lambda\lambda'}, \Lambda)$, we define $\boldsymbol{k}' = (k'_\lambda): \boldsymbol{B}' \to \boldsymbol{I}$, by putting $k'_\lambda = r_{\lambda\rho(\lambda)} k_{\rho(\lambda)}: B'_\lambda \to I_\lambda$. By Lemma 11.2, all h_μ are monomorphisms. Therefore, by the same theorem, $\boldsymbol{h}'$ is a monomorphism of Mod^{Λ}. Since $\boldsymbol{I}$ is injective, one obtains a morphism $\boldsymbol{g}': \boldsymbol{A}' \to \boldsymbol{I}$ such that $\boldsymbol{g}'\boldsymbol{h}' = \boldsymbol{k}'$. Note that, for every $\mu \in M$, one has $k'_\mu = k_\mu$ and $h'_\mu = h_\mu$. Therefore, $\boldsymbol{k}'|M = \boldsymbol{k}, \boldsymbol{h}'|M = \boldsymbol{h}$. Consequently, the restriction $\boldsymbol{g} = \boldsymbol{g}'|M: \boldsymbol{A} \to \boldsymbol{I}|M$ satisfies the desired condition $\boldsymbol{gh} = \boldsymbol{k}$. □

Bibliographic notes

The cofinality theorem (Theorem 14.9), which is the main result of this section, was obtained in (Mitchell 1973). Mitchell's proof is an appropriate generalization of the Cartan – Eilenberg mapping theorem (see Chapter VIII, Theorem 3.1 of (Cartan, Eilenberg 1956)). The proof-supporting material on tensor products (Section 14.1) appeared in (Mitchell 1972). An important step in Mitchell's proof is the generalization to direct systems of a theorem on modules, which asserts that flat modules are colimits of free modules of finite type (Govorov 1965), (Lazard 1969). Our proof avoids reference to flat systems by producing an elementary proof of the fact that the augmented chain complex $(\boldsymbol{L}, \eta)$ is acyclic (Lemma 14.7). This proof is taken from (Mardešić, Prasolov 1998). The chain complex $\boldsymbol{L}$ plays an important role in 20 and appeared already in (Roos 1961). In the special case of injective mappings

$\phi: M \to \Lambda$, the cofinality theorem (Remark 14.18) was first announced in (Roos 1961) and proved in (Nöbeling 1961) and (Jensen 1972). Theorem 14.14 is from (Mitchell 1973). Corollary 14.17 is from (Mardešić 1996a).

15. Higher limits on the category pro-Mod

In previous sections of this chapter we have considered the derived limits $\lim^n$ as functors from the category Mod^{Λ} to the category Mod. In the first subsection of this section we will extend $\lim^n$ to a functor pro-Mod $\to$ Mod. In the next subsection we will show that the main properties of $\lim^n$ are preserved by this extension.

15.1 $\lim^n$ as a functor on pro-Mod

We will first extend the functors $\lim^n\colon \mathsf{Mod}^{\Lambda} \to \mathsf{Mod}$ to functors $\lim^n$: inv-Mod$\to$Mod, $n \geq 0$. The objects of inv-Mod are inverse systems of R-modules $\boldsymbol{X} = (X_\lambda, p_{\lambda\lambda'}, \Lambda)$, where the index sets Λ range over all preordered sets, and the morphisms $\boldsymbol{f} = (f, f_\mu)\colon \boldsymbol{X} \to \boldsymbol{Y} = (Y_\mu, p_{\mu\mu'}, M)$ consist of an increasing function $f\colon M \to \Lambda$ and of homomorphisms $f_\mu\colon X_{f(\mu)} \to Y_\mu$, $\mu \in M$, which satisfy (1.1.6). Since we have already defined the modules $\lim^n \boldsymbol{X}$, $n \geq 0$, it remains to define the homomorphisms $\lim^n \boldsymbol{f}\colon \lim^n \boldsymbol{X} \to \lim^n \boldsymbol{Y}$, induced by a morphism $\boldsymbol{f}\colon \boldsymbol{X} \to \boldsymbol{Y}$ of inv-Mod.

Recall that $\lim^n \boldsymbol{X}$ is the n-th cohomology module of the cochain complex $K(\boldsymbol{X})$, which consists of modules $K^n(\boldsymbol{X})$, defined by (11.5.2), and of coboundary operators $\delta^n\colon K^{n-1} \to K^n$, defined by (11.5.3). We first associate with $\boldsymbol{f}$ a mapping of cochain complexes $\boldsymbol{f}^{\#}\colon K(\boldsymbol{X}) \to K(\boldsymbol{Y})$, given by homomorphisms $f^n\colon K^n(\boldsymbol{X}) \to K^n(\boldsymbol{Y})$, $n \geq 0$. If $c \in K^n(\boldsymbol{X})$, we define the cochain $f^n c \in K^n(\boldsymbol{Y})$ by putting

$$(f^n c)(\boldsymbol{\mu}) = f_{\mu_0} c(f(\boldsymbol{\mu})), \tag{1}$$

where $f(\boldsymbol{\mu}) = (f(\mu_0), \dots, f(\mu_n))$, for $\boldsymbol{\mu} = (\mu_0, \dots, \mu_n) \in M_n$. Since $f\colon M \to \Lambda$ is increasing, $f(\boldsymbol{\mu}) \in \Lambda_n$, and thus, $c(f(\boldsymbol{\mu}))$ is a well-defined element of $X_{f(\mu_0)}$. Consequently, $f_{\mu_0} c(f(\boldsymbol{\mu})) \in Y_{\mu_0}$, which makes $f^n c$ a well-defined cochain belonging to $K^n(\boldsymbol{Y})$.

LEMMA 15.1. *$\boldsymbol{f}^{\#}\colon K(\boldsymbol{X}) \to K(\boldsymbol{Y})$ is a mapping of cochain complexes, i.e.,*

$$f^n \delta^n = \delta^n f^{n-1}, \; n \geq 1. \tag{2}$$

Proof. For $\boldsymbol{\mu} = (\mu_0, \ldots, \mu_n) \in M_n$ and a cochain $c \in K^{n-1}(\boldsymbol{X})$, the coboundary formula (11.5.3) and (1) yield

$$(f^n \delta^n c)(\boldsymbol{\mu}) = f_{\mu_0} p_{f(\mu_0)f(\mu_1)} c(d^0 f(\boldsymbol{\mu})) + \sum_{j=1}^{j=n} (-1)^j f_{\mu_0} c(d^j f(\boldsymbol{\mu})), \tag{3}$$

$$(\delta^n f^{n-1} c)(\boldsymbol{\mu}) = q_{\mu_0\mu_1} f_{\mu_1} c(f(d^0 \boldsymbol{\mu})) + \sum_{j=1}^{j=n} (-1)^j f_{\mu_0} c(f(d^j \boldsymbol{\mu})). \tag{4}$$

Since $f(d^j \boldsymbol{\mu}) = f(\mu_0, \ldots, \widehat{\mu_j}, \ldots, \mu_n) = d^j f(\boldsymbol{\mu})$, $f_{\mu_0} p_{f(\mu_0)f(\mu_1)} = q_{\mu_0\mu_1} f_{\mu_1}$, the right sides of (3) and (4) coincide. □

The desired homomorphism $\lim^n \boldsymbol{f} \colon \lim^n \boldsymbol{X} \to \lim^n \boldsymbol{Y}$ is, by definition, the homomorphism of cohomology modules, induced by the cochain mapping $\boldsymbol{f}^{\#} \colon K(\boldsymbol{X}) \to K(\boldsymbol{Y})$.

REMARK 15.2. In the special case when $\Lambda = M$ and $\boldsymbol{f}$ is a level morphism, i.e., $f = \mathrm{id}$, formula (1) reduces to (11.5.8). Therefore, the present definition of $\lim^n$ extends the one of 15.1.

That $\lim^n$: inv-Mod → Mod is a functor, follows from the next lemma.

LEMMA 15.3. *If $\boldsymbol{f} \colon \boldsymbol{X} \to \boldsymbol{Y}$ and $\boldsymbol{g} \colon \boldsymbol{Y} \to \boldsymbol{Z} = (Z_\nu, r_{\nu\nu'}, N)$ are morphisms of* inv-Mod, *then*

$$\boldsymbol{g}^{\#} \boldsymbol{f}^{\#} = (\boldsymbol{g}\boldsymbol{f})^{\#} \tag{5}$$

and thus,

$$\lim{}^n \boldsymbol{g} \, \lim{}^n \boldsymbol{f} = \lim{}^n (\boldsymbol{g}\boldsymbol{f}). \tag{6}$$

Proof. Let $\boldsymbol{h} = \boldsymbol{g}\boldsymbol{f}$ and let $\boldsymbol{f}^{\#}$, $\boldsymbol{g}^{\#}$ and $\boldsymbol{h}^{\#}$ consist of homomorphisms $f^n \colon K^n(\boldsymbol{X}) \to K^n(\boldsymbol{Y})$, $g^n \colon K^n(\boldsymbol{Y}) \to K^n(\boldsymbol{Z})$ and $h^n \colon K^n(\boldsymbol{X}) \to K^n(\boldsymbol{Z})$, respectively. Then $\boldsymbol{g}^{\#} \boldsymbol{f}^{\#}$ consists of homomorphisms $g^n f^n \colon K^n(\boldsymbol{X}) \to K^n(\boldsymbol{Z})$ and we must show that

$$g^n f^n = h^n. \tag{7}$$

If $c \in K^n(\boldsymbol{X})$, put $c' = f^n c \in K^n(\boldsymbol{Y})$. Then, for $\boldsymbol{\nu} \in N_n$, one has

$$(g^n f^n c)(\boldsymbol{\nu}) = (g^n c')(\boldsymbol{\nu}) = g_{\nu_0} c'(g(\boldsymbol{\nu})). \tag{8}$$

However,

$$c'(g(\boldsymbol{\nu})) = (f^n c)(g(\boldsymbol{\nu})) = f_{g(\nu_0)} c(fg(\boldsymbol{\nu})), \tag{9}$$

and thus,

$$(g^n f^n c)(\boldsymbol{\nu}) = g_{\nu_0} f_{g(\nu_0)} c(fg(\boldsymbol{\nu})). \tag{10}$$

On the other hand, by (1.1.8), $\boldsymbol{h}$ is given by the function $h = fg$ and by the homomorphisms $h_{\nu_0} = g_{\nu_0} f_{g(\nu_0)}$. Hence, also

$$(h^n c)(\boldsymbol{\nu}) = h_{\nu_0} c(h(\boldsymbol{\nu})) = g_{\nu_0} f_{g(\nu_0)} c(fg(\boldsymbol{\nu})). \;\square \tag{11}$$

In extending $\lim^n$ further to a functor $\lim^n$: pro-Mod→ Mod, we assume that all systems are indexed by directed cofinite sets. We need the following lemma.

LEMMA 15.4. *If $\boldsymbol{f}, \boldsymbol{f}' : \boldsymbol{X} \to \boldsymbol{Y}$ are congruent morphisms of systems, the induced cochain mappings $\boldsymbol{f}^{\#}, \boldsymbol{f}'^{\#} : K(\boldsymbol{X}) \to K(\boldsymbol{Y})$ are homotopic and thus, induce the same homomorphism of cohomology modules, i.e.,* $\lim^n \boldsymbol{f} = \lim^n \boldsymbol{f}'$.

Proof. There is no loss of generality in assuming that $\boldsymbol{f}'$ is a shift of $\boldsymbol{f}$, i.e., $\boldsymbol{f} = (f, f_{\mu_0})$, $\boldsymbol{f}' = (f', f'_{\mu_0})$, where $f' \geq f$ and $f'_{\mu_0} = f_{\mu_0} p_{f(\mu_0) f'(\mu_0)}$. We must exhibit a sequence of homomorphisms $F^n : K^n(\boldsymbol{X}) \to K^{n-1}(\boldsymbol{Y})$, $n \geq 0$, such that

$$\delta^n F^n + F^{n+1} \delta^{n+1} = f'^n - f^n. \tag{12}$$

For $n = 0$, we interpret (12) to mean $F^1 \delta^1 = f'^0 - f^0$.

To facilitate the writing of explicit formulae for F^n, we introduce the following notation. If $\boldsymbol{\mu} = (\mu_0, \dots, \mu_{n-1}) \in M_{n-1}$, $n \geq 1$ and $0 \leq k \leq n-1$, let

$$\langle \boldsymbol{\mu} | k \rangle = (f(\mu_0), \dots, f(\mu_k), f'(\mu_k), \dots, f'(\mu_{n-1})) \in \Lambda_n. \tag{13}$$

Note that, for $0 \leq j \leq n$,

$$\langle d^j \boldsymbol{\mu} | k \rangle = \begin{cases} d^j \langle \boldsymbol{\mu} | k+1 \rangle, & 0 \leq j \leq k, \\ d^{j+1} \langle \boldsymbol{\mu} | k \rangle, & k+1 \leq j \leq n. \end{cases} \tag{14}$$

Moreover,

$$d^{k+1} \langle \boldsymbol{\mu} | k \rangle = d^{k+1} \langle \boldsymbol{\mu} | k+1 \rangle. \tag{15}$$

If $c \in K^n(\boldsymbol{X})$, $n \geq 1$, we define the cochain $F^n c \in K^{n-1}(\boldsymbol{Y})$, by putting

$$(F^n c)(\boldsymbol{\mu}) = \sum_{k=0}^{n-1} (-1)^k f_{\mu_0} c \langle \boldsymbol{\mu} | k \rangle, \tag{16}$$

where $c\langle \boldsymbol{\mu} | k \rangle$ is a shorter notation for $c(\langle \boldsymbol{\mu} | k \rangle)$. Note that $\langle \boldsymbol{\mu} | k \rangle \in \Lambda_n$, $c\langle \boldsymbol{\mu} | k \rangle \in X_{f(\mu_0)}$ and $f_{\mu_0} c \langle \boldsymbol{\mu} | k \rangle \in Y_{\mu_0}$, which shows that $F^n c$ is indeed an $(n-1)$-cochain of $K(\boldsymbol{Y})$.

In order to verify (12), consider a cochain $c \in K^n(\boldsymbol{X})$ and a multiindex $\boldsymbol{\mu} \in M_n$. For $n \geq 1$, by the boundary formula (11.5.3) and (16),

$$\begin{aligned} (\delta^{n-1} F^n c)(\boldsymbol{\mu}) &= \textstyle\sum_{k=0}^{n-1} (-1)^k q_{\mu_0 \mu_1} f_{\mu_1} c \langle d^0 \boldsymbol{\mu} | k \rangle \\ &\quad + \textstyle\sum_{k=0}^{n-1} \sum_{j=1}^{n} (-1)^{j+k} f_{\mu_0} c \langle d^j \boldsymbol{\mu} | k \rangle. \end{aligned} \tag{17}$$

On the other hand, for $n \geq 0$,

$$(F^{n+1} \delta^n c)(\boldsymbol{\mu}) = f_{\mu_0}((\delta^n c)\langle \boldsymbol{\mu} | 0 \rangle) + \sum_{k=1}^{n} (-1)^k f_{\mu_0}((\delta^n c)\langle \boldsymbol{\mu} | k \rangle). \tag{18}$$

Now note that $d^0 \langle \boldsymbol{\mu} | 0 \rangle = d^0(f(\mu_0), f'(\mu_0), \dots, f'(\mu_n)) = f'(\boldsymbol{\mu})$ and thus, by (1),

$$f_{\mu_0} p_{f(\mu_0) f(\mu_1)} c(d^0 \langle \boldsymbol{\mu} | 0 \rangle) = f'_{\mu_0} c(f'(\boldsymbol{\mu})) = (f'^n c)(\boldsymbol{\mu}). \tag{19}$$

Therefore, the first term of the right side of (18) equals

$$f_{\mu_0}((\delta^n c)\langle \boldsymbol{\mu}|0\rangle) = (f'^n c)(\boldsymbol{\mu}) + \sum_{j=1}^{n+1}(-1)^j f_{\mu_0} c(d^j\langle \boldsymbol{\mu}|0\rangle). \tag{20}$$

For $k \geq 1$, the first two terms of $\langle \boldsymbol{\mu}|k\rangle$ are $f(\mu_0), f(\mu_1)$ and thus,

$$\begin{aligned}(-1)^k f_{\mu_0}(\delta^n c\langle \boldsymbol{\mu}|k\rangle) &= (-1)^k f_{\mu_0} p_{f(\mu_0)f(\mu_1)} c(d^0\langle \boldsymbol{\mu}|k\rangle) \\ &+ \textstyle\sum_{j=1}^{n+1}(-1)^{j+k} f_{\mu_0} c(d^j\langle \boldsymbol{\mu}|k\rangle).\end{aligned} \tag{21}$$

Summing up (20) with (21), for $k = 1, \dots, n$, one obtains

$$\begin{aligned}(F^{n+1}\delta^n c)(\boldsymbol{\mu}) &= (f'^n c)(\boldsymbol{\mu}) + \textstyle\sum_{k=1}^{n}(-1)^k f_{\mu_0} p_{f(\mu_0)f(\mu_1)} c(d^0\langle \boldsymbol{\mu}|k\rangle) \\ &+ \textstyle\sum_{k=0}^{n}\sum_{j=1}^{n+1}(-1)^{j+k} f_{\mu_0} c(d^j\langle \boldsymbol{\mu}|k\rangle).\end{aligned} \tag{22}$$

Consequently, the left side of (12) is obtained by summing up (17) and (22). Let us first note that the single sums S_1 and S_2 in (17) and (22) add up to 0. To see this, take into account (1.1.6), change in S_2 the index k to $k+1$ and use $d^0\langle \boldsymbol{\mu}|k+1\rangle = \langle d^0\boldsymbol{\mu}|k\rangle$, which is just (14), for $j = 0$.

To prove (12), it now remains to show that the two double sums D_1 and D_2 in (17) and (22) add up to $-f^n(c(\boldsymbol{\mu}))$. Decompose D_1 in two double sums D_1', D_1'', by putting in the first one the summands with $1 \leq j \leq k$ and in the second one the summands with $k+1 \leq j \leq n$. Apply again (14) and then, change the summation indices $k+1$ to k in D_1' and $j+1$ to j in D_1''. One obtains

$$D_1 = \left(\sum_{k=2}^{n}\sum_{j=1}^{k-1} + \sum_{k=0}^{n-1}\sum_{j=k+2}^{n+1}\right)(-1)^{j+k-1} f_{\mu_0} c(d^j\langle \boldsymbol{\mu}|k\rangle). \tag{23}$$

Note that each of the terms in (23) appears in the double sum D_2 with the opposite sign and therefore, cancels in $D_1 + D_2$. In the latter sum remain only the terms $(-1)^{j+k} f_{\mu_0} c(d^j\langle \boldsymbol{\mu}|k\rangle)$ of D_2, where (k,j) is either of the form (k,k), $1 \leq k \leq n$, or of the form $(k, k+1)$, $0 \leq k \leq n$. Consequently,

$$D_1 + D_2 = \sum_{k=1}^{n} f_{\mu_0} c(d^k\langle \boldsymbol{\mu}|k\rangle) - \sum_{k=0}^{n-1} f_{\mu_0} c(d^{k+1}\langle \boldsymbol{\mu}|k\rangle) - f_{\mu_0} c(d^{n+1}\langle \boldsymbol{\mu}|n\rangle). \tag{24}$$

Changing the index k to $k-1$ in the second of these sums and using (15) (for $k-1$), one concludes that the two sums cancel. It only remains the term $-f_{\mu_0} c(d^{n+1}\langle \boldsymbol{\mu}|n\rangle) = -f_{\mu_0} c(f(\boldsymbol{\mu})) = -(f^n c)(\boldsymbol{\mu})$, as desired. □

For a morphism $[\boldsymbol{f}]\colon \boldsymbol{X} \to \boldsymbol{Y}$ of pro-Mod, we define $\lim^n[\boldsymbol{f}]$ by

$$\lim{}^n[\boldsymbol{f}] = \lim{}^n \boldsymbol{f}. \tag{25}$$

$\lim^n[\boldsymbol{f}]$ is well defined, because, by Lemma 15.4, it depends on the congruence class $[\boldsymbol{f}]$ and not on its representative $\boldsymbol{f}$.

THEOREM 15.5. $\lim^n$: pro-Mod $\to$ Mod *is an additive functor. It extends the functor* $\lim^n$: $\mathsf{Mod}^{\Lambda} \to$ Mod.

Proof. $\lim^n([\boldsymbol{g}][\boldsymbol{f}]) = \lim^n[\boldsymbol{g}]\lim^n[\boldsymbol{f}]$. However, $[\boldsymbol{g}][\boldsymbol{f}] = [\boldsymbol{gf}]$, and by Lemma 15.3, $\lim^n[\boldsymbol{gf}] = \lim^n(\boldsymbol{gf}) = \lim^n\boldsymbol{g}\ \lim^n\boldsymbol{f} = \lim^n[\boldsymbol{g}]\lim^n[\boldsymbol{f}]$. An analogous argument shows that $\lim^n([\boldsymbol{g}]+[\boldsymbol{f}]) = \lim^n[\boldsymbol{g}] + \lim^n[\boldsymbol{f}]$. □

REMARK 15.6. In the case of inverse systems of modules over directed sets, which are not cofinite, one uses a more general type of morphisms $[\boldsymbol{f}]: \boldsymbol{X} \to \boldsymbol{Y}$ (see Remark 1.5), based on index functions which need not be increasing. Using the construction mentioned in Remarks 6.28 and 6.30, it is possible to extend the definition of $\lim^n[\boldsymbol{f}]$ to such morphisms as well. In particular, if the index sets are ordered, one associates with every morphism $\boldsymbol{f}: \boldsymbol{X} \to \boldsymbol{Y}$, a morphism of inv-Top between cofinite systems $\boldsymbol{f}^*: \boldsymbol{X}^* \to \boldsymbol{Y}^*$. Moreover, one associates with $\boldsymbol{X}$ and $\boldsymbol{Y}$ certain morphisms $\boldsymbol{u}_{\boldsymbol{X}}: \boldsymbol{X} \to \boldsymbol{X}^*$, $\boldsymbol{v}_{\boldsymbol{Y}}: \boldsymbol{Y} \to \boldsymbol{Y}^*$, which make the following diagram commutative.

$$\begin{array}{ccc} \boldsymbol{X} & \xrightarrow{\ \boldsymbol{f}\ } & \boldsymbol{Y} \\ {\scriptstyle \boldsymbol{u}_{\boldsymbol{X}}}\downarrow & & \downarrow{\scriptstyle \boldsymbol{u}_{\boldsymbol{Y}}} \\ \boldsymbol{X}^* & \xrightarrow[\ \boldsymbol{f}^*\ ]{} & \boldsymbol{Y}^* \end{array} \tag{26}$$

Moreover, the induced homomorphisms $\lim^n\boldsymbol{u}_{\boldsymbol{X}}$, $\lim^n\boldsymbol{v}_{\boldsymbol{Y}}$ are isomorphisms. It is then natural to define $\lim^n\boldsymbol{f}$ by the formula,

$$\lim{}^n\boldsymbol{f} = (\lim{}^n\boldsymbol{u}_{\boldsymbol{Y}})^{-1}\lim{}^n\boldsymbol{f}^*\lim{}^n\boldsymbol{u}_{\boldsymbol{X}}. \tag{27}$$

Similar constructions, based on Remark 6.28, enable one to replace arbitrary systems of modules by systems over ordered sets.

15.2 Properties of $\lim^n$ on pro-Mod

An essential tool in showing that also in pro-Mod $\lim^n$ preserves its main properties from 11.3 is the following *reindexing lemma.*

LEMMA 15.7. *Let* $[\boldsymbol{f}]: \boldsymbol{X} \to \boldsymbol{Y}$ *be a morphism in* pro-Mod. *Then there exist a cofinite directed set* N *and a commutative diagram*

$$\begin{array}{ccc} \boldsymbol{X} & \xrightarrow{\ [\boldsymbol{f}]\ } & \boldsymbol{Y} \\ {\scriptstyle [\boldsymbol{i}]}\downarrow & & \downarrow{\scriptstyle [\boldsymbol{j}]} \\ \boldsymbol{X}' & \xrightarrow[\ [\boldsymbol{f}']\ ]{} & \boldsymbol{Y}', \end{array} \tag{1}$$

where $[\boldsymbol{i}]$ *and* $[\boldsymbol{j}]$ *are isomorphisms in* pro-Mod *and* $\boldsymbol{f}'$ *is a morphism of* Mod^N.

For a proof see, e.g., (Mardešić, Segal 1982), I.1.3, Theorem 3.

THEOREM 15.8. *The category* pro - Mod *is abelian.*

Proof. The category pro - Mod has zero-objects. Such an object is the rudimentary system $\mathbf{0} = \{0\}$. Generally, $\boldsymbol{X} = (X_\lambda, p_{\lambda\lambda'}, \Lambda)$ is a zero object if and only if every $\lambda \in \Lambda$ admits a $\lambda' \geq \lambda$ such that $p_{\lambda\lambda'} = 0$ (see e.g., (Mardešić, Segal 1982), II.2.3, Theorem 7). A zero-morphism is a morphism, which factors through a zero-object. If $\boldsymbol{Y} = (Y_\mu, q_{\mu\mu'}, M)$ and $\boldsymbol{f} = (f, f_\mu): \boldsymbol{X} \to \boldsymbol{Y}$ is a morphism of systems, then $[\boldsymbol{f}] = \mathbf{0}$ if and only if every $\mu \in M$ admits a $\lambda \geq f(\mu)$ such that $f_\mu p_{f(\mu)\lambda} = 0$, i.e., $\boldsymbol{f}$ admits a shift $\boldsymbol{f}' = (f', f'_\mu)$ such that $f'_\mu = 0$, for every $\mu \in M$.

To see that every morphism $[\boldsymbol{f}]: \boldsymbol{X} \to \boldsymbol{Y}$ of pro - Mod admits a kernel, apply Lemma 15.7 to $[\boldsymbol{f}]$. It suffices to show that $[\boldsymbol{f}']$ has a kernel $[\boldsymbol{u}']: \boldsymbol{U}' \to \boldsymbol{X}'$, because then $[\boldsymbol{i}]^{-1}[\boldsymbol{u}']$ will be a kernel of $[\boldsymbol{f}]$. Similarly, to show that every monomorphism $[\boldsymbol{f}]$ is a kernel, it suffices to show that $[\boldsymbol{f}']$ is a kernel of a morphism $[\boldsymbol{v}']: \boldsymbol{Y}' \to \boldsymbol{V}'$, because then $[\boldsymbol{f}]$ is a kernel of $[\boldsymbol{v}'][\boldsymbol{j}]$. Note that whenever $[\boldsymbol{f}]$ is a monomorphism, then also $[\boldsymbol{f}']$ is a monomorphism. We have thus reduced the task to the case of morphisms $[\boldsymbol{f}]: \boldsymbol{X} \to \boldsymbol{Y}$, which have a level-preserving representative $\boldsymbol{f} = (f_\lambda)$.

For $\lambda \in \Lambda$, let $u_\lambda: U_\lambda \to X_\lambda$ be the kernel of the homomorphism $f_\lambda: X_\lambda \to Y_\lambda$ and let $u_{\lambda\lambda'}: U_{\lambda'} \to U_\lambda$ be the homomorphism induced by $p_{\lambda\lambda'}$. Then $\boldsymbol{U} = (U_\lambda, u_{\lambda\lambda'}, \Lambda) \in \mathsf{Mod}^\Lambda$ and $\boldsymbol{u} = (u_\lambda): \boldsymbol{U} \to \boldsymbol{X}$ is a level morphism. It is now easy to show that $[\boldsymbol{u}]: \boldsymbol{U} \to \boldsymbol{X}$ is a kernel of $[\boldsymbol{f}]$ (see e.g., (Mardešić, Segal 1982), II.2.3, Theorem 8).

Now assume that $\boldsymbol{f}$ is level-preserving and $[\boldsymbol{f}]$ is a monomorphism. For $\lambda \in \Lambda$, put $V_\lambda = Y_\lambda / f_\lambda(X_\lambda)$ and let $v_{\lambda\lambda'}: V_{\lambda'} \to V_\lambda$ be the homomorphism induced by $q_{\lambda\lambda'}$. Then $\boldsymbol{V} = (V_\lambda, v_{\lambda\lambda'}, \Lambda)$ is an inverse system. Moreover, denoting by $v_\lambda: Y_\lambda \to V_\lambda$ the quotient homomorhism, one obtains a morphism $\boldsymbol{v} = (v_\lambda): \boldsymbol{Y} \to \boldsymbol{V}$ of Mod^Λ. We claim that $[\boldsymbol{f}]$ is a kernel of $[\boldsymbol{v}]$. To prove this assertion, consider the inverse system $\boldsymbol{T} = (T_\lambda, t_{\lambda\lambda'}, \Lambda)$, where $T_\lambda = f_\lambda(X_\lambda)$ and $t_{\lambda\lambda'} = q_{\lambda\lambda'}|T_{\lambda'}$. Let $\boldsymbol{w} = (w_\lambda): \boldsymbol{T} \to \boldsymbol{Y}$ be the morphism consisting of inclusions $w_\lambda: T_\lambda \to Y_\lambda$, and let $\boldsymbol{t} = (t_\lambda): \boldsymbol{X} \to \boldsymbol{T}$ be the morphism given by homomorphisms $t_\lambda: X_\lambda \to T_\lambda$, which satisfy $w_\lambda t_\lambda = f_\lambda$. Clearly, $\boldsymbol{wt} = \boldsymbol{f}$, and thus, also

$$[\boldsymbol{w}][\boldsymbol{t}] = [\boldsymbol{f}]. \tag{2}$$

Since w_λ is the kernel of v_λ, one concludes as above that $[\boldsymbol{w}]$ is the kernel of $[\boldsymbol{v}]$. Hence, in view of (2), our assertion will be proved if we show that $[\boldsymbol{t}]$ is an isomorphism. First note that every t_λ is surjective. This implies that $[\boldsymbol{t}]$ is an epimorphism (see e.g., (Mardešić, Segal 1982), II. 2.1, Corollary 3). On the other hand, (2) and the assumption that $[\boldsymbol{f}]$ is a monomorphism imply that also $[\boldsymbol{t}]$ is a monomorphism. However, it is known that pro - Mod is a balanced category, i.e., bimorphisms are isomorphisms (see (Mardešić, Segal 1982), II.2.2, Theorem 6). This completes the proof of the assertion that every monomorphism of pro - Mod is a kernel.

Similar arguments show that every morphism in pro - Mod has a cokernel and every epimorphism is a cokernel. It suffices to consider the case when $\boldsymbol{f} = (f_\lambda): \boldsymbol{X} \to \boldsymbol{Y}$ is a level morphism. Let $\boldsymbol{V}$ and $\boldsymbol{v}$ be as above. Let us show that $[\boldsymbol{v}]$ is a cokernel of $[\boldsymbol{f}]$. Since $[\boldsymbol{v}][\boldsymbol{f}] = [\boldsymbol{0}]$, it remains to show that, whenever a morphism $\boldsymbol{g}: \boldsymbol{Y} \to \boldsymbol{Z}$ satisfies $[\boldsymbol{g}][\boldsymbol{f}] = [\boldsymbol{0}]$, then there exists a unique $[\boldsymbol{h}]: \boldsymbol{V} \to \boldsymbol{Z}$ such that $[\boldsymbol{g}] = [\boldsymbol{h}][\boldsymbol{v}]$. If $\boldsymbol{Z} = (Z_\nu, r_{\nu\nu'}, N)$ and $\boldsymbol{g} = (g, g_\nu)$, then there exists an increasing function $g': N \to \Lambda$ such that $g'(\nu) \geq g(\nu)$ and $g_\nu f_{g(\nu)} p_{g(\nu)g'(\nu)} = g_\nu q_{g(\nu)g'(\nu)} f_{g'(\nu)} = 0$. Therefore g' and the homomorphisms $g'_\nu = g_\nu q_{g(\nu)g'(\nu)}$ define a shift $\boldsymbol{g}': \boldsymbol{Y} \to \boldsymbol{Z}$ of $\boldsymbol{g}$ such that $g'_\nu f_{g'(\nu)} = 0$. Clearly, the homomorphisms g'_ν induce homomorphisms $h_\nu: V_{g'(\nu)} \to Z_\nu$ such that $g'_\nu = h_\nu v_{h(\nu)}$, where $h: N \to \Lambda$ is given by $h(\nu) = g'(\nu)$. It is readily seen that h and h_ν determine a morphism $\boldsymbol{h}: \boldsymbol{V} \to \boldsymbol{Z}$ such that $\boldsymbol{g}' = \boldsymbol{h}\boldsymbol{v}$. Since $[\boldsymbol{g}'] = [\boldsymbol{g}]$, we conclude that indeed, $[\boldsymbol{g}] = [\boldsymbol{h}][\boldsymbol{v}]$. To prove uniqueness of $[\boldsymbol{h}]$, consider another morphism $\boldsymbol{h}': \boldsymbol{V} \to \boldsymbol{Z}$ such that $[\boldsymbol{g}] = [\boldsymbol{h}'][\boldsymbol{v}]$. It is easy to define morphisms $\boldsymbol{k} = (k, k_\nu), \boldsymbol{k}' = (k, k'_\nu): \boldsymbol{V} \to \boldsymbol{Z}$, which are shifts of $\boldsymbol{h}$ and $\boldsymbol{h}'$, respectively, by the same function $k \geq h, h'$, and are such that $k_\nu v_{k(\nu)} = k'_\nu v_{k(\nu)}$. Since $v_{k(\nu)}$ is an epimorphism, it follows that $k_\nu = k'_\nu$, for all $\nu \in N$, and thus, $\boldsymbol{k} = \boldsymbol{k}'$. It now follows that $[\boldsymbol{h}] = [\boldsymbol{k}] = [\boldsymbol{k}'] = [\boldsymbol{h}']$.

Now assume that $\boldsymbol{f}$ is level-preserving and $[\boldsymbol{f}]$ is an epimorphism. Choose $\boldsymbol{U}$ and $\boldsymbol{u}$ as above. We claim that $[\boldsymbol{f}]$ is a cokernel of $[\boldsymbol{u}]$. Let $\boldsymbol{S} = (S_\lambda, s_{\lambda\lambda'}, \Lambda)$ be the system defined by $S_\lambda = X_\lambda / u_\lambda(U_\lambda)$ and by the homomorphisms $s_{\lambda\lambda'}$, induced by $p_{\lambda\lambda'}$. Moreover, let $\boldsymbol{r} = (r_\lambda): \boldsymbol{X} \to \boldsymbol{S}$ and $\boldsymbol{s} = (s_\lambda): \boldsymbol{S} \to \boldsymbol{Y}$ be level morphisms, where r_λ are the quotient homomorphisms, and s_λ are the homomorphisms induced by f_λ. Clearly, $\boldsymbol{s}\boldsymbol{r} = \boldsymbol{f}$. Moreover, by the above argument, $[\boldsymbol{r}]$ is the cokernel of $[\boldsymbol{u}]$. Therefore, it suffices to show that $[\boldsymbol{s}]$ is an isomorphism, because this will imply that also $[\boldsymbol{f}]$ is a cokernel of $[\boldsymbol{u}]$. However, the homomorphisms s_λ are injective, which implies that $[\boldsymbol{s}]$ is a monomorphism (see (Mardešić, Segal 1982), II.2.1, Corollary 1). On the other hand, since $[\boldsymbol{f}]$ is an epimorphism and $[\boldsymbol{s}][\boldsymbol{r}] = [\boldsymbol{f}]$, one concludes that $[\boldsymbol{s}]$ is also an epimorphism, hence, $[\boldsymbol{s}]$ is indeed an isomorphism.

To complete the proof of Theorem 15.8, it remains to show the existence of finite products and coproducts. It suffices to consider the case of two systems $\boldsymbol{A} = (A_\lambda, a_{\lambda\lambda'}, \Lambda)$ and $\boldsymbol{B} = (B_\mu, b_{\mu\mu'}, M)$. Let $\Lambda \times M$ be the direct product, ordered by putting $(\lambda, \mu) \leq (\lambda', \mu')$, provided $\lambda \leq \lambda'$ and $\mu \leq \mu'$. Then $\boldsymbol{A} \times \boldsymbol{B} = (A_\lambda \times B_\mu, a_{\lambda\lambda'} \times b_{\mu\mu'}, \Lambda \times M)$ is also an inverse system. Choose a fixed $\mu \in M$ and define a function $p^A: \Lambda \to \Lambda \times M$ by putting $p^A(\lambda) = (\lambda, \mu)$. Let $p^A_\lambda: A_\lambda \times B_\mu \to A_\lambda$ be the first projection. Clearly, (p^A, p^A_λ) is a morphism $\boldsymbol{p}^A: \boldsymbol{A} \times \boldsymbol{B} \to \boldsymbol{A}$. It depends on μ, but its congruence class $[\boldsymbol{p}^A]: \boldsymbol{A} \times \boldsymbol{B} \to \boldsymbol{A}$ does not. Similarly, using the second projection, we define a morphism $[\boldsymbol{p}^B]: \boldsymbol{A} \times \boldsymbol{B} \to \boldsymbol{B}$. Then $\boldsymbol{A} \times \boldsymbol{B}$ and the morphisms $[\boldsymbol{p}^A], [\boldsymbol{p}^B]$ form a direct product in pro - Mod.

Indeed, let $\boldsymbol{C} = (C_\gamma, c_{\gamma\gamma'}, \Gamma)$ be a system and let $\boldsymbol{f}^A = (f^A, f^A_\lambda): \boldsymbol{C} \to \boldsymbol{A}$ and $\boldsymbol{f}^B = (f^B, f^B_\lambda): \boldsymbol{C} \to \boldsymbol{B}$ be morphisms. Consider a function $f: \Lambda \times M \to \Gamma$, satisfying $f(\lambda, \mu) \geq f^A(\lambda), f^B(\mu)$. Then consider the homomorphisms $f_{(\lambda,\mu)}\colon C_{f(\lambda,\mu)} \to A_\lambda \times B_\mu$, given by $f_{(\lambda,\mu)} = (f^A_\lambda p_{f^A(\lambda) f(\lambda,\mu)}) \times (f^B_\mu p_{f^B(\mu) f(\lambda,\mu)})$. It is easy to verify that f and $f_{(\lambda,\mu)}$ form a morphism $\boldsymbol{f}\colon \boldsymbol{C} \to \boldsymbol{A} \times \boldsymbol{B}$. If $\boldsymbol{p}^A$ is the representative of $[\boldsymbol{p}^A]$, obtained by fixing a given μ, then $\boldsymbol{p}^A \boldsymbol{f}$ is given by the function fp^A and by the homomorphisms $p^A_\lambda f_{p^A(\lambda)} = f^A_\lambda p_{f^A(\lambda) f(\lambda,\mu)}$ and we see that $\boldsymbol{p}^A \boldsymbol{f}$ is the shift of $\boldsymbol{f}^A$, by the function $\lambda \mapsto f(\lambda,\mu)$. Consequetly, $[\boldsymbol{p}^A][\boldsymbol{f}] = [\boldsymbol{f}^A]$. Analogously, $[\boldsymbol{p}^B][\boldsymbol{f}] = [\boldsymbol{f}^B]$. Uniqueness of $[\boldsymbol{f}]$ is also easily verified. The proof of the existence of finite coproducts is similar. □

In the set of morphisms $\boldsymbol{X} \to \boldsymbol{Y}$ one defines addition as follows. For morphisms $[\boldsymbol{f}], [\boldsymbol{g}]\colon \boldsymbol{X} \to \boldsymbol{Y}$, one can always choose representatives $\boldsymbol{f} = (f, f_\mu), \boldsymbol{g} = (g, g_\mu)$ such that $f = g$. Then $(f, f_\mu + g_\mu)$ is a morphism $\boldsymbol{f} + \boldsymbol{g}\colon \boldsymbol{X} \to \boldsymbol{Y}$ of inv-Mod. Moreover, if $\boldsymbol{f}' = (f', f'_\mu), \boldsymbol{g}' = (g', g'_\mu)$ is another choice of representatives with $f' = g'$, then $\boldsymbol{f}' + \boldsymbol{g}' \equiv \boldsymbol{f} + \boldsymbol{g}$. Consequently, $[\boldsymbol{f}] + [\boldsymbol{g}]$ is well defined by putting $[\boldsymbol{f}] + [\boldsymbol{g}] = [\boldsymbol{f} + \boldsymbol{g}]$. It is easy to see that in this way pro-Mod $(\boldsymbol{X}, \boldsymbol{Y})$ becomes an abelian group. Both distributive laws are valid (see (11.1.6) and (11.1.7)).

The next theorem states the main property of the functors $\lim^n$ on pro-Mod and corresponds to Theorem 11.32.

THEOREM 15.9. *Let*

$$E = (\mathbf{0} \to \boldsymbol{X} \overset{[\boldsymbol{f}]}{\to} \boldsymbol{Y} \overset{[\boldsymbol{g}]}{\to} \boldsymbol{Z} \to \mathbf{0}) \tag{3}$$

be a short exact sequence in pro-Mod. *Then there exist homomorphisms* $\theta^n_E\colon \lim^n \boldsymbol{Z} \to \lim^{n+1} \boldsymbol{X}$ *such that the following sequence of modules is exact*

$$\ldots \to \lim{}^n \boldsymbol{X} \overset{\lim^n[\boldsymbol{f}]}{\longrightarrow} \lim{}^n \boldsymbol{Y} \overset{\lim^n[\boldsymbol{g}]}{\longrightarrow} \lim{}^n \boldsymbol{Z} \overset{\theta^n_E}{\longrightarrow} \lim{}^{n+1} \boldsymbol{X} \to \ldots. \tag{4}$$

Moreover, if

$$F = (\mathbf{0} \to \boldsymbol{U} \overset{[\boldsymbol{s}]}{\to} \boldsymbol{V} \overset{[\boldsymbol{t}]}{\to} \boldsymbol{W} \to \mathbf{0}) \tag{5}$$

is another short exact sequence and $r = ([\boldsymbol{u}], [\boldsymbol{v}], [\boldsymbol{w}])\colon E \to F$ *is a morphism of short exact sequences, i.e., a commutative diagram*

$$\begin{array}{ccccccccc}
\mathbf{0} & \longrightarrow & \boldsymbol{X} & \overset{[\boldsymbol{f}]}{\longrightarrow} & \boldsymbol{Y} & \overset{[\boldsymbol{g}]}{\longrightarrow} & \boldsymbol{Z} & \longrightarrow & \mathbf{0} \\
 & & \downarrow{\scriptstyle [\boldsymbol{u}]} & & \downarrow{\scriptstyle [\boldsymbol{v}]} & & \downarrow{\scriptstyle [\boldsymbol{w}]} & & \\
\mathbf{0} & \longrightarrow & \boldsymbol{U} & \underset{[\boldsymbol{s}]}{\longrightarrow} & \boldsymbol{V} & \underset{[\boldsymbol{t}]}{\longrightarrow} & \boldsymbol{W} & \longrightarrow & \mathbf{0},
\end{array} \tag{6}$$

then, for every $n \geq 0$, *also commutes the diagram*

$$\begin{array}{ccc} \lim^n \boldsymbol{Z} & \xrightarrow{\theta_E^n} & \lim^{n+1}\boldsymbol{X} \\ \downarrow{\scriptstyle \lim^n[\boldsymbol{w}]} & & \downarrow{\scriptstyle \lim^{n+1}[\boldsymbol{u}]} \\ \lim^n \boldsymbol{W} & \xrightarrow[\theta_F^n]{} & \lim^{n+1}\boldsymbol{U}\,. \end{array} \tag{7}$$

The proof of Theorem 15.9 requires the following lemma.

LEMMA 15.10. *If* (3) *is a short exact sequence in* pro - Mod, *then there exists a cofinite directed set* Π, *a short exact sequence in* Mod$^{\Pi}$

$$E' = (\mathbf{0} \to \boldsymbol{X}' \xrightarrow{\boldsymbol{f}'} \boldsymbol{Y}' \xrightarrow{\boldsymbol{g}'} \boldsymbol{Z}' \to \mathbf{0}) \tag{8}$$

and a commutative diagram with exact rows

$$\begin{array}{ccccccccc} \mathbf{0} & \longrightarrow & \boldsymbol{X} & \xrightarrow{[\boldsymbol{f}]} & \boldsymbol{Y} & \xrightarrow{[\boldsymbol{g}]} & \boldsymbol{Z} & \longrightarrow & \mathbf{0} \\ & & \downarrow{\scriptstyle [\boldsymbol{i}]} & & \downarrow{\scriptstyle [\boldsymbol{j}]} & & \downarrow{\scriptstyle [\boldsymbol{k}]} & & \\ \mathbf{0} & \longrightarrow & \boldsymbol{X}' & \xrightarrow[{[\boldsymbol{f}']}]{} & \boldsymbol{Y}' & \xrightarrow[{[\boldsymbol{g}']}]{} & \boldsymbol{Z}' & \longrightarrow & \mathbf{0}\,, \end{array} \tag{9}$$

where $[\boldsymbol{i}]$, $[\boldsymbol{j}]$ *and* $[\boldsymbol{k}]$ *are isomorphisms in* pro - Mod.

Proof. Applying Lemma 15.7 to $[\boldsymbol{g}]: \boldsymbol{Y} \to \boldsymbol{Z}$, we obtain a cofinite directed set Π and the right rectangle in diagram (9), where $\boldsymbol{g}' = (g'_\pi)$ is a morphism from Mod$^{\Pi}$. There is no loss of generality in assuming that every $g'_\pi: Y'_\pi \to Z'_\pi$ is surjective. Indeed, if this is not the case, consider a system $\boldsymbol{Z}'' = (Z''_\pi, r''_{\pi\pi'}, \Pi)$, where $Z''_\pi = g'_\pi(Y'_\pi)$ and $r''_{\pi\pi'}$ is induced by $r'_{\pi\pi'}$. Also consider the morphism $\boldsymbol{k}': \boldsymbol{Z}'' \to \boldsymbol{Z}'$, given by inclusions $Z''_\pi \to Z'_\pi$, and the morphism $\boldsymbol{g}'': \boldsymbol{Y}' \to \boldsymbol{Z}''$, given by homomorphisms $g''_\pi: Y'_\pi \to Z''_\pi$, which are induced by g'_π. Clearly, $\boldsymbol{k}'\boldsymbol{g}'' = \boldsymbol{g}'$. Since all k'_π are injections, it follows that $[\boldsymbol{k}']$ is a monomorphism in pro - Mod. On the other hand, $[\boldsymbol{g}]$ is an epimorphism, which implies that also $[\boldsymbol{g}']$ and $[\boldsymbol{k}']$ are epimorphisms. Hence, $[\boldsymbol{k}']$ is an isomorphism and one can replace $[\boldsymbol{k}]$ by $[\boldsymbol{k}']^{-1}[\boldsymbol{k}]$.

Let the system $\boldsymbol{X}' = (X'_\pi, p'_{\pi\pi'}, \Pi)$ consist of modules $X'_\pi = (g'_\pi)^{-1}(0)$ and homomorphisms $p'_{\pi\pi'} = q'_{\pi\pi'}|X'_{\pi'}$ and let $\boldsymbol{f}': \boldsymbol{X}' \to \boldsymbol{Y}'$ be given by inclusions $f'_\pi: X'_\pi \to Y'_\pi$. Clearly, (8) is an exact sequence. Moreover, $[\boldsymbol{f}']$ is a kernel of $[\boldsymbol{g}']$, $[\boldsymbol{f}']$ is a monomorphism and $[\boldsymbol{g}']$ is an epimorphism, which implies that the second row in (9) is also exact. Since $[\boldsymbol{j}]$ and $[\boldsymbol{k}]$ are isomorphisms, it readily follows that $[\boldsymbol{j}]^{-1}[\boldsymbol{f}']$ is a kernel of $[\boldsymbol{g}]$. On the other hand, $[\boldsymbol{f}]$ is also a kernel of $[\boldsymbol{g}]$. Consequently, by uniqueness of the kernels, there exists a unique isomorphism $[\boldsymbol{i}]: \boldsymbol{X} \to \boldsymbol{X}'$, such that the left rectangle in diagram (9) also commutes. □

To $(E', [\boldsymbol{i}], [\boldsymbol{j}], [\boldsymbol{k}])$ as in Lemma 15.10, we refer as to a *cofinite level presentation* of the sequence (3). Theorem 11.32 associates with the sequence (8) homomorphisms $\theta^n_{E'}\colon \lim^n \boldsymbol{Z}' \to \lim^{n+1} \boldsymbol{X}'$, $n \geq 0$, which make the following sequence exact.

$$\ldots \to \lim^n \boldsymbol{X} \overset{\lim^n \boldsymbol{f}'}{\longrightarrow} \lim^n \boldsymbol{Y} \overset{\lim^n \boldsymbol{g}'}{\longrightarrow} \lim^n \boldsymbol{Z} \overset{\theta^n_{E'}}{\longrightarrow} \lim^{n+1} \boldsymbol{X} \to \ldots . \tag{10}$$

Since $\lim^n$ are functors on pro - Mod, (9) yields the following commutative diagram (solid arrows).

$$\begin{array}{ccccccccc}
\longrightarrow & \lim^n \boldsymbol{X} & \xrightarrow{\lim^n [\boldsymbol{f}]} & \lim^n \boldsymbol{Y} & \xrightarrow{\lim^n [\boldsymbol{g}]} & \lim^n \boldsymbol{Z} & \overset{\theta^n_E}{\cdots\cdots\rightarrow} & \lim^{n+1} \boldsymbol{X} & \longrightarrow \\
 & \downarrow{\scriptstyle \lim^n [\boldsymbol{i}]} & & \downarrow{\scriptstyle \lim^n [\boldsymbol{j}]} & & \downarrow{\scriptstyle \lim^n [\boldsymbol{k}]} & & \downarrow{\scriptstyle \lim^{n+1} [\boldsymbol{i}]} & \\
\longrightarrow & \lim^n \boldsymbol{X}' & \xrightarrow[\lim^n \boldsymbol{f}']{} & \lim^n \boldsymbol{Y}' & \xrightarrow[\lim^n \boldsymbol{g}']{} & \lim^n \boldsymbol{Z}' & \xrightarrow[\theta^n_{E'}]{} & \lim^{n+1} \boldsymbol{X}' & \longrightarrow .
\end{array} \tag{11}$$

The second row of this diagram is the exact sequence (10) and the vertical arrows are isomorphisms. We complete the diagram by adding homomorphisms $\theta^n_E\colon \lim^n \boldsymbol{Z} \to \lim^{n+1} \boldsymbol{X}$, defined by

$$\theta^n_E = (\lim^{n+1} [\boldsymbol{i}])^{-1}\, \theta^n_{E'}\, (\lim^n [\boldsymbol{k}]). \tag{12}$$

Clearly, the first row of (11) becomes exact. By definition, this is the desired sequence (4).

We still need to show that the homomorphisms θ^n_E do not depend on the choice of the cofinite level presentations of $\boldsymbol{E}$. We first prove two more lemmas.

LEMMA 15.11. *Let*

$$E = (\boldsymbol{0} \to \boldsymbol{X} \overset{f}{\to} \boldsymbol{Y} \overset{g}{\to} \boldsymbol{Z} \to \boldsymbol{0}), \tag{13}$$

$$F = (\boldsymbol{0} \to \boldsymbol{U} \overset{s}{\to} \boldsymbol{V} \overset{t}{\to} \boldsymbol{W} \to \boldsymbol{0}) \tag{14}$$

be sequences of level morphisms in Mod^{Π} *and* Mod^{P}, *respectively, and let* $[\boldsymbol{u}]$, $[\boldsymbol{v}]$ *and* $[\boldsymbol{w}]$ *be morphisms of* pro - Mod, *which make diagram* (6) *commutative. Then there exist representatives* $\boldsymbol{u}$, $\boldsymbol{v}$, $\boldsymbol{w}$ *of* $[\boldsymbol{u}], [\boldsymbol{v}], [\boldsymbol{w}]$, *which make the following diagram in* inv - Top *commutative.*

$$\begin{array}{ccccccccc}
\boldsymbol{0} & \longrightarrow & \boldsymbol{X} & \xrightarrow{\boldsymbol{f}} & \boldsymbol{Y} & \xrightarrow{\boldsymbol{g}} & \boldsymbol{Z} & \longrightarrow & \boldsymbol{0} \\
 & & \downarrow{\scriptstyle \boldsymbol{u}} & & \downarrow{\scriptstyle \boldsymbol{v}} & & \downarrow{\scriptstyle \boldsymbol{w}} & & \\
\boldsymbol{0} & \longrightarrow & \boldsymbol{U} & \xrightarrow[\boldsymbol{s}]{} & \boldsymbol{V} & \xrightarrow[\boldsymbol{t}]{} & \boldsymbol{W} & \longrightarrow & \boldsymbol{0} .
\end{array} \tag{15}$$

Proof. Choose arbitrary representatives $\boldsymbol{u}', \boldsymbol{v}'$ and $\boldsymbol{w}'$ of $[\boldsymbol{u}]$, $[\boldsymbol{v}]$ and $[\boldsymbol{w}]$, respectively. Then, $\boldsymbol{s}\boldsymbol{u}' \equiv \boldsymbol{v}'\boldsymbol{f}$, implies the existence of an increasing function $\phi' : P \to \Pi$ such that $\phi' \geq u', v'$ and

$$s_\rho u'_\rho p_{u'(\rho)\phi'(\rho)} = v'_\rho f_{v'(\rho)} p_{v'(\rho)\phi'(\rho)} = v'_\rho q_{v'(\rho)\phi'(\rho)} f_{\phi'(\rho)}. \tag{16}$$

Now denote by $\boldsymbol{u}$ and $\boldsymbol{v}$ the shifts of $\boldsymbol{u}'$ and $\boldsymbol{v}'$, respectively, by the function ϕ' (or by any increasing function $\phi \geq \phi'$). Clearly, $\boldsymbol{s}\boldsymbol{u} = \boldsymbol{v}\boldsymbol{f}$. The same argument applies to $\boldsymbol{v}', \boldsymbol{w}'$ and yields shifts $\boldsymbol{v}$ and $\boldsymbol{w}$ of $\boldsymbol{v}'$ and $\boldsymbol{w}'$, respectively, by an increasing function ϕ''. Moreover, $\boldsymbol{t}\boldsymbol{v} = \boldsymbol{w}\boldsymbol{g}$. There is no loss of generality in assuming that $\phi' = \phi''$ and thus, one obtains in both cases the same morphism $\boldsymbol{v}$, i.e., one obtains the commutative diagram (15). □

The next lemma strengthens the second part of Theorem 11.32 and reads as follows.

LEMMA 15.12. *Let* (15) *be a commutative diagram, whose rows E and F consist of level morphisms of* inv-Mod, *indexed by Π and P respectively, and whose vertical arrows are morphisms of* inv-Mod. *If the rows are exact and $\theta^n_E : \lim^n \boldsymbol{Z} \to \lim^{n+1} \boldsymbol{X}$, $\theta^n_F : \lim^n \boldsymbol{W} \to \lim^{n+1} \boldsymbol{U}$ are the induced homomorphisms, then diagram* (7) *commutes.*

Proof. By Lemma 15.3, diagram (15) yields a commutative diagram of cochain complexes and cochain mappings.

$$\begin{array}{ccccccccc}
0 & \longrightarrow & K(\boldsymbol{X}) & \xrightarrow{\boldsymbol{f}^\#} & K(\boldsymbol{Y}) & \xrightarrow{\boldsymbol{g}^\#} & K(\boldsymbol{Z}) & \longrightarrow & 0 \\
 & & \downarrow \boldsymbol{u}^\# & & \downarrow \boldsymbol{v}^\# & & \downarrow \boldsymbol{w}^\# & & \\
0 & \longrightarrow & K(\boldsymbol{U}) & \xrightarrow[\boldsymbol{s}^\#]{} & K(\boldsymbol{V}) & \xrightarrow[\boldsymbol{t}^\#]{} & K(\boldsymbol{W}) & \longrightarrow & 0.
\end{array} \tag{17}$$

By Lemma 11.42, the rows in this diagram are exact. Therefore, application of cohomology yields a morphism between the long cohomology sequences of these two rows. Since these sequences coincide with the sequences (4), for the rows of (6), one concludes that the corresponding diagram (7) is indeed commutative. □

LEMMA 15.13. *The homomorphisms θ^n_E, given by* (12), *do not depend on the choice of the cofinite level presentations of the sequence E.*

Proof. Assume that $(E', [\boldsymbol{i}], [\boldsymbol{j}], [\boldsymbol{k}])$ is a cofinite level presentation of the sequence E, while

$$E'' = (\boldsymbol{0} \to \boldsymbol{X}'' \xrightarrow{\boldsymbol{f}''} \boldsymbol{Y}'' \xrightarrow{\boldsymbol{g}''} \boldsymbol{Z}'' \to \boldsymbol{0}), \tag{18}$$

$[\boldsymbol{i}'] : \boldsymbol{X} \to \boldsymbol{X}''$, $[\boldsymbol{j}'] : \boldsymbol{Y} \to \boldsymbol{Y}''$ and $[\boldsymbol{k}'] : \boldsymbol{Z} \to \boldsymbol{Z}''$ form another level presentation of E. In view of (12), we must show that

$$(\lim^{n+1}[\boldsymbol{i}])^{-1}\,\theta^n_{E'}\,(\lim^n[\boldsymbol{k}]) = (\lim^{n+1}[\boldsymbol{i}'])^{-1}\,\theta^n_{E''}\,(\lim^n[\boldsymbol{k}']). \tag{19}$$

Putting

$$[\boldsymbol{u}] = [\boldsymbol{i}'][\boldsymbol{i}]^{-1},\ [\boldsymbol{v}] = [\boldsymbol{j}'][\boldsymbol{j}]^{-1},\ [\boldsymbol{w}] = [\boldsymbol{k}'][\boldsymbol{k}]^{-1}, \tag{20}$$

one obtains a commutative diagram:

$$\begin{array}{ccccccccc}
\mathbf{0} & \longrightarrow & \boldsymbol{X}' & \xrightarrow{[\boldsymbol{f}']} & \boldsymbol{Y}' & \xrightarrow{[\boldsymbol{g}']} & \boldsymbol{Z}' & \longrightarrow & \mathbf{0} \\
 & & \downarrow{\scriptstyle [\boldsymbol{u}]} & & \downarrow{\scriptstyle [\boldsymbol{v}]} & & \downarrow{\scriptstyle [\boldsymbol{w}]} & & \\
\mathbf{0} & \longrightarrow & \boldsymbol{X}'' & \xrightarrow[{[\boldsymbol{f}'']}]{} & \boldsymbol{Y}'' & \xrightarrow[{[\boldsymbol{g}'']}]{} & \boldsymbol{Z}'' & \longrightarrow & \mathbf{0}\,.
\end{array} \tag{21}$$

Applying Lemma 15.11, we infer that there are representatives $\boldsymbol{u}\colon \boldsymbol{X}' \to \boldsymbol{X}''$, $\boldsymbol{v}\colon \boldsymbol{Y}' \to \boldsymbol{Y}''$ and $\boldsymbol{w}\colon \boldsymbol{Z}' \to \boldsymbol{Z}''$ of the classes $[\boldsymbol{u}], [\boldsymbol{v}]$ and $[\boldsymbol{w}]$, respectively, such that commutes the diagram obtained from (21), by replacing all arrows by the corresponding representatives $\boldsymbol{f}'$, $\boldsymbol{g}'$, $\boldsymbol{f}''$,$\boldsymbol{g}''$,$\boldsymbol{u}$,$\boldsymbol{v}$,$\boldsymbol{w}$. To this diagram we then apply Lemma 15.12 and thus, obtain the commutative diagram

$$\begin{array}{ccc}
\lim^n \boldsymbol{Z}' & \xrightarrow{\theta^n_{E'}} & \lim^{n+1}\boldsymbol{X}' \\
{\scriptstyle \lim^n[\boldsymbol{w}]}\downarrow & & \downarrow{\scriptstyle \lim^{n+1}[\boldsymbol{u}]} \\
\lim^n \boldsymbol{Z}'' & \xrightarrow[\theta^n_{E''}]{} & \lim^{n+1}\boldsymbol{X}''\,,
\end{array} \tag{22}$$

i.e., we obtain the equality

$$\lim^{n+1}[\boldsymbol{u}]\theta^n_{E'} = \theta^n_{E''}\lim^n[\boldsymbol{w}]. \tag{23}$$

However, (20) transforms (23) to the desired relation (19). □

To prove the second assertion of Theorem 15.9, consider cofinite level presentations $(E', [\boldsymbol{i}], [\boldsymbol{j}], [\boldsymbol{k}])$ of (3) and $(F', [\boldsymbol{i}'], [\boldsymbol{j}'], [\boldsymbol{k}'])$ of (5). Clearly, there exist unique morphisms $[\boldsymbol{u}'], [\boldsymbol{v}'], [\boldsymbol{w}']$, such that

$$[\boldsymbol{u}'][\boldsymbol{i}] = [\boldsymbol{i}'][\boldsymbol{u}],\ [\boldsymbol{v}'][\boldsymbol{j}] = [\boldsymbol{j}'][\boldsymbol{v}],\ [\boldsymbol{w}'][\boldsymbol{k}] = [\boldsymbol{k}'][\boldsymbol{w}]. \tag{24}$$

One readily concludes that $[\boldsymbol{v}'][\boldsymbol{f}'][\boldsymbol{i}] = [\boldsymbol{s}'][\boldsymbol{u}'][\boldsymbol{i}]$. Since $[\boldsymbol{i}]$ is an isomorphism, it follows that

$$[\boldsymbol{v}'][\boldsymbol{f}'] = [\boldsymbol{s}'][\boldsymbol{u}']. \tag{25}$$

Similarly,

$$[\boldsymbol{w}'][\boldsymbol{g}'] = [\boldsymbol{t}'][\boldsymbol{v}'], \tag{26}$$

i.e., the following diagram commutes.

$$\begin{array}{ccccccccc} 0 & \longrightarrow & X' & \xrightarrow{[f']} & Y' & \xrightarrow{[g']} & Z' & \longrightarrow & 0 \\ & & \downarrow{[u']} & & \downarrow{[v']} & & \downarrow{[w']} & & \\ 0 & \longrightarrow & U' & \xrightarrow[{[s']}]{} & V' & \xrightarrow[{[t']}]{} & W' & \longrightarrow & 0. \end{array} \tag{27}$$

Let $\theta^n_{E'}$ and $\theta^n_{F'}$ be the homomorphisms induced by the exact sequences $E' = (\mathbf{0} \to \boldsymbol{X}' \to \boldsymbol{Y}' \to \boldsymbol{Z}' \to \mathbf{0})$ and $F' = (\mathbf{0} \to \boldsymbol{U}' \to \boldsymbol{V}' \to \boldsymbol{W}' \to \mathbf{0})$. Applying subsequently Lemmas 15.11 and 15.12 to diagram (27), one concludes that

$$\theta^n_{F'}(\lim{}^n[\boldsymbol{w}']) = (\lim{}^{n+1}[\boldsymbol{u}'])\theta^n_{E'}. \tag{28}$$

On the other hand, by definition,

$$\theta^n_{E'}(\lim{}^n[\boldsymbol{k}]) = (\lim{}^{n+1}[\boldsymbol{i}])\theta^n_{E}, \tag{29}$$

$$\theta^n_{F'}(\lim{}^n[\boldsymbol{k}']) = (\lim{}^{n+1}[\boldsymbol{i}'])\theta^n_{F}. \tag{30}$$

Moreover, by (24),

$$\lim{}^{n+1}[\boldsymbol{u}']\lim{}^{n+1}[\boldsymbol{i}] = \lim{}^{n+1}[\boldsymbol{i}']\lim{}^{n+1}[\boldsymbol{u}], \tag{31}$$

$$\lim{}^{n}[\boldsymbol{w}']\lim{}^{n}[\boldsymbol{k}] = \lim{}^{n}[\boldsymbol{k}']\lim{}^{n}[\boldsymbol{w}]. \tag{32}$$

Now (28)–(32) prove that

$$(\lim{}^{n+1}[\boldsymbol{i}'])(\lim{}^{n+1}[\boldsymbol{u}])\theta^n_E = (\lim{}^{n+1}[\boldsymbol{i}'])\theta^n_F(\lim{}^n[\boldsymbol{w}]). \tag{33}$$

Finally, since $\lim^{n+1}[\boldsymbol{i}']$ is an isomorphism, (33) establishes the desired commutativity of diagram (7). □

We now strengthen Theorem 15.9 to the following result.

THEOREM 15.14. *The functors* $\lim^n$: pro-Mod $\to$ Mod, *together with the connecting homomorphisms* θ^n, $n = 0, 1, \ldots$, *form a universal connected sequence of functors.*

In view of Theorem 11.37, in order to prove Theorem 15.14, it suffices to prove the following result.

THEOREM 15.15. *The category* pro-Mod *has enough injective objects. For every injective object* J *of* pro-Mod *and every* $n \geq 1$, $\lim^n \boldsymbol{J} = 0$.

To prove this theorem we first establish two lemmas.

LEMMA 15.16. *Let* $\boldsymbol{h}: \boldsymbol{B} \to \boldsymbol{A}$ *be a morphism of* Mod^Λ, *which induces a monomorphism* $[\boldsymbol{h}]$ *of* pro-Mod. *If* $\boldsymbol{I}$ *is an injective object of* Mod^M *and* $\boldsymbol{k}: \boldsymbol{B} \to \boldsymbol{I}$ *is a morphism of* inv-Mod, *then there exists a morphism* $\boldsymbol{g}: \boldsymbol{A} \to \boldsymbol{I}$ *of* inv-Mod *such that* $\boldsymbol{gh} \equiv \boldsymbol{k}$.

Proof. Let $\boldsymbol{A} = (A_\lambda, a_{\lambda\lambda'}, \Lambda)$, $\boldsymbol{B} = (B_\lambda, b_{\lambda\lambda'}, \Lambda)$, $\boldsymbol{I} = (I_\mu, q_{\mu\mu'}, M)$ and let $\boldsymbol{h} = (h_\mu)$ and $\boldsymbol{k} = (k, k_\mu)$. Put $\boldsymbol{A}' = \mathrm{Im}(\boldsymbol{A})$, i.e., $\boldsymbol{A}' = (A'_\lambda, a'_{\lambda\lambda'}, \Lambda)$, where $A'_\lambda = h_\lambda(A_\lambda)$ and $a'_{\lambda\lambda'}$ are the appropriate restrictions of $a_{\lambda\lambda'}$, and let $\boldsymbol{h}' = (h'_\lambda)\colon \boldsymbol{B} \to \boldsymbol{A}'$ be the level morphism induced by $\boldsymbol{h}$, i.e., $h'_\lambda\colon B_\lambda \to A'_\lambda$ is the homomorphism induced by h_λ. Let $\boldsymbol{i} = (i_\lambda)\colon \boldsymbol{A}' \to \boldsymbol{A}$ be the morphism given by inclusions $i_\lambda\colon A'_\lambda \to A_\lambda$. Note that $i_\lambda h'_\lambda = h_\lambda$ and thus, $\boldsymbol{ih}' = \boldsymbol{h}$.

To prove the assertion, it suffices to exhibit a morphism $\boldsymbol{g}' = (g, g'_\mu)\colon \boldsymbol{A}' \to \boldsymbol{I}$ such that

$$\boldsymbol{g}'\boldsymbol{h}' \equiv \boldsymbol{k}. \tag{34}$$

Indeed, let $\boldsymbol{C} = (C_\mu, c_{\mu\mu'}, M)$ and $\boldsymbol{D} = (D_\mu, d_{\mu\mu'}, M)$ be systems from Mod^M, defined by putting $C_\mu = A'_{g(\mu)}$, $c_{\mu\mu'} = a'_{g(\mu)g(\mu')}$ and $D_\mu = A_{g(\mu)}$, $d_{\mu\mu'} = a_{g(\mu)g(\mu')}$, respectively, and let $\boldsymbol{j} = (j_\mu)\colon \boldsymbol{C} \to \boldsymbol{D}$ be given by $j_\mu = i_{g(\mu)}$. Note that every j_μ is a monomorphism of modules and therefore, $\boldsymbol{j}$ is a monomorphism from Mod^M. Furthermore, $\boldsymbol{g}'\colon \boldsymbol{A}' \to \boldsymbol{I}$ yields a level morphism $\overline{\boldsymbol{g}}' = (g'_\mu)\colon \boldsymbol{C} \to \boldsymbol{I}$, because $g'_\mu\colon C_\mu \to I_\mu$ and $q_{\mu\mu'}g'_{\mu'} = g'_\mu c_{\mu\mu'}$, for $\mu \le \mu'$. Since $\boldsymbol{I}$ is injective in Mod^M, we conclude that there is a morphism $\overline{\boldsymbol{g}} = (g_\mu)\colon \boldsymbol{D} \to \boldsymbol{I}$ from Mod^M, such that $\overline{\boldsymbol{g}}' = \overline{\boldsymbol{g}}\boldsymbol{j}$, i.e., $g'_\mu = g_\mu j_\mu = g_\mu i_{g'(\mu)}$. Since $g_\mu d_{\mu\mu'} = q_{\mu\mu'} g_{\mu'}$, it follows that $\boldsymbol{g} = (g, g_\mu)\colon \boldsymbol{A} \to \boldsymbol{I}$ is a morphism of inv - Mod such that $\boldsymbol{g}' = \boldsymbol{gi}$. Consequently, $\boldsymbol{k} \equiv \boldsymbol{g}'\boldsymbol{h}' = \boldsymbol{gih}' = \boldsymbol{gh}$.

We will now construct $\boldsymbol{g}'$. Since $[\boldsymbol{h}]$ is a monomorphism of pro - Mod, every $\lambda \in \Lambda$ admits a $\lambda' \ge \lambda$ such that $\mathrm{Ker}(h_{\lambda'}) \subseteq \mathrm{Ker}(b_{\lambda\lambda'})$ (see (Mardešić, Segal 1982), II.2.1, Theorem 1). Consequently (due to the cofiniteness of M), there exists and increasing function $g\colon M \to \Lambda$, $g \ge k$, such that

$$\mathrm{Ker}(h_{g(\mu)}) \subseteq \mathrm{Ker}(b_{k(\mu)g(\mu)}),\ \ \mu \in M. \tag{35}$$

To define g'_μ note that, by (35), $k_\mu b_{k(\mu)g(\mu)}(\mathrm{Ker}(h_{g(\mu)})) = 0$. Consequently, there is a unique homomorphism $\tilde{k}_\mu\colon B_{g(\mu)}/\mathrm{Ker}(h_{g(\mu)}) \to I_\mu$ such that

$$k_\mu b_{k(\mu)g(\mu)} = \tilde{k}_\mu r_\mu, \tag{36}$$

where $r_\mu\colon B_{g(\mu)} \to B_{g(\mu)}/\mathrm{Ker}(h_{g(\mu)})$ is the quotient homomorphism. Now note that $h_{g(\mu)}$ induces an isomorphism $\tilde{h}_\mu\colon B_{g(\mu)}/\mathrm{Ker}(h_{g(\mu)}) \to A'_{g(\mu')}$ such that

$$\tilde{h}_\mu r_\mu = h'_{g(\mu)}. \tag{37}$$

Consequently, there is a unique homomorphism $g'_\mu\colon A'_{g(\mu')} \to A_\mu$ such that

$$\tilde{k}_\mu = g'_\mu \tilde{h}_\mu. \tag{38}$$

Clearly, (36), (38) and (37) imply

$$k_\mu b_{k(\mu)g(\mu)} = g'_\mu h'_{g(\mu)}. \tag{39}$$

Let us show that $\boldsymbol{g}' = (g, g'_\mu)$ is indeed a morphism of inv - Mod. If $\mu \le \mu'$, by (39), we see that

$$
\begin{aligned}
&q_{\mu\mu'} g'_{\mu'} h'_{g(\mu')} = \\
&q_{\mu\mu'} k_{\mu'} b_{k(\mu')g(\mu')} = k_\mu b_{k(\mu)k(\mu')} b_{k(\mu')g(\mu')} = \\
&k_\mu b_{k(\mu)g(\mu)} b_{g(\mu)g(\mu')} = g'_\mu h'_{g(\mu)} b_{g(\mu)g(\mu')} = \\
&g'_\mu a_{g(\mu)g(\mu')} h'_{g(\mu')}.
\end{aligned} \tag{40}
$$

Since $h'_{g(\mu')}$ is an epimorphism, it follows that indeed,

$$q_{\mu\mu'} g'_{\mu'} = g_{\mu'} a_{g(\mu)g(\mu')}. \tag{41}$$

Finally, (39) shows that $\boldsymbol{k} \equiv \boldsymbol{g}'\boldsymbol{h}'$. □

LEMMA 15.17. *If $\boldsymbol{I}$ is an injective object of* Mod^M, *then $\boldsymbol{I}$ is also an injective object of* pro - Mod.

Proof. Let $\boldsymbol{h}: \boldsymbol{B} \to \boldsymbol{A}$ and $\boldsymbol{k}: \boldsymbol{B} \to \boldsymbol{I}$ be morphisms of inv - Mod such that $[\boldsymbol{h}]$ is a monomorphism of pro - Mod. By Lemma 15.7, there exist morphisms of $\boldsymbol{i}: \boldsymbol{B} \to \boldsymbol{B}'$, $\boldsymbol{j}: \boldsymbol{A} \to \boldsymbol{A}'$, $\boldsymbol{h}': \boldsymbol{B}' \to \boldsymbol{A}'$ such that

$$[\boldsymbol{j}][\boldsymbol{h}] = [\boldsymbol{h}'][\boldsymbol{i}], \tag{42}$$

$[\boldsymbol{i}]$ and $[\boldsymbol{j}]$ are isomorphisms of pro - Mod and $\boldsymbol{h}'$ is a level morphism, i.e., it belongs to Mod^M, for some directed set M. Since $[\boldsymbol{i}]$ is an isomorphism, one can find a morphism $\boldsymbol{k}': \boldsymbol{B}' \to \boldsymbol{I}$ such that $[\boldsymbol{k}'][\boldsymbol{i}] = [\boldsymbol{k}]$. Applying Lemma 15.16 to $\boldsymbol{h}'$ and $\boldsymbol{k}'$, we obtain a morphism $\boldsymbol{g}': \boldsymbol{B}' \to \boldsymbol{I}$ such that $[\boldsymbol{g}'][\boldsymbol{h}'] = [\boldsymbol{k}']$. Define $\boldsymbol{g}: \boldsymbol{A} \to \boldsymbol{I}$ by putting $\boldsymbol{g} = \boldsymbol{g}'\boldsymbol{j}$. Then $[\boldsymbol{g}][\boldsymbol{h}] = [\boldsymbol{k}]$, which proves that $\boldsymbol{I}$ is indeed an injective object of pro - Mod. □

Proof of Theorem 15.15. Let $\boldsymbol{X}$ be an object of pro - Mod, indexed by Λ. By Theorem 11.18, there exists a system $\boldsymbol{I}$, which is an injective object of Mod^Λ and there exists an $\boldsymbol{f} = (f_\lambda): \boldsymbol{X} \to \boldsymbol{I}$, which is a monomorphism of Mod^Λ. By Lemma 11.2, every f_λ is a monomorphism. This implies that $[\boldsymbol{f}]: \boldsymbol{X} \to \boldsymbol{I}$ is a monomorphism of pro - Mod (see (Mardešić, Segal 1982), II.2.1, Corollary 1). However, by Lemma 15.17, $\boldsymbol{I}$ is also an injective object of pro - Mod, which proves that the latter category has enough injective objects.

To prove the second assertion of Theorem 15.15, assume that $\boldsymbol{J}$ is an injective object of pro - Mod. If $\boldsymbol{J}$ is indexed by Λ, one finds (as above) a monomorphism $[\boldsymbol{f}]: \boldsymbol{J} \to \boldsymbol{I}$, where $\boldsymbol{I}$ is injective in Mod^Λ and thus, by Lemma 11.31, $\lim^n \boldsymbol{I} = 0$, for $n \geq 1$. Since $\boldsymbol{J}$ is injective, there exists a morphism $[\boldsymbol{g}]: \boldsymbol{I} \to \boldsymbol{J}$ such that $[\boldsymbol{g}][\boldsymbol{f}] = \mathrm{id}$. Since $\lim^n$ is a functor on pro - Mod (Theorem 15.5), one concludes that $\lim^n[\boldsymbol{g}]\lim^n[\boldsymbol{f}] = \mathrm{id}$. However, $\lim^n[\boldsymbol{f}] = 0$ and thus, the identity on $\lim^n \boldsymbol{J}$ equals zero, i.e., $\lim^n \boldsymbol{J} = 0$. □

Bibliographic notes

The definition of the cochain mapping $\boldsymbol{f}^\#$ in 15.1, as well as Lemma 15.4, are special cases of a more general situation, concerning strong homology, and considered already in (Lisica, Mardešić 1985d, 1985e). The extension of $\lim^n$

to systems over index sets which are not cofinite was considered in detail in (Watanabe 1991b), which in turn is a special case of the more general situation encountered in strong homology (Mardešić 1987). In particular, these papers give explicit formulae for the homotopy inverse of the cochain mappings $\boldsymbol{u}^{\#}_{\boldsymbol{X}}$. Various elementary facts concerning the category pro-Ab can be found in (Mardešić, Segal 1982). The reindexing lemma 15.7 appears already in (Artin, Mazur 1969). The assertion that pro-$\mathcal{A}$ is abelian, whenever $\mathcal{A}$ is an abelian category, is stated in (Verdier 1965).

IV. HOMOLOGY GROUPS

16. Homology pro-groups

The first subsection is devoted to the definition of (ordinary) homology pro-groups $H_m(\boldsymbol{X};G)$ of an inverse system of spaces $\boldsymbol{X}$. Their limits are the Čech homology groups $\check{H}_m(\boldsymbol{X};G)$. The higher derived limits $\lim^r H_m(\boldsymbol{X};G)$ are also well defined. The second subsection is devoted to the construction of examples which show that, in general, these limits are non-trivial. This fact has important consequences for the strong homology groups, defined in sections 17 and 18.

16.1 Homology pro-groups and Čech homology

An *inverse system of chain complexes* $\boldsymbol{C} = (C_\lambda, p_{\lambda\lambda'}, \Lambda)$, also called a *pro-chain complex*, is an inverse system consisting of chain complexes C_λ and chain mappings $p_{\lambda\lambda'}: C_{\lambda'} \to C_\lambda$. We will assume that C_λ consists of abelian groups C_λ^n, $n \in \mathbb{Z}$, and of boundary operators $\partial_\lambda^n: C_\lambda^n \to C_\lambda^{n-1}$, while $p_{\lambda\lambda'}$ consists of homomorphisms $p_{\lambda\lambda'}^n: C_{\lambda'}^n \to C_\lambda^n$, such that the following diagram commutes.

$$\begin{array}{ccc} C_\lambda^n & \xleftarrow{\;p_{\lambda\lambda'}^n\;} & C_{\lambda'}^n \\ \downarrow{\scriptstyle \partial_\lambda^n} & & \downarrow{\scriptstyle \partial_{\lambda'}^n} \\ C_\lambda^{n-1} & \xleftarrow[\;p_{\lambda\lambda'}^{n-1}\;]{} & C_{\lambda'}^{n-1} \, . \end{array} \tag{1}$$

A *morphism* $\boldsymbol{f}: \boldsymbol{C} \to \boldsymbol{D} = (D_\mu, q_{\mu\mu'}, M)$ *of pro-chain complexes* consists of an increasing function $f: M \to \Lambda$ and of chain mappings $f_\mu: C_{f(\mu)} \to D_\mu$ such that

$$f_\mu p_{f(\mu)f(\mu')} = q_{\mu\mu'} f_{\mu'}, \ \mu \le \mu'. \tag{2}$$

Recall that f_μ consists of homomorphisms $f_\mu^n: C_{f(\mu)}^n \to D_\mu^n$ such that

$$\partial_\mu^n f_\mu^n = f_\mu^{n-1} \partial_{f(\mu)}^n . \tag{3}$$

Pro-chain complexes and their morphisms form the *category of pro-chain complexes*, denoted by inv - Chn.

Application of the homology functor H_n to a pro-chain complex $\boldsymbol{C}$ yields an inverse system of abelian groups $H_n(\boldsymbol{C}) = (H_n(C_\lambda), p_{\lambda\lambda'*}, \Lambda)$, called the n-th *homology pro-group* of $\boldsymbol{C}$, $n \in \mathbb{Z}$. A morphism of pro-chain complexes $\boldsymbol{f}: \boldsymbol{C} \to \boldsymbol{D}$ induces a morphism of the homology pro-groups $\boldsymbol{f}_* = H_n(\boldsymbol{f}): H_n(\boldsymbol{C}) \to H_n(\boldsymbol{D})$, given by f and the homomorphisms $f_{\mu*}: H_n(C_{f(\mu)}) \to H_n(D_\mu)$. Clearly, H_n: inv-Chn$\to$ inv-Ab is a functor.

With every inverse system of spaces $\boldsymbol{X} = (X_\lambda, p_{\lambda\lambda'}, \Lambda)$ one can associate an inverse system of chain complexes $S(\boldsymbol{X})$, called the *singular pro-complex* of the system $\boldsymbol{X}$. It is the pro-chain complex $\boldsymbol{C} = (C_\lambda, q_{\lambda\lambda'}, \Lambda)$, where $C_\lambda = S(X_\lambda)$ is the singular complex of the space X_λ, while $q_{\lambda\lambda'} = S(p_{\lambda\lambda'}): S(X_{\lambda'}) \to S(X_\lambda)$ is the induced chain mapping. The induced homology pro-group $H_n(S(\boldsymbol{X}))$ is denoted by $H_n(\boldsymbol{X})$ and is called the n-th *homology pro-group* of the system $\boldsymbol{X}$, $n \geq 0$. Clearly, $H_n(\boldsymbol{X}) = (H_n(X_\lambda), p_{\lambda\lambda'*}, \Lambda)$, where $H_n(X_\lambda)$ is the *singular homology group* (integer coefficients) of X_λ and $p_{\lambda\lambda'*}$ is the homomorphism induced by $p_{\lambda\lambda'}: X_{\lambda'} \to X_\lambda$. If $\boldsymbol{f} = (f, f_\mu): \boldsymbol{X} \to \boldsymbol{Y}$ is a mapping of systems, then $S(\boldsymbol{f}) = (f, S(f_\mu)): S(\boldsymbol{X}) \to S(\boldsymbol{Y})$ is a morphism of pro-chain complexes. Consequently, $S(\boldsymbol{f})$ induces morphisms of homology pro-groups $\boldsymbol{f}_* = H_n(\boldsymbol{f}): H_n(\boldsymbol{X}) \to H_n(\boldsymbol{Y})$, $n \geq 0$. It is readily seen that in this way one obtains a sequence of functors H_n: pro-Top$\to$ Ab, $n \geq 0$.

The above definition can be extended to inverse systems in the homotopy category H(Top). If $[\boldsymbol{X}] = (X_\lambda, [p_{\lambda\lambda'}], \Lambda)$ is such a system, then different representatives $p_{\lambda\lambda'}$ of the same homotopy class $[p_{\lambda\lambda'}]$ induce chain mappings $p_{\lambda\lambda'\#}: S(X_{\lambda'}) \to S(X_\lambda)$, which are chain homotopic and thus, they induce the same homomorphism $p_{\lambda\lambda'*}: H_n(S(X_{\lambda'})) \to H_n(S(X_\lambda))$, denoted by $[p_{\lambda\lambda'}]_*$. Consequently, $H_n[\boldsymbol{X}] = (H_n(X_\lambda), [p_{\lambda\lambda'}]_*, \Lambda)$ is a well-defined abelian pro-group. The same arguments apply to morphisms $[\boldsymbol{f}] = (f, [f_\mu])$ of pro-H(Top).

To define *homology pro-groups of* $\boldsymbol{X}$ *with coefficients* in an abelian group G, denoted by $H_n(\boldsymbol{X}; G)$, one takes the homology pro-groups of the pro-chain complex $S(\boldsymbol{X}; G) = S(\boldsymbol{X}) \otimes G$, which consists of the chain complexes $S(X_\lambda) \otimes G$ and the chain mappings $p_{\lambda\lambda'\#} \otimes 1$. One defines $H_n([\boldsymbol{X}]; G)$ analogously.

Homology pro-groups of a space X are defined as follows. One associates with X a polyhedral resolution $\boldsymbol{p}: X \to \boldsymbol{X}$ (its existence was established in Theorem 6.22). The n-th *homology pro-group of the space* X is the homology pro-group $H_n(\boldsymbol{X}; G)$. More generally, instead of a resolution $\boldsymbol{p}$, one can take a polyhedral H(Top)-expansion $[\boldsymbol{p}]: X \to [\boldsymbol{X}]$ of X and define the n-th homology pro-group of X as $H_n([\boldsymbol{X}]; G)$. Recall that $[\boldsymbol{p}]$ consists of an inverse system $[\boldsymbol{X}] = (X_\lambda, [p_{\lambda\lambda'}], \Lambda)$ in H(Top) and of a collection $[\boldsymbol{p}]$ of morphisms $[p_\lambda]: X \to X_\lambda$, which satisfy $[p_{\lambda\lambda'}][p_{\lambda'}] = [p_\lambda]$, for $\lambda \leq \lambda'$, as well as the Morita conditions (M1) and (M2) (see 7.1). Also recall that the homotopy functor H transforms every resolution into an H(Top)-expansion (see Theorem 7.6 and Remark 7.2). If $[\boldsymbol{p}']: X \to [\boldsymbol{X}']$ is another polyhedral H(Top)-expansion of X, then by ((Mardešić, Segal 1982), I.4.1), there exists a unique isomorphism

$[\boldsymbol{i}]: [\boldsymbol{X}] \to [\boldsymbol{X}']$ of pro-H(Top) such that $[\boldsymbol{i}][\boldsymbol{p}] = [\boldsymbol{p}']$. This isomorphism induces a natural isomorphism in pro-Ab, which identifies the homology pro-groups $H_n([\boldsymbol{X}];G)$ and $H_n([\boldsymbol{X}'];G)$. Therefore, $H_n(\boldsymbol{X};G) \approx H_n([\boldsymbol{X}];G)$ is a well-defined abelian pro-group, called the n-th *homology pro-group of the space* X *with coefficients in* G. We will denote it by $\boldsymbol{H}_n(X;G)$. The notation pro-$H_n(X;G)$ is also used. In the above definitions, instead of polyhedral systems, one can also use the more general systems consisting of spaces having the homotopy type of polyhedra.

In general, a mapping $f: X \to Y$ does not induce a mapping between the corresponding polyhedral resolutions. However, it does induce a morphism $[\boldsymbol{f}] = (f, [f_\mu]): [\boldsymbol{X}] \to [\boldsymbol{Y}]$ of pro-H(Top), such that $[\boldsymbol{q}][f] = [\boldsymbol{f}][\boldsymbol{p}]$ (see (Mardešić, Segal 1982), I.2.3). Moreover, the morphism $[\boldsymbol{f}]$ induces a morphism of homology pro-groups $H_n[\boldsymbol{f}]: H_n([\boldsymbol{X}];G) \to H_n([\boldsymbol{Y}];G)$, $n \geq 0$. This morphism depends only on the homotopy class $[f]$ of f and can therefore, be denoted by $[f]_* = \boldsymbol{H}_n[f]$. In this way the n-th pro-homology becomes a functor H(Top)$\to$ pro-Ab. The inverse limit of the homology pro-groups $\boldsymbol{H}_n(X;G)$ is, by definition, the n-th *Čech homology group* of the space X, denoted by $\check{H}_n(X;G)$.

The *Čech expansion* of a space X is a particular polyhedral H(Top)-expansion $[\boldsymbol{p}]: X \to [\boldsymbol{X}]$. The index set Λ consists of all normal open coverings λ of X and $\lambda \leq \lambda'$, provided λ' refines λ. The term X_λ is the nerve of the covering λ, endowed with the CW-topology. For $\lambda \leq \lambda'$, $p_{\lambda\lambda'}: X_{\lambda'} \to X_\lambda$ is a simplicial mapping, which sends a vertex $V' \in \lambda'$ to some vertex $V \in \lambda$, for which $V' \subseteq V$. The mapping $p_{\lambda\lambda'}$ may not be unique, but its homotopy class $[p_{\lambda\lambda'}]$ is unique. For every $\lambda \in \Lambda$, $p_\lambda: X \to X_\lambda$ is a canonical mapping, subordinated to the covering λ. Again, p_λ may not be unique, but its homotopy class $[p_\lambda]$ is unique. $[\boldsymbol{X}] = (X_\lambda, [p_{\lambda\lambda'}], \Lambda)$ is an inverse system in pro-H(Top) and $[\boldsymbol{p}]: X \to [\boldsymbol{X}]$ is a polyhedral expansion of X (see (Mardešić, Segal 1982) I.4.2 and Appendix 1.3 or (Morita 1975b)). Clearly, the Čech homology group associated with the Čech expansion, coincides with the classical definition of the Čech homology group. Therefore, the definition of Čech homology, based on a resolution or on an H(Top)-expansion coincides with the classical definition.

16.2 Higher limits of homology pro-groups

Since the homology pro-groups $H_n(\boldsymbol{C})$ of a pro-chain complex $\boldsymbol{C}$ are well defined, so are their higher limits $\lim^m H_n(\boldsymbol{C})$. In particular, application to the pro-chain complex $S(\boldsymbol{X};G)$ yields higher limits $\lim^m H_n(\boldsymbol{X};G)$ of systems $\boldsymbol{X}$ in pro-H(Top). Since a morphism $[\boldsymbol{f}] = (f, [f_\mu]): \boldsymbol{X} \to \boldsymbol{Y}$ of pro-H(Top) induces morphisms $H_n[\boldsymbol{f}]: H_n(\boldsymbol{X};G) \to H_n(\boldsymbol{Y};G)$ of pro-Ab, one obtains induced homomorphisms $\lim^m H_n[\boldsymbol{f}]: \lim^m H_n(\boldsymbol{X};G) \to \lim^m H_n(\boldsymbol{Y};G)$. One also defines $\lim^m \boldsymbol{H}_n(X;G)$, for a space X, as $\lim^m H_n([\boldsymbol{X}];G)$, where $[\boldsymbol{X}]$ is a polyhedral homotopy expansion of X. Since $\lim^m H_n$ is a functor on pro-Ab

(Theorem 15.5), one concludes that $\lim^m H_n$: H(Top) $\to$ Ab is a well-defined functor.

EXAMPLE 16.1. It is easy to exhibit examples of compact metric spaces X with $\lim^1 \boldsymbol{H}_n(X;\mathbb{Z}) \neq 0$. Indeed, let X be the *dyadic solenoid*, i.e., the limit of the inverse sequence

$$\boldsymbol{X} = (S^1 \xleftarrow{2} S^1 \xleftarrow{2} S^1 \longleftarrow \ldots), \tag{1}$$

where $S^1 = \{z \in \mathbb{C}\colon |z| = 1\}$ and 2 denotes the mapping $z \mapsto z^2$. Since X and S^1 are compact spaces, the natural projections define a polyhedral resolution $\boldsymbol{p}\colon X \to \boldsymbol{X}$ of X. Therefore, $\boldsymbol{H}_1(X;\mathbb{Z}) = H_1(\boldsymbol{X};\mathbb{Z})$ is the pro-group

$$\boldsymbol{H} = (\mathbb{Z} \xleftarrow{2} \mathbb{Z} \xleftarrow{2} \mathbb{Z} \longleftarrow \ldots), \tag{2}$$

where 2 now denotes multiplication by 2. Using the explicit formulae for $\lim^1$ (see 11.6), it is easy to see that $\lim^1 \boldsymbol{H} \neq 0$. This also follows from Example 11.23, because the vanishing of $\lim^1 \boldsymbol{H}$ would imply that the level morphism $\boldsymbol{f}\colon \boldsymbol{H} \to \boldsymbol{H}'' = (\mathbb{Z}/3 \xleftarrow{2} \mathbb{Z}/3 \xleftarrow{2} \ldots)$, given by the quotient homomorphisms $\mathbb{Z} \to \mathbb{Z}/3$, induces a surjection $\lim \boldsymbol{f}$, which is not the case. Hence, $\lim^1 \boldsymbol{H}_1(X;\mathbb{Z}) = \lim^1 \boldsymbol{H} \neq 0$. Note that the Čech group $\check{H}_1(X;\mathbb{Z}) = 0$. To obtain examples where $\lim^1 \boldsymbol{H}_n(X;\mathbb{Z}) \neq 0$, for $n > 1$, it suffices to take iterated suspensions of (1).

REMARK 16.2. It follows from Theorem 11.52, that

$$\lim^m \boldsymbol{H}_n(X;G) = 0, \tag{3}$$

for compact metric spaces X and $m \geq 2$. Generally, $\lim^m \boldsymbol{H} = 0$, for $m \geq 2$ and abelian pro-groups $\boldsymbol{H}$, consisting of finitely generated groups (see [Jensen 1972], Remark on p. 65). Therefore, (3) holds also for compact Hausdorff spaces X, $m \geq 2$ and finitely generated coefficient groups G. A more general result is proved in 21.

In the non-compact case, higher derived limits of homology pro-groups are in general non-trivial. We will now construct examples, which demonstrate this fact.

EXAMPLE 16.3. Let $m, n \geq 1$, be fixed integers and let Λ be a directed set. We define an inverse system $\boldsymbol{X} = \boldsymbol{X}(m,n,\Lambda)$, consisting of CW-complexes as follows. Let $(B^m, *)$ denote the standard m-cell with a base-point $*$, chosen on its boundary $\partial B^m = S^{m-1}$. For every $\lambda \in \Lambda$ and every multiindex $\boldsymbol{\lambda} = (\lambda_0, \ldots, \lambda_n) \in \Lambda_n$, let

$$X^\lambda_{\boldsymbol{\lambda}} = \begin{cases} S^m, & \lambda \leq \lambda_0, \\ B^m, & \lambda \not\leq \lambda_0. \end{cases} \tag{4}$$

Let X_λ be the wedge (weak topology) of the collection of all $X^\lambda_{\boldsymbol{\lambda}}$, $\boldsymbol{\lambda} \in \Lambda_n$,

$$X_\lambda = \bigvee_{\boldsymbol{\lambda}} X_{\boldsymbol{\lambda}}^{\lambda}. \tag{5}$$

For $\lambda \le \lambda'$, let $p_{\lambda\lambda'}: X_{\lambda'} \to X_\lambda$ be the mapping

$$p_{\lambda\lambda'} = \bigvee_{\boldsymbol{\lambda}} p_{\boldsymbol{\lambda}}^{\lambda\lambda'}, \tag{6}$$

where $p_{\boldsymbol{\lambda}}^{\lambda\lambda'}: X_{\boldsymbol{\lambda}}^{\lambda'} \to X_{\boldsymbol{\lambda}}^{\lambda}$ is defined as follows. If $\lambda' \le \lambda_0$, or if $\lambda' \not\le \lambda_0$ and $\lambda \not\le \lambda_0$, then $p_{\boldsymbol{\lambda}}^{\lambda\lambda'}$ is the identity mapping on S^m and B^m, respectively. In the only remaining case, i.e., when $\lambda' \not\le \lambda_0$ and $\lambda \le \lambda_0$, one has $X_{\boldsymbol{\lambda}}^{\lambda'} = B^m$ and $X_{\boldsymbol{\lambda}}^{\lambda} = S^m$. Let $p_{\boldsymbol{\lambda}}^{\lambda\lambda'} = \phi$, where $\phi: B^m \to S^m$ is the mapping, which collapses the boundary of B^m to the base-point $* \in S^m$. It is, readily seen that $\boldsymbol{X} = (X_\lambda, p_{\lambda\lambda'}, \Lambda)$ is a polyhedral inverse system.

LEMMA 16.4. *If the cofinality* $\mathrm{cof}(\Lambda) = \aleph_{n-1}$, *then* $\boldsymbol{X} = \boldsymbol{X}(m,n,\Lambda)$ *has the property that*

$$\lim{}^n H_m(\boldsymbol{X};\mathbb{Z}) \neq 0. \tag{7}$$

Proof. By (4) and (5), one has

$$H_m(X_\lambda;\mathbb{Z}) = \bigoplus_{\lambda \le \boldsymbol{\lambda}} \mathbb{Z}. \tag{8}$$

Moreover, the restriction of $p_{\boldsymbol{\lambda}}^{\lambda\lambda'}$ to $X_{\boldsymbol{\lambda}}^{\lambda'}$ induces the identity homomorphism between the corresponding summands $\mathbb{Z}$. Therefore, $p_{\lambda\lambda' *}: H_m(X_{\lambda'};\mathbb{Z}) \to H_m(X_\lambda;\mathbb{Z})$ is the inclusion homomorphism. Comparing this with the standard projective resolution of $\boldsymbol{\Delta}(\Lambda)$ (see (12.2.14)), we conclude that $H_m(\boldsymbol{X};\mathbb{Z})$ coincides with the n-th term $\boldsymbol{P}_n$ of this resolution, i.e.,

$$H_m(\boldsymbol{X};\mathbb{Z}) = \boldsymbol{P}_n\,. \tag{9}$$

However, since $\mathrm{cof}(\Lambda) = \aleph_{n-1}$, we know, by Corollary 13.12, that $\lim^n \boldsymbol{P}_n \neq 0$. Therefore, (7) is a consequence of (9). □

EXAMPLE 16.5. For a directed set Λ having no terminal element and for integers $m, n \ge 1$, we define a space $X = X(m,n,\Lambda)$ and mappings $p_\lambda: X \to X_\lambda$, $\lambda \in \Lambda$, as follows. As a set,

$$X = \bigvee_{\boldsymbol{\lambda}} X_{\boldsymbol{\lambda}}, \tag{10}$$

where all summands are m-balls, i.e., $X_{\boldsymbol{\lambda}} = B^m$, for all mulitiindices $\boldsymbol{\lambda} \in \Lambda_n$. The mappings p_λ are given by

$$p_\lambda = \bigvee_{\boldsymbol{\lambda}} p_{\boldsymbol{\lambda}}^{\lambda}, \tag{11}$$

where $p_{\boldsymbol{\lambda}}^{\lambda}: X_{\boldsymbol{\lambda}} \to X_{\boldsymbol{\lambda}}^{\lambda}$ is given by

$$p_{\boldsymbol{\lambda}}^{\lambda} = \begin{cases} \phi, & \lambda \leq \lambda_0, \\ \text{id}, & \lambda \not\leq \lambda_0. \end{cases} \tag{12}$$

X is topologized by taking for a basis $\mathcal{B}$ of its topology the collection of all sets $(p_\lambda)^{-1}(V_\lambda)$, where $\lambda \in \Lambda$ and V_λ is an open subset of X_λ. It is readily seen that $\boldsymbol{p} = (p_\lambda): X \to \boldsymbol{X}$ is a mapping of systems.

LEMMA 16.6. *The leaves $X_{\boldsymbol{\lambda}}$ inherit from X the usual topology of B^m and are closed subsets of X. The mappings p_λ satisfy condition* (B2) *and this stronger form of condition* (B1)*:*

($\overline{\text{B1}}$). *For an arbitrary open covering $\mathcal{U}$ of X, there exist a $\lambda \in \Lambda$ and an open covering $\mathcal{U}_\lambda$ of X_λ, such that $p_\lambda^{-1}(\mathcal{U}_\lambda)$ refines $\mathcal{U}$.*
Consequently, $\mathbf{p} = (p_\lambda): X \to \boldsymbol{X}$ is a polyhedral resolution of X. Moreover, the space X is paracompact.

Proof. For any $\lambda \in \Lambda$ and $V \subseteq X_\lambda$ open, one has

$$p_\lambda^{-1}(V) \cap X_{\boldsymbol{\lambda}} = (p_{\boldsymbol{\lambda}}^{\lambda})^{-1}(V \cap X_{\boldsymbol{\lambda}}^{\lambda}). \tag{13}$$

Since $V \cap X_{\boldsymbol{\lambda}}^{\lambda}$ is an open subset of $X_{\boldsymbol{\lambda}}^{\lambda}$ and $p_{\boldsymbol{\lambda}}^{\lambda}: X_{\boldsymbol{\lambda}} \to X_{\boldsymbol{\lambda}}^{\lambda}$ is continuous with respect to the usual topology of $X_{\boldsymbol{\lambda}}$, one concludes that the open sets of $X_{\boldsymbol{\lambda}}$, in the topology inherited from X, are also open in the usual topology. Conversely, let $U \subseteq X_{\boldsymbol{\lambda}}$ be open in the usual topology. Choose a $\lambda \in \Lambda$, for which $\lambda \not\leq \lambda_0$. Such λ exist, because λ_0 is not a terminal element. Note that $p_{\boldsymbol{\lambda}}^{\lambda} = \text{id}$. Therefore, U is also an open set of the leaf $X_{\boldsymbol{\lambda}}^{\lambda}$. Consequently, one can choose an open set V in X_λ such that $V \cap X_{\boldsymbol{\lambda}}^{\lambda} = U$. Then, by (13),

$$p_\lambda^{-1}(V) \cap X_{\boldsymbol{\lambda}} = V \cap X_{\boldsymbol{\lambda}}^{\lambda} = U, \tag{14}$$

which shows that U is open also in the topology inherited from X. Also notice that, for every multiindex $\boldsymbol{\lambda} \in \Lambda_n$, the open leaf $X_{\boldsymbol{\lambda}} \backslash \{*\}$ is an open subset of X, because it coincides with the set $p_\lambda^{-1}(X_{\boldsymbol{\lambda}}^{\lambda} \backslash \{*\})$. It is now obvious that every leaf $X_{\boldsymbol{\lambda}}$ is a closed subset of X.

In order to establish ($\overline{\text{B1}}$), consider an open covering $\mathcal{U}$ of X. There is no loss of generality in assuming that $\mathcal{U} \subseteq \mathcal{B}$. At least one member $U_* \in \mathcal{U}$ contains the base-point $* \in X$. Since $U_* \in \mathcal{B}$, there exists a $\lambda_* \in \Lambda$ and an open set V_* in $X_{\lambda*}$ such that $U_* = p_{\lambda*}^{-1}(V_*)$. Note that

$$* = p_{\lambda*}(*) \in p_{\lambda*}(U_*) \subseteq V_*. \tag{15}$$

Now consider any multiindex $\boldsymbol{\lambda} = (\lambda_0, \ldots, \lambda_n) \in \Lambda_n$. Let us first assume that $\lambda_* \not\leq \lambda_0$ and therefore, $X_{\boldsymbol{\lambda}} = X_{\boldsymbol{\lambda}}^{\lambda_*} = B^m$ and $p_{\boldsymbol{\lambda}}^{\lambda_*} = \text{id}$. For any $U \in \mathcal{U}$, which differs from U_*, consider the set U', obtained from $U \cap X_{\boldsymbol{\lambda}}$ by removing the base-point $*$. Every set U' is contained in $X_{\boldsymbol{\lambda}} \backslash \{*\}$ and is open on the leaf $X_{\boldsymbol{\lambda}}$. Therefore, it is an open subset of $X_{\boldsymbol{\lambda}}^{\lambda_*} \backslash \{*\}$, hence, also of X^{λ_*}. Consequently, the sets U' form a collection $\mathcal{V}_{\boldsymbol{\lambda}}$ of open subsets of X^{λ_*}, which covers $X_{\boldsymbol{\lambda}}^{\lambda_*} \backslash V_*$ and has the property that $p_{\lambda_*}^{-1}(\mathcal{V}_{\boldsymbol{\lambda}})$ refines $\mathcal{U}$. Now consider the case when $\lambda_* \leq \lambda_0$. Then, $X_{\boldsymbol{\lambda}} = B^m$, $X_{\boldsymbol{\lambda}}^{\lambda_*} = S^m$ and $p_{\boldsymbol{\lambda}}^{\lambda_*} = \phi$. For any $U \in \mathcal{U}$, which differs from U_*, consider the set U', obtained from $U \cap X_\lambda$ by removing the boundary of X_λ. Every set U' is contained in the interior of the m-ball X_λ and is open on the leaf X_λ. Since ϕ maps the interior of this leaf onto $X_\lambda^{\lambda_*} \backslash \{*\} = S^m \backslash \{*\}$ homeomorphically, it follows that the sets $\phi(U')$ form a collection $\mathcal{V}_\lambda$ of open subsets of $X_\lambda^{\lambda_*} \backslash \{*\}$ and thus, also of X^{λ_*}. Moreover, $\mathcal{V}_\lambda$ covers $X_\lambda^{\lambda_*} \backslash V_*$ and $p_{\lambda_*}^{-1}(\mathcal{V}_\lambda)$ refines $\mathcal{U}$. It is now clear that all the collections $\mathcal{V}_\lambda$, together with the open set V, form an open covering $\mathcal{V}$ of X_{λ_*}, which has the desired property that $p_{\lambda_*}^{-1}(\mathcal{V})$ refines $\mathcal{U}$.

The mappings $p_\lambda : X \to X_\lambda$ are surjections, which makes condition (B2) obviously true. Consequently, Theorem 6.7 implies that $\boldsymbol{p} = (p_\lambda) : X \to \boldsymbol{X}$ is a polyhedral resolution of X.

To prove that X is a Hausdorff space, it suffices to show that, for any pair of distinct points $x, x' \in X$, there exists a $\lambda \in \Lambda$ such that $p_\lambda(x) \neq p_\lambda(x')$. Assume that, $x \in X_\lambda$, $x' \in X_{\lambda'}$ and choose and index $\lambda \in \Lambda$ such that $\lambda \nleq \lambda_0$ and $\lambda \nleq \lambda_0'$. Such λ exist, because neither λ_0 nor λ_0' are terminal elements. Therefore, there exist elements $\mu, \mu' \in \Lambda$ such that $\mu_0 \nleq \lambda_0$ and $\mu' \nleq \lambda_0'$. Clearly, any $\lambda \geq \mu, \mu'$ has the desired property. Now note that for such a λ, both mappings p_λ^λ and $p_{\lambda'}^\lambda$ are identity mappings, and the assertion is obviously true. To prove that X is paracompact, consider any open covering $\mathcal{U}$ of X. By property $(\overline{\text{B1}})$, there is a $\lambda \in \Lambda$ and an open covering $\mathcal{V}$ of X_λ such that $p^{-1}(\mathcal{V})$ refines $\mathcal{U}$. Since polyhedra are paracompact, $\mathcal{V}$ admits a locally finite refinement $\mathcal{W}$. Consequently, $p^{-1}(\mathcal{W})$ is a locally finite refinement of $p^{-1}(\mathcal{V})$, and thus, also of $\mathcal{U}$. □

REMARK 16.7. Paracompact spaces are topologically complete and for such spaces resolutions are always inverse limits (Theorem 6.16). Therefore, X is the inverse limit of $\boldsymbol{X}$.

An immediate consequence of Lemmas 16.4 and 16.6 is the following theorem.

THEOREM 16.8. *If $m, n \geq 1$ and Λ is a directed set of cofinality $\operatorname{cof}(\Lambda) = \aleph_{n-1}$, then $X = X(m,n,\Lambda)$ is a paracompact space with the property that the n-th derived limit of its m-th homology pro-group with integer coefficients does not vanish, i.e.,*

$$\lim{}^n \boldsymbol{H}_m(X;\mathbb{Z}) \neq 0. \tag{16}$$

REMARK 16.9. The topology of X is not the weak topology of a wedge of m-balls, because this would imply that X is a polyhedron. However, polyhedra have rudimentary polyhedral resolutions. Therefore, their homology pro-groups are also rudimentary. Clearly, for $\Lambda = *$, $\operatorname{hd}(\Lambda) = 0$ and therefore, by Theorem 13.3, all limits $\lim^n$ vanish, for $n \geq 1$.

The next lemma shows that, for cofinite uncountable Λ, $X(m,n,\Lambda)$ is not a cluster of m-balls.

LEMMA 16.10. *If Λ is cofinite and uncountable, then every compact subset K of the space $X = X(m,n,\Lambda)$ is contained in the union of a finite collection of leaves of X.*

Proof. Assume that there exist a sequence of different multiindices $\boldsymbol{\lambda}^i = (\lambda_0^i, \dots, \lambda_n^i) \in \Lambda_n$ and a sequence of points $x^i \in K \cap (X_{\boldsymbol{\lambda}^i} \backslash \{*\})$, $i = 1, 2, \dots$. Since Λ is cofinite, for every i the set M^i of all predecessors of λ_0^i is finite. Therefore, $M = \cup_i M^i$ is also a countable set. Since the cardinal card$(\Lambda) > \aleph_0$, there exists a $\lambda \in \Lambda \backslash M$. Clearly, for every i, $\lambda \not\leq \lambda_0^i$. Consequently, $(X^\lambda_{\boldsymbol{\lambda}^i}, *) = (B^m *) = (X_{\boldsymbol{\lambda}^i}, *)$ and $p^\lambda_{\boldsymbol{\lambda}^i} = \mathrm{id}$, for all i. Therefore, the points $p_\lambda(x^i) = p^\lambda_{\boldsymbol{\lambda}^i}(x^i) = x^i$ belong to a sequence of different leaves $X^\lambda_{\boldsymbol{\lambda}^i} \backslash \{*\}$ of X_λ. This is a contradiction, because all the points $p_\lambda(x^i)$ belong to the compact set $p_\lambda(K)$, which by the CW-topology of X_λ, must be contained in the union of a finite collection of leaves of X_λ. □

Bibliographic notes

The Čech homology groups have their origins in the classical papers (Alexandroff 1927), (Vietoris 1927), (Čech 1932), where they were defined for compact metric and for compact Hausdorff spaces, respectively. C.H. Dowker defined Čech cohomology groups using arbitrary open coverings (Dowker 1947). The definition of Čech cohomology for arbitrary topological spaces (based on normal coverings) appeared in (Morita 1975b). The definition of homology pro-groups and the Čech groups as their inverse limits developed after the advent of shape theory (see the Bibliographic notes in (Mardešić, Segal 1982)). The space $X(m,n,\Lambda)$ and Theorem 16.8 are taken from (Mardešić 1996a).

17. Strong homology groups of systems

In this section we define strong homology groups $\overline{H}_m(\boldsymbol{X};G)$ of inverse systems of spaces $\boldsymbol{X}$. They are homology groups of the total complex of a certain bicomplex, whose boundary operators are the boundary operators ∂ of singular complexes and the Nöbeling – Roos operators δ of abelian pro-groups (defined in 11.5). We also define a sequence of homology groups $\overline{H}_m^{(r)}(\boldsymbol{X};G)$, $r = 0, 1, \ldots$, of *height* r. They approximate the strong homology group $\overline{H}_m(\boldsymbol{X};G)$, which can be viewed as homology of height ∞. Homology of height 0 coincides with Čech homology and gives the coarsest approximation. The precise relation between the homology groups $\overline{H}_m^{(r)}(\boldsymbol{X};G)$, for different heights r, as well as the relation between their limit and the strong group $\overline{H}_m(\boldsymbol{X};G)$, are given by certain exact sequences (the Miminoshvili sequences). These sequences enable us to show that mappings $\boldsymbol{f}\colon \boldsymbol{X} \to \boldsymbol{Y}$, which induce isomorphisms of homology pro-groups also induce isomorphisms of strong homology groups.

17.1 Strong homology of pro-chain complexes

We first associate with every pro-chain complex $\boldsymbol{C} = (C_\lambda, p_{\lambda\lambda'}, \Lambda)$ a *bicomplex* (*double complex*) $K(\boldsymbol{C}) = (K^{ns}, \partial^{ns}, \delta^{ns})$, $n, s \in \mathbb{Z}$ as follows. For a fixed $n \in \mathbb{Z}$, $\boldsymbol{C}^n = (C^n_\lambda, p^n_{\lambda\lambda'}, \Lambda)$ is an inverse system of abelian groups. We associate with this system the cochain complex $K(\boldsymbol{C}^n)$, defined in 11.5. It consists of abelian groups $K^s(\boldsymbol{C}^n)$, which we now denote by $K^{ns}(\boldsymbol{C})$ or just K^{ns}, and of homomorphisms $\delta^{ns}(\boldsymbol{C}) = \delta^{ns}\colon K^{n,s-1} \to K^{ns}$. According to (11.5.2) and (11.5.3),

$$K^{ns} = \prod_{\boldsymbol{\lambda} \in \Lambda_s} C^n_{\lambda_0}, \tag{1}$$

for $s \geq 0$ and $K^{ns} = 0$, for $s < 0$. Moreover, for $c \in K^{n,s-1}$, $s \geq 1$, $\delta^{ns}c \in K^{ns}$ is given by

$$(\delta^{ns}c)(\boldsymbol{\lambda}) = p^n_{\lambda_0\lambda_1} c(d^0\boldsymbol{\lambda}) + \sum_{j=1}^{s} (-1)^j c(d^j\boldsymbol{\lambda}), \;\; \boldsymbol{\lambda} \in \Lambda_s. \tag{2}$$

The boundary operators $\partial^n_{\lambda_0}: C^n_{\lambda_0} \to C^{n-1}_{\lambda_0}$ of the chain complex C_{λ_0} determine operators $\partial^{ns}: K^{ns} \to K^{n-1,s}$, defined by

$$(\partial^{ns} c)(\boldsymbol{\lambda}) = \partial^n_{\lambda_0}(c(\boldsymbol{\lambda})), \tag{3}$$

where $c \in K^{ns}$ and $\boldsymbol{\lambda} \in \Lambda_s$.

LEMMA 17.1. $K(\boldsymbol{C}) = (K^{ns}, \partial^{ns}, \delta^{ns})$, $n, s \in \mathbb{Z}$, *is a bicomplex, i.e.,*

$$\partial^{n-1,s}\partial^{ns} = 0, \;\; \delta^{n,s+1}\delta^{ns} = 0, \tag{4}$$

$$\delta^{n-1,s}\partial^{n,s-1} = \partial^{ns}\delta^{ns}. \tag{5}$$

Note that (5) means commutativity of the following diagram.

$$\begin{array}{ccc} K^{n,s-1} & \xrightarrow{\delta^{ns}} & K^{ns} \\ \downarrow{\scriptstyle \partial^{n,s-1}} & & \downarrow{\scriptstyle \partial^{ns}} \\ K^{n-1,s-1} & \xrightarrow[\delta^{n-1,s}]{} & K^{n-1,s}. \end{array} \tag{6}$$

Proof. The first relation in (4) is an immediate consequence of (3) and the fact that $\partial^{n-1}_{\lambda_0}\partial^n_{\lambda_0} = 0$. The second relation in (4) was proved in Lemma 11.39. In order to verify (5), let $c \in K^{n,s-1}$ and $\boldsymbol{\lambda} \in \Lambda_s$. By (3) and (16.1.1), one has

$$(p^{n-1}_{\lambda_0\lambda_1}\partial^{n,s-1}c)(d^0\boldsymbol{\lambda}) = (p^{n-1}_{\lambda_0\lambda_1}\partial^n_{\lambda_1}c)(d^0\boldsymbol{\lambda}) = (\partial^n_{\lambda_0}p^n_{\lambda_0\lambda_1}c)(d^0\boldsymbol{\lambda}), \tag{7}$$

$$(\partial^{n,s-1}c)(d^j\boldsymbol{\lambda}) = (\partial^n_{\lambda_0}c)(d^j\boldsymbol{\lambda}), \;\; j > 0. \tag{8}$$

Therefore, by (2),

$$(\delta^{n-1,s}\partial^{n,s-1}c)(\boldsymbol{\lambda}) = (\partial^n_{\lambda_0}\delta^{ns}c)(\boldsymbol{\lambda}) = (\partial^{ns}\delta^{ns}c)(\boldsymbol{\lambda}). \; \square \tag{9}$$

We will now associate with the bicomplex $K(\boldsymbol{C}) = (K^{ns}, \partial^{ns}, \delta^{ns})$ a *total chain complex* $T(\boldsymbol{C}) = T = (T^m, d^m)$. The term T^m, $m \in \mathbb{Z}$, is determined by the terms K^{ns} of the bicomplex, lying on the m-th diagonal, i.e., the set of all pairs (n, s), for which $n - s = m$ (in (10) the diagonals are represented by solid lines).

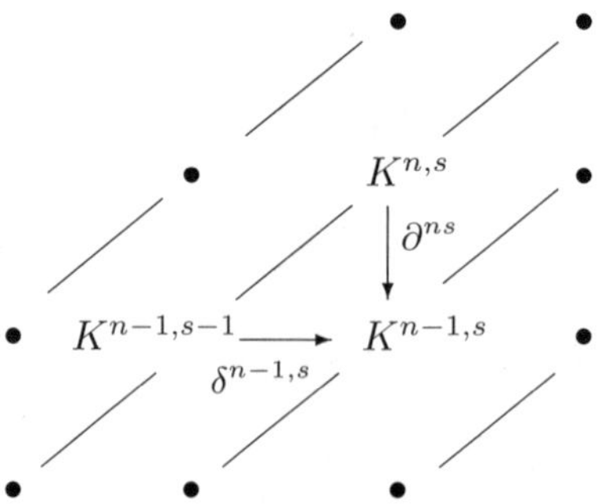

(10)

In general, a bicomplex yields two types of total complexes. In the first type (used more often), T^m is the direct sum of the terms K^{ns}, lying on the m-th diagonal. In the second type, T^m is the direct product of these terms (see e.g., (Hilton, Stammbach 1971), Ch. V.1). For our purposes we need the second type. Therefore, we put

$$T^m = \prod_{s=0}^{\infty} K^{m+s,s} = \prod_{s=0}^{\infty} \prod_{\boldsymbol{\lambda} \in \Lambda_s} C_{\lambda_0}^{m+s}, \quad m \in \mathbb{Z}. \tag{11}$$

The elements $c \in T^m$ will be called *strong m-chains* of $\boldsymbol{C}$. They are sequences $c = (c_0, c_1, \ldots)$, where $c_s \in K^{m+s,s}$ is a function, which assigns to every $\boldsymbol{\lambda} = (\lambda_0, \ldots, \lambda_s) \in \Lambda_s$ an $(m+s)$-chain $c_s(\boldsymbol{\lambda})$ of C_{λ_0}, i.e., $c_s(\boldsymbol{\lambda}) \in C_{\lambda_0}^{m+s}$. In particular, c_0 assigns to every $\lambda_0 \in \Lambda$ an m-chain of the complex C_{λ_0}, c_1 assigns to every pair $\lambda_0 \leq \lambda_1$ an $(m+1)$-chain of C_{λ_0}, etc. A strong chain $c \in T^m$ can also be viewed as a function c, defined on all multiindices $\boldsymbol{\lambda} \in \Lambda_s$, $s \geq 0$, whose values $c(\boldsymbol{\lambda}) = c_s(\boldsymbol{\lambda}) \in C_{\lambda_0}^{m+s}$. Addition in T^m is defined in the obvious way, i.e., $(c + c')(\boldsymbol{\lambda}) = c(\boldsymbol{\lambda}) + c'(\boldsymbol{\lambda})$.

To define the *boundary operator* $d^m \colon T^m \to T^{m-1}$, we define two operators $\partial^m \colon T^m \to T^{m-1}$ and $\delta^m \colon T^m \to T^{m-1}$ and then put

$$d^m c = \partial^m c + (-1)^m \delta^m c, \quad c \in T^m. \tag{12}$$

By definition,

$$(\partial^m c)_s = \partial^{m+s,s} c_s, \quad s \geq 0, \tag{13}$$

$$(\delta^m c)_s = \delta^{m+s-1,s} c_{s-1}, \quad s \geq 1. \tag{14}$$

Combining formulae (13) and (14) with (3) and (2), we see that, for $c \in T^m$ and $\boldsymbol{\lambda} \in \Lambda_s$,

$$(\partial^m c)(\boldsymbol{\lambda}) = \partial_{\lambda_0}^{m+s}(c(\boldsymbol{\lambda})), \tag{15}$$

$$(\delta^m c)(\boldsymbol{\lambda}) = p_{\lambda_0 \lambda_1}^{m+s} c(d^0 \boldsymbol{\lambda}) + \sum_{j=1}^{s} (-1)^j c(d^j \boldsymbol{\lambda}). \tag{16}$$

LEMMA 17.2. *$T = (T^m, d^m)$ is a chain complex.*

Proof. By (12), we see that

$$d^{m-1} d^m c = \partial^{m-1} \partial^m c + (-1)^m (\partial^{m-1} \delta^m c - \delta^{m-1} \partial^m c) - \delta^{m-1} \delta^m c. \tag{17}$$

Now note that,

$$(\partial^{m-1} \partial^m c)_s = \partial^{m-1+s,s} \partial^{m+s,s} c_s = 0. \tag{18}$$

Similarly,

$$(\delta^{m-1} \delta^m c)_s = \delta^{m+s-2,s} \delta^{m+s-1,s} c_{s-2} = 0. \tag{19}$$

Finally, by (5),

$$(\partial^{m-1}\delta^m c)_s - (\delta^{m-1}\partial^m c)_s = \\ \partial^{m+s-1,s}\delta^{m+s-1,s}c_{s-1} - \delta^{m-2+s,s}\partial^{m+s-1,s-1}c_{s-1} = 0. \square \tag{20}$$

We now define the m-th *strong homology group* $\overline{H}_m(\boldsymbol{C})$ of a pro-chain complex $\boldsymbol{C}$ as the m-th homology group of the total chain complex $T(\boldsymbol{C})$, i.e.,

$$\overline{H}_m(\boldsymbol{C}) = H_m(T(\boldsymbol{C})). \tag{21}$$

REMARK 17.3. $K(\boldsymbol{C})$ is a chain-cochain bicomplex, because its operators are of degree $(-1,0)$ and $(0,1)$, respectively. The usual definition requires anticommutativity of diagram (6) in which case the total boundary operator is obtained by summing up the two partial boundary operators. The difference between our definition and the standard one is inessential and it can be avoided by endowing the operators δ^{ns} with a sign $(-1)^{n+s+1}$.

REMARK 17.4. To an abelian pro-group $\boldsymbol{X} = (X_\lambda, p_{\lambda\lambda'}, \Lambda)$ one can associate a pro-chain complex $\boldsymbol{C} = (C_\lambda, q_{\lambda\lambda'}, \Lambda)$, by putting $C^0_\lambda = X_\lambda, q^0_{\lambda\lambda'} = p_{\lambda\lambda'}$, i.e., $\boldsymbol{C}^0 = \boldsymbol{X}$, while $C^n_\lambda = 0$, for $n \neq 0$. Then, $K^{ns} = 0$, for $n \neq 0$, and thus, $T^{-m} = K^{0m}$, for $m \geq 0$, and $T^{-m} = 0$, for $m < 0$. Moreover, since all $\partial^{ns} = 0$, one concludes that $d^{-m}: T^{-m} \to T^{-(m+1)}$ coincides with $(-1)^m\delta^{0,m+1}: K^{0m} \to K^{0,m+1}$. Comparing (1) and (2) with (11.5.2) and (11.5.3), we see that the chain complexes (T^m, d^m) and $K(\boldsymbol{X})$ from 11.5 have the same homology. More precisely, the strong homology $\overline{H}_{-m}(\boldsymbol{C})$ coincides with $\lim^m \boldsymbol{X}$, $m \geq 0$. Consequently, strong homology groups of pro-chain complexes generalizes higher derived limits of pro-groups.

A *morphism of pro-chain complexes* $\boldsymbol{f} = (f, f_\mu): \boldsymbol{C} \to \boldsymbol{D}$ induces a chain mapping $f^\# = T(\boldsymbol{f}): T(\boldsymbol{C}) \to T(\boldsymbol{D})$, defined as follows. For every $n \in \mathbb{Z}$, $\boldsymbol{C}^n = (C^n, p^n_{\lambda\lambda'}, \Lambda)$ and $\boldsymbol{D}^n = (D^n, q^n_{\mu\mu'}, M)$ are abelian pro-groups and $\boldsymbol{f}^n = (f, f^n_\mu): \boldsymbol{C}^n \to \boldsymbol{D}^n$ is a morphism of inv-Ab, because (16.1.2) implies

$$f^n_\mu p^n_{f(\mu)f(\mu')} = q^n_{\mu\mu'} f^n_{\mu'}, \ \mu \leq \mu'. \tag{22}$$

According to 15.1, $\boldsymbol{f}^n$ induces a cochain mapping $(f^n)^\#: K(\boldsymbol{C}^n) \to K(\boldsymbol{D}^n)$. Since the cochain complexes $K(\boldsymbol{C}^n)$ and $K(\boldsymbol{D}^n)$ consist of abelian groups $K^{ns}(\boldsymbol{C})$ and $K^{ns}(\boldsymbol{D})$, $(f^n)^\#$ consists of homomorphisms $f^{ns}: K^{ns}(\boldsymbol{C}^n) \to K^{ns}(\boldsymbol{D}^n)$. By (15.1.1), these homomorphisms are given by

$$(f^{ns}c)(\boldsymbol{\mu}) = f^n_{\mu_0} c(f(\boldsymbol{\mu})), \tag{23}$$

where $c \in K^{ns}(\boldsymbol{C})$ and $\boldsymbol{\mu} = (\mu_0, \dots, \mu_s) \in M_s$. The homomorphisms $f^{m+s,s}$ induce a homomorphism $f^m = \prod_{s=0}^\infty f^{m+s,s}: T^m(\boldsymbol{C}) \to T^m(\boldsymbol{D})$, which maps $c = (c_s) \in T^m(\boldsymbol{C})$ to $f^m(c) = c'$, where $c'_s = f^{m+s,s}(c_s)$, i.e.,

$$(f^m(c))_s = f^{m+s,s}(c_s). \tag{24}$$

Hence, for $c \in T^m(\boldsymbol{C})$ and $\boldsymbol{\mu} \in M_s$, one has $(f^m c)(\boldsymbol{\mu}) = (f^m(c))_s(\boldsymbol{\mu}) = f^{m+s,s}(c_s)(\boldsymbol{\mu})$. Consequently, (23) yields

$$(f^m c)(\boldsymbol{\mu}) = f^{m+s}_{\mu_0} c(f(\boldsymbol{\mu})). \tag{25}$$

LEMMA 17.5. *The homomorphisms* $f^m: T^m(\boldsymbol{C}) \to T^m(\boldsymbol{D})$, $m \in \mathbb{Z}$, *define a chain mapping* $f^\# = T(\boldsymbol{f}): T(\boldsymbol{C}) \to T(\boldsymbol{D})$, *i.e.,*

$$f^{m-1} d^m = d^m f^m. \tag{26}$$

Proof. In view of (12), it suffices to show that

$$f^{m-1} \partial^m = \partial^m f^m, \tag{27}$$

$$f^{m-1} \delta^m = \delta^m f^m. \tag{28}$$

By (24) and (13), for every $c \in T^m(\boldsymbol{C})$ and $s \geq 0$, one has

$$(f^{m-1} \partial^m c)_s = f^{m+s-1,s}((\partial^m c)_s) = f^{m+s-1,s} \partial^{m+s,s} c_s, \tag{29}$$

$$(\partial^m f^m c)_s = \partial^{m+s,s}((f^m c)_s) = \partial^{m+s,s} f^{m+s,s} c_s. \tag{30}$$

Moreover, by (23) and (3), for $\boldsymbol{\mu} \in M_s$,

$$(f^{m+s-1,s} \partial^{m+s,s} c_s)(\boldsymbol{\mu}) = f_{\mu_0}^{m+s-1} \partial_{f(\mu_0)}^{m+s} (c_s(f(\boldsymbol{\mu})), \tag{31}$$

$$(\partial^{m+s,s} f^{m+s,s} c_s)(\boldsymbol{\mu}) = \partial_{\mu_0}^{m+s} f_{\mu_0}^{m+s} c_s(f(\boldsymbol{\mu})). \tag{32}$$

However, since $f_{\mu_0}: C_{f(\mu_0)} \to D_{\mu_0}$ is a chain mapping, one has

$$f_{\mu_0}^{n-1} \partial_{f(\mu_0)}^n = \partial_{\mu_0}^n f_{\mu_0}^n, \tag{33}$$

which shows that the right sides of (29) and (30) coincide.

The validity of (28) is a consequence of Lemma (15.1). More precisely, by (25) and (14),

$$(f^{m-1} \delta^m c)_s = f^{m-1+s,s}((\delta^m c)_s) = f^{m-1+s,s} \delta^{m+s-1,s}(c_{s-1}), \tag{34}$$

$$(\delta^m f^m c)_s = \delta^{m+s-1,s}((f^m c)_{s-1}) = \delta^{m+s-1,s} f^{m+s-1,s-1}(c_{s-1}). \tag{35}$$

Therefore, (28) follows from

$$f^{m-1+s,s} \delta^{m+s-1,s} = \delta^{m+s-1,s} f^{m+s-1,s-1}, \tag{36}$$

which is just (15.1.2), for the morphism $\boldsymbol{f}^{m+s-1}: \boldsymbol{C}^{m+s-1} \to \boldsymbol{D}^{m+s-1}$. □

Taking into account (21), it is clear how to define the homomorphism $f_* = \overline{H}_m(\boldsymbol{f}): \overline{H}_m(\boldsymbol{C}) \to \overline{H}_m(\boldsymbol{D})$, induced by a morphism of pro-chain complexes $\boldsymbol{f}: \boldsymbol{C} \to \boldsymbol{D}$. This is just the homomorphism induced by the chain mapping $f^\# = T(\boldsymbol{f}): T(\boldsymbol{C}) \to T(\boldsymbol{D})$.

We define *strong homology groups* $\overline{H}_m(\boldsymbol{X})$ of an inverse system of spaces $\boldsymbol{X}$ as follows. According to 16.1, $\boldsymbol{X} = (X_\lambda, p_{\lambda\lambda'}, \Lambda)$ determines its singular pro-chain complex $S(\boldsymbol{X}) = (S(X_\lambda), S(p_{\lambda\lambda'}), \Lambda)$. We then put

$$\overline{H}_m(\boldsymbol{X}) = \overline{H}_m(S(\boldsymbol{X})). \tag{37}$$

If $\boldsymbol{f}: \boldsymbol{X} \to \boldsymbol{Y}$ is a mapping of inverse systems of spaces, then $\boldsymbol{f}$ induces a morphism of the corresponding singular pro-chain complexes $S(\boldsymbol{f}): S(\boldsymbol{X}) \to$

$S(\boldsymbol{Y})$ and the latter determines a homomorphism of strong homology groups $\overline{H}_m(S(\boldsymbol{f})): \overline{H}_m(\boldsymbol{S}(\boldsymbol{X})) \to \overline{H}_m(S(\boldsymbol{Y}))$. One defines $f_* = \overline{H}_m(\boldsymbol{f}): \overline{H}_m(\boldsymbol{X}) \to \overline{H}_m(\boldsymbol{Y})$ by putting $f_* = \overline{H}_m(S(\boldsymbol{f}))$.

Similarly, if G is an abelian group, one considers the pro-chain complex $S(\boldsymbol{X}) \otimes G$ (defined in 16.1) and one puts

$$\overline{H}_m(\boldsymbol{X}; G) = \overline{H}_m(S(\boldsymbol{X}) \otimes G). \tag{38}$$

Moreover, a mapping $\boldsymbol{f} = (f, f_{\mu_0}): \boldsymbol{X} \to \boldsymbol{Y}$ induces a morphism $S(\boldsymbol{X}) \otimes G \to S(\boldsymbol{Y}) \otimes G$ of inv-Chn and the latter determines a homomorphism $f_*: \overline{H}_m(\boldsymbol{X}; G) \to \overline{H}_m(\boldsymbol{Y}; G)$. In this way one obtains the functor $\overline{H}_m(\,.\,; G)$: inv-Top $\to$ Ab.

Strong homology groups are also functors in their second variable G. Recall that $\overline{H}_m(\boldsymbol{X}; G) = \overline{H}_m(\boldsymbol{C})$, where $\boldsymbol{C}$ is the pro-chain complex $\boldsymbol{C} = S(\boldsymbol{X}) \otimes G$ (see (38)). A homomorphism $\phi: G \to G''$ induces a morphism of pro-chain complexes $1 \otimes \phi: S(\boldsymbol{X}) \otimes G \to S(\boldsymbol{X}) \otimes G''$ and the latter induces a chain mapping $T(1 \otimes \phi)$ of the corresponding total complexes, which in turn induces homomorphisms of the corresponding strong homology groups $\phi_*: \overline{H}_m(\boldsymbol{X}; G) \to \overline{H}_m(\boldsymbol{X}; G'')$. Clearly, these homomorphisms are functorial. The following theorem will be used in 21.3.

THEOREM 17.6. *Every short exact sequence of abelian groups*

$$0 \to G' \to G \to G'' \to 0 \tag{39}$$

induces a natural exact sequence

$$\ldots \to \overline{H}_{m+1}(X; G'') \to \overline{H}_m(X; G') \to \overline{H}_m(X; G) \to \overline{H}_m(X; G'') \to . \tag{40}$$

Proof. Since the groups of singular chains $S_m(X_{\lambda_0})$, $m \in \mathbb{Z}$, are free abelian groups, (39) implies exactness of the induced sequences, $s \geq 0$,

$$0 \to S_{m+s}(X_{\lambda_0}) \otimes G' \to S_{m+s}(X_{\lambda_0}) \otimes G \to S_{m+s}(X_{\lambda_0}) \otimes G'' \to 0. \tag{41}$$

Taking direct products of copies of (41) over all multiindices $\boldsymbol{\lambda} \in \Lambda_s$ and all $s \geq 0$, one obtains exact sequences of groups, $m \in \mathbb{Z}$,

$$0 \to T^m(S(\boldsymbol{X}) \otimes G') \to T^m(S(\boldsymbol{X}) \otimes G) \to T^m(S(\boldsymbol{X}) \otimes G'') \to 0. \tag{42}$$

Consequently, we also have an exact sequence of chain complexes

$$0 \to T(S(\boldsymbol{X}) \otimes G') \to T(S(\boldsymbol{X}) \otimes G) \to T(S(\boldsymbol{X}) \otimes G'') \to 0. \tag{43}$$

The induced long exact sequence of homology groups is just (40).

To prove naturality of the sequence (40), first recall that a mapping between cofinite systems $\boldsymbol{f}: \boldsymbol{X} \to \boldsymbol{Y} = (Y_\mu, q_{\mu\mu'}, M)$ consists of an increasing function $f: M \to \Lambda$ and of mappings $f_{\mu_0}: X_{f(\mu_0)} \to Y_{\mu_0}$ such that

$q_{\mu\mu'} f_{\mu'} = f_\mu p_{f(\mu)f(\mu')}$, $\mu \leq \mu'$. The induced homomorphism $f_*: \overline{H}_m(\boldsymbol{X}; G) \to \overline{H}_m(\boldsymbol{Y}; G)$ is induced by a chain mapping $f_\#: T(S(\boldsymbol{X}) \otimes G) \to T(S(\boldsymbol{X}) \otimes G)$, given by the formula

$$f_\# c(\mu_0, \dots, \mu_s) = f_{\mu_0 \#} c(f(\mu_0), ..., f(\mu_s)). \tag{44}$$

The commutativity of the following diagram is easily verified.

$$\begin{array}{ccc} T^m(S(\boldsymbol{X}) \otimes G') & \xrightarrow{\phi_\#} & T^m(S(\boldsymbol{X}) \otimes G) \\ {\scriptstyle \boldsymbol{f}_\#}\downarrow & & \downarrow{\scriptstyle \boldsymbol{f}_\#} \\ T^m(S(\boldsymbol{Y}) \otimes G') & \xrightarrow[\phi_\#]{} & T^m(S(\boldsymbol{Y})G) \; . \end{array} \tag{45}$$

As an immediate consequence one obtains the commutativity of the corresponding diagram of strong homology groups of $\boldsymbol{X}$ and $\boldsymbol{Y}$. Moreover, since every mapping $f: X \to Y$ between topological spaces admits a cofinite polyhedral resolution $(\boldsymbol{p}, \boldsymbol{q}, \boldsymbol{f})$ (see Theorem 6.22), one also obtains a commutative diagram of strong homology groups of spaces.

$$\begin{array}{ccc} \overline{H}_m(X; G') & \xrightarrow{\phi_*} & \overline{H}_m(X; G) \\ {\scriptstyle f_*}\downarrow & & \downarrow{\scriptstyle f_*} \\ \overline{H}_m(Y; G') & \xrightarrow[\phi_*]{} & \overline{H}_m(Y; G) \; . \end{array} \tag{46}$$

It is now clear that the exact sequence (40) is natural with respect to mappings of X. □

To define the strong homology groups $\overline{H}^{(r)}(\boldsymbol{C})$, for pro-chain complexes $\boldsymbol{C}$ and $r \geq 0$, we proceed as follows. We first consider a chain complex $T^{(r)} = T^{(r)}(\boldsymbol{C})$. It is obtained by truncating the chain complex $T = T(\boldsymbol{C})$ (see 16.2) at height r. More precisely, $T^{(r)}$ consists of abelian groups $T^{(r)m}$, given by

$$T^{(r)m} = \prod_{s=0}^{r} K^{m+s,s}, \tag{47}$$

where $K^{n,s}$ is defined by (1). The boundary operators d^m are defined as before, i.e., by formulae (12), (13) and (14). However, one now restricts s to $s \leq r$.

Let $\pi^{(r)m}: T^{(r)m} \to T^{(r-1)m}$, $r \geq 1$, be the natural projection, i.e., let $\pi^{(r)m}(c_0, \dots, c_s) = (c_0, \dots, c_{s-1})$. The homomorphisms $\pi^{(r)m}$, $m \in \mathbb{Z}$, determine a chain mapping $\pi^{(r)}: T^{(r)} \to T^{(r-1)}$, which induces homomorphisms of homology groups $\pi^{(r)}_m = H_m(\pi^{(r)}): H_m(T^{(r)}) \to H_m(T^{(r-1)})$. Now the *strong homology group of height* $r \geq 0$ is defined, by putting

$$\overline{H}_m^{(r)}(\boldsymbol{C}) = \pi_m^{(r+1)}(H_m(T^{(r+1)}(\boldsymbol{C}))),\ r \geq 0. \tag{48}$$

For $r < 0$, one puts $\overline{H}_m^{(r)}(\boldsymbol{C}) = 0$. Note that $\overline{H}_m^{(r)}(\boldsymbol{C}) \subseteq H_m(T^{(r)})$ and thus,

$$\pi_m^{(r)}(\overline{H}_m^{(r)}(\boldsymbol{C})) \subseteq \overline{H}_m^{(r-1)}(\boldsymbol{C}). \tag{49}$$

Consequently, the restriction

$$\overline{\pi}_m^{(r)} = \pi_m^{(r)} | \overline{H}_m^{(r)}(\boldsymbol{C}) \tag{50}$$

is a homomorphism

$$\overline{\pi}_m^{(r)} \colon \overline{H}_m^{(r)}(\boldsymbol{C}) \to \overline{H}_m^{(r-1)}(\boldsymbol{C}), \tag{51}$$

which relates the m-th strong homology groups of consecutive heights. For $r = 0$, the following lemma holds.

LEMMA 17.7. *The homology group $\overline{H}_m^{(0)}(\boldsymbol{C})$ is isomorphic to the Čech group $\check{H}_m(\boldsymbol{C}) = \lim H_m(\boldsymbol{C})$.*

Proof. If z_0 is an m-cycle of $T^{(0)m}$ and $\lambda_0 \in \Lambda$, then $z_0(\lambda_0) \in C_{\lambda_0}^m$ and $\partial_{\lambda_0}^m(z_0(\lambda_0)) = 0$, so that $z_0(\lambda_0)$ is an m-cycle of C_{λ_0}. Consequently, its homology class $[z_0(\lambda_0)] \in H_m(C_{\lambda_0})$ is well defined and so is $h(z_0) \in \prod_{\lambda_0} H_m(C_{\lambda_0})$, where

$$(h(z_0))(\lambda_0) = [z_0(\lambda_0)]. \tag{52}$$

We have thus defined a homomorphism $h \colon Z_m(T^{(0)}) \to \prod_{\lambda_0} H_m(C_{\lambda_0})$, where $Z_m(T^{(0)})$ denotes the abelian group of all m-cycle of $T^{(0)}$. If $z_0 = d^{m+1}x_0$ is a boundary, then $z_0(\lambda_0) = \partial_{\lambda_0}^{m+1}x_0(\lambda_0)$ and thus, $[z_0(\lambda_0)] = 0$, which shows that $h(z_0) = 0$. Consequently, h induces a homomorphism $k \colon H_m(T^{(0)}) \to \prod_{\lambda_0} H_m(C_{\lambda_0})$.

We now consider the restriction of k to $\pi_m^{(1)} H_m(T^{(1)}) \subseteq H_m(T^{(0)})$ and we show that it assumes values in $\lim H_m(\boldsymbol{C}) \subseteq \prod_{\lambda_0} H_m(C_{\lambda_0})$. Indeed, let $\zeta = \pi_m^{(1)}\zeta'$, where $\zeta' \in H_m(T^{(1)})$. If $z' = (z_0', z_1') \in Z_m(T^{(1)})$ is a cycle from the homology class ζ', then $z_0' = \pi^{(1)}(z')$ is a cycle from the homology class ζ. By (12)–(14),

$$\partial_{\lambda_0}^m(z_0'(\lambda_0)) = 0, \tag{53}$$

$$p_{\lambda_0\lambda_1}^m z_0'(\lambda_1) - z_0'(\lambda_0) = (-1)^{m+1}\partial_{\lambda_0}^{m+1}(z_1'(\lambda_0, \lambda_1)). \tag{54}$$

Consequently, the homology classes $[z_0'(\lambda_0)] \in H_m(T^{(1)})$ have the property that

$$p_{\lambda_0\lambda_1 *}^m [z_0'(\lambda_1)] = [z_0'(\lambda_0)], \tag{55}$$

and thus, $k(\zeta) = ([z_0'(\lambda_0)]) \in \lim\ H_m(\boldsymbol{C})$. Since $\overline{H}_m^{(0)}(\boldsymbol{C}) = \pi_m^{(1)} H_m(T^{(1)}) \subseteq H_m(T^{(0)})$, we see that k is a homomorphism $k \colon \overline{H}_m^{(0)}(\boldsymbol{C}) \to \lim H_m(\boldsymbol{C})$.

Now assume that $k(\zeta) = 0$, i.e., $h(z)(\lambda_0) = [z_0(\lambda_0)] = 0$, for all λ_0. Then there exist $(m+1)$-chains $x_{\lambda_0} \in C^{m+1}_{\lambda_0}$ such that $z_0(\lambda_0) = \partial^{m+1}_{\lambda_0} x_{\lambda_0}$. Putting $x(\lambda_0) = x_{\lambda_0}$, we obtain a chain $x \in T^{(0)m+1}$ such that $z_0 = d^m x$. Consequently, $\zeta = [z_0] = 0$, which shows that k is a monomorphism.

To show that k is an epimorphism, note that an element of $\lim H_m(\boldsymbol{C})$ is of the form $([z_{\lambda_0}])$, where z_{λ_0} is an m-cycle of C_{λ_0} and $p^m_{\lambda_0\lambda_1 *}[z_{\lambda_1}] = [z_{\lambda_0}]$, for $\lambda_0 \leq \lambda_1$. Consequently, there exist $(m+1)$-chains $x_{\lambda_0\lambda_1} \in C^{m+1}_{\lambda_0}$ such that

$$p^m_{\lambda_0\lambda_1}(z_{\lambda_1}) - z_{\lambda_0} = (-1)^{m+1}\partial^m_{\lambda_0} x_{\lambda_0\lambda_1}. \tag{56}$$

Define an $(m+1)$-chain $z' = (z_0, z_1) \in T^{(1)m+1}$, by putting $z_0(\lambda_0) = z_{\lambda_0}$, $z_1(\lambda_0, \lambda_1) = x_{\lambda_0\lambda_1}$. Then (56) shows that $d^{m+1}z' = 0$, i.e., $z' \in Z_{m+1}(T^{(1)})$. Moreover, $\pi^{(1)}(z') = z_0$ and therefore, $\zeta = \pi^{(1)}_m \zeta' \in \overline{H}^{(0)}_m(\boldsymbol{C})$ has the property that $k(\zeta) = h(z_0) = [z_0(\lambda_0)] = ([z_{\lambda_0}])$. □

A morphism of pro-chain complexes $\boldsymbol{f} = (f, f_\mu)\colon \boldsymbol{C} \to \boldsymbol{D}$ induces a homomorphism $\overline{f}^{(r)}_m\colon \overline{H}^{(r)}_m(\boldsymbol{C}) \to \overline{H}^{(r)}_m(\boldsymbol{D})$, defined as follows. Put $f^{(r)m} = \prod_{s=0}^r f^{m+s,s}\colon T^{(r)m}(\boldsymbol{C}) \to T^{(r)m}(\boldsymbol{D})$, where f^{ns}, $0 \leq s \leq r$, is defined as before, by (23). The proof of Lemma 17.5 applies and shows that the homomorphisms $f^{(r)m}$ form a chain mapping $T^{(r)}(\boldsymbol{f})\colon T^{(r)}(\boldsymbol{C}) \to T^{(r)}(\boldsymbol{D})$. Clearly, $T^{(r)}(\boldsymbol{f})$ induces homomorphisms $f^{(r)}_m\colon H_m(T^{(r)}(\boldsymbol{C})) \to H_m(T^{(r)}(\boldsymbol{D}))$. Now note that

$$\pi^{(r)} f^{(r)m} = f^{(r-1)m}\pi^{(r)}. \tag{57}$$

Indeed, if $c = (c_0, \dots, c_r) \in T^{(r)m}(\boldsymbol{C})$, where $c_s \in K^{m+s,s}$, $s \leq r$, then $\pi^{(r)} f^{(r)m}(c) = \pi^{(r)}(c'_0, \dots, c'_r) = (c'_0, \dots, c'_{r-1})$, where $c'_s = f^{m+s,s}(c_s)$. On the other hand, $f^{(r-1)m}\pi^{(r)}(c) = f^{(r-1)m}(c_0, \dots, c_{r-1}) = (c'_0, \dots, c'_{r-1})$. Clearly, (57) implies

$$\pi^{(r)}_m f^{(r)}_m = f^{(r-1)}_m \pi^{(r)}_m. \tag{58}$$

Since $f^{(r+1)}_m H_m(T^{(r+1)}(\boldsymbol{C})) \subseteq H_m(T^{(r+1)}(\boldsymbol{D}))$, (48) (for $\boldsymbol{C}$ and $\boldsymbol{D}$) and (58) yield

$$f^{(r)}_m(\overline{H}^{(r)}_m(\boldsymbol{C})) \subseteq \overline{H}^{(r)}_m(\boldsymbol{D}). \tag{59}$$

The restriction $f^{(r)}_m|\overline{H}^{(r)}_m(\boldsymbol{C})$ is the desired homomorphism $\overline{f}^{(r)}_m : \overline{H}^{(r)}_m(\boldsymbol{C}) \to \overline{H}^{(r)}_m(\boldsymbol{D})$. Note that $\boldsymbol{h} = \boldsymbol{g}\boldsymbol{f}$ implies $\overline{h}^{(r)}_m = \overline{g}^{(r)}_m \overline{f}^{(r)}_m$ and $\boldsymbol{f} = \mathrm{id}$ implies $\overline{f}^{(r)}_m = \mathrm{id}$.

The pro-groups $\overline{H}^{(r)}_m(\boldsymbol{X})$, for systems of spaces $\boldsymbol{X}$, as well as the homomorphisms $f_* = \overline{H}^{(r)}_m(\boldsymbol{f})\colon \overline{H}^{(r)}_m(\boldsymbol{X}) \to \overline{H}^{(r)}_m(\boldsymbol{Y})$, for mappings $\boldsymbol{f}\colon \boldsymbol{X} \to \boldsymbol{Y}$, are obtained by first applying the singular complex functor S. In the case of groups $\overline{H}^{(r)}_m(\boldsymbol{X}; G)$ with coefficients in G, one applies the functor $S \otimes G$.

17.2 The first Miminoshvili sequence

In this subsection we study the homomorphism $\overline{\pi}_m^{(r)}: \overline{H}_m^{(r)}(\boldsymbol{C}) \to \overline{H}_m^{(r-1)}(\boldsymbol{C})$ between two strong homology groups of consecutive heights, defined in 17.1. The main result is the following theorem.

THEOREM 17.8. *For every pro-chain complex $\boldsymbol{C}$ and $m \in \mathbb{Z}$, there exists an exact sequence*

$$\begin{aligned} \ldots \to \overline{H}_{m+1}^{(r-2)}(\boldsymbol{C}) \to \quad & \lim\nolimits^r H_{m+r}(\boldsymbol{C}) \to \overline{H}_m^{(r)}(\boldsymbol{C}) \xrightarrow{\overline{\pi}_m^{(r)}} \overline{H}_m^{(r-1)}(\boldsymbol{C}) \to \\ & \lim\nolimits^{r+1} H_{m+r}(\boldsymbol{C}) \to \ldots\ , \end{aligned} \tag{1}$$

which is natural with respect to morphisms of pro-chain complexes.

Since $\overline{H}_{m+1}^{(r-2)}(\boldsymbol{C}) = 0$, for $r \leq 1$, the sequence (1) begins with

$$0 \to \lim\nolimits^1 H_{m+1}(\boldsymbol{C}) \to \overline{H}_m^{(1)}(\boldsymbol{C}) \overset{\overline{\pi}_m^{(1)}}{\to} \overline{H}_m^{(0)}(\boldsymbol{C}) \to \ldots\ . \tag{2}$$

We refer to (1) as the *first sequence of Miminoshvili*, , because it was first announced in (Miminoshvili 1984).

We will obtain (1) as the derived sequence of an exact couple. Therefore, we first recall the relevant definitions and facts on exact couples (see e.g., VIII.1 of (Hilton, Stammbach 1971)). A *bigraded* R-module is the direct sum $D = \oplus D_m^r$ of a collection of R-modules D_m^r, indexed by pairs $(r,m) \in \mathbb{Z} \times \mathbb{Z}$. A *homomorphism* $f: D \to E$ of bigraded modules of *bidegree* $(s,t) \in \mathbb{Z} \times \mathbb{Z}$ is a homomorphism $f = \oplus f_m^r: D \to E$, where $f_m^r: D_m^r \to E_{m+t}^{r+s}$. An *exact couple* of bigraded modules consists of two bigraded modules $D = \oplus D_m^r$, $E = \oplus E_m^r$ and of three homomorphisms $i: D \to D, j: D \to E$, $k: E \to D$ of given bidegrees, which make the following diagram exact at every vertex.

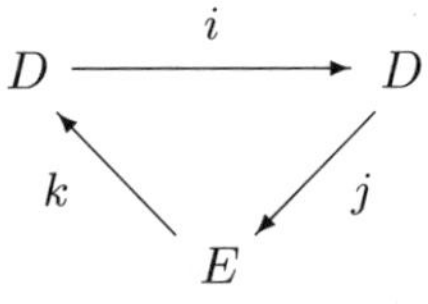

(3)

If the bidegrees of i, j and k are $(-1,0)$, $(1,-1)$ and $(0,0)$, respectively, then exactness means that

$$\operatorname{Im}(j_m^r) = \operatorname{Ker}(k_{m-1}^{r+1}), \tag{4}$$

$$\operatorname{Im}(k_m^r) = \operatorname{Ker}(i_m^r), \tag{5}$$

$$\operatorname{Im}(i_m^r) = \operatorname{Ker}(j_m^{r-1}). \tag{6}$$

With an exact couple is associated its *derived exact couple* (D', E', i', j', k')

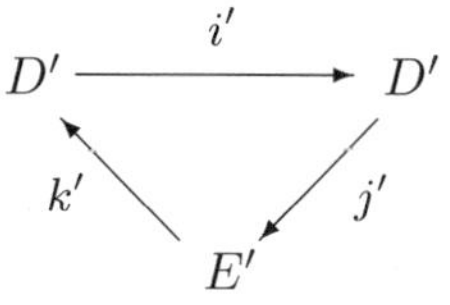

(7)

If the bidegrees of i, j, k are as above, than the derived couple is defined as follows. $D'^r_m = i^{r+1}_m(D^{r+1}_m) \subseteq D^r_m$ and $i'^r_m = i^r_m | D'^r_m \colon D'^r_m \to i^r_m(D^r_m) = D'^{r-1}_m$. Hence, $i' \colon D' \to D'$ is of bidegree $(-1, 0)$. Note that $\partial = jk \colon E \to E$ has the property that $\partial^2 = 0$. Therefore, (E, ∂) is a bigraded differential module. Let E' be its bigraded homology module $E' = H(E, \partial)$. It consists of modules E'^r_m, formed by homology classes $[z]$, where the cycle $z \in E^r_m$. To define j', consider any $d' \in D'^r_m$. There exists an element $d \in D^{r+1}_m$ such that $d' = i^{r+1}_m(d)$. Then $j^r_m(d) \in E^{r+2}_{m-1}$ is a cycle of (E, ∂), because $jk(j(d)) = 0$. Consequently, its homology class $[j^r_m(d)] \in E'^{r+2}_{m-1}$ is defined. Moreover, this class depends only on d'. Indeed, if $d_1 \in D$ is another element satisfying $i(d_1) = d'$, then $i(d_1 - d) = 0$ and by exactness, there exists an element $e \in E$ such that $d_1 - d = k(e)$. Consequently, $j(d_1) - j(d) = (jk)(e)$, which proves that $[j(d_1)] = [j(d)]$. Therefore, putting $j'^r_m(d') = [j^r_m(d)] \in E'^{r+2}_{m-1}$, one obtains a well-defined homomorphism $j' \colon D' \to E'$ of bidegree $(2, -1)$. To define k', consider an $e' \in E'^r_m$ and let $e \in E^r_m$ be a cycle which belongs to the class e', i.e., $e' = [e]$. Put $k'^r_m(e') = k^r_m(e) \in D^r_m$. Since $j(k(e)) = \partial(e) = 0$, exactness yields an element $d \in D^{r+1}_m$ such that $k^r_m(e) = i^{r+1}_m(d) \in D'^r_m$. If $e_1 \in E$ is another representative of e', then $e_1 - e = jk(e_2)$, for some $e_2 \in E$. Therefore, $k(e_1) - k(e) = 0$, which shows that $k^r_m(e)$ depends only on the class e'. Hence, $k' \colon E' \to D'$ is a well-defined homomorphism of bidegree $(0, 0)$.

LEMMA 17.9. *If the homomorphisms i, j, k of an exact couple (D, E, i, j, k) have bidegrees $(-1, 0)$, $(1, -1)$, $(0, 0)$, then the derived couple (D', E', i', j', k') is exact and its homomorphisms i', j', k' have bidegrees $(-1, 0)$, $(2, -1)$, $(0, 0)$. Consequently, for every $m \in \mathbb{Z}$, the following long sequence is exact.*

$$\ldots \to D'^{r-2}_{m+1} \xrightarrow{j'} E'^r_m \xrightarrow{k'} D'^r_m \xrightarrow{i'} D'^{r-1}_m \xrightarrow{j'} E'^{r+1}_{m-1} \to \ldots \,. \tag{8}$$

Proof. For $d' \in D'$, there exists a $d \in D$ such that $d' = i(d)$. Then $j'(d') = [j(d)]$ and therefore, $k'j'(d') = kj(d) = 0$, which shows that $\mathrm{Im}(j') \subseteq \mathrm{Ker}\,(k')$. Conversely, assume that $e' \in \mathrm{Ker}\,(k')$. Let $e \in E$ be such that $e' = [e]$. Then $k(e) = k'(e') = 0$. By exactness of (3), there exists a $d \in D$ such that $j(d) = e$. Therefore, for $d' = i(d) \in D'$, one obtains $j'(d') = [j(d)] = [e] = e'$, which proves that $\mathrm{Ker}\,(k') \subseteq \mathrm{Im}(j')$.

If $e' \in E'$, there exists a cycle $e \in E$ such that $e' = [e]$. Therefore, $i'k'(e') = i'(k[e]) = (ik)[e] = 0$, because $ik = 0$. This shows that $\mathrm{Im}(k') \subseteq \mathrm{Ker}\,(i')$. Conversely, assume that $d' \in \mathrm{Ker}\,(i')$. Then there exists a $d \in D$ such that $d' = i(d)$. Moreover, $i(d') = i'(d') = 0$. By exactness of (3), there exists an $e \in E$ such that $k(e) = d'$. Note that $\partial(e) = jk(e) = j(d') = ji(d) = 0$, so that the homology class $e' = [e] \in E'$ is well defined. Then $k'(e') = k(e) = d'$, which proves that $\mathrm{Ker}\,(i') \subseteq \mathrm{Im}(k')$.

Finally, consider any $d' \in D'$. There exists a $d \in D$ such that $d' = i(d)$. Therefore, $j'i'(d') = j'i(d') = [j(d')] = [ji(d)] = 0$, because $ji = 0$. This shows that $\mathrm{Im}(i') \subseteq \mathrm{Ker}\,(j')$. Conversely, assume that $d' \in \mathrm{Ker}\,(j')$. Then there exists a $d \in D$ such that $d' = i(d)$. Moreover, $[j(d)] = j'(d') = 0$, which implies the existence of an $e \in E$ such that $j(d) = \partial(e) = jk(e)$. Now $j(d - k(e)) = 0$ implies the existence of a $d_1 \in D$ such that $d - k(e) = i(d_1)$. Clearly, $d_1' = i(d_1) \in D'$. Moreover, $i'(d_1') = i(d_1') = i(d) - ik(e) = i(d) = d'$, because $ik = 0$. □

A *homomorphism* $h = (f, g)\colon (D, E, i, j, k) \to (\tilde{D}, \tilde{E}, \tilde{\imath}, \tilde{\jmath}, \tilde{k})$ *of exact couples* consists of homomorphisms $f\colon D \to \tilde{D}$, $g\colon E \to \tilde{E}$ of bidegree (0,0) such that

$$fk = \tilde{k}g, \;\; fi = \tilde{\imath}f, \;\; gj = \tilde{\jmath}f, \tag{9}$$

i.e., the following diagram commutes.

$$\begin{array}{ccccccc} E & \xrightarrow{k} & D & \xrightarrow{i} & D & \xrightarrow{j} & E \\ {\scriptstyle g}\downarrow & & {\scriptstyle f}\downarrow & & {\scriptstyle f}\downarrow & & \downarrow{\scriptstyle g} \\ \tilde{E} & \xrightarrow[\tilde{k}]{} & \tilde{D} & \xrightarrow[\tilde{\imath}]{} & \tilde{D} & \xrightarrow[\tilde{\jmath}]{} & \tilde{E}\,. \end{array} \tag{10}$$

LEMMA 17.10. *A homomorphism of exact couples* $h = (f, g)\colon (D, E, i, j, k) \to (\tilde{D}, \tilde{E}, \tilde{\imath}, \tilde{\jmath}, \tilde{k})$ *induces a homomorphism between the corresponding derived couples* $h' = (f', g')\colon (D', E', i', j', k') \to (\tilde{D}', \tilde{E}', \tilde{\imath}', \tilde{\jmath}', \tilde{k}')$. *By definition,* $f' = f|D'$. *Moreover, if* $e' = [e] \in E'$, *where* $e \in E$ *and* $\partial(e) = 0$, *then* $g'(e') = [g(e)] \in \tilde{E}'$.

Proof. First note that $f(D') = fi(D) = \tilde{\imath}f(D) \subseteq \tilde{\imath}(\tilde{D}) = \tilde{D}'$, so that $f' = f|D'$ is a homomorphism $f'\colon D' \to \tilde{D}'$. Also note that $\partial = jk$ and $\tilde{\partial} = \tilde{\jmath}\tilde{k}$ satisfy $g\partial = \tilde{\partial}g$. Therefore, for $e \in E$ and $\partial(e) = 0$, one has $\tilde{\partial}g(e) = g\partial(e) = 0$, so that $[g(e)] \in \tilde{E}'$. If $e' = [e] = [e_1]$, then there exists an $e_2 \in E$ such that $e_1 - e = \partial(e_2)$ and therefore, $g(e_1) - g(e) = g\partial(e_2) = \tilde{\partial}g(e_2)$, so that $[g(e_1)] = [g(e)]$. Consequently, $[g(e)]$ is a well-defined element $g'(e') \in \tilde{E}'$, which depends only on $e' = [e] \in E'$.

The verification of the analogues of (9) is straightforward. Indeed, for $e' \in E'$, one has $\tilde{k}'g'(e') = \tilde{k}'[g(e)] = \tilde{k}g(e) = fk(e) = f'k(e) = f'k'(e')$. For $d' \in D'$, one has $\tilde{\imath}'f'(d') = \tilde{\imath}'f(d') = \tilde{\imath}f(d')fi(d') = fi'(d') = f'i'(d')$. Finally, $\tilde{\jmath}'f'(d') = \tilde{\jmath}'f(d') = [\tilde{\jmath}f(d')] = [gj(d')] = g'[j(d')] = g'j'(d')$. □

Proof of Theorem 17.8. Let $\boldsymbol{C}$ be a pro-chain complex of abelian groups. For a given integer $r \geq 0$, consider the chain mapping $\pi^{(r)}\colon T^{(r)}(\boldsymbol{C}) \to T^{(r-1)}(\boldsymbol{C})$, defined in 17.1. Recall that in dimension m, $\pi^{(r)}$ is given by the natural projection

$$\pi^{(r)m}\colon \prod_{s=0}^{r} K^{m+s,s} \to \prod_{s=0}^{r-1} K^{m+s,s}. \tag{11}$$

Also consider the chain complex $M^{(r)}(\boldsymbol{C})$, which consists of abelian groups $M^{(r)m}(\boldsymbol{C}) = K^{m+r,r}$ and of boundary operators, which are restrictions of $d^m: T^{(r)m}(\boldsymbol{C}) \to T^{(r)m-1}(\boldsymbol{C})$ to $K^{m+r,r}$, i.e., coincide with the homomorphisms $\partial^{m+r,r}: K^{m+r,r} \to K^{m+r-1,r}$. Let $u^{(r)}: M^{(r)}(\boldsymbol{C}) \to T^{(r)}(\boldsymbol{C})$ be the chain mapping given by natural inclusions $u^{(r)m}: M^{(r)m}(\boldsymbol{C}) \to T^{(r)m}(\boldsymbol{C})$, i.e., for $z \in M^{(r)m}(\boldsymbol{C})$ and $\boldsymbol{\lambda} \in \Lambda_s$, $0 \le s \le r$, let

$$(u^{(r)m}(z))(\boldsymbol{\lambda}) = \begin{cases} 0, & 0 \le s < r, \\ z(\boldsymbol{\lambda}), & s = r. \end{cases} \tag{12}$$

Since $\pi^{(r)m}$ is a surjection, whose kernel equals $K^{m+r,r}$, the following sequence of chain complexes is exact.

$$0 \to M^{(r)}(\boldsymbol{C}) \overset{u^{(r)}}{\to} T^{(r)}(\boldsymbol{C}) \overset{\pi^{(r)}}{\to} T^{(r-1)}(\boldsymbol{C}) \to 0. \tag{13}$$

For every $r \ge 0$, the sequence (13) induces a long exact sequence of homology groups,

$$\begin{aligned} \ldots \to H_{m+1}(T^{(r)}) \overset{\Delta^{(r)}_{m+1}}{\longrightarrow} H_m(M^{(r)}) \overset{u^{(r)}_m}{\longrightarrow} H_m(T^{(r)}) \overset{\pi^{(r)}_m}{\longrightarrow} H_m(T^{(r-1)}) \overset{\Delta^{(r-1)}_m}{\longrightarrow} \\ H_{m-1}(M^{(r)}) \to \ldots . \end{aligned} \tag{14}$$

The sequences (14), $m \in \mathbb{Z}$, can be interpreted as an exact couple of bigraded R-modules (D, E, i, j, k), by putting

$$D^r_m = H_m(T^{(r)}(\boldsymbol{C})), \; E^r_m = H_m(M^{(r)}(\boldsymbol{C})), \tag{15}$$

$$i^r_m = \pi^{(r)}_m, \; j^r_m = \Delta^{(r)}_m, \; k^r_m = u^{(r)}_m. \tag{16}$$

Note that the bidegrees of i, j, k are (-1,0), (1,-1) and (0,0), respectively. Therefore, by Lemma 17.9, the derived couple consists of exact sequences (8), $m \in \mathbb{Z}$. To complete the proof of Theorem 17.8, it suffices to show that (8) is of the form (1), i.e., it suffices to prove the following equalities.

$$D'^r_m = \overline{H}^{(r)}_m(\boldsymbol{C}), \tag{17}$$

$$E'^r_m = \lim{}^r H_{m+r}(\boldsymbol{C}), \tag{18}$$

$$i'^r_m = \overline{\pi}^{(r)}_m. \tag{19}$$

By the definition of D'^r_m and by (16), (15), (7.1.48), $D'^r_m = i^{r+1}_m(D^{r+1}_m) = \pi^{(r+1)}_m(H_m(T^{(r+1)}(\boldsymbol{C})) = \overline{H}^{(r)}_m(\boldsymbol{C})$. Moreover, by the definition of i'^r_m and by (16), (17) and (17.1.50), one obtains $i'^r_m = i^r_m|D'^r_m = \pi^{(r)}_m|\overline{H}^{(r)}_m(\boldsymbol{C}) = \overline{\pi}^{(r)}_m$.

In order to prove (18), consider the inverse system of groups $\boldsymbol{X} = H_{m+r}(\boldsymbol{C})$ and recall that

$$\lim{}^s \boldsymbol{X} = H^s K(\boldsymbol{X}), \tag{20}$$

where $K(\boldsymbol{X}) = (K^s(\boldsymbol{X}), \delta^{s+1})$ is the cochain complex, defined in 11.5. Consequently,

$$K^s(\boldsymbol{X}) = \prod_{\boldsymbol{\lambda}\in\Lambda_s} X_{\lambda_0} = \prod_{\boldsymbol{\lambda}\in\Lambda_s} H_{m+r}(C_{\lambda_0}), \tag{21}$$

and $\delta^{s+1}\colon K^s(\boldsymbol{X}) \to K^{s+1}(\boldsymbol{X})$ is given by

$$(\delta^{s+1}[z])(\boldsymbol{\lambda}) = p^{m+s}_{\lambda_0\lambda_1 *}([z](d^0\boldsymbol{\lambda})) + \sum_{j=1}^{s+1}(-1)^j[z](d^j\boldsymbol{\lambda}), \tag{22}$$

where $[z] \in K^s(\boldsymbol{X})$, $\boldsymbol{\lambda} \in \Lambda_{s+1}$. Hence, by (20),

$$\lim{}^r H_{m+r}(\boldsymbol{C}) = \operatorname{Ker}(\delta^{r+1})/\operatorname{Im}(\delta^r). \tag{23}$$

On the other hand,

$$E'^r_m = \operatorname{Ker}(j^r_m k^r_m)/\operatorname{Im}(j^{r-1}_{m+1}k^{r-1}_{m+1}). \tag{24}$$

Therefore, to prove (18), it suffices to prove that the mappings δ^{r+1} and $j^r_m k^r_m$ coincide up to a sign. Actually, we will show that

$$j^r_m k^r_m = (-1)^m \delta^{r+1}. \tag{25}$$

By (17.1.1) and (17.1.3),

$$M^{(r)m}(\boldsymbol{C}) = K^{m+r,r} = \prod_{\boldsymbol{\lambda}\in\Lambda_r} C^{m+r}_{\lambda_0}, \tag{26}$$

$$\partial^{m+r,r} = \prod_{\boldsymbol{\lambda}\in\Lambda_r} \partial^{m+r}_{\lambda_0}, \tag{27}$$

and thus, by (15),

$$E^r_m = \prod_{\boldsymbol{\lambda}\in\Lambda_r} H_{m+r}(C_{\lambda_0}). \tag{28}$$

(21) and (28) show that the domains and the codomains of the mappings δ^{r+1} and $j^r_m k^r_m$ coincide.

In order to prove (25), note that every element of $E^r_m = H_m(M^{(r)}(\boldsymbol{C}))$ is the homology class $[z] \in \prod_{\boldsymbol{\lambda}\in\Lambda_r} H_{m+r}(C_{\lambda_0})$ of a cycle $z \in M^{(r)m}(\boldsymbol{C})$, where

$$[z](\boldsymbol{\lambda}) = [z(\boldsymbol{\lambda})],\ \boldsymbol{\lambda} \in \Lambda_r. \tag{29}$$

By (16), $k^r_m[z] = [u^{(r)m}(z)]$. To determine $j^r_m k^r_m[z] = \Delta^{(r)m}[u^{(r)m}(z)]$, choose a chain $y \in T^{(r+1),m}(\boldsymbol{C})$ such that $\pi^{(r+1),m}(y) = u^{(r)m}(z)$, i.e.,

$$y(\boldsymbol{\lambda}) = (u^{(r)m}(z))(\boldsymbol{\lambda}),\ \lambda \in \Lambda_s,\ 0 \le s \le r. \tag{30}$$

Note that (30) and (12) imply

$$y(\boldsymbol{\lambda}) = z(\boldsymbol{\lambda}),\ \lambda \in \Lambda_r. \tag{31}$$

By exactness of (13) there is a unique $x \in M^{(r+1)m-1} = K^{m+r,r+1}$ such that $d^m y = u^{(r+1),m-1}(x)$. Since z and $u^{(r)m}z$ are cycles, by the well-known construction of the connecting homomorphism $\Delta^{(r)m}$, x is an $(m-1)$-cycle of $M^{(r+1)}(\boldsymbol{C})$ and $\Delta^{(r)m}[u^{(r)m}(z)]$ is its homology class $[x] \in H_{m-1}(M^{(r+1)}(\boldsymbol{C}))$. By (17.1.12),

$$x(\boldsymbol{\lambda}) = (d^m y)(\boldsymbol{\lambda}) = (\partial^m y)(\boldsymbol{\lambda}) + (-1)^m(\delta^m y)(\boldsymbol{\lambda}),\ \boldsymbol{\lambda} \in \Lambda_{r+1}. \tag{32}$$

Now consider the chain $w \in M^{(r+1)m}$, defined by $w(\boldsymbol{\lambda}) = y(\boldsymbol{\lambda})$, for $\boldsymbol{\lambda} \in \Lambda_{r+1}$. Note that

$$(\partial^m y)(\boldsymbol{\lambda}) = \partial_{\lambda_0}^{m+r+1}(y(\boldsymbol{\lambda})) = \partial_{\lambda_0}^{m+r+1}(w(\boldsymbol{\lambda})). \tag{33}$$

Since $w(\boldsymbol{\lambda}) \in C_{\lambda_0}^{m+r+1}$, (33) shows that $[(\partial^m y)(\boldsymbol{\lambda})] \in H_{m+r}(C_{\lambda_0})$ vanishes. On the other hand, since $(\delta^m y)(\boldsymbol{\lambda}) = (\delta^{m+r,r+1}y)(\boldsymbol{\lambda})$, (17.1.2) yields

$$(\delta^m y)(\boldsymbol{\lambda}) = p_{\lambda_0\lambda_1}^{m+r}(y(d^0\boldsymbol{\lambda})) + \sum_{j=1}^{r+1}(-1)^j y(d^j\boldsymbol{\lambda}),\ \boldsymbol{\lambda} \in \Lambda_{r+1}. \tag{34}$$

However, (31) shows that $y(d^j\boldsymbol{\lambda}) = z(d^j\boldsymbol{\lambda})$. Using this in (34) and passing to homology classes in $H_{m+r}(C_{\lambda_0})$, we conclude that

$$[(\delta^m y)(\boldsymbol{\lambda})] = p_{\lambda_0\lambda_1 *}^{m+r}[z(d^0\boldsymbol{\lambda})] + \sum_{j=1}^{r+1}(-1)^m[z(d^j\boldsymbol{\lambda})],\ \lambda \in \Lambda_{r+1}. \tag{35}$$

Substituting in (35) $[z(d^j\boldsymbol{\lambda})]$ by $[z](d^j\boldsymbol{\lambda})$ (see (29)) and comparing the result with (22), we conclude that the right side of (35) equals $(\delta^{r+1}[z])(\boldsymbol{\lambda})$. Since also $[x](\boldsymbol{\lambda}) = [x(\boldsymbol{\lambda})]$ and $[(\partial^m y)(\boldsymbol{\lambda})] = 0$, passing to homology in (32) yields the desired result $[x](\boldsymbol{\lambda}) = (\delta^{r+1}[z])(\boldsymbol{\lambda})$, for $\boldsymbol{\lambda} \in \Lambda_{r+1}$, i.e., $[x] = \delta^{r+1}[z]$.

To complete the proof we still need to prove naturality with respect to morphisms of pro-chain complexes $\boldsymbol{f} = (f, f_\mu)\colon \boldsymbol{C} \to \boldsymbol{D}$. These morphisms induce chain mappings $f^{(r)}\colon T^{(r)}(\boldsymbol{C}) \to T^{(r)}(\boldsymbol{D})$, defined by (17.1.23). Moreover, one has chain mappings $g^{(r)}\colon M^{(r)}(\boldsymbol{C}) \to M^{(r)}(\boldsymbol{D})$, defined by (15.1.1). It is readily seen that the following diagram of chain complexes commutes.

$$\begin{array}{ccccccccc}
0 & \longrightarrow & M^{(r)}(\boldsymbol{C}) & \xrightarrow{u} & T^{(r)}(\boldsymbol{C}) & \xrightarrow{\pi} & T^{(r-1)}(\boldsymbol{C}) & \longrightarrow & 0 \\
 & & \downarrow{\scriptstyle g^{(r)}} & & \downarrow{\scriptstyle f^{(r)}} & & \downarrow{\scriptstyle f^{(r-1)}} & & \\
0 & \longrightarrow & M^{(r)}(\boldsymbol{D}) & \xrightarrow[u]{} & T^{(r)}(\boldsymbol{D}) & \xrightarrow[\pi]{} & T^{(r-1)}(\boldsymbol{D}) & \longrightarrow & 0.
\end{array} \tag{36}$$

Passing to homology, one obtains a homomorphism between the exact sequence (14) for $\boldsymbol{C}$ and the corresponding sequence for $\boldsymbol{D}$. This can be interpreted as a homomorphism of the exact couple (D, E, i, j, k) for $\boldsymbol{C}$ to the corresponding exact couple for $\boldsymbol{D}$. Passing to derived couples, one obtains a commutative diagram (10), which represents a homomorphism of the long exact sequences (14), for $\boldsymbol{C}$ and $\boldsymbol{D}$, respectively. Since these sequences coincide with (1), for $\boldsymbol{C}$ and $\boldsymbol{D}$, one has the desired commutative diagram, which expresses the naturality of (1). □

17.3 The second Miminoshvili sequence

With every pro-chain complex $\boldsymbol{C}$ and $m \in \mathbb{Z}$, one can associate the inverse sequence of abelian groups

$$\overline{H}_m^{(*)}(\boldsymbol{C}) = (\overline{H}_m^{(0)}(\boldsymbol{C}) \leftarrow \ldots \leftarrow \overline{H}_m^{(r)}(\boldsymbol{C}) \overset{\overline{\pi}_m^{(r+1)}}{\longleftarrow} \overline{H}_m^{(r+1)}(\boldsymbol{C}) \leftarrow \ldots). \quad (1)$$

Its terms and bonding homomorphisms were defined in 17.1. The natural projections $\pi^{(r\infty)m}\colon T^m(\boldsymbol{C}) \to T^{(r)m}(\boldsymbol{C})$ determine a chain mapping $\pi^{(r\infty)}\colon T(\boldsymbol{C}) \to T^{(r)}(\boldsymbol{C})$, which induces homomorphisms of the corresponding homology groups $\pi_m^{(r\infty)}\colon \overline{H}_m(\boldsymbol{C}) = H_m(T(\boldsymbol{C})) \to H_m(T^{(r)}(\boldsymbol{C}))$. Clearly,

$$\pi_m^{(r\infty)} = \pi_m^{(r+1)} \pi_m^{(r+1,\infty)}. \quad (2)$$

Therefore,

$$\pi_m^{(r\infty)}(\overline{H}_m(\boldsymbol{C})) \subseteq \pi_m^{(r+1)}(H_m(T^{(r+1)}(\boldsymbol{C}))) = \overline{H}_m^{(r)}(\boldsymbol{C}). \quad (3)$$

Consequently, one can view $\pi_m^{(r\infty)}$ as a homomorphism $\overline{\pi}_m^{(r\infty)}\colon \overline{H}_m(\boldsymbol{C}) \to \overline{H}_m^{(r)}(\boldsymbol{C})$. Clearly,

$$\overline{\pi}_m^{(r)\infty} = \overline{\pi}_m^{(r+1)} \overline{\pi}_m^{(r+1)\infty}. \quad (4)$$

Hence, the homomorphisms $\overline{\pi}_m^{(r)\infty}$ induce a homomorphism $\overline{\pi}_m^{\infty}\colon \overline{H}_m(\boldsymbol{C}) \to \lim \overline{H}_m^{(*)}(\boldsymbol{C})$. The next theorem asserts that $\overline{\pi}_m^{\infty}$ is an epimorphism and it describes its kernel.

THEOREM 17.11. *For every pro-chain complex $\boldsymbol{C}$ and $m \in \mathbb{Z}$, there exists an exact sequence*

$$0 \to \lim{}^1 \overline{H}_{m+1}^{(*)}(\boldsymbol{C}) \to \overline{H}_m(\boldsymbol{C}) \xrightarrow{\overline{\pi}_m^{\infty}} \lim \overline{H}_m^{(*)}(\boldsymbol{C}) \to 0, \quad (5)$$

which is natural with respect to morphisms of pro-chain complexes.

To (5) we refer as to the *second sequence of Miminoshvili.* The proof of Theorem 17.11 is based on a general lemma, which we will now state and prove (see (Massey 1978), Appendix, Theorem A19).

Let $\boldsymbol{K} = (K_i, p_{ii'}, \mathbb{N})$ be an inverse sequence of chain complexes $K_i = (K_i^n, \partial_i^n)$ and chain mappings $p_{ii'}\colon K_{i'} \to K_i$, which consist of homomorphisms $p_{ii'}^n\colon K_{i'}^n \to K_i^n$. For n fixed, $\boldsymbol{K}^n = (K_i^n, p_{ii'}^n, \mathbb{N})$ is an inverse sequence of abelian groups with limit $K_\infty^n = \lim \boldsymbol{K}^n$ and projections $p_i^n\colon K_\infty^n \to K_i^n$. Moreover, the boundary operators ∂_i^n induce a mapping $\boldsymbol{\partial}^n\colon \boldsymbol{K}^n \to \boldsymbol{K}^{n-1}$, whose limit is a boundary operator $\partial_\infty^n\colon K_\infty^n \to K_\infty^{n-1}$. In this way one obtains a chain complex $\lim \boldsymbol{K} = K_\infty = (K_\infty^n, \partial_\infty^n)$ as well as chain mappings $p_i\colon K_\infty \to K_i$, given by the homomorphisms p_i^n. Note that $p_{ii'}^n p_{i'}^n = p_i^n$, for $i \le i'$. Passing to homology, for each $n \in \mathbb{Z}$, one obtains an inverse sequence of homology groups $H_n(\boldsymbol{K}) = (H_n(K_i), p_{ii'*}, \mathbb{N})$ and a sequence of homomorphisms $p_{i*}^n\colon H_n(\lim \boldsymbol{K}) \to H_n(K_i)$, which satisfy $p_{ii'*}^n p_{i'*}^n = p_{i*}^n$, for $i \le i'$, and thus, induces a homomorphism $p_*^n\colon H_n(\lim \boldsymbol{K}) \to \lim H_n(\boldsymbol{K})$.

LEMMA 17.12. *If $\boldsymbol{K} = (K_i, p_{ii'}, \mathbb{N})$ is an inverse sequence of chain complexes, which satisfies the condition*

$$\lim{}^1 \boldsymbol{K}^n = 0, \tag{6}$$

for all $n \in \mathbb{Z}$, then for every $m \in \mathbb{Z}$, there exists an exact sequence

$$0 \to \lim{}^1 H_{m+1}(\boldsymbol{K}) \to H_m(\lim \boldsymbol{K}) \xrightarrow{p_*^m} \lim\,(H_m(\boldsymbol{K})) \to 0, \tag{7}$$

which is natural with respect to level morphisms of pro-chain complexes.

Proof. Denote by $Z^n(K_i)$ the subgroup of the cycles of K_i^n. Since the homomorphisms $p_{ii'}^n$ map $Z^n(K_{i'})$ to $Z^n(K_i)$, the groups $Z^n(K_i)$ and the restrictions of $p_{ii'}^n$ define an inverse sequence $Z^n(\boldsymbol{K})$. Analogously, we define the groups $B^n(K_i)$ of the boundaries of K_i^n and the inverse sequence $B^n(\boldsymbol{K})$, consisting of those groups. Note that the quotient homomorphisms $q_i^n \colon Z^n(K_i) \to H^n(K_i)$ define a morphism $\boldsymbol{q}^n \colon Z^n(\boldsymbol{K}) \to H_n(\boldsymbol{K})$. Clearly, the following sequences are exact

$$0 \to Z^n(\boldsymbol{K}) \xrightarrow{\boldsymbol{i}^n} \boldsymbol{K}^n \xrightarrow{\boldsymbol{\partial}^n} B^{n-1}(\boldsymbol{K}) \to 0, \tag{8}$$

$$0 \to B^n(\boldsymbol{K}) \xrightarrow{\boldsymbol{j}^n} Z^n(\boldsymbol{K}) \xrightarrow{\boldsymbol{q}^n} H_n(\boldsymbol{K}) \to 0, \tag{9}$$

where $\boldsymbol{i}$ and $\boldsymbol{j}$ are morphisms given by inclusions. Now apply Corollary 11.51 to (8) and take into account (6). One obtains an exact sequence

$$0 \to \lim\, Z^n(\boldsymbol{K}) \xrightarrow{i^n} \lim\, \boldsymbol{K}^n \xrightarrow{\partial^n} \lim\, B^{n-1}(\boldsymbol{K}) \xrightarrow{\theta} \lim{}^1\, Z^n(\boldsymbol{K}) \to 0, \tag{10}$$

as well as the equality

$$\lim{}^1\, B^{n-1}(\boldsymbol{K}) = 0. \tag{11}$$

Applying the same corollary to (9) and taking into account (11) in dimension n, one obtains an exact sequence

$$0 \to \lim\, B^n(\boldsymbol{K}) \xrightarrow{j^n} \lim\, Z^n(\boldsymbol{K}) \xrightarrow{q^n} \lim\, H_n(\boldsymbol{K}) \to 0, \tag{12}$$

as well as the conclusion that

$$\lim{}^1 \boldsymbol{q}^n \colon \lim{}^1\, Z^n(\boldsymbol{K}) \to \lim{}^1\, H_n(\boldsymbol{K}) \tag{13}$$

is an isomorphism. In (10) and (12), $i^n = \lim \boldsymbol{i}^n$, $j^n = \lim \boldsymbol{j}^n$, $q^n = \lim \boldsymbol{q}^n$. Note that i^n and j^n are monomorphisms and therefore, can be viewed as inclusions.

Next note that

$$B^{n-1}(\lim \boldsymbol{K}) \subseteq \lim\, B^{n-1}(\boldsymbol{K}). \tag{14}$$

Indeed, every element $b = (b_i) \in B^{n-1}(\lim \boldsymbol{K})$ is of the form $b = \partial_\infty^n c$, where $c \in K_\infty^n$. Since $K_\infty^n = \lim \boldsymbol{K}^n$, we see that c is a sequence (c_i), where

$c_i \in K_i^n$ is such that $p_{ii'}^n(c_{i'}) = c_i$. Consequently, $b_i = \partial_i^n c_i = \partial_i^n p_{ii'}^n(c_{i'}) = p_{ii'}^{n-1}\partial_{i'}^n(c_{i'}) = p_{ii'}^{n-1}(b_{i'})$. This shows that indeed, $b \in \lim B^{n-1}(\boldsymbol{K})$.

Also note that, by (12) in dimension $n-1$,

$$\lim B^{n-1}(\boldsymbol{K}) \subseteq \lim Z^{n-1}(\boldsymbol{K}). \tag{15}$$

Clearly, (15) and (14) yield an exact sequence

$$0 \to \frac{\lim B^{n-1}(\boldsymbol{K})}{B^{n-1}(\lim \boldsymbol{K})} \xrightarrow{\alpha} \frac{\lim Z^{n-1}(\boldsymbol{K})}{B^{n-1}(\lim \boldsymbol{K})} \xrightarrow{\beta} \frac{\lim Z^{n-1}(\boldsymbol{K})}{\lim B^{n-1}(\boldsymbol{K})} \to 0, \tag{16}$$

where α is induced by the inclusion (15) and β is induced by the identity homomorphism of the numerators. We will show that (16) can be identified with (7), for $m = n-1$, and thus, the latter sequence must also be exact.

Indeed, from the definition of the chain complex $\lim \boldsymbol{K}$, it follows immediately that

$$\lim Z^{n-1}(\boldsymbol{K}) = Z^{n-1}(\lim \boldsymbol{K}). \tag{17}$$

Therefore, the second term of (16) equals $H_{n-1}(\lim \boldsymbol{K})$. Exactness of the sequence (12) in dimension $n-1$ shows that q^{n-1} induces an isomorphism between the third term of (16) and $\lim H_{n-1}(\boldsymbol{K})$. Moreover, the composition of β with this isomorphism coincides with the homomorphism q_*^{n-1} from (7).

We next show that there is an isomorphism between the first term of (16) and $\lim^1 H_n(\boldsymbol{K})$, which is the first term of (7), for $m = n-1$. Indeed, since $\partial_\infty^n(\lim \boldsymbol{K}^n) = B^{n-1}(\lim \boldsymbol{K})$, (10) shows that θ induces an isomorphism $\tilde{\theta}$ between the first term of (16) and the group $\lim^1 Z^n(\boldsymbol{K})$. Furthermore, by (13), $\lim^1 \boldsymbol{q}^n$ is an isomorphism between the latter group and $\lim^1 H_n(\boldsymbol{K})$. Clearly, there is a unique homomorphism $\tilde{\alpha}\colon \lim^1 H_n(\boldsymbol{K}) \to H_{n-1}(\lim \boldsymbol{K})$ such that

$$\tilde{\alpha}(\lim{}^1 \boldsymbol{q}^n)\tilde{\theta} = \alpha. \tag{18}$$

We have thus, exhibited an isomorphism between the sequences (16) and (7), for $m = n-1$. Since the sequence (16) is exact, the same must be true of the sequence (7), for $m = n-1$.

To complete the proof of Lemma 17.12, we still need to prove naturality of the sequence (7). Let $\boldsymbol{f} = (f_i)\colon \boldsymbol{K} \to \boldsymbol{L} = (L_i, q_{ii'}, \mathbb{N})$ be a level morphism of pro-chain complexes, which satisfies condition (6). Note that the chain mappings $f_i\colon K_i \to L_i$ induce homomorphisms $f_{i*}\colon H_{n-1}(K_i) \to H_{n-1}(L_i)$, which form a level morphism of pro-groups $\boldsymbol{f}_*\colon H_{n-1}(\boldsymbol{K}) \to H_{n-1}(\boldsymbol{L})$. Clearly, $\boldsymbol{f}_*$ induces homomorphisms $\lim \boldsymbol{f}_*\colon \lim H_{n-1}(\boldsymbol{K}) \to \lim H_{n-1}(\boldsymbol{L})$. Analogously, we obtain homomorphisms $\lim^1 \boldsymbol{f}_*\colon \lim^1 H_n(\boldsymbol{K}) \to \lim^1 H_n(\boldsymbol{L})$. Moreover, the chain mappings $f_i\colon K_i \to L_i$ define a limit chain mapping $\lim \boldsymbol{f}\colon \lim \boldsymbol{K} \to \lim \boldsymbol{L}$, which induces a homorphism $(\lim \boldsymbol{f})_*\colon H_{n-1}(\lim \boldsymbol{K}) \to H_{n-1}(\lim L)$. We must show that the following diagram commutes.

$$\begin{array}{ccccc}
\lim^1 H_n(\boldsymbol{K}) & \longrightarrow & H_{n-1}(\lim \boldsymbol{K}) & \xrightarrow{q_*^{n-1}} & \lim H_{n-1}(\boldsymbol{K}) \\
\lim^1 \boldsymbol{f}_* \big\downarrow & & \big\downarrow (\lim \boldsymbol{f})_* & & \big\downarrow \lim \boldsymbol{f}_* \\
\lim^1 H_n(\boldsymbol{L}) & \longrightarrow & H_{n-1}(\lim \boldsymbol{L}) & \xrightarrow[q_*^{n-1}]{} & \lim H_{n-1}(\boldsymbol{L}) \,.
\end{array} \tag{19}$$

A straightforward verification shows that the right rectangle in (19) commutes, i.e., q_*^{n-1} is natural. In order to prove commutativity of the left rectangle of diagram (19), i.e., to prove that also the homomorphism $\tilde{\alpha}$ is natural, it suffices to show that all three homomorphisms α, $\lim^1 \boldsymbol{q}^n$ and $\tilde{\theta}$ from (18) are natural.

First note that the chain mappings $f_i: K_i \to L_i$ map $B^{n-1}(K_i)$ to $B^{n-1}(L_i)$ and determine a morphism $\boldsymbol{f}: B^{n-1}(\boldsymbol{K}) \to B^{n-1}(\boldsymbol{L})$, which induces a homomorphism $\lim \boldsymbol{f}: \lim(B^{n-1}(\boldsymbol{K})) \to \lim(B^{n-1}(\boldsymbol{L}))$. It is straightforward to verify that $\lim \boldsymbol{f}(B^{n-1}(\boldsymbol{K})) \subseteq B^{n-1}(\boldsymbol{L})$. Therefore, $\lim \boldsymbol{f}$ induces a homomorphism

$$\lim(B^{n-1}(\boldsymbol{K}))/B^{n-1}(\boldsymbol{K}) \to \lim(B^{n-1}(\boldsymbol{L}))/B^{n-1}(\boldsymbol{L}) \tag{20}$$

between the first terms of (16) for $\boldsymbol{K}$ and $\boldsymbol{L}$, respectively. Similarly, $\boldsymbol{f}$ induces a homomorphism between the second terms of the corresponding sequences (16). Naturality of α means commutativity of the following diagram.

$$\begin{array}{ccc}
\dfrac{\lim B^{n-1}(\boldsymbol{K})}{B^{n-1}(\lim \boldsymbol{K})} & \xrightarrow{\alpha} & \dfrac{\lim Z^{n-1}(\boldsymbol{K})}{B^{n-1}(\lim \boldsymbol{K})} \\
\big\downarrow & & \big\downarrow \\
\dfrac{\lim B^{n-1}(\boldsymbol{L})}{B^{n-1}(\lim \boldsymbol{L})} & \xrightarrow[\alpha]{} & \dfrac{\lim Z^{n-1}(\boldsymbol{L})}{B^{n-1}(\lim \boldsymbol{L})} \,.
\end{array} \tag{21}$$

Since the homomorphism α was induced by inclusion, commutativity of (21) is readily verified.

To prove naturality of $\lim^1 \boldsymbol{q}^n$, it suffices to note that, by the definition of f_{i*}, one has $f_{i*}q_i^n = q_i^n f_i$, for each i. Therefore, $\boldsymbol{f}_* \boldsymbol{q}^n = \boldsymbol{q}^n \boldsymbol{f}$. Since $\lim^1$ is a functor, one obtains the desired relation $\lim^1 \boldsymbol{f}_* \lim^1 \boldsymbol{q}^n = \lim^1 \boldsymbol{q}^n \lim^1 \boldsymbol{f}$. Finally, to establish naturality of $\tilde{\theta}$, it suffices to show naturality of θ. Note that $\boldsymbol{f}: \boldsymbol{K} \to \boldsymbol{L}$ induces a level morphism $\boldsymbol{f}^n: \boldsymbol{K}^n \to \boldsymbol{L}^n$, which induces a morphism between the exact sequence (8) and the corresponding sequence for L. The latter induces a morphism between the corresponding long exact sequences (10), which implies the desired naturality of θ. □

REMARK 17.13. Condition (6) from Lemma 17.12 is satisfied for every n, whenever all the bonding homomorphisms $p_{ii'}^n$ of $\boldsymbol{K}^n$ are surjective. Indeed, in this case, for any sequence (ξ_i), $\xi_i \in K_i^n$, one can construct, by induction, a sequence (u_i), $u_i \in K_i^n$ such that $p_{ii+1}^n(u_{i+1}) = \xi_i + u_i$. However, by 11.6, the latter condition shows that $\lim^1 (\boldsymbol{K}^n) = 0$. A more general condition, which insures (6), is the Mittag – Leffler condition (see (Mardešić, Segal 1982), II.6.2, Theorem 10).

Proof of Theorem 17.11. Let $\boldsymbol{C}$ be a pro-chain complex and let $m \in \mathbb{Z}$. Consider the inverse sequence $H_m(T^{(*)}(\boldsymbol{C}))$, defined by

$$H_m(T^{(0)}(\boldsymbol{C})) \leftarrow \ldots \leftarrow H_m(T^{(r)}(\boldsymbol{C})) \xleftarrow{\pi_m^{r+1}} H_m(T^{(r+1)}(\boldsymbol{C})) \leftarrow \ldots. \tag{22}$$

We will first prove that there exists an exact sequence

$$0 \to \lim{}^1 H_{m+1}(T^{(*)}(\boldsymbol{C})) \to \overline{H}_m(\boldsymbol{C}) \xrightarrow{\pi_m} \lim H_m(T^{(*)}(\boldsymbol{C})) \to 0, \tag{23}$$

which is natural with respect to morphisms of pro-chain complexes.

Consider the inverse sequence of chain complexes

$$T^{(*)}(\boldsymbol{C}) = (T^{(0)}(\boldsymbol{C}) \leftarrow \ldots \leftarrow T^{(r)}(\boldsymbol{C}) \xleftarrow{\pi^{(r+1)}} T^{(r+1)}(\boldsymbol{C}) \leftarrow \ldots) \tag{24}$$

and note that the chain mappings $\pi^{(r+1)}$ consist of projections to direct factors, which are surjections. Therefore, by Remark 17.13, Lemma 17.12 is applicable to the inverse sequence of chain complexes $\boldsymbol{K} = T^{(*)}(\boldsymbol{C})$. The obtained short exact sequence (7) coincides with (23). Indeed, $H_m(\boldsymbol{K})$ becomes $H_m(T^{(*)}(\boldsymbol{C}))$. On the other hand, by (17.1.47) and (17.2.11),

$$\lim\,(T^{(r)m}(\boldsymbol{C})) = \lim \prod_{s=0}^{r} K^{m+s,s} = \prod_{s=0}^{\infty} K^{m+s,s} = T^m(\boldsymbol{C}). \tag{25}$$

Hence, $\lim \boldsymbol{K}$ becomes $T(\boldsymbol{C})$, while $H_m(\lim \boldsymbol{K})$ becomes $H_m(T(\boldsymbol{C})) = \overline{H}_m(\boldsymbol{C})$. Moreover, every morphism of pro-chain complexes $\boldsymbol{f}: \boldsymbol{C} \to \boldsymbol{D}$ induces a level morphism of inverse sequences $T^{(*)}(\boldsymbol{f}): T^{(*)}(\boldsymbol{C}) \to T^{(*)}(\boldsymbol{D})$. Naturality of (23) follows from the naturality of (7).

It remains to show that the sequences (5) and (23) are naturally isomorphic. First note that

$$\overline{H}_m^{(r)}(\boldsymbol{C}) \subseteq H_m(T^{(r)}\boldsymbol{C}). \tag{26}$$

The inclusions $u^{(r)}$, given by (26), define a level morphism of inverse sequences $\boldsymbol{u}: \overline{H}_m^{(*)}(\boldsymbol{C}) \to H_m(T^{(*)}(\boldsymbol{C}))$, because the following diagram commutes (solid arrows).

$$\begin{array}{ccc} \overline{H}_m^{(r)}(\boldsymbol{C}) & \xleftarrow{\pi^{(r)}} & \overline{H}_m^{(r+1)}(\boldsymbol{C}) \\ \downarrow u^{(r)} & \overset{\pi^{(r)}}{\nwarrow} & \downarrow u^{(r+1)} \\ H_m(T^{(r)}\boldsymbol{C}) & \xleftarrow[\pi^{(r)}]{} & H_m(T^{(r+1)}(\boldsymbol{C})). \end{array} \tag{27}$$

Clearly, $\boldsymbol{u}$ induces homomorphisms $\lim \boldsymbol{u}: \lim \overline{H}_m^{(*)}(\boldsymbol{C}) \to \lim H_m(T^{(*)}(\boldsymbol{C}))$ and $\lim^1 \boldsymbol{u}: \lim^1 \overline{H}_{m+1}(\boldsymbol{C}) \to \lim^1 H_{m+1}(T^{(*)}(\boldsymbol{C}))$, and make the following diagram commutative.

$$\begin{array}{ccccc}
\lim^1 \overline{H}^{(*)}_{m+1}(\boldsymbol{C}) & \longrightarrow & \overline{H}_m(\boldsymbol{C}) & \xrightarrow{\overline{\pi}_m} & \lim \overline{H}^{(*)}_m(\boldsymbol{C}) \\
\downarrow{\scriptstyle \lim^1 \boldsymbol{u}} & & \downarrow{\scriptstyle \mathrm{id}} & & \downarrow{\scriptstyle \lim \boldsymbol{u}} \\
\lim^1 H_{m+1}(T^{(*)}(\boldsymbol{C})) & \longrightarrow & \overline{H}_m(\boldsymbol{C}) & \xrightarrow[\pi_m]{} & \lim H_m(T^{(*)}(\boldsymbol{C})) \, .
\end{array} \tag{28}$$

In order to show that $\lim^1 \boldsymbol{u}$ and $\lim \boldsymbol{u}$ are isomorphisms, it suffices to show that $\boldsymbol{u}$ is an isomorphism. This is indeed the case, because diagram (27) remains commutative if one inserts the diagonal $\pi^{(r)}: H_m(T^{(r+1)}(\boldsymbol{C})) \to \overline{H}^{(r)}_m(\boldsymbol{C})$. Therefore, Morita's lemma applies (see (Mardešić, Segal 1982), II, Theorem 2.2.5) and yields the desired conclusion. □

COROLLARY 17.14. *If $m \in \mathbb{Z}$ and $\boldsymbol{C}$ is a pro-chain complex having the property that*

$$\lim^r H_{m+r}(\boldsymbol{C}) = 0, \ r \geq 2, \tag{29}$$

$$\lim^{r+1} H_{m+r}(\boldsymbol{C}) = 0, \ r \geq 1, \tag{30}$$

and for all sufficiently large r,

$$\lim^r H_{m+1+r}(\boldsymbol{C}) = 0, \tag{31}$$

then there exists a natural exact sequence

$$0 \to \lim^1 H_{m+1}(\boldsymbol{C}) \to \overline{H}^{(1)}_m(\boldsymbol{C}) \to \check{H}_m(\boldsymbol{C}) \to 0. \tag{32}$$

Moreover, there exists a natural isomorphism

$$\overline{\pi}_m : \overline{H}_m(\boldsymbol{C}) \to \overline{H}^{(1)}_m(\boldsymbol{C}). \tag{33}$$

Proof. By Theorem 17.8, $\overline{\pi}^{(r)}_m : \overline{H}^{(r)}_m(\boldsymbol{C}) \to \overline{H}^{(r-1)}_m(\boldsymbol{C})$ is an isomorphism, for $r \geq 2$, and an epimorphism, for $r = 1$. In particular, since $\overline{H}^{(0)}_m(\boldsymbol{C}) \approx \check{H}_m(\boldsymbol{C})$, one has a natural exact sequence

$$0 \to \lim^1 H_{m+1}(\boldsymbol{C}) \to \overline{H}^{(1)}_m(\boldsymbol{C}) \to \check{H}_m(\boldsymbol{C}) \to 0. \tag{34}$$

Moreover, the inverse sequence (1) stabilizes eventually. Consequently, the isomorphisms $\overline{\pi}^{(r)}_m$ induce and isomorphism $\overline{\pi}^{(r,\infty)}_m : \lim \overline{H}^{(*)}_m(\boldsymbol{C}) \to \overline{H}^{(1)}_m(\boldsymbol{C})$. Moreover, since also $\overline{H}^{(*)}_{m+1}(\boldsymbol{C})$ is stable, $\lim^1 \overline{H}^{(*)}_{m+1}(\boldsymbol{C}) = 0$. Hence, Theorem 17.11 proves that $\overline{\pi}_m : \overline{H}_m(\boldsymbol{C}) \to \lim \overline{H}^{(*)}_m(\boldsymbol{C})$ is also an isomorphism. The desired isomorphism (33) is obtained by composing these two isomorphisms. □

REMARK 17.15. If X is a compact metric space, there exists an inverse sequence of compact polyhedra $\boldsymbol{X}$ such that $X = \lim \boldsymbol{X}$. The singular pro-complex $\boldsymbol{C} = S(\boldsymbol{X}) \otimes G$ satisfies conditions (29), (30) and (31), because $H_n(\boldsymbol{C})$ is a tower of abelian groups. Consequently, Corollary 17.14 establishes the well-known *Milnor exact sequence* for Steenrod homology of metric compacta (Milnor 1960).

$$0 \to \lim{}^1 \boldsymbol{H}_{m+1}(X;G) \to \overline{H}_m(X;G) \to \check{H}_m(X;G) \to 0. \tag{35}$$

For an alternative direct proof see (Mardešić 1991c). For a generalization of Milnor's sequence to homology of inverse systems of compacta see (Mdzinarishvili 1984)

17.4 Isomorphism theorems for strong homology

The first result in this subsection is a theorem relating isomorphisms of homology pro-groups to isomorphisms of strong homology groups.

THEOREM 17.16. *Let $\boldsymbol{f}: \boldsymbol{C} \to \boldsymbol{D}$ be a morphism of pro-chain complexes. If $\boldsymbol{f}$ induces an isomorphism of homology pro-groups $H_m(\boldsymbol{f}): H_m(\boldsymbol{C}) \to H_m(\boldsymbol{D})$, for all $m \in \mathbb{Z}$, then $\overline{H}_m(\boldsymbol{f}): \overline{H}_m(\boldsymbol{C}) \to \overline{H}_m(\boldsymbol{D})$ is an isomorphism of strong homology groups, for all m.*

Since $\lim^n$ is a functor on the category of pro-groups (see Theorem 15.5), an isomorphism of homology pro-groups $H_m(\boldsymbol{C}) \to H_m(\boldsymbol{D})$ induces an isomorphism of their derived limits $\lim^r H_m(\boldsymbol{C}) \to \lim^r H_m(\boldsymbol{D})$. Therefore, Theorem 17.16 is an immediate consequence of the following theorem.

THEOREM 17.17. *Let $\boldsymbol{f}: \boldsymbol{C} \to \boldsymbol{D}$ be a morphism of pro-chain complexes such that $\lim^r H_m(\boldsymbol{f}): \lim^r H_m(\boldsymbol{C}) \to \lim^r H_m(\boldsymbol{D})$ is an isomorphism, for every $r \geq 0$ and every $m \in \mathbb{Z}$. Then $\overline{H}_m(\boldsymbol{f}): \overline{H}_m(\boldsymbol{C}) \to \overline{H}_m(\boldsymbol{D})$ is an isomorphism of strong homology groups, for every $m \in \mathbb{Z}$.*

Proof. We first show that $\boldsymbol{f}$ induces isomorphisms $\overline{H}^{(r)}_m(\boldsymbol{f}): \overline{H}^{(r)}_m(\boldsymbol{C}) \to \overline{H}^{(r)}_m(\boldsymbol{D})$ of strong homology groups of height r, for all $r, m \in \mathbb{Z}$. The proof is by induction on r. If $r < 0$, both groups vanish. To prove the induction step, one considers the commutative diagram, which expresses naturality with respect to $\boldsymbol{f}$ of the first Miminoshvili sequence. The corresponding terms of the exact sequences (17.2.1), for $\boldsymbol{C}$ and $\boldsymbol{D}$, respectively, are related by isomorphisms $\overline{H}^{(r-i)}_{m+i-1}(\boldsymbol{C}) \to \overline{H}^{(r-i)}_{m+i-1}(\boldsymbol{D})$ and $\lim^{(r+i-1)} H_{m+r}(\boldsymbol{C}) \to \lim^{(r+i-1)} H_{m+r}(\boldsymbol{C})$, for $i = 1, 2$. Consequently, the five lemma shows that also $\overline{H}^{(r)}_m(\boldsymbol{f})$ is an isomorphism.

Clearly, the isomorphisms $\overline{H}^{(r)}_m(\boldsymbol{f}): \overline{H}^{(r)}_m(\boldsymbol{C}) \to \overline{H}^{(r)}_m(\boldsymbol{D})$, $r \geq 0$, induce a level isomorphism of inverse sequences $\overline{H}^{(*)}_m(\boldsymbol{C}) \to \overline{H}^{(*)}_m(\boldsymbol{D})$. In turn this isomorphism induces an isomorphism of the corresponding limits $\lim \overline{H}^{(*)}_m(\boldsymbol{C}) \to \lim \overline{H}^{(*)}_m(\boldsymbol{D})$. The same is true of the sequences $\overline{H}^{(*)}_{m+1}(\boldsymbol{C})$ and $\overline{H}^{(*)}_{m+1}(\boldsymbol{D})$ and their first derived limits. Now the naturality of the second Miminoshvili sequence enables one to conclude that the induced homomorphism $\overline{H}_m(\boldsymbol{C}) \to \overline{H}_m(\boldsymbol{D})$ is indeed an isomorphism. □

As in 14.2, an increasing function $\phi: M \to \Lambda$ between directed sets assigns to every pro-chain complex $\boldsymbol{C} = (C_\lambda, p_{\lambda\lambda'}, \Lambda)$ an induced pro-chain complex $\boldsymbol{C}^* = (C^*_\mu, p^*_{\mu\mu'}, M)$, where $C^*_\mu = C_{\phi(\mu)}, p^*_{\mu\mu'} = p_{\phi(\mu)\phi(\mu')}$. Moreover, there is an induced morphism of pro-chain complexes $\boldsymbol{\phi} = (\phi, \phi_\mu): \boldsymbol{C} \to \boldsymbol{C}^*$, where $\phi_\mu: C_{\phi(\mu)} \to C^*_\mu$ is the identity on $C_{\phi(\mu)}$. In the special case when $M \subseteq \Lambda$, $\boldsymbol{C}^*$ is the restriction of $\boldsymbol{C}$ to M.

COROLLARY 17.18. *If $\phi: M \to \Lambda$ is cofinal in Λ, then $\boldsymbol{\phi}$ induces isomorphisms of Čech homology groups $\check{H}_m(\boldsymbol{C}) \to \check{H}_m(\boldsymbol{C}^*)$ and strong homology groups $\overline{H}_m(\boldsymbol{C}) \to \overline{H}_m(\boldsymbol{C}^*)$. In particular, Čech and strong homology groups are invariant under restrictions to cofinal subsystems.*

Proof. The first assertion is an immediate consequence of Lemma 14.10. The second assertion follows from Theorem 14.9 and Theorem 17.17. □

COROLLARY 17.19. *Let $\boldsymbol{f} = (f_\lambda): \boldsymbol{C} \to \boldsymbol{D}$ be a level morphism of pro-chain complexes. If every $f_\lambda: C_\lambda \to D_\lambda$ induces an isomorphism of homology groups $f_{\lambda *}: H_m(C_\lambda) \to H_m(D_\lambda)$, then $\boldsymbol{f}$ induces an isomorphism of homology pro-groups $H_m(\boldsymbol{C}) \to H_m(\boldsymbol{D})$, hence, it also induces isomorphisms of Čech and strong homology groups.*

Proof. The first assertion is obvious and the second one follows from Theorem 17.16. □

COROLLARY 17.20. *Let $\boldsymbol{K} = (K_\lambda, p_{\lambda\lambda'}, \Lambda)$ be an inverse system of simplicial complexes and simplicial mappings and let $\boldsymbol{C} = (C_\lambda, p_{\lambda\lambda'}, \Lambda)$ be the induced pro-chain complex, where C_λ is the ordered (oriented) chain complex of K_λ and $p_{\lambda\lambda'}: C_{\lambda'} \to C_\lambda$ is the induced chain mapping. Furthermore, let $\boldsymbol{X} = (X_\lambda, p_{\lambda\lambda'}, \Lambda)$ be the inverse system consisting of the geometric realizations $X_\lambda = |K_\lambda|$ and induced mappings $p_{\lambda\lambda'}: X_{\lambda'} \to X_\lambda$. Then, for every abelian group G, there are natural isomorphisms between the homology pro-groups $H_m(\boldsymbol{C}; G) \to H_m(\boldsymbol{X}; G)$. Moreover, there are natural isomorphisms between the Čech homology groups $\check{H}_m(\boldsymbol{C}; G) \to \check{H}_m(\boldsymbol{X}; G)$, Čech cohomology groups $\check{H}^m(\boldsymbol{X}; G) \to \check{H}^m(\boldsymbol{C}; G)$ and strong homology groups $\overline{H}_m(\boldsymbol{C}; G) \to \overline{H}_m(\boldsymbol{X}; G)$.*

Proof. $H_m(\boldsymbol{C}; G) = H_m(\boldsymbol{C} \otimes G)$ and $H_m(\boldsymbol{X}; G) = H_m(S(\boldsymbol{X}) \otimes G)$, where S denotes the singular complex. To prove the first assertion note that there are natural chain mappings $f_\lambda: C(X_\lambda) \to S(X_\lambda)$, which induce isomorphisms of homology groups. According to the standard universal coefficient theorem (see e.g., (Munkres 1984), Theorem 55.1), there exist a natural exact sequence,

$$0 \to H_m(C_\lambda) \otimes G \to H_m(C_\lambda; G) \to \mathrm{Tor}\,(H_{m-1}(C_\lambda), G) \to 0, \qquad (1)$$

as well as an analogous sequence for $S(X_\lambda)$. This implies that the chain mappings $f_\lambda \otimes 1: C(X_\lambda) \otimes G \to S(X_\lambda) \otimes G$ induce isomorphisms of homology groups $H_m(C_\lambda; G) \to H_m(S(X_\lambda); G)$. Therefore, the pro-chain mapping $\boldsymbol{f} \otimes 1$ induces an isomorphism of homology pro-groups $H_m(\boldsymbol{C}; G) \to H_m(\boldsymbol{X}; G)$. Clearly, this isomorphism induces an isomorphism of the corresponding Čech homology groups. Moreover, by Corollary 17.19, $\boldsymbol{f}$ also induces an isomorphism of the corresponding strong homology groups. The argument for Čech cohomology groups is analogous to the one for Čech homology groups. The only difference is that instead of (1), one uses the universal coefficient formula (see e.g., (Munkres 1984), Theorem 53.1)

$$0 \to \mathrm{Ext}\,(H_{m-1}(C_\lambda), G) \to H^m(C_\lambda; G) \to \mathrm{Hom}\,(H_m(C_\lambda), G) \to 0. \ \square \qquad (2)$$

In 19 we will need a generalization of Theorem 17.17, which involves $\boldsymbol{C}$ and a collection of pro-chain complexes $\boldsymbol{D}^\alpha$, $\alpha \in A$. If A is a singleton, the next theorem reduces to Theorem 17.16.

THEOREM 17.21. *Let* $\boldsymbol{C}$ *be a pro-chain complex and let* $\boldsymbol{D}^\alpha$, $\alpha \in A$, *be a collection of pro-chain complexes. If* $\boldsymbol{f}^\alpha: \boldsymbol{C} \to \boldsymbol{D}^\alpha$, $\alpha \in A$, *is a collection of morphisms of pro-chain complexes, which induces isomorphisms* $\lim^r H_m(\boldsymbol{C}) \to \prod_{\alpha \in A} \lim^r H_m(\boldsymbol{D}^\alpha)$, *for all* $r \geq 0$ *and all* $m \in \mathbb{Z}$, *then it also induces isomorphisms* $\overline{H}_m(\boldsymbol{C}) \to \prod_{\alpha \in A} \overline{H}_m(\boldsymbol{D}^\alpha)$, *for all* m.

Proof. The proof is similar to the proof of Theorem 17.16. Each $\boldsymbol{f}^\alpha$ induces a morphism between the first Miminoshvili sequences for $\boldsymbol{C}$ and $\boldsymbol{D}^\alpha$ respectively. Since a direct product of exact sequences is exact, one obtains a commutative diagram with exact rows, whose first row is the sequence for $\boldsymbol{C}$ and the second row is the direct product of the sequences for $\boldsymbol{D}^\alpha$, $\alpha \in A$. The assumptions of the theorem and the induction hypotheses prove that the morphisms $\boldsymbol{f}^\alpha$ induce isomorphisms $\overline{H}_m^{(r)}(\boldsymbol{C}) \to \prod_\alpha \overline{H}_m^{(r)}(\boldsymbol{D}^\alpha)$, for all $r \geq 0$, hence also an isomorphism between the inverse sequence

$$\overline{H}_m^{(*)}(\boldsymbol{C}) = (\overline{H}_m^{(0)}(\boldsymbol{C}) \leftarrow \ldots \leftarrow \overline{H}_m^{(r)}(\boldsymbol{C}) \leftarrow \ldots), \qquad (3)$$

$$\prod_\alpha \overline{H}_m^{(*)}(\boldsymbol{D}^\alpha) = (\prod_\alpha \overline{H}_m^{(0)}(\boldsymbol{D}^\alpha) \leftarrow \ldots \leftarrow \prod_\alpha \overline{H}_m^{(r)}(\boldsymbol{D}^\alpha) \leftarrow \ldots). \qquad (4)$$

Clearly, the latter isomorphism induces isomorphisms of the corresponding limits and the first derived limits, respectively.

On the other hand, by the naturality of the second Miminoshvili sequence with respect to the morphisms $\boldsymbol{f}^\alpha: \boldsymbol{C} \to \boldsymbol{D}^\alpha$, the following diagram is commutative and has exact rows.

$$\begin{array}{ccccccccc} 0 & \longrightarrow & \lim^1 \overline{H}^{(*)}_{m+1}(\boldsymbol{C}) & \longrightarrow & \overline{H}_m(\boldsymbol{C}) & \longrightarrow & \lim \overline{H}^{(*)}_m(\boldsymbol{C}) & \longrightarrow & 0 \\ & & \downarrow & & \downarrow & & \downarrow & & \\ 0 & \rightarrow & \prod_\alpha \lim^1 \overline{H}^{(*)}_{m+1}(\boldsymbol{D}^\alpha) & \rightarrow & \prod_\alpha \overline{H}_m(\boldsymbol{D}^\alpha) & \rightarrow & \prod_\alpha \lim \overline{H}^{(*)}_m(\boldsymbol{D}^\alpha) & \rightarrow & 0\,. \end{array} \tag{5}$$

Therefore, in order to prove that the middle vertical arrow is an isomorphism, it suffices to show that the two remaining vertical arrows are isomorphisms. To achieve this, consider the following diagram

$$\begin{array}{ccc} \lim^r \overline{H}^{(*)}_{m+1}(\boldsymbol{C}) & & \\ \downarrow & \searrow & \\ \prod_\alpha \lim^r \overline{H}^{(*)}_{m+r}(\boldsymbol{D}^\alpha) & \longrightarrow & \lim^r \prod_\alpha \overline{H}^{(*)}_m(\boldsymbol{D}^\alpha) \end{array} \quad . \tag{6}$$

for $r = 0, 1$. By Corollary 12.15, the diagram is commutative and the horizontal arrow is an isomorphism. Since we proved above that also the skew arrow is an isomorphism, we obtain the desired conclusion that the vertical arrow is indeed an isomorphism. □

Bibliographic notes

T. Porter has studied pro-chain complexes $\boldsymbol{C}$ already in (Porter 1977). In particular, he associated with $\boldsymbol{C}$ a chain complex, which differs from $T(\boldsymbol{X})$ only in the choice of signs. This construction was rediscovered in (Miminoshvili 1984) and (Lisica, Mardešić 1983, 1985d) and applied to the definition of strong homology groups. Since this homology uses the total complex of a bicomplex, in (Mdzinarishvili 1986b) it was named *total homology*. The groups of finite height $\overline{H}^{(r)}_m(\boldsymbol{X})$ were introduced in (Miminoshvili 1984), where the exact sequences of 17.2 and 17.3 were announced. Proofs appeared in (Mdiznarishvili 1986b), (Mardešić, Prasolov 1988), (Mardešić, Miminoshvili 1990) and (Miminoshvili 1991). Exact couples were first used in the proofs in (Mardešić, Prasolov 1988). A detailed proof of the naturality of the Miminoshvili sequences was given in (Mardešić, Miminoshvili 1990). It is well known that, by iterating the construction of the derived exact couple, one obtains a spectral sequence (see e.g., (Hilton, Stammbach 1971), Ch. VIII). Application of this procedure to the exact couple described in 17.2, yields a spectral sequence with initial terms $E_2^{st} = \lim^s H_{-t}(\boldsymbol{C})$, which converges towards the strong homology group $\overline{H}_{-s-t}(\boldsymbol{C})$, provided $\lim^1 E_r^{st} = 0$ (Prasolov 1989).

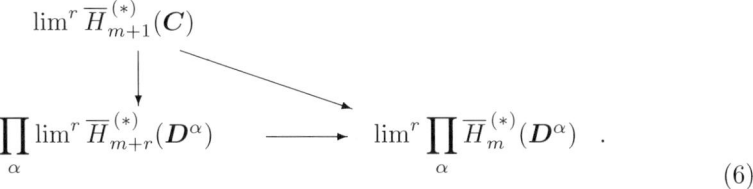

18. Strong homology on CH(pro-Top)

In 17 we have defined strong homology groups $\overline{H}_m(\boldsymbol{X};G)$ of an inverse system of spaces. In this section, with every coherent mapping $\boldsymbol{f}:\boldsymbol{X}\to\boldsymbol{Y}$, we associate homomorphisms $f_*:\overline{H}_m(\boldsymbol{X};G)\to\overline{H}_m(\boldsymbol{Y};G)$, which depend only on the homotopy class $[\boldsymbol{f}]$. In this way $\overline{H}_m(\,.\,;G)$, $m\in\mathbb{Z}$, become functors from the coherent homotopy category CH(Top) to the category Ab of abelian groups.

18.1 Chain mappings induced by coherent mappings

In 17.1 we defined the strong homology groups $\overline{H}_m(\boldsymbol{X};G)$ of an inverse system of spaces $\boldsymbol{X}$ as well as the homomorphisms $f_*=\overline{H}_m(\boldsymbol{f}):\overline{H}_m(\boldsymbol{X};G)\to\overline{H}_m(\boldsymbol{Y};G)$, induced by a mapping $\boldsymbol{f}:\boldsymbol{X}\to\boldsymbol{Y}$. In this subsection we will generalize the latter construction to coherent mappings $\boldsymbol{f}:\boldsymbol{X}\to\boldsymbol{Y}$.

We first recall some facts, concerning products of singular chains, needed in the sequel. We will denote by Δ^i, not only the standard i-simplex Δ^i with vertices $e_0,\dots,e_i$ (see 1.2), but also its singular i-chain, given by the identity mapping id: $\Delta^i\to\Delta^i$. Note that the boundary of the chain Δ^i, $i\geq 1$, is the singular $(i-1)$-chain

$$\partial\Delta^i=\sum_{j=0}^{i}(-1)^j d_j\Delta^{i-1}, \tag{1}$$

where d_j are the face operators $d_j:\Delta^{i-1}\to\Delta^i$, defined in 1.2; $\partial\Delta^0=0$.

On the product space $\Delta^p\times\Delta^q$, we define a singular $(p+q)$-chain, also denoted by $\Delta^p\times\Delta^q$, as follows. Consider the set V of all pairs $e_{ij}=(e_i,e_j)\in\Delta^p\times\Delta^q$, $0\leq i\leq p$, $0\leq j\leq q$, and order it by putting $e_{ij}\leq e_{i'j'}$, provided $i\leq i'$ and $j\leq j'$. It is easy to see that every linearly ordered subset of $p+q+1$ distinct points of V spans an ordered $(p+q)$-simplex $s=(e_{00},\dots,e_{ij},\dots,e_{pq})$, which can be viewed as a singular $(p+q)$-simplex of $\Delta^p\times\Delta^q$. Moreover, these simplices and their faces form a triangulation K of the space $\Delta^p\times\Delta^q$ (see, e.g., (Eilenberg, Steenrod 1952), Lemma 8.9).

To each $(p+q)$-simplex s of the triangulation K one can assign a unique value $\varepsilon(s)\in\{-1,1\}$ in such a way that $\varepsilon(s_0)=1$, for the *leading simplex* $s_0=$

$(e_{00},\ldots,e_{p0},e_{p1},\ldots,e_{pq})$, and $\varepsilon(s') = -\varepsilon(s)$, whenever the simplices s, s' are *adjacent*, i.e., differ in one vertex only. To see that such a choice is possible, for a given $(p+q)$-simplex $s = (e_{00},\ldots,e_{ij},\ldots,e_{pq})$ and $k \in \{1,\ldots,p\}$, let $\gamma(s,k)+1$ be the number of times when $i = k$. E.g., for the leading simplex s_0, $\gamma(s_0,k)+1 = 1$, for $1 \le k \le p-1$ and $\gamma(s_0,p)+1 = q+1$. Then put

$$\gamma(s) = pq + \sum_{k=1}^{p} k\gamma(s,k), \tag{2}$$

$$\varepsilon(s) = (-1)^{\gamma(s)}. \tag{3}$$

Clearly, $\gamma(s_0) = 2pq$ and thus, $\varepsilon(s_0) = 1$. Moreover, if s, s' are adjacent, i.e., if they are of the form $s = (\ldots,e_{ij},e_{i+1,j},\ldots)$, $s' = (\ldots,e_{ij},e_{i,j+1},\ldots)$, then $\gamma(s',i) = \gamma(s,i)+1$, $\gamma(s',i+1) = \gamma(s,i+1)-1$, while $\gamma(s',k) = \gamma(s,k)$, for $k \ne i, i+1$. Consequently, $\gamma(s)$ and $\gamma(s')$ have different parity and thus, $\varepsilon(s) \ne \varepsilon(s')$. Uniqueness of $\varepsilon(s)$ is a consequence of the fact that any $(p+q)$-simplex s of K can be connected to s_0 by a chain of $(p+q)$-simplices $s_0,\ldots,s_i,s_{i+1},\ldots,s$, where s_i and s_{i+1} are adjacent. Since $\varepsilon(s_0) = 1$ and $\varepsilon(s_{i+1}) = -\varepsilon(s_i)$, $\varepsilon(s)$ is completely determined. Finally, we define $\Delta^p \times \Delta^q$, by putting

$$\Delta^p \times \Delta^q = \sum \varepsilon(s)s, \tag{4}$$

where summation is taken over all $(p+q)$-simplices of K.

It is now easy to generalize the described construction and, for spaces X and Y, define the product $x \times y \in S^{p+q}(X \times Y) \otimes G$ of two singular chains $x \in S^p(X) \otimes G$, $y \in S^q(Y) = S^q(Y) \otimes \mathbb{Z}$. Assume that $x = \sum_i g_i a_i$, $y = \sum_j m_j b_j$, where $a_i \colon \Delta^p \to X$, $b_j \colon \Delta^q \to Y$ are singular simplices and $g_i \in G$, $m_j \in \mathbb{Z}$, $i = 1,\ldots,m$, $j = 1,\ldots,n$. Then

$$x \times y = \sum_{i,j} m_j g_i (a_i \times b_j)^{\#}(\Delta^p \times \Delta^q) \in S^{p+q}(X \times Y), \tag{5}$$

where $(a_i \times b_j)^{\#}$ is the chain mapping induced by the mapping $a_i \times b_j \colon \Delta^p \times \Delta^q \to X \times Y$.

LEMMA 18.1. *For singular chains* $x \in S^p(X), y \in S^q(Y), z \in S^r(Z)$,

$$x \times (y+z) = x \times y + x \times z,\ (x+y) \times z = x \times z + y \times z, \tag{6}$$

$$(x \times y) \times z = x \times (y \times z). \tag{7}$$

If $\tau \colon X \times Y \to Y \times X$ *is the involution which permutes coordinates, then*

$$\tau^{\#}(x \times y) = (-1)^{pq} y \times x. \tag{8}$$

$$\partial(x \times y) = (\partial x) \times y + (-1)^p x \times \partial y, \tag{9}$$

Moreover, if $f \colon X \to X'$ *and* $g \colon Y \to Y'$ *are mappings, then*

$$(f \times g)^{\#}(x \times y) = f^{\#}(x) \times g^{\#}(y). \tag{10}$$

Proof. The distribution laws (6) follow immediately from the definition. To prove (7), it suffices to show that $(\Delta^p \times \Delta^q) \times \Delta^r = \Delta^p \times (\Delta^q \times \Delta^r)$. Put $e_{ijk} = (c_{ij}, e_k) = (e_i, e_{jk})$ and note that both products are sums of $(p+q+r)$-simplices of the form $t = (\ldots, e_{ijk}, \ldots)$ with coefficients ± 1. In both products the leading simplex $(e_{000}, \ldots, e_{p00}, \ldots, e_{pq0}, \ldots, e_{pqr})$ assumes the value 1. Hence, it suffices to verify that in both products adjacent simplices s, s' assume opposite values. Three cases can occur. In the first case $s = (\ldots, e_{ijk}, e_{i+1,jk}, e_{i+1,j+1,k}, \ldots)$ and $s' = (\ldots, e_{ijk}, e_{i,j+1,k}, e_{i+1,j+1,k}, \ldots)$. Note that in this case the simplices $t = (\ldots, e_{ij}, e_{i+1,j}, e_{i+1,j+1}, \ldots)$ and $t' = (\ldots, e_{ij}, e_{i,j+1}, e_{i+1,j+1}, \ldots)$ are adjacent in $\Delta^p \times \Delta^q$ and therefore, occur with opposite coefficients. However, s and s' have the same coefficients in $t \times \Delta^q$ and $t' \times \Delta^q$, because $\gamma(s,l) = \gamma(s',l)$, for all l. Hence, s and s' appear with opposite coefficients in $(\Delta^p \times \Delta^q) \times \Delta^r$. On the other hand, $u = (\ldots, e_{jk}, e_{j+1,k}, \ldots)$ belongs to $\Delta^q \times \Delta^r$ and s, s' are adjacent in $\Delta^p \times u$, hence, they have opposite coefficients also in $\Delta^p \times (\Delta^q \times \Delta^r)$. Similar arguments apply when $s = (\ldots, e_{ijk}, e_{i+1,jk}, e_{i+1,j,k+1}, \ldots), s' = (\ldots, e_{ijk}, e_{i,j,k+1}, e_{i+1,j,k+1}, \ldots)$ or $s = (\ldots, e_{ijk}, e_{i,j+1,k}, e_{i,j+1,k+1}, \ldots), s' = (\ldots, e_{ijk}, e_{i,j,k+1}, e_{i,j+1,k+1}, \ldots)$.

To prove (8), note that τ maps every $(p+q)$-simplex s of the triangulation K of $\Delta^p \times \Delta^q$ to a simplex $t = \tau(s)$ of the corresponding triangulation L of $\Delta^q \times \Delta^p$. The $(q+p)$-chain $\Delta^q \times \Delta^p$ has the form $\sum \eta(t)t$, where $t = \tau(s)$ and $\eta(t)$ is defined by rules analogous to those used in the definiton of $\varepsilon(s)$. Now consider the simplex $s^0 \in K$, given by $s^0 = (e_{00}, \ldots, e_{0q}, e_{1q}, \ldots, e_{pq})$. For this simplex $\varepsilon(s^0) = (-1)^{pq}$. Moroever, $\tau^\#(s^0) = (e_{00}, \ldots, e_{q0}, e_{q1}, \ldots, e_{qp})$ is the leading simplex t_0 of the triangulation L of $\Delta^q \times \Delta^p$ and thus, $\eta(\tau^\#(s^0)) = 1$. Consequently, $\varepsilon(s) = (-1)^{pq}\eta(\tau(s))$, for this special simplex. However, if s, s' are adjacent simplices of K, then $\tau(s), \tau(s')$ are adjacent simplices of L and therefore, the above formula holds for all $(p+1)$-simplices $s \in K$ and one concludes that

$$\begin{aligned} \tau^\#(\Delta^p \times \Delta^q) = \quad & \textstyle\sum \varepsilon(s)\tau^\#(s) = (-1)^{pq} \sum \eta(\tau^\#(s))\tau^\#(s) = \\ & (-1)^{pq} \textstyle\sum \eta(t)t = (-1)^{pq}(\Delta^q \times \Delta^p). \end{aligned} \tag{11}$$

To prove (9), it suffices to consider the special case when $x = \Delta^p$, $y = \Delta^q$ and $z = \Delta^r$. In this case both sides of (9) are linear combinations of $(p+q-1)$-simplices t of K and it suffices to show that the corresponding coefficients coincide. If t is of the form $(\ldots, e_{ij}, e_{i+1,j+1}, \ldots)$, it does not belong to the boundary of $\Delta^p \times \Delta^q$ and thus, does not appear in the right side of (9). On the other hand, t appears only in ∂s and $\partial s'$, for $s = (\ldots, e_{ij}, e_{i+1,j}, e_{i+1,j+1}, \ldots)$ and $s' = (\ldots, e_{ij}, e_{i,j+1}, e_{i+1,j+1}, \ldots)$, in both cases with the same coefficient. Since s and s' are adjacent, $\varepsilon(s) = -\varepsilon(s')$ and the two terms cancel in $\partial(\Delta^p \times \Delta^q)$.

Now consider the simplex $t = (e_{00}, \ldots, e_{p-1,0}, e_{p-1,1}, \ldots, e_{p-1,q})$. Since $\Delta^{p-1} = (e_0, \ldots, e_{p-1})$ appears in $\partial\Delta^p$ with coefficient $(-1)^p$ and t is the leading $(p+q-1)$-simplex in the triangulation K of $\Delta^{p-1} \times \Delta^q$, we conclude that t appears in $\partial\Delta^p \times \Delta^q$, hence also in the right side of (9), with coefficient $(-1)^p$. On the other hand, t appears only in the boundary of

$s = (e_{00}, \ldots, e_{p-1,0}, e_{p-1,1}, \ldots, e_{p-1,q}, e_{pq})$, the corresponding coefficient being $(-1)^{p+q}$. Since $\gamma(s) = 2pq - q$, we conclude that t appears in the left side of (9) also with coefficient $(-1)^p$. Using the fact that adjacent simplices have opposite signs, one readily concludes that coefficients in both sides of (9) coincide, for $(p+q-1)$-simplices t, which belong to $\Delta^{p-1} \times \Delta^q$. The same argument applies to simplices t, which belong to $d_i\Delta^p = (e_0, \ldots, \widehat{e_i}, \ldots, e_p)$, hence, to simplices t belonging to $\partial\Delta^p \times \Delta^q$.

Similarly, to verify validity of (9), for simplices t belonging to $\Delta^p \times \partial\Delta^q$, it suffices to determine the coefficients of $t = (e_{00}, \ldots, e_{p,0}, e_{p,1}, \ldots, e_{p,q-1})$ in both sides of (9). Since $\Delta^{q-1} = (e_0, \ldots, e_{q-1})$ appears in $\partial\Delta^q$ with coefficient $(-1)^q$ and t is the leading $(p+q-1)$-simplex in the triangulation K of $\Delta^p \times \Delta^{q-1}$, we conclude that t appears in $\Delta^p \times \partial\Delta^q$ with coefficient $(-1)^q$, hence, it apears in the right side of (9) with coefficient $(-1)^{p+q}$. On the other hand, t appears only in the boundary of $s = (e_{00}, \ldots, e_{p,0}, e_{p,1}, \ldots, e_{p,q-1}, e_{pq})$, the corresponding coefficient being $(-1)^{p+q}$. Since $\gamma(s) = 2pq$, we conclude that t appears in the left side of (9) also with coefficient $(-1)^{p+q}$.

To obtain (10), it suffices to consider the case of singular simplices $a \in S^p(X), b \in S^q(Y)$. By definition,

$$(f \times g)^\#(a \times b) = (f \times g)^\#(a \times b)^\#(\Delta^p \times \Delta^q) = (fa \times gb)^\#(\Delta^p \times \Delta^q). \quad (12)$$

On the other hand,

$$f^\#(a) \times g^\#(b) = fa \times gb = (fa \times gb)^\#(\Delta^p \times \Delta^q). \ \square \quad (13)$$

REMARK 18.2. In the above lemma some of the singular chains can have coefficients in G and the remaining chains must have integer coefficients. E.g., in (7), only one of the three chains x, y, z can have coefficients in G.

Now consider two inverse systems of topological spaces $\boldsymbol{X} = (X_\lambda, p_{\lambda\lambda'} \Lambda)$, $\boldsymbol{Y} = (Y_\mu, p_{\mu\mu'}, M)$ and a coherent mapping $\boldsymbol{f}: \boldsymbol{X} \to \boldsymbol{Y}$. Recall that $\boldsymbol{f}$ is given by an increasing function $f: M \to \Lambda$, and by mappings $f_{\boldsymbol{\mu}}: X_{f(\mu_p)} \times \Delta^p \to Y_{\mu_0}$, defined for every multiindex $\boldsymbol{\mu} = (\mu_0, \ldots, \mu_p)$, $p \geq 0$, satisfying the coherence conditions (1.2.7) and (1.2.8). We will first define an induced chain mapping $f^\#: TS(\boldsymbol{X}; G) \to TS(\boldsymbol{Y}; G)$ between the total complexes (see 17.1) of the singular pro-chain complexes $S(\boldsymbol{X}; G)$ and $S(\boldsymbol{Y}; G)$ (see 16.1). Since strong homology $\overline{H}_*(\boldsymbol{X}; G)$ is the homology of $TS(\boldsymbol{X}; G)$ (see (17.1.21)), we will obtain in this way the *induced homomorphism of strong homology groups* $f_*: \overline{H}_*(\boldsymbol{X}; G) \to \overline{H}_*(\boldsymbol{Y}; G)$.

Recall that the m-chains of $TS(\boldsymbol{X}; G)$, $m \in \mathbb{Z}$, i.e., the strong m-chains of $\boldsymbol{X}$ with coefficients in G, are functions c, defined on multiindices $\boldsymbol{\lambda} = (\lambda_0, \ldots, \lambda_s)$ of Λ, $s \geq 0$, whose values $c(\boldsymbol{\lambda})$ are singular $(m+s)$-chains of the space X_{λ_0} with coefficients in G (see 17.1). If $\boldsymbol{\mu} = (\mu_0, \ldots, \mu_s) \in M_s$ is a multiindex of M of length s, then $f(\mu_0) \leq \ldots \leq f(\mu_s)$. Consequently, for every i, $0 \leq i \leq s$, $(f(\mu_i), \ldots, f(\mu_s))$ is a multiindex of Λ of length $s - i$. Hence, if c is an m-chain of $TS(\boldsymbol{X}; G)$, then $c(f(\mu_i), \ldots, f(\mu_s))$ is a well-defined singular $(m+s-i)$-chain of $X_{f(\mu_i)}$ and $c(f(\mu_i), \ldots, f(\mu_s)) \times \Delta^i$ is a singular $(m+s)$-chain of $X_{f(\mu_i)} \times \Delta^i$. Since $f_{\mu_0 \dots \mu_i}$ maps $X_{f(\mu_i)} \times \Delta^i$ to Y_{μ_0}, it induces a chain mapping $(f_{\mu_0 \dots \mu_i})^{\#}: S(X_{f(\mu_i)} \times \Delta^i; G) \to S(Y_{\mu_0}; G)$. Consequently, $(f_{\mu_0 \dots \mu_i})^{m+s}(c(f(\mu_i), \dots, f(\mu_s)) \times \Delta^i)$ is a singular $(m+s)$-chain of Y_{μ_0} with coefficients in G. It is then clear that the following formula defines an m-chain $f^m(c)$ of $TS(\boldsymbol{Y}; G)$.

$$(f^m c)(\boldsymbol{\mu}) = \sum_{i=0}^{s} (-1)^{si} f^{\#}_{\mu_0 \dots \mu_i}(c(f(\mu_i), \dots, f(\mu_s)) \times \Delta^i), \ \boldsymbol{\mu} \in M_s. \tag{14}$$

Here $f^{\#}_{\mu_0 \dots \mu_i}$ is a shorter notation for $(f_{\mu_0 \dots \mu_i})^{m+s}$, where the dimensional index is not specified. For simplicity, in the sequel, we will often omit G from the notation.

LEMMA 18.3. *The homomorphisms* $f^m: T^m S(\boldsymbol{X}) \to T^m S(\boldsymbol{Y}), m \in \mathbb{Z}$, *given by* (14), *define a chain mapping* $f^{\#} = TS(\boldsymbol{f}): TS(\boldsymbol{X}) \to TS(\boldsymbol{Y})$.

Proof. We must show that, for any strong m-chain c of $\boldsymbol{X}$, and $\boldsymbol{\mu} = (\mu_0, \dots, \mu_s) \in M_s$, one has

$$(d(f^{\#} c))(\boldsymbol{\mu}) = (f^{\#}(dc))(\boldsymbol{\mu}). \tag{15}$$

The case $s = 0$ is obvious, because in that case $d = \partial$ and

$$(\partial(fc))(\mu_0) = \partial(f_{\mu_0}(c(\mu_0))) = f_{\mu_0}(\partial(c(\mu_0))) = f_{\mu_0}(dc(\mu_0)). \tag{16}$$

Now assume that $s \geq 1$. In this case the verification of (15) is also straightforward, but requires some care. This is why we describe it in detail.

By (17.1.12),

$$(d(f^{\#} c))(\boldsymbol{\mu}) = (\partial(f^{\#} c))(\boldsymbol{\mu}) + (-1)^m (\delta(f^{\#} c))(\boldsymbol{\mu}), \tag{17}$$

However, by (17.1.15), $(\partial(f^{\#} c))(\boldsymbol{\mu}) = \partial(f^{\#} c(\boldsymbol{\mu}))$ and thus, by (14),

$$(\partial(f^{\#} c))(\boldsymbol{\mu}) = \sum_{j=0}^{s} (-1)^{sj} \partial f^{\#}_{\mu_0 \dots \mu_j}(c(f(\mu_j), \dots, f(\mu_s)) \times \Delta^j). \tag{18}$$

Since $f^{\#}_{\mu_0 \dots \mu_j}$ is a chain mapping, it commutes with the boundary operator ∂. Taking this into account and using (9), we conclude that

$$\begin{aligned}(\partial(f^{\#} c))(\boldsymbol{\mu}) &= \sum_{j=0}^{s} (-1)^{sj} f^{\#}_{\mu_0 \dots \mu_j}(\partial c(f(\mu_j), \dots, f(\mu_s)) \times \Delta^j) + \\ &\sum_{j=1}^{s} (-1)^{sj+m+s-j} f^{\#}_{\mu_0 \dots \mu_j}(c(f(\mu_j), \dots, f(\mu_s)) \times \partial(\Delta^j)).\end{aligned} \tag{19}$$

Furthermore, for $j \geq 1$, (1), (10) and (1.2.9) imply

$$f^{\#}_{\mu_0\dots\mu_j}(c(f(\mu_j),\dots,f(\mu_s))\times\partial(\Delta^j)) =$$
$$\textstyle\sum_{i=0}^{j}(-1)^i f^{\#}_{\mu_0\dots\mu_j}(c(f(\mu_j),\dots,f(\mu_s))\times d_i\Delta^{j-1}) = \tag{20}$$
$$\textstyle\sum_{i=0}^{j}(-1)^i(qf_{d^i(\mu_0,\dots,\mu_j)}(p\times 1))^{\#}(c(f(\mu_j),\dots,f(\mu_s))\times\Delta^{j-1}).$$

Substituting (20) to (19), we obtain

$$\partial(f^{\#}c)(\boldsymbol{\mu}) = \sum_{j=0}^{s}(-1)^{sj}f^{\#}_{\mu_0\dots\mu_j}(\partial c(f(\mu_j),\dots,f(\mu_s))\times\Delta^j) +$$
$$\sum_{j=1}^{s}\sum_{i=0}^{j-1}(-1)^{sj+m+s-j+i}qf^{\#}_{d^i(\mu_0,\dots,\mu_j)}(c(f(\mu_j),\dots,f(\mu_s))\times\Delta^{j-1}) + \tag{21}$$
$$\sum_{j=1}^{s}(-1)^{sj+m+s}(f_{\mu_0\dots\mu_{j-1}}(p\times 1))^{\#}(c(f(\mu_j),\dots,f(\mu_s))\times\Delta^{j-1}).$$

Furthermore, in view of (17.1.16) and (14),

$$(\delta(f^{\#}c))(\boldsymbol{\mu}) = \sum_{i=0}^{s}(-1)^i q^{\#}(f^{\#}c)(d^i\mu) =$$
$$\sum_{i=0}^{s}\sum_{j=0}^{i-1}(-1)^{i+(s-1)j}f^{\#}_{\mu_0\dots\mu_j}(c(f(\mu_j),\dots,\widehat{f(\mu_i)},\dots,f(\mu_s))\times\Delta^j) + \tag{22}$$
$$\sum_{i=0}^{s}\sum_{j=i+1}^{s}(-1)^{i+(s-1)(j-1)}qf^{\#}_{d^i(\mu_0\dots\mu_j)}(c(f(\mu_j),\dots,f(\mu_s))\times\Delta^{j-1}).$$

Since the summation indices in the double sum in (21) and the second double sum in (22) range over the same set, after substitution in (17), these double sums cancel and one obtains

$$(d(f^{\#}c))(\boldsymbol{\mu}) = \sum_{j=0}^{s}(-1)^{sj}f^{\#}_{\mu_0\dots\mu_j}(\partial c(f(\mu_j),\dots,f(\mu_s))\times\Delta^j) +$$
$$\sum_{j=1}^{s}(-1)^{sj+m+s}(f_{\mu_0,\dots,\mu_{j-1}}(p\times 1))^{\#}(c(f(\mu_j),\dots,f(\mu_s))\times\Delta^{j-1}) + \tag{23}$$
$$\sum_{i=0}^{s}\sum_{j=0}^{i-1}(-1)^{m+i+(s-1)j}f^{\#}_{\mu_0\dots\mu_j}(c(f(\mu_j),\dots,\widehat{f(\mu_i)},\dots,f(\mu_s))\times\Delta^j).$$

On the other hand,

$$(f^{\#}(\partial c))(\boldsymbol{\mu}) = \sum_{j=0}^{s}(-1)^{sj}f^{\#}_{\mu_0\dots\mu_j}(\partial c(f(\mu_j),\dots,f(\mu_s))\times\Delta^j), \tag{24}$$

$$
(f^{\#}(\delta c))(\boldsymbol{\mu}) = \sum_{j=0}^{s}(-1)^{sj} f^{\#}_{\mu_0\dots\mu_j}((\delta c)(f(\mu_j),\dots,f(\mu_s)\times\Delta^j) = \\
\sum_{j=0}^{s}\sum_{i=j}^{s}(-1)^{sj+i-j} f^{\#}_{\mu_0\dots\mu_j}(c(f(\mu_j),\dots,\widehat{f(\mu_i)},\dots,f(\mu_s))\times\Delta^j)). \tag{25}
$$

Consequently,

$$
(f^{\#}(dc))(\boldsymbol{\mu}) = \sum_{j=0}^{s}(-1)^{sj} f^{\#}_{\mu_0\dots\mu_j}(\partial c(f(\mu_j),\dots,f(\mu_s))\times\Delta^j) + \\
\sum_{j=0}^{s}\sum_{i=j+1}^{s}(-1)^{m+sj+i-j} f^{\#}_{\mu_0\dots\mu_j}(c(f(\mu_j),\dots,\widehat{f(\mu_i)},\dots,f(\mu_s))\times\Delta^j)) + \\
\sum_{j=0}^{s-1}(-1)^{m+sj}(f_{\mu_0\dots\mu_j}(p\times 1))^{\#}(c(f(\mu_{j+1}),\dots,f(\mu_s))\times\Delta^j)). \tag{26}
$$

Now note that the double sums in (23) and (26) range over the same set and therefore coincide. The first simple sums in the two relations also coincides. Finally, the remaining simple sum of (23) becomes the remaining simple sum of (26) if one replaces the summation index j by $j-1$. Consequently, the right sides of (23) and (26) coincide. □

18.2 Chain mappings induced by congruence classes

In this subsection we consider homomorphisms of strong homology groups induced by congruent coherent mappings. The lemma which follows is important in our development of strong homology. Although the proof is in principle a straightforward computation, unfortunately, it is rather tedious. Much care is needed to detect the various pairs of terms which cancel.

LEMMA 18.4. *Let $\boldsymbol{f},\boldsymbol{g}\colon \boldsymbol{X}\to\boldsymbol{Y}$ be congruent coherent mappings. Then the induced chain mappings $f^{\#},g^{\#}\colon TS(\boldsymbol{X};G)\to TS(\boldsymbol{Y};G)$ are chain homotopic and thus, induce the same homomorphisms of strong homology groups.*

Proof. It suffices to consider the case when $\boldsymbol{g}$ is a shift of $\boldsymbol{f}$, i.e., when $\boldsymbol{f}$ and $\boldsymbol{g}$ are given by increasing functions $f,g\colon M\to\Lambda$ such that $f\le g$ and

$$
g_{\boldsymbol{\mu}} = f_{\boldsymbol{\mu}}(p_{f(\mu_s)g(\mu_s)}\times 1)\colon X_{g(\mu_s)}\times\Delta^s\to Y_{\mu_0},\ \ \boldsymbol{\mu}\in M_s, \tag{1}
$$

(see (1.2.54)). The proof will be completed if we find a chain homotopy H between the chain complexes $TS(\boldsymbol{X};G)$ and $TS(\boldsymbol{Y};G)$, which connects $f^{\#}$ to $g^{\#}$. In defining the needed homomorphisms $H^m\colon T^m(S(\boldsymbol{X};G))\to T^{m+1}(S(\boldsymbol{Y};G))$ we will use the notation

$$
\langle j|\boldsymbol{\mu}|k\rangle = (f(\mu_j),\dots,f(\mu_k),g(\mu_k),\dots,g(\mu_s)), \tag{2}
$$

where $\boldsymbol{\mu} \in M_s$ and $0 \le j \le k \le s$. For a strong m-chain c of $S(\boldsymbol{X};G)$, we define the strong $(m+1)$-chain $H^m c$ of $S(\boldsymbol{Y};G)$, by putting

$$(H^m c)(\boldsymbol{\mu}) = (-1)^{m+1} \sum_{j=0}^{s} \sum_{k=j}^{s} (-1)^{sj+j+k} f^{\#}_{\mu_0 \dots \mu_j}(c\langle j|\boldsymbol{\mu}|k\rangle \times \Delta^j), \ \boldsymbol{\mu} \in M_s. \tag{3}$$

The length of $\langle j|\boldsymbol{\mu}|k\rangle$ is $s-j+1$, therefore, $c\langle j|\boldsymbol{\mu}|k\rangle$ is an $(m+s-j+1)$-chain and $(H^m c)(\boldsymbol{\mu})$ is indeed an $(m+1+s)$-chain of Y_{μ_0}. Clearly, $H^m : T^m S(\boldsymbol{X};G) \to T^{m+1} S(\boldsymbol{Y};G)$ is a homomorphism. Hence, it remains to show that

$$(d(H^m c))(\boldsymbol{\mu}) + (H^{m-1}(dc))(\boldsymbol{\mu}) = (g^m c)(\boldsymbol{\mu}) - (f^m c)(\boldsymbol{\mu}). \tag{4}$$

If $s = 0$,

$$(d^m(Hc))(\mu_0) = \partial((H^m c)(\mu_0)) = (-1)^{m+1} \partial f^{\#}_{\mu_0} c(f(\mu_0), g(\mu_0)). \tag{5}$$

On the other hand,

$$\begin{aligned}&(H^{m-1}(dc))(\mu_0) = (-1)^m f^{\#}_{\mu_0}(dc)(f(\mu_0), g(\mu_0)) = \\ &(-1)^m f^{\#}_{\mu_0} \partial c(f(\mu_0), g(\mu_0)) + (f^{\#}_{\mu_0} \delta c(f(\mu_0), g(\mu_0)).\end{aligned} \tag{6}$$

However,

$$\begin{aligned} f^{\#}_{\mu_0} \delta c(f(\mu_0), g(\mu_0)) &= f^{\#}_{\mu_0} p^{\#}_{f(\mu_0) g(\mu_0)} c(g(\mu_0)) - f^{\#}_{\mu_0} c(f(\mu_0)) \\ &= g^{\#} c(\mu_0) - f^{\#} c(\mu_0) \end{aligned} \tag{7}$$

and (4) holds.

We now assume that $s \ge 1$. Then

$$(d(H^m c))(\boldsymbol{\mu}) = (\partial(H^m c))(\boldsymbol{\mu}) + (-1)^{m+1} (\delta(H^m c))(\boldsymbol{\mu}). \tag{8}$$

By (3) and (18.1.9),

$$\begin{aligned}(\partial(H^m c))(\boldsymbol{\mu}) = (-1)^{m+1} &\sum_{j=0}^{s} \sum_{k=j}^{s} (-1)^{sj+j+k} f^{\#}_{\mu_0 \dots \mu_j}(\partial c\langle j|\mu|k\rangle \times \Delta^j) + \\ &\sum_{j=1}^{s} \sum_{k=j}^{s} (-1)^{sj+s+k} f^{\#}_{\mu_0 \dots \mu_j}(c\langle j|\mu|k\rangle \times \partial \Delta^j).\end{aligned} \tag{9}$$

Using (18.1.1) and the coherence conditions for $\boldsymbol{f}$, we see that the second double sum in (9) assumes the form

$$\sum_{j=1}^{s} \sum_{k=j}^{s} \sum_{i=0}^{j} (-1)^{sj+s+k+i} (q f_{d^i(\mu_0 \dots \mu_j)}(p \times 1))^{\#}(c\langle j|\mu|k\rangle \times \Delta^{j-1}). \tag{10}$$

Furthermore,

$$(-1)^{m+1}(\delta(H^m c))(\boldsymbol{\mu}) = (-1)^{m+1}\sum_{i=0}^{s}(-1)^i q^{\#}(H^m c)(d^i\boldsymbol{\mu}) = \sum_{i=0}^{s}\sum_{j=0}^{s-1}\sum_{k=j}^{s-1}(-1)^{sj+k+i}(qf_{\langle d^i\mu|j\rangle})^{\#}(c\langle j|d^i\boldsymbol{\mu}|k\rangle \times \Delta^j), \tag{11}$$

where $\langle d^i\boldsymbol{\mu}|j\rangle$ denotes the initial part of $d^i\boldsymbol{\mu}$ of length j. Clearly,

$$\langle d^i\boldsymbol{\mu}|j\rangle = \begin{cases} (\mu_0,\ldots,\mu_j), & 0\le j\le i-1\le s-1, \\ d^i(\mu_0,\ldots,\mu_{j+1}), & 0\le i\le j\le s-1. \end{cases} \tag{12}$$

The value of $\langle j|d^i\boldsymbol{\mu}|k\rangle$ depends on the mutual position of j,i,k. Indeed, the value is $(f(\mu_j),\ldots,f(\mu_k),g(\mu_k),\ldots,\widehat{g(\mu_i)},\ldots,g(\mu_s))$, for $0\le j\le i-1$, $j\le k\le i-1$, it is $(f(\mu_j),\ldots,\widehat{f(\mu_i)},\ldots,f(\mu_{k+1}),g(\mu_{k+1}),\ldots,g(\mu_s))$, for $0\le j\le i-1$, $i\le k\le s-1$, and it is $(f(\mu_{j+1}),\ldots,f(\mu_{k+1}),g(\mu_{k+1}),\ldots,g(\mu_s))$, for $i\le j\le s-1$, $j\le k\le s-1$. Therefore,

$$\langle j|d^i\boldsymbol{\mu}|k\rangle = \begin{cases} d^{i-j+1}\langle j|\boldsymbol{\mu}|k\rangle, & 0\le j\le i-1, \quad j\le k\le i-1, \\ d^{i-j}\langle j|\boldsymbol{\mu}|k+1\rangle, & 0\le j\le i-1, \quad i\le k\le s-1, \\ \langle j+1|\boldsymbol{\mu}|k+1\rangle, & i\le j\le s-1, \quad j\le k\le s-1. \end{cases} \tag{13}$$

Decomposing the triple sum in (11) according to the conditions, which appear in (13), we obtain

$$\begin{aligned}(-1)^{m+1}&(\delta(H^m c))(\boldsymbol{\mu}) = \\ &\sum_{i=1}^{s}\sum_{j=0}^{i-1}\sum_{k=j}^{i-1}(-1)^{sj+k+i}(qf_{\mu_0\cdots\mu_j})^{\#}(c(d^{i-j+1}\langle j|\boldsymbol{\mu}|k\rangle)\times\Delta^j) + \\ &\sum_{i=1}^{s-1}\sum_{j=0}^{i-1}\sum_{k=i}^{s-1}(-1)^{sj+k+i}(qf_{\mu_0\cdots\mu_j})^{\#}(c(d^{i-j}\langle j|\boldsymbol{\mu}|k+1\rangle)\times\Delta^j) + \\ &\sum_{i=0}^{s-1}\sum_{j=i}^{s-1}\sum_{k=j}^{s-1}(-1)^{sj+k+i}(qf_{d^i(\mu_0\cdots\mu_{j+1})})^{\#}(c(\langle j+1|\boldsymbol{\mu}|k+1\rangle)\times\Delta^j).\end{aligned} \tag{14}$$

Replacing in the third triple sum of (14) $j+1$ by j and $k+1$ by k, this sum assumes the form

$$\sum_{i=0}^{s-1}\sum_{j=i+1}^{s}\sum_{k=j}^{s}(-1)^{sj+s+k+i+1}(qf_{d^i(\mu_0,\ldots,\mu_j)})^{\#}(c(\langle j|\boldsymbol{\mu}|k\rangle)\times\Delta^{j-1}). \tag{15}$$

Note that the signs of the corresponding terms in (10) and (15) are opposite. Moreover, the range B of the summation indices in (15) can also be described by the following set of inequalities.

$$1\le j\le s,\ j\le k\le s,\ 0\le i\le j-1. \tag{16}$$

Comparing B with the range A of the indices (i,j,k) in (10), one sees that $B\subseteq A$ and $A\backslash B$ is given by

$$1 \le j \le s,\ j \le k \le s,\ i = j. \tag{17}$$

Moreover, $i < j$ for $(i, j, k) \in B$, and thus, in (10), $p \times 1$ reduces to identity. Consequently, the two triple sums add up to

$$\sum_{j=1}^{s} \sum_{k=j}^{s} (-1)^{sj+s+j+k} (f_{\mu_0 \ldots \mu_{j-1}} (p \times 1))^{\#} (c\langle j|\boldsymbol{\mu}|k\rangle \times \Delta^{j-1}). \tag{18}$$

Here the factor q has been ommitted because $i = j > 0$. Finally, if one replaces $j - 1$ by j in (18) and $k + 1$ by k in the second triple sum of (14), one concludes that

$$\begin{aligned}
(d(H^m c))(\boldsymbol{\mu}) &= (-1)^{m+1} \sum_{j=0}^{s} \sum_{k=j}^{s} (-1)^{sj+j+k} f^{\#}_{\mu_0 \ldots \mu_j} (\partial c\langle j|\boldsymbol{\mu}|k\rangle \times \Delta^j) + \\
&\sum_{j=0}^{s-1} \sum_{k=j+1}^{s} (-1)^{sj+j+k+1} (f_{\mu_0 \ldots \mu_j} (p \times 1))^{\#} (c\langle j+1|\boldsymbol{\mu}|k\rangle \times \Delta^j) + \\
&\sum_{i=1}^{s} \sum_{j=0}^{i-1} \sum_{k=j}^{i-1} (-1)^{sj+k+i} (q f_{\mu_0 \ldots \mu_j})^{\#} (c(d^{i-j+1}\langle j|\boldsymbol{\mu}|k\rangle) \times \Delta^j) + \\
&\sum_{i=1}^{s-1} \sum_{j=0}^{i-1} \sum_{k=i+1}^{s} (-1)^{sj+k+i+1} (q f_{\mu_0 \ldots \mu_j})^{\#} (c(d^{i-j}\langle j|\boldsymbol{\mu}|k\rangle) \times \Delta^j).
\end{aligned} \tag{19}$$

On the other hand,

$$\begin{aligned}
(H^{m-1}(dc))(\boldsymbol{\mu}) &= (-1)^m \sum_{j=0}^{s} \sum_{k=j}^{s} (-1)^{sj+j+k} f^{\#}_{\mu_0 \ldots \mu_j} (\partial c\langle j|\boldsymbol{\mu}|k\rangle \times \Delta^j) + \\
&\sum_{j=0}^{s} \sum_{k=j}^{s} \sum_{i=0}^{s-j+1} (-1)^{sj+j+k+i} (f_{\mu_0 \ldots \mu_j} (p \times 1))^{\#} (c(d^i\langle j|\boldsymbol{\mu}|k\rangle) \times \Delta^j).
\end{aligned} \tag{20}$$

Now note that

$$\langle j|\boldsymbol{\mu}|k\rangle = \begin{cases} (f(\mu_j), \ldots, f(\mu_k), g(\mu_k), \ldots, g(\mu_s)), & 1 \le j+1 \le k \le s, \\ (f(\mu_j), g(\mu_j), \ldots, g(\mu_s)), & 0 \le k = j \le s, \end{cases} \tag{21}$$

and thus,

$$d^0\langle j|\boldsymbol{\mu}|k\rangle = \begin{cases} \langle j+1|\boldsymbol{\mu}|k\rangle, & j+1 \le k \\ (g(\mu_j), \ldots, g(\mu_s)), & k = j. \end{cases} \tag{22}$$

Consequently, the part of the triple sum of (20), which corresponds to $i = 0$, equals

$$\begin{aligned}
&\sum_{j=0}^{s-1} \sum_{k=j+1}^{s} (-1)^{sj+j+k} (f_{\mu_0 \ldots \mu_j} (p \times 1))^{\#} (c(d^0\langle j|\boldsymbol{\mu}|k\rangle) \times \Delta^j) + \\
&\sum_{j=0}^{s} (-1)^{sj} (f_{\mu_0 \ldots \mu_j} (p \times 1))^{\#} (c(d^0\langle j|\boldsymbol{\mu}|j\rangle) \times \Delta^j)
\end{aligned} \tag{23}$$

and (20) assumes the form

$$(H^{m-1}(dc))(\boldsymbol{\mu}) = (-1)^m \sum_{j=0}^{s}\sum_{k=j}^{s}(-1)^{sj+j+k} f^{\#}_{\mu_0\ldots\mu_j}(\partial c\langle j|\boldsymbol{\mu}|k\rangle \times \Delta^j) +$$
$$\sum_{j=0}^{s-1}\sum_{k=j+1}^{s}(-1)^{sj+j+k}(f_{\mu_0\ldots\mu_j}(p\times 1))^{\#}(c\langle j+1|\boldsymbol{\mu}|k\rangle \times \Delta^j) +$$
$$\sum_{j=0}^{s}(-1)^{sj}(f_{\mu_0\ldots\mu_j}(p\times 1))^{\#}(c(g(\mu_j),\ldots,g(\mu_s))\times \Delta^j) +$$
$$\sum_{j=0}^{s}\sum_{k=j}^{s}\sum_{i=1}^{s-j+1}(-1)^{sj+j+k+i} f^{\#}_{\mu_0\ldots\mu_j}(c(d^i\langle j|\boldsymbol{\mu}|k\rangle)\times \Delta^j). \tag{24}$$

We now add up (19) and (24) to obtain $(d(H^m c))(\boldsymbol{\mu}) + (H^{m-1}(dc))(\boldsymbol{\mu})$. Note that the first double sums cancel and so do the next double sums. Furthermore, since $\boldsymbol{g}$ is a shift of $\boldsymbol{f}$,

$$\begin{aligned}(f_{\mu_0\ldots\mu_j}(p\times 1))^{\#}(c(g(\mu_j),\ldots,g(\mu_s))\times \Delta^j) &=\\ g^{\#}_{\mu_0\ldots\mu_j}(c(g(\mu_j),\ldots,g(\mu_s))\times \Delta^j)\end{aligned} \tag{25}$$

and therefore, the third term of (24) equals $(g^{\#}c)(\boldsymbol{\mu})$. Hence, in order to verify (4), it remains to show that the triple sum in (24) and the two triple sums in (19) add up to $-(f^{\#}c)(\boldsymbol{\mu})$. To show this, we decompose the range $[1, s-j+1]$ of i in three disjoint subintervals $[1, k-j-1]$, $[k-j, k-j+1]$ and $[k-j+2, s-j+1]$. Accordingly, we decompose the triple sum of (24) in three terms

$$\sum_{j=0}^{s-2}\sum_{k=j+2}^{s}\sum_{i=1}^{k-j-1} + \sum_{j=0}^{s}\sum_{k=j}^{s}\sum_{i=k-j}^{k-j+1} + \sum_{j=0}^{s-1}\sum_{k=j}^{s-1}\sum_{i=k-j+2}^{s-j+1}. \tag{26}$$

The first of these terms cancels the last term of (19). To see this, first note that in the latter term the order of summation can be rearranged and the sum assumes the form

$$\sum_{j=0}^{s-2}\sum_{k=j+2}^{s}\sum_{i=j+1}^{k-1}(-1)^{sj+k+i+1}(qf_{\mu_0\ldots\mu_j})^{\#}(c(d^{i-j}\langle j|\boldsymbol{\mu}|k\rangle)\times \Delta^j). \tag{27}$$

To obtain the desired conclusion, it now suffices to replace, in (27), i by $i+j$. Similarly, the third term in (26) cancels the next to the last term in (19). Indeed, the latter term can also be written in the form

$$\sum_{j=0}^{s-1}\sum_{k=j}^{s-1}\sum_{i=k+1}^{s}(-1)^{sj+k+i}(qf_{\mu_0\ldots\mu_j})^{\#}(c(d^{i-j+1}\langle j|\boldsymbol{\mu}|k\rangle)\times \Delta^j) \tag{28}$$

and it suffices to replace i by $i+j-1$.

The remaining middle sum in (26) can be written in the following form.

$$\begin{aligned}&\sum_{j=0}^{s}(-1)^{sj+1}f^{\#}_{\mu_0\dots\mu_j}(c(d^1\langle j|\boldsymbol{\mu}|j\rangle)\times\Delta^j)+\\&\sum_{j=0}^{s-1}\sum_{k=j+1}^{s}(-1)^{sj}f^{\#}_{\mu_0\dots\mu_j}(c(d^{k-j}\langle j|\boldsymbol{\mu}|k\rangle)\times\Delta^j)+\\&\sum_{j=0}^{s-1}\sum_{k=j+1}^{s}(-1)^{sj+1}f^{\#}_{\mu_0\dots\mu_j}(c(d^{k-j+1}\langle j|\boldsymbol{\mu}|k\rangle)\times\Delta^j).\end{aligned}\tag{29}$$

Now note that, for $j+1\le k\le s-1$, one has

$$\begin{aligned}d^{k-j+1}\langle j|\boldsymbol{\mu}|k\rangle &= (f(\mu_j),\dots,f(\mu_k),g(\mu_{k+1}),\dots,g(\mu_s))\\&= d^{k-j+1}\langle j|\boldsymbol{\mu}|k+1\rangle.\end{aligned}\tag{30}$$

and for $j+1\le k=s$, one has

$$d^{s-j+1}\langle j|\boldsymbol{\mu}|s\rangle=(f(\mu_j),\dots,f(\mu_s)).\tag{31}$$

Therefore, the last term of (29) assumes the form

$$\begin{aligned}&\sum_{j=0}^{s-1}\sum_{k=j+1}^{s-1}(-1)^{sj+1}f^{\#}_{\mu_0\dots\mu_j}(c(d^{k-j+1}\langle j|\boldsymbol{\mu}|k+1\rangle)\times\Delta^j)+\\&\sum_{j=0}^{s-1}(-1)^{sj+1}f^{\#}_{\mu_0\dots\mu_j}(f(\mu_j),\dots,f(\mu_s)).\end{aligned}\tag{32}$$

If in the double sum of (32) we replace $k+1$ by k, we see that this sum cancels the second term of (29), except for the part corresponding to $k=j+1$. Consequently, (29) assumes the form

$$\begin{aligned}&\sum_{j=0}^{s}(-1)^{sj+1}f^{\#}_{\mu_0\dots\mu_j}(c(d^1\langle j|\boldsymbol{\mu}|j\rangle)\times\Delta^j)+\\&\sum_{j=0}^{s-1}(-1)^{sj}f^{\#}_{\mu_0\dots\mu_j}(c(d^1\langle j|\boldsymbol{\mu}|j+1\rangle)\times\Delta^j)+\\&\sum_{j=0}^{s-1}(-1)^{sj+1}f^{\#}_{\mu_0\dots\mu_j}(f(\mu_j),\dots,f(\mu_s)).\end{aligned}\tag{33}$$

Now note that, for $0\le j\le s-1$, one has

$$d^1\langle j|\boldsymbol{\mu}|j\rangle=(f(\mu_j),g(\mu_{j+1}),\dots,g(\mu_s))=d^1\langle j|\boldsymbol{\mu}|j+1\rangle.\tag{34}$$

Moreover, $d^1\langle s|\boldsymbol{\mu}|s\rangle=f(\mu_s)$. Therefore, the second term in (33) cancels all of the first term except its last summand $(-1)^{ss+1}f^{\#}_{\mu_0\dots\mu_s}(c(d^1\langle s|\boldsymbol{\mu}|s\rangle)\times\Delta^s)=(-1)^{ss+1}f^{\#}_{\mu_0\dots\mu_s}(c(f(\mu_s))\times\Delta^s)$. However, this summand and the last sum in (33) add up to $-(f^{\#}(c))(\boldsymbol{\mu})$. □

REMARK 18.5. One can show that our choice of signs in formulae (18.1.14) and (3) is the only choice, which warrants cancellation and equality of terms in the presented proofs of Lemmas 18.3 and 18.4

18.3 Chain mappings induced by homotopy classes

In this subsection we will show that the homomorphisms of strong homology groups $f_*: \overline{H}_m(\boldsymbol{X}) \to \overline{H}_m(\boldsymbol{Y})$, induced by a coherent mapping $\boldsymbol{f}: \boldsymbol{X} \to \boldsymbol{Y}$, depend only on its homotopy class $[\boldsymbol{f}]$. More precisely the following lemma holds.

LEMMA 18.6. *If $\boldsymbol{f}, \boldsymbol{f}': \boldsymbol{X} \to \boldsymbol{Y}$ are coherent mappings which are connected by a coherent homotopy $\boldsymbol{F}: \boldsymbol{X} \times I \to \boldsymbol{Y}$, then the induced chain mappings $f^\#, f'^\#: TS(\boldsymbol{X}; G) \to TS(\boldsymbol{Y}; G)$ are chain homotopic and thus, they induce the same homomorphisms of strong homology groups.*

Proof. Let $\boldsymbol{f}, \boldsymbol{f}'$ and $\boldsymbol{F}$ be given by increasing functions f, f' and F, where $f, f' \le F$, and by mappings $f_{\boldsymbol{\mu}}, f'_{\boldsymbol{\mu}}$ and $F_{\boldsymbol{\mu}}$, respectively. Let $\boldsymbol{g}, \boldsymbol{g}': \boldsymbol{X} \to \boldsymbol{Y}$ be shifts of $\boldsymbol{f}, \boldsymbol{f}'$ by the function F, given by mappings $g_{\boldsymbol{\mu}}, g'_{\boldsymbol{\mu}}$, $\boldsymbol{\mu} \in M_n$. Then, for $(x, t) \in X_{F(\mu_n)} \times I \times \Delta^n$, one has

$$F_{\boldsymbol{\mu}}(x, 0, t) = f_{\boldsymbol{\mu}}(p_{f(\mu_n)F(\mu_n)}(x), t) = g_{\boldsymbol{\mu}}(x, t), \tag{1}$$

$$F_{\boldsymbol{\mu}}(x, 1, t) = f'_{\boldsymbol{\mu}}(p_{f'(\mu_n)F(\mu_n)}(x), t) = g'_{\boldsymbol{\mu}}(x, t). \tag{2}$$

By Lemma 18.4, the induced chain mappings $f^\#$ and $g^\#$ are homotopic and so are $f'^\#$ and $g'^\#$. Hence, it suffices to show that $g^\#$ and $g'^\#$ are homotopic chain mappings, i.e., there exists a chain homotopy H between the chain complexes $TS(\boldsymbol{X}; G)$ and $TS(\boldsymbol{Y}; G)$, which connects $g'^\#$ to $g^\#$. For a strong m-chain c of $S(\boldsymbol{X}; G)$, we define the strong $(m+1)$-chain $H^m c$ of $S(\boldsymbol{Y}; G)$, by putting, for $\boldsymbol{\mu} \in M_s$,

$$(H^m c)(\boldsymbol{\mu}) = \\ (-1)^{m+1} \sum_{j=0}^{s} (-1)^{sj+s+j} F^\#_{\mu_0 \dots \mu_j}(c(F(\mu_j), \dots, F(\mu_s)) \times I \times \Delta^j). \tag{3}$$

Here I also denotes the singular 1-chain of I, given by the homeomorphism $\Delta^1 \to I$, which maps $(1-u, u)$ to u. Hence, $c(F(\mu_j), \dots, F(\mu_s)) \times I \times \Delta^j$ is a singular $(m+1+s)$-chain of $X_{\mu_j} \times I \times \Delta^j$. Since $F_{\mu_0 \dots \mu_j}: X_{\mu_j} \times I \times \Delta^j \to Y_{\mu_0}$, we see that $F^\#_{\mu_0 \dots \mu_j}(c(F(\mu_j), \dots, F(\mu_s)) \times I \times \Delta^j)$ is a singular $(m+1+s)$-chain of Y_{μ_0}, and $(H^m c)(\boldsymbol{\mu})$ is indeed a strong $(m+1)$-chain of $S(\boldsymbol{Y}; G)$. It remains to show that

$$(d(H^m c))(\boldsymbol{\mu}) + (H^{m-1}(dc))(\boldsymbol{\mu}) = (g^m(c))(\boldsymbol{\mu}) - (g'^m(c))(\boldsymbol{\mu}), \tag{4}$$

Omitting the verification in the rather trivial case $s = 0$, we proceed to verify (4), for $s \ge 1$. In this case

$$(d(H^m c))(\boldsymbol{\mu}) = \partial(H^m c)(\boldsymbol{\mu}) + (-1)^{m+1}(\delta(H^m c))(\boldsymbol{\mu}). \tag{5}$$

Moreover,

$$\begin{aligned}
&\partial(H^m c)(\boldsymbol{\mu}) = \\
&(-1)^{m+1}\sum_{j=0}^{s}(-1)^{sj+s+j}F^{\#}_{\mu_0\ldots\mu_j}(\partial c(F(\mu_j),\ldots,F(\mu_s))\times I\times\Delta^j) + \\
&\sum_{j=0}^{s}(-1)^{sj+1}F^{\#}_{\mu_0\ldots\mu_j}(c(F(\mu_j),\ldots,F(\mu_s))\times\partial I\times\Delta^j) + \\
&\sum_{j=1}^{s}(-1)^{sj}F^{\#}_{\mu_0\ldots\mu_j}(c(F(\mu_j),\ldots,F(\mu_s))\times I\times\partial(\Delta^j)).
\end{aligned} \tag{6}$$

Since $\partial I = (1) - (0)$, formulae (1), (2) and (18.1.14) show that the second term of (6) equals

$$(g^{\#}(c))(\boldsymbol{\mu}) - (g'^{\#}(c))(\boldsymbol{\mu}). \tag{7}$$

We now apply the boundary formula (18.1.1) and the coherence conditions for $\boldsymbol{F}$ to the last term of (6). We then replace j by $j+1$ and split off the terms with $i = j+1$. It turns out that the last term of (6) equals

$$\begin{aligned}
&\sum_{j=0}^{s-1}\sum_{i=0}^{j}(-1)^{sj+s+i}q^{\#}F^{\#}_{d^i(\mu_0,\ldots,\mu_{j+1})}(c(F(\mu_{j+1}),\ldots,F(\mu_s))\times I\times\Delta^j)+ \\
&\sum_{j=0}^{s-1}(-1)^{sj+s+j+1}(F_{(\mu_0\ldots\mu_j)}(p\times 1))^{\#}(c(F(\mu_{j+1}),\ldots,F(\mu_s))\times I\times(\Delta^j)).
\end{aligned} \tag{8}$$

Next, take into account the fact that

$$(-1)^{m+1}(\delta H^m c)(\boldsymbol{\mu}) = (-1)^{m+1}\sum_{i=0}^{s}(-1)^i q^{\#}(H^m c)(d^i\boldsymbol{\mu}). \tag{9}$$

Distinguishing the cases $0 \le j \le i-1$ and $i \le j \le s-1$, the right side of (9) assumes the form

$$\begin{aligned}
&\sum_{i=1}^{s}\sum_{j=0}^{i-1}(-1)^{sj+s+i+1}q^{\#}F^{\#}_{\mu_0\ldots\mu_j}(c(F(\mu_j),\ldots,\widehat{F(\mu_i)},\ldots,F(\mu_s))\times I\times\Delta^j)+ \\
&\sum_{i=0}^{s-1}\sum_{j=i}^{s-1}(-1)^{sj+s+i+1}q^{\#}F^{\#}_{d^i(\mu_0,\ldots,\mu_{j+1})}(c(F(\mu_{j+1}),\ldots,F(\mu_s))\times I\times\Delta^j).
\end{aligned} \tag{10}$$

Now note that the range of the indices in the first term of (8) can also be described by the inequalities

$$0 \le i \le s-1,\ i \le j \le s-1. \tag{11}$$

Therefore, the sum of the first term of (8) and the second term of (10) vanishes. Consequently, one has

$$\begin{aligned}
&(d(H^m c))(\boldsymbol{\mu}) = (g^{\#}(c))(\boldsymbol{\mu}) - (g'^{\#}(c))(\boldsymbol{\mu})+\\
&(-1)^{m+1}\sum_{j=0}^{s}(-1)^{sj+s+j}F_{\mu_0\ldots\mu_j\#}(\partial c(F(\mu_j),\ldots,F(\mu_s))\times I\times\Delta^j)+\\
&\sum_{j=0}^{s-1}(-1)^{sj+s+j+1}(F_{\mu_0\ldots\mu_j}(p\times 1))^{\#}(c(F(\mu_{j+1}),\ldots,F(\mu_s))\times I\times\Delta^j)+\\
&\sum_{i=1}^{s}\sum_{j=0}^{i-1}(-1)^{sj+s+i+1}q^{\#}F^{\#}_{\mu_0\ldots\mu_j}(c(F(\mu_j),\ldots,\widehat{F(\mu_i)},\ldots,F(\mu_s))\times I\times\Delta^j).
\end{aligned} \tag{12}$$

On the other hand,

$$\begin{aligned}
&(H^{m-1}(dc))(\boldsymbol{\mu}) =\\
&(-1)^m\sum_{j=0}^{s}(-1)^{sj+s+j}F^{\#}_{\mu_0\ldots\mu_j}((dc)(F(\mu_j),\ldots,F(\mu_s))\times I\times\Delta^j).
\end{aligned} \tag{13}$$

However,

$$\begin{aligned}
&(dc)(F(\mu_j),\ldots,F(\mu_s)) = \partial c(F(\mu_j),\ldots,F(\mu_s)) +\\
&(-1)^m\sum_{i=j}^{s}(-1)^{j+i}p^{\#}(c(F(\mu_j),\ldots,\widehat{F(\mu_i)},\ldots,F(\mu_s))).
\end{aligned} \tag{14}$$

Consequently,

$$\begin{aligned}
&(H^{m-1}(dc))(\boldsymbol{\mu}) =\\
&(-1)^m\sum_{j=0}^{s}(-1)^{sj+s+j}F^{\#}_{\mu_0\ldots\mu_j}(\partial c(F(\mu_j),...,F(\mu_s))\times I\times\Delta^j) +\\
&\sum_{j=0}^{s-1}\sum_{i=j}^{s}(-1)^{sj+s+i}(F_{\mu_0\ldots\mu_j}(p\times 1))^{\#}(c(F(\mu_j),...,\widehat{F(\mu_i)},...,F(\mu_s))\times I\times\Delta^j)
\end{aligned} \tag{15}$$

We now sum up (12) and (15). The second term of (12) and the first term of (15) cancel. Moreover, the range of the indices in the double sum of (12) can also be described by the conditions

$$0\le j\le s-1,\ j+1\le i\le s. \tag{16}$$

Therefore, this double sum cancels all terms in the double sum of (15), except terms for which $i=j$. Consequently, the two double sums add up to

$$\sum_{j=0}^{s-1}(-1)^{sj+s+j}(F_{\mu_0\ldots\mu_j}(p\times 1))^{\#}(c(F(\mu_{j+1}),\ldots,F(\mu_s))\times I\times\Delta^j). \tag{17}$$

However, (17) cancels the third term of (12). □

18.4 Chain mappings induced by composition

The purpose of this subsection is to establish the following lemma.

LEMMA 18.7. *If $\boldsymbol{f}: \boldsymbol{X} \to \boldsymbol{Y}$ and $\boldsymbol{g}: \boldsymbol{Y} \to \boldsymbol{Z}$ are coherent mappings, then the composition $g^{\#} f^{\#}$ is chain homotopic to the chain mapping $h^{\#}$, induced by the composition $\boldsymbol{h} = \boldsymbol{g}\boldsymbol{f}$.*

To prove the lemma we need some preparation. In defining composition of coherent mappings (see 1.3), we decomposed the standard simplex Δ^n in subpolyhedra P_i^n, $0 \le i \le n$. Then we considered mappings $a_i^n: P_i^n \to \Delta^{n-i}, b_i^n: P_i^n \to \Delta^i$ which determined an affine homeomorphism $c_i^n: P_i^n \to \Delta^{n-i} \times \Delta^i$. Let $\tau_i^n: \Delta^{n-i} \times \Delta^i \to \Delta^i \times \Delta^{n-i}$ be the involution, which interchanges the two factors, and let $c_i'^n: \Delta^i \times \Delta^{n-i} \to P_i^n$ denote the inverse of $\tau_i^n c_i^n$. Viewing $\Delta^i \times \Delta^{n-i}$ as a singular n-chain of the space $\Delta^i \times \Delta^{n-i}$, it follows that $c_i'^{n\#}(\Delta^i \times \Delta^{n-i})$ is a well-defined singular n-chain of P_i^n, for which we also use the notation P_i^n. We then define a singular n-chain P^n of Δ^n by putting

$$P^n = \sum_{i=0}^{n} P_i^n. \tag{1}$$

P^n can be viewed as a subdivision of the singular simplex Δ^n.

LEMMA 18.8. *The boundary of the singular chain P^n satisfies the equality*

$$\partial P^n = \sum_{j=0}^{n} (-1)^j (d_j^n)^{\#}(P^{n-1}), \tag{2}$$

where $d_j^n: \Delta^{n-1} \to \Delta^n$, $0 \le j \le n$, are the standard face operators (see 1.2).

Using the explicit formulae for a_i^n and b_i^n (see (1.3.3) and (1.3.4)), it is easy to obtain an explicit formula for $c_i'^n$. If $u = (u_0, ..., u_i) \in \Delta^i$ and $v = (v_0, ..., v_{n-i}) \in \Delta^{n-i}$, then

$$c_i'^n(u, v) = (\frac{1}{2}u_0, ..., \frac{1}{2}u_{i-1}, \frac{1}{2}(u_i + v_0), \frac{1}{2}v_1, ..., \frac{1}{2}v_{n-i}). \tag{3}$$

Now a straightforward computation shows that

$$\begin{aligned}
c_i'^n(1 \times d_j^{n-i}) &= d_{i+j}^n c_i'^{n-1}, && 0 \le i \le n-1, \quad 1 \le j \le n-i, \\
c_{i+1}'^n(d_j^{i+1} \times 1) &= d_j^n c_i'^{n-1}, && 0 \le i \le n-1, \quad 0 \le j \le i, \\
c_{i+1}'^n(d_{i+1}^{i+1} \times 1) &= c_i'^n(1 \times d_0^{n-i}), && 0 \le i \le n-1.
\end{aligned} \tag{4}$$

By (1) and by the definition of P_i^n, one has

$$
\begin{aligned}
&\partial P^n = \sum_{i=0}^{n} \partial (c_i'^n)^\#(\Delta^i \times \Delta^{n-i}) = \\
&\sum_{i=1}^{n} (c_i'^n)^\#(\partial \Delta^i \times \Delta^{n-i}) + \sum_{i=0}^{n-1} (-1)^i (c_i'^n)^\#(\Delta^i \times \partial \Delta^{n-i}) = \\
&\sum_{i=1}^{n} \sum_{j=0}^{i} (-1)^j (c_i'^n(d_j^i \times 1))^\#(\Delta^{i-1} \times \Delta^{n-i}) + \\
&\sum_{i=0}^{n-1} \sum_{j=0}^{n-i} (-1)^{i+j} (c_i'^n(1 \times d_j^{n-i}))^\#(\Delta^i \times \Delta^{n-i-1}).
\end{aligned} \tag{5}
$$

If in the first double sum of (5) we replace $i-1$ by i and then, split off the terms for which $j = i+1$, this double sum assumes the form

$$
\begin{aligned}
&\sum_{i=0}^{n-1} \sum_{j=0}^{i} (-1)^j (c_{i+1}'^n(d_j^{i+1} \times 1))^\#(\Delta^i \times \Delta^{n-i-1}) + \\
&\sum_{i=0}^{n-1} (-1)^{i+1} (c_{i+1}'^n(d_{i+1}^{i+1} \times 1))^\#(\Delta^i \times \Delta^{n-i-1}).
\end{aligned} \tag{6}
$$

On the other hand, if in the second double sum of (5), we split off the terms for which $j = 0$, this double sum assumes the form

$$
\begin{aligned}
&\sum_{i=0}^{n-1} \sum_{j=1}^{n-i} (-1)^{i+j} (c_i'^n(1 \times d_j^{n-i}))^\#(\Delta^i \times \Delta^{n-i-1}) + \\
&\sum_{i=0}^{n-1} (-1)^i (c_i'^n(1 \times d_0^{n-i}))^\#(\Delta^i \times \Delta^{n-i-1}).
\end{aligned} \tag{7}
$$

Taking into account the third of the equalities in (4), we see that the two single sums in (6) and (7) add up to zero. Therefore, the two remaining equalities of (4) show that

$$
\begin{aligned}
\partial P^n = &\sum_{i=0}^{n-1} \sum_{j=0}^{i} (-1)^j (d_j^n c_i'^{n-1})^\#(\Delta^i \times \Delta^{n-i-1}) + \\
&\sum_{i=0}^{n-1} \sum_{j=1}^{n-i} (-1)^{i+j} (d_{i+j}^n c_i'^{n-1})^\#(\Delta^i \times \Delta^{n-i-1}).
\end{aligned} \tag{8}
$$

Finally, if in the second sum of (8) we replace j by $j - i$, we obtain

$$
\partial P^n = \sum_{i=0}^{n-1} \sum_{j=0}^{n} (-1)^j (d_j^n c_i'^{n-1})^\#(\Delta^i \times \Delta^{n-i-1}). \tag{9}
$$

However, by the definition of P^n and P_i^n, one has

$$\begin{aligned}\sum_{j=0}^{n}(-1)^j(d_j^n)^\#(P^{n-1}) &= \sum_{j=0}^{n}\sum_{i=0}^{n-1}(-1)^j(d_j^n)^\#(P_i^{n-1}) \\ &= \sum_{i=0}^{n-1}\sum_{j=0}^{n}(-1)^j(d_j^n c_i'^{n-1})^\#(\Delta^i \times \Delta^{n-i-1}),\end{aligned} \tag{10}$$

which establishes the desired formula (2). □

With every coherent mapping $\boldsymbol{f}: \boldsymbol{X} \to \boldsymbol{Y}$ we have associated in 18.1 an induced chain mapping $f^\#: TS(\boldsymbol{X}; G) \to TS(\boldsymbol{Y}; G)$. We now associate with $\boldsymbol{f}$ another chain mapping $f_P^\#: TS(\boldsymbol{X}; G) \to TS(\boldsymbol{Y}; G)$. The definition differs from the previous one (see (18.1.14)) only in that Δ^i is replaced by its subdivision P^i. Hence, if c is a strong m-chain of $S(\boldsymbol{X}; G)$, then

$$(f_P^m(c))(\boldsymbol{\mu}) = \sum_{i=0}^{s}(-1)^{si} f_{\mu_0\ldots\mu_i}^\#(c(f(\mu_i), \ldots, f(\mu_s)) \times P^i), \ \boldsymbol{\mu} \in M_s. \tag{11}$$

Since P^i is a singular i-chain of Δ^i, one concludes that $c(f(\mu_i), ..., f(\mu_s)) \times P^i$ is a singular $(m+s)$-chain of $X_{f(\mu_i)} \times \Delta^i$ and thus, $(f_P^m c)(\boldsymbol{\mu})$ is a singular $(m+s)$-chain of Y_{μ_0}, which proves that $f_P^m(c)$ is indeed a strong m-chain of $S(\boldsymbol{Y}; G)$.

LEMMA 18.9. *The homomorphisms f_P^m satisfy*

$$d(f_P^m(c)) = f_P^{m-1}(dc) \tag{12}$$

and thus, $f_P: TS(\boldsymbol{X}; G) \to TS(\boldsymbol{Y}; G)$ is a chain mapping.

Proof. The proof of (12) proceeds just as the proof of Lemma 18.3, the only difference being that the usual boundary formula for Δ^i (see (18.1.1)) is replaced by formula (2). □

The next step needed to prove Lemma 18.7. is the following lemma.

LEMMA 18.10. *The chain mappings $f^\#, f_P^\#: TS(\boldsymbol{X}; G) \to TS(\boldsymbol{Y}; G)$ are chain homotopic.*

Proof. We must define a chain homotopy H, which connects $f^\#$ to $f_P^\#$, i.e., we need homomorphisms $H^m: T^m S(\boldsymbol{X}; G) \to T^{m+1} S(\boldsymbol{Y}; G)$, $m \in \mathbb{Z}$, which satisfy

$$(d(H^m c))(\boldsymbol{\mu}) + (H^{m-1}(dc))(\boldsymbol{\mu}) = (f^m c)(\boldsymbol{\mu}) - (f_P^m(c))(\boldsymbol{\mu}), \tag{13}$$

for every strong m-chain c of $S(\boldsymbol{X}; G)$ and every $\boldsymbol{\mu} \in M_s$.

In order to define H, we need the *cone of a singular chain* x (see e.g., (Dold 1972), p. 34). If K is a convex subset of $\mathbb{R}^m$, and $v \in K$ is an arbitrary point, then the cone construction assigns to every singular i-simplex $s: \Delta^i \to K$ the $(i+1)$-simplex $vs: \Delta^{i+1} \to K$ of K, defined as the mapping

$$vs(t_0, ..., t_{i+1}) = \begin{cases} v, & t_0 = 1, \\ t_0 v + (1-t_0) s(\frac{t_1}{1-t_0}, ..., \frac{t_{i+1}}{1-t_0}), & t_0 > 1. \end{cases} \tag{14}$$

It is readily seen that

$$(vs)d_0 = s, \tag{15}$$

$$(vs)\, d_{j+1} = \begin{cases} v(sd_j), & 0 < i,\ 0 < j \le i, \\ v, & i = 0,\ j = 1. \end{cases} \tag{16}$$

For a singular i-chain $x = \sum_j m_j s_j$ in K, one puts $vx = \sum_j m_j(vs_j)$. Clearly, $c \mapsto vx$ is a homomorphism $S^i(K) \to S^{i+1}(K)$. Moreover, using (15) and (16), it is easy to see that

$$\partial(vx) = \begin{cases} x - v(\partial x), & i > 0, \\ x - \kappa(x)v, & i = 0, \end{cases} \tag{17}$$

where $\kappa(x) = \sum_j m_j$.

By induction on $i \ge 0$, we now define singular i-chains D^i of $\mathrm{Bd}(I \times \Delta^i)$. For each i, we choose a point $v^i \in I \times \Delta^i \subseteq \mathbb{R}^{i+2}$ and we put

$$D^i = \begin{cases} 1 \times \Delta^0 - 0 \times \Delta^0, & i = 0, \\ (1 \times P^i) - (0 \times \Delta^i) - \displaystyle\sum_{j=0}^{i} (-1)^j (1 \times d^i_j)^\# v^{i-1} D^{i-1}, & i \ge 1. \end{cases} \tag{18}$$

We now show, by induction on $i \ge 0$, that $\partial D^i = 0$. This is obvious, for $i = 0$. Therefore, we assume that $i \ge 1$. By (18) and (17), for $i \ge 2$, we have

$$\partial D^i = (1 \times \partial P^i) - (0 \times \partial \Delta^i) - \sum_{j=0}^{i} (-1)^j (1 \times d^i_j)^\# D^{i-1}. \tag{19}$$

This holds also for $i = 1$, because $\kappa(D^0) = 0$. Using (18.1.1), (2) and (18) applied to D^{i-1}, one obtains

$$\partial D^i = \sum_{j=0}^{i} \sum_{k=0}^{i-1} (-1)^{j+k} (1 \times d^i_j d^{i-1}_k)^\# v^{i-2} D^{i-2}. \tag{20}$$

However, this double sum equals zero because

$$d^i_j d^{i-1}_k = d^j_k d^{i-1}_{j-1},\ 0 \le k < j \le i, \tag{21}$$

(see (1.2.16)).

We now define the desired chain homotopy H, by putting

$$(H^m c)(\boldsymbol{\mu}) = \\ (-1)^{m+1} \sum_{j=0}^{s} (-1)^{sj+s+j} f^\#_{\mu_0 \dots \mu_j}(c(f(\mu_j), ..., f(\mu_s)) \times \pi^\#(v^j D^j)), \tag{22}$$

where c is a strong m-chain of $S(\boldsymbol{X};G)$, $\boldsymbol{\mu} \in M_s$, and π denotes the second projection $I \times \Delta^j \to \Delta^j$. The verification of (13) is similar to the verification of (18.3.4) with $f, f_{\mu_0 \dots \mu_j}$ and $\pi^\#(v^j D^j)$ playing the role of $F, F_{\mu_0 \dots \mu_j}$ and $I \times \Delta^j$, respectively. Omitting the verification of (13) in the trivial case $s = 0$, we assume that $s \geq 1$.

Since $\partial D^j = 0$, (22) implies

$$\begin{aligned} &\partial(H^m c)(\boldsymbol{\mu}) = \\ &(-1)^{m+1} \sum_{j=0}^{s} (-1)^{sj+s+j} f^\#_{\mu_0 \dots \mu_j} (\partial c(f(\mu_j), \dots, f(\mu_s)) \times \pi^\#(v^j D^j)) + \\ &\sum_{j=0}^{s} (-1)^{sj+1} f^\#_{\mu_0 \dots \mu_j} (c(f(\mu_j), \dots, f(\mu_s)) \times \pi^\#(D^j)). \end{aligned} \tag{23}$$

Now note that $(\pi^j)(s \times 1_{\Delta^j}) = 1_{\Delta^j}$, for any $s \in I$, and thus, $\pi^\#(1 \times P^j) = P^j$, $\pi^\#(0 \times \Delta^j) = \Delta^j$. Moreover, it is easy to see that

$$\pi^j(1 \times d_i^j) = d_i^j \pi^{j-1}, \; 0 \leq i \leq j. \tag{24}$$

Therefore, by (18), (11) and (18.1.14), the second sum of (23) assumes the form

$$\begin{aligned} &(f^m(c))(\boldsymbol{\mu}) - (f^m_P(c))(\boldsymbol{\mu}) + \\ &\sum_{j=1}^{s} \sum_{i=0}^{j} (-1)^{sj+i} f^\#_{\mu_0 \dots \mu_j} (c(f(\mu_j), \dots, f(\mu_s)) \times (d_i^j \pi^{j-1})^\#(v^{j-1} D^{j-1})). \end{aligned} \tag{25}$$

Using the coherence properties of $f_{\boldsymbol{\mu}}$, replacing the summation index $j - 1$ by j and splitting off the terms with $i = j + 1$, one concludes that the double sum in (25) equals

$$\begin{aligned} &\sum_{j=0}^{s-1} \sum_{i=0}^{j} (-1)^{sj+s+i} (q f_{d^i(\mu_0, \dots, \mu_{j+1})})^\# (c(f(\mu_{j+1}), \dots, f(\mu_s)) \times \pi^\#(v^j D^j)) + \\ &\sum_{j=0}^{s-1} (-1)^{sj+s+j+1} (f_{\mu_0 \dots \mu_j} (p \times 1))^\# (c(f(\mu_{j+1}), \dots, f(\mu_s)) \times \pi^\#(v^j D^j)). \end{aligned} \tag{26}$$

Next note that $(-1)^{m+1}(\delta D c)(\boldsymbol{\mu})$ assumes the form

$$\begin{aligned} &\sum_{i=1}^{s} \sum_{j=0}^{i-1} (-1)^{sj+s+i+1} (f_{\mu_0 \dots \mu_j})^\# (c(f(\mu_j), \dots, \widehat{f(\mu_i)}, \dots, f(\mu_s)) \times \pi^\#(v^j D^j)) + \\ &\sum_{i=0}^{s-1} \sum_{j=i}^{s-1} (-1)^{sj+s+i+1} (q f_{d^i(\mu_0, \dots, \mu_{j+1})})^\# (c(f(\mu_{j+1}), \dots, f(\mu_s)) \times \pi^\#(v^j D^j)). \end{aligned} \tag{27}$$

As in the proof of Lemma 18.6, the double sum in (26) and the second double sum in (27) add up to zero. Consequently,

$$\begin{aligned}&(d(H^m c))(\boldsymbol{\mu}) = (f^m(c))(\boldsymbol{\mu}) - (f_P^m(c))(\boldsymbol{\mu}) + \\ &(-1)^{m+1}\sum_{j=0}^{s}(-1)^{sj+s+j} f^{\#}_{\mu_0\dots\mu_j}(\partial c(f(\mu_j),\dots,f(\mu_s)) \times \pi^{\#}(v^j D^j)) + \\ &\sum_{j=0}^{s-1}(-1)^{sj+s+j+1}(f_{\mu_0\dots\mu_j}(p\times 1))^{\#}(c(f(\mu_{j+1}),\dots,f(\mu_s)) \times \pi^{\#}(v^j D^j)) + \\ &\sum_{i=1}^{s}\sum_{j=0}^{i-1}(-1)^{sj+s+i+1} f^{\#}_{\mu_0\dots\mu_j}(c(f(\mu_j),\dots,\widehat{f(\mu_i)},\dots,f(\mu_s)) \times \pi^{\#}(v^j D^j)).\end{aligned} \tag{28}$$

On the other hand,

$$\begin{aligned}&(H^{m-1}(dc))(\boldsymbol{\mu}) = \\ &(-1)^m\sum_{j=0}^{s}(-1)^{sj+s+j} f^{\#}_{\mu_0\dots\mu_j}(\partial c(f(\mu_j),\dots,f(\mu_s)) \times \pi^{\#}(v^j D^j)) + \\ &\sum_{j=0}^{s-1}\sum_{i=j}^{s}(-1)^{sj+s+i} f^{\#}_{\mu_0\dots\mu_j}(p^{\#}\times 1)(c(f(\mu_j),\dots,\widehat{f(\mu_i)},\dots,f(\mu_s)) \times \pi^{\#}(v^j D^j))\end{aligned} \tag{29}$$

We now add up (28) and (29). The second term of (28) and the first term of (29) cancel. Moreover, the range of the indices in the double sum of (28) can also be described by conditions (18.3.16). Therefore, this double sum cancels all terms in the double sum of (29), except terms for which $i = j$. Consequently, the two double sums add up to

$$\sum_{j=0}^{s-1}(-1)^{sj+s+j}(f_{\mu_0\dots\mu_j}(p\times 1))^{\#}(c(f(\mu_{j+1}),\dots,f(\mu_s)) \times \pi^{\#}(v^j D^j)). \tag{30}$$

Finally, (30) cancels the third term of (28). □

Proof of Lemma 18.7. Let $\boldsymbol{f}$ and $\boldsymbol{g}$ be coherent mappings and let $\boldsymbol{h} = \boldsymbol{g}\boldsymbol{f}$ be their composition. We will prove that

$$g^{\#} f^{\#} = h_P^{\#}, \tag{31}$$

where $h_P^{\#}$ is the chain mapping associated with $\boldsymbol{h}$ by formula (11) for $\boldsymbol{h}$. Since, by Lemma 18.10, $h_P^{\#}$ and $h^{\#}$ are homotopic chain mappings, this will yield the desired assertion that $g^{\#} f^{\#}$ and $h^{\#}$ are homotopic chain mappings.

Let $\boldsymbol{f}$ and $\boldsymbol{g}$ be given by increasing functions f, g and by mappings f_{μ}, g_{μ}, respectively. By (18.1.14) applied to $\boldsymbol{g}$, for a strong m-chain c of $S(\boldsymbol{X}; G)$ and $\boldsymbol{\nu} \in N_s$, one has

$$((g^m f^m)c)(\boldsymbol{\nu}) = \sum_{j=0}^{s}(-1)^{sj} g^{\#}_{\nu_0\dots\nu_j}((f^m c)(g(\nu_j),\dots,g(\nu_s)) \times \Delta^j). \tag{32}$$

One determines $(f^m c)(g(\nu_j), ..., g(\nu_s))$, by applying the same formula to $\boldsymbol{f}$. Then, (32) becomes

$$((g^m f^m)c)(\boldsymbol{\nu}) = \sum_{j=0}^{s}\sum_{i=j}^{s}(-1)^{si+ji+j} g^\#_{\nu_0\ldots\nu_j}(f^\#_{g(\nu_j)\ldots g(\nu_i)}(z \times \Delta^{i-j}) \times \Delta^j)), \tag{33}$$

where

$$z = c(fg(\nu_i), ..., fg(\nu_s)) \in S^{m+s-i}(X_{fg(\nu_i)}; G). \tag{34}$$

On the other hand, the composition $\boldsymbol{h} = \boldsymbol{gf}$ is given by the increasing function $h = gf$ and by mappings h_ν, defined by (1.3.5). Therefore, by (11) and (1), $h^\#_P$ is given by

$$(h^m_P(c))(\boldsymbol{\nu}) = \sum_{i=0}^{s}\sum_{j=0}^{i}(-1)^{si} h^\#_{\nu_0\ldots\nu_i}(z \times P^i_j). \tag{35}$$

Note that summation in the double sums of (33) and (35) ranges over the same set. Therefore, (31) will be proved if we show that

$$(-1)^{ji+j} g^\#_{\nu_0\ldots\nu_j}(f^\#_{g(\nu_j)\ldots g(\nu_i)}(z \times \Delta^{i-j})) \times \Delta^j) = h^\#_{\nu_0\ldots\nu_i}(z \times P^i_j), \tag{36}$$

for $\boldsymbol{\nu} \in N_s$, $j \le i \le s$, and for every chain $z \in S^{m+s-i}(X_{h(\nu_i)}; G)$.

By the composition formula for coherent mappings (see (1.3.5)),

$$h_{\nu_0\ldots\nu_i}|(X_{h(\nu_i)} \times P^i_j) = g_{\nu_0\ldots\nu_j}(f_{g(\nu_j)\ldots g(\nu_i)} \times 1)(1 \times c^i_j). \tag{37}$$

Therefore,

$$h^\#_{\nu_0\ldots\nu_i}(z \times P^i_j) = g^\#_{\nu_0\ldots\nu_j}(f_{g(\nu_j)\ldots g(\nu_i)} \times 1)^\#(1 \times c^i_j)^\#(z \times P^i_j). \tag{38}$$

Now note that, by (18.1.10),

$$(1 \times c^i_j)^\#(z \times P^i_j) = z \times (c^i_j)^\#(P^i_j). \tag{39}$$

Also recall that, by definition, $P^i_j = (c'^i_j)^\#(\Delta^j \times \Delta^{i-j})$. Moreover, $c'^i_j \colon \Delta^j \times \Delta^{i-j} \to P^i_j$ is the inverse of $\tau^i_j c^i_j \colon P^i_j \to \Delta^j \times \Delta^{i-j}$, where $\tau^i_j \colon \Delta^{i-j} \times \Delta^j \to \Delta^j \times \Delta^{i-j}$ is the involution which permutes coordinates. Therefore,

$$(c^i_j)^\#(P^i_j) = (c^i_j c'^i_j)^\#(\Delta^j \times \Delta^{i-j}) = (\tau^i_j)^\#(\Delta^j \times \Delta^{i-j}) \tag{40}$$

Taking into account (8), one concludes that

$$(c^i_j)^\#(P^i_j) = (-1)^{ij+j}\Delta^{i-j} \times \Delta^j. \tag{41}$$

If we now substitute (41) to (39), we obtain

$$(1 \times c^i_j)^\#(z \times P^i_j) = (-1)^{ij+j}(z \times \Delta^{i-j} \times \Delta^j). \tag{42}$$

Finally, the desired relation (36) follows by substituting (42) to (38). □

18.5 Induced chain mappings and the coherence functor

In 1.4 we have associated with every mapping of systems $\boldsymbol{f}$ a coherent mapping $C(\boldsymbol{f})$. In this subsection, we establish the following lemma.

LEMMA 18.11. *If $\boldsymbol{f}: \boldsymbol{X} \to \boldsymbol{Y}$ is a mapping and $\boldsymbol{g} = C(\boldsymbol{f}): \boldsymbol{X} \to \boldsymbol{Y}$, then the induced chain mappings $f^{\#}, g^{\#}: TS(\boldsymbol{X}; G) \to TS(\boldsymbol{Y}; G)$ are homotopic and thus, induce the the same homomorphisms of strong homology groups. In particular, $C(\mathbf{1}_{\boldsymbol{X}})$ induces the identity homomorphisms* $\mathrm{id}: \overline{H}_m(\boldsymbol{X}; G) \to \overline{H}_m(\boldsymbol{X}; G)$.

Proof. Let $\boldsymbol{f}$ be given by an increasing function $f: M \to \Lambda$ and by mappings $f_{\mu_0}: X_{f(\mu_0)} \to Y_{\mu_0}$. According to 1.4, $\boldsymbol{g} = C(\boldsymbol{f})$ is given by the function $g = f$ and by the mappings

$$g_{\boldsymbol{\mu}} = f_{\mu_0} p_{f(\mu_0) f(\mu_s)} \pi' = q_{\mu_0 \mu_s} f_{\mu_s} \pi', \tag{1}$$

where π' denotes the first projection $X_{f(\mu_s)} \times \Delta^s \to X_{f(\mu_s)}$. According to (17.1.25), for a strong m-chain c of $S(\boldsymbol{X}; G)$, one has

$$(f^m(c))(\boldsymbol{\mu}) = f_{\mu_0}^{m+s}(c(f(\mu_0), \dots, f(\mu_s))). \tag{2}$$

On the other hand, by (18.1.14),

$$\begin{gathered}(g^m(c))(\boldsymbol{\mu}) = \sum_{i=0}^{s} (-1)^{si} g^{\#}_{\mu_0 \dots \mu_i} (c(f(\mu_i, \dots, f(\mu_s)) \times \Delta^i) = \\ \sum_{i=0}^{s} (-1)^{si} (q_{\mu_0 \mu_i} f_{\mu_i} \pi')^{\#} (c(f(\mu_i, \dots, f(\mu_s)) \times \Delta^i).\end{gathered} \tag{3}$$

We need a chain homotopy H, which connects $g^{\#}$ to $f^{\#}$, i.e., satisfies

$$(d(H^m c))(\boldsymbol{\mu}) + (H^{m-1}(dc))(\boldsymbol{\mu}) = (f^m(c))(\boldsymbol{\mu}) - (g^m(c))(\boldsymbol{\mu}). \tag{4}$$

We will show that such a chain homotopy is given by

$$\begin{gathered}(-1)^{m+1} (H^m c)(\boldsymbol{\mu}) = \\ \sum_{j=0}^{s} (-1)^{sj+s+j} (q_{\mu_0 \mu_j} f_{\mu_j} \pi')^{\#} (c(f(\mu_j), \dots, f(\mu_s)) \times \Delta^{j+1}).\end{gathered} \tag{5}$$

Omitting the easy case $s = 0$, we will verify (4), for $s \geq 1$. First note that

$$\begin{gathered}\partial(H^m c)(\boldsymbol{\mu}) = \\ (-1)^{m+1} \sum_{j=0}^{s} (-1)^{sj+s+j} (q_{\mu_0 \mu_j} f_{\mu_j} \pi')^{\#} (\partial c(f(\mu_j), \dots, f(\mu_s)) \times \Delta^{j+1}) + \\ \sum_{j=0}^{s} (-1)^{sj+1} (q_{\mu_0 \mu_j} f_{\mu_j} \pi')^{\#} (c(f(\mu_j), \dots, f(\mu_s)) \times \partial \Delta^{j+1}).\end{gathered} \tag{6}$$

Now apply (18.1.1) and note that $\pi'(1 \times d_i^{j+1}) = \pi'$ does not depend on i. Therefore, the second sum in (6) equals

$$\sum_{j=0}^{s}(-1)^{sj+1}\varepsilon(j+1)(q_{\mu_0\mu_j} f_{\mu_j}\pi')^{\#}(c(f(\mu_j),...,f(\mu_s)) \times \Delta^j), \tag{7}$$

where $\varepsilon(j) = \sum_{i=0}^{j}(-1)^i$.

On the other hand,

$$(-1)^{m+1}(\delta(H^m c))(\boldsymbol{\mu}) = (-1)^{m+1}\sum_{i=0}^{s}(-1)^i q^{\#}(H^m c)(d^i(\underline{H}_m(\boldsymbol{X};G))) =$$
$$\sum_{i=0}^{s}\sum_{j=0}^{i-1}(-1)^{sj+s+i+1}(qf_{\mu_j}\pi')^{\#}(c(f(\mu_j),...,\widehat{f(\mu_i)},...,f(\mu_s)) \times \Delta^{j+1}) +$$
$$\sum_{i=0}^{s-1}\sum_{j=i+1}^{s}(-1)^{sj+i+1}(qf_{\mu_j}\pi')^{\#}(c(f(\mu_j),...,f(\mu_s)) \times \Delta^j). \tag{8}$$

Now note that

$$\{(i,j)|\, 0 \le i \le s-1, i+1 \le j \le s\} = \{(i,j)|\, 1 \le j \le s, 0 \le i \le j-1\}. \tag{9}$$

Therefore, interchanging the order of summation, the second double sum of (8) equals

$$\sum_{j=1}^{s}(-1)^{sj+1}\epsilon(j-1)(qf_{\mu_j}\pi')^{\#}(c(f(\mu_j),...,f(\mu_s)) \times \Delta^j). \tag{10}$$

In (7) we can omit the term $j = 0$, because $\varepsilon(1) = 0$. Since $\varepsilon(j+1) + \varepsilon(j-1) = 1 + (-1)^{j+1}$, we conclude that the sum of (7) and (10) equals

$$\begin{aligned} &- \sum_{j=1}^{s}(-1)^{sj}(q_{\mu_0\mu_j} f_{\mu_j}\pi')^{\#}(c(f(\mu_j),...,f(\mu_s)) \times \Delta^j) \\ &+ \sum_{j=1}^{s}(-1)^{sj+j}(q_{\mu_0\mu_j} f_{\mu_j}\pi')^{\#}(c(f(\mu_j),...,f(\mu_s)) \times \Delta^j). \end{aligned} \tag{11}$$

Taking into account (2), (3) and the fact that $q_{\mu_0\mu_j} f_{\mu_j} = f_{\mu_0} p_{f(\mu_0)f(\mu_j)}$, one sees that the first sum in (11) equals $(f^m(c))(\boldsymbol{\mu}) - (g^m(c))(\boldsymbol{\mu})$. Consequently,

$$\begin{aligned}
&(d(H^m c))(\boldsymbol{\mu}) = (f^m(c))(\boldsymbol{\mu}) - (g^m(c))(\boldsymbol{\mu}) + \\
&(-1)^{m+1}\sum_{j=0}^{s}(-1)^{sj+s+j}(q_{\mu_0\mu_j}f_{\mu_j}\pi')^{\#}(\partial c(f(\mu_j),...,f(\mu_s))\times \Delta^{j+1}) + \\
&\sum_{i=1}^{s}\sum_{j=0}^{i-1}(-1)^{sj+s+i+1}(q_{\mu_0\mu_j}f_{\mu_j}\pi')^{\#}(c(f(\mu_j),...,\widehat{f(\mu_i)},...,f(\mu_s))\times \Delta^{j+1}) + \\
&\sum_{j=1}^{s}(-1)^{sj+j}(q_{\mu_0\mu_j}f_{\mu_j}\pi')^{\#}(c(f(\mu_j),...,f(\mu_s))\times \Delta^{j}).
\end{aligned} \tag{12}$$

On the other hand,

$$\begin{aligned}
&(H^{m-1}(dc))(\boldsymbol{\mu}) = \\
&(-1)^{m}\sum_{j=0}^{s}(-1)^{sj+s+j}(q_{\mu_0\mu_j}f_{\mu_j}\pi')^{\#}((dc)(f(\mu_j),...,f(\mu_s))\times \Delta^{j+1}) + \\
&(-1)^{m}\sum_{j=0}^{s}(-1)^{sj+s+j}(q_{\mu_0\mu_j}f_{\mu_j}\pi')^{\#}((\partial c)(f(\mu_j),...,f(\mu_s)) + \\
&\sum_{j=0}^{s-1}\sum_{i=j}^{s}(-1)^{sj+s+i}(q_{\mu_0\mu_j}f_{\mu_j}p\pi')^{\#}c(f(\mu_j),...,\widehat{f(\mu_i)},...,f(\mu_s))\times \Delta^{j+1}).
\end{aligned} \tag{13}$$

We now add up (12) and (13). The first sums cancel. The range of the double sum in (12) can also be described by $0 \le j \le s-1,\ j+1 \le i \le s$. Therefore, the double sum of (12) cancels all the terms of the double sum of (13), except for the terms with $i = j$, which yield

$$\sum_{j=0}^{s-1}(-1)^{sj+s+j}(q_{\mu_0\mu_j}f_{\mu_j}p_{\mu_j\mu_{j+1}}\pi')^{\#}c(f(\mu_{j+1}),...,f(\mu_s))\times \Delta^{j+1}). \tag{14}$$

Replacing $j+1$ by j in the latter sum we obtain

$$\sum_{j=1}^{s}(-1)^{sj+j+1}(q_{\mu_0\mu_j}f_{\mu_j}\pi')^{\#}c(f(\mu_j),...,f(\mu_s))\times \Delta^{j}). \tag{15}$$

However, this sum cancels the last term of (12). This completes the proof of the first assertion. The second assertion is an immediate consequence of the first assertion and the fact that the mapping $\mathbf{1}_X$ induces the identity homomorphism (see 17.1). □

The above lemmas enable us to define the *strong homology functors* on CH(pro-Top). Indeed, if $[\boldsymbol{f}]: \boldsymbol{X} \to \boldsymbol{Y}$ is a morphism of CH(pro-Top), i.e., a homotopy class of coherents mappings $\boldsymbol{f}: \boldsymbol{X} \to \boldsymbol{Y}$, we assign to $[\boldsymbol{f}]$ the homomorphism $f_*: \overline{H}_m(\boldsymbol{X};G) \to \overline{H}_m(\boldsymbol{Y};G)$, induced by the chain mapping $f^{\#}: TS(\boldsymbol{X};G) \to TS(\boldsymbol{Y};G)$. By Lemma 18.6, f_* depends only on the homotopy class $[\boldsymbol{f}]$. Therefore, $[\boldsymbol{f}]_* = f_*$ are well-defined homomorphisms $\overline{H}_m(\boldsymbol{X};G) \to \overline{H}_m(\boldsymbol{Y};G)$.

THEOREM 18.12. *The function which associates with* $\boldsymbol{X}$ *the strong homology group* $\overline{H}_m(\boldsymbol{X};G)$ *and associates with* $[\boldsymbol{f}]: \boldsymbol{X} \to \boldsymbol{Y}$ *the homomorphism* $f_*: \overline{H}_m(\boldsymbol{X};G) \to \overline{H}_m(\boldsymbol{Y};G)$ *is a functor* $\overline{H}_m(\,.\,;G)$: CH(pro-Top) $\to$ Ab.

Proof. Let $[\boldsymbol{f}]: \boldsymbol{X} \to \boldsymbol{Y}$ and $[\boldsymbol{g}]: \boldsymbol{Y} \to \boldsymbol{Z}$ be morphisms of CH(pro-Top). It follows from Lemma 18.7 that $([\boldsymbol{g}][\boldsymbol{f}])_* = [\boldsymbol{g}]_*[\boldsymbol{f}]_*$. Moreover, for the identity morphism $[C(\mathbf{1}_{\boldsymbol{X}})]$ of CH(pro-Top), Lemma 18.11 implies $[C(\mathbf{1}_{\boldsymbol{X}})]_* = \mathrm{id}$. □

Bibliographic notes

This section is based on (Lisica, Mardešić 1985d, 1985e). The choice of signs in the explicit formulae used in these papers differs from the present choice. This is due to the fact that, in expressions like $X_\lambda \times \Delta^n$, the factors were interchanged. While this ordering is irrelevant for strong shape, it affects homology, because chain multiplication is commutative only up to a sign.

19. Strong homology of spaces

In this section we define strong homology groups of spaces and prove that they are invariants of strong shape. The construction of these groups uses in an essential way strong expansions and coherent homotopy. The definition is also extended to pairs of spaces and it is shown that, for a very broad class of pairs (normal pairs), strong homology groups satisfy all the *Eilenberg – Steenrod axioms*, including exactness. We also establish some specific properties of strong homology, in particular, strong excision and the cluster property. For nice spaces, e.g., for polyhedra, strong homology coincides with singular homology.

19.1 Strong homology groups of spaces

In 17.1 we have defined strong homology groups $\overline{H}_m(\boldsymbol{X};G)$ of an inverse system of spaces. Moreover, in 18, with every coherent mapping $\boldsymbol{f}:\boldsymbol{X}\to\boldsymbol{Y}$ we have associated homomorphisms $f_*:\overline{H}_m(\boldsymbol{X};G)\to\overline{H}_m(\boldsymbol{Y};G)$. We will now define *strong homology groups of spaces* $\overline{H}_m(X;G)$ and homomorphisms of such groups $F_*:\overline{H}_m(X;G)\to\overline{H}_m(Y;G)$, induced by strong shape morphisms $F:X\to Y$.

Given a space X, choose a cofinite strong HPol - expansion $\boldsymbol{p}:X\to\boldsymbol{X}$ of X, e.g., a cofinite polyhedral or ANR - resolution. Existence of such expansions is insured by Theorems 6.22, 6.23, 7.6 and Lemmas 6.29, 6.31. Then put

$$\overline{H}_m(X;G)=\overline{H}_m(\boldsymbol{X};G). \tag{1}$$

These groups are completely determined by X, m and G and do not depend on the choice of the expansion $\boldsymbol{p}$. Indeed, if $\boldsymbol{p}':X\to\boldsymbol{X}'$ is another cofinite strong HPol - expansion, then by Theorem 8.1, there exists a unique morphism $[\boldsymbol{i}_{\boldsymbol{X}\boldsymbol{X}'}]:\boldsymbol{X}\to\boldsymbol{X}'$ of CH(pro - Top) such that $[\boldsymbol{i}_{\boldsymbol{X}\boldsymbol{X}'}][C(\boldsymbol{p})]=[C(\boldsymbol{p}')]$. Similarly, there is a unique morphism $[\boldsymbol{i}_{\boldsymbol{X}'\boldsymbol{X}}]:\boldsymbol{X}'\to\boldsymbol{X}$ such that $[\boldsymbol{i}_{\boldsymbol{X}'\boldsymbol{X}}][C(\boldsymbol{p}')]=[C(\boldsymbol{p})]$. By uniqueness, $[\boldsymbol{i}_{\boldsymbol{X}'\boldsymbol{X}}][\boldsymbol{i}_{\boldsymbol{X}\boldsymbol{X}'}]=\mathrm{id}$, $[\boldsymbol{i}_{\boldsymbol{X}\boldsymbol{X}'}][\boldsymbol{i}_{\boldsymbol{X}'\boldsymbol{X}}]=\mathrm{id}$, which shows that $[\boldsymbol{i}_{\boldsymbol{X}\boldsymbol{X}'}]$ is an isomorphism in CH(pro - Top). Therefore, by Theorem 18.12, $[\boldsymbol{i}_{\boldsymbol{X}\boldsymbol{X}'}]$ induces an isomorphism $[\boldsymbol{i}_{\boldsymbol{X}\boldsymbol{X}'}]_*$, which identifies the groups $\overline{H}_m(\boldsymbol{X};G)$ and $\overline{H}_m(\boldsymbol{X}';G)$.

Now let $F: X \to Y$ be a strong shape morphism. According to 8.2, F is represented by a triple $(\boldsymbol{p}, \boldsymbol{q}, [\boldsymbol{f}])$, where $\boldsymbol{p}: X \to \boldsymbol{X}$, $\boldsymbol{q}: Y \to \boldsymbol{Y}$ are cofinite strong HPol-expansions of X and Y, respectively and $[\boldsymbol{f}]: \boldsymbol{X} \to \boldsymbol{Y}$ is a morphism of CH(pro-Top). Consider the homomorphisms $f_*: \overline{H}_m(\boldsymbol{X}; G) \to \overline{H}_m(\boldsymbol{Y}; G)$, induced by the morphism $[\boldsymbol{f}]$. If F is also represented by a triple $(\boldsymbol{p}', \boldsymbol{q}', [\boldsymbol{f}'])$, where $\boldsymbol{p}': X \to \boldsymbol{X}'$, $\boldsymbol{q}': Y \to \boldsymbol{Y}'$, $[\boldsymbol{f}']: \boldsymbol{X}' \to \boldsymbol{Y}'$, then $[\boldsymbol{f}'][\boldsymbol{i}_{\boldsymbol{X}\boldsymbol{X}'}] = [\boldsymbol{f}][\boldsymbol{i}_{\boldsymbol{Y}\boldsymbol{Y}'}]$ and therefore, $[\boldsymbol{f}']_*[\boldsymbol{i}_{\boldsymbol{X}\boldsymbol{X}'}]_* = [\boldsymbol{f}]_*[\boldsymbol{i}_{\boldsymbol{Y}\boldsymbol{Y}'}]_*$, which shows that the isomorphisms, which identify the homology groups $\overline{H}_m(\boldsymbol{X}; G)$ with $\overline{H}_m(\boldsymbol{X}'; G)$ and $\overline{H}_m(\boldsymbol{Y}; G)$ with $\overline{H}_m(\boldsymbol{Y}'; G)$, also identify the homomorphisms $[\boldsymbol{f}]_*$ with $[\boldsymbol{f}']_*$ and thus, yield a well-defined homomorphism $F_* = f_*: \overline{H}_m(\boldsymbol{X}; G) \to \overline{H}_m(\boldsymbol{Y}; G)$. We refer to $F_* = [\boldsymbol{f}]_*$ as to the *homomorphism induced by the strong shape morphism* F. Clearly, $(GF)_* = G_* F_*$ and $1_{X*} = \mathrm{id}$. Consequently, we have the following theorem.

THEOREM 19.1. *The function, which with a space X associates the strong homology group $\overline{H}_m(X; G)$ and with a strong shape morphism $F: X \to Y$ associates the homomorphism $F_*: \overline{H}_m(X; G) \to \overline{H}_m(Y; G)$, is a functor $\overline{H}_m(\,.\,; G)$: SSh(Top) $\to$ Ab.*

Proof. The assertion is an immediate consequence of the definitions and of Theorem 18.12. □

Every mapping $f: X \to Y$ induces a homomorphism of strong homology groups $f_* = F_*: \overline{H}_m(X; G) \to \overline{H}_m(Y; G)$, where F_* is the homomorphism induced by the strong shape morphism

$$F = \overline{S}[f]: X \to Y. \tag{2}$$

Here $[f]$ denotes the homotopy class of f and $\overline{S}$ is the strong shape functor $\overline{S}$:H(Top)$\to$ SSh(Top) (see 8.2). An immediate consequence of this definition and of Theorem 19.1 is the following result.

COROLLARY 19.2. *The function which to a space X assigns the strong homology group $\overline{H}_m(X; G)$ and to a mapping $f: X \to Y$ assigns the induced homomorphism $f_* = \overline{H}_m(f): \overline{H}_m(X; G) \to \overline{H}_m(Y; G)$ is a functor $\overline{H}_m(\,.\,; G)$: Top $\to$ Ab. If the mappings are homotopic, $f \simeq f'$, then $f_* = f'_*$. If the mappings induce the same morphism of strong shape, then also $f_* = f'_*$.*

We now define *relative strong homology groups* $\overline{H}_m(X, A; G)$, i.e., *strong homology groups of pairs* of spaces (X, A), $A \subseteq X$. The construction is analogous to the one used in the absolute case, i.e., in the case of spaces X. Associate with (X, A) a cofinite strong HPol_2-expansion $\boldsymbol{p}: (X, A) \to (\boldsymbol{X}, \boldsymbol{A}) = ((X_\lambda, A_\lambda), p_{\lambda\lambda'}, \Lambda)$ (see Remark 7.20), e.g., a cofinite strong ANR_2-resolution (see 6.5). Also consider the systems $\boldsymbol{X} = (X_\lambda, p_{\lambda\lambda'}, \Lambda)$, $\boldsymbol{A} = (A_\lambda, p_{\lambda\lambda'}, \Lambda)$ and the natural inclusion $\boldsymbol{i}: \boldsymbol{A} \to \boldsymbol{X}$. Clearly, $\boldsymbol{i}$ embeds the singular pro-chain complex $S(\boldsymbol{A})$ as a pro-chain subcomplex of $S(\boldsymbol{X})$. Let $S(\boldsymbol{X}, \boldsymbol{A}) = S(\boldsymbol{X})/S(\boldsymbol{A})$ be the corresponding quotient pro-chain complex, which consists of quotients

$S(X_\lambda)/S(A_\lambda)$. Consider the pro-chain complex $S(\boldsymbol{X},\boldsymbol{A};G) = S(\boldsymbol{X},\boldsymbol{A}) \otimes G$. Finally, define $\overline{H}_m(X,A;G)$ as the homology group of the total complex $T(S(\boldsymbol{X},\boldsymbol{A};G))$ (see 17.1).

Similarly, if $F\colon (X,A) \to (Y,B)$ is a strong shape morphism of pairs of spaces (see 8.2), given by a triple $(\boldsymbol{p},\boldsymbol{q},\boldsymbol{f})$, where $\boldsymbol{p},\boldsymbol{q}$ are cofinite strong HPol_2-expansions of (X,A) and (Y,B), respectively, and $\boldsymbol{f}\colon (\boldsymbol{X},\boldsymbol{A}) \to (\boldsymbol{Y},\boldsymbol{B})$ is a coherent mapping for pairs (see 2.3), then $\boldsymbol{f}$ induces a morphism $\boldsymbol{f}^\#$ of pro-chain complexes $S(\boldsymbol{X},\boldsymbol{A}) \otimes G \to S(\boldsymbol{Y},\boldsymbol{B}) \otimes G$, which in turn induces a chain mapping of the corresponding total complexes and thus, induces homomorphisms of the corresponding homology groups $f_*\colon \overline{H}_m(X,A;G) \to \overline{H}_m((Y,B);G)$. These homomorphisms are independent of the particular choice of the expansions and yield functors $\mathrm{SSh}(\mathsf{Top}_2) \to \mathsf{Ab}$ (for more details see (Mardešić 1991c)).

Strong homology groups of pointed spaces $\overline{H}_m(X,*;G)$ are defined by viewing $(X,*)$ as a special pair of spaces.

REMARK 19.3. In full analogy with $\overline{H}_m(X,*;G)$, one can also define *strong homotopy groups* $\overline{\pi}_m(X,*)$ of pointed topological spaces. Various general results proved for strong homology groups remain valid also for strong homotopy groups. In particular, this is true for groups in dimensions $m \geq 2$, because homotopy groups in these dimensions are abelian groups. For more details on the groups $\overline{\pi}_m(X,*)$ see (Mohorianu 1997).

REMARK 19.4. For a pair of spaces (X,A), also consider the pair $(X,\overline{A})$, where $\overline{A}$ denotes the closure of A. Let $i\colon (X,A) \to (X,\overline{A})$ be the corresponding inclusion map. Then i induces an isomorphism of strong homology groups $i_*\colon \overline{H}_m(X,A) \to \overline{H}_m(X,\overline{A})$. This is an immediate consequence of the fact that, whenever $\boldsymbol{p}\colon (X,A) \to (\boldsymbol{X},\boldsymbol{A})$ is a polyhedral or an ANR-resolution of (X,A) (more generally, a resolution where A_λ is closed in X_λ), then $\boldsymbol{p}$ can be viewed as a resolution of $(X,\overline{A})$. Therefore, $\overline{H}_m(X,\overline{A};G) = \overline{H}_m(\boldsymbol{X},\boldsymbol{A};G) = \overline{H}_m(X,A;G)$.

An essential feature of strong homology is the validity of the *exactness axiom*, i.e., the validity of the following assertion.

THEOREM 19.5. *If (X,A) is a normal pair of spaces, there exist functorial homomorphisms $\partial\colon \overline{H}_m(X,A;G) \to \overline{H}_{m-1}(A;G)$ such that the following sequence is exact,*

$$\ldots \to \overline{H}_m(A;G) \xrightarrow{i_*} \overline{H}_m(X;G) \xrightarrow{j_*} \overline{H}_m(X,A;G) \xrightarrow{\partial} \overline{H}_{m-1}(A;G) \to \ldots, \tag{3}$$

here i_* and j_* are induced by natural inclusions.

Recall that in collectionwise normal spaces X, in particular in paracompact spaces X, all pairs (X,A) with A closed are normal pairs (see e.g., (Mardešić, Segal 1982), I.6.5). Therefore, the restrictions in Theorem 19.5 are very mild.

The proof uses the following lemma.

LEMMA 19.6. *For every inverse system of pairs $(\boldsymbol{X}, \boldsymbol{A})$ and every abelian group G,*

$$0 \to T^m S(\boldsymbol{A}; G) \xrightarrow{i^\#} T^m S(\boldsymbol{X}; G) \xrightarrow{j^\#} T^m S(\boldsymbol{X}, \boldsymbol{A}; G) \to 0, \tag{4}$$

is an exact sequence of chain complexes.

Proof. For every $\lambda_0 \in \Lambda$, one has the exact sequence of groups,

$$0 \to S^{m+s}(A_{\lambda_0}) \xrightarrow{i} S^{m+s}(X_{\lambda_0}) \xrightarrow{j} S^{m+s}(X_{\lambda_0})/S^{m+s}(A_{\lambda_0}) \to 0. \tag{5}$$

Since $S^{m+s}(A_{\lambda_0})$ is a free abelian group, (5) implies exactness of the sequence

$$0 \to S^{m+s}(A_{\lambda_0}) \otimes G \xrightarrow{i\otimes 1} S^{m+s}(X_{\lambda_0}) \otimes G \xrightarrow{j\otimes 1} S^{m+s}(X_{\lambda_0}, A_{\lambda_0}) \otimes G \to 0. \tag{6}$$

Taking direct products of copies of (6), for $\boldsymbol{\lambda} \in \Lambda_s$ and $s \geq 0$, one obtains the desired exact sequence (4). □

Proof of Theorem 19.5. Let (X, A) be a normal pair of spaces. Then there exists a polyhedral resolution $\boldsymbol{p}: (X, A) \to (\boldsymbol{X}, \boldsymbol{A})$ such that $\boldsymbol{p}: X \to \boldsymbol{X}$ and $\boldsymbol{p}: A \to \boldsymbol{A}$ are also polyhedral resolutions (see Remark 6.39 or (Mardešić, Segal 1982), I.6.5). Lemma 19.6, applied to $(\boldsymbol{X}, \boldsymbol{A})$ yields a short exact sequence of chain complexes (4). However, every such sequence generates a long exact sequence of homology groups. To complete the proof it suffices to show that the obtained sequence coincides with the sequence (3). This is indeed the case, because the strong groups of $\boldsymbol{A}$, $\boldsymbol{X}$ and $(\boldsymbol{X}, \boldsymbol{A})$ are the homology groups of the chain complexes $T^m S(\boldsymbol{A}; G)$, $T^m S(\boldsymbol{X}; G)$ and $T^m S(\boldsymbol{X}, \boldsymbol{A}; G)$, respectively. □

EXAMPLE 19.7. The following example shows that Theorem 19.5 does not hold for arbitrary pairs (X, A). Let $X = B^m$ be the m-cell, $m \geq 2$, with boundary $\partial B^m = S^{m-1}$. Choose a point $* \in S^{m-1}$ and put $A = \partial B^m \setminus \{*\}$. By Remark 19.4, $\overline{H}_m(X, A; \mathbb{Z}) \approx \overline{H}_m(X, \overline{A}; \mathbb{Z})$. However, $(X, \overline{A}) = (B^m, S^{m-1})$. We will show in 19.4 that for polyhedral pairs, strong homology coincides with singular homology. Therefore, $\overline{H}_m(X, A) = H_m(B^m, S^{m-1}) \approx \mathbb{Z}$. Moreover, $\overline{H}_m(X) = H_m(B^m) = 0$ and $\overline{H}_{m-1}(A) = 0$, because $A \approx \mathbb{R}^{m-1}$. Consequently, the sequence (3) cannot be exact.

EXAMPLE 19.8. It is well known that Čech homology is not exact, even for compact metric pairs. To give a specific example, consider the unit disc $B^2 = \{z \in \mathbb{C}: |z| \leq 1\}$ and its boundary $S^1 = \partial B^2$. Let $f: (B^2, S^1) \to (B^2, S^1)$ be the mapping given by $f(z) = z^3$. Let P^2 be the projective plane obtained from B^2 by identifying pairs of points $z, -z$, where $z \in S^1$. The quotient mapping $q: B^2 \to P^2$ maps S^1 to the projective line $P^1 \approx S^1$. Consider the inverse sequence $(\boldsymbol{X}, \boldsymbol{A}) = ((X_n, A_n), p_{nn'}, \mathbb{N})$, where all $(X_n, A_n) = (P^2, P^1)$, $p_{nn'} = p^{n'-n}$ and $p: (P^2, P^1) \to (P^2, P^1)$ is the mapping induced by f. Let $(X, A) = \lim(\boldsymbol{X}, \boldsymbol{A})$. The first Čech homology groups with integer coefficients assume the values $\check{H}_1(A) = \check{H}_1(X, A) = 0$ and $\check{H}_1(X) \approx \mathbb{Z}/2$. Therefore, the sequence $\check{H}_1(A) \to \check{H}_1(X) \to \check{H}_1(X, A)$ cannot be exact (for more details see e.g., (Mardešić, Segal 1982), II.3.1, Example 2).

19.2 Strong excision property

The goal of this subsection is to prove that strong homology groups have the *strong excision property*, i.e., the following theorem holds.

THEOREM 19.9. *Let (X, A) be a pair of topological spaces, $A \neq \emptyset$, and let $q: (X, A) \to (X, *)$ be the quotient mapping which collapses A to a single point $*$. Then the induced homomorphisms $q_*: \overline{H}_m(X, A; G) \to \overline{H}_m(X/A, *; G)$ are isomorphisms, for all $m \in \mathbb{Z}$.*

The proof uses the following lemma on resolutions.

LEMMA 19.10. *Let $\boldsymbol{p} = (p_\lambda): (X, A) \to (\boldsymbol{X}, \boldsymbol{A}) = ((X_\lambda, A_\lambda), p_{\lambda\lambda'}, \Lambda)$ be a resolution of the pair (X, A), $A \neq \emptyset$. If $X' = X/A$, $X'_\lambda = X_\lambda/A_\lambda$ and $p'_{\lambda\lambda'}: (X'_{\lambda'}, *) \to (X'_\lambda, *)$, $\lambda \leq \lambda'$, $p'_\lambda: (X', *) \to (X'_\lambda, *)$ are mappings induced by $p_{\lambda\lambda'}$ and p_λ, respectively, then $(\boldsymbol{X}', *) = ((X'_\lambda, *), p'_{\lambda\lambda'}, \Lambda)$ is an inverse system of pointed spaces and $\boldsymbol{p}' = (p'_\lambda): (X', *) \to (\boldsymbol{X}', *)$ is a resolution. Moreover, if $q_\lambda: X_\lambda \to X'_\lambda$, $\lambda \in \Lambda$, and $q: X \to X'$ are the respective quotient mappings, then $\boldsymbol{q} = (q_\lambda): (\boldsymbol{X}, \boldsymbol{A}) \to (\boldsymbol{X}', *)$ is a level mapping and $\boldsymbol{qp} = \boldsymbol{p}'q$.*

Proof. That $(\boldsymbol{X}', *)$ is an inverse system, $\boldsymbol{q}$ is a level mapping and $\boldsymbol{qp} = \boldsymbol{p}'q$, immediately follows from the definitions. It remains to prove that $\boldsymbol{p}'$ is a resolution, i.e., $\boldsymbol{p}'$ has properties (B1) and (B2) (see 6.2).

Verification of (B1). For a normal covering $\mathcal{U}' \in \mathrm{Cov}(X')$, choose a normal star-refinement $\mathcal{U}'_1 \in \mathrm{Cov}(X')$,

$$\mathrm{St}(\mathcal{U}'_1) \prec \mathcal{U}'. \tag{1}$$

Application of (B1) for $\boldsymbol{p}$ to

$$\mathcal{U}_1 = q^{-1}(\mathcal{U}'_1) \tag{2}$$

yields a $\mu \in \Lambda$ and a $\mathcal{U}_\mu \in \mathrm{Cov}(X_\mu)$ such that

$$p_\mu^{-1}(\mathcal{U}_\mu) \prec \mathcal{U}_1. \tag{3}$$

Choose $\mathcal{U}_{\mu 1} \in \mathrm{Cov}(X_\mu)$ such that

$$\mathrm{St}(\mathcal{U}_{\mu 1}) \prec \mathcal{U}_\mu. \tag{4}$$

By property (B2)* for $\boldsymbol{p}$ (see 6.5), there exists a $\lambda \geq \mu$ such that

$$p_{\mu\lambda}(A_\lambda) \subseteq \mathrm{St}(p_\mu(A), \mathcal{U}_{\mu 1}). \tag{5}$$

Define $\mathcal{U}_\lambda \in \mathrm{Cov}(X_\lambda)$, by putting

$$\mathcal{U}_\lambda = (p_{\mu\lambda})^{-1}(\mathcal{U}_{\mu 1}). \tag{6}$$

Finally, let $\mathcal{U}'_\lambda$ be the open covering of X'_λ, which consists of all sets $U'_\lambda = q_\lambda(U_\lambda)$, where $U_\lambda \in \mathcal{U}_\lambda$ and $U_\lambda \cap A_\lambda = \emptyset$, and of the set $U'_* = q_\lambda(U_*)$, where $U_* = \mathrm{St}(A_\lambda, \mathcal{U}_\lambda)$. We will show that

$$(p'_\lambda)^{-1}(\mathcal{U}'_\lambda) \prec \mathcal{U}'. \tag{7}$$

First consider members $U'_\lambda = q_\lambda(U_\lambda)$ of $\mathcal{U}'_\lambda$, where $U_\lambda \in \mathcal{U}_\lambda$, $U_\lambda \cap A_\lambda = \emptyset$. By (6), U_λ is of the form $(p_{\mu\lambda})^{-1}(U_{\mu 1})$, where $U_{\mu 1} \in \mathcal{U}_{\mu 1}$. By (4), there exists a member $U_\mu \in \mathcal{U}_\mu$ such that $U_{\mu 1} \subseteq U_\mu$ and thus,

$$p_\lambda^{-1}(U_\lambda) \subseteq p_\mu^{-1}(U_\mu). \tag{8}$$

By (3), there exists a member $U_1 \in \mathcal{U}_1$ such that $p_\mu^{-1}(U_\mu) \subseteq U_1$. Moreover, by (2), there exists a member $U'_1 \in \mathcal{U}'_1$ such that $U_1 = q^{-1}(U'_1)$ and thus,

$$qp_\lambda^{-1}(U_\lambda) \subseteq U'_1. \tag{9}$$

If we show that

$$(p'_\lambda)^{-1}(U'_\lambda) \subseteq qp_\lambda^{-1}(U_\lambda), \tag{10}$$

(9) and (1) will imply

$$(p'_\lambda)^{-1}(U'_\lambda) \subseteq U', \tag{11}$$

for some $U' \in \mathcal{U}'$.

To prove (10), consider a point $x' \in (p'_\lambda)^{-1}(U'_\lambda)$. Since $U'_\lambda = q_\lambda(U_\lambda)$, there exists a point $x_\lambda \in U_\lambda$ such that $p'_\lambda(x') = q_\lambda(x_\lambda)$ and since q is a surjection, there exists a point $x \in X$ such that $q(x) = x'$. Taking into account that $q_\lambda p_\lambda = p'_\lambda q$, one concludes that $q_\lambda p_\lambda(x) = q_\lambda(x_\lambda)$. Note that $x_\lambda \in U_\lambda$ and $U_\lambda \cap A_\lambda = \emptyset$ implies $q_\lambda(x_\lambda) \neq *$ and therefore, $p_\lambda(x) \notin A_\lambda$. Since $q_\lambda | X_\lambda \backslash A_\lambda$ is injective, one concludes that $p_\lambda(x) = x_\lambda \in U_\lambda$. This shows that $x \in p_\lambda^{-1}(U_\lambda)$ and $x' = q(x) \in qp_\lambda^{-1}(U_\lambda)$, which establishes (10).

To complete the proof of (7), it remains to exhibit a member $U' \in \mathcal{U}'$ such that

$$(p'_\lambda)^{-1}(q_\lambda(U_*)) \subseteq U'. \tag{12}$$

First note that (6), (5) and (4) imply

$$p_{\mu\lambda}(U_*) \subseteq p_{\mu\lambda}(\mathrm{St}(A_\lambda, \mathcal{U}_\lambda) \subseteq \mathrm{St}(p_\mu(A), \mathrm{St}(\mathcal{U}_{\mu 1})) \subseteq \mathrm{St}(p_\mu(A), \mathcal{U}_\mu), \tag{13}$$

which yields

$$p_\lambda^{-1}(U_*) \subseteq p_\lambda^{-1} p_{\mu\lambda}{}^{-1}(\mathrm{St}(p_\mu(A), \mathcal{U}_\mu) = p_\mu^{-1}(\mathrm{St}(p_\mu(A), \mathcal{U}_\mu)). \tag{14}$$

Now note that

$$p_{\mu}^{-1}(\mathrm{St}(p_{\mu}(A),\mathcal{U}_{\mu}))\subseteq \mathrm{St}(A,p_{\mu}^{-1}(\mathcal{U}_{\mu}))\subseteq \mathrm{St}(A,\mathcal{U}_1)=\mathrm{St}(A,q^{-1}(\mathcal{U}_1')). \quad (15)$$

Therefore, (14) implies

$$p_{\lambda}^{-1}(U_*)\subseteq \mathrm{St}(A,q^{-1}(\mathcal{U}_1')). \quad (16)$$

Taking into account the fact that q and q_λ are surjections and $q_\lambda p_\lambda = p'_\lambda q$, one concludes that

$$(p'_\lambda)^{-1}q_\lambda(U_*) = qq^{-1}(p'_\lambda)^{-1}q_\lambda(U_*) = qp_\lambda^{-1}q_\lambda^{-1}q_\lambda(U_*) = qp_\lambda^{-1}(U_*). \quad (17)$$

Note that (16) and (17) yield

$$(p'_\lambda)^{-1}q_\lambda(U_*)\subseteq q(\mathrm{St}(A,q^{-1}(\mathcal{U}_1'))). \quad (18)$$

By (2) and (1), there exists a $U'\in\mathcal{U}'$ such that

$$q(\mathrm{St}(A,q^{-1}(\mathcal{U}_1')))\subseteq \mathrm{St}(*,\mathcal{U}_1')\subseteq U'. \quad (19)$$

Now, (18) and (19) yield the desired relation (12).

It is readily seen that the above described construction, which to $\mathcal{U}_\lambda$ assigned $\mathcal{U}'_\lambda$, preserves star-refinements. Since $\mathcal{U}_\lambda$ was a normal covering, it admits a sequence of consecutive star-refinements. Consequently, $\mathcal{U}'_\lambda$ also admits such a sequence, which shows that $\mathcal{U}'_\lambda$ is a normal covering (see e.g., (Mardešić, Segal 1982), Appendix 1.3, Remark 1).

Verification of (B2). For $\lambda\in\Lambda$ and $\mathcal{U}'_\lambda\in \mathrm{Cov}(X'_\lambda)$, let

$$\mathcal{U}_\lambda = q_\lambda^{-1}(\mathcal{U}'_\lambda). \quad (20)$$

By (B2) for $\boldsymbol{p}$, there exists a $\mu\geq\lambda$ such that

$$p_{\lambda\mu}(X_\mu)\subseteq \mathrm{St}(p_\lambda(X),\mathcal{U}_\lambda). \quad (21)$$

Let us show that

$$p'_{\lambda\mu}(X'_\mu)\subseteq \mathrm{St}(p'_\lambda(X'),\mathcal{U}'_\lambda). \quad (22)$$

Note that

$$p'_{\lambda\mu}(X'_\mu) = p'_{\lambda\mu}q_\mu(X_\mu) = q_\lambda p_{\lambda\mu}(X_\mu), \quad (23)$$

$$p'_\lambda(X') = p'_\lambda q(X) = q_\lambda p_\lambda(X). \quad (24)$$

Therefore, (22) is equivalent to

$$q_\lambda p_{\lambda\mu}(X_\mu)\subseteq \mathrm{St}(q_\lambda p_\lambda(X),\mathcal{U}'_\lambda). \quad (25)$$

However, (21) and (20) imply (25), because

$$q_\lambda p_{\lambda\mu}(X_\mu)\subseteq q_\lambda(\mathrm{St}(p_\lambda(X),\mathcal{U}_\lambda))\subseteq \mathrm{St}(q_\lambda p_\lambda(X),\mathcal{U}'_\lambda).\ \square \quad (26)$$

Proof of Theorem 19.9. For the pair (X,A) choose a cofinite polyhedral resolution $\boldsymbol{p}\colon (X,A)\to(\boldsymbol{X},\boldsymbol{A})$. Let $(\boldsymbol{X}',*)$, $\boldsymbol{p}'$, $\boldsymbol{q}$ and q be as in Lemma 19.10. Then $\boldsymbol{p}'$ is a cofinite resolution. Since every $X'_\lambda = X_\lambda/A_\lambda$ is a CW-complex (even a polyhedron), $\boldsymbol{p}'$ is a CW - resolution. Moreover, the quotient mappings $q_\lambda: (X_\lambda, A_\lambda) \to (X'_\lambda, *)$ induce isomorphisms of singular homology groups $q_{\lambda *}: H_m(X_\lambda, A_\lambda; G) \to H_m(X'_\lambda, *; G)$ (see e.g., (Dold 1972), V, Corollary 4.4). Consequently, the level mapping $\boldsymbol{q}: (\boldsymbol{X}, \boldsymbol{A}) \to (\boldsymbol{X}, *)$ induces an isomorphism of homology pro-groups $\boldsymbol{q}_*: H_m(\boldsymbol{X}, \boldsymbol{A}; G) \to H_m(\boldsymbol{X}, *; G)$. Now, an application of Theorem 17.16 yields the conclusion that $\boldsymbol{q}_*$ induces an isomorphism of strong homology groups $\overline{H}_m(\boldsymbol{X}, \boldsymbol{A}; G) \to \overline{H}_m(\boldsymbol{X}, *; G)$. However, by definition, the latter coincides with $q_*: \overline{H}_m(X, A; G) \to \overline{H}_m(X, *; G)$. □

REMARK 19.11. In the above proof one cannot replace polyhedral resolutions by ANR - resolutions, because the quotient space of an ANR - pair need not be an ANR. E.g., for the ANR - pair $(\mathbb{R}, \mathbb{Z})$, the quotient space $\mathbb{R}/\mathbb{Z}$ is not metrizable, because the point $* = q_\lambda(\mathbb{Z})$ does not have a countable basis of neighborhoods.

REMARK 19.12. Čech homology groups also have the strong excision property. As in the proof of Theorem 19.9, one concludes that $\boldsymbol{q}_*: H_m(\boldsymbol{X}, \boldsymbol{A}; G) \to H_m(\boldsymbol{X}, *; G)$ is an isomorphism of pro-groups. Since lim is a functor on inv - Ab, the desired conclusion follows.

EXAMPLE 19.13. Singular homology groups do not have the strong excision property. For instance, if A is the Warsaw circle and X is the corresponding "disc", then $H_2(X; \mathbb{Z}) = 0$, $H_1(A; \mathbb{Z}) = 0$ and thus, by exactness, $H_2(X, A; \mathbb{Z}) = 0$. However, $H_2(X/A; \mathbb{Z}) \approx \mathbb{Z}$, because X/A is homeomorphic to S^2.

A mapping of pairs $f: (X, A) \to (Y, B)$ is called a *relative homeomorphism* if $f: X \to Y$ is a quotient mapping and the restriction $f|X \backslash A$ is a homeomorphism $X \backslash A \to Y \backslash B$ (see e.g., (Whitehead 1978), Ch.I.§5)).

COROLLARY 19.14. *A relative homeomorphism* $f: (X, A) \to (Y, B)$ *induces isomorphisms of groups* $f_*: \overline{H}_m(X, A; G) \to \overline{H}_m(Y, B; G)$.

Proof. Since $f: X \to Y$, $q_X: X \to X/A$ and $q_Y: Y \to Y/B$ are quotient mappings and the induced mapping $f': X/A \to Y/B$ satisfies the equality $f' q_X = q_Y f$, it follows that also f' is a quotient mapping. Moreover, f' is a bijection, because $f|X \backslash A$ is a bijection $X \backslash A \to Y \backslash B$. Therefore, f' is a homeomorphism and thus, it induces isomorphisms f'_* of strong homology groups. By Theorem 19.9, q_X and q_Y also induce isomorphisms of strong homology groups. Since $f'_* q_{X*} = q_{Y*} f_*$, we conclude that f_* is also an isomorphism of groups. □

For a pair of space (X, A) and a subset $U \subseteq A$ the inclusion mapping $i: (X \backslash U, A \backslash U) \to (X, A)$ is well defined and is called the *excision mapping.*

THEOREM 19.15. *If* $U \subseteq A$ *is an open subset of* X *and* $A \backslash U \neq \emptyset$, *then the excision mapping* $i: (X \backslash U, A \backslash U) \to (X, A)$ *induces isomorphisms of strong homology groups.*

Proof. Let $q: (X, A) \to (X/A, *)$, $q_U: (X\backslash U, A\backslash U) \to ((X\backslash U)/(A\backslash U), *)$ be the respective quotient mappings. Clearly, $i: (X\backslash U, A\backslash U) \to (X, A)$ induces a mapping $\tilde{\imath}: ((X\backslash U)/(A\backslash U), *) \to (X/A, *)$ such that $qi = \tilde{\imath} q_U$. The induced homomorphisms between the corresponding strong homology groups satisfy the equality

$$q_* i_* = \tilde{\imath}_* q_{U*}. \tag{27}$$

By Theorem 19.9, q_* and q_{U*} are isomorphisms. Therefore, to complete the proof, it suffices to prove that also $\tilde{\imath}_*$ is an isomorphism. However, this follows from the next lemma, which asserts that $\tilde{\imath}$ is a homeomorphism.

LEMMA 19.16. *Let (X, A), U and i be as in Theorem 19.15. Then the induced mapping $\tilde{\imath}: ((X\backslash U)/(A\backslash U), *) \to (X/A, *)$ is a homeomorphism.*

Proof. We define the inverse $\tilde{\jmath}: (X/A, *) \to ((X\backslash U)/(A\backslash U), *)$ of $\tilde{\imath}$ as follows. For $y = * \in X/A$, put $\jmath(*) = *$. For $y \neq *$, i.e., $y = q(x)$, where $x \in X\backslash A$, put $\tilde{\jmath}(y) = q_U(x)$. Note that $x \notin A$ implies $x \notin U$, because $U \subseteq A$, and thus, x is uniquely determined by y. It is readily seen that $\tilde{\jmath}$ is indeed the inverse of $\tilde{\imath}$. It remains to prove its continuity. Continuity at points $y \neq *$ is obvious. Let us show that $\tilde{\jmath}$ is continuous also at the point $*$.

Let $\tilde{V}$ be an open neighborhood of $*$ in $(X\backslash U)/(A\backslash U)$. Then $V = q_U^{-1}(\tilde{V})$ is an open set in $X\backslash U$, for which

$$A\backslash U \subseteq V. \tag{28}$$

Choose an open set W_1 from X such that

$$W_1 \cap (X\backslash U) = V \tag{29}$$

and consider the set $W = W_1 \cup U$. Since U is open in X, so is W. Note that (28) and (29) imply

$$A\backslash U \subseteq W_1 \cap (X\backslash U) = W \cap (X\backslash U). \tag{30}$$

Since $W \cap U = U$ and $U \subseteq A$, we see that (30) implies

$$A = (A\backslash U) \cup U \subseteq (W \cap (X\backslash U)) \cup (W \cap U) = W. \tag{31}$$

It follows from (31) that $q(W)$ is an open neighborhood of $*$ in X/A. To complete the proof it now suffices to show that

$$\tilde{\jmath}(q(W)) \subseteq \tilde{V}. \tag{32}$$

Let $x \in W$. If $x \in A$, then $q(x) = *$ and thus, $\tilde{\jmath}(q(x)) = \tilde{\jmath}(*) = * \in \tilde{V}$. If $x \in W\backslash A$, then $x \in W\backslash U$, because $U \subseteq A$. Consequently, $x \in W_1 \cap (X\backslash U) = V$. However, in this case, $\tilde{\jmath}(q(x)) = q_U(x) \in q_U(V) \subseteq \tilde{V}$. □

EXAMPLE 19.17. The following example shows that the assertion of Theorem 19.15 does not hold if one omits the assumptions that U is open and $A\backslash U \neq \emptyset$. Let $X = S^1 \vee S^1$ and let $U = A$ be one of the two circles S^1. Then $(X\backslash U, A\backslash U) \approx (\mathbb{R}, \emptyset)$ and thus, $\overline{H}_1(X\backslash U, A\backslash U; \mathbb{Z}) = 0$. However, $X/A \approx S^1$ and thus, $\overline{H}_1(X, A; \mathbb{Z}) \approx \overline{H}_1(X/A; *) \approx \mathbb{Z}$.

REMARK 19.18. The Eilenberg – Steenrod *excision axiom* asserts that the inclusion mapping $i\colon (X\backslash U, A\backslash U) \to (X, A)$ induces isomorphisms of homology groups provided $\mathrm{Cl}(U) \subseteq \mathrm{Int}(U)$. It is well known that singular homology satisfies this axiom (see e.g., (Dold 1972), III, Corollary 7.4).

19.3 Strong homology of clusters

Let $(X^\alpha, *)$, $\alpha \in A$, be a collection of pointed spaces. Their *cluster* (also called *bouquet*) $X = \bigvee_{\alpha \in A} X^\alpha$ is defined as the subspace of the direct product $\prod_{\alpha\in A} X^\alpha$, which consists of points $x = (x_\alpha)$, where at most one of the coordinates $x^\alpha \in X^\alpha$ differs from the base point $* \in X^\alpha$. The cluster X has a natural base-point, the point $*$ all of whose coordinates equal $*$. For every $\alpha \in A$, there is a well-defined mapping $f^\alpha\colon (X, *) \to (X^\alpha, *)$, which maps X^α by identity and maps all other leaves $X^{\alpha'}$, $\alpha' \neq \alpha$, to the base-point $*$. The mappings f^α induce homomorphisms $f^\alpha_*\colon \overline{H}_m(X, *; G) \to \overline{H}(X^\alpha, *; G)$, which in turn determine a homomorphism

$$f_*\colon \overline{H}_m(X, *; G) \to \prod_{\alpha\in A} \overline{H}(X^\alpha, *; G). \tag{1}$$

THEOREM 19.19. *For every collection of pointed spaces $(X^\alpha, *)$, $\alpha \in A$, f_* from* (1) *is an isomorphism.*

The proof uses a particular resolution of the cluster X, which we will now describe. For $\alpha \in A$, let $\boldsymbol{p}^\alpha = (p^\alpha_\beta)$ be a mapping of the pointed space $(X^\alpha, *)$ to a cofinite system of pointed spaces $(\boldsymbol{X}^\alpha, *) = ((X^\alpha_\beta, *), p^\alpha_{\beta\beta'}, B^\alpha)$. We require that $B^\alpha \cap B^{\alpha'} = \emptyset$, for $\alpha \neq \alpha'$. Let Λ be the set of all finite non-empty sets $\lambda = \{\beta_1, \dots, \beta_k\}$ such that there exist different elements $\alpha_1, \dots, \alpha_k \in A$, for which $\beta_i \in B^{\alpha_i}$. Note that every $\lambda \in \Lambda$ determines the finite set $A(\lambda) = \{\alpha_1, \dots, \alpha_k\} \subseteq A$ as well as the indices $\lambda(\alpha_i) = \beta_i \in B^{\alpha_i}$, $i = 1, \dots, k$. For $\lambda, \lambda' \in \Lambda$, put $\lambda \leq \lambda'$, provided $A(\lambda) \subseteq A(\lambda')$ and $\lambda(\alpha) \leq \lambda'(\alpha)$, whenever $\alpha \in A(\lambda)$. Clearly, $(\Lambda, \leq)$ is a cofinite directed set. For $\lambda \in \Lambda$, put

$$(X_\lambda, *) = \bigvee_{\alpha\in A(\lambda)} (X^\alpha_{\lambda(\alpha)}, *) = (X^{\alpha_1}_{\beta_1}, *) \vee \ldots \vee (X^{\alpha_k}_{\beta_k}, *). \tag{2}$$

Clearly, if all X^α_β are polyhedra, then so is X_λ.

For $\lambda \leq \lambda'$, define $p_{\lambda\lambda'}\colon (X_{\lambda'}, *) \to (X_\lambda, *)$, by putting

$$p_{\lambda\lambda'}|X^\alpha_{\lambda'(\alpha)} = \begin{cases} p^\alpha_{\lambda(\alpha)\lambda'(\alpha)}, & \alpha \in A(\lambda), \\ *, & \alpha \in A(\lambda')\backslash A(\lambda). \end{cases} \tag{3}$$

Also define mappings $p_\lambda: (X, *) \to (X_\lambda, *)$, by putting

$$p_\lambda|X^\alpha = \begin{cases} p^\alpha_{\lambda(\alpha)}, & \alpha \in A(\lambda), \\ *, & \alpha \in A\backslash A(\lambda). \end{cases} \tag{4}$$

Then $(\boldsymbol{X}, *) = ((X_\lambda, *), p_{\lambda\lambda'}, \Lambda)$ is an inverse system and $\boldsymbol{p} = (p_\lambda): (X, *) \to (\boldsymbol{X}, *)$ is a mapping of systems.

LEMMA 19.20. *If all $\boldsymbol{p}^\alpha: (X^\alpha, *) \to (\boldsymbol{X}^\alpha, *)$, $\alpha \in A$, are cofinite resolutions, then also $\boldsymbol{p}: (X, *) \to (\boldsymbol{X}, *)$ is a cofinite resolution.*

Proof. We first verify property (B1). For $\mathcal{U} \in \mathrm{Cov}(X)$, choose $\mathcal{V} \in \mathrm{Cov}(X)$ so that

$$\mathrm{St}(\mathcal{V}) \prec \mathcal{U}. \tag{5}$$

Choose a member $U_* \in \mathcal{U}$ such that

$$\mathrm{St}(*, \mathcal{V}) \subseteq U_*. \tag{6}$$

Clearly, U_* contains all X^α, except for a finite set of indices $\{\alpha_1, \dots, \alpha_k\}$. By (B1) for $\boldsymbol{p}^{\alpha_i}$, there exist a $\beta_i \in B^{\alpha_i}$ and a $\mathcal{U}^i \in \mathrm{Cov}(X^{\alpha_i}_{\beta_i})$ such that

$$p^{\alpha_i\,-1}_{\beta_i}(\mathcal{U}^i) \prec \mathcal{V}|X^{\alpha_i}. \tag{7}$$

Put $\lambda = \{\beta_1, \dots, \beta_k\} \in \Lambda$ and note that $A(\lambda) = \{\alpha_1, \dots, \alpha_k\}$, $\lambda(\alpha_i) = \beta_i$. Consider X_λ and its covering $\mathcal{U}_\lambda$, which consists of all the sets $U_\lambda \in \mathcal{U}^1 \cup \dots \cup \mathcal{U}^k$ with $* \notin U_\lambda$ and of the set W, which is the union of all the members U of $\mathcal{U}^1 \cup \dots \cup \mathcal{U}^k$ containing the point $*$. It is easy to see that $\mathcal{U}_\lambda \in \mathrm{Cov}(X_\lambda)$.

We claim that

$$p_\lambda^{-1}(\mathcal{U}_\lambda) \prec \mathcal{U}. \tag{8}$$

Indeed, if $U_\lambda \in \mathcal{U}^i$ and $* \notin U_\lambda$, by (7),

$$p_\lambda^{-1}(U_\lambda) = (p^{\alpha_i}_{\beta_i})^{-1}(U_\lambda) \subseteq V \subseteq U, \tag{9}$$

for some $V \in \mathcal{V}$ and some $U \in \mathcal{U}$. On the other hand, if $U \in \mathcal{U}^i$ and $* \in U$, by (7),

$$(p_\lambda|X^{\alpha_i})^{-1}(U) = (p^{\alpha_i}_{\beta_i})^{-1}(U_\lambda) \subseteq V, \tag{10}$$

for some $V \in \mathcal{V}$, which contains $*$. Therefore,

$$(p_\lambda|X^{\alpha_i})^{-1}(W) \subseteq \mathrm{St}(*, \mathcal{V}) \subseteq U_*,\ i = 1, \dots, k. \tag{11}$$

Since also

$$(p_\lambda|X^\alpha)^{-1}(W) = X^\alpha \subseteq U_*,\ \alpha \in A\backslash\{\alpha_1, \dots, \alpha_k\}, \tag{12}$$

we see that

$$p_\lambda^{-1}(W) = \bigcup_{\alpha\in A} (p_\lambda|X^\alpha)^{-1}(W) \subseteq U_* \in \mathcal{U}, \tag{13}$$

which proves assertion (8).

To verify (B2), consider any $\lambda \in \Lambda$ and $\mathcal{U}_\lambda \in \mathrm{Cov}(X_\lambda)$ and let $A(\lambda) = \{\alpha_1, \dots, \alpha_k\}$. Since $\mathcal{U}^i = \mathcal{U}_\lambda|X^{\alpha_i}_{\lambda(\alpha_i)} \in \mathrm{Cov}(X^{\alpha_i}_{\lambda(\alpha_i)})$, $i = 1, \dots, k$, there exists a $\beta_i' \in B^{\alpha_i}$, $\beta_i' \geq \lambda(\alpha_i)$, such that

$$p^\alpha_{\lambda(\alpha_i)\beta_i'}(X^{\alpha_i}_{\beta_i'}) \subseteq \mathrm{St}(p^\alpha_{\lambda(\alpha_i)}(X^{\alpha_i}), \mathcal{U}^i). \tag{14}$$

Put $\lambda' = \{\beta_1', \dots, \beta_k'\}$. Then $\lambda \leq \lambda'$ and

$$p_{\lambda\lambda'}(X_{\lambda'}) \subseteq \mathrm{St}(p_\lambda(X), \mathcal{U}_\lambda). \ \square \tag{15}$$

Proof of Theorem 19.19. For every $\alpha \in A$ choose a cofinite polyhedral resolution $\boldsymbol{p}^\alpha\colon (X^\alpha, *) \to (\boldsymbol{X}^\alpha, *)$. Put $D^\alpha_\beta = S(X^\alpha_\beta, *)\otimes G$ and let $q^\alpha_{\beta\beta'}\colon D^\alpha_{\beta'} \to D^\alpha_\beta$ denote the chain mapping induced by $p^\alpha_{\beta\beta'}$. Then $\boldsymbol{D}^\alpha = (D^\alpha_\beta, q^\alpha_{\beta\beta'}, B^\alpha)$ is a pro-chain complex and

$$\overline{H}_m(X^\alpha, *; G) = \overline{H}_m(\boldsymbol{D}^\alpha). \tag{16}$$

Moreover, put $C_\lambda = S(X_\lambda, *) \otimes G$ and let $p_{\lambda\lambda'}\colon C_{\lambda'} \to C_\lambda$, $\lambda \leq \lambda'$, denote the chain mapping induced by $p_{\lambda\lambda'}\colon (X_{\lambda'}, *) \to (X_\lambda, *)$. Then $\boldsymbol{C} = (C_\lambda, p_{\lambda\lambda'}, \Lambda)$ is a pro-chain complex. Since, by Lemma 19.20, $\boldsymbol{p}\colon (X, *) \to (\boldsymbol{X}, *)$ is a cofinite polyhedral resolution, it follows that

$$\overline{H}_m(X, *; G) = \overline{H}_m(\boldsymbol{C}). \tag{17}$$

Let $\boldsymbol{f}^\alpha\colon (\boldsymbol{X}, *) \to (\boldsymbol{X}^\alpha, *)$ be the mapping of systems, given by $f^\alpha\colon B^\alpha \to \Lambda$ and by $f^\alpha_\beta\colon (X_{f^\alpha(\beta)}, *) \to (X^\alpha_\beta, *)$, where $f^\alpha(\beta) = \{\beta\} \in \Lambda$ and f^α_β is the identity mapping of $X_{f^\alpha(\beta)} = X_{\{\beta\}} = X^\alpha_\beta$. Note that $\boldsymbol{f}^\alpha\boldsymbol{p} = \boldsymbol{p}^\alpha f^\alpha$, because $p_{\{\beta\}} = p^\alpha_\beta f^\alpha$. Clearly, $\boldsymbol{f}^\alpha$ induces a morphism of pro-chain complexes $\boldsymbol{f}^\alpha\colon \boldsymbol{C} \to \boldsymbol{D}^\alpha$, given by $f^\alpha\colon B^\alpha \to \Lambda$ and by the identity chain mappings $f^\alpha_\beta\colon C_{\{\beta\}} \to D^\alpha_\beta$. Moreover, the induced homomorphism of homology groups $f^\alpha_*\colon \overline{H}_m(\boldsymbol{C}) = \overline{H}_m(\boldsymbol{D}^\alpha)$ is the homomorphism induced by the mapping $f^\alpha\colon X \to X^\alpha$. The forthcoming lemma enables us to apply Theorem 17.21 to the chain mapping $\boldsymbol{f}^\alpha\colon \boldsymbol{C} \to \boldsymbol{D}^\alpha$ and conclude that the induced homomorphism

$$f_*\colon \overline{H}_m(\boldsymbol{C}) = \prod_{\alpha\in A} \overline{H}_m(\boldsymbol{D}^\alpha) \tag{18}$$

is an isomorphism. However, by the above arguments, (18) coincides with the homomorphism f_* of Theorem 19.19. $\square$

LEMMA 19.21. *The homomorphisms*

$$f^r\colon \lim{}^r H_m(\boldsymbol{C}) \to \prod_{\alpha\in A} \lim{}^r H_m(\boldsymbol{D}^\alpha), \tag{19}$$

induced by the morphisms of pro-chain complexes $\boldsymbol{f}^\alpha\colon \boldsymbol{C} \to \boldsymbol{D}^\alpha$, $\alpha \in A$, *are isomorphisms.*

Proof. Abbreviate $H_m(D^\alpha_\beta)$ to H^α_β and $H_m(\boldsymbol{D}^\alpha)$ to $\boldsymbol{H}^\alpha = (H^\alpha_\beta, q^\alpha_{\beta\beta'}, B^\alpha)$. Also let G_λ stand for $H_m(C_\lambda)$, so that $\boldsymbol{G} = (G_\lambda, p_{\lambda\lambda'}, \Lambda) = H_m(\boldsymbol{C})$. By (2), X_λ is a finite wedge of polyhedra $(X^\alpha_{\lambda(\alpha)}, *)$. Therefore, its (singular) homology group $H_m(C_\lambda)$ is the direct sum of the corresponding homology groups of summands, i.e.,

$$G_\lambda = \bigoplus_{\alpha\in A(\lambda)} H^\alpha_{\lambda(\alpha)} \subseteq \prod_{\alpha\in A} H^\alpha_{\lambda(\alpha)}. \tag{20}$$

Moreover, the homomorphisms $p_{\lambda\lambda'}: G_{\lambda'} \to G_\lambda$ are given by

$$p_{\lambda\lambda'}|H^\alpha_{\lambda'(\alpha)} = \begin{cases} q^\alpha_{\lambda(\alpha)\lambda'(\alpha)}, & \alpha \in A(\lambda), \\ 0, & a \in A(\lambda') \setminus A(\lambda). \end{cases} \tag{21}$$

In new notation $\boldsymbol{f}^\alpha_*: H_m(\boldsymbol{C}) \to H_m(\boldsymbol{D}^\alpha)$ becomes $\boldsymbol{f}^\alpha_*: \boldsymbol{G} \to \boldsymbol{H}^\alpha$ and is given by $f^\alpha(\beta) = \{\beta\}$ and by the identity homomorphisms $f^\alpha_\beta: G_{\{\beta\}} \to H^\alpha_\beta$, $\beta \in B^\alpha$.

The morphism $\boldsymbol{f}^\alpha_*$ induces a cochain mapping $f^\alpha: K(\boldsymbol{G}) \to K(\boldsymbol{H}^\alpha)$, which in turn induces homomorphisms of the corresponding cohomology groups $f^{\alpha r}_*: \lim^r \boldsymbol{G} \to \lim^r \boldsymbol{H}^\alpha$ (see 11.5). By definition, if x is an r-cochain of $K(\boldsymbol{G})$, then $f^{\alpha r}(x)$ is the r-cochain y^α of $K(\boldsymbol{H}^\alpha)$, given by

$$y^\alpha(\beta_0, \dots, \beta_r) = x(\{\beta_0\}, \dots, \{\beta_r\}) \in G_{\{\beta_0\}} = H^\alpha_{\beta_0}. \tag{22}$$

$\prod_A K(\boldsymbol{H}^\alpha)$ itself can be viewed as a cochain complex. Its coboundary $\prod_A \delta$ operates coordinatewise, i.e., operates on each factor $K(\boldsymbol{H}^\alpha)$ as the corresponding coboundary operator δ. The cochain mappings $f^\alpha: K(\boldsymbol{G}) \to K(\boldsymbol{H}^\alpha)$ determine a cochain mapping $f: K(\boldsymbol{G}) \to \prod_A K(\boldsymbol{H}^\alpha)$. By definition, f^r maps an r-cochain x of $K(\boldsymbol{G})$ to the r-cochain $f^r(x) = y$ of $\prod_A K(\boldsymbol{H}^\alpha)$, where $y = (y^\alpha)$ and $y^\alpha = f^{\alpha r}(x)$, i.e., it is given by (22). Note that $\prod_A \lim^r \boldsymbol{H}^\alpha$ can be identified with $\lim^r \prod_A \boldsymbol{H}^\alpha$ and the induced homomorphism $f^r: \lim^r \boldsymbol{G} \to \prod_A \boldsymbol{H}^\alpha$ becomes the homomorphism of cohomology groups, induced by the cochain mapping $f: K(\boldsymbol{G}) \to \prod_A K(\boldsymbol{H}^\alpha)$ (see Corollary 12.15). Therefore, the assertion of Lemma 19.21 will be proved if we exhibit a cochain mapping $g: \prod_A K(\boldsymbol{H}^\alpha) \to K(\boldsymbol{G})$ such that

$$fg = \mathrm{id}, \tag{23}$$

$$gf \simeq \mathrm{id}. \tag{24}$$

For $y = (y^\alpha) \in \prod_A K^r(\boldsymbol{H}^\alpha)$, we define $x = g^r(y) \in K^r(\boldsymbol{G})$, by putting

$$x(\boldsymbol{\lambda}) = \sum_{\alpha\in A(\lambda_0)} y^\alpha(\boldsymbol{\lambda}(\alpha)), \tag{25}$$

where $\boldsymbol{\lambda} = (\lambda_0, \dots, \lambda_r)$ and $\boldsymbol{\lambda}(\alpha) = (\lambda_0(\alpha), \dots, \lambda_r(\alpha))$. Note that $y^\alpha(\boldsymbol{\lambda}(\alpha)) \in H^\alpha_{\lambda_0(\alpha)}$ and thus, $x(\boldsymbol{\lambda}) \in G_{\lambda_0}$. It is easy to verify that $\delta^r g^{r-1} = g^r \delta^r$. Indeed, for $\boldsymbol{\lambda} \in \Lambda_r$,

$$\begin{aligned}(\delta^r(g^{r-1}(y))(\boldsymbol{\lambda}) &= p_{\lambda_0\lambda_1}(x(d^0(\boldsymbol{\lambda})) + \textstyle\sum_{j=1}^r(-1)^j x(d^j\boldsymbol{\lambda}) = \\ &\textstyle\sum_{\alpha\in A(\lambda_1)} p_{\lambda_0\lambda_1}(y^\alpha(d^0(\boldsymbol{\lambda}(\alpha)))) + \sum_{j=1}^r \sum_{A(\lambda_0)}(-1)^j y^\alpha(d^j(\boldsymbol{\lambda}(\alpha)).\end{aligned} \tag{26}$$

However, (21) impies that $p_{\lambda_0\lambda_1}$ maps $y^\alpha(d^0(\boldsymbol{\lambda}(\alpha)))$ to 0, for $\alpha \in A(\lambda_1)\backslash A(\lambda_0)$ and thus, the above sum over $A(\lambda_1)$ reduces to the corresponding sum over $A(\lambda_0)$. It is then readily seen that

$$(\delta^r(g^{r-1}(y)))(\boldsymbol{\lambda}) = \sum_{A(\lambda_0)} (\delta^r y^\alpha)(\boldsymbol{\lambda}(\alpha)). \tag{27}$$

On the other hand,

$$(g^r(\delta^r(y)))(\boldsymbol{\lambda}) = \sum_{A(\lambda_0)} (\delta^r y)^\alpha(\boldsymbol{\lambda}(\alpha)). \tag{28}$$

However, by definition, $(\delta^r y)^\alpha = \delta^r y^\alpha$ and thus, the right sides of (27) and (28) coincide, which shows that, indeed, $g^r\delta^r(y) = \delta^r g^{r-1}(y)$.

To verify (23), put $g(y) = x$ and $f(x) = z$. Then, by (22), $z^\alpha(\boldsymbol{\beta}) = x(\{\beta_0\},\dots,\{\beta_r\})$, for $\boldsymbol{\beta} = (\beta_0,\dots,\beta_r)$. As $\beta_i \in B^\alpha$, for $\lambda_0 = \{\beta_0\},\dots,\lambda_r = \{\beta_r\}$, we have $A(\lambda_0) = \alpha$, $\lambda_0(\alpha) = \beta_0,\dots,\lambda_r(\alpha) = \beta_r$ and thus, by (25), $x(\{\beta_0\},\dots,\{\beta_r\}) = y^\alpha(\boldsymbol{\beta})$, which shows that $z^\alpha = y^\alpha$ and thus, $z = y$, i.e., $(fg)(y) = y$.

Let us now compute gf. Put $f(z) = y$ and $g(y) = x$. Then, $x(\boldsymbol{\lambda})$ is given by (25), where by (22), $y^\alpha(\boldsymbol{\lambda}(\alpha)) = z^\alpha(\{\lambda_0(\alpha)\},\dots,\{\lambda_r(\alpha)\})$ and thus,

$$(gf(z))(\boldsymbol{\lambda}) = \sum_{\alpha\in A(\lambda_0)} z(\{\lambda_0(\alpha)\},\dots,\{\lambda_r(\alpha)\}). \tag{29}$$

We shall now define a cochain homotopy D on $K(\boldsymbol{G})$, i.e., a sequence of homomorphisms $D^r\colon K^r(\boldsymbol{G}) \to K^{r-1}(\boldsymbol{G})$ such that

$$D^{r+1}\delta + \delta D^r = 1 - gf. \tag{30}$$

For $z \in K^{r+1}(\boldsymbol{G})$ and $\boldsymbol{\lambda} \in \Lambda_r$, put

$$(D^{r+1}(z))(\boldsymbol{\lambda}) = \sum_{\alpha\in A(\lambda_0)} \sum_{i=0}^r (-1)^i z^\alpha(\{\lambda_0(\alpha)\},\dots,\{\lambda_i(\alpha)\},\lambda_i,\dots,\lambda_r). \tag{31}$$

We then have

$$(D^{r+1}\delta^r(z))(\boldsymbol{\lambda}) = \sum_{\alpha\in A(\lambda_0)} u^\alpha(\boldsymbol{\lambda}), \tag{32}$$

where

$$
\begin{aligned}
u^\alpha(\boldsymbol{\lambda}) = z^\alpha(\boldsymbol{\lambda}) \;+\; & \sum_{j=0}^{r}(-1)^{j+1} z^\alpha(\{\lambda_0(\alpha)\}, \lambda_0, \ldots, \widehat{\lambda_j}, \ldots, \lambda_r) + \\
& \sum_{i=1}^{r} (-1)^i p^\alpha_{\lambda_0(\alpha)\lambda_1(\alpha)}(z^\alpha(\{\lambda_1(\alpha)\}, \ldots, \{\lambda_i(\alpha)\}, \lambda_i, \ldots, \lambda_r)) + \\
& \sum_{i=1}^{r}\sum_{j=1}^{i}(-1)^{i+j} z^\alpha(\{\lambda_0(\alpha)\}, \ldots, \widehat{\{\lambda_j(\alpha)\}}, \ldots, \{\lambda_i(\alpha)\}, \lambda_i, \ldots, \lambda_r) + \\
& \sum_{i=1}^{r}\sum_{j=i}^{r}(-1)^{i+j+1} z^\alpha(\{\lambda_0(\alpha)\}, \ldots, \{\lambda_i(\alpha)\}, \lambda_i, \ldots, \widehat{\lambda_j}, \ldots, \lambda_r).
\end{aligned} \tag{33}
$$

On the other hand,

$$
(\delta^r D^r(x))(\boldsymbol{\lambda}) = \sum_{\alpha \in A(\lambda_0)} v^\alpha(\boldsymbol{\lambda}), \tag{34}
$$

where

$$
\begin{aligned}
v^\alpha(\boldsymbol{\lambda}) = & \sum_{i=1}^{r}(-1)^{i-1} p^\alpha_{\lambda_0(\alpha)\lambda_1(\alpha)}(z^\alpha(\{\lambda_1(\alpha)\}, \ldots, \{\lambda_i(\alpha)\}, \lambda_i, \ldots, \lambda_r)) + \\
& \sum_{j=1}^{r}\sum_{i=0}^{j-1}(-1)^{i+j} z^\alpha(\{\lambda_0(\alpha)\}, \ldots, \{\lambda_i(\alpha)\}, \lambda_i, \ldots, \widehat{\lambda_j}, \ldots, \lambda_r)) + \\
& \sum_{j=1}^{r-1}\sum_{i=j+1}^{r} (-1)^{i+j+1} z^\alpha(\{\lambda_0(\alpha)\}, \ldots, \widehat{\{\lambda_j(\alpha)\}}, \ldots, \{\lambda_i(\alpha)\}, \lambda_i, \ldots, \lambda_r)).
\end{aligned} \tag{35}
$$

If we add up (33) and (35), the simple sum of (35) cancels the second simple sum of (33). Note that

$$
\{(i,j)\colon 2 \le i \le r, 1 \le j \le i-1\} = \{(i,j)\colon 1 \le j \le r-1, j+1 \le i \le r\}, \tag{36}
$$

$$
\{(i,j)\colon 1 \le i \le r-1, i+1 \le j \le r\} = \{(i,j)\colon 1 \le j \le r, 1 \le i \le j-1\}. \tag{37}
$$

Therefore, the second double sum of (35) cancells all members of the first double sum of (33), except for terms obtained for $j = i$ and $i = 1$, which form the sum

$$
\sum_{i=1}^{r} z^\alpha(\{\lambda_0(\alpha)\} \ldots, \widehat{\{\lambda_i(\alpha)\}}, \lambda_i \ldots, \lambda_r). \tag{38}
$$

Similarly, summing up the two remaining double sums, most terms cancell. From the first double sum of (35) there remain terms for $i = 0$. They form the sum

$$
\sum_{i=1}^{r}(-1)^j z^\alpha(\{\lambda_0(\alpha)\}, \lambda_0, \ldots, \widehat{\lambda_j}, \ldots, \lambda_r) \tag{39}
$$

which cancels the first simple sum of (33), except for its first term, which equals

$$- z^\alpha(\{\lambda_0(\alpha)\}, \lambda_0, \ldots, \lambda_r). \tag{40}$$

From the double sum of (33), there remain terms for $j = i$ and $i = r$, which form the sum

$$-\sum_{i=1}^{r} z^\alpha(\{\lambda_0(\alpha)\}, \ldots, \{\lambda_i(\alpha)\}\widehat{\lambda_i}, \ldots, \lambda_r). \tag{41}$$

Now note that the sequence $(\{\lambda_0(\alpha), \ldots, \widehat{\{\lambda_i(\alpha)\}}, \lambda_i, \ldots, \lambda_r)$ coincides with the sequence $(\{\lambda_0(\alpha)\}, \ldots, \{\lambda_{i-1}(\alpha)\}, \widehat{\lambda_{i-1}}, \ldots, \lambda_r))$. Therefore, in summing up (38), (40) and (41) all terms cancel, except for the last term of (41), which is $-z^\alpha(\{\lambda_0(\alpha)\}, \ldots, \{\lambda_r(\alpha)\})$. We have thus obtained the formula

$$u^\alpha(\boldsymbol{\lambda}) + v^\alpha(\boldsymbol{\lambda}) = z^\alpha(\boldsymbol{\lambda}) - \sum_{j=1}^{r} z^\alpha(\{\lambda_0(\alpha)\}, \ldots, \{\lambda_r(\alpha)\}). \tag{42}$$

Summing up (42) over $A(\lambda_0)$ and taking into account (29), we finally obtain the desired equation,

$$(D^{r+1}\delta^r(z))(\boldsymbol{\lambda}) + (\delta^r D^r(z))(\boldsymbol{\lambda}) = z(\boldsymbol{\lambda}) - (gf(z))(\boldsymbol{\lambda}). \ \square \tag{43}$$

REMARK 19.22. If X^α, $\alpha \in A$, is a collection of spaces, one can consider their coproduct (disjoint sum) $X = \coprod_A X^\alpha$ and the natural inclusions $i^\alpha\colon X^\alpha \to X$. Clearly, these mappings induce homomorphisms of strong homology groups $i^\alpha_*\colon \overline{H}_m(X^\alpha; G) \to \overline{H}_m(X; G)$. In turn, these homomorphisms induce a homomorphism $i_*\colon \bigoplus_A \overline{H}_m(X^\alpha; G) \to \overline{H}_m(X; G)$. It is natural to ask whether i_* is an isomorphism. The first counter-example has been exhibited in (Mardešić, Prasolov 1988). The proof that the example (which is a separable metric space) has the desired property reduces to showing that a certain abelian pro-group $\boldsymbol{A}$ has non-trivial first derived limit $\lim^1 \boldsymbol{A} \neq 0$. This was shown to be equivalent to a set-theoretic assertion, proved using the continuum hypothesis. Later it was shown in (Dow, Simon, Vaughan 1989) that this assertion is undecidable in the ZFC set theory.

Recently, A.V. Prasolov has exhibited a paracompact counter-example which is valid in every set-theoretic model compatible with the ZFC axioms. The space X is the coproduct of a collection of $\aleph_0$ copies X^α of the space $X(m+2, 0, \Lambda)$, where Λ is a directed set of cofinality $\mathrm{cof}(\Lambda) = \aleph_1$ (see Example 16.3). Then, for every abelian group $G \neq 0$, $i_*\colon \bigoplus \overline{H}_m(X^\alpha; G) \to \overline{H}_m(X; G)$ fails to be an epimorphism, for $m \geq -2$ (Prasolov 1998a, 1998b).

19.4 Strong homology and dimension

A space X is said to have *shape dimension* $\mathrm{sd}\, X \leq n$ provided it admits an expansion $[\boldsymbol{p}]\colon X \to [\boldsymbol{X}] = (X_\lambda, [p_{\lambda\lambda'}], \Lambda)$ in pro-H(Top) such that every term X_λ is a polyhedron of dimension $\dim X_\lambda \leq n$. In particular, if a space X is the resolution of a system of polyhedra of dimension $\leq n$, then $\mathrm{sd}\, X \leq n$. It is known that $\mathrm{sd}\, X \leq \dim X$, where $\dim X$ denotes covering dimension, based on normal coverings (see (Mardešić, Segal 1982), II.1.1, Theorem 3).

The main result of this subsection is the following simple theorem.

THEOREM 19.23. *Let X be a space with finite shape dimension $\mathrm{sd}\, X = n$. Then the strong homology group $\overline{H}_m(X;G) = 0$, for $m > n$. Moreover, $\overline{H}_n(X;G) \approx \check{H}_n(X;G)$.*

In the proof we need the following lemma.

LEMMA 19.24. *If $m > \mathrm{sd}\, X$, then the homology pro-group $\boldsymbol{H}_m(X;G) = 0$. Hence, also the Čech group $\check{H}_m(X;G) = 0$.*

Proof. Let $\mathrm{sd}\, X = n$. Then there exists an expansion $[\boldsymbol{p}]: X \to [\boldsymbol{X}]$ in pro-H(Top), where X_λ are polyhedra of dimension $\dim X_\lambda \leq n$. In dimensions which exceed the dimension of a polyhedron, its homology groups vanish and thus, $H_m(X_\lambda;G) = 0$, for $m > n$. Consequently, the pro-group $\boldsymbol{H}_m(X;G) = H_m([\boldsymbol{X}];G) = 0$ and $\check{H}_m(X;G) = \lim \boldsymbol{H}_m(X;G) = 0$. □

Proof of Theorem 19.23. Choose a cofinite strong HPol-expansion $\boldsymbol{p}: X \to \boldsymbol{X}$ of X, e.g., a cofinite polyhedral resolution. Then $\overline{H}_m(X;G) = \overline{H}_m(\boldsymbol{X};G)$ (see (19.1.1)). Consider the pro-chain complex $\boldsymbol{C} = S(\boldsymbol{X}) \otimes G$ and recall that $\overline{H}_p(\boldsymbol{X};G) = \overline{H}_p(\boldsymbol{C})$ (see (17.1.38)). Lemma 19.24 shows that $H_p(\boldsymbol{C}) = 0$, for $p > n$. Clearly, this implies $\lim^s H_{m+r}(\boldsymbol{C}) = \lim^s H_{m+r+1}(\boldsymbol{C}) = 0$, for $m \geq n$, $r \geq 1$ and any $s \geq 0$. Therefore, by Corollary 17.14, there is a natural isomorphism $\overline{H}_m(\boldsymbol{C}) \to \overline{H}_m^{(1)}(\boldsymbol{C})$ and a natural exact sequence

$$0 \to \lim{}^1 H_{m+1}(\boldsymbol{C}) \to \overline{H}_m^{(1)}(\boldsymbol{C}) \to \check{H}_m(\boldsymbol{C}) \to 0. \tag{1}$$

Since $\lim^1 H_{m+1}(\boldsymbol{C}) = 0$, we conclude that there is a natural isomorphism $\overline{H}_m(\boldsymbol{C}) \to \check{H}_m(\boldsymbol{C})$. If $m > n$, then $H_m(\boldsymbol{C}) = 0$ implies $\check{H}_m(\boldsymbol{C}) = \lim H_m(\boldsymbol{C}) = 0$ and thus, also $\overline{H}_m(\boldsymbol{C}) = 0$. If $m = n$, we obtain the desired isomorphism $\overline{H}_n(X;G) \approx \check{H}_n(X;G)$. □

REMARK 19.25. There exist examples of metric continua X of finite dimension $\dim X = n$ such that, for some values $m > n$ and rational coefficients $\mathbb{Q}$, the *singular* homology groups $H_m(X;\mathbb{Q}) \neq 0$ (Barratt, Milnor 1962).

REMARK 19.26. In (Bauer 1976) a new type of homology groups for topological spaces X, called shape homology groups, was introduced. A further study of these groups was carried out in (Koyama 1984a). These groups can be defined as the singular homology groups of the homotopy limit of a polyhedral resolution of X. It was shown in (Koyama 1984b) that there exists a 2-dimensional metric continuum X such that the third Bauer shape homology group with rational coefficients is non-trivial. Hence, this homology does not have the property of Theorem 19.23 and therefore, it differs from strong homology.

19.5 Strong homology of polyhedra

The purpose of this subsection is to prove the following result.

THEOREM 19.27. *If a space X has the homotopy type of a polyhedron, then its strong homology group $\overline{H}_m(X;G)$ is isomorphic to its singular homology group $H_m(X;G)$. The analogous statement also holds for relative groups.*

An immediate consequence of Theorem 19.27 is the following corollary, which shows that strong homology groups also satisfy the Eilenberg – Steenrod *dimension axiom.*

COROLLARY 19.28. *For a space consisting of a single point $*$, the strong homology group $\overline{H}_m(*;G)$ vanishes, for $m \neq 0$, and equals G, for $m = 0$.*

Proof of Theorem 19.27. According to the definition of strong homology groups $\overline{H}_m(X;G) = \overline{H}_m(\boldsymbol{X};G)$, where $\boldsymbol{X}$ is any cofinite strong HPol-expansion of X. Since X belogs to HPol, the rudimentary system $\boldsymbol{X} = (X)$ is such an expansion. Therefore, the proof will be completed if we show that, for a rudimentary system $\boldsymbol{X}$, which consists of a single space X, the strong homology group $\overline{H}_m(\boldsymbol{X};G)$ is isomorphic to the singular group $H_m(X;G)$. However, this assertion is a special case of the following lemma.

LEMMA 19.29. *Let $\boldsymbol{C}$ be a rudimentary pro-chain complex, i.e., $\boldsymbol{C}$ consists of a single chain complex C. Then the strong homology group $\overline{H}_m(\boldsymbol{C})$ is isomorphic to the homology group $H_m(C)$.*

The assertion of Lemma 19.29 is not totally obvious because of the presence of degenerate multiindices of the form $\boldsymbol{\lambda} = (\lambda, \dots, \lambda)$. However, in defining strong homology groups, restriction to non-degenerate multiindices does not affect the groups as we will now show.

Denote by $\hat{\Lambda}_s$ the set of all non-degenerate multiindices $\boldsymbol{\lambda} = (\lambda_0, \dots, \lambda_s)$, i.e., multiindices of length s such that $\lambda_0 < \lambda_1 < \dots < \lambda_s$. With every pro-chain complex $\boldsymbol{C}$ we associate a chain complex $\hat{T}(\boldsymbol{C})$, which is a subcomplex of the chain complex $T(\boldsymbol{C})$, considered in 17.1. It consists of subgroups $\hat{T}^m(\boldsymbol{C}) \subseteq T^m(\boldsymbol{C})$, which in turn consist of functions c, defined on non-degenerate multiindices $\boldsymbol{\lambda} \in \hat{\Lambda}_s \subseteq \Lambda_s$, $s \geq 0$, with values $c(\boldsymbol{\lambda}) \in C^{m+s}_{\lambda_0}$. The boundary operators $\hat{d}^m \colon \hat{T}^m \to \hat{T}^{m-1}$ are restrictions of the boundary operators $d^m \colon T^m \to T^{m-1}$. Let $i = (i^m) \colon T(\boldsymbol{C}) \to \hat{T}(\boldsymbol{C})$ be the chain mapping, which to an m-chain $c \in T^m$ assigns its restriction to non-degenerate multiindices, i.e.,

$$(i^m c)(\boldsymbol{\lambda}) = c(\boldsymbol{\lambda}), \ \boldsymbol{\lambda} \in \hat{\Lambda}_s, \ s \geq 0. \tag{1}$$

LEMMA 19.30. *The chain mapping $i \colon T(\boldsymbol{C}) \to \hat{T}(\boldsymbol{C})$ is a chain equivalence. Consequently, the induced homomorphism $i_* \colon H_m(T(\boldsymbol{C})) \to H_m(\hat{T}(\boldsymbol{C}))$ is an isomorphism, for every $m \in \mathbb{Z}$.*

Proof. It suffices to exhibit a chain mapping $r: \hat{T}(\boldsymbol{C}) \to T(\boldsymbol{C})$ such that

$$ir = \mathrm{id}, \; ri \simeq \mathrm{id}. \tag{2}$$

We define $r^m: \hat{T}^m \to T^m$ by putting

$$(r^m c)(\boldsymbol{\lambda}) = \begin{cases} c(\boldsymbol{\lambda}), & \boldsymbol{\lambda} \in \hat{\Lambda}_s, \\ 0, & \boldsymbol{\lambda} \in \Lambda_s \setminus \hat{\Lambda}_s. \end{cases} \tag{3}$$

r is a chain mapping, because, for $c \in \hat{T}^m$,

$$(dr^m c)(\boldsymbol{\lambda}) = (r^m \hat{d} c)(\boldsymbol{\lambda}), \; \boldsymbol{\lambda} \in \hat{\Lambda}_s. \tag{4}$$

This is obvious if $\boldsymbol{\lambda} \in \hat{\Lambda}_s$, because both sides of (4) equal $dc(\boldsymbol{\lambda})$. Now assume that $\boldsymbol{\lambda} \in \Lambda_s \setminus \hat{\Lambda}_s$, i.e., $\boldsymbol{\lambda}$ is degenerate. Clearly, $r^m \hat{d} c(\boldsymbol{\lambda}) = 0$ and it remains to show that also $dr^m c(\boldsymbol{\lambda}) = 0$. Since $\boldsymbol{\lambda}$ is degenerate, there exists an i, $0 \le i \le s-1$ such that $\lambda_i = \lambda_{i+1}$. Therefore, the multiindices $d^j \boldsymbol{\lambda}$, $0 \le j \le s$, for $j \ne i, i+1$ are also degenerate. Hence, by (3), $r^m c(\boldsymbol{\lambda}) = 0$ and $r^m c(d^j \boldsymbol{\lambda}) = 0$, for $j \ne i, i+1$. Consequently, the boundary formulae (17.1.12), (17.1.15) and (17.1.16) show that

$$(-1)^m dr^m c(\boldsymbol{\lambda}) = (-1)^i p r^m c(d^i \boldsymbol{\lambda}) + (-1)^{i+1} p r^m c(d^{i+1} \boldsymbol{\lambda}). \tag{5}$$

However, $\lambda_i = \lambda_{i+1}$ implies $d^i \boldsymbol{\lambda} = d^{i+1} \boldsymbol{\lambda}$ and the two terms in (5) cancel.

The assertion $ir = \mathrm{id}$ is obviously fulfilled. In order to show that $ri \simeq \mathrm{id}$, we define a chain homotopy H of the chain complex $T(\boldsymbol{C})$ to itself. It is given by homomorphisms $H^m: T^m(\boldsymbol{C}) \to T^{m+1}(\boldsymbol{C})$, where

$$(H^m c)(\boldsymbol{\lambda}) = \begin{cases} 0, & \boldsymbol{\lambda} \in \hat{\Lambda}_s, \\ (-1)^{m+k(\boldsymbol{\lambda})} c(s^{k(\boldsymbol{\lambda})} \boldsymbol{\lambda}), & \boldsymbol{\lambda} \in \Lambda_s \setminus \hat{\Lambda}_s. \end{cases} \tag{6}$$

Here, for $\boldsymbol{\lambda} \in \Lambda_s$ degenerate, $k(\boldsymbol{\lambda})$ denotes the smallest index k, $0 \le k < s$, such that $\lambda_k = \lambda_{k+1}$ and $s^k: \Lambda_s \to \Lambda_{s+1}$ is the k-th degeneracy operator, defined in 1.2.

Let us show that H is indeed a chain homotopy, i.e., for $c \in T^m(\boldsymbol{C})$, one has

$$(dH^m c)(\boldsymbol{\lambda}) + (H^{m-1} dc)(\boldsymbol{\lambda}) = (r^m i^m c)(\boldsymbol{\lambda}) - c(\boldsymbol{\lambda}), \; \boldsymbol{\lambda} \in \Lambda_s. \tag{7}$$

If $\boldsymbol{\lambda} \in \hat{\Lambda}_s$, then $(r^m i^m c)(\boldsymbol{\lambda}) = c(\boldsymbol{\lambda})$ and $(H^{m-1} dc)(\boldsymbol{\lambda}) = 0$. Moreover, $(dH^m c)(\boldsymbol{\lambda}) = 0$, because $\boldsymbol{\lambda} \in \hat{\Lambda}_s$ implies $d^j \boldsymbol{\lambda} \in \hat{\Lambda}_s$ and the assertion follows from the boundary formulae and (6).

Now assume that $\boldsymbol{\lambda}$ is degenerate. By the boundary formulae (17.1.12), (17.1.15), (17.1.16) and by (6), we see that

$$\begin{aligned}(dH^m c)(\boldsymbol{\lambda}) = \; & (-1)^{m+k(\boldsymbol{\lambda})} \partial c(s^{k(\boldsymbol{\lambda})} \boldsymbol{\lambda}) + \\ & \textstyle\sum_{j=0}^{s} (-1)^{j+1+k(d^j \boldsymbol{\lambda})} p_\# c(s^{k(d^j \boldsymbol{\lambda})} d^j \boldsymbol{\lambda}).\end{aligned} \tag{8}$$

We decompose the sum in (8) in three sums as follows.

$$\sum_{j=0}^{s} = \sum_{j=0}^{k-1} + \sum_{j=k}^{k+1} + \sum_{j=k+2}^{s}, \tag{9}$$

where $k = k(\boldsymbol{\lambda})$. The middle sum vanishes, because $\lambda_k = \lambda_{k+1}$ implies $d^k\boldsymbol{\lambda} = d^{k+1}\boldsymbol{\lambda}$. Furthermore,

$$k(d^j\boldsymbol{\lambda}) = \begin{cases} k-1, & 0 \le j < k, \\ k, & k+1 < j \le s. \end{cases} \tag{10}$$

Therefore, by (1.2.21),

$$s^{k(d^j\boldsymbol{\lambda})} d^j \boldsymbol{\lambda} = \begin{cases} d^j s^k \boldsymbol{\lambda}, & 0 \le j < k, \\ d^{j+1} s^k, & k+1 < j \le s. \end{cases} \tag{11}$$

If we substitute the values from (10) and (11) into (9) and in the last sum replace $j+1$ by j, we obtain

$$\begin{aligned}(dH^m c)(\boldsymbol{\lambda}) = (-1)^{m+k}\partial c(s^k\boldsymbol{\lambda}) \quad &+ \textstyle\sum_{j=0}^{k-1}(-1)^{j+k} p^\# c(d^j s^k \boldsymbol{\lambda}) \\ &+ \textstyle\sum_{j=k+3}^{s+1}(-1)^{j+k} p^\# c(d^j s^k \boldsymbol{\lambda}).\end{aligned} \tag{12}$$

On the other hand,

$$(H^{m-1}dc)(\boldsymbol{\lambda}) = (-1)^{m-1+k}\partial c(s^k\boldsymbol{\lambda}) + (-1)^m \sum_{j=0}^{s} (-1)^{j+k+1} p^\# c(d^j s^k \boldsymbol{\lambda}). \tag{13}$$

If we sum up (12) and (13), all terms cancel, except the terms of (13) for $j = k, k+1, k+2$, and we obtain

$$\begin{aligned}&(dH^m c)(\boldsymbol{\lambda}) + (H^{m-1}dc)(\boldsymbol{\lambda}) = -p^\# c(d^k s^k \boldsymbol{\lambda}) + \\ &p^\# c(d^{k+1} s^k \boldsymbol{\lambda}) - p^\# c(d^{k+2} s^k \boldsymbol{\lambda}).\end{aligned} \tag{14}$$

Since $s^k\boldsymbol{\lambda} = (\lambda_0, \dots, \lambda_k, \lambda_k, \lambda_{k+1}, \dots, \lambda_s)$ and $\lambda_k = \lambda_{k+1}$, one has $d^k s^k \boldsymbol{\lambda} = d^{k+1} s^k \boldsymbol{\lambda} = d^{k+2} s^k \boldsymbol{\lambda} = \boldsymbol{\lambda}$ (in this case $p = \mathrm{id}$). Consequently, the right side of (14) equals $-c(\boldsymbol{\lambda})$. Since $r^m i^m c(\boldsymbol{\lambda}) = 0$, we have obtained the desired relation (7). □

Proof. Let $\boldsymbol{C}$ be a rudimentary pro-chain complex, i.e., $\boldsymbol{C}$ is a chain complex C. By Lemma 19.30, $\overline{H}_m(\boldsymbol{C}) = H_m(T(\boldsymbol{C})) \approx H_m(\hat{T}(\boldsymbol{C}))$. It thus suffices to note that $\hat{T}(\boldsymbol{C}) = C$. Since Λ is a singleton $\{\lambda\}$, $\boldsymbol{\lambda} = (\lambda)$ is the only nondegenerate multiindex in Λ. Hence, $\hat{T}^m(\boldsymbol{C}) = C^m$. □

Taking into account Theorems 19.5, 19.15, 19.27 and Corollary 19.28, one readily obtains the following result.

COROLLARY 19.31. *For pairs of topological spaces (X, A), where A is normally embedded in X, in particular, for pairs where X is paracompact and A is closed, strong homology groups $\overline{H}_m((X, A); G)$ satisfy all the Eilenberg – Steenrod axioms.*

REMARK 19.32. In 1960 J. Milnor proved that, for a given coefficient group G, on the category of metric compacta there is a unique homology theory, which satisfies all the Eilenberg – Steenrod axioms, has the strong excision property and has the (countable) cluster property. In this way he obtained an axiomatic characterization of strong (Steenrod) homology groups for metric compacta (Milnor 1960).

REMARK 19.33. The strong homology groups, as defined in this chapter, are *unreduced* homology groups. One can also define *reduced strong homology groups*. By definition, the m-th reduced group of a space X is the kernel of the homomorphism $\overline{H}_m(X;G) \to \overline{H}_m(*;G)$, induced by the mapping $X \to *$. In dimensions $m \neq 0$, the group coincides with the unreduced one and, for $m = 0$, the unreduced group is obtained by adding to the reduced group a direct summand G.

19.6 Strong homology of metric compacta

In this subsection we show that strong homology groups of metric compacta X admit a simple description. This is due to the fact that strong homology groups of towers of chain complexes $\boldsymbol{C} = (C_i, p_{ii'}, \mathbb{N})$ admit a simple description. Indeed, in the case of towers the chain complex $T(\boldsymbol{C})$ can be replaced by a homotopy equivalent chain complex $\overline{T} = \overline{T}(\boldsymbol{C})$, defined as follows.

The group of m-chains $\overline{T}^m$ of $\overline{T}$ consists of functions c, defined on singletons (i) and on pairs $(i, i+1)$, $i \in \mathbb{N}$, where $c(i) = c_i \in C_i^m$ and $c(i, i+1) = c_{i,i+1} \in C_i^{m+1}$. The boundary operator $\overline{d}\colon \overline{C}^m \to \overline{C}^{m-1}$ is defined as in 17.1, i.e., by putting

$$(\overline{d}c)_i = \partial c_i, \tag{1}$$

$$(\overline{d}c)_{i,i+1} = \partial c_{i,i+1} + (-1)^m (p_{ii+1}(c_{i+1}) - c_i). \tag{2}$$

LEMMA 19.34. *If $\boldsymbol{C}$ is a tower of chain complexes, the chain mapping $k\colon T(\boldsymbol{C}) \to \overline{T}(\boldsymbol{C})$, defined by restricting the chains of $T(\boldsymbol{C})$ to multiindices (i) and $(i, i+1)$, is a homotopy equivalence. Consequently, $k_*\colon H_m(T(\boldsymbol{C})) \to H_m(\overline{T}(\boldsymbol{C}))$ is an isomorphism of groups.*

Proof. By Lemma 19.30, we know that the restriction to non-degenerate multiindices $i\colon T(\boldsymbol{C}) \to \hat{T}(\boldsymbol{C})$ is a homotopy equivalence. Let $j\colon \hat{T}(\boldsymbol{C}) \to \overline{T}(\boldsymbol{C})$ denote the chain mapping obtained by restricting the chains of $\hat{T}(\boldsymbol{C})$ to multiindices (i) and $(i, i+1)$. Clearly, $k = ji$ and the lemma will be proved if we show that j is also a homotopy equivalence. To prove this assertion, we first define a chain mapping $r\colon \overline{T}(\boldsymbol{C}) \to \hat{T}(\boldsymbol{C})$. If $c \in \overline{T} = \overline{T}(\boldsymbol{C})$, we put

$$(rc)_i = c_i,\ i \in \mathbb{N}, \tag{3}$$

$$(rc)_{i,i+k} = \sum_{j=0}^{k-1} p_{ii+j}(c_{i+j,i+j+1}),\ k \geq 1,\ i \in \mathbb{N}, \tag{4}$$

$$(rc)_{\boldsymbol{i}} = 0,\ \boldsymbol{i} \in \hat{\mathbb{N}}_n,\ n \geq 2. \tag{5}$$

To see that r is a chain mapping, we must show that, for every $c \in \overline{T}^m$, one has

$$(\hat{d}rc)_{\boldsymbol{i}} = (r\overline{d}c)_{\boldsymbol{i}},\ \boldsymbol{i} \in \hat{\mathbb{N}}_n, n \geq 0. \tag{6}$$

If $n = 0$, i.e., if $\boldsymbol{i} = (i)$ is a singleton, (6) holds, because both sides equal ∂c_i. Now assume that $n = 1$, i.e., $\boldsymbol{i} = (i, i+k)$, $k \geq 1$. Then by (4), (12), (15) and (16),

$$\begin{aligned}(\hat{d}rc)_{i,i+k} &= \partial((rc)_{i,i+k}) + (-1)^m(p_{ii+k}((rc)_{i+k}) - (rc)_i) \\ &= \textstyle\sum_{j=0}^{k-1} p_{ii+j}\partial(c_{i+j,i+j+1}) + (-1)^m(p_{ii+k}(c_{i+k}) - c_i).\end{aligned} \tag{7}$$

On the other hand,

$$\begin{aligned}&(r\overline{d}c)_{i,i+k} = \textstyle\sum_{j=0}^{k-1} p_{ii+j}(\partial c_{i+j,i+j+1}) + \\ &\textstyle\sum_{j=0}^{k-1}(-1)^m p_{ii+j+1}(c_{i+j+1}) - \sum_{j=0}^{k-1}(-1)^m p_{ii+j}(c_{i+j}).\end{aligned} \tag{8}$$

The first terms of (7) and (8) coincide. Moreover, in the second row of (8), most terms cancel and the row equals $(-1)^m(p_{ii+k}(c_{i+k}) - c_i)$. Consequently, (6) holds in this case too.

Now assume that $n \geq 2$. Then (5) implies $(r\overline{d}c)_{\boldsymbol{i}} = 0$, and we must show that also $(\hat{d}rc)_{\boldsymbol{i}} = 0$. Obviously, this is the case if $n \geq 3$, because then $(rc)_{\boldsymbol{i}} = 0$ and $(rc)_{d^j\boldsymbol{i}} = 0$. It remains to prove the assertion in the case $n = 2$, i.e., when $\boldsymbol{i} = (i, i+k, i+k+l)$ and $k \geq 1$, $l \geq 1$. Note that,

$$(-1)^m(\hat{d}rc)_{i,i+k,i+k+l} = p_{ii+k}(rc)_{i+k,i+k+l} - (rc)_{i,i+k+l} + (rc)_{i,i+k}. \tag{9}$$

However, it readily follows from (4) that the right side of (9) equals 0. This completes the proof of (6).

Formulae (3) and (4) immediately yield

$$jr = \mathrm{id}. \tag{10}$$

To complete the proof of Lemma 19.34 we still need a chain homotopy D, which connects id to rj.

We define the homomorphisms $D: \hat{T}^m \to \hat{T}^{m+1}$ as follows. If $\boldsymbol{i} = (i)$ is a singleton, we put

$$(Dc)_i = 0. \tag{11}$$

If $\boldsymbol{i} = (i_0, \ldots, i_{n-1}, i_n) \in \hat{\mathbb{N}}_n$, $n \geq 1$, we introduce the abbreviation $s(\boldsymbol{i}) = i_n - i_{n-1}$. Since $\boldsymbol{i}$ is non-degenerate, one always has $s(\boldsymbol{i}) \geq 1$. If $s(\boldsymbol{i}) = 1$, we put

$$(Dc)_{\boldsymbol{i}} = 0. \tag{12}$$

However, if $s(\boldsymbol{i}) \geq 2$, we consider the non-degenerate multiindices $u_j(\boldsymbol{i}) \in \hat{\mathbb{N}}^{n+1}$, given by

$$u_j(\boldsymbol{i}) = (i_0, \ldots, i_{n-1}, i_{n-1}+j, i_{n-1}+j+1), \tag{13}$$

for $1 \leq j \leq s(\boldsymbol{i})-1$. Then, we put

$$(Dc)_{\boldsymbol{i}} = (-1)^{m+n} \sum_{j=1}^{s(\boldsymbol{i})-1} c_{u_j(\boldsymbol{i})}. \tag{14}$$

Let us show that indeed,

$$(\hat{d}Dc)_{\boldsymbol{i}} + (D\hat{d}c)_{\boldsymbol{i}} = (rc)_{\boldsymbol{i}} - c_{\boldsymbol{i}}. \tag{15}$$

If $n = 0$, (15) holds, because $(\hat{d}Dc)_i = 0 = (D\hat{d}c)_i$ and $(rc)_i = c_i$. If $n = 1$, and $\boldsymbol{i} = (i, i+k)$, $k \geq 1$, then $s(\boldsymbol{i}) = k$. In the case $k = 1$, $(\hat{d}Dc)_{\boldsymbol{i}} = 0 = (D\hat{d}c)_{\boldsymbol{i}}$ and $(rc)_{\boldsymbol{i}} = c_{\boldsymbol{i}}$ and thus, (15) holds again. Now assume that $k \geq 2$. In this case $u_j(\boldsymbol{i}) = (i, i+j, i+j+1)$, $1 \leq j \leq k-1$, and therefore,

$$(\hat{d}Dc)_{\boldsymbol{i}} = (-1)^{m+1} \sum_{j=1}^{k-1} \partial c_{u_j(\boldsymbol{i})}, \tag{16}$$

$$\begin{aligned} (D\hat{d}c)_{\boldsymbol{i}} = \quad & (-1)^m \textstyle\sum_{j=1}^{k-1} \partial c_{u_j(\boldsymbol{i})} + \sum_{j=1}^{k-1} p_{ii+j} c_{i+j,i+j+1} - \\ & \textstyle\sum_{j=1}^{k-1} c_{i,i+j+1} + \sum_{j=1}^{k-1} c_{i,i+j}. \end{aligned} \tag{17}$$

In summing up (16) with (17), the first sums cancel. Moreover, most terms in the second row of (17) cancel each other and what remains equals $c_{i,i+1} - c_{i,i+k}$. However, by (4), this expression added to the second sum of (17) yields $(rc)_{i,i+k} - c_{i,i+k}$, as desired.

Now assume that $n \geq 2$. In view of (5), we must show that

$$(\hat{d}Dc)_{\boldsymbol{i}} + (D\hat{d}c)_{\boldsymbol{i}} = -c_{\boldsymbol{i}}. \tag{18}$$

We first consider the case when $s(\boldsymbol{i}) = 1$. In this case, by (12), $(D\hat{d}c)_{\boldsymbol{i}} = 0$ and we must show that

$$(\hat{d}Dc)_{\boldsymbol{i}} = -c_{\boldsymbol{i}}. \tag{19}$$

By the boundary formula,

$$(\hat{d}Dc)_{\boldsymbol{i}} = \partial(Dc)_{\boldsymbol{i}} + (-1)^{m+1} \sum_{j=0}^{n} (-1)^j p(Dc)_{d^j \boldsymbol{i}}. \tag{20}$$

However, $(Dc)_{\boldsymbol{i}} = 0$. Moreover, for $0 \leq j \leq n-2$, $s(d^j \boldsymbol{i}) = 1$ and thus, $(Dc)_{d^j \boldsymbol{i}} = 0$. Consequently,

$$(\hat{d}Dc)_{\boldsymbol{i}} = (-1)^{m+n}(Dc)_{d^{n-1}\boldsymbol{i}} + (-1)^{m+n+1}(Dc)_{d^n \boldsymbol{i}}. \tag{21}$$

If $s' = s(d^n \boldsymbol{i}) = 1$, then $s'' = s(d^{n-1}\boldsymbol{i}) = 2$ and $u_1(d^{n-1}\boldsymbol{i}) = \boldsymbol{i}$. Consequently, $(Dc)_{d^n \boldsymbol{i}} = 0$ and $(Dc)_{d^{n-1}\boldsymbol{i}} = (-1)^{m+n+1}c_{\boldsymbol{i}}$ and thus, $(\hat{d}Dc)_{\boldsymbol{i}} = -c_{\boldsymbol{i}}$. Formula (18) also holds when $s' = s(d^n \boldsymbol{i}) \geq 2$. Indeed, $s'' > s' \geq 2$ and

$$u_j(d^n \boldsymbol{i}) = u_j(d^{n-1}\boldsymbol{i}),\ 1 \leq j \leq s' - 1. \tag{22}$$

Therefore, all terms of $(-1)^{m+n}((Dc)_{d^{n-1}\boldsymbol{i}} - (Dc)_{d^n \boldsymbol{i}})$, given by (14), cancel except for the last term of $(-1)^{m+n}(Dc)_{d^{n-1}\boldsymbol{i}}$, which equals $-c_{\boldsymbol{i}}$, because $s'' - 1 = i_{n-1} - i_{n-2}$ and $i_{n-1} + 1 = i_n$.

We now consider the remaining case $s(\boldsymbol{i}) \geq 2$. In this case $(rc)_{\boldsymbol{i}} = 0$ and

$$\begin{aligned}&(D\hat{d}c)_{\boldsymbol{i}} = (-1)^{m+n-1}\textstyle\sum_{j=1}^{s(\boldsymbol{i})-1} \partial c_{u_j(\boldsymbol{i})} + \\ &(-1)^{n-1}\textstyle\sum_{j=1}^{s(\boldsymbol{i})-1} p(c_{d^0 u_j(\boldsymbol{i})}) + (-1)^{n-1}\textstyle\sum_{j=1}^{s(\boldsymbol{i})-1}\sum_{k=1}^{n+1}(-1)^k c_{d^k u_j(\boldsymbol{i})}.\end{aligned} \tag{23}$$

Note that, for $0 \leq k \leq n-2$, one has $s(d^k \boldsymbol{i}) = s(\boldsymbol{i}) \geq 2$ and thus, $(Dc)_{\boldsymbol{i}}$ and $(Dc)_{d^k \boldsymbol{i}}$ are given by (14). If also $1 \leq j \leq s(\boldsymbol{i}) - 1$, one has $u_j(d^k \boldsymbol{i}) = d^k u_j(\boldsymbol{i})$. Therefore, (14) yields

$$\begin{aligned}&(\hat{d}Dc)_{\boldsymbol{i}} = (-1)^{m+n}\textstyle\sum_{j=1}^{s(\boldsymbol{i})-1} \partial c_{u_j(\boldsymbol{i})} + \\ &(-1)^n \textstyle\sum_{j=1}^{s(\boldsymbol{i})-1} p(c_{d^k u_j(\boldsymbol{i})}) + (-1)^n \sum_{k=1}^{n-2}\sum_{j=1}^{s(\boldsymbol{i})-1}(-1)^k c_{d^k u_j(\boldsymbol{i})} + \\ &(-1)^{m+n}((Dc)_{d^{n-1}\boldsymbol{i}} - (Dc)_{d^n \boldsymbol{i}}).\end{aligned} \tag{24}$$

If we sum up (23) and (24), many terms cancel, and one obtains

$$\begin{aligned}&(D\hat{d}c)_{\boldsymbol{i}} + (\hat{d}Dc)_{\boldsymbol{i}} = (-1)^{m+n}((Dc)_{d^{n-1}\boldsymbol{i}} - (Dc)_{d^n \boldsymbol{i}}) + \\ &\textstyle\sum_{j=1}^{s(\boldsymbol{i})-1} c_{d^{n-1}u_j(\boldsymbol{i})} - \sum_{j=1}^{s(\boldsymbol{i})-1} c_{d^n u_j(\boldsymbol{i})} + \sum_{j=1}^{s(\boldsymbol{i})-1} c_{d^{n+1}u_j(\boldsymbol{i})}.\end{aligned} \tag{25}$$

Now note that, for $1 \leq j \leq s(\boldsymbol{i}) - 1$, one has $d^n u_j(\boldsymbol{i}) = d^{n+1}u_j(\boldsymbol{i})$. Therefore, in the last two sums all terms cancel, except for two terms, which equal $-c_{\boldsymbol{i}}$ and $c_{d^{n+1}u_1(\boldsymbol{i})}$, respectively. Consequently,

$$\begin{aligned}&(D\hat{d}c)_{\boldsymbol{i}} + (\hat{d}Dc)_{\boldsymbol{i}} = (-1)^{m+n}((Dc)_{d^{n-1}\boldsymbol{i}} - (Dc)_{d^n \boldsymbol{i}}) + \\ &\textstyle\sum_{j=1}^{s(\boldsymbol{i})-1} c_{d^{n-1}u_j(\boldsymbol{i})} + (c_{d^{n+1}u_1(\boldsymbol{i})} - c_{\boldsymbol{i}}).\end{aligned} \tag{26}$$

Since $s'' \geq s(\boldsymbol{i}) \geq 2$, (14) yields

$$(Dc)_{d^{n-1}\boldsymbol{i}} = (-1)^{m+n-1}\sum_{j=1}^{s''-1} c_{u_j(d^{n-1}\boldsymbol{i})}. \tag{27}$$

In the case when $s' = 1$, i.e., $i_{n-1} = i_{n-2} + 1$, we see that $s(d^n \boldsymbol{i}) = 1$ and $s'' - 1 = s(\boldsymbol{i})$. Moreover, $d^{n-1}u_j(\boldsymbol{i}) = u_{j+1}(d^{n-1}\boldsymbol{i})$, for $1 \leq j \leq s(\boldsymbol{i}) - 1$ Therefore, $(Dc)_{d^n \boldsymbol{i}} = 0$ and (27) substituted in (26) cancels the sum in (26) with the exception of the first term $-c_{u_1(d^{n-1}\boldsymbol{i})}$, which in this case equals $c_{d^{n+1}u_1(\boldsymbol{i})}$. Consequently, (26) assumes the desired form

$$(D\hat{d}c)_i + (\hat{d}Dc)_i = -c_i. \tag{28}$$

Finally, assume that $s' \geq 2$. In this case, (14) yields

$$(Dc)_{d^n i} = (-1)^{m+n-1} \sum_{j=1}^{s'-1} c_{u_j(d^n i)}. \tag{29}$$

We also know that (22) holds, for $1 \leq j \leq s'' - 1$. Therefore, (27) and (29) yield

$$(-1)^{m+n}((Dc)_{d^{n-1}i} - (Dc)_{d^n i}) = -\sum_{j=s'}^{s''-1} c_{u_j(d^{n-1}i)}. \tag{30}$$

Now note that

$$u_{s'+k}(d^{n-1}\boldsymbol{i}) = d^{n-1}u_k(\boldsymbol{i}),\ 1 \leq k \leq s(\boldsymbol{i}) - 1. \tag{31}$$

Therefore, after substituting (30) in (26), the sum in (26) cancels the right side of (30) with the exception of its first term $-c_{u_{s'}(d^{n-1}i)}$. However, this term equals $-c_{d^{n+1}u_1(i)}$ and it cancels the corresponding term in (26), thus yielding the desired relation (18). □

An immediate consequence of Lemma 19.34 is the following teorem.

THEOREM 19.35. *If X is a metric compact space and $\boldsymbol{X} = (X_i, p_{ii+1}, \mathbb{N})$ is an inverse sequence of compact polyhedra (compact* ANR*'s) such that $X = \lim \boldsymbol{X}$, the strong homology group $\overline{H}_m(X;G)$ is the m-th homology group of the chain complex $\overline{T}(S(\boldsymbol{X}) \otimes G))$.*

Bibliographic notes

Strong homology groups have a long history. Their origins can be traced back to the papers (Kolmogoroff 1936), (Chogoshvili 1940) and especially, (Steenrod 1940). In the latter paper N.E. Steenrod introduced strong homology groups of metric compacta and proved that for closed subsets $X \subseteq S^n$ and for cohomology groups of the complement $S^n \setminus X$, the Alexander duality law holds, i.e., $\overline{H}_m(X) \approx \check{H}^{n-m-1}(X)$. Failure of this law for Čech homology was the main reason for introducing strong homology. Originally, Steenrod defined his homology group $\overline{H}_m(X;G)$ as the $(m+1)$-st homology group with *infinite cycles* of the contractible telescope CTel $\boldsymbol{X}$ of an inverse system of compact polyhedra with $\lim \boldsymbol{X} = X$ (Steenrod 1940). K.A. Sitnikov gave a different description of strong homology groups of metric compacta (Sitnikov 1951,1954). Our description of $\overline{H}_m(X;G)$, given in Theorem 19.35, essentially follows Sitnikov's description and it was the starting point of the Lisica – Mardešić approach to strong homology. That the Sitnikov groups are isomorphic to the *Steenrod homology groups* was shown in (Sklyarenko 1969).

Over the last 50 years a large literature on the subject accumulated. Many different approaches and results, referring to various classes of spaces, were considered. We mention here only some of the relevant papers: (Bauer 1984,1987), (Berikashvili 1984), (Borel, Moore 1960), (Cordier 1987), (Deheuvels 1962), (Dydak, Nowak 1991), (Inassaridze 1972, 1991), (Inassaridze, Mdzinarishvili 1980), (Koyama 1984a, 1984b), (Kuz'minov, Shvedov 1974, 1975), (Lisica, Mardešić 1983, 1984a, 1985d, 1985e, 1985f, 1986), (Lisitsa 1977, 1983a, 1985), (Mardešić 1987a, 1996b), (Mardešić, Miminoshvili 1990), (Mardešić, Prasolov 1988, 1998), (Mardešić, Watanabe 1988), (Massey 1978), (Mdzinarishvili 1965, 1972, 1978, 1981, 1986a, 1986b), (Milnor 1960, 1962), (Miminoshvili 1984, 1991), (Petkova 1973), (Prasolov 1989), (Saneblidze 1983a, 1983b, 1992), (Sklyarenko 1969, 1971, 1979, 1989a, 1989b, 1995), (Watanabe 1987b). Especially informative are Sklyarenko's survey articles.

The appearence of strong shape theory led to a new approach to strong homology, which we followed in our exposition. In particular, we base the construction of strong homology groups, their strong shape invariance and exactness on (Lisica, Mardešić 1983, 1985d, 1985e, 1985f). Theorem 19.9 and 19.19 are from (Mardešić, Miminoshvili 1990). Weaker versions of these theorems were obtained in (Dydak, Nowak 1991) and (Watanabe 1987b). A weaker version of Theorem 19.15 appeared in (Lisica, Mardešić 1985f). Theorem 19.23 is from (Mardešić, Watanabe 1988). Theorems 19.27 and 19.35 are from (Lisica, Mardešić 1985d).

20. Spectral sequences. Abelian groups

The proof of the key result of the next section (Theorem 21.6) uses in an essential way the Roos spectral sequence and its consequences, which we describe in this section (see 20.3). In order to make the text as self-contained as possible, we develop general techniques of spectral sequences in subsections 20.1 and 20.2. In 20.4 we discuss pure extensions of abelian groups and in 20.5 we establish the needed results from the theory of abelian groups.

20.1 The spectral sequence of a filtered complex

A (decreasing) *filtration of a module* M is a sequence F of submodules $M_p = F_p(M) \subseteq M,\ p \in \mathbb{Z}$, such that

$$\ldots \supseteq M_p \supseteq M_{p+1} \supseteq \ldots, \tag{1}$$

$$\cup_p M_p = M. \tag{2}$$

The filtration F is *regular* provided $M_p = 0$, for all sufficiently large p. Every *filtered module* (M, F) determines a graded module, denoted by $G(M, F)$ and called the *associated graded module* of (M, F). It consists of all the quotients $F_p(M)/F_{p+1}(M)$.

A (decreasing) *filtration of a cochain complex* $C = (C^n, \delta)$ is a sequence F of subcomplexes $C_p = F_p(C) = (C_p^n, \delta)$ of C such that, for each $n \in \mathbb{Z}$, the modules $C_p^n,\ p \in \mathbb{Z}$, form a filtration of the module C^n. The filtration of C is *regular* provided the filtrations of all C^n are regular.

A (regular) filtration F of a cochain complex C induces on each of the cohomology modules $H^n(C)$ a (regular) filtration F, defined by

$$F_p(H^n(C)) = i_p(H^n\,(C_p)), \tag{3}$$

where i_p is the homomorphism induced by the inclusion $C_p \to C$. Indeed, if $i_{pp+1}\colon H^n(C_{p+1}) \to H^n(C_p)$ is the homomorphism induced by the inclusion $C_{p+1} \to C_p$, then $i_p i_{pp+1} = i_{p+1}$ and thus,

$$F_{p+1}(H^n(C)) = i_p i_{pp+1}(H^n\,(C_{p+1})) \subseteq F_p(H^n(C)). \tag{4}$$

Moreover,

$$\cup_p F_p(H^n(C)) = H^n(C), \tag{5}$$

because every n-cocycle of $C = \cup C_p$ is also an n-cocycle of C_p, for some $p \in \mathbb{Z}$. If F is regular on C, then, for any n, $C_p^n = 0$, provided p is sufficiently large. Clearly, for such p also $H^n(C_p) = 0$ and $F_p(H^n(C)) = 0$.

The filtration F on $H^n(C)$, induced by the filtration F on C, determines the *associated bigraded cohomology module* of (C, F), which is denoted by $G(H^*(C), F)$. This is a *bigraded module*, i.e., a collection of modules indexed by pairs of integers. It consists of the modules

$$F_p(H^n(C))/F_{p+1}(H^n(C)),\ p, q \in \mathbb{Z}, \tag{6}$$

and contains valuable information on the cohomology modules $H^n(C)$, $n \in \mathbb{Z}$. However, it does not determine them completely, as the next example shows.

EXAMPLE 20.1. Let $0 \to A \to B \to D \to 0$ and $0 \to A \to B' \to D \to 0$ be two exact sequences of abelian groups with B' not isomorphic to B. E.g., such are the sequences $0 \to \mathbb{Z} \xrightarrow{2} \mathbb{Z} \to \mathbb{Z}/2 \to 0$ and $0 \to \mathbb{Z} \to \mathbb{Z} \oplus \mathbb{Z}/2 \to \mathbb{Z}/2 \to 0$. Let C be the cochain complex defined by $C^n = 0$, for $n \neq 0$, $C^0 = B$, so that $H^n(C) = 0$, for $n \neq 0$, and $H^0(C) = B$. Let F be the filtration on C, defined by $C_p^n = 0$, for $n \neq 0$, and $C_p^0 = B, A, 0$, for $p < 0$, $p = 0$ and $p > 0$, respectively. Then $F_p(H^n(C)) = 0$, for $n \neq 0$, and $F_p(H^0(C)) = B, A, 0$, for $p < 0$, $p = 0$ and $p > 0$, respectively. Consequently, the terms of the bigraded module of cohomology of C equal 0, for $n \neq 0$, and equal $0, B/A, A, 0$, for $p < -1, p = -1, p = 0$ and $p > 0$, respectively. Let C' and F' be defined analogously, using the second exact sequence instead of the first one. The bigraded cohomology modules of C and C' are isomorphic, because $B/A \approx B'/A$. Nevertheless, the cohomologies of C and C' differ, because $H^0(C) = B$ and $H^0(C') = B'$ are not isomorphic.

With every regularly filtered cochain complex one can associate a spectral sequence, which converges towards $G(H^*(C), F)$. To define spectral sequences, we need the notion of a *bigraded differential module* (E, d) (see Fig. 20.1). This is a bigraded module $E = (E^{pq})$, endowed with a *differential* d of *bidegree* $(r, -r+1)$, for some $r \in \mathbb{Z}$, i.e., a collection of homomorphisms

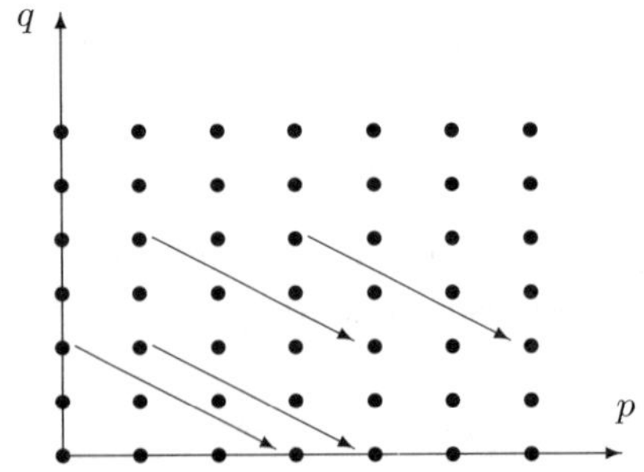

Fig. 20.1. Bigraded differential module of bidegree $(3, -2)$

$$d^{pq}: E^{pq} \to E^{p+r,q-r+1}, \tag{7}$$

such that $(d)^2 = 0$, i.e., the compositions

$$E^{p-r,q+r-1} \xrightarrow{d} E^{pq} \xrightarrow{d} E^{p+r,q-r+1} \tag{8}$$

equal 0. Consequently, $E^{p+ir,q-ir+i}$ and $d^{p+ir,q-ir+i}$, $i \in \mathbb{Z}$, form a cochain complex. The quotient $\mathrm{Ker}(d^{pq})/\mathrm{Im}(d^{p-r,q+r-1})$ is the cohomology of that complex at $i = 0$ and we denote it by $H^{pq}(E)$. The modules $H^{pq}(E), p, q \in \mathbb{Z}$, form a bigraded module $(H^{pq}(E))$, called the *bigraded cohomology module* of (E, d).

A (cohomology) *spectral sequence* is a sequence E of bigraded differential modules (E_r, d_r), $r \geq k$, with differentials d_r of bidegrees $(r, -r+1)$, together with (canonical) isomorphisms, which identify $H^{pq}(E_r)$ with E^{pq}_{r+1}. The term E_k is the *initial term* of the spectral sequence. Clearly, E_r and d_r determine E_{r+1}, but not the differential d_{r+1}. A spectral sequence is called *convergent* provided each pair (p, q) admits an $r(p, q) \geq k$, such that $d^{pq}_r = 0$, for $r \geq r(p, q)$. In this case, $E^{pq}_r = \mathrm{Ker}(d^{pq}_r)$ and thus, the quotient homomorphism $\mathrm{Ker}(d^{pq}_r) \to \mathrm{Ker}(d^{pq}_r)/\mathrm{Im}(d^{p-r,q+r-1}_r) = H^{pq}(E_r)$, composed with the canonical isomorphism $H^{pq}(E_r) \to E^{pq}_{r+1}$, yields an epimorphism $E^{pq}_r \to E^{pq}_{r+1}$, $r \geq r(p, q)$. In this way one obtains a direct sequence

$$E^{pq}_{r(p,q)} \to \cdots \to E^{pq}_r \to E^{pq}_{r+1} \cdots, \tag{9}$$

whose colimit is denoted by E^{pq}_∞. We say that the (convergent) spectral sequence (E_r, d_r) *converges towards* the bigraded module (E^{pq}_∞). If the homomorphisms in (9) are isomorphisms for sufficiently large r, one speaks of *strong convergence.* This is the case for *positive spectral sequences* (also called *first quadrant sequences*). These are sequences, where for some index $r \geq k$ (hence, also for all larger indices) $E^{pq}_r = 0$, whenever either $p < 0$ or $q < 0$. Indeed, for $r > q+1, k-1$, one has $E^{p+r,q-r+1}_r = 0$ and thus, $d^{pq}_r = 0$. Furthermore, for $r > p, k-1$, one has $E^{p-r,q+r-1}_r = 0$ and thus, also $d^{p-r,q+r-1}_r = 0$. Consequently, $E^{pq}_r = H^{pq}(E_r) \to E^{pq}_{r+1}$ is an isomorphism.

A *mapping of spectral sequences* $f: E \to E' = (E'_r, d'_r)$ consists of homomorphisms $f_r: E_r \to E'_r$ of bigraded differential modules, which are compatible with the canonical isomorphisms $H^{pq}(E_r) \to E^{pq}_{r+1}$ and $H^{pq}(E'_r) \to E'^{pq}_{r+1}$. Note that f_r consists of homomorphisms $f^{pq}_r: E^{pq}_r \to E'^{pq}_r$, which commute with the differentials, i.e., $d'^{pq}_r f^{pq}_r = f^{p+r,q-r+1}_r d^{p+r,q-r+1}_r$. Clearly, in the case of a convergent sequence, f induces homomorphisms $f_\infty: E_\infty \to E'_\infty$.

With every filtered cochain complex $C = (C^n, \delta)$ one associates a cohomology spectral sequence (E_r, d_r) as follows. Let Z^{pq}_r and B^{pq}_r, $r \in \mathbb{Z}$, denote the submodules of $C^{p+q}_p = F_p(C^{p+q})$, defined by

$$Z^{pq}_r = \{x \in C^{p+q}_p \mid \delta x \in C^{p+q+1}_{p+r}\}. \tag{10}$$

$$B^{pq}_r = \{x \in C^{p+q}_p \mid (\exists y \in C^{p+q-1}_{p-r})\ x = \delta y\}. \tag{11}$$

Note that

$$B_r^{pq} = \delta Z_r^{p-r,q+r-1}. \tag{12}$$

One refers to p as the *degree of filtration*, to q as the *complementary degree* and to $p+q$ as the *total degree* or *dimension*. It is readily seen that

$$\ldots \subseteq B_r^{pq} \subseteq B_{r+1}^{pq} \subseteq \ldots \subseteq Z_{r+1}^{pq} \subseteq Z_r^{pq} \subseteq \ldots . \tag{13}$$

One also defines submodules Z_∞^{pq} and B_∞^{pq} of C_p^{p+q}, by putting

$$Z_\infty^{pq} = \cap_r Z_r^{pq}, \tag{14}$$

$$B_\infty^{pq} = \cup_r B_r^{pq}. \tag{15}$$

Since $\cup_r C_{p-r}^{p+q-1} = C^{p+q-1}$, the submodule B_∞^{pq} consists of all $(p+q)$-cochains of C_p, which are coboundaries in C, i.e.,

$$B_\infty^{pq} = \{x \in C_p^{p+q} |\ (\exists y \in C^{p+q-1})\ x = \delta y\}. \tag{16}$$

Similarly, for regular filtrations, Z_∞^{pq} consists of all $(p+q)$-cocycles of C_p, i.e.,

$$Z_\infty^{pq} = \{x \in C_p^{p+q} |\ \delta x = 0\}. \tag{17}$$

This is so because $x \in Z_r^{pq}$ implies $\delta x \in C_{p+r}^{p+q+1}$ and $C_{p+r}^{p+q+1} = 0$, for all sufficiently large r.

Now put

$$E_r^{pq} = Z_r^{pq}/(B_{r-1}^{pq} + Z_{r-1}^{p+1,q-1}). \tag{18}$$

This module is well defined, because $C_{p+1}^{p+q} \subseteq C_p^{p+q}$ implies

$$Z_{r-1}^{p+1,q-1} \subseteq Z_r^{pq}. \tag{19}$$

On the bigraded module $E_r = (E_r^{pq})$, the coboundary operator δ induces a differential $d_r = (d_r^{pq})$ of bidegree $(r, -r+1)$. Indeed, by (18),

$$E_r^{p+r,q-r+1} = Z_r^{p+r,q-r+1}/(B_{r-1}^{p+r,q-r+1} + Z_{r-1}^{p+r+1,q-r}), \tag{20}$$

and by (12),

$$\delta Z_r^{pq} = B_r^{p+r,q-r+1} \subseteq Z_r^{p+r,q-r+1}. \tag{21}$$

Therefore, δ maps the numerator of (18) to the numerator of (20). Analogously, δ maps the denominator of (18) to the denominator of (20), because $\delta B_{r-1}^{pq} = 0$ and

$$\delta Z_{r-1}^{p+1,q-1} = B_{r-1}^{p+r,q-r+1}. \tag{22}$$

Consequently, δ induces a homomorphism $d_r^{pq} \colon E_r^{pq} \to E_r^{p+r,q-r+1}$.

THEOREM 20.2. *There exist canonical isomorphisms $H^{pq}(E_r) \to E_{r+1}^{pq}$, which make (E_r, d_r) a spectral sequence, called the* spectral sequence of the filtered complex (C, F).

Proof. We will first establish the following equalities.

$$\operatorname{Ker}(d_r^{pq}) = \frac{Z_{r+1}^{pq} + Z_{r-1}^{p+1,q-1}}{B_{r-1}^{pq} + Z_{r-1}^{p+1,q-1}}, \tag{23}$$

$$\operatorname{Im}(d_r^{p-r,q+r-1}) = \frac{B_r^{pq} + Z_{r-1}^{p+1,q-1}}{B_{r-1}^{pq} + Z_{r-1}^{p+1,q-1}}. \tag{24}$$

By (13) and (19), the numerator of the right side of (23) is contained in Z_r^{pq}. Moreover, by (21) and (13),

$$\delta Z_{r+1}^{pq} \subseteq Z_{r+1}^{p+r+1,q-r} \subseteq Z_{r-1}^{p+r+1,q-r}. \tag{25}$$

This and (20) show that the right side of (23) is contained in $\operatorname{Ker}(d_r^{pq})$. Conversely, if $x \in Z_r^{pq} \subseteq C_p^{p+q}$ represents an element of $\operatorname{Ker}(d_r^{pq})$, then (20), (12) and (19) show that $\delta x = \delta y + z$, where $y \in Z_{r-1}^{p+1,q-1} \subseteq Z_r^{pq} \subseteq C_p^{p+q}$, $z \in Z_{r-1}^{p+r+1,q-r} \subseteq C_{p+r+1}^{p+q+1}$. Putting $u = x - y$, one concludes that $u \in C_p^{p+q}$, $\delta u = z \in C_{p+r+1}^{p+q+1}$ and thus, $u \in Z_{r+1}^{pq}$. Consequently, $x = u + y \in Z_{r+1}^{pq} + Z_{r-1}^{p+1,q-1}$.

In order to prove (24), first note that every element from the left side of (24) is of the form $d_r[x] = [\delta x]$, where $x \in Z_r^{p-r,q+r-1}$. Since $\delta x \in \delta Z_r^{p-r,q+r-1}$, (12) shows that $\delta x \in B_r^{pq}$ and thus, $[\delta x]$ belongs to the right side of (24). Conversely, since $Z_{r-1}^{p+1,q-1}$ is contained in the denominator of the right side of (24), every element of that side has a representative $y \in B_r^{pq}$. By (12), there exists an element $x \in Z_r^{p-r,q+r-1}$ such that $y = \delta x$. Consequently, by (18), $[x] \in E_r^{p-r,q+r-1}$ has the property that $d_r[x] = [y]$.

Since $H^{pq}(E_r)$ is the quotient of the left sides of (23) and (24), one concludes that

$$H^{pq}(E_r) \approx \frac{Z_{r+1}^{pq} + Z_{r-1}^{p+1,q-1}}{B_r^{pq} + Z_{r-1}^{p+1,q-1}} \approx \frac{Z_{r+1}^{pq}}{Z_{r+1}^{pq} \cap (B_r^{pq} + Z_{r-1}^{p+1,q-1})}. \tag{26}$$

Now, $B_r^{pq} \subseteq Z_{r+1}^{pq}$ and

$$Z_{r+1}^{pq} \cap Z_{r-1}^{p+1,q-1} = Z_r^{p+1,q-1}, \tag{27}$$

imply

$$H^{pq}(E_r) \approx \frac{Z_{r+1}^{pq}}{(B_r^{pq} + Z_r^{p+1,q-1})} = E_{r+1}^{pq} \tag{28}$$

and one obtains the desired canonical isomorphism $H^{pq}(E_r) \to E_{r+1}^{pq}$. □

THEOREM 20.3. *If (E_r, d_r) is the spectral sequence of a filtered complex (C, F), then*

$$E_0^{pq} = C_p^{p+q} / C_{p+1}^{p+q} \tag{29}$$

and the homomorphisms d_0^{pq} are induced by $\delta\colon C_p^{p+q} \to C_p^{p+q+1}$. Moreover,

$$E_1^{pq} = H^{p+q}(C_p / C_{p+1}) \tag{30}$$

and the homomorphisms d_1^{pq} coincide with the connecting homomorphisms $H^{p+q}(C_p/C_{p+1}) \to H^{p+q+1}(C_{p+1}/C_{p+2})$, associated with the exact sequence of cochain complexes

$$0 \to C_{p+1}/C_{p+2} \to C_p/C_{p+2} \to C_p/C_{p+1} \to 0. \tag{31}$$

Proof. First note that $Z_0^{pq} = C_p^{p+q}$, because $\delta(C_p^{p+q}) \subseteq C_p^{p+q+1}$. Furthermore, $Z_{-1}^{p+1,q-1} = C_{p+1}^{p+q}$ and $B_{-1}^{pq} \subseteq C_{p+1}^{p+q}$. Therefore, (18) for $r = 0$ implies (29). Clearly, $d_0\colon E_0^{pq} \to E_0^{pq}$ is induced by $\delta\colon C_p^{p+q} \to C_p^{p+q+1}$. In order to prove (30), note that the quotient complex C_p/C_{p+1} consists of modules C_p^n/C_{p+1}^n and the diferential $C_p^n/C_{p+1}^n \to C_p^{n+1}/C_{p+1}^{n+1}$ is induced by $\delta\colon C_p^n \to C_p^{n+1}$. Consequently, E_1^{pq} can be identified with $H^{pq}(E_0) = H^{p+q}(C_p/C_{p+1})$, which establishes (30). To determine the differential $d_1^{pq}\colon E_1^{pq} \approx H^{p+q}(C_p/C_{p+1}) \to H^{p+1+q}(C_{p+1}/C_{p+2}) \approx E_1^{p+1,q}$, consider an element $\alpha \in H^{p+q}(C_p/C_{p+1})$. Then α is represented by a cochain $z \in C_p^{p+q}$ with $\delta z \in C_{p+1}^{p+q+1}$. Consequently, $z \in Z_1^{pq}$ and $d_1(\alpha) = [\delta z] \in H^{p+1,q}(E_0) \approx E_1^{p+1,q}$. However, this is just the way one defines the connecting homomorphism in the cohomology sequence associated with (31). □

THEOREM 20.4. *If (C, F) is a regularly filtered cochain complex, then its spectral sequence (E_r, d_r) converges towards the bigraded cohomology module associated with (C, F), i.e., for every pair of integers (p, q),*

$$E_\infty^{pq} \approx F_p(H^{p+q}(C))/F_{p+1}(H^{p+q}(C)). \tag{32}$$

Proof. By regularity, for every pair (p, q) there exists an integer $s(p, q)$ such that $C_s^{p+q+1} = 0$, for $s \geq s(p, q)$. Therefore,

$$Z_r^{p+r,q-r+1} \subseteq C_{p+r}^{p+q+1} = 0, \ r \geq s(p,q) - p. \tag{33}$$

Hence, also $E_r^{p+r,q-r+1} = 0$ and $d_r^{pq} = 0$, which shows that (E_r, d_r) is a convergent spectral sequence. To determine its limit, consider the exact sequence determined by (18),

$$0 \to (B_{r-1}^{pq} + Z_{r-1}^{p+1,q-1}) \to Z_r^{pq} \to E_r^{pq} \to 0. \tag{34}$$

Note that $Z_r^{pq} = Z_\infty^{pq}$, for r sufficiently large. This is so because $x \in Z_r^{pq}$ implies $x \in C_p^{p+q}$ and $\delta x \in C_{p+r}^{p+q+1}$. However, $C_{p+r}^{p+q+1} = 0$, for sufficiently large r, so that $\delta x = 0$. Hence, $x \in Z_\infty^{pq}$. Consequently, for sufficiently large r, the sequence (34) assumes the form

$$0 \to (B_{r-1}^{pq} + Z_\infty^{p+1,q-1}) \to Z_\infty^{pq} \to E_r^{pq} \to 0. \tag{35}$$

Using the inclusion mappings $B^{pq}_{r-1} \to B^{pq}_r$ and the homomorphisms $E^{pq}_r \to E^{pq}_{r+1}$ from (9), one obtains a direct sequence of exact sequences (35). Passing to the colimit, one obtains the sequence

$$0 \to (B^{pq}_\infty + Z^{p+1,q-1}_\infty) \to Z^{pq}_\infty \to E^{pq}_\infty \to 0, \tag{36}$$

which is also exact, by Lemma 14.2. Consequently, we have proved that

$$E^{pq}_\infty \approx Z^{pq}_\infty/(B^{pq}_\infty + Z^{p+1,q-1}_\infty). \tag{37}$$

It remains to show that

$$Z^{pq}_\infty/(B^{pq}_\infty + Z^{p+1,q-1}_\infty) \approx F_p(H^{p+q}(C))/F_{p+1}(H^{p+q}(C)). \tag{38}$$

Composing the homomorphism $Z^{pq}_\infty \to H^{p+q}(C_p)$, which maps every cocycle $x \in Z^{pq}_\infty$ to its cohomology class $[x] \in H^{p+q}(C_p)$, with $i_p\colon H^{p+q}(C_p) \to H^{p+q}(C)$, one obtains a homomorphism $Z^{pq}_\infty \to F_p(H^{p+q}(C))$. Every $x \in B^{pq}_\infty$ is a coboundary in C and therefore, $i_p[x] = 0$. Moreover, $x \in Z^{p+1,q-1}_\infty$ implies $x \in C^{p+q}_{p+1}$, $\delta x = 0$ and thus, $[x] \in i_{pp+1}(H^{p+q}(C_{p+1}))$, which in turn implies $i_p[x] \in i_{p+1}(H^{p+q}(C_{p+1})) = F_{p+1}(H^{p+q}(C))$. Consequently, we obtain a homomorphism of the left side of (38) to the right side of (38). This homomorphism is surjective, because $Z^{pq}_\infty \to H^{p+q}(C_p)$ is surjective. To see that this homomorphism is also injective, assume that $x \in Z^{pq}_\infty$ and $i_p[x] \in F_{p+1}(H^{p+q}(C)) = i_{p+1}H^{p+q}(C_{p+1})$. Then there exists a $(p+q)$-cocycle $y \in C_{p+1}$ such that $i_p[x] = i_{p+1}[y] = i_p i_{pp+1}[y]$. Clearly, $y \in Z^{p+1,q-1}_\infty$. Moreover, x and y are cohomologous cocycles of C, i.e., $x = y + \delta z$, where $z \in C^{p+q-1}$. Since $\delta z = x - y \in C^{p+q}_p$, we conclude that $\delta z \in B^{pq}_\infty$. Consequently, $x \in B^{pq}_\infty + Z^{p+1,q-1}_\infty$, i.e., x represents the class 0 of the left side of (38). □

A *mapping of filtered cochain complexes* (C, F) to (C', F') is a mapping of cochain complexes $f\colon C \to C'$, which preserves filtration, i.e., $f(F_p(C)) \subseteq F'_p(C')$. By (10), (11) and (18), it is readily seen that f induces homomorphisms $f^{pq}_r\colon E^{pq}_r \to E^{pq}_r$, which form a mapping $f\colon E \to E'$ between the spectral sequences of the given filtered complexes.

Convergent spectral sequences are particularly useful, provided for some r, say $r = 2$, the modules E^{pq}_r vanish for most pairs (p, q). The next theorem is devoted to such a case.

THEOREM 20.5. *Let $M = (M^n)$ be a graded module and let each M^n be endowed with a regular filtration F. Let (E_r, d_r), $r \geq 2$, be a spectral sequence, which converges towards the associated bigraded module of (M, F), i.e.,*

$$E^{pq}_\infty \approx F_p(M^{p+q})/F_{p+1}(M^{p+q}). \tag{39}$$

If $E^{pq}_2 = 0$, whenever $q \neq 0, 1$, then, for each $n \in \mathbb{Z}$, there exists an exact sequence

$$E^{n0}_2 \xrightarrow{\varepsilon_n} M^n \xrightarrow{\eta_n} E^{n-1,1}_2 \xrightarrow{d_2} E^{n+1,0}_2 \xrightarrow{\varepsilon_{n+1}} M^{n+1}. \tag{40}$$

If also $E_2^{p1} = 0$, *for all* p, *then the spectral sequence* (E_r, d_r) *is called degenerate and*

$$E_2^{n0} \approx M^n. \tag{41}$$

Proof. Since $E_{r+1}^{pq} \approx H^{pq}(E_r) = \mathrm{Ker}\,(d_r^{pq})/\mathrm{Im}\,(d_r^{p-r,q+r-1})$ and $\mathrm{Ker}\,(d_r^{pq}) \subseteq E_r^{pq}$, one concludes, by induction on r, that all non-trivial terms of E_r, for $r \geq 2$, lie on the lines $q = 0$ and $q = 1$. Hence, also $E_\infty^{pq} = 0$, for $q \neq 0, 1$. Consequently, by (39),

$$F_{n-q}(M^n) = F_{n-q+1}(M^n), \tag{42}$$

for all $n \in \mathbb{Z}$ and $q \neq 0, 1$. Substituting $q = 2, 3, \ldots$ in (42), one obtains

$$\ldots = F_{n-2}(M^n) = F_{n-1}(M^n),\ n \in \mathbb{Z}, \tag{43}$$

and substituting $q = -1, -2, \ldots$, one obtains

$$F_{n+1}(M^n) = F_{n+2}(M^n) = \ldots,\ n \in \mathbb{Z}. \tag{44}$$

Since F is a decreasing filtration of M^n, (43) and $M^n = \cup_p F_p(M^n)$ imply

$$M^n = F_{n-1}(M^n),\ n \in \mathbb{Z}. \tag{45}$$

Furthermore, the regularity of F and (44) show that

$$F_{n+1}(M^n) = F_{n+2}(M^n) = \ldots = 0,\ n \in \mathbb{Z}. \tag{46}$$

By (39), (46) and (45), one now concludes that

$$E_\infty^{n0} \approx F_n(M^n)/F_{n+1}(M^n) = F_n(M^n),\ n \in \mathbb{Z}, \tag{47}$$

$$E_\infty^{n-1,1} \approx F_{n-1}(M^n)/F_n(M^n) = M^n/F_n(M^n),\ n \in \mathbb{Z}. \tag{48}$$

On the other hand, for $r \geq 3$ and arbitrary p, q, one has $d_r^{pq} = 0$, because either the domain or the codomain of d_r^{pq} equals 0 (for $r = 3$, see Fig. 20.2). Therefore, $\mathrm{Ker}(d_r^{pq}) = E_r^{pq}$ and $\mathrm{Im}(d_r^{p-r,q+r-1}) = 0$, which implies that the homomorphisms $E_r^{pq} = H^{pq}(E_r) \to E_{r+1}^{pq}$ are isomorphisms. Consequently,

$$E_r^{pq} \approx E_\infty^{pq},\ r \geq 3,\ p, q \in \mathbb{Z}. \tag{49}$$

In particular, one has

$$E_3^{n0} \approx E_\infty^{n0}, \tag{50}$$

$$E_3^{n-1,1} \approx E_\infty^{n-1,1}. \tag{51}$$

Clearly, (47), (48), (50) and (51) yield an exact sequence

$$0 \to E_3^{n0} \xrightarrow{\alpha_n} M^n \xrightarrow{\beta_n} E_3^{n-1,1} \to 0,\ n \in \mathbb{Z}. \tag{52}$$

One also has an exact sequence

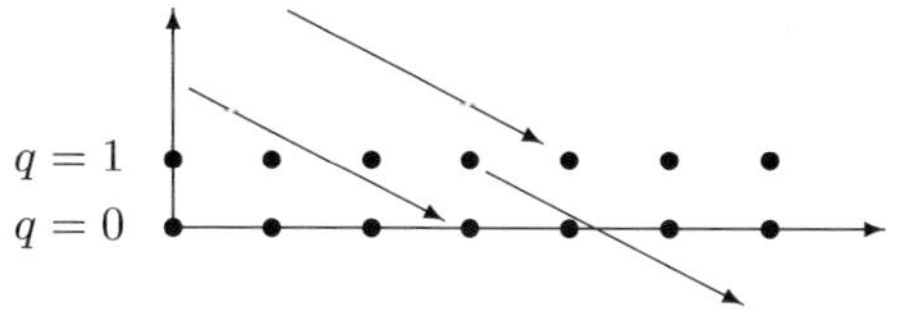

Fig. 20.2. The bigraded differential module (E_3, d_3)

$$0 \to E_3^{n-1,1} \xrightarrow{\gamma_n} E_2^{n-1,1} \xrightarrow{d_2} E_2^{n+1,0}, \quad n \in \mathbb{Z}. \tag{53}$$

Indeed, $E_3^{n-1,1} \approx H^{n-1,1}(E_2) = \mathrm{Ker}\,(d_2^{n-1,1})$, because $E_2^{n-3,2} = 0$ and thus, also $\mathrm{Im}(d_2^{n-3,2}) = 0$. Finally, we also have an exact sequence

$$E_2^{n-1,1} \xrightarrow{d_2} E_2^{n+1,0} \xrightarrow{\delta_{n+1}} E_3^{n+1,0} \to 0, \quad n \in \mathbb{Z}. \tag{54}$$

Indeed, $E_3^{n+1,0} \approx H^{n+1,0}(E_2) = E_2^{n+1,0}/\mathrm{Im}\,(d_2^{n-1,1})$, because $E_2^{n+3,-1} = 0$ and thus, $Z_2^{n+1,0} = E_2^{n+1,0}$.

Combining the exact sequences (52), (53) and (54), one obtains diagram (55), where α_n, γ_n and α_{n+1}, are monomorphisms, while δ_n, β_n and δ_{n+1} are epimorphisms. Putting $\varepsilon_n = \alpha_n\delta_n$ and $\eta_n = \gamma_n\beta_n$ and using the exactness of the sequences (52), (53) and (54), one readily obtains the desired exact sequence (40).

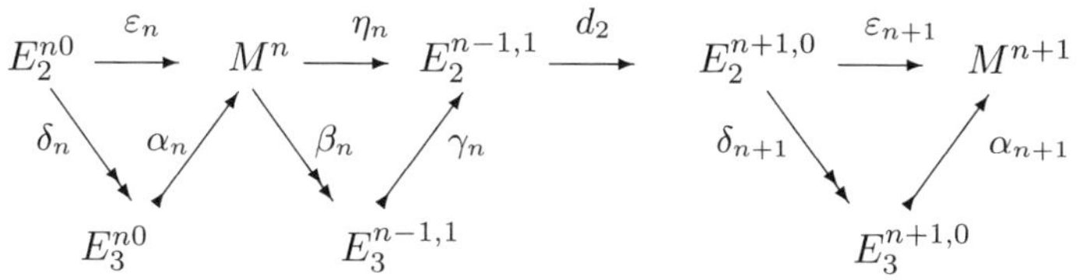

(55)

Now assume that also $E_2^{n1} = 0$, for all $n \in \mathbb{Z}$. By (40), we have an exact sequence

$$E_2^{n-2,1} \to E_2^{n0} \to M^n \to E_2^{n-1,1}, \quad n \in \mathbb{Z}. \tag{56}$$

Since $E_2^{n-2,1} = E_2^{n-1,1} = 0$, one obtains the desired conclusion $E_2^{n0} \approx M^n$, $n \in \mathbb{Z}$. □

20.2 The spectral sequences of a bicomplex

A *cochain bicomplex* of modules $C = (C, \delta', \delta'')$ is a bigraded module (C^{pq}) endowed with two differentials δ' and δ'' of bidegrees $(1,0)$ and $(0,1)$, respectively, such that $\delta'\delta'' + \delta''\delta' = 0$. A *mapping of cochain bicomplexes* is a mapping of bigraded modules $f: C \to D$, which commutes with the coboundary operators, i.e., $\delta' f = f\delta', \delta'' f = f\delta''$. With every cochain bicomplex C is associated a *total cochain complex* of the first type $T(\boldsymbol{C}) = (C^n, \delta)$. It is defined by

$$C^n = \bigoplus_{p+q=n} C^{pq},\ n \in \mathbb{Z}, \tag{1}$$

$$\delta = \delta' + \delta'' \colon C^n \to C^{n+1}. \tag{2}$$

Clearly, a mapping of cochain bicomplexes $f \colon C \to D$ induces a mapping of total complexes $f \colon T(C) \to T(D)$.

Beside the cohomology $H^n(C)$ of the total complex $T(C)$, one also associates with C *iterated cohomology modules* $H'^p(H''^q(C))$ and $H''^q(H'^p(C))$, defined as follows. For a fixed index p, the bicomplex C determines a cochain complex

$$C^{p*} = (\ldots \to C^{p,q-1} \xrightarrow{\delta''} C^{pq} \xrightarrow{\delta''} C^{p,q+1} \to \ldots). \tag{3}$$

By definition, its q-th cohomology module equals

$$H^q(C^{p*}) = \frac{\operatorname{Ker}(\delta'' \colon C^{pq} \to C^{p,q+1})}{\operatorname{Im}(\delta'' \colon C^{p,q-1} \to C^{pq})}. \tag{4}$$

The differential δ' induces homomorphisms $\delta' \colon H^q(C^{p*}) \to H^q(C^{p+1,*})$. Indeed, the anticommutativity of the diagram

$$\begin{array}{ccc} C^{p,q+1} & \xrightarrow{\delta'} & C^{p+1,q+1} \\ \delta'' \uparrow & & \uparrow \delta'' \\ C^{pq} & \xrightarrow[\delta']{} & C^{p+1,q} \end{array} \tag{5}$$

implies that δ' maps $\operatorname{Ker}(\delta'' \colon C^{pq} \to C^{p,q+1})$ to $\operatorname{Ker}(\delta'' \colon C^{p+1,q} \to C^{p+1,q+1})$. The same diagram for $q-1$ implies that δ' maps $\operatorname{Im}(\delta'' \colon C^{p,q-1} \to C^{p,q})$ to $\operatorname{Im}(\delta'' \colon C^{p+1,q-1} \to C^{p+1,q})$. Since $(\delta')^2 = 0$, one sees that

$$\ldots \to H^q(C^{p-1,*}) \xrightarrow{\delta'} H^q(C^{p,*}) \xrightarrow{\delta'} H^q(C^{p+1,*}) \to \ldots \tag{6}$$

is a cochain complex. One denotes it by $H''^q(C)$ to emphasize that its terms are cohomologies with respect to δ''. Finally, $H'^p(H''^q(C))$ is the p-th cohomology module of (6). The module $H''^q H'^p(C)$ is defined analogously.

On the total complex $T(C) = (C^n, \delta)$ one defines two (decreasing) filtrations F', F''. The *first filtration* is given by the subcomplexes $C'_p = F'_p(C) = (C'^n_p, \delta)$, where

$$C'^n_p = \bigoplus_{i \geq p} C^{i,n-i},\ p \in \mathbb{Z}. \tag{7}$$

Note that C'_p is a subcomplex of $C = (C^n, \delta)$. Indeed, $C'^n_p \subseteq C^n$. Moreover, $\delta'(C^{i,n-i}) \subseteq C^{i+1,n-i}$ and $\delta''(C^{i,n-i}) \subseteq C^{i,n+1-i}$ belong to C'^{n+1}_p, for $i \geq p$. Also note that the modules C'^n_p, $p \in \mathbb{Z}$, form a decreasing filtration of C^n. The *second filtration* is defined by

$$C''^n_q = \bigoplus_{j \geq q} C^{n-j,j}, \quad q \in \mathbb{Z}. \tag{8}$$

If $(C^{pq}, \delta', \delta'')$ is a first quadrant bicomplex, then for any $n \in \mathbb{Z}$, $p > n$ implies $C'^n_p = 0$, because $p \geq i$ implies $n - i < 0$, hence also $C^{i,n-i} = 0$. Consequently, the first filtration of a first quadrant bicomplex is regular. An analogous argument shows that also the second filtration is regular.

For a mapping of cochain bicomplexes $f: C \to D$, $f(F'_p(C)) \subseteq F'_p(D)$ and $f(F''_p(C)) \subseteq F''_p(D)$. Therefore, for both filtrations of C, f is a mapping of filtered complexes.

THEOREM 20.6. *Let $C = (C^{pq}, \delta', \delta'')$ be a cochain bicomplex. The spectral sequence (E'_r, d'_r), associated with the first filtration F' of the total complex $T(C) = (C^n, \delta)$ of C, has the second term*

$$E'^{pq}_2 \approx H'^p(H''^q(C)). \tag{9}$$

Analogously, the spectral sequence (E''_r, d''_r) associated with the second filtration F'' of $T(C)$ has the second term

$$E''^{pq}_2 \approx H''^q(H'^p(C)). \tag{10}$$

For a mapping of cochain bicomplexes $f: C \to D$, for both filtrations of C, f is a mapping of filtered complexes and thus, it induces mappings of the corresponding spectral sequences $f': E'(C) \to E'(D)$ and $f'': E''(C) \to E''(D)$. If one identifies $E'^{pq}_2(C)$ with $H'^p(H''^q(C))$ and $E'^{pq}_2(D)$ with $H'^p(H''^q(D))$ according to (9), $f': E'^{pq}_2(C) \to E'^{pq}_2(D)$ gets identified with the homomorphism $f': H'^p(H''^q(C)) \to H'^p(H''^q(D))$, naturally induced by $f: C \to D$.

Proof. By (20.1.29),

$$E'^{pq}_0 = C'^{p+q}_p / C'^{p+q}_{p+1}, \tag{11}$$

and the differential $d^{pq}_0: E'^{pq}_0 \to E'^{p,q+1}_0$ is induced by $\delta = \delta' + \delta''$. However, $\delta'(C'^{p+q}_p) \subseteq C'^{p+q+1}_{p+1}$, which is the denominator of the expression (11), for $(p, q+1)$. Hence, δ' induces the homomorphism 0 and d^{pq}_0 is induced by $\delta'': C^{pq} \to C^{p,q+1}$. Consequently,

$$E'^{pq}_1 = H^{p+q}(C'_p / C'_{p+1}) = H''^q(C^{p,*}). \tag{12}$$

By Theorem 20.3, the operator $d^{pq}_1: E'^{pq}_1 \to E'^{p+1,q}_1$ is the connecting homomorphism $H''^q(C^{p,*}) \to H''^{q+1}(C^{p+1,*})$, which belongs to the exact sequence of cochain complexes

$$0 \to C'_{p+1}/C'_{p+2} \to C'_p/C'_{p+2} \to C'_p/C'_{p+1} \to 0. \tag{13}$$

However, this sequence coincides with the sequence

$$0 \to C^{p+1,*} \to C^{p,*} \oplus C^{p+1,*} \to C^{p,*} \to 0, \tag{14}$$

because

$$(C'_p/C'_{p+2})^n = C'^n_p/C'^n_{p+2} = C^{p,n-p} \oplus C^{p+1,n-p-1}. \tag{15}$$

Moreover, the differentials on $C^{p,*}$ and $C^{p+1,*}$ coincide with δ'', while the differential on the middle term of (14) is given by $\delta = \delta' \oplus \delta'' \colon C^{pq} \to C^{p+1,q} \oplus C^{p,q+1}$. Consequently, $d_1^{pq} \colon E_1'^{pq} = H''^q(C^{p,*}) \to H''^q(C^{p+1,*}) = E_1'^{p+1,q}$ maps a class $\alpha \in H''^q(C^{p,*})$ to the class of δz, where $z \in C^{pq}$ is a cocycle from α. Since $\delta'' z = 0$, one has $\delta z = \delta' z \in C^{p+1,q}$, which shows that d_1^{pq} is induced by $\delta' \colon C^{p,*} \to C^{p+1,*}$. This establishes (9). The proofs of the remaining assertions are either analogous or straightforward. □

20.3 The Roos spectral sequence

For a direct system of modules $\boldsymbol{A} = (A^\lambda, a^{\lambda\lambda'}, \Lambda)$, $\mathrm{Hom}\,(\boldsymbol{A}, G)$ denotes the inverse system

$$\mathrm{Hom}\,(\boldsymbol{A}, G) = (\mathrm{Hom}\,(A^\lambda, G), \mathrm{Hom}\,(a^{\lambda\lambda'}, 1), \Lambda). \tag{1}$$

Similarly,

$$\mathrm{Ext}^n\,(\boldsymbol{A}, G) = (\mathrm{Ext}^n\,(A^\lambda, G), \mathrm{Ext}^n\,(a^{\lambda\lambda'}, 1), \Lambda). \tag{2}$$

The main purpose of this subsection is to prove the existence of a spectral sequence with properties stated in the theorem which follows. This sequence was discovered by J.-E. Roos and we call it the *Roos spectral sequence* (Roos 1961).

THEOREM 20.7. *For a direct system of modules* $\boldsymbol{A} = (A^\lambda, a^{\lambda\lambda'}, \Lambda)$ *and an arbitrary module* G, *there exists a strongly convergent spectral sequence* (E_r, d_r), $r \geq 0$, *and there exist regular filtrations* F *of the modules*

$$M^n = \mathrm{Ext}^n\,(\mathrm{colim}\,\boldsymbol{A}, G),\ \ n \geq 0, \tag{3}$$

such that

$$E_2^{pq} \approx \lim{}^p\,\mathrm{Ext}^q\,(\boldsymbol{A}, G), \tag{4}$$

$$E_\infty^{pq} \approx F_p(M^{p+q})/F_{p+1}(M^{p+q}). \tag{5}$$

The proof of the above theorem is obtained by constructing a certain bicomplex $C = (C, \delta', \delta'')$ to which we refer as to the *Roos bicomplex of the direct system* $\boldsymbol{A}$. The desired spectral sequence is then the spectral sequence (E'_r, d'_r) associated with the first filtration F' of the total complex $T(C)$. The sequence (E''_r, d''_r) associated with the second filtration F'' of $T(C)$ is degenerate and is needed in the proof of (5).

Proof of Theorem 20.7. First recall that in 14.1 we have associated with $\boldsymbol{A}$ a chain complex $L = (L_n, \partial)$ and a homomorphism $\eta \colon L_0 \to \mathrm{colim}\boldsymbol{A}$. The modules L_n, the boundary operators $\partial \colon L_n \to L_{n-1}$ and η were given by

(14.1.17), (14.1.18) and (14.1.19), respectively. It was proved in Lemma 14.7 that the following sequence is exact.

$$0 \leftarrow \operatorname{colim} \boldsymbol{A} \xleftarrow{\eta} L_0 \xleftarrow{\partial} L_1 \xleftarrow{\partial} L_2 \leftarrow \dots. \tag{6}$$

Choose an injective resolution of the module G,

$$0 \to G \xrightarrow{\delta} I^0 \xrightarrow{\delta} I^1 \dots. \tag{7}$$

Then form a first quadrant cochain bicomplex $(C^{pq}, \delta', \delta'')$. By definition,

$$C^{pq} = \operatorname{Hom}(L_p, I^q) \tag{8}$$

and the differentials $\delta' \colon C^{pq} \to C^{p+1,q}$ and $\delta'' \colon C^{pq} \to C^{p,q+1}$ are induced by $\partial \colon L_p \to L_{p-1}$ and $(-1)^p\delta \colon I^q \to I^{q+1}$, respectively. Note that $\delta'^2 = 0$, $\delta''^2 = 0$, because $\partial^2 = 0$, $\delta^2 = 0$. Moreover, for $\alpha \in \operatorname{Hom}(L_p, I^q)$, one has $\delta'\delta''(\alpha) = (-1)^p \delta\alpha\partial$, $\delta''\delta'(\alpha) = (-1)^{p+1}\delta\alpha\partial$ and thus, $\delta'\delta'' + \delta''\delta' = 0$. The two filtrations F', F'' of the total complex $T(C)$ yield convergent spectral sequences (E'_r, d'_r) and (E''_r, d''_r), respectively. By Theorems 20.6 and 20.5,

$$E_2'^{pq} \approx H'^p(H''^q(C)), \tag{9}$$

$$E_\infty'^{pq} \approx F'_p(H^{p+q}(T(C)))/F'_{p+1}(H^{p+q}(T(C))), \tag{10}$$

where F' now denotes the induced filtrations on the modules $H^n(T(C))$.

To determine $H'^p(H''^q(C))$, recall that $H^q(C^{p,*})$ is the q-th cohomology of the cochain complex

$$C^{p,*} = (\ldots \to \operatorname{Hom}(L_p, I^q) \xrightarrow{\delta''} \operatorname{Hom}(L_p, I^{q+1}) \to \ldots). \tag{11}$$

Hence, by the definition of Ext^q,

$$H^q(C^{p,*}) = \operatorname{Ext}^q(L_p, G). \tag{12}$$

The cochain complex $H''^q(C)$ consists of the modules $H^q(C^{p,*})$ and the operator $\delta' \colon H^q(C^{p,*}) \to H^q(C^{p+1,*})$, which is induced by $\delta' \colon C^{pq} \to C^{p+1,q}$. Consequently, it is given by

$$\delta' = \operatorname{Ext}^q(\partial, 1) \colon \operatorname{Ext}^q(L_p, G) \to \operatorname{Ext}^q(L_{p+1}, G). \tag{13}$$

To determine the p-th cohomology $H'^p(H''^q(C))$ of $H''^q(C)$, we use the following lemma.

LEMMA 20.8. *The cochain complex $H''^q(C)$ is isomorphic to the cochain complex $K(\boldsymbol{X})$, for $\boldsymbol{X} = \operatorname{Ext}^q(\boldsymbol{A}, G)$, defined in* 11.5.

The explicit description of $\lim^p \boldsymbol{X}$ as the p-th cohomology of $K(\boldsymbol{X})$ (see Corollary 11.47) shows that

$$E_2'^{pq} \approx H'^p(H''^q(C)) = \lim^p \operatorname{Ext}^q(\boldsymbol{A}, G). \tag{14}$$

Now consider the sequence (E_r'', d_r'') associated with the second filtration F'' of $T(C)$. By Theorem 20.6, one has

$$E_2''^{pq} \approx H''^q(H'^p(C)), \tag{15}$$

where F'' now denotes the induced filtration on the modules $H^n(T(C))$. Note that $H^p(C^{*,q})$ is the p-th cohomology of the cochain complex

$$0 \to \operatorname{Hom}(L_0, I^q) \to \ldots \to \operatorname{Hom}(L_p, I^q) \xrightarrow{\delta'} \operatorname{Hom}(L_{p+1}, I^q) \to \ldots. \tag{16}$$

Since I^q is an injective module, $\operatorname{Hom}(\,.\,, I^q)$ is an exact contravariant functor. Therefore, exactness of (6) implies exactness of

$$0 \to \operatorname{Hom}(\operatorname{colim} \boldsymbol{A}, I^q) \to \operatorname{Hom}(L_0, I^q) \xrightarrow{\delta'} \operatorname{Hom}(L_1, I^q) \to \ldots. \tag{17}$$

Comparing (16) and (17), one concludes that

$$H^p(C^{*,q}) = \begin{cases} \operatorname{Hom}(\operatorname{colim} \boldsymbol{A}, I^q), & p = 0, \\ 0, & p \neq 0. \end{cases} \tag{18}$$

Recall that the cochain complex $H'^p(C)$ consists of the modules $H^p(C^{*,q})$. Its coboundary operator $\delta'' \colon H^p(C^{*,q}) \to H^p(C^{*,q+1})$ is induced by $\delta'' \colon C^{pq} \to C^{p,q+1}$. Hence, for $p = 0$, it is given by

$$\operatorname{Hom}(1, \delta'') \colon \operatorname{Hom}(\operatorname{colim} \boldsymbol{A}, I^q) \to \operatorname{Hom}(\operatorname{colim} \boldsymbol{A}, I^{q+1}). \tag{19}$$

Consequently, by (15),

$$E_2''^{pq} = \begin{cases} \operatorname{Ext}^q(\operatorname{colim} \boldsymbol{A}, G), & p = 0 \\ 0, & p \neq 0, \end{cases} \tag{20}$$

Formula (20) shows that the sequence (E_r'', d_r'') is degenerate (notice that the roles of p and q have been interchanged). Therefore, Theorem 20.5 applies and yields

$$E_2''^{0n} \approx H^n(T(C)). \tag{21}$$

Hence, by (20),

$$H^n(T(C)) \approx \operatorname{Ext}^n(\operatorname{colim} \boldsymbol{A}, G). \tag{22}$$

F' can be viewed as a filtration of $M^n = \operatorname{Ext}^n(\operatorname{colim} \boldsymbol{A}, G)$ and thus, (10) and (22) show that (E_r', d_r') and F' have the required properties. □

Proof of Lemma 20.8. For a family of modules (A^λ) and an arbitrary module G, consider the homomorphism $\Phi = \prod_\lambda \operatorname{Hom}(i_\lambda, 1) \colon \operatorname{Hom}(\bigoplus_\lambda A^\lambda, G) \to \prod_\lambda \operatorname{Hom}(A^\lambda, G)$, where $i_\lambda \colon A^\lambda \to \bigoplus_\lambda A^\lambda$ are natural inclusions. Φ is natural with respect to G. Moreover, it is an isomorphism with inverse Ψ, given by

$\Psi = \bigoplus_\lambda p_\lambda = \prod_\lambda \mathrm{Hom}(A^\lambda, G) \to \bigoplus_\lambda A^\lambda$, where $p_\lambda : \prod_\lambda \mathrm{Hom}(A^\lambda, G) \to \mathrm{Hom}(A^\lambda, G)$ denotes the natural projection. The isomorphism Φ induces isomorphisms $\Phi^q : \mathrm{Ext}^q(\bigoplus_\lambda A^\lambda, G) \to \prod_\lambda \mathrm{Ext}^q(A^\lambda, G)$, $q \geq 1$, defined as follows (see e.g., (Cartan, Eilenberg 1956), Chapter V, Proposition 9.4). Consider any injective resolution of G, say (7). It induces a commutative diagram

$$\begin{array}{ccccccc} 0 & \longrightarrow & \mathrm{Hom}(\bigoplus_\lambda A^\lambda, I^0) & \xrightarrow{\mathrm{Hom}(1,\delta)} & \mathrm{Hom}(\bigoplus_\lambda A^\lambda, I^1) & \longrightarrow & \dots \\ & & \Phi \big\downarrow & & \big\downarrow \Phi & & \\ 0 & \longrightarrow & \prod_\lambda \mathrm{Hom}(A^\lambda, I^0) & \xrightarrow[\prod_\lambda \mathrm{Hom}(1,\delta)]{} & \prod_\lambda \mathrm{Hom}(A^\lambda, I^1) & \longrightarrow & \dots, \end{array} \tag{23}$$

whose vertical arrows are isomorphisms. Hence, Φ induces an isomorphism of cochain complexes $\Phi : \mathrm{Hom}(\bigoplus_\lambda A^\lambda, \boldsymbol{I}) \to \prod_\lambda \mathrm{Hom}(A^\lambda, \boldsymbol{I})$. Clearly, this isomorphism of complexes induces isomorphisms Φ^q of the corresponding cohomology groups. By the definition of Ext^q, the q-th cohomology of the first row equals $\mathrm{Ext}^q(\bigoplus_\lambda A^\lambda, G)$, while the q-th cohomology of the second row equals $\prod_\lambda \mathrm{Ext}^q(A^\lambda, G)$. Hence, $\Phi^q : \mathrm{Ext}^q(\bigoplus_\lambda A^\lambda, G) \to \prod_\lambda \mathrm{Ext}^q(A^\lambda, G)$ is an isomorphism.

In particular, for the family $(A^{\lambda_0}, \boldsymbol{\lambda})$, where $\boldsymbol{\lambda} = (\lambda_0, \dots, \lambda_p) \in \Lambda_p$, one obtains isomorphisms

$$\Phi : \mathrm{Hom}\,(\bigoplus_{\boldsymbol{\lambda} \in \Lambda_p} A^{\lambda_0}, G) \longrightarrow \prod_{\boldsymbol{\lambda} \in \Lambda_p} \mathrm{Hom}\,(A^{\lambda_0}, G); \tag{24}$$

and induced isomorphisms

$$\Phi^q : \mathrm{Ext}^q\,(\bigoplus_{\boldsymbol{\lambda} \in \Lambda_p} A^{\lambda_0}, G) \longrightarrow \prod_{\boldsymbol{\lambda} \in \Lambda_p} \mathrm{Ext}^q\,(A^{\lambda_0}, G),\ q \geq 1, \tag{25}$$

i.e.,

$$\Phi^q : \mathrm{Ext}^q\,(L_p, G) \to K^p(\mathrm{Ext}^q\,(\boldsymbol{A}, G)). \tag{26}$$

To complete the proof it remains to verify commutativity of the diagram

$$\begin{array}{ccc} \mathrm{Ext}^q\,(L_{p-1}, G) & \xrightarrow{\mathrm{Ext}^q\,(\partial, 1)} & \mathrm{Ext}^q\,(L_p, G) \\ \Phi^q \big\downarrow & & \big\downarrow \Phi^q \\ K^{p-1}\,(\mathrm{Ext}^q\,(\boldsymbol{A}, G)) & \xrightarrow[\delta]{} & K^p\,(\mathrm{Ext}^q\,(\boldsymbol{A}, G)). \end{array} \tag{27}$$

To achieve this, it suffices to show commutativity of the diagram

$$\begin{array}{ccc} \mathrm{Hom}\,(L_{p-1}, I^q) & \xrightarrow{\mathrm{Hom}\,(\partial,1)} & \mathrm{Hom}\,(L_p, I^q) \\ \Big\downarrow \Phi & & \Big\downarrow \Phi \\ K^{p-1}\,(\mathrm{Hom}\,(\boldsymbol{A}, I^q)) & \xrightarrow[\delta]{} & K^p\,(\mathrm{Hom}\,(\boldsymbol{A}, I^q)) \,, \end{array} \tag{28}$$

because this would prove commutativity of the corresponding diagram of mappings of cochain complexes, whose q-th cohomology modules form the vertices of diagram (27).

By definition, Φ maps a homomorphism $\alpha\colon L_{p-1} \to I^q$ to the element $\beta \in K^{p-1}\,(\mathrm{Hom}\,(\boldsymbol{A}, I^q))$, where $\beta(\lambda_0, \ldots, \lambda_{p-1})\colon A^{\lambda_0} \to I^q$ is the homomorphism, which maps $y \in A^{\lambda_0}$ to $\alpha(i_{\lambda_0 \ldots \lambda_{p-1}}(y)) \in I^q$. Therefore, $\delta\alpha(\lambda_0, \ldots, \lambda_p)$ maps y to

$$\alpha(i_{\lambda_1 \ldots \lambda_p}(a^{\lambda_0\lambda_1}(y))) + \sum_{j=1}^{p} (-1)^j \alpha(i_{\lambda_0 \ldots \widehat{\lambda_j} \ldots \lambda_p}(y)). \tag{29}$$

However, using formula (14.1.18) for ∂, it is readily seen that the same result is obtained following the other path in the diagram. □

We will now assume that the ground ring R of our modules is $\mathbb{Z}$, i.e., we will specialize Theorem 20.7 to the case of abelian groups.

THEOREM 20.9. *Let* $\boldsymbol{A} = (A^\lambda, a^{\lambda\lambda'}, \Lambda)$ *be a direct system of abelian groups and let* G *be an arbitrary abelian group. Then there exist isomorphisms*

$$\mathrm{Hom}\,(\mathrm{colim}\,\boldsymbol{A}, G) \approx \lim \mathrm{Hom}\,(\boldsymbol{A}, G), \tag{30}$$

$$\lim{}^p\, \mathrm{Ext}\,(\boldsymbol{A}, G) \approx \lim{}^{p+2}\, \mathrm{Hom}\,(\boldsymbol{A}, G), \;\; p \geq 1. \tag{31}$$

Moreover, there exists an exact sequence

$$\begin{gathered} 0 \to \lim{}^1 \mathrm{Hom}\,(\boldsymbol{A}, G) \to \mathrm{Ext}\,(\mathrm{colim}\,\boldsymbol{A}, G) \xrightarrow{\pi} \lim \mathrm{Ext}\,(\boldsymbol{A}, G) \to \\ \lim{}^2 \mathrm{Hom}\,(\boldsymbol{A}, G) \to 0, \end{gathered} \tag{32}$$

where the homomorphism $\pi\colon \mathrm{Ext}\,(\mathrm{colim}\,\boldsymbol{A}, G) \to \lim \mathrm{Ext}\,(\boldsymbol{A}, G)$ *is induced by the canonical homomorphisms* $a^\lambda\colon A^\lambda \to \mathrm{colim}\,\boldsymbol{A}$.

Proof. Every abelian group A is the image of a free abelian group F_0. Therefore, there exists a short exact sequence

$$0 \to A \xleftarrow{e} F_0 \leftarrow F_1 \leftarrow 0, \tag{33}$$

where $F_1 = \mathrm{Ker}(e) \subseteq F_0$. However, it is well known that a subgroup of a free abelian group is itself a free abelian group (see e.g., (Fuchs 1970), Theorem 14.5). Since free abelian groups are projective, one concludes that (33) is a short projective resolution of A and thus, $\mathrm{Ext}^q\,(A, G) = 0$, for $q \geq 2$.

Consequently, for abelian groups, the spectral sequence (E_r, d_r) of Theorem 20.7 has the property that $E_2^{pq} \neq 0$ can occur only for $q = 0$ or $q = 1$. This enables us to use the exact sequence (20.1.40) of Theorem 20.5. We conclude that, for $n \geq 2$, there is an exact sequence

$$0 \to \lim^{n-1} \operatorname{Ext}(\boldsymbol{A}, G) \to \lim^{n+1} \operatorname{Hom}(\boldsymbol{A}, G) \to 0, \tag{34}$$

which yields (31). For $n = 0$, one obtains an exact sequence

$$0 \to \lim \operatorname{Hom}(\boldsymbol{A}, G) \to \operatorname{Hom}(\operatorname{colim} \boldsymbol{A}, G) \to 0, \tag{35}$$

which yields (30). For $n = 1$, (20.1.40) becomes the exact sequence (32).

It remains to show that $\eta_1 \colon M^1 \to E_2^{01}$ coincides with the homomorphism π, induced by the canonical homomorphisms $a^\lambda \colon A^\lambda \to \operatorname{colim} \boldsymbol{A}$. To prove this, for an arbitrary $\lambda \in \Lambda$, consider the bicomplex $D = D(\lambda) = (D^{pq}, \delta', \delta'')$, where $D^{pq} = 0$, except for $D^{0q} = \operatorname{Hom}(A^\lambda, I^q)$, $q = 0, 1$. Moreover, $\delta'' \colon D^{00} \to D^{0,1}$ is given by $\operatorname{Hom}(1, \delta)$, while all other δ'' and all δ' equal 0. As before, the spectral sequence reduces to the lines $q = 0, 1$. Therefore, one can use again Theorem 20.5 and obtain an exact sequence (20.1.40). We will first show that in this case the homomorphism $\eta_1 \colon M^1 \to E_2^{01}$ coincides with $\mathrm{id} \colon \operatorname{Ext}(A^\lambda, G) \to \operatorname{Ext}(A^\lambda, G)$.

Indeed, using (20.1.10), (20.1.11), (20.1.18), (20.1.23) and (20.1.24), it is readily seen that the spectral sequence E of the first filtration of the bicomplex $D(\lambda)$ has the property that the canonical isomorphisms $H^{0q}(E_r) \to E_{r+1}^{0q}$, $r \geq 1$, are identities. Therefore, the canonical homomorphisms $E_r^{0q} = H^{0q}(E_r) \to E_{r+1}^{0q}$ and $E_r^{0q} \to E_\infty^{0q}$, $r \geq 1$, are identity isomorphisms. In particular, in the present case, the homomorphisms $\gamma_1 \colon E_3^{01} \to E_2^{01}$ from (20.1.51) and $E_3^{01} \to E_\infty^{01}$ are identity isomorphisms. Furthermore, for $M^1 = H^1(T(D))$, one has $F_1(M^1) = 0$, because $D_1 = 0$. Hence, $M^1 \to M^1/F_1(M^1)$ is the identity. To show that $\beta_1 \colon M^1 \to E_3^{01}$ from (20.1.50) is also the identity, it suffices to note that the canonical homomorphism $E_\infty^{01} \to M^1/F_1(M^1)$ is the identity. Indeed, $D_0 = D$ and therefore, the inclusion $D_0 \to D$ induces the identity homomorphism $i_0 \colon H^1(D_0) \to H^1(D)$. Moreover, by (20.1.16), (20.1.17) and (20.1.36), $E_\infty^{01} = Z_\infty^{01}/B_\infty^{01}$. However, $Z_\infty^{01} = \operatorname{Ker}(\delta \colon D^1 \to D^2)$ and $B_\infty^{01} = \operatorname{Im}(\delta \colon D^0 \to D^1)$, so that the homomorphism $E_\infty^{01} \to H^1(T(D))$ is the identity isomorphism. Consequently, $\beta_1 = \mathrm{id}$ and $\eta_1 = \gamma_1 \beta_1 \colon M^1 \to E_2^{01}$ are the identity isomomorphisms. Finally, since the total differential $D^0 \to D^1$ coincides with $\delta'' \colon \operatorname{Hom}(A^\lambda, I^0) \to \operatorname{Hom}(A^\lambda, I^1)$, one concludes that $H^1(T(D)) = \operatorname{Ext}(A^\lambda, G)$ and the assertion is proved.

We now define a mapping of bicomplexes $f \colon C \to D(\lambda)$, where C is the bicomplex used in the construction of the Roos sequence. By definition, $f^{0q} \colon C^{0q} \to D^{0q}$ is the homomorphism $\operatorname{Hom}(L_0, I^q) \to \operatorname{Hom}(A^\lambda, I^q)$, induced by the natural inclusion $A^\lambda \to \oplus_{\lambda_0} A^{\lambda_0} = L_0$. The mapping f induces a mapping between the corresponding spectral sequences and thus, it also induces a mapping between the corresponding exact sequences (20.1.40) (for $n = 1$). This yields a commutative diagram

$$\begin{array}{ccc} \operatorname{Ext}(\operatorname{colim}\boldsymbol{A}, G) & \xrightarrow{\pi} & \lim \operatorname{Ext}(\boldsymbol{A}, G) \\ \rho \downarrow & & \downarrow \\ \operatorname{Ext}(A^\lambda, G) & \xrightarrow[\text{id}]{} & \operatorname{Ext}(A^\lambda, G)\,. \end{array} \tag{36}$$

The homomorphism ρ is induced by id: $A^\lambda \to A^\lambda$, i.e., by the natural homomorphisms $A^\lambda \to \operatorname{colim}\boldsymbol{A}$. The homomorphism $\lim \operatorname{Ext}(\boldsymbol{A}, G) \to \operatorname{Ext}(A^\lambda, G)$ sends the element $(\varepsilon^{\lambda_0})$ to ε^λ, because it is also induced by id: $A^\lambda \to A^\lambda$. Consequently, if $\pi(\varepsilon) = (\varepsilon^{\lambda_0})$, then by (53), $\varepsilon^\lambda = \rho(\varepsilon)$. This shows that π is precisely the homomorphism described in Theorem 20.9. □

20.4 Pure extension functors Pext^n

In this subsection we apply relative homological algebra to the case of purely exact sequences (see, e.g., (Butler, Horrocks 1961)). A short exact sequence of modules

$$0 \to X' \xrightarrow{u'} X \xrightarrow{u} X'' \to 0 \tag{1}$$

is said to be *purely exact* provided, for every module Y, the induced sequence

$$0 \to X' \otimes Y \xrightarrow{u' \otimes 1} X \otimes Y \xrightarrow{u \otimes 1} X'' \otimes Y \to 0 \tag{2}$$

is exact. Since $\otimes Y$ is a right exact functor, the above definition is equivalent to the assertion that the monomorphism $X' \to X$ induces a monomorphism $X' \otimes Y \to X \otimes Y$, for every module Y.

EXAMPLE 20.10. If the exact sequence (1) splits, i.e., if X' is a direct summand of X, then (1) is purely exact.

REMARK 20.11. If (1) is a purely exact sequence, then for every module Y, the sequence (2) is also purely exact. Indeed, applying $\otimes Z$ to (2) is the same as applying $\otimes(Y \otimes Z)$ to (1).

A long exact sequence of modules

$$\ldots \leftarrow X_{n-1} \xleftarrow{d_n} X_n \xleftarrow{d_{n+1}} X_{n+1} \leftarrow \ldots \tag{3}$$

is said to be *purely exact* provided the short exact sequences

$$0 \leftarrow \operatorname{Im}(d_n) \xleftarrow{d_n} X_n \xleftarrow{\supseteq} \operatorname{Ker}(d_n) \leftarrow 0 \tag{4}$$

are purely exact.

A module P is said to be *purely projective*, provided for any purely exact sequence (1), every homomorphism $f: P \to X''$ admits a homomorphism $g: P \to X$ such that $ug = f$.

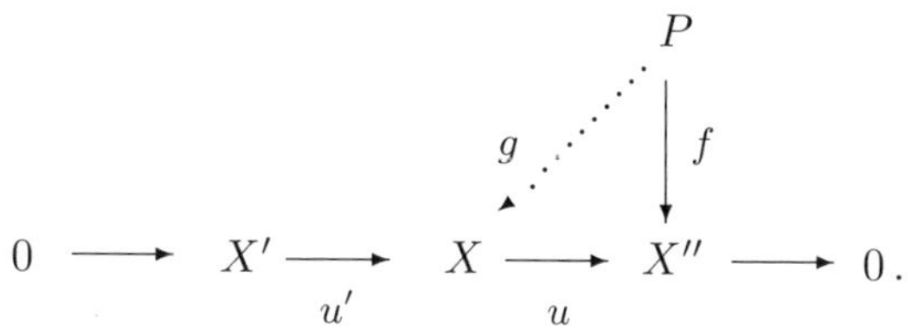

(5)

Dually, a module I is said to be *purely injective*, provided for any purely exact sequence (1), every homomorphism $f: X' \to I$ admits a homomorphism $g: X \to I$ such that $gu' = f$.

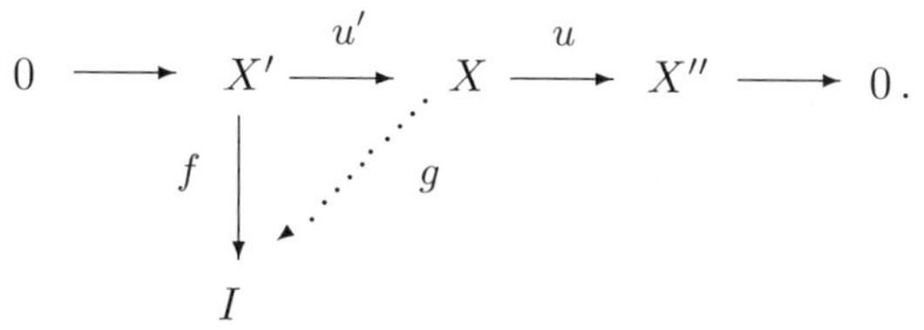

(6)

A *purely projective resolution* of a module A is a purely exact sequence

$$0 \leftarrow A \xleftarrow{\varepsilon} P_0 \xleftarrow{d_1} P_1 \xleftarrow{d_2} P_2 \leftarrow \ldots \tag{7}$$

such that the modules P_n are purely projective. One defines *purely injective resolutions* analogously.

To conclude that every R-module admits purely projective and purely injective resolutions, it suffices to know that there are enough purely projective and enough purely injective R-modules. This means that, for every R-module A, there exist purely exact sequences

$$0 \leftarrow X \xleftarrow{e} P_0 \xleftarrow{u_1} R_0 \leftarrow 0, \tag{8}$$

$$0 \rightarrow X \xrightarrow{e} I^0 \xrightarrow{v_1} J^0 \rightarrow 0, \tag{9}$$

where P_0 is purely projective and I^0 is purely injective.

If a ring R is such that there are enough purely projective and enough purely injective modules, then one can repeat the arguments which led to the definition of the functor Ext^n, using purely exact sequences instead of exact sequences and purely projective (purely injective) modules instead of projective (injective) modules. One thus obtains new functors Pext^n, instead of the functors Ext^n. In particular, to obtain the module $\mathrm{Pext}^n(A, B)$, one chooses a purely projective resolution $0 \leftarrow A \leftarrow P_0 \leftarrow P_1 \leftarrow \ldots$ of A. Then $\mathrm{Pext}^n(A, B)$ is the n-th homology module of the cochain complex $0 \to \mathrm{Hom}(P_0, B) \to \mathrm{Hom}(P_1, B) \to \ldots$. The same module $\mathrm{Pext}^n(A, B)$ is also obtained by choosing a purely injective resolution $0 \to B \to I^0 \to I^1 \to \ldots$ of B and then, taking the n-th homology module of the cochain complex $0 \to \mathrm{Hom}(A, I^0) \to \mathrm{Hom}(A, I^1) \to \ldots$. Modifying appropriately the proofs of theorems for Ext^n, one obtains analogous theorems for Pext^n. In particular, a short purely exact sequence of modules

$$0 \to X' \to X \to X'' \to 0 \tag{10}$$

yields an exact sequence

$$\begin{aligned} &0 \to \mathrm{Hom}\,(A, X') \to \mathrm{Hom}\,(A, X) \to \mathrm{Hom}\,(A, X'') \to \mathrm{Pext}^1(A, X') \to \dots \\ &\mathrm{Pext}^n(A, X') \to \mathrm{Pext}^n(A, X) \to \mathrm{Pext}^n(A, X'') \to \mathrm{Pext}^{n+1}(A, X') \to \dots . \end{aligned} \tag{11}$$

Similarly, a short purely exact sequence

$$0 \to A' \to A \to A'' \to 0 \tag{12}$$

yields an exact sequence

$$\begin{aligned} &0 \to \mathrm{Hom}\,(A'', X) \to \mathrm{Hom}\,(A, X) \to \mathrm{Hom}\,(A', X) \to \mathrm{Pext}^1(A'', X) \to \dots \\ &\mathrm{Pext}^n(A'', X) \to \mathrm{Pext}^n(A, X) \to \mathrm{Pext}^n(A', X) \to \mathrm{Pext}^{n+1}(A'', X) \to \dots . \end{aligned} \tag{13}$$

Furthermore, $\mathrm{Pext}^n(A, X) = 0$, whenever A is purely projective or X is purely injective.

The following theorem is the analogue of Theorem 20.7.

THEOREM 20.12. *Let R be a ring such that R-modules admit enough purely projective and enough purely injective modules. Then, for a direct system of modules $\boldsymbol{A} = (A^\lambda, a^{\lambda\lambda'}, \Lambda)$ and an arbitrary module G, there exists a strongly convergent spectral sequence (E_r, d_r), $r \geq 0$, and there exist regular filtrations F of the modules*

$$M^n = \mathrm{Pext}^n\,(\mathrm{colim}\,\boldsymbol{A}, G),\ n \in \mathbb{Z}, \tag{14}$$

such that

$$E_2^{pq} \approx \lim{}^p\, \mathrm{Pext}^q\,(\boldsymbol{A}, G), \tag{15}$$

$$E_\infty^{pq} \approx F_p(M^{p+q})/F_{p+1}(M^{p+q}). \tag{16}$$

In Lemma 14.7 we proved exactness of the sequence

$$0 \leftarrow \mathrm{colim}\boldsymbol{A} \overset{\eta}{\longleftarrow} L_0 \overset{\partial}{\longleftarrow} L_1 \overset{\partial}{\longleftarrow} L_2 \leftarrow \dots . \tag{17}$$

In order to prove Theorem 20.12, we need the following modification.

LEMMA 20.13. *The sequence* (17) *is purely exact.*

Proof. By the exactness of (17), the following short sequences are exact.

$$\begin{aligned} &0 \leftarrow \mathrm{colim}\boldsymbol{A} \leftarrow L_0 \leftarrow \partial L_1 \leftarrow 0, \\ &\dots\dots\dots\dots\dots\dots\dots\dots \\ &0 \leftarrow \partial L_n \leftarrow L_n \leftarrow \partial L_{n+1} \leftarrow 0, \\ &\dots\dots\dots\dots\dots\dots\dots\dots . \end{aligned} \tag{18}$$

We must show that they are even purely exact. According the proof of Lemma 14.7, these sequences are colimits of direct systems of exact sequences

$$\begin{aligned} &0 \leftarrow A^\mu \leftarrow L_0^\mu \leftarrow \partial(L_1^\mu) \leftarrow 0, \\ &\cdots\cdots\cdots\cdots\cdots\cdots\cdots\cdots\cdots \\ &0 \leftarrow \partial(L_n^\mu) \leftarrow L_n^\mu \leftarrow \partial(L_{n+1}^\mu) \leftarrow 0, \\ &\cdots\cdots\cdots\cdots\cdots\cdots\cdots\cdots\cdots . \end{aligned} \tag{19}$$

Since the functor $\otimes Y$ commutes with colimits, it suffices to show that these short sequences are purely exact. Indeed, in the proof of Lemma 14.7, we exhibited homomorphisms $\varepsilon^\mu \colon A^\mu \to L_0^\mu$ and $c_n^\mu \colon L_n^\mu \to L_{n+1}^\mu$, $n \geq 0$, such that ε^n is a right inverse of $L_0^\mu \to A^\mu$ and $(-1)^{n+1}\partial c_n^\mu : L_n^\mu \to \partial(L_{n+1}^\mu)$ is a left inverse of $\partial(L_{n+1}^\mu) \to L_n^\mu$ (see (14.1.29) and (14.1.30)). Consequently, the sequences (19) split and thus, they are purely exact. □

Proof of Theorem 20.12. We associate with $\boldsymbol{A}$ the same chain complex L and the same homomorphism $\eta \colon L_0 \to \operatorname{colim}\boldsymbol{A}$ as in 20.3. We then choose a purely injective resolution (20.3.7) of G. A bicomplex $C = (C^{pq}, \delta', \delta'')$ is defined as before and so are the spectral sequences $(E'_r, d'_r), (E''_r, d''_r)$. By definition of Pext^n, formula (20.3.12) becomes $H^q(C^{p,*}) = \operatorname{Pext}^q(L_p, G)$. Moreover, the analogue of Lemma 20.8 now asserts that the cochain complex $H''^q(C)$ is isomorphic to the cochain complex $K(\boldsymbol{X})$, where $\boldsymbol{X} = \operatorname{Pext}^q(\boldsymbol{A}, G)$. Therefore, the explicit description of $\lim^p \boldsymbol{X}$ yields $E'^{pq}_2 \approx \lim^p \operatorname{Pext}^q(\boldsymbol{A}, G)$. Since the module I^q is purely injective, $\operatorname{Pext}^1(A, I^q) = 0$, for every module A. Therefore, (13) shows that the functor $\operatorname{Hom}(\,.\,, I^q)$ transforms purely exact sequences into exact sequences. However, by Lemma 20.13, the sequence (17) is purely exact and thus, the sequence (20.3.17) is exact. As before, one concludes that (20.3.18) holds. Moreover, (20.3.20) remains valid if one replaces Ext^q by Pext^q. Applying Theorem 20.5 as before, one obtains (20.3.21) and the analogue of (20.3.22) with Ext^n replaced by Pext^n. Hence, F' is a filtration of $\operatorname{Pext}^n(\operatorname{colim}\boldsymbol{A}, G)$ with desired properties. □

As in 20.3, we would now like to specialize the ring R to the case $R = \mathbb{Z}$, i.e., to the case of abelian groups. In order to be able to apply Theorem 20.12, we need the following facts.

THEOREM 20.14. *In the category of abelian groups there are enough purely projective and enough purely injective groups.*

To obtain the analogue of Theorem 20.9, we also need the following result.

THEOREM 20.15. *Every abelian group A admits a short purely projective resolution*

$$0 \leftarrow A \leftarrow P_0 \leftarrow P_1 \leftarrow 0. \tag{20}$$

Consequently, for an arbitrary abelian group G, $\operatorname{Pext}^n(A, G) = 0$, *for* $n \geq 2$.

Theorems 20.14 and 20.15, whose proofs are given in the next subsection (see Remarks 20.24 and 20.43), enable us to repeat the arguments given in 20.3 and thus, obtain the following theorem.

THEOREM 20.16. *Let* $\boldsymbol{A}$ *be a direct system of abelian groups and let* G *be an arbitrary abelian group. Then there exist isomorphisms*

$$\operatorname{Hom}(\operatorname{colim} \boldsymbol{A}, G) \approx \lim \operatorname{Hom}(\boldsymbol{A}, G), \tag{21}$$

$$\lim^p \operatorname{Pext}(\boldsymbol{A}, G) \approx \lim^{p+2} \operatorname{Hom}(\boldsymbol{A}, G), \ p \geq 1. \tag{22}$$

Moreover, there exists an exact sequence

$$\begin{gathered} 0 \to \lim^1 \operatorname{Hom}(\boldsymbol{A}, G) \to \operatorname{Pext}(\operatorname{colim} \boldsymbol{A}, G) \to \lim \operatorname{Pext}(\boldsymbol{A}, G) \to \\ \lim^2 \operatorname{Hom}(\boldsymbol{A}, G) \to 0. \end{gathered} \tag{23}$$

Having in view applications to pro-homology of compact Hausdorff spaces, we will now consider direct systems $\boldsymbol{A}$ of finitely generated abelian groups and obtain the result which plays the key role in the next section.

THEOREM 20.17. *Let* $\boldsymbol{A}$ *be a direct system of finitely generated abelian groups. Then*

$$\lim^r \operatorname{Ext}(\boldsymbol{A}, G) = 0, \ \text{for } r \geq 1, \tag{24}$$

$$\lim^r \operatorname{Hom}(\boldsymbol{A}, G) = 0, \ \text{for } r \geq 2, \tag{25}$$

$$\lim^1 \operatorname{Hom}(\boldsymbol{A}, G) \approx \operatorname{Pext}(\operatorname{colim} \boldsymbol{A}, G). \tag{26}$$

Moreover, there is an exact sequence

$$0 \to \lim^1 \operatorname{Hom}(\boldsymbol{A}, G) \to \operatorname{Ext}(\operatorname{colim} \boldsymbol{A}, G) \xrightarrow{\pi} \lim \operatorname{Ext}(\boldsymbol{A}, G) \to 0, \tag{27}$$

where π *is induced by the canonical homomorphisms* $A^\lambda \to \operatorname{colim} \boldsymbol{A}$.

To prove Theorem 20.17, we need several previously proved results, as well as the following fact (which is a corollary of Theorem 20.19, proved in the next subsection).

COROLLARY 20.18. *Every finitely generated abelian group* A *is purely projective and therefore,* $\operatorname{Pext}(A, G) = 0$, *for every abelian group* G.

Proof of Theorem 20.17. Since $\boldsymbol{A}$ consists of finitely generated abelian groups, Corollary 20.18 yields $\operatorname{Pext}(\boldsymbol{A}, G) = 0$ and by (22), $\lim^r \operatorname{Hom}(\boldsymbol{A}, G) = 0$, for $r \geq 3$. Moreover, the exact sequence (23) yields $\lim^2 \operatorname{Hom}(\boldsymbol{A}, G) = 0$, as well as (26). Finally, one obtains (24) and (27), by applying Theorem 20.9. □

20.5 Some theorems on abelian groups

The purpose of this subsection is to prove Theorems 20.14 and 20.15 as well as Corollary 20.18 and thus, complete the proofs of Theorems 20.16 and 20.17. It is well known that every finitely generated abelian group is a direct sum of finitely many cyclic groups (see e.g., (Fuchs 1970), Theorem 15.5). Therefore, Corollary 20.18 is an immediate consequence of the following theorem.

THEOREM 20.19. *An abelian group is purely projective if and only if it is the direct sum of a collection of cyclic groups.*

The proofs of Theorems 20.14, 20.15 and 20.19 require a deeper analysis of purity in abelian groups. We will first characterize purely exact sequences of abelian groups, i.e., pure subgroups (see e.g., (Fuchs 1970), Theorem 60.4).

LEMMA 20.20. *A subgroup $X' \subseteq X$ of an abelian group X is a pure subgroup of X if and only if, every element $x' \in X'$, divisible in X by an integer $m \in \mathbb{Z}$, is also divisible in X' by m.*

Proof. Let X' be a subgroup of X. It is readily seen that the condition stated in the lemma is equivalent to the assertion that the induced homomorphism $X'/mX' \to X/mX$ is a monomorphism. However, the latter homomorphism is easily identified with $X' \otimes (\mathbb{Z}/m) \to X \otimes (\mathbb{Z}/m)$, which proves the necessity of the condition of Lemma 20.20. To prove sufficiency, note that $X' \otimes \mathbb{Z} \to X \otimes \mathbb{Z}$ can be identified with $X' \to X$. Therefore, if the condition from the lemma is fulfilled, then $X' \otimes Y \to X \otimes Y$ is a monomorphism, for $Y = \mathbb{Z}$ and $Y = \mathbb{Z}/m$. Since every finitely generated abelian group Y is a direct sum of finitely many copies $\mathbb{Z}$ and $\mathbb{Z}/m$, it follows that $X' \otimes Y \to X \otimes Y$ is a monomorphism, for such groups as well. Finally, an arbitrary abelian group Y is the colimit of the direct system $\boldsymbol{Y}$ of its finitely generated abelian subgroups. Therefore, the assertion that $X' \otimes Y \to X \otimes Y$ is a monomorphism follows by applying Lemma 14.2 and the elementary fact that $\operatorname{colim}(X \otimes \boldsymbol{Y})$ is naturally isomorphic to $X \otimes \operatorname{colim} \boldsymbol{Y}$. □

EXAMPLE 20.21. A simple example of a subgroup, which is not pure, is the subgroup $X' = \{[0], [2]\} \approx \mathbb{Z}/2$ of the cyclic group $X = \{[0], [1], [2], [3]\} = \mathbb{Z}/4$. Clearly, $x = [1]$ is a solution of the equation $2x = [2] \in X$, but this equation has no solutions in X'.

REMARK 20.22. For a characterization of pure submodules over arbitrary rings see ((Cohn 1959), Theorem 2.4).

The sufficiency part of Theorem 20.19 is covered by the following lemma.

LEMMA 20.23. *Every direct sum of cyclic groups is purely projective.*

Proof. It is easy to see that a direct sum of purely projective abelian groups is itself purely projective. The free cyclic group $\mathbb{Z}$ is projective and thus, also purely projective. Therefore, it suffices to prove that every finite cyclic group $\mathbb{Z}/m$ is purely projective. Let $f\colon \mathbb{Z}/m \to X''$ be a homomorphism. Denote by $q\colon \mathbb{Z} \to \mathbb{Z}/m$ the quotient homomorphism. Since $\mathbb{Z}$ is free, hence also projective, there exists a homomorphism $h\colon \mathbb{Z} \to X$ such that $uh = fq$. Consider the element $x = h(1) \in X$ and note that $u(mx) = fq(m) = 0$. By exactness of (20.4.1), $mx \in u'(X')$. One concludes, by Lemma 20.20, that there exists an element $x' \in X'$ such that $mx = mu'(x')$. Now define a homomorphism $k\colon \mathbb{Z} \to X$ by putting $k(1) = x - u'(x')$. Note that $m(x - u'(x')) = 0$. Therefore, k induces a homomorphism $g\colon \mathbb{Z}/m \to X$ such that $gq = k$. Clearly, $ug = f$ because $ug[1] = uk(1) = u(x) - uu'(x') = u(x) = uh(1) = f[1]$. □

REMARK 20.24. Lemmas 20.27 and 20.23 prove that in the category of abelian groups there are enough purely projective groups. Note that Theorem 20.15 is a much stronger statement.

The *necessity part* of Theorem 20.19 is an immediate consequence of the lemma and of the theorem which follow.

LEMMA 20.25. *Every purely projective group is a subgroup (actually a direct summand) of a direct sum of cyclic groups.*

THEOREM 20.26. *If A is a direct sum of cyclic groups, then every subgroup B of A is also a direct sum of cyclic groups.*

To establish Lemma 20.25, we need the following lemma.

LEMMA 20.27. *For every abelian group A there exist a direct sum of cyclic groups P and an epimorphism $u\colon P \to A$ such that* $\mathrm{Ker}(u)$ *is a pure subgroup of P.*

Proof. For every element $a \in A$, let $\langle a \rangle$ denote the cyclic subgroup of A generated by a. Consider the direct sum $P = \bigoplus_{a\in A}\langle a \rangle$ and the homomorphism $u\colon P \to A$, which maps each summand $\langle a \rangle \subseteq P$ to A by the natural inclusion. Clearly, u is an epimorphism. Let us show that the kernel $\mathrm{Ker}(u)$ is a pure subgroup of P. Indeed, assume that $m \in \mathbb{Z}$, $b \in P$ and $mb \in \mathrm{Ker}(u)$, i.e., $mu(b) = 0$. We must find an element $c \in \mathrm{Ker}(u)$ such that $mc = mb$. Notice that $u(b) = a \in A$ and therefore, $u(a) = a$. Consequently, $u(b-a) = 0$, i.e., $c = b - a \in \mathrm{Ker}(u)$. Moreover, $mc = mb - ma = mb - mu(b) = mb$. □

Proof of Lemma 20.25. Let A be a purely projective group. By Lemma 20.27, there exists a purely exact sequence (20.4.1), where $X'' = A$ and $X = P$ is a direct sum of cyclic groups. Taking for $f\colon A \to A$ the identity homomorphism, one obtains a homomorphism $g\colon A \to P$, such that $ug = \mathrm{id}$. Consequently, the sequence (20.4.1) splits and thus, A is a direct summand of P. □

Theorem 20.26 is a well-known result in infinite abelian groups, due to L.Ya. Kulikov (see (Kaplansky 1954), Theorem 13 or (Fuchs 1970), Theorem 18.1). Its proof requires some preparation. We first recall some notions from group theory. A group is *periodic* if every element $a \in A$ is of finite order $o(a)$. A periodic group is *bounded* if the orders of its elements are bounded. A periodic group is a p-*group*, where p is a given prime number, provided the orders of its elements are powers of p, i.e., of the form p^i. In a p-group A the *height* $h(a)$ of an element $a \in A\backslash\{0\}$ is the maximal integer r such that a is divisible by p^r. If there is no such integer, i.e., if a is divisible by all powers p^i, then $h(a) = \infty$. If $B \subseteq A$ is a subgroup of A and $b \in B$, we denote the height of b in B by $h_B(b)$ and its height in A by $h_A(b)$. Note that $h_B(b) \leq h_A(b)$. We denote by $A[n]$ the subgroup of elements $a \in A$, for which $na = 0$. If X is a subset of an abelian group A, we denote by $\langle X \rangle$ the subgroup of A generated by X.

We will now prove a lemma, which covers a special case of Theorem 20.26.

LEMMA 20.28. *If a p-group A is a direct sum of cyclic groups, then so is each of its subgroups $B \subseteq A$.*

The proof of Lemma 20.28 is based on a series of further lemmas.

LEMMA 20.29. *Let A be a p-group and let $B \subseteq A$ be a subgroup all of whose elements $b \in B$ have finite height $h_B(b) < \infty$. If $h_B(b) = h_A(b)$, for all elements $b \in B[p]$, then B is a pure subgroup of A.*

Proof. We will first prove that $h_B(b) = h_A(b)$, for all elements $b \in B$. The proof is by induction on k, where $o(b) = p^k$. For $k = 1$, the assertion holds by assumption. Now assume that it holds for elements whose order is p^k, where $k \geq 1$. Let $b \in B$ be an element of order p^{k+1}. Since $o(pb) = p^k$, the induction hypothesis implies that $h_A(pb) = h_B(pb)$ has a finite value r. Therefore, there exists an element $b' \in B \subseteq A$ such that $pb = p^r b'$. Clearly, $r - 1 \leq h_B(p^{r-1}b') \leq h_A(p^{r-1}b')$. On the other hand, if $p^i x = p^{r-1}b'$, for some $x \in A$, then $p^{i+1}x = p^r b' = pb$, which implies $i+1 \leq r$, i.e., $i \leq r-1$. Consequently, $h_A(p^{r-1}b') \leq i \leq r-1$ and we see that $h_B(p^{r-1}b') = h_A(p^{r-1}b') = r-1$. Note that $p^n x = b$ implies $p^{n+1}x = pb$ and since $h_B(pb) = h_A(pb) = r$, we conclude that $n+1 \leq r$, i.e., $n \leq r-1$. Consequently, both heights $h_B(b)$ and $h_A(b)$ cannot exceed $r-1$. Now define $b'' \in B$ by putting $b = p^{r-1}b' + b''$. Since $pb'' = 0$, by assumption, $h_A(b'') = h_B(b'')$ has a finite value s. If in a p-group two elements x_1, x_2 have different heights $r_1 \neq r_2$, then it is clear that the height of their sum $x_1 + x_2$ equals $\min\{r_1, r_2\}$. Therefore, if $s \neq r-1$, one concludes that $h_B(b) = h_A(b) = \min\{r-1, s\}$. If $s = r-1$, then $b = p^{r-1}b' + b''$ implies $h_B(b) \geq r-1$ and $h_A(b) \geq r-1$. Since we have already shown that $h_B(b) \leq r-1$ and $h_A(b) \leq r-1$, one concludes that in this case $h_B(b) = h_A(b) = r-1$.

To conclude the proof, we must show that, whenever an element $b \in B$ is divisible in A by some integer n, then it is divisible by n already in B. Let

$o(b) = p^k$ and let $n = mp^i$, $i \geq 0$, where m and p are relatively prime. Then there exist integers r, s such that $1 = rp^k + sm$ and thus, $b = msb$. Since b is divisible in A by n, hence, also by p^i, one concludes that $sb \in B$ is also divisible in A by p^i. Consequently, the height $h_A(sb) \geq i$. Since $h_B(sb) = h_A(sb)$, one concludes that $h_B(sb) \geq i$. Consequently, there exists an element $b' \in B$ such that $sb = p^i b'$ and thus, $b = msb = nb'$. □

LEMMA 20.30. *Let A be a p-primary group and $G \subseteq A$ a pure subgroup, whose elements have finite height in G. Let $x \in A \backslash G$ be an element of order $o(x) = p$ and height $h_A(x) = r < \infty$ and let $h_A(a + x) \leq h_A(x)$, for every $a \in G \cap A[p]$. If $z \in A$ is an element such that $x = p^r z$, then the group $H = \langle G \cup \{z\} \rangle$ is a pure subgroup of A, $G \cap \langle z \rangle = 0$ and $H \approx G \oplus \langle z \rangle$.*

Proof. Let us first show that $G \cap \langle z \rangle = 0$ and thus, $H \approx G \oplus \langle z \rangle$. Assume the contrary, i.e., that $G \cap \langle z \rangle \neq 0$. Then $G \cap \langle z \rangle$ would be a non-trivial subgroup of $\langle z \rangle$. Since A is a p-group, $\langle z \rangle$ is a cyclic group of order p^k, for some integer k. However, every non-trivial subgroup of such a cyclic group, hence also $G \cap \langle z \rangle$, is one of the subgroups $\langle p^i z \rangle$, where $0 \leq i \leq k-1$. Since $p^{r+1} z = px = 0$, one concludes that $k-1 \leq r$. Therefore, $x = p^r z = p^{r-i} p^i z \in \langle p^i z \rangle = G \cap \langle z \rangle \subseteq G$, which is a contradiction.

Now note that the direct sum of two groups having elements of finite height has itself elements of finite height. G has elements of finite height by assumption and $\langle z \rangle \approx \mathbb{Z}/p^k$, being a finite group, also has elements of finite height. Consequently, the group $H \approx G \oplus \langle z \rangle$ has also elements of finite height. Therefore, we can use Lemma 20.30 and prove that H is pure in A, by showing that $h_H(u) = h_A(u)$, for every element $u \in H[p]$.

Clearly, u is of the form $u = a + mz$, where $a \in G$, $m \in \mathbb{Z}$. Note that $pa = 0$, because $pa + mpz = pu = 0$ and $G \cap \langle z \rangle = 0$. We first consider the case $m = 0$, i.e., $u = a \in G[p]$. Then $h_G(a) = h_A(a)$, because G is a pure subgroup of A. However, $G \subseteq H \subseteq A$ implies $h_G(a) \leq h_H(a) \leq h_A(a)$ and thus, $h_G(a) = h_H(a) = h_A(a)$. Next we consider the case $u = a + x = a + p^r z$. Clearly, $H \subseteq A$ implies $h_H(x) \leq h_A(x) = r$. On the other hand, $x = p^r z$ and $z \in H$ imply $h_H(x) \geq r$, hence, $h_H(x) = h_A(x) = r$. We already know that $h_H(a) = h_A(a)$. If $h_A(a) \neq r$, then $h_H(a + x) = \min\{h_H(a), h_H(x)\}$ and also $h_A(a + x) = \min\{h_A(a), h_A(x)\}$, therefore, $h_H(a + x) = h_A(a + x)$. Now assume that $h_A(a) = r$ and thus, $h_H(a) = r$. Since also $h_H(x) = r$, we conclude that $h_H(a + x) \geq r$. Therefore, $h_A(a + x) \geq h_H(a + x)$ implies $h_A(a + x) \geq r$, which together with the assumption $h_A(a + x) \leq r$ yields $h_A(a + x) = h_H(a + x) = r$.

Next we show that $h_A(a + nx) = h_H(a + nx)$, whenever n is relatively prime to p. Indeed, in that case there exist integers r, s such that $1 = rn + sp$ and therefore, $x = rnx$ and $a = rna$. If $a + nx = p^i u$, for some $u \in A$, then $ra + x = ra + rnx = p^i ru$. Since $ra \in G \cap A[p]$, one has $h_H(ra + x) = h_A(ra + x)$. Therefore, there exists a $v \in H$ such that $ra + x = p^i v$ and thus, $a + nx = rna + nx = p^i nv$, which shows that $h_H(a + nx) \geq i$. All this proves that $h_H(a+nx) \geq h_A(a+nx)$. On the other hand, $H \subseteq A$ implies $h_H(a+nx) \leq h_A(a+nx)$ and thus, $h_H(a+nx) = h_A(a+nx)$.

We now consider the general case $u = a + mz$. We write m in the form $m = np^i$, where $i \geq 0$ and n is relatively prime to p. We can assume that $i \leq r$, because $i > r$ implies $mz = np^i z = np^{i-r}x = 0$ and thus, $u = a \in G \cap A[p]$, which refers to the case we have already settled. The case $i < r$, cannot occur, because $pu = pa + np^{i+1}z = 0$ and $pa = 0$ would imply $np^{i+1}z = 0$, hence, also $np^r z = nx = 0$. Since the order of x is p, p would have to be a divisor of n. The only remaining case is $u = a + np^r z = a + nx$, where n and p are relatively prime. However, we have already handled that case. □

A subset X of an abelian group A is said to be *independent* whenever $x_1, \ldots, x_r \in X$, $n_1, \ldots, n_r \in \mathbb{Z}$ and $n_1x_1 + \ldots + n_rx_r = 0$ imply $n_1x_1 = \ldots = n_rx_r = 0$. Clearly, a group A is generated by an independent set X if and only if A is the direct sum of the cyclic subgroups $\langle x \rangle$, $x \in X$.

LEMMA 20.31. *Let A be a p-primary group and let $B \subseteq A[p]$ be a subgroup which has the property that the heights $h_A(b)$ of its elements $b \in B$ are bounded. Let $X \subseteq A$ be an independent subset such that the subgroup $\langle X \rangle$ is pure in A and $\langle X \rangle \cap A[p] \subseteq B$. Then there exists an independent subset $Y \subseteq A$ such that $X \subseteq Y$, the subgroup $\langle Y \rangle$ is pure in A and $\langle Y \rangle \cap A[p] = B$.*

Proof. Let $\mathcal{Y}$ denote the collection of all independent subsets $Y \subseteq A$ such that $X \subseteq Y$, the subgroup $\langle Y \rangle$ is pure in A and $\langle Y \rangle \cap A[p] \subseteq B$. $\mathcal{Y}$ is not empty because it contains $\langle X \rangle$. Order $\mathcal{Y}$ by inclusion $\subseteq$. It is readily seen that the union of any totally ordered subcollection of $\mathcal{Y}$ belongs to $\mathcal{Y}$. Therefore, Zorn's lemma proves the existence of maximal elements $Y \in \mathcal{Y}$. To complete the proof it suffices to show that $\langle Y \rangle \cap A[p] = B$.

Assume to the contrary that there exists an element $b \in B \backslash \langle Y \rangle$. Consider all elements of the form $a + b$, where $a \in \langle Y \rangle \cap A[p]$. Since all such elements belong to B and the elements of B have bounded heights in A, there exists such an element $x = a_0 + b$ of maximal height. Note that $h_A(a+x) \leq h_A(x)$, for $a \in \langle Y \rangle \cap A[p]$, because $a + x = (a + a_0) + b$ and $a + a_0 \in \langle Y \rangle \cap A[p]$,. Also note that $B \subseteq A$ implies $h_B(x) \leq h_A(x)$. Therefore, $h_B(x) = r$, for some $r < \infty$. Consequently, there exists an element $z \in B$ such that $x = p^r z$. Note that $x \in B \subseteq A[p]$ and therefore, $px = 0$. Moreover, $x \notin \langle Y \rangle$, because $a_0 \in \langle Y \rangle$ and $x \in \langle Y \rangle$ would imply $b \in \langle Y \rangle$. Also note that $\langle Y \rangle$ is a direct sum of finite cyclic groups and therefore, every element $u \in \langle Y \rangle$ has finite height $h_{\langle Y \rangle}(u) < \infty$. Indeed, u belongs to the sum C of finitely many direct summands of $\langle Y \rangle$. As a direct summand, C is pure in $\langle Y \rangle$, hence, the height of u in $\langle Y \rangle$ equals its height in C. However, C is a finite group and therefore, $h_C(u) < \infty$. All this enables us to apply Lemma 20.30 to $G = \langle Y \rangle \subseteq A$, x and z and conclude that $Z = Y \cup \{z\}$ is an independent subset of A such that $\langle Z \rangle$ is a pure subgroup of A. Finally, $\langle Y \rangle \cap A[p] \subseteq B$ and $z \in B$ imply $\langle Z \rangle \cap A[p] \subseteq B$ and thus, $Z \in \mathcal{Y}$. However, since $z \notin Y$, this contradicts the maximality property of Y. □

LEMMA 20.32. *Let A be a p-primary group and let $A_1 \subseteq A_2 \subseteq \dots$ be an increasing sequence of subgroups, whose union equals $A[p]$. If the elements of each subgroup A_i have bounded heights in A, then A is a direct sum of cyclic groups.*

Proof. Using Lemma 20.31, one constructs, by induction, an increasing sequence of independent subsets $X_1 \subseteq X_2 \subseteq \dots$ of A such that $\langle X_i \rangle$ is a pure subgroup of A and $\langle X_i \rangle \cap A[p] = A_i$. The initial set X_1 is obtained by applying the lemma to $A_1 \subseteq A$ and to the empty subset of A. Let $X = \cup X_i$. Clearly, the set X is an independent set and thus, $\langle X \rangle$ is a direct sum of cyclic groups. Therefore, it suffices to prove that $\langle X \rangle = A$.

Since $\langle X \rangle = \cup \langle X_i \rangle$ and the subgroups $\langle X_i \rangle$ are pure, so is also $\langle X \rangle$. Moreover, $A[p] \subseteq \langle X \rangle$. We will now prove, by induction on the order p^n of elements $a \in A$, that every a belongs to $\langle X \rangle$, i.e., $A = \langle X \rangle$. If $n = 1$, then $a \in A[p] \subseteq \langle X \rangle$. Now assume that the assertion has been proved for n and that $o(a) = p^{n+1}$. Note that $o(pa) = p^n$ and thus, $pa \in \langle X \rangle$. Since, $\langle X \rangle$ is pure, there exists an element $x \in \langle X \rangle$ such that $pa = px$. Therefore, $a - x \in A[p] \subseteq \langle X \rangle$. Consequently, also $a \in \langle X \rangle$. □

Proof of Lemma 20.28. Let A be the direct sum of a collection of cyclic p-groups C_λ. Denote by A_n the sum of all C_λ of order p^i, where $i \leq n$ and note that $A_1 \subseteq A_2 \subseteq \dots$ and $A = \cup A_n$. Now put $B_n = A_n \cap B[p]$. Clearly, $B_1 \subseteq B_2 \subseteq \dots$ and $\cup B_n = B[p]$. To complete the proof, it suffices to show that the elements of B_n have bounded heights in B. Indeed, if this is the case, one can apply Lemma 20.32 and conclude that B is a direct sum of cyclic groups. In fact we will show that the elements of $A_n \supseteq B_n$ have bounded heights even in A. As a direct summand of A, the subgroup A_n is pure in A and therefore, $h_{A_n}(a) = h_A(a)$, for every element $a \in A_n$. Hence, it suffices to prove that the elements of A_n have bounded heights in A_n. This is indeed the case, because $p^r A_n = 0$, for $r \geq n$ and the height of an element $a \in A_n \setminus \{0\}$ must be $< r$. □

Proof of Theorem 20.26. Let $T \subseteq A$ denote the *torsion subgroup* of the group A, i.e., the subgroup formed by all elements of finite order. Clearly, $B \cap T$ is the torsion subgroup of B and $B/(B \cap T)$ is a subgroup of A/T. If $A = \oplus A_\lambda$, where A_λ are cyclic groups, then $T_\lambda = T \cap A_\lambda$ are the torsion subgroups of A_λ, $T = \oplus T_\lambda$ and $A/T \approx \oplus(A_\lambda/T_\lambda)$. Now note that $T_\lambda = 0$, if A_λ is infinite cyclic, and thus, $A_\lambda/T_\lambda \approx \mathbb{Z}$. On the other hand, $T_\lambda = A_\lambda$ if A_λ is finite cyclic, and thus, $A_\lambda/T_\lambda = 0$. Consequently, A/T is a free cyclic group and its subgroup $B/(B \cap T)$ must also be free. Therefore, the exact sequence

$$0 \to B \cap T \to B \to B/(B \cap T) \to 0 \tag{1}$$

splits, i.e.,

$$B \approx (B \cap T) \oplus B/(B \cap T). \tag{2}$$

It thus suffices to prove that $B \cap T$ is a direct sum of cyclic groups. Since T is a direct sum of finite cyclic groups T_λ, we have reduced the problem to the case when A is a direct sum of finite cyclic groups, i.e., when A is a periodic group.

We now use the well-known fact that every periodic group A is the direct sum $A = \oplus A_p$ of its p-primary components A_p, i.e., subgroups whose orders are powers of a given prime number p. In particular, $B = \oplus B_p$ and it suffices to prove that every group B_p is a direct sum of cyclic groups. However, A_p is readily obtained from the direct decomposition of A in cyclic subgroups, by collecting together the cyclic summands, whose orders are powers of a given p. Since $B_p \subseteq A_p$ and A_p is a direct sum of cyclic groups, we have reduced the problem to the case when A is a p-group and $B \subseteq A$. However, Lemma 20.28 asserts that, in this case, B is indeed a direct sum of cyclic groups. □

Proof of Theorem 20.15. For every abelian group A, Lemma 20.27 yields a short purely exact sequence $0 \leftarrow A \leftarrow P_0 \leftarrow P_1 \leftarrow 0$, where P_0 is a direct sum of cyclic groups. By Theorem 20.26, P_1 is also a direct sum of cyclic groups. Consequently, by Theorem 20.19, P_0 and P_1 are both purely projective groups and the sequence is a purely projective resolution of A, as required by Theorem 20.15. □

To obtain a characterization theorem for purely injective abelian groups, one needs cocyclic groups. An abelian group A is a *cocyclic group* if it admits an element $c \in A$, called the *cogenerator*, which has the property that every homomorphism $\phi: A \to B$, for which $\phi(c) \neq 0$, is necessarily a monomorphism. Clearly, an abelian group is cocyclic if and only if the intersection of all of its non-trivial subgroups is non-trivial. Every non-trivial element of this minimal non-trivial subgroup is a cogenerator.

EXAMPLE 20.33. The cyclic group $\mathbb{Z}/p^k$ of order p^k, where p is a prime and $k \geq 1$, is a cocyclic group. Indeed, if c is a generator of $\mathbb{Z}/p^k$, then the sequence of cyclic subgroups $0 \subset \langle p^{k-1}c\rangle \subset \ldots \subset \langle pc\rangle \subset \langle c\rangle = \mathbb{Z}/p^k$ contains all subroups of $\mathbb{Z}/p^k$. Hence, $\langle p^{k-1}c\rangle$ is the minimal non-trivial subgroup of $\mathbb{Z}/p^k$. Another example of a cocyclic group is the *quasicyclic group* $\mathbb{Z}/p^\infty$. To define it, one considers the increasing sequence of cyclic groups $0 \subset \mathbb{Z}/p \subset \mathbb{Z}/p^2 \subset \ldots$ with generators $c_k \in \mathbb{Z}/p^k$ such that $pc_{k+1} = c_k$. Then $\mathbb{Z}/p^\infty$ is the colimit of this sequence. One can show that every subgroups of $\mathbb{Z}/p^\infty$ is a member of the sequence $0 \subset \langle c_1\rangle \subset \langle c_2\rangle \subset \ldots$ (see e.g., (Fuchs 1970), Ch.I, §3). Consequently, $\langle c_1\rangle$ is the smallest non-trivial subgroup of $\mathbb{Z}/p^\infty$.

LEMMA 20.34. *Every cocylic group C is isomorphic to one of the cyclic groups $\mathbb{Z}/p^k$ or to the quasicyclic group $\mathbb{Z}/p^\infty$.*

Proof. Let $c \in C$ be a cogenerator of the cocyclic group C. Then the cyclic subgroup $\langle c\rangle$ generated by c is the minimal non-trivial subgroup of C. Therefore, the order $o(c)$ must be a prime number p. Since c belongs to every non-trivial subgroup of C, one concludes that C cannot contain neither elements of infinite order nor elements whose order is divisible by a prime $q \neq p$. Consequently, C is a p-group, i.e., the order of every element of C is a power of p. We will prove, by induction on n, that C contains at most one subgroup C_n of order p^n and this subgroup is a cyclic group with generator c_n. For $n = 0$, this is the trivial subgroup. Assume that the assertion holds for n and let A, B be two subgroups of order p^{n+1}. Choose elements $a \in A\backslash C_n$, $b \in B\backslash C_n$. Clearly, the orders $o(a) = o(b) = p^{n+1}$. Therefore, there exist integers r, s such that $pa = rc_n$, $pb = sc_n$ and r, s are relatively prime to p. Choose integers r', s' so that $rr' \equiv 1 \equiv ss' (\bmod p^n)$. Then r', s' are relatively prime to p and for $a' = r'a$, $b' = s'b$, one obtains $\langle a' \rangle = \langle a \rangle$, $pa' = c_n$, $\langle b' \rangle = \langle b \rangle$, $pb' = c_n$. Consequently, $p(a' - b') = 0$ and there exists an integer t such that $a' - b' = tc_n$, i.e., $a' = b' + tpb'$, $b' = a' - tpa'$. However, this implies that the elements a' and b' generate the same cyclic group, i.e., $\langle a' \rangle = \langle b' \rangle$. Hence, $A = \langle a \rangle = \langle b \rangle = B$. Consequently, the group C is the union of an increasing sequence of subgroups $C_n \approx \mathbb{Z}/p^n$. If this sequence is finite, one concludes that $C \approx \mathbb{Z}/p^k$, for some integer k. If this sequence is infinite, then $C \approx \mathbb{Z}/p^\infty$. □

THEOREM 20.35. *An abelian group A is purely injective if and only if it is a direct summand of a direct product of cocyclic groups.*

The sufficiency part of Theorem 20.35 is covered by the following lemma.

LEMMA 20.36. *Every direct product of cocyclic groups is purely injective.*

In the proof of this lemma we need another theorem of Kulikov, here stated as the following lemma (see e.g., (Kaplansky 1954), Theorem 7 or (Fuchs 1970), Theorem 27.5).

LEMMA 20.37. *Let $B \subseteq A$ be a pure subgroup of the abelian group A. If B is bounded, then it is a direct summand of A.*

Proof of Lemma 20.36. It is easy to see that a direct product of purely injective abelian groups is itself purely injective. Therefore, it suffices to consider the case of a cocyclic group C. Consider the purely exact sequence (20.4.1). There is no loss of generality in assuming that X' is a pure subgroup of X and u' is inclusion. Let $f: X' \to C$ be a homomorphism. We must exhibit a homomorphism $g: X \to C$, which extends f. A subgroup of a cocylic group is again a cocyclic group, because non-trivial subgroups of $\mathbb{Z}/p^k$ are groups isomorphic to $\mathbb{Z}/p^i$, $1 \leq i \leq k$, and non-trivial proper subgroups of $\mathbb{Z}/p^\infty$ are isomorphic to groups $\mathbb{Z}/p^i$, $i \geq 1$. Therefore, there is no loss of generality in assuming that $f: X' \to C$ is an epimorphism. Since the quasicyclic group $\mathbb{Z}/\infty$ is divisible and divisible groups are injective, g exists if $C = \mathbb{Z}/p^\infty$. It remains to consider the case when C is isomorphic to the cyclic group $\mathbb{Z}/p^k$. Consider the subgroup $Y = \mathrm{Ker}(f) \subseteq X'$. Clearly, $f = hq'$, where $q': X' \to X'/Y$ is the quotient homomorphism and $h: X'/Y \to C$ is an isomorphism. To complete the proof, it suffices to show that $X'/Y \subseteq X/Y$ is a direct summand of X/Y. Indeed, if this is the case, one considers the quotient homomorphism $q: X \to X/Y$ and a retraction $r: X/Y \to X'/Y$. Then the homomorphism $hrq: X \to C$ is the desired extension of f.

To prove that X'/Y is a direct summand of X/Y, note that $X'/Y \approx C \approx \mathbb{Z}/p^k$ is a finite group. Therefore, there exists an integer n such that $n(X'/Y) = 0$. Consequently, $nX' \subseteq Y$ and $Y/nX' \subseteq X'/nX'$. Let us show that X'/nX' is a pure subgroup of X/nX'. Indeed, if $x' \in X'$, $x \in X$ and the class $[x'] = n[x]$, for some integer n, then there exists an element $y \in nX'$ such that $x' - y = nx$. Since $x' - y \in X'$ and X' is pure in X, one concludes that there exists an element $y' \in X'$ such that $x' - y = ny' \in nX'$. Since $[y] = 0$, one concludes that indeed, $[x'] = n[y']$. Now note that the group X'/nX' is bounded, because $n(X'/nX') = 0$. Therefore, Lemma 20.37 applies and yields the conclusion that X'/nX' is a direct summand of X/nX'. If R is a subgroup of X/nX' such that $X/nX' = X'/nX' \oplus R$, then division by Y/nX' yields the desired conclusion that $X/Y \approx X'/Y \oplus R$. □

To prove Lemma 20.37, we need some further lemmas. The first one is a well-know theorem, due to H. Prüfer (see e.g., (Kaplansky 1954), Theorem 6 or (Fuchs 1970), Theorem 17.2).

LEMMA 20.38. *Every bounded abelian group A is a direct sum of cyclic groups.*

Proof. As a bounded group, A is periodic and thus, the direct sum of its p-primary components. Therefore, it suffices to prove the lemma under the assumption that A is a p-group. Since A is bounded, there exists an integer n such that $p^n A = 0$. Therefore, for every element $a \in A \backslash \{0\}$, the height $h_A(a) < n$. Therefore, one can apply Lemma 20.32 to the sequence $A_i = A[p]$, $i = 1, 2, \ldots$, and conclude that A is a direct sum of cyclic groups. □

LEMMA 20.39. *Let A be an abelian group and let $B \subseteq A$ be a pure subgroup. For every element $[y] \in A/B$, there exists an element $x \in A$ such that $[x] = [y]$ and the order $o(x)$ of x equals the order $o[y]$ of $[y]$.*

Proof. Let n be the order of $[y] \in A/B$. Then $ny \in B$. Since B is pure in A, there exists an element $b \in B$ such that $nb = ny$. Put $x = y - b$. Clearly, $[x] = [y]$. Moreover, $nx = 0$. On the other hand, $kx = 0$ implies $ky = kb$ and thus, $k[y] = k[b] = 0$. Consequently, k must be a multiple of n, which shows that $o(x) = n$. □

LEMMA 20.40. *Let B be a pure subgroup of an abelian group A. If A/B is a direct sum of cyclic groups, then B is a direct summand of A.*

Proof. Let $A/B = \oplus_{\lambda \in \Lambda} C_\lambda$, where each C_λ is a cyclic group. For each $\lambda \in \Lambda$, choose a generator $[y_\lambda]$ of C_λ. By Lemma 20.39, there exist elements $x_\lambda \in A$, such that $[x_\lambda] = [y_\lambda]$ and $o(x_\lambda) = o[y_\lambda]$, for each $\lambda \in \Lambda$. Denote by D the subgroup of A generated by all the elements x_λ, $\lambda \in \Lambda$. Let us show that $A = B \oplus D$. If $a \in A$, then $[a] = \sum_i n_i [x_{\lambda_i}]$. Therefore, $b = a - \sum_i n_i x_{\lambda_i} \in B$. Since $d = \sum_i n_i x_{\lambda_i} \in D$, one concludes that $a = b + d \in B + D$. Now assume that $a \in B \cap D$. Then a is of the form $a = \sum_i n_i x_{\lambda_i}$. Consequently, $\sum_i n_i[y_{\lambda_i}] = [a] = 0$. However, this implies $n_i[y_{\lambda_i}] = 0$, for each i. If $o[y_{\lambda_i}] = \infty$, then $n_i = 0$. If $o[y_{\lambda_i}] = m$ is finite, then n_i is a multiple of m. Since also $o(x_{\lambda_i}) = m$, one concludes that $n_i x_i = 0$. Consequently, $a = 0$, i.e., $B \cap D = 0$. □

Proof of Lemma 20.37. Since B is a bounded group, there exists an integer n such that $nB = 0$. Let us first show that $(B + nA)/nA$ is a pure subgroup of A/nA. Indeed, if $a \in A$, $m \in \mathbb{Z}$ and $m[a] \in (B + nA)/nA$, then there are elements $b \in B$ and $a' \in A$ such that $ma = b + na'$. If r is the greatest common divisor of m and n, then there exist integers $\lambda, \mu \in \mathbb{Z}$ such that $r = \lambda m + \mu n$. Consequently, $b \in rA$. Since B is pure in A, there exists an element $b' \in B$ such that $b = rb' = \lambda mb' + \mu nb'$. However, $nb' = 0$, because $nB = 0$. Consequently, $[b] = \lambda m[b'] \in m(B + nA)/nA$, which proves the assertion. Now notice that $n(A/nA) = 0$, which shows that the group A/nA is bounded. However, by Lemma 20.38, every bounded group is a direct sum of cyclic groups. One can now apply Lemma 20.40 to conclude that $(B+nA)/nA$ is a direct summand of A/nA.

To complete the proof, consider a complementary direct summand $R \subseteq A/nA$ of $(B + nA)/nA$ and let $q\colon A \to A/nA$ denote the quotient homomorphism. Then put $C = q^{-1}(R) \subseteq A$. Clearly, $nA \subseteq C$ and therefore, $B + C = (B + nA) + C = A$. Moreover, $B \cap C \subseteq (B + nA) \cap C = nA$ and thus, $B \cap C \subseteq B \cap nA$. Since B is a pure subgroup of A, it follows that $B \cap nA \subseteq nB = 0$. Consequently, $A = B \oplus C$. □

LEMMA 20.41. *Every abelian group A embeds as a pure subgroup in a direct product of cocyclic groups.*

In the proof we need the following lemma.

LEMMA 20.42. *For every abelian group A and element $a \in A \backslash \{0\}$, there exist a cocyclic group C and an epimorphism $f\colon A \to C$ such that $f(a) \neq 0$.*

Proof. Let $\mathcal{M}$ be the collection of all subgroups of A, which do not contain a. Order $\mathcal{M}$ by inclusion $\subseteq$. Clearly, the union of every totally ordered set from $\mathcal{M}$ belongs to $\mathcal{M}$. Therefore, Zorn's lemma applies and yields a maximal subgroup $M \in \mathcal{M}$, which has the property that $a \notin M$. Now put $C = A/M$ and let $f\colon A \to C$ be the quotient homomorphism. Clearly, $c = f(a) \neq 0$. Moreover, c is a cogenerator of C, which makes C a cocyclic group. Indeed, let $\phi\colon C \to Y$ be a homomorphism such that $\phi(c) \neq 0$, i.e., $\phi f(a) \neq 0$. Consequently, $a \notin N = f^{-1}(\mathrm{Ker}(\phi))$ and thus, $N \in \mathcal{M}$. Since $M \subseteq N$, one concludes that $M = N$, hence, $\mathrm{Ker}(\phi) = f(M) = 0$, which proves that ϕ is indeed a monomorphism. □

Proof of Lemma 20.41. Let $\{M_\lambda | \lambda \in \Lambda\}$ be the set of all subgroups $M_\lambda \subseteq A$, which have the property that $B_\lambda = A/M_\lambda$ is a cocyclic group and let $f_\lambda\colon A \to B_\lambda$ denote the corresponding quotient homomorphisms. Then the homomorphism $f = \prod_\lambda f_\lambda\colon A \to B = \prod_\lambda B_\lambda$ is a monomorphism. Indeed, for every $a \in A$, $a \neq 0$, Lemma 20.42 yields a subgroup $M_\lambda \subseteq A$ such that B_λ is cocyclic and $f_\lambda(a) \neq 0$. Therefore, $f(a) \neq 0$. It remains to show that $f(A)$ is a pure subgroup of B.

Let $b \in B$, $a \in A$ and $n \in \mathbb{Z}$ be such that $nb = f(a)$. We claim that there exists an $a' \in A$ such that $a = na'$ and therefore, $nb = nf(a') \in nf(A)$. Assume to the contrary that $a \notin nA$. Using Zorn's lemma, one can construct a subgroup $M \subseteq A$, which is maximal with respect to the properties $nA \subseteq M$ and $a \notin M$. Since A/M is a cocylic group, M must be one of the subgroups M_λ. Hence, $A/M = B_\lambda$ and the quotient homomorphism $A \to A/M$ coincides with the homomorphism f_λ. Note that $f(a) = nb \in nB$ implies $f_\lambda(a) \in nB_\lambda$. However, $nA \subseteq M_\lambda$ implies $nB_\lambda = 0$ and thus, $f_\lambda(a) = 0$. On the other hand, $a \notin M_\lambda$ implies $f_\lambda(a) \neq 0$, which is a contradiction. □

Proof of Theorem 20.35. A direct summand of a purely injective group is itself purely injective. Therefore, sufficiency of the condition follows from Lemma 20.36. To prove necessity, assume that A is a purely injective group. By Lemma 20.41, there exists a purely exact sequence (20.4.1), where $X' = A$ and $X = I$ is a direct product of cocyclic groups, hence, a purely injective group. Putting $f = \mathrm{id}\colon A \to A$, one obtains a homomorphism $g\colon I \to A$, such that $gu' = \mathrm{id}$. Consequently, the sequence splits and A is a direct summand of I. □

REMARK 20.43. Lemmas 20.41 and 20.36 prove that in the category of abelian groups there are enough purely injective groups.

We conclude this subsection by proving the analogue of Theorem 20.15.

THEOREM 20.44. *Every abelian group G admits a short purely injective resolution*

$$0 \to G \to I^0 \to I^1 \to 0. \tag{3}$$

Proof. By Remark 20.43, there exists a purely exact sequence (3), where I^0 is purely injective. To complete the proof, it suffices to show that I^1 is also purely injective. Let us first show that $\mathrm{Pext}(A, I^1) = 0$, for every abelian group A. Indeed, the pure exactness of (3) implies exactness of the sequence

$$\ldots \to \mathrm{Pext}\,(A, I^0) \to \mathrm{Pext}\,(A, I^1) \to \mathrm{Pext}^2(A, G) \to \ldots. \tag{4}$$

Moreover, $\mathrm{Pext}\,(A, I^0) = 0$, because I^0 is purely injective, and $\mathrm{Pext}^2(A, G) = 0$, by Theorem 20.15. Now consider a purely exact sequence (20.4.1). It induces an exact sequence

$$\ldots \to \mathrm{Hom}\,(X, I^1) \to \mathrm{Hom}\,(X', I^1) \to \mathrm{Pext}\,(X'', I^1) \to \ldots. \tag{5}$$

Since $\mathrm{Pext}\,(X'', I^1) = 0$, we conclude that $\mathrm{Hom}\,(X, I^1) \to \mathrm{Hom}\,(X', I^1)$ is an epimorphism. Consequently, every homomorphism $f\colon X' \to I^1$ extends to a homomorphism $g\colon X \to I^1$, which proves that I^1 is indeed a purely injective group. □

Bibliographic notes

Results on spectral sequences in subsections 20.1 and 20.2 are standard. Our exposition is a more detailed version of the one given in (Godement 1958). The exposition of subsections 20.3 and 20.4 is based on (Mardešić, Prasolov 1998). The Roos spectral sequence and Theorem 20.7 were announced in (Roos 1961). An outline of the proof was given in (Jensen 1972), where Theorem 20.12 is also mentioned. Theorem 20.9 appears in (Kuz'minov 1971) (see the proof of his Theorem 5) and in (Yosimura 1972/73b) (see his section 1.4). In both papers the result was derived using the Roos spectral sequence. A different proof was given in (Huber, Meier 1978). It is based on a result from (Nöbeling 1961). The exposition of results on abelian groups follows (Fuchs 1970) and (Kaplansky 1954), where one can also find information concerning the original sources of the results.

21. Strong homology of compact spaces

This section is devoted to strong homology of compact Hausdorff spaces. In subsection 21.1 we recall the universal coefficient theorem for compact polyhedra. In 21.2 we establish the all-important fact that the higher derived limits $\lim^r H_m(\boldsymbol{X}; G)$ of an inverse system $\boldsymbol{X}$ of compact polyhedra vanish, for $r \geq 2$ (Theorem 21.6). It is the proof of this fact that requires the machinery developed in 20. We also prove the Milnor exact sequence (Theorem 21.9). In 21.3 we prove the universal coefficient theorem for compact Hausdorff spaces (Theorem 21.15). In view of the axiomatic characterization of strong homology (Berikashvili 1984), our strong homology of compact Hausdorff spaces coincides with that of several other authors. In 21.4 we obtain a large commutative diagram (see (21.4.1)), which embodies the just mentioned results and contains additional information (Theorem 21.18). The last subsection is devoted to strong homology with compact supports.

21.1 Universal coefficients for compact polyhedra

The following theorem is known as the *universal coefficient theorem* for compact polyhedra (see e.g., (Munkres 1984), Corollary 56.4).

THEOREM 21.1. *Let X be a compact polyhedron and let G be an arbitrary abelian group. Then for every $m \geq 0$, there exists an exact sequence*

$$0 \to \operatorname{Ext}(H^{m+1}(X), G) \to H_m(X; G) \to \operatorname{Hom}(H^m(X), G) \to 0, \qquad (1)$$

where $H^m(X)$ denotes the m-th cohomology group of X with integer coefficients $\mathbb{Z}$. The sequence is natural with respect to (continuous) mappings $f: X \to Y$.

EXAMPLE 21.2. Compactness of X is an essential assumption. To obtain a non-compact counter-example, consider the wedge $X = \bigvee_i X_i$ of a sequence of copies X_i of the projective plane P^2 and let $G = \mathbb{Z}/2$, $m = 2$. Clearly, $H^3(X) = 0, H_2(X; \mathbb{Z}/2) = \oplus(\mathbb{Z}/2)$ and $H^2(X) = \prod(\mathbb{Z}/2)$. Therefore, (1) would imply

$$\operatorname{Hom}(\prod(\mathbb{Z}/2), \mathbb{Z}/2) \approx \oplus(\mathbb{Z}/2). \qquad (2)$$

However, the group $\oplus(\mathbb{Z}/2)$ is countable, while $\operatorname{Hom}(\prod(\mathbb{Z}/2), \mathbb{Z}/2)$ is not. Indeed, there are uncountably many homomorphisms $\alpha: \oplus(\mathbb{Z}/2) \to \mathbb{Z}/2$. Since $\mathbb{Z}/2$ is a field, $\oplus(\mathbb{Z}/2)$ is a subspace of the vector space $\prod(\mathbb{Z}/2)$ over $\mathbb{Z}/2$. Therefore, every homomorphism α extends to a homomorphism $\beta: \prod(\mathbb{Z}/2) \to \mathbb{Z}/2$.

We will derive Theorem 21.1 from its algebraic analogue, which reads as follows.

LEMMA 21.3. *Let C be a chain complex, which consists of finitely generated free abelian groups and let G be an arbitrary abelian group. Then for every $m \geq 0$, there exists an exact sequence*

$$0 \to \operatorname{Ext}(H^{m+1}(C), G) \to H_m(C; G) \to \operatorname{Hom}(H^m(C), G) \to 0, \qquad (3)$$

which is natural with respect to chain mappings $f: C \to D$.

Proof of Theorem 21.1. Every compact polyhedron X admits a finite triangulation K, i.e., a finite simplicial complex K such that its carrier $|K| = X$. By definition, $H_m(X; G) = H_m(C(K); G)$, where $C(K)$ is the chain complex consisting of all oriented simplices of K. Consequently, $C(K)$ is free and finitely generated. Similarly, $H^m(X) = H^m(C(K))$. Therefore, (3) implies (1). To prove naturality, consider a mapping $f: X \to Y$ between compact polyhedra and let K and L be their respective triangulations. Then the homomorphism $f_*: H_m(X; G) = H_m(C(K); G) \to H_m(C(L); G) = H_m(Y; G)$, induced by $f: |K| \to |L|$, is by definition the homomorphism induced by the composition of two chain mappings $\sigma: C(K) \to C(K')$ and $\phi: C(K') \to C(L)$, where K' is any subdivision of K, which admits a simplicial approximation $\varphi: K' \to L$ of f, ϕ is induced by φ and σ is the subdivision operator (see e.g., (Munkres 1984), Chapter II,§17). It is well known that f_* does not depend on the choice of K' and ϕ. Similarly, $f^*: H^m(Y) = H^m(C(L)) \to H^m(C(K)) = H^m(X)$ is induced by the same chain mapping $\phi\sigma: C(K) \to C(L)$. Therefore, naturality of (3) for $\phi\sigma$ implies naturality of (3) for f, i.e., commutativity of the diagram

$$\begin{array}{ccccccccc} 0 & \to & \operatorname{Ext}(H^{m+1}(X), G) & \to & H_m(X; G) & \to & \operatorname{Hom}(H^m(X), G) & \to & 0 \\ & & \downarrow \scriptstyle\operatorname{Ext}(f^*, 1) & & \downarrow \scriptstyle f_* & & \downarrow \scriptstyle\operatorname{Hom}(f^*, 1) & & \\ 0 & \to & \operatorname{Ext}(H^{m+1}(Y), G) & \to & H_m(Y; G) & \to & \operatorname{Hom}(H^m(Y), G) & \to & 0. \square \end{array} \qquad (4)$$

We will derive Lemma 21.3 from the standard universal coefficient theorem for cohomology of free chain complexes, which reads as follows (see e.g., (Mac Lane 1963), III, Theorem 4.1 or (Munkres 1984), Theorem 53.1).

LEMMA 21.4. *Let C be a chain complex consisting of free abelian groups and let G be an arbitrary abelian group. Then for every $m \geq 0$, there exists an exact sequence*

$$0 \to \mathrm{Ext}\,(H_{m-1}(C);G) \to H^m(C;G) \to \mathrm{Hom}\,(H_m(C);G) \to 0, \tag{5}$$

which is natural with respect to chain mappings $f: C \to D$.

In order to prove Lemma 21.3, beside Lemma 21.4, we also need the next lemma.

LEMMA 21.5. *Let C be a chain complex and G an abelian group. Then there exists a unique chain mapping $\Phi: C \otimes G \to \mathrm{Hom}\,(\mathrm{Hom}\,(C,\mathbb{Z}),G)$ such that*

$$(\Phi(c \otimes g))(\alpha) = \alpha(c)g, \tag{6}$$

where $m \in \mathbb{Z}$, $c \in C_m$, $g \in G$, $\alpha \in \mathrm{Hom}\,(C_m,\mathbb{Z})$. The chain mapping Φ is natural with respect to chain mappings $f: C \to D$. If C is a free finitely generated chain complex, then $\mathrm{Hom}\,(C,\mathbb{Z})$ is a free finitely generated cochain complex and Φ is an isomorphism of chain complexes.

Proof. First define bilinear functions $\Phi: C_m \times G \to \mathrm{Hom}\,(\mathrm{Hom}\,(C_m,\mathbb{Z}),G)$, $m \in \mathbb{Z}$, by putting

$$(\Phi(c,g))(\alpha) = \alpha(c)g. \tag{7}$$

Clearly, they induce homomorphisms $\Phi: C_m \otimes G \to \mathrm{Hom}\,(\mathrm{Hom}\,(C_m,\mathbb{Z}),G)$, which satisfy (6). To see that these homomorphisms form a chain mapping, it suffices to verify that $d\Phi(c \otimes g) = \Phi(\partial c \otimes g)$, where d and ∂ denote the boundary operators in the chain complexes $\mathrm{Hom}\,(\mathrm{Hom}\,(C,\mathbb{Z}),G)$ and C, respectively. If δ denotes the coboundary operator in the cochain complex $\mathrm{Hom}\,(C,\mathbb{Z})$, then for $\alpha \in \mathrm{Hom}\,(C_{m-1},\mathbb{Z})$, one has

$$(d\Phi(c\otimes g))(\alpha) = (\Phi(c\otimes g))(\delta\alpha) = (\delta\alpha)(c)g = (\alpha(\partial c))g = (\Phi(\partial c\otimes g))(\alpha). \tag{8}$$

The uniqueness of Φ is obvious.

To prove naturality of Φ with respect to a chain mapping $f: C \to D$, one must prove the commutativity of the following diagrams

$$\begin{array}{ccc} C_m \otimes G & \xrightarrow{\ \Phi\ } & \mathrm{Hom}\,(\mathrm{Hom}\,(C_m,\mathbb{Z}),G) \\ {\scriptstyle f\otimes 1}\downarrow & & \downarrow{\scriptstyle \mathrm{Hom}\,(\mathrm{Hom}\,(f,1),1)} \\ D_m \otimes G & \xrightarrow[\ \Phi\ ]{} & \mathrm{Hom}\,(\mathrm{Hom}\,(D_m,\mathbb{Z}),G)\,. \end{array} \tag{9}$$

Clearly, for $c \in C_m, g \in G$ and $\beta \in \mathrm{Hom}(D_m,\mathbb{Z})$, one has

$$\begin{gathered} \mathrm{Hom}\,(\mathrm{Hom}\,(f,1),1)\Phi(c \otimes g)(\beta) = \Phi(c \otimes g)\mathrm{Hom}\,(f,1)(\beta) = \\ \Phi(c \otimes g)(\beta f) = (\beta f)(c)g = \Phi(f(c) \otimes g)(\beta) = \Phi(f \otimes 1)(c \otimes g)(\beta). \end{gathered} \tag{10}$$

We now assume that C_m is free and has a finite basis $\{c_1,\dots,c_k\}$. Let $\alpha_1,\dots,\alpha_k \in \mathrm{Hom}\,(C_m,\mathbb{Z})$ be homomorphisms characterized by the requirement

$$\alpha_i(c_j) = \begin{cases} 1, & i = j \\ 0, & i \neq j. \end{cases} \tag{11}$$

It is readily seen that $\{\alpha_1, \ldots, \alpha_k\}$ is a basis of $\mathrm{Hom}\,(C_m, \mathbb{Z})$ and thus, the latter module is also free and finitely generated. We now define a homomorphism $\Psi\colon \mathrm{Hom}\,(\mathrm{Hom}\,(C_m, \mathbb{Z}), G) \to C_m \otimes G$ by assigning to every homomorphism $\varphi\colon \mathrm{Hom}\,(C_m, \mathbb{Z}) \to G$ the value

$$\Psi(\varphi) = \sum_{i=1}^{k} (c_i \otimes \varphi(\alpha_i)) \in C \otimes G. \tag{12}$$

It remains to show that Ψ is the inverse of Φ. Indeed, one has

$$\Psi\Phi(c_j \otimes g) = \sum_{i=1}^{k} c_i \otimes \Phi(c_j \otimes g)(\alpha_i) = \sum_{i=1}^{k} c_i \otimes \alpha_i(c_j) g = c_j \otimes g. \tag{13}$$

Conversely, to show that also $\Phi\Psi = \mathrm{id}$, it suffices to verify that $\Phi(\Psi(\varphi))(\alpha_j) = \varphi(\alpha_j)$, for every homomorphism $\varphi\colon \mathrm{Hom}\,(C, \mathbb{Z}) \to G$ and every $j \in \{1, \ldots, k\}$. Indeed,

$$\Phi(\Psi(\varphi))(\alpha_j) = \sum_{i=1}^{k} \Phi(c_i \otimes \varphi(\alpha_i))(\alpha_j) = \sum_{i-1}^{k} \alpha_j(c_i)\varphi(\alpha_i) = \varphi(\alpha_j). \;\square \tag{14}$$

Proof of Lemma 21.3. By Lemma 21.5, $\mathrm{Hom}\,(C, \mathbb{Z})$ is a free finitely generated cochain complex with coboundary $\delta = \mathrm{Hom}\,(\partial, 1)\colon \mathrm{Hom}\,(C^m, \mathbb{Z}) \to \mathrm{Hom}\,(C^{m+1}, \mathbb{Z})$. Clearly, putting $K_m = \mathrm{Hom}\,(C^{-m}, \mathbb{Z})$, $d = \delta\colon K_{m+1} \to K_m$, one obtains a free finitely generated chain complex $K = (K_m, d)$. Therefore, Lemma 21.4, applied to K and $-m$, yields a natural exact sequence

$$0 \to \mathrm{Ext}\,(H_{-m-1}(K); G) \to H^{-m}(K; G) \to \mathrm{Hom}\,(H_{-m}(K); G) \to 0. \tag{15}$$

By definition,

$$H^m(C) = \frac{\mathrm{Ker}\,(\delta\colon \mathrm{Hom}\,(C_m, \mathbb{Z}) \to \mathrm{Hom}\,(C_{m+1}, \mathbb{Z}))}{\mathrm{Im}\,(\delta\colon \mathrm{Hom}\,(C_{m-1}, \mathbb{Z}) \to \mathrm{Hom}\,(C_m, \mathbb{Z}))} \tag{16}$$

and thus, $H^m(C) = H_{-m}(K)$. Similarly, $H^{m+1}(C) = H_{-m-1}(K)$. Moreover, Lemma 21.5 yields a natural isomorphism of chain complexes $\Phi\colon C \otimes G \to \mathrm{Hom}\,(\mathrm{Hom}\,(C, \mathbb{Z}), G)$. One concludes that Φ induces natural isomorphisms between the m-th homology group of $C \otimes G$, which by definition equals $H_m(C; G)$, and the m-th homology group of $\mathrm{Hom}\,(\mathrm{Hom}\,(C, \mathbb{Z}), G)$, i.e.,

$$\frac{\mathrm{Ker}\,(d\colon \mathrm{Hom}\,(K_{-m}, G) \to \mathrm{Hom}\,(K_{-m+1}, G))}{\mathrm{Im}\,(d\colon \mathrm{Hom}\,(K_{-m-1}, G) \to \mathrm{Hom}\,(K_{-m}, G))} = H^{-m}(K, G). \tag{17}$$

Consequently, the natural sequence (15) yields a natural sequence (3). $\square$

21.2 Homology of compact spaces

THEOREM 21.6. *Let* $\boldsymbol{X}$ *be an inverse system of compact polyhedra. Then, for every abelian group* G *and every integer* $m \geq 0$,

$$\lim^r H_m(\boldsymbol{X}; G) = 0, \tag{1}$$

for $r \geq 2$. *Moreover,*

$$\lim^1 H_m(\boldsymbol{X}; G) \approx \lim^1 \mathrm{Hom}\,(H^m(\boldsymbol{X}), G), \tag{2}$$

and there exists an exact sequence

$$\begin{aligned} 0 \to \lim \mathrm{Ext}\,(H^{m+1}\,(\boldsymbol{X}), G) \quad &\xrightarrow{\beta} \lim H_m\,(\boldsymbol{X}; G) \\ &\xrightarrow{\alpha} \lim \mathrm{Hom}\,(H^m(\boldsymbol{X}), G) \to 0. \end{aligned} \tag{3}$$

Recall that every compact Hausdorff space X admits an inverse system of compact polyhedra $\boldsymbol{X} = (X_\lambda, p_{\lambda\lambda'}, \Lambda)$ such that $X = \lim \boldsymbol{X}$ (see e.g., (Mardešić, Segal 1982), I.5.2, Theorem 7) and $H_m(\boldsymbol{X}; G)$ is the homology pro-group $\boldsymbol{H}_m\,(X; G)$ (see 16.1).

REMARK 21.7. Recall that we proved in Theorem 16.8 that, for every $m \geq 1$ and every $r \geq 1$, there exists a paracompact space X and a polyhedral resolution $\boldsymbol{p}: X \to \boldsymbol{X}$ such that $\lim^r \boldsymbol{H}_m(X; \mathbb{Z}) = \lim^r H_m(\boldsymbol{X}; \mathbb{Z}) \neq 0$.

REMARK 21.8. For inverse systems $\boldsymbol{H}$ of finitely generated abelian groups, $\lim^r \boldsymbol{H} = 0$, whenever $r \geq 2$ (see (Jensen 1972), Remark on p. 65). This result implies (1) in the special case when the coefficient group G is finitely generated, because in this case the groups $H_m(X_\lambda; G)$ are finitely generated. Note that there exist inverse systems $\boldsymbol{H}$ of finitely generated free modules over Noetherian rings, for which $\lim^2 \boldsymbol{H} \neq 0$. This follows from Theorem 4.3 of (Gruson, Jensen 1981) and Proposition 5 of (Jensen 1977).

Proof of Theorem 21.6. First note that $H^m(\boldsymbol{X}) = (H^m(X_\lambda), p^*_{\lambda\lambda'}, \Lambda)$ is a direct system of abelian groups. Next note that, by the naturality of the exact sequence (21.1.1), for any pair $\lambda \leq \lambda'$, one has a commutative diagram (21.1.4) with $X = X_{\lambda'}, Y = X_\lambda$ and $f = p_{\lambda\lambda'}$. Therefore, one also has an exact sequence of inverse systems

$$0 \to \mathrm{Ext}\,(H^{m+1}(\boldsymbol{X}), G) \to H_m\,(\boldsymbol{X}; G) \to \mathrm{Hom}\,(H^m(\boldsymbol{X}), G) \to 0. \tag{4}$$

However, (4) induces a long exact sequence of abelian groups

$$\begin{aligned} 0 &\to \lim \mathrm{Ext}\,(H^{m+1}\,(\boldsymbol{X}), G) \to \lim H_m\,(\boldsymbol{X}; G) \to \lim \mathrm{Hom}\,(H^m(\boldsymbol{X}), G) \\ &\to \lim^1 \mathrm{Ext}\,(H^{m+1}\,(\boldsymbol{X}), G) \to \lim^1 H_m\,(\boldsymbol{X}; G) \to \lim^1 \mathrm{Hom}\,(H^m(\boldsymbol{X}), G) \\ &\to \lim^2 \mathrm{Ext}\,(H^{m+1}\,(\boldsymbol{X}), G) \to \lim^2 H_m\,(\boldsymbol{X}; G) \to \lim^2 \mathrm{Hom}\,(H^m(\boldsymbol{X}), G) \\ &\cdots\cdots\cdots\cdots\cdots\cdots\cdots\cdots\cdots\cdots\cdots\cdots\cdots\cdots \end{aligned} \tag{5}$$

Since $H^m(\boldsymbol{X})$ and $H^{m+1}(\boldsymbol{X})$ are direct systems of finitely generated abelian groups, Theorem 20.17 applies and one obtains (1), for $r \geq 2$. Moreover, for $r = 1$ and $r = 0$, one obtains the isomorphism (2) and the exact sequence (3), respectively. □

THEOREM 21.9. *For every compact Hausdorff space X and every abelian group G there exists a natural exact sequence*

$$0 \to \lim{}^1 \boldsymbol{H}_{m+1}(X;G) \xrightarrow{\phi} \overline{H}_m(X;G) \xrightarrow{\chi} \check{H}_m(X;G) \to 0. \tag{6}$$

Proof. Let $\boldsymbol{p}\colon X \to \boldsymbol{X}$ be an inverse system of compact polyhedra with $\lim \boldsymbol{X} = X$ and let $\boldsymbol{C}$ be the pro-chain complex $S(\boldsymbol{X}) \otimes G$, where S is the singular complex functor. Since $\lim^r H_n(\boldsymbol{C}) = \lim^r H_n(\boldsymbol{X};G) = 0$, for $r \geq 2$ and all $n \in \mathbb{Z}$, Corollary 17.14 applies to $\boldsymbol{C}$ and yields the desired conclusion. □

REMARK 21.10. For compact metric spaces, Theorem 21.9 was proved before (see Remark 17.15).

THEOREM 21.11. *For any compact Hausdorff space X and any abelian group G, in negative dimensions $m < 0$, the strong homology group*

$$\overline{H}_m(X;G) = 0. \tag{7}$$

Proof. Choose an inverse system of compact polyhedra $\boldsymbol{X} = (X_\lambda, p_{\lambda\lambda'}, \Lambda)$ having the property that $\lim \boldsymbol{X} = X$. Clearly, $H_m(X_\lambda;G) = 0$, for $m < 0$, therefore, also $\lim H_m(\boldsymbol{X};G) = 0$, for $m < 0$. Using (6), one concludes that (7) holds, for $m < -1$, and also

$$\overline{H}_{-1}(X;G) \approx \lim{}^1 H_0(\boldsymbol{X};G). \tag{8}$$

Taking into account (2), one also concludes that

$$\overline{H}_{-1}(X;G) \approx \lim{}^1 \mathrm{Hom}\,(H^0(\boldsymbol{X}),G). \tag{9}$$

If K_λ is a triangulation of X_λ, then $H^0\,(X_\lambda) = H^0\,(K_\lambda)$. Since K_λ is a finite simplicial complex, the groups $C^m(K_\lambda)$ of oriented m-cochains of K_λ are finitely generated, hence, so are the subgroups $Z^m(K_\lambda)$ of m-cocycles and the homology groups $H^0\,(K_\lambda)$. For $m = 0$, the group of boundaries equals zero and therefore, $H^0(K_\lambda) = Z^0(K_\lambda)$ is also a finitely generated free abelian groups. This enables us to apply Theorem 20.17 to the direct system $\boldsymbol{A} = H^0\,(\boldsymbol{X})$. In particular, the exact sequence (20.4.27) yields an exact sequence

$$0 \to \lim{}^1 \mathrm{Hom}\,(H^0\,(\boldsymbol{X}),G) \to \mathrm{Ext}\,(\mathrm{colim}\, H^0\,(\boldsymbol{X}),G). \tag{10}$$

Since $\mathrm{colim}\, H^0(\boldsymbol{X})$ is the Čech cohomology group $\check{H}^0(X)$, the proof will be completed if we prove the following lemma.

LEMMA 21.12. *For every compact Hasdorff space X, the Čech cohomology group $\check{H}^0(X)$ is a free abelian group and thus,*

$$\operatorname{Ext}(\check{H}^0(X), G) = 0. \tag{11}$$

We first establish the following simple lemma.

LEMMA 21.13. *For every compact Hasdorff space X, the Čech cohomology group with integer coefficients $\check{H}^0(X)$ is isomorphic to the group $C(X,\mathbb{Z})$ of all (continuous) mappings $f: X \to \mathbb{Z}$.*

Proof. Choose an inverse system of compact polyhedra $\boldsymbol{X} = (X_\lambda, p_{\lambda\lambda'}, \Lambda)$ such that $\boldsymbol{p} = (p_\lambda): X \to \boldsymbol{X}$ is an inverse limit. Then $\check{H}^0(X) = \operatorname{colim} H^0(\boldsymbol{X})$ and every element of $\check{H}^0(X)$ is a class $\alpha = [\alpha_\lambda] \in \operatorname{colim} H^0(\boldsymbol{X})$, where $\alpha_\lambda \in H^0(X_\lambda)$, for some $\lambda \in \Lambda$. By the standard universal coefficient theorem, $H^0(X_\lambda) \approx \operatorname{Hom}(H_0(X_\lambda), \mathbb{Z})$ (Lemma 21.4). Therefore, α_λ can be viewed as a homomorphism $\alpha_\lambda: H_0(X_\lambda) \to \mathbb{Z}$. On the other hand, $H_0(X_\lambda)$ is a finitely generated free abelian group, whose generators can be identified with the components $C_1, \ldots, C_k$ of X_λ. Therefore, α_λ is completely determined by the values $n_i = \alpha_\lambda(C_i) \in \mathbb{Z}$, $i = 1, \ldots, k$. Let ϕ_λ be the mapping $X_\lambda \to \mathbb{Z}$, given by $\phi_\lambda | C_i = n_i$. If $\lambda \leq \lambda'$ and $\alpha_{\lambda'} = p^*_{\lambda\lambda'}(\alpha_\lambda)$, then $\phi_\lambda p_{\lambda\lambda'} = \phi_{\lambda'}$ and thus, $\phi_\lambda p_\lambda = \phi_{\lambda'} p_{\lambda'}$. Consequently, the mapping $f = \phi_\lambda p_\lambda: X \to \mathbb{Z}$ depends only on the class $\alpha \in \operatorname{colim} H^0(\boldsymbol{X})$ and $\Phi(\alpha) = \phi p_\lambda$ defines a mapping $\Phi: \check{H}^0(X) \to C(X,\mathbb{Z})$. It is readily seen that Φ is a homomorphism. In fact, it is an isomorphism. Indeed, if $f: X \to \mathbb{Z}$ is a mapping, by property (M1) (see 7.1), there exist a $\lambda \in \Lambda$ and a mapping $\phi: X_\lambda \to \mathbb{Z}$ such that the mappings ϕp_λ and f are homotopic. Since $\mathbb{Z}$ is discrete, this implies that $\phi p_\lambda = f$. Clearly, the mapping ϕ must be constant on every component of X_λ and therefore, it determines a homomorphism $\alpha: H_0(X_\lambda) \to \mathbb{Z}$, hence, also an element $\alpha \in H^0(X_\lambda)$ and a class $[\alpha] \in \operatorname{colim} H^0(\boldsymbol{X}) \approx \check{H}^0(X)$. It is readily seen that the homomorphism $\Psi: C(X,\mathbb{Z}) \to \check{H}^0(X)$, defined in this way, is the inverse of Φ. □

Proof of Lemma 21.12. We associate with every set X the abelian group $F(X)$ of all functions $f: X \to \mathbb{Z}$ having finite images $f(X)$. Clearly, if X is a compact space and $f \in C(X,\mathbb{Z})$, then $f \in F(X)$, i.e., $C(X,\mathbb{Z})$ is a subgroup of the group $F(X)$. Since every subgroup of a free abelian group is itself a free abelian group, Lemma 21.12 is an immediate consequence of Lemma 21.13 and of the following theorem (Nöbeling 1968).

LEMMA 21.14. *For every set X the group $F(X)$ of functions $f: X \to \mathbb{Z}$ assuming only a finite set of values, is a free abelian group.*

21.3 Universal coefficients for compact spaces

In this subsection we will prove the *universal coefficient theorem* for strong homology of compact Hausdorff spaces.

THEOREM 21.15. *For every compact Hausdorff space X, abelian group G and integer $m \in \mathbb{Z}$, there exists an exact sequence*

$$0 \to \mathrm{Ext}\,(\check{H}^{m+1}(X), G) \xrightarrow{\psi} \overline{H}_m(X;G) \xrightarrow{\theta} \mathrm{Hom}\,(\check{H}^m(X), G) \to 0, \qquad (1)$$

which is natural in both variables.

Proof. Every abelian group G embeds in a divisible group. Moreover, images of divisible groups are divisible and divisible groups are injective. Consequently, there exists a short injective resolution

$$0 \to G \xrightarrow{u} I^0 \xrightarrow{v} I^1 \to 0. \qquad (2)$$

By Theorem 17.6, (2) induces an exact sequence

$$\ldots \to \overline{H}_{m+1}(X;I^1) \to \overline{H}_m(X;G) \to \overline{H}_m(X;I^0) \to \overline{H}_m(X;I^1) \to \ldots . \qquad (3)$$

This long exact sequence breaks up into short exact sequences

$$0 \to A_{m+1} \to \overline{H}_m(X;G) \to B_m \to 0, \;\; m \in \mathbb{Z}, \qquad (4)$$

where

$$A_m = \mathrm{Coker}\,(\overline{H}_m(X;I^0) \to \overline{H}_m(X;I^1)), \qquad (5)$$

$$B_m = \mathrm{Ker}\,(\overline{H}_m(X;I^0) \to \overline{H}_m(X;I^1)). \qquad (6)$$

We will show that the sequence (4) coincides with the sequence (1) from Theorem 21.15, i.e., that there exist isomorphisms

$$A_m \approx \mathrm{Ext}\,(\check{H}^m(X), G), \qquad (7)$$

$$B_m \approx \mathrm{Hom}\,(\check{H}^m(X), G). \qquad (8)$$

Let us first note that

$$\mathrm{Ext}\,(\check{H}^m(X), I^0) = 0, \qquad (9)$$

because I^0 is injective. Therefore, application of the functor $\mathrm{Hom}\,(\check{H}^m(X), .)$ to (2) yields an exact sequence

$$\begin{aligned} 0 \to \mathrm{Hom}\,(\check{H}^m(X), G) \to \;& \mathrm{Hom}\,(\check{H}^m(X), I^0) \to \mathrm{Hom}\,(\check{H}^m(X), I^1) \to \\ & \mathrm{Ext}\,(\check{H}^m(X), G) \to 0, \end{aligned} \qquad (10)$$

which shows that

$$\mathrm{Coker}\,(\mathrm{Hom}\,(\check{H}^m(X), I^0) \to \mathrm{Hom}\,(\check{H}^m(X), I^1)) \approx \mathrm{Ext}\,(\check{H}^m(X), G), \qquad (11)$$

$$\operatorname{Ker}(\operatorname{Hom}(\check{H}^m(X), I^0) \to \operatorname{Hom}(\check{H}^m(X), I^1)) \approx \operatorname{Hom}(\check{H}^m(X), G). \quad (12)$$

To establish the existence of isomorphisms (7) and (8), it suffices to establish the existence of isomorphisms

$$\overline{H}_m(X, I^k) \approx \operatorname{Hom}(\check{H}^m(X), I^k),\ k = 0, 1, \quad (13)$$

which make the following diagram commutative.

$$\begin{array}{ccc} \overline{H}_m(X; I^0) & \longrightarrow & \overline{H}_m(X; I^1) \\ \approx \downarrow & & \downarrow \approx \\ \operatorname{Hom}(\check{H}^m(X), I^0) & \longrightarrow & \operatorname{Hom}(\check{H}^m(X), I^1). \end{array} \quad (14)$$

We define the isomorphism (13) as the composition of three isomorphisms. The first one is the homomorphism $\chi\colon \overline{H}_m(X, I^k) \to \check{H}_m(X; I^k)$ from Theorem 21.9 (for $G = I^k$). To see that it is an isomorphism, it suffices to show that

$$\lim{}^1 H_{m+1}(\boldsymbol{X}; I^k) = 0,\ k = 0, 1. \quad (15)$$

First note that Theorem 21.6 yields an isomorphism

$$\lim{}^1 H_{m+1}(\boldsymbol{X}; I^k) \approx \lim{}^1 \operatorname{Hom}(H^{m+1}(\boldsymbol{X}), I^k),\ k = 0, 1. \quad (16)$$

Next note that I^k is injective, hence also purely injective. Consequently, one has $\operatorname{Pext}(\check{H}^{m+1}(X), I^k) = 0$. Now, the exact sequence (20.4.23), applied to $\boldsymbol{A} = H^{m+1}(\boldsymbol{X})$ and $G = I^k$ yields

$$\lim{}^1 \operatorname{Hom}(H^{m+1}(\boldsymbol{X}), I^k) = 0,\ k = 0, 1. \quad (17)$$

The second isomorphism, used in the definition of (13), is the homomorphism $\alpha\colon \check{H}_m(X; I^k) \to \lim \operatorname{Hom}(H^m(\boldsymbol{X}), I^k)$ from (21.2.3) (for $G = I^k$). Since I^k is injective, $\operatorname{Ext}(H^{m+1}(\boldsymbol{X}); I^k) = 0$, and therefore, the exact sequence (21.2.3) implies that α is indeed an isomorphism. The third isomorphism is the isomorphism $\lim \operatorname{Hom}(H^m(\boldsymbol{X}), I^k) \to \operatorname{Hom}(\check{H}^m(X), I^k)$ from Theorem 20.9.

Naturality of the exact sequence (1) is an immediate consequence of the way it was constructed and of the naturality of the sequence (17.1.40). □

REMARK 21.16. For polyhedra strong homology coincides with singular homology (see Theorem 19.27). Therefore, Theorem 21.15 generalizes the universal coefficient theorem for compact polyhedra (see Theorem 21.1).

REMARK 21.17. In 1980 N.A. Berikashvili gave a simple axiomatic characterization of the strong homology groups on compact Hausdorff spaces. His axioms consist of the Eilenberg – Steenrod axioms and of the assertion that the sequence (1) of Theorem 21.15 is exact (Berikashvili 1980, 1984). Consequently, Theorem 21.15 implies that our strong homology groups coincide with the homology groups constructed by various other authors, e.g., with the groups considered in (Massey 1978). A direct proof that our groups coincide with the latter ones was given in (Günther 1992c).

21.4 A filtration of the strong homology group

The next theorem describes a natural four term filtration of the strong homology group $\overline{H}_m(X;G)$ of a compact Hausdorff space X as well as the corresponding associated graded group.

THEOREM 21.18. *For every compact Hausdorff space X, abelian group G and integer $m \in \mathbb{Z}$, the strong homology group $\overline{H}_m(X;G)$ has a natural filtration $0 = F_0 \subseteq F_1 \subseteq F_2 \subseteq F_3 = \overline{H}_m(X;G)$ having the following properties:*

(*i*) $F_1 \approx \operatorname{Pext}(\check{H}^{m+1}(X), G) \approx \lim^1 H_{m+1}(\boldsymbol{X};G) \approx \lim^1 \operatorname{Hom}(H^{m+1}(\boldsymbol{X}), G)$,

(*ii*) $F_2 \approx \operatorname{Ext}(\check{H}^{m+1}(X), G)$,

(*iii*) $F_2/F_1 \approx \lim \operatorname{Ext}(H^{m+1}(\boldsymbol{X}), G)$,

(*iv*) $F_3/F_1 \approx \check{H}_m(X;G)$,

(*v*) $F_3/F_2 \approx \operatorname{Hom}(\check{H}^m(X), G) \approx \lim \operatorname{Hom}(H^m(\boldsymbol{X}), G)$.

(*vi*) *The composition*

$$\operatorname{Pext}(\check{H}^{m+1}(X), G) \overset{\approx}{\to} F_1 \hookrightarrow F_2 \overset{\approx}{\to} \operatorname{Ext}(\check{H}^{m+1}(X), G),$$

formed by the isomorphisms from (i) and (ii) and by the inclusion $F_1 \hookrightarrow F_2$, equals the natural inclusion $\operatorname{Pext}(\check{H}^{m+1}(X), G) \hookrightarrow \operatorname{Ext}(\check{H}^{m+1}(X), G)$.

Proof. We begin the proof by considering a large commutative diagram, which contains much information on the various groups related to the strong homology groups of Hausdorff compact spaces.

$$\begin{array}{ccccccccc}
 & & 0 & & 0 & & & & \\
 & & \downarrow & & \downarrow & & & & \\
0 & \to & \operatorname{Pext}(\check{H}^{m+1}(X), G) & \underset{\approx}{\overset{\kappa}{\to}} & \lim^1 H_{m+1}(\boldsymbol{X};G) & \to & 0 & & \\
 & & \iota\downarrow & & \phi\downarrow & & \downarrow & & \\
0 & \to & \operatorname{Ext}(\check{H}^{m+1}(X), G) & \overset{\psi}{\to} & \overline{H}_m(X;G) & \overset{\theta}{\to} & \operatorname{Hom}(\check{H}^m(X), G) & \to & 0 \\
 & & \pi\downarrow & & \chi\downarrow & & \approx\downarrow\gamma & & \\
0 & \to & \lim \operatorname{Ext}(H^{m+1}(\boldsymbol{X}), G) & \underset{\beta}{\to} & \check{H}_m(X;G) & \underset{\alpha}{\to} & \lim \operatorname{Hom}(H^m(\boldsymbol{X}), G) & \to & 0 \\
 & & \downarrow & & \downarrow & & \downarrow & & \\
 & & 0 & & 0 & & 0 & &
\end{array} \qquad (1)$$

The homomorphisms $\psi, \theta, \beta, \alpha$ have already been defined in the exact sequences of Theorems 21.15 and 21.6, which appear in the diagram as the two lower rows. Applying naturality of the exact sequence (21.3.1) to the mapping $p_\lambda : X \to X_\lambda$, one obtains the following commutative diagram.

$$\begin{array}{ccccccccc}
0 & \to & \operatorname{Ext}(\check{H}^{m+1}(X),G) & \to & \overline{H}_m(X;G) & \to & \operatorname{Hom}(\check{H}_m(X),G) & \to & 0 \\
 & & \downarrow & & \downarrow & & \downarrow & & \\
0 & \to & \operatorname{Ext}(H^{m+1}(X_\lambda),G) & \to & H^m(X_\lambda;G) & \to & \operatorname{Hom}(H^m(X_\lambda),G) & \to & 0.
\end{array} \tag{2}$$

Note that X_λ is a compact polyhedron and thus, $\overline{H}_m(X_\lambda;G) = H_m(X_\lambda;G)$ and $\check{H}^m(X_\lambda;G) = H^m(X_\lambda;G)$. Passing to the limit in diagram (2), one obtains the mappings π, χ, γ, which together with the last two rows of diagram (1) form a commutative diagram. Here χ coincides with the homomorphism from the exact sequence (21.2.6), which makes the corresponding vertical line of (1) also exact. It is an elementary fact that $\gamma\colon \operatorname{Hom}(\operatorname{colim}\boldsymbol{A}, G) \to \lim \operatorname{Hom}(\boldsymbol{A}, G)$ is an isomorphism, for any direct system of modules $\boldsymbol{A}$. Finally, for $\boldsymbol{A} = H^{m+1}(\boldsymbol{X})$, the exact sequence of Theorem 20.17 assumes the form

$$\begin{aligned}
0 \to \lim{}^1 \operatorname{Hom}(H^{m+1}(\boldsymbol{X}),G) &\to \operatorname{Ext}(\check{H}^{m+1}(\boldsymbol{X}),G) \\
&\xrightarrow{\pi} \lim \operatorname{Ext}(H^{m+1}(\boldsymbol{X}),G) \to 0,
\end{aligned} \tag{3}$$

where π is induced by the canonical homomorphisms $a^\lambda\colon A^\lambda \to \operatorname{colim}\boldsymbol{A}$.

In order to obtain the homomorphism ι, consider a short purely injective resolution (see Theorem 20.44).

$$0 \to G \xrightarrow{p} J^0 \xrightarrow{q} J^1 \to 0. \tag{4}$$

A standard argument from homological algebra shows that there exists a mapping of exact sequences

$$\begin{array}{ccccccccc}
0 & \to & G & \xrightarrow{p} & J^0 & \xrightarrow{q} & J^1 & \to & 0 \\
 & & =\downarrow & & \downarrow\sigma & & \downarrow\tau & & \\
0 & \to & G & \xrightarrow{u} & I^0 & \xrightarrow{v} & I^1 & \to & 0,
\end{array} \tag{5}$$

where the second row is as in (21.3.2).

Viewed as a cochain mapping, (σ, τ) is unique up to cochain homotopy. For any direct system $\boldsymbol{A}$, (5) induces a mapping from the pure version of the Roos bicomplex of $\boldsymbol{A}$ to the Roos bicomplex of $\boldsymbol{A}$, which in turn induces a mapping between the corresponding spectral sequences (see 20.4 and 20.3). Hence, it also induces a mapping between the induced exact sequences (20.4.23) and (20.4.27). If $\boldsymbol{A}$ is a system of finitely generated abelian groups, one has $\lim^2 \operatorname{Hom}(\boldsymbol{A}, G) = 0$ (see Theorem 20.17) and the induced mapping assumes the form of a commutative diagram.

$$\begin{array}{ccccccccc}
0 & \to & \lim{}^1\operatorname{Hom}(\boldsymbol{A},G) & \to & \operatorname{Pext}(\operatorname{colim}\boldsymbol{A},G) & \to & \lim \operatorname{Pext}(\boldsymbol{A},G) & \to & 0 \\
 & & =\downarrow & & \downarrow\iota & & \downarrow & & \\
0 & \to & \lim{}^1\operatorname{Hom}(\boldsymbol{A},G) & \to & \operatorname{Ext}(\operatorname{colim}\boldsymbol{A},G) & \xrightarrow{\pi} & \lim \operatorname{Ext}(\boldsymbol{A},G) & \to & 0.
\end{array} \tag{6}$$

In particular, this is the case for $\boldsymbol{A} = H^{m+1}(\boldsymbol{X})$. Note that in this case also $\operatorname{Pext}(\boldsymbol{A}, G) = 0$ and thus, $\lim \operatorname{Pext}(\boldsymbol{A}, G) = 0$. Consequently, the first row of (6) yields an isomorphism

$$\lim{}^1 \operatorname{Hom}(H^{m+1}(\boldsymbol{X}), G) \approx \operatorname{Pext}(\check{H}^{m+1}(X), G). \tag{7}$$

Moreover, (6) shows that the following sequence is exact.

$$\begin{aligned} 0 \to \operatorname{Pext}(\check{H}^{m+1}(X), G) &\xrightarrow{\iota} \operatorname{Ext}(\check{H}^{m+1}(X), G) \\ &\xrightarrow{\pi} \lim \operatorname{Ext}(H^{m+1}(\boldsymbol{X}), G) \to 0. \end{aligned} \tag{8}$$

This sequence is the first vertical line in (1).

It is now easy to conclude, from the already established properties of diagram (1), that there exists a unique homomorphism $\kappa\colon \operatorname{Pext}(\check{H}^{m+1}(X), G) \to \lim^1 H_{m+1}(\boldsymbol{X}; G)$, which makes diagram (1) commutative. Moreover, κ is an isomorphism.

We define the subgroups F_1 and F_2 of $\overline{H}_m(X; G)$, by putting

$$F_1 = \operatorname{Ker}(\chi), \tag{9}$$

$$F_2 = \operatorname{Ker}(\theta). \tag{10}$$

Since χ and θ are natural mappings, F_1 and F_2 are natural subgroups of $\overline{H}_m(X; G)$. Due to the fact that γ is an isomorphism, the commutativity of (1) implies

$$F_2 = \operatorname{Ker}(\gamma\theta) = \operatorname{Ker}(\alpha\chi). \tag{11}$$

Therefore, $F_1 \subseteq F_2$. The assertion (iv) is an immediate consequence of the exactness of the Milnor sequence. Exactness of the middle row of (1) shows that the restriction $\psi_F\colon \operatorname{Ext}(\check{H}^{m+1}(X), G) \to F_2$ of ψ to F_2 is an isomorphism, which establishes (ii). It also shows that $F_3/F_2 \approx \operatorname{Hom}(\check{H}^m(X), G)$. This and $\operatorname{Hom}(\check{H}^m(X), G) \approx \lim \operatorname{Hom}(H^m(\boldsymbol{X}), G)$ establishes (v). To prove (iii), note that χ is an epimorphism. Therefore, it induces an epimorphism $\chi\colon \operatorname{Ker}(\alpha\chi) \to \operatorname{Ker}(\alpha)$, whose kernel is $\operatorname{Ker}(\chi) \subseteq \operatorname{Ker}(\alpha\chi)$. Consequently,

$$\operatorname{Ker}(\alpha\chi)/\operatorname{Ker}(\chi) \approx \operatorname{Ker}(\alpha). \tag{12}$$

Now, (9), (11) and (12) establish (iii). To prove (i), first note that the existence of an isomorphism $\lim^1 H_m(\boldsymbol{X}; G) \approx \lim^1 \operatorname{Hom} H^m(\boldsymbol{X}, G)$ was established in Theorem 21.6. Next note that $F_1 = \operatorname{Im}(\phi)$ and thus, the restriction ϕ_F of ϕ to $F_1 \subseteq \overline{H}_m(X; G)$ is an isomorphism. Since κ is an isomorphism, it follows that also $\phi_F\kappa\colon \operatorname{Pext}(\check{H}^{m+1}(X), G) \to F_1$ is an isomorphism and (i) holds. Moreover, the composition of the isomorphism $(\phi_F\kappa)$, of the inclusion $F_1 \hookrightarrow F_2$ and of the isomorphism $(\psi_F)^{-1}\colon F_2 \to \operatorname{Ext}(\check{H}^{m+1}(X), G)$ coincides with ι. Consequently, to prove (vi) and complete the proof of Theorem 21.18, it suffices to prove the next lemma.

LEMMA 21.19. *For every A, $\iota\colon \operatorname{Pext}(A, G) \to \operatorname{Ext}(A, G)$ coincides with the canonical embedding $i\colon \operatorname{Pext}(A, G) \hookrightarrow \operatorname{Ext}(A, G)$.*

Proof. To define the canonical embedding $i\colon \mathrm{Pext}\,(A,G) \to \mathrm{Ext}\,(A,G)$, we need a different (and well-known) interpretation of the groups $\mathrm{Ext}\,(A,G)$ and $\mathrm{Pext}\,(A,G)$ (see (Mac Lane 1963), Chapter III). It involves equivalence classes of exact (purely exact) sequences

$$0 \to G \xrightarrow{g} B \xrightarrow{b} A \to 0. \tag{13}$$

Two such sequences are considered equivalent provided there exists a commutative diagram

$$\begin{array}{ccccccccc} 0 & \to & G & \xrightarrow{g} & B & \xrightarrow{b} & A & \to & 0 \\ & & =\downarrow & & \downarrow & & \downarrow= & & \\ 0 & \to & G & \xrightarrow{g'} & B' & \xrightarrow{b'} & A & \to & 0. \end{array} \tag{14}$$

Note that, due to the five lemma, $B' \to B$ must be an isomorphism.

To obtain the desired interpretation, consider a short injective resolution

$$0 \to G \xrightarrow{u} I^0 \xrightarrow{v} I^1 \to 0. \tag{15}$$

Since $\mathrm{Ext}\,(A, I^0) = 0$, (15) implies

$$\mathrm{Ext}\,(A,G) \approx \mathrm{Coker}\,(\mathrm{Hom}(A,I^0) \to \mathrm{Hom}(A,I^1)). \tag{16}$$

With every homomorphism $\alpha \in \mathrm{Hom}(A,I^1)$ associate a commutative diagram with exact rows

$$\begin{array}{ccccccccc} 0 & \to & G & \xrightarrow{g} & B & \xrightarrow{b} & A & \to & 0 \\ & & =\downarrow & & \downarrow\beta & & \downarrow\alpha & & \\ 0 & \to & G & \xrightarrow{u} & I^0 & \xrightarrow{v} & I^1 & \to & 0. \end{array} \tag{17}$$

To obtain such a diagram it suffices to take for B, β, b the pull-back of the homomorphisms v, α. Viewing B as the set $\{(x,a) \in I^0 \times A) | v(x) = \alpha(a)\}$, one defines $g\colon G \to B$ by putting $g(y) = (u(y), 0)$. Conversely, one can show that in a commutative diagram with exact rows (17), the right-hand square is a pull-back diagram (see e.g., (Hilton, Stammbach 1971), III, Lemma 1.3). Using properties of the pull-back, it is easy to see that any two sequences (13), which embed as the upper row in a digram (17), belong to the same equivalence class.

If $\alpha, \alpha' \in \mathrm{Hom}(A, I^1)$ are such that $\alpha' - \alpha = v\gamma$, for some $\gamma \in \mathrm{Hom}(A, I^0)$, i.e., if they represent the same element of (16), then the exact sequence (13), which forms the first row of (17), and the exact sequence

$$0 \to G \xrightarrow{g'} B' \xrightarrow{b'} A \to 0, \tag{18}$$

which froms the first row of the diagram corresponding to (17) for α', are equivalent extensions. The needed isomorphism $B \to B'$ is given by $(x,a) \mapsto (x + \gamma(a), a)$. In this way every equivalence class of exact sequences (13) is associated with some element of $\operatorname{Ext}(A, G)$, because the sequences (13) and (15) can be embedded in a commutative diagram (17). The established correspondence between elements of $\operatorname{Ext}(A, G)$ and equivalence classes of exact sequences (13) yields the desired interpretation of $\operatorname{Ext}(A, G)$ in terms of extensions of G by A.

One can now repeat the above reasoning for purely exact sequences and Pext by considering a short purely injective resolution (4) of G. Since J^0 is purely injective, $\operatorname{Pext}(A, J^0) = 0$ and therefore,

$$\operatorname{Pext}(A, C) \approx \operatorname{Coker}(\operatorname{Hom}(A, J^0) \to \operatorname{Hom}(A, J^1)). \tag{19}$$

As before, using the pull-back construction, one associates with every homomorphism $\alpha \in \operatorname{Hom}(A, J^1)$ a commutative diagram

$$\begin{array}{ccccccccc} 0 & \to & G & \xrightarrow{g} & B & \xrightarrow{b} & A & \to & 0 \\ & & =\downarrow & & \downarrow\beta & & \downarrow\alpha & & \\ 0 & \to & G & \xrightarrow{p} & J^0 & \xrightarrow{q} & J^1 & \to & 0. \end{array} \tag{20}$$

Since $p(G)$ is purely embedded in J^0, one readily concludes that $g(G)$ is purely embedded in B. Consequently, the exact sequence $0 \to G \to B \to A \to 0$ is purely exact. As before, the described construction establishes the desired bijection between $\operatorname{Pext}(G, A)$ and the equivalence classes of pure extensions of G by A.

Using the described interpretation one can also introduce the group structure in the sets of equivalence classes of extensions and pure extensions, respectively. For an explicit description of the group structures see (Mac Lane 1963), Chapter III,§2). The described interpretation enables us to define the canonical embedding $i\colon \operatorname{Pext}(A, G) \to \operatorname{Ext}(A, G)$. One simply views every purely exact sequence (4) as an exact sequence, i.e., one forgets its purity. Therefore, to prove Lemma 21.19, we must show that the homomorphism $\iota\colon \operatorname{Pext}(A, G) \to \operatorname{Ext}(A, G)$, induced by (5), corresponds to the above described canonical embedding i.

By definition, $\iota\colon \operatorname{Pext}(G, A) \to \operatorname{Ext}(G, A)$ is induced by the commutative diagram (5). To an element $\varepsilon \in \operatorname{Pext}(G, A)$, determined by a homomorphism $\alpha\colon A \to J^1$ (see (19)), corresponds a purely exact sequence $0 \to G \to B \to A \to 0$, which fits into diagram (20). On the other hand, $\iota(\varepsilon)$ is given by $\tau\alpha\colon A \to I^1$. Now notice that the juxtaposition of the diagrams (20) and (5) yields a diagram of type (17), whose last vertical arrow is $\tau\alpha\colon A \to I^1$, while its top row is again the sequence $0 \to G \to B \to A \to 0$. Therefore, this purely exact sequence, viewed as an exact sequence, corresponds to $i(\varepsilon)$. In other words, if one interprets $\operatorname{Pext}(G, A)$ as the set of equivalence classes of pure extensions of G by A and one interprets $\operatorname{Ext}(G, A)$ as the set of equivalence classes of extensions of G by A, then the mapping $\iota\colon \operatorname{Pext}(G, A) \to \operatorname{Ext}(G, A)$ just forgets the purity of the extensions belonging to $\operatorname{Pext}(G, A)$. $\square$

21.5 Strong homology with compact supports

Strong homology groups with compact supports are defined in the usual way. If X is a topological space, one considers the direct system $\boldsymbol{C} = (C_\mu, i_{\mu\mu'}, M)$, where C_μ ranges over all compact subsets of X and $i_{\mu\mu'}: C_\mu \to C_{\mu'}$ are inclusion mappings. Applying the strong homology functor $\overline{H}_m(\,.\,;G)$, one obtains a direct system of abelian groups $\overline{H}_m(\boldsymbol{C};G)$. The m-th *strong homology group with compact supports* $\overline{H}^c_m(X;G)$ is the colimit of $\overline{H}_m(\boldsymbol{C};G)$. Clearly, the inclusions $C_\mu \subseteq X$ induce homomorphisms $\overline{H}_m(C_\mu;G) \to \overline{H}_m(X;G)$, which then induce a homomorphism

$$\overline{H}^c_m(X;G) \to \overline{H}_m(X;G). \tag{1}$$

We will show, by examples, that the groups $\overline{H}^c_m(X;G)$ and $\overline{H}_m(X;G)$ need not be isomorphic, hence, (1) need not be an isomorphism. In 16.2 we exhibited a paracompact space $X = X(m,n,\Lambda)$, which depends on integers $m,n \geq 1$, and on an infinite directed set Λ. The space X is the limit of a resolution $\boldsymbol{p}: X \to \boldsymbol{X} = \boldsymbol{X}(m,n,\Lambda)$, consisting of m-dimensional polyhedra. Therefore, $\dim X \leq m$. Since X contains copies of the m-cell B^m, one also has $\dim X \geq m$, and thus, $\dim X = m$. In this subsection we will prove that, for a suitable choice of m, n, p and Λ, the groups $\overline{H}_p(X;\mathbb{Z})$ and $\overline{H}^c_p(X;\mathbb{Z})$ differ. The proof requires several lemmas. We first prove a lemma concerning Čech homology of the spaces $X(m,n,\Lambda)$.

LEMMA 21.20. *For the space $X = X(m,n,\Lambda)$ and $p > 0$, the Čech homology group*

$$\check{H}_p(X;G) = 0. \tag{2}$$

Proof. First recall that

$$\check{H}_p(X;G) = \lim H_p(\boldsymbol{X};G). \tag{3}$$

By (16.2.5),

$$H_p(X_\lambda;G) = \bigoplus_{\boldsymbol{\lambda}} H_p(X^\lambda_{\boldsymbol{\lambda}};G). \tag{4}$$

For $p \neq m$, $H_p(S^m,*) = H_p(B^m,*) = 0$ and thus, by (16.2.4), $H_p(X_\lambda;G) = 0$, hence, also $H_p(\boldsymbol{X};G) = \boldsymbol{0}$. Consequently, (2) holds. If $p = m$, then $H_m(S^m;G) \approx G$, $H_m(B^m;G) = 0$ and thus, (4) implies

$$H_m(X_\lambda;G) = \bigoplus_{\lambda \leq \boldsymbol{\lambda}} G. \tag{5}$$

Moreover, by (16.2.6), $p_{\lambda\lambda'*}: H_m(X_{\lambda'};G) \to H_m(X_\lambda;G)$, $\lambda \leq \lambda'$, is the natural inclusion. Let $\alpha = (\alpha_\lambda) \in \lim H_m(X_\lambda;G)$ and let $\alpha_\lambda \in H_m(X_\lambda;G)$ be contained in the subgroup of $H_m(X_\lambda;G)$, obtained by summing up copies of G over a finite collection of sequences $\lambda^i_0 \leq \ldots \leq \lambda^i_n$, $i = 1, \ldots, k$. Choose

$\lambda' \in \Lambda$ so that $\lambda' > \lambda, \lambda_0^1, \ldots, \lambda_0^k$. Then the corresponding summands do not appear in $H_m(X_{\lambda'}; G)$. Therefore, $\alpha_\lambda = p_{\lambda\lambda'}(\alpha_{\lambda'}) = 0$, which shows that $\alpha = 0$, i.e., (2) holds also for $p = m$. □

LEMMA 21.21. *If* $\boldsymbol{X}$ *is an inverse system of* CW - *complexes of bounded dimension, then for all p,*

$$\overline{H}_p(\boldsymbol{X}; G) \approx \lim \overline{H}_p^{(*)}(\boldsymbol{X}; G), \tag{6}$$

where $\overline{H}_p^{(*)}(\boldsymbol{X}; G)$ *denotes the inverse sequence*

$$\overline{H}_p^{(0)}(\boldsymbol{X}; G) \leftarrow \ldots \leftarrow \overline{H}_p^{(r-1)}(\boldsymbol{X}; G) \leftarrow \overline{H}_p^{(r)}(\boldsymbol{X}; G) \leftarrow \ldots. \tag{7}$$

Proof. For a given p and any sufficiently large r, the number $p + 1 + r$ exceeds the dimension of all $X_\lambda, \lambda \in \Lambda$. Therefore, $H_{p+1+r}(\boldsymbol{X}; G) = 0$ and

$$\lim^{r+1} H_{p+1+r}(\boldsymbol{X}; G) = 0. \tag{8}$$

It now follows from the first Miminoshvili sequence (see (17.2.1)) that, for sufficiently large r,

$$\overline{H}_{p+1}^{(r)}(\boldsymbol{X}; G) \to \overline{H}_{p+1}^{(r-1)}(\boldsymbol{X}; G) \tag{9}$$

is an epimorphism. Therefore, the sequence $\overline{H}_{p+1}^{(*)}(\boldsymbol{X}; G)$ has the Mittag – Leffler property and its first derived limit vanishes (see, e.g., (Mardešić, Segal 1982), II.6.2, Theorem 10). Now, the second Miminoshvili sequence (see (17.3.5)) yields the desired conclusion (6). □

LEMMA 21.22. *If* $\boldsymbol{X} = (X_\lambda, p_{\lambda\lambda'}, \Lambda)$ *is an inverse system of spaces indexed by a directed set* Λ *of cofinality* $\mathrm{cof}(\Lambda) = \aleph_{n-1}$, $n \geq 1$, *then, for any p and any* $r \geq n$, *there exists an epimorphism*

$$\lim \overline{H}_p^{(*)}(\boldsymbol{X}; G) \to \overline{H}_p^{(r-1)}(\boldsymbol{X}; G). \tag{10}$$

Proof. By Goblot's theorem (Theorem 13.9), for any inverse system of abelian groups $\boldsymbol{H}$, indexed by a set Λ of cofinality $\mathrm{cof}(\Lambda) = \aleph_{n-1}$, $n \geq 1$, the derived limit $\lim^s \boldsymbol{H} = 0$, provided $s \geq n + 1$. Therefore, by the first Miminoshvili sequence (see (17.2.1)), one concludes that

$$\overline{H}_p^{(r)}(\boldsymbol{X}; G) \to \overline{H}_p^{(r-1)}(\boldsymbol{X}; G). \tag{11}$$

is an epimorphism, for $r \geq n$. Consequently, all bonding homomorphisms in the sequence

$$\overline{H}_p^{(n-1)}(\boldsymbol{X}; G) \leftarrow \overline{H}_p^{(n)}(\boldsymbol{X}; G) \leftarrow \ldots \tag{12}$$

are epimorphisms and thus, the projections of the limit group $\lim \overline{H}_p^{(*)}(\boldsymbol{X}; G)$ to any of the terms of (12) is also an epimorphism. □

THEOREM 21.23. *Let $X = X(m, 2, \Lambda)$, where $m \geq 2$ and* $\mathrm{cof}(\Lambda) = \aleph_1$. *Then the strong homology group*

$$\overline{H}_{m-2}(X; \mathbb{Z}) \neq 0. \tag{13}$$

Proof. By Lemma 17.7, the strong homology group $\overline{H}^{(0)}_{m-1}(\boldsymbol{X}; \mathbb{Z})$ is isomorphic to the Čech homology group $\check{H}_{m-1}(X; \mathbb{Z})$. Therefore, by Lemma 21.20, $\overline{H}^{(0)}_{m-1}(\boldsymbol{X}; \mathbb{Z}) = 0$. Now, the first Miminoshvili sequence (for $r = 2$ and m replaced by $m - 2$) yields an exact sequence

$$0 \to \lim{}^2 H_m(\boldsymbol{X}; \mathbb{Z}) \to \overline{H}^2_{m-2}(\boldsymbol{X}; \mathbb{Z}). \tag{14}$$

It was proved in Lemma 16.4 that

$$\lim{}^2 H_m(\boldsymbol{X}; \mathbb{Z}) \neq 0. \tag{15}$$

Consequently,

$$\overline{H}^{(2)}_{m-2}(\boldsymbol{X}; \mathbb{Z}) \neq 0. \tag{16}$$

By Lemma 21.22 (for $n = 2, p = m - 2$ and $r = 3$), there is an epimorphism

$$\lim \overline{H}^{(*)}_{m-2}(\boldsymbol{X}; \mathbb{Z}) \to \overline{H}^2_{m-2}(\boldsymbol{X}; \mathbb{Z}). \tag{17}$$

Therefore, (16) implies that $\lim \overline{H}^{(*)}_{m-2}(\boldsymbol{X}; \mathbb{Z}) \neq 0$. Now (13) follows, by applying Lemma 21.21. □

REMARK 21.24. In the proof of Lemma 21.20 it was shown that $H_p(\boldsymbol{X}; \mathbb{Z}) = 0$, for $p > 0$ and $p \neq m$. In particular, $H_{m-1}(\boldsymbol{X}; \mathbb{Z}) = 0$, for $m \geq 2$. Hence, also $\lim{}^1 H_{m-1}(\boldsymbol{X}; \mathbb{Z}) = 0$. Moreover, by Lemma 21.20, for $m \geq 3$, also $\overline{H}^{(0)}_{m-2}(\boldsymbol{X}; \mathbb{Z}) = 0$. Therefore, the beginning of the first Miminoshvili sequence (see (17.2.2)) shows that $\overline{H}^{(1)}_{m-2}(\boldsymbol{X}; \mathbb{Z}) = 0$. Consequently, (16) shows that the strong homology groups of height 2 can differ from those of height 0 and height 1.

THEOREM 21.25. *Let $X = X(m, n, \Lambda)$. If Λ is cofinite and uncountable, then the strong homology group*

$$\overline{H}^c_m(X; G) = 0, \ m \neq 0. \tag{18}$$

Proof. It was proved in Lemma 16.10 that, for Λ cofinite and uncountable, every compact subset $C \subseteq X$ is contained in a subspace $C' \subseteq X$, which is the wedge of a finite collection of m - cells. Since C' is a compact contractible polyhedron $\overline{H}^c_m(C'; G) = H_m(C'; G) = 0$ and (18) follows. □

Combining Theorems 21.23 and 21.25, we obtain the following corollary.

COROLLARY 21.26. *If Λ is cofinite,* $\mathrm{cof}(\Lambda) = \aleph_1$ *and* $m \geq 3$*, then for* $X = X(m, n, \Lambda)$*, one has*

$$\overline{H}_{m-2}(X;\mathbb{Z}) \neq 0, \tag{19}$$

$$\overline{H}^{c}_{m-2}(X;\mathbb{Z}) = 0. \tag{20}$$

It is now easy to improve the preceding corollary to the following result.

COROLLARY 21.27. *There exists a paracompact space X, all of whose strong homology groups with compact supports and all Čech groups vanish in dimensions $p \neq 0$. Nevertheless, in dimensions $p \geq -1$, all strong homology groups with integer coefficients are non-trivial.*

Proof. Let Λ be a cofinite directed set with $\mathrm{cof}(\Lambda) = \aleph_1$. For each $m \geq 1$, put $X_m = X(m, 2, \Lambda)$ and let X be the topological sum (coproduct)

$$X = \coprod_{m \geq 1} X_m. \tag{21}$$

X is an infinite-dimensional paracompact space. Since $\overline{H}_p(X_m; G) = 0$, for each m, and the Čech homology is additive (see (Mardešić, Prasolov 1988), §7), the Čech groups of X also vanish.

In order to prove that the groups with compact supports vanish, notice that every compact subset $C \subseteq X$ is the finite sum of a collection of compact subsets $C_m \subseteq X_m$, $m = 1, \ldots, k$. Since strong homology is additive with respect to finite sums, one concludes that

$$\overline{H}_p(C; G) = \bigoplus_{m=1}^{k} \overline{H}_p(C_m; G). \tag{22}$$

However, by Theorem 21.25, every class $a_m \in \overline{H}_p(C_m; G)$ admits a compact set C'_m, such that $C_m \subseteq C'_m \subseteq X_m$ and the homomorphism $\overline{H}_p(C_m; G) \to \overline{H}_p(C'_m; G)$ annihilates α_m. Therefore, every class α from $\overline{H}_p(C; G)$ is annihilated by the homomorphism induced by $C \subseteq C'_1 \cup \ldots \cup C'_m$. Finally, since X_m is a retract of X, $\overline{H}_{m-2}(X_m; \mathbb{Z})$ is a direct summand of $\overline{H}_{m-2}(X; \mathbb{Z})$. Consequently, by Theorem 21.23, $\overline{H}_{m-2}(X; \mathbb{Z}) \neq 0$, for all $m \geq 1$. □

REMARK 21.28. The first example showing that strong homology groups differ from groups with compact supports was exhibited in (Mardešić, Prasolov 1988). It had the advantage of being rather simple (finite-dimensional separable metric). However, the proof depended on showing that the abelian pro-group $\boldsymbol{A}$, mentioned in Remark 19.22, has $\lim^1 \boldsymbol{A} \neq 0$, an assertion proved using the continuum hypothesis. The first example, which uses only the ZFC axioms, was obtained in (Günther 1992c). Günther considered the pair (X, A), where X is the set of all countable ordinals and A is the subset of the limit ordinals of X. He proved that A is closed and normally embedded in X and

$$\overline{H}_0^c(X, A; \mathbb{Z}/2) \to \overline{H}_0(X, A; \mathbb{Z}/2) \tag{23}$$

is not an isomorphism. Notice that in Günther's example X is not paracompact. Moreover, dim $X = 0$ and therefore, the strong homology groups $\overline{H}_0(X, A; \mathbb{Z}/2)$ coincide with the Čech groups (Mardešić, Watanabe 1988).

PROBLEM. Is there a separable metric space X whose strong homology groups and strong homology groups with compact supports differ ?

Since the continuum hypothesis is compatible with the ZFC axioms, in ZFC set theory one cannot give a negative answer to this problem.

REMARK 21.29. That Čech homology and Čech homology with compact supports differ was shown in (Mishchenko 1953).

REMARK 21.30. Günther used the pair (X, A), described in Remark 21.28, to show that strong homology groups with compact supports are not invariants of strong shape (see (Günther 1992c), Theorem 8).

REMARK 21.31. Strong homology groups in negative dimensions need not vanish. This phenomenon was first demonstrated in (Mardešić, Prasolov 1988). However, the proof depended on the continuum hypothesis. Recently A.V. Prasolov noticed (private communication) that, for every $r \geq 1$ and $m \geq 1$, the paracompact space $X = X(m, m+r, \Lambda)$ with $\mathrm{cof}(\Lambda) = \aleph_{m+r-1}$ has dim $X = m$ and $\overline{H}_{-r}(X; \mathbb{Z}) \neq 0$. The proof uses only the ZFC set theory. By Theorem 21.11, the phenomenon cannot occur in compact spaces.

Bibliographic notes

Theorem 21.6 was first proved in (Kuz'minov 1971) (see Theorem 6). The same proof with all the non-trivial details and supporting material appeared in (Mardešić, Prasolov 1998) and our exposition follows the one in the latter paper. The exact sequence (21.2.6) is often called the *Milnor sequence*, because J. Milnor proved it in the case of metric compacta. Theorem 21.18 giving the filtration of the strong groups is also from (Mardešić, Prasolov 1998). All results in subsection 21.5 are from (Mardešić 1996b).

22. Generalized strong homology

This section is a brief outline of results obtained in generalized strong homology. For proofs we refer to the literature.

Homology theories satisfying all the axioms of Eilenberg and Steenrod are called *ordinary*. On compact CW-complexes an ordinary homology theory (h_n, ∂) is completely determined by its coefficient group $G = h_0(*)$, where $*$ denotes the space consisting of a single point. Recall that the dimension axiom requires that $h_n(*) = 0$, for all $n \neq 0$. Omission of this axiom yields the notion of an *extraordinary* or *generalized homology* theory. The most important examples of generalized homology theories are bordism groups and K-homology groups.

There is a general method for generating generalized homology theories on compact CW-complexes. It is based on the notion of a spectrum (Whitehead, G.W. 1962). Groups of non-compact CW-complexes are usually defined as colimits of homology groups of compact subcomplexes.

A *spectrum* (also called a CW-spectrum) E consists of a sequence of pointed CW-complexes E_m, $m \geq 0$, and of a sequence of pointed mappings $\gamma_m : \Sigma E_m \to E_{m+1}$, called *structure mappings*. Here Σ denotes the (reduced) suspension functor. Recall that, by definition, the *smash product* $X \wedge Y$ of two pointed spaces X and Y is the quotient space $(X \times Y)/(X \vee Y)$ and $\Sigma X = X \wedge S^1$. A spectrum E is an *Ω- spectrum* provided the adjoint mappings $\delta_n : E_n \to \Omega E_{n+1}$ of the structure mappings γ_n are homotopy equivalences. Here Ω denotes the loop-space functor, $\Omega X = X^{S^1}$. In the literature one encounters different definitions and versions of spectra (and prespectra). Especially useful are Kan spectra, whose terms are Kan complexes. Application of the singular complex functor S to a spectrum E yields a Kan spectrum. Conversely, application of the realization functor R converts a Kan spectrum to a CW-spectrum.

Clearly, every pointed mapping $\varphi : S^{n+k} \to E_k$ determines its suspension $\Sigma\varphi : \Sigma S^{n+k} \to \Sigma E_k$. Composing $\Sigma\varphi$ with $\gamma_k : \Sigma E_k \to E_{k+1}$, one obtains a mapping $S^{n+k+1} \to E_{k+1}$. For every n, the described procedure yields a homomorphism of homotopy groups,

$$\gamma_{k,k+1} : \pi_{n+k}(E_k) \to \pi_{n+k+1}(E_{k+1}). \tag{1}$$

Consequently, $(\pi_{n+k}(E_k), \gamma_{k,k+1})$ is a direct sequence of groups. Its colimit is, by definition, the *homotopy group $\pi_n(E)$ of the spectrum E*,

$$\pi_n(E) = \operatorname{colim}_k \pi_{n+k}(E_k). \tag{2}$$

For a pointed compact CW-complex X the CW-complexes $X \wedge E_n$ and the mappings $1 \wedge \gamma_n: \Sigma(X \wedge E_n) = X \wedge \Sigma E_n \to X \wedge E_{n+1}$ define a spectrum $X \wedge E$. By definition, its homotopy groups are the *reduced homology groups* of X with *coefficients in the spectrum* E,

$$\tilde{E}_n(X) = \pi_n(X \wedge E). \tag{3}$$

Moreover, a pointed mapping $f: X \to Y$ induces mappings $f \wedge 1: X \wedge E_n \to Y \wedge E_n$, which in turn induce homomorphisms of homology groups $f_n: \tilde{E}_n(X) \to \tilde{E}_n(Y)$.

If $\psi: S^{n+k} \to X \wedge E_k$ is a mapping, its suspension $\Sigma\psi$ can also be interpreted as a mapping $S^{n+k+1} \to \Sigma X \wedge E_k$. Therefore, one obtains homomorphisms $\pi_{n+k}(X \wedge E_k) \to \pi_{n+1+k}(\Sigma X \wedge E_k)$. It is easy to see that these homomorphisms induce an isomorphism $\sigma_n: \tilde{E}_n(X) \to \tilde{E}_{n+1}(\Sigma X)$, called the *suspension isomorphism.* The relative groups are defined by putting $E_n(X, A) = \tilde{E}_n(X/A)$. The suspension isomorphism σ_{n-1} induces the boundary operator $\partial: E_n(X, A) \to \tilde{E}_{n-1}(X)$. Indeed, since the cone CA is contractible, the mapping $X \cup CA \to X \cup CA/CA \approx X/A$ is a homotopy equivalence and thus, it induces an isomorphism of homology groups $\tilde{E}_n(X \cup CA) \to \tilde{E}_n(X/A)$. Consequently, it suffices to define a homomorphism $\partial: \tilde{E}_n(X \cup CA) \to \tilde{E}_{n-1}(A)$. By definition, this is the composition of the homomorphism $\tilde{E}_n(X \cup CA) \to \tilde{E}_n(\Sigma A)$, induced by the quotient mapping $X \cup CA \to (X \cup CA)/X \approx \Sigma A$, and the inverse of the suspension isomorphism $\sigma_{n-1}: \tilde{E}_{n-1}(A) \to \tilde{E}_n(\Sigma A)$.

The described procedure yields all generalized homology theories on compact CW-complexes. In particular, the *Eilenberg – Mac Lane spectrum* $K(G)$ yields the standard (singular) homology groups $H_n(X; G)$ of compact CW-complexes X. The terms of this spectrum are the *Eilenberg – Mac Lane complexes* $E_n = K(G, n)$. They are CW-complexes, characterized up to natural homotopy equivalence, by the requirement that $\pi_k(K(G, n)) = 0$, for $k \neq n$, while $\pi_n(K(G, n)) = G$. To define the structure mappings $\gamma_n: \Sigma K(G, n) \to K(G, n+1)$, note that in general $\pi_k(\Omega Y) \approx \pi_{k+1}(Y)$. Therefore, $\Omega K(G, n+1)$ is an Eilenberg – Mac Lane space of type (G, n). Moreover, $\Omega K(G, n+1)$ has the homotopy type of a CW-complex. It follows that there is a homotopy equivalence $\delta_n: K(G, n) \to \Omega K(G, n+1)$. By definition, $\gamma_n: \Sigma K(G, n) \to K(G, n+1)$ is the adjoint of the mapping δ_n.

Bordism groups are obtained using the appropriate *Thom spectrum* . In the case of the orthogonal group O, $E_n = MO(n)$, where $MO(n)$ is the Thom complex of the universal n-plane bundle ξ_n over the classifying space $BO(n)$. The structure mapping $\Sigma MO(n) \to MO(n+1)$ is obtained by noticing that $\Sigma MO(n) = MO(n) \wedge S^1$ is the Thom complex of $\xi_n \oplus \varepsilon^1$, where ε^1 denotes the trivial line bundle over a point. Therefore, the natural bundle mapping $\xi_n \oplus \varepsilon^1 \to \xi_{n+1}$ induces a mapping between the corresponding Thom complexes.

Complex K - homology groups are induced by the BU - *spectrum*. In this spectrum $E_{2n-1} = U$, $E_{2n} = BU$. The structure mappings $\Sigma E_{2n-1} \to E_{2n}$ are induced by a well-known homotopy equivalence $\Omega BU(n) \approx U(n)$ and the structure mappings $\Sigma E_{2n} \to E_{2n+1}$ are given by the Bott periodicity theorem.

The first appearance of a generalized strong homology theory beyond the realm of CW - complexes was an extension of the complex K - homology groups to metric compacta. It originated in 1973, when L.G. Brown, R.G. Douglas and P.A. Fillmore were studying essentially normal operators on an infinite-dimensional separable complex Hilbert space H (Brown, Douglas, Fillmore 1973a, 1973b, 1977). Let $\mathcal{L}$ denote the C^* - algebra of bounded linear operators $T: H \to H$ and let $\mathcal{K} \subseteq \mathcal{L}$ be the subalgebra of compact operators. It is well known that $\mathcal{K}$ is a two-sided ideal and therefore, the quotient $\mathcal{L}/\mathcal{K}$ is a well-defined C^* - algebra, called the Calkin algebra. Let $\pi: \mathcal{L} \to \mathcal{L}/\mathcal{K}$ denote the quotient homomorphism. Recall that a bounded operator $T: H \to H$ is normal if $T^*T = TT^*$ and it is essentially normal if $T^*T - TT^* \in \mathcal{K}$, i.e., if $\pi(T)$ is normal in the Calkin algebra. Also recall that the essential spectrum $\sigma_e(T)$ of an essentially normal operator T coincides with the spectrum $\sigma(\pi(T))$ of the normal element $\pi(T) \in \mathcal{L}/\mathcal{K}$. The initial question considered by the above mentioned authors was to find conditions when an essentially normal operator is the sum of a normal operator and a compact operator.

To answer this question Brown, Douglas and Fillmore associated with every compact metric space X an abelian group $\mathcal{E}xt(X)$. It consists of the equivalence classes of exact sequences of C^* - algebras of the form

$$0 \to \mathcal{K} \to \mathcal{E} \to C(X) \to 0, \tag{4}$$

where $C(X)$ is the algebra of continuous complex-valued functions on X and $\mathcal{E}$ is a subalgebra of $\mathcal{L}$, which contains $\mathcal{K}$ and the identity operator $I: H \to H$. If $T \in \mathcal{L}$ and $\mathcal{E}(T)$ denotes the subalgebra of $\mathcal{L}$ generated by $\mathcal{K}$, T and I, then π induces an exact sequence of the form

$$0 \to \mathcal{K} \to \mathcal{E}(T) \to C^*(\pi(T), 1) \to 0, \tag{5}$$

where $C^*(\pi(T), 1)$ is the subalgebra of $\mathcal{L}/\mathcal{K}$ generated by $\pi(T)$ and the unit element 1. Clearly, the algebra $C^*(\pi(T), 1)$ is commutative. Therefore, if T is essentially normal, the spectral theory of commutative C^* - algebras shows that $C^*(\pi(T), 1) \approx C(\sigma(\pi(T))) = C(\sigma_e(T))$. Consequently, putting $X = \sigma_e(T)$, (5) assumes the form (4), for $\mathcal{E} = \mathcal{E}(T)$, and thus, determines an element of the group $\mathcal{E}xt(X)$. The vanishing of this element is a necessary and sufficient condition for T to be the sum of a normal and a compact operator.

For a compactum $X \subseteq \mathbb{C}$, Brown, Douglas and Fillmore proved that $\mathcal{E}xt(X)$ is isomorphic to the reduced cohomology group $\tilde{H}^0(\mathbb{C}\backslash X)$. In order to study $\mathcal{E}xt(X)$, for more general metric compacta X, they introduced higher groups, $E_n(X) = \mathcal{E}xt(X)$, for n odd, and $E_n(X) = \mathcal{E}xt(\Sigma X)$, for n even, as well as homomorphisms induced by mappings $f\colon X \to Y$ and suspension isomorphisms $\sigma_n\colon E_n(X) \to E_{n+1}(\Sigma X)$. Then they proved that these data form a generalized homology theory for metric compacta, which on compact CW-complexes coincides with the reduced homology theory with coefficients in the BU-spectrum, i.e., with complex K-theory. In particular, the relation $\mathcal{E}xt(X) \approx \tilde{H}^0(\mathbb{C}\backslash X)$, for compact $X \subseteq \boldsymbol{C}$, is just the Alexander duality law for that particular generalized homology theory.

Subsequently, J. Kaminker and C. Schochet exhibited a spectral sequence of the Atiyah – Hirzebruch type which, for finite-dimensional X, relates Steenrod homology to the Brown – Douglas – Fillmore homology. In particular, the term E^2_{pq} equals the reduced Steenrod homology group $\overline{H}_p(X; \tilde{E}_q(*))$ (Kaminker, Schochet 1975, 1977). The next step was done in (Kahn, Kaminker, Schochet 1977) and consisted in the construction of a generalized homology for pointed metric compacta, which on compact CW-complexes coincides with a given generalized homology theory. In view of the above described relation between generalized homology of compact CW-complexes and homology with coefficients in a spectrum, it was natural to define this homology as a homology with coefficients in a spectrum E. In fact, if X is a pointed metric compactum X, let $F(X,S)$ be its function spectrum. The n-th term of this spectrum is the space $F(X,S^n) = (S^n)^X$ of pointed mappings $X \to S^n$ and the structure mapping $\gamma_n\colon \Sigma F(X,S^n) \to F(X,S^{n+1})$ sends $\phi \wedge t \in \Sigma F(X,S^n) = (S^n)^X \wedge S^1$ to the mapping $\psi \in F(X,S^{n+1}) = (S^{n+1})^X$, where $\psi(x) = \phi(x) \wedge t \in S^n \wedge S^1 = S^{n+1}$. Since the terms of $F(X,S)$ need not be CW-complexes, one replaces $F(X,S)$ with an equivalent spectrum $F(X)$, formed by CW-complexes. By definition,

$$E_k(X) = [F(X), E]_k, \tag{6}$$

where, for spectra F and E, $[F,E]_k$ denotes the set of all morphisms of degree k in the Boardman homotopy category of spectra (Boardman 1970). The obtained homology theory is a generalized homology theory, which satisfies the strong excision axiom and the cluster axiom (Kahn, Kaminker, Schochet 1977). Such a theory is called a *strong homology theory* and F.W. Bauer showed that it is completely determined by its restriction to compact CW-complexes (Bauer 1987). D.A. Edwards and H.M. Hastings gave a different construction of the same homology (Edwards, Hastings 1976). For a pointed metric compactum X their homology groups with coefficients in a spectrum E are give by the formula

$$E_n(X) = \pi_n(\operatorname{holim} S(RV(X) \wedge E)). \tag{7}$$

Here $V(X)$ denotes the Vietoris complex of X and holim denotes the homotopy limit of a simplicial spectrum (Bousfield, Kan 1972).

In a series of papers beginning with 1976, F.W. Bauer defined homology groups of pointed spaces with coefficients in a spectrum E by the formula

$$\overline{E}_n(X) = \operatorname{colim}_k \overline{\pi}_{n+k}(X \overline{\wedge} E_k), \tag{8}$$

where $\overline{\pi}_m(Y)$ is the m-th strong shape group of Y, i.e., the group of pointed strong shape morphisms $S^m \to Y$ (Bauer 1976, 1984, 1987) and $\overline{\wedge}$ is a suitable modification of $\wedge$. His generalized homology factors through the strong shape category. In general, it does not satisfy the cluster axiom on metric compacta and therefore, it is not a strong homology theory in the above sense. However, in the case of the Eilenberg – Mac Lane spectrum, Bauer's homology does satisfy the cluster axiom. Hence, for shape connected metric compacta it coincides with ordinary strong homology (Steenrod homology).

In 1986 M.A. Batanin approached the problem of defining strong shape invariant generalized strong homology theories for arbitrary spaces by first constructing strong homology groups with coefficients in a spectrum E on the coherent homotopy category CH(pro-Pol)≈Ho(pro-Pol). He then replaced spaces by their polyhedral resolutions (Batanin 1986). This program was further developed by A.V. Prasolov, who used strong ANR-expansions of spaces (Prasolov to appear). In particular, Prasolov showed that, in the case of the Eilenberg – Mac Lane spectrum $E = K(G)$, the obtained homology coincides with (ordinary) strong homology, as defined in the present book. We will now briefly describe the basic definitions and results of the Batanin – Prasolov theory.

An inverse system $\boldsymbol{E} = (E_\lambda, f_{\lambda\lambda'}, \Lambda)$ of spectra is an inverse system in the category of spectra. In the case of a system of Ω-spectra, its homotopy limit $\operatorname{holim}\boldsymbol{E}$ is defined as the spectrum whose n-th term is the homotopy limit of the inverse system $\boldsymbol{E}_n$, formed by the n-th terms of the spectra E_λ (Thomason 1985). To define the structure mappings, one considers the mappings $\delta_{\lambda n}: E_{\lambda n} \to \Omega E_{\lambda n+1}$, adjoint to the structure mappings $\gamma_{\lambda n}: \Sigma E_{\lambda n} \to E_{\lambda n+1}$ of E_λ. These mappings induce a mapping $\delta_n: \operatorname{holim}\boldsymbol{E}_n \to \operatorname{holim}\Omega\boldsymbol{E}_{n+1}$. However, holim and Ω commute and therefore, we obtain a mapping $\delta_n: \operatorname{holim}\boldsymbol{E}_n \to \Omega(\operatorname{holim}\boldsymbol{E}_{n+1})$. Its adjoint is the desired structure mapping $\gamma_n: \Sigma(\operatorname{holim}\boldsymbol{E}_n) \to \operatorname{holim}\boldsymbol{E}_{n+1}$.

If $\boldsymbol{P} = (P_\lambda, p_{\lambda\lambda'}, \Lambda)$ is an inverse system of plyhedra and E is a spectrum, one defines an inverse system of spectra $\boldsymbol{P} \wedge E$. Its terms are the spectra $P_\lambda \wedge E$ and its bonding mappings are the mappings of spectra $p_{\lambda\lambda'} \wedge 1$. If this spectrum is an Ω-spectrum, one can take its homotopy limit. Finally, we need a functor ω which to every spectrum E assigns an Ω-spectrum $\omega(E)$, which is equivalent to E in the Boardman homotopy category of spectra (Boardman 1970). The spectrum $\omega(E)$ consists of spaces $\omega(E_n)$ and structure mappings $\omega(\gamma_n): \Sigma\omega(E_n) \to \omega(E_{n+1})$, defined as follows. One considers the direct sequence

$$E_n \xrightarrow{\delta_n} \Omega E_{n+1} \xrightarrow{\Omega\delta_{n+1}} \Omega^2 E_{n+2} \to \dots, \tag{9}$$

where $\delta_n: E_n \to \Omega E_{n+1}$ is the adjoint of $\gamma_n: \Sigma E_n \to E_{n+1}$. By definition, $\omega(E_n)$ is the homotopy colimit (telescope) of the sequence (9). The mapping $\omega(\gamma_n)$ is induced by the mappings $\Sigma\Omega^k E_{n+k} \to \Omega^{k-1} E_{n+k}$, which are adjoint to the identity mappings on $\Omega^k E_{n+k}$.

According to the Batanin – Prasolov definition, the unreduced strong homology groups $\overline{E}(X)$ with coefficients in the spectrum E are given by the formula

$$\overline{E}(X) = \pi_n (\operatorname{holim} \omega((\boldsymbol{P} \cup *) \wedge E)), \tag{10}$$

where $\boldsymbol{P}$ is a strong ANR-expansion of X. The reduced groups of a pointed space $(X, *)$ are then obtained by putting $\tilde{E}_n(X,*) = \operatorname{Coker}(E_n(*)) \to E_n(X)$.

Prasolov proved that his generalized homology groups have all the desired properties of a strong homology theory. This includes the strong shape invariance, the strong excision axiom and an Atiyah – Hirzebruch spectral sequence, whose E^2-term is given by the ordinary strong homology. Finally, Prasolov's theory works not just for spaces but also for inverse systems of topological spaces.

In a series of papers beginning in 1989, Bauer developed generalized homology theories h_n for arbitrary spaces using the approach of chain functors. The basic idea in this approach is that homology groups $h_n(X, A)$ of a pair of spaces must be of the form $H_n(C_*(X, A))$, where $C_*(X, A)$ is a chain complex associated with the pair (X, A) and H_n denotes the usual homology of a chain complex. The usual condition $C_*(X, A) = C_*(X)/C_*(A)$ is too restrictive because it can yield only ordinary homology groups or their direct sums. Therefore, Bauer devised the more sophisticated notion of a chain functor $\boldsymbol{C}_*$. It consists of two functors $C_*, C'_* \colon \mathsf{Top} \to \mathsf{Ch}$ to the category Ch of chain complexes, of a subfunctor $l \colon C'_* \to C_*$, of natural inclusions $i' \colon C_*(A) \subseteq C'_*(X, A)$ and of (unnatural) chain mappings $\kappa \colon C_*(X) \to C'_*(X, A)$, $\phi \colon C'_*(X, A) \to C_*(X)$. All these data are subject to a number of conditions, which make it possible to define a boundary operator $\partial \colon h_n(X, A) \to h_{n-1}(A)$ (Bauer 1989). Bauer proves that his homology groups are strong shape invariants and he characterizes them by means of a continuity condition at the level of chains.

References

Alder, M.D. (1974): Inverse limits of simplicial complexes. Compositio Math. **29**, 1–7

Alexandroff, P. (1926): Simpliziale Approximationen in der allgemeinen Topologie. Math. Ann. **96**, 489–511

Alexandroff, P. (1927): Une définition des nombres de Betti pour un ensemble fermé quelconque. C. R. Acad. Sci. Paris **184**, 317–319

Alexandroff, P. (1929): Untersuchungen über Gestalt und Lage abgeschlossener Mengen beliebiger Dimension. Ann. of Math. **30**, 101–187

Aleksandrov, P.S., Pasynkov, B.A. (1973): Introduction to dimension theory. Nauka, Moscow (Russian)

Anderson, R.D. (1967): On topological infinite deficiency. Michigan J. Math. **14**, 365–383

Artin, M., Mazur, B. (1969): Etale homotopy. Lecture Notes in Math. **100**. Springer, Berlin Heidelberg New York

Bacon, P. 1975: Continuous functors. General Topology Appl. **5**, 321–331

Barratt, M.G., Milnor, J. (1962): An example of anomalous singular homology. Proc. Amer. Math. Soc. **13**, 293–297

Batanin, M.A. (1986): On the coherent prohomotopy category of Lisica-Mardešić and generalized Steenrod homology theories. USSR Academy of Sci., Siberian Section, Novosibirsk, preprint, pp. 1–23 (Russian)

Batanin, M.A. (1993): Coherent categories with respect to monads and coherent prohomotopy theory. Cahiers Topol. Géom. Diff. Categor. **34**, No.4, 279–304

Bauer, F.W. (1976): A shape theory with singular homology. Pacific J. Math. **64**, 25–65

Bauer, F.W. (1978): Some relations between shape constructions. Cahiers Topol. Géom. Diff. **19**, 337–367

Bauer, F.W. (1984): Duality in manifolds. Ann. Mat. Pura Appl. (4) **136**, 241–302

Bauer, F.W. (1987): Extensions of generalized homology theories. Pacific J. Math. **128**, 25–61

Bauer, F.W. (1989): Generalized homology theories and chain complexes. Ann. Mat. Pura Appl. **155**, 143–191

Bauer, F.W. (1991): A strong shape theoretical version of a result due to E. Lima. Topology Appl. **40**, 17–21

Bauer, F.W. (1994a): Extensions of chain functors. Comm. Algebra **22**, no.9, 3331–3417

Bauer, F.W. (1994b): The existence of strong homology theory. Comm. Algebra **22**, no.9, 3419–3432

Bauer, F.W. (1994c): The universality of strong homology theory. Comm. Algebra **22**, no.9, 3433–3447

Bauer, F.W. (1995a): Bordism theories and chain complexes. J. Pure Appl. Algebra **102**, 251–272

Bauer, F.W. (1995b): A strong shape theory admitting an S-dual. Topology Appl. **62**, 207–232

Bauer, F.W. (1997): A strong shape theory with S-duality. Fund. Math. **154**, 37–56

Bauer, F.W., Dugundji, J. (1969): Categorical homotopy and fibrations. Trans. Amer. Math. Soc. **140**, 239–256

Berikashvili, N.A. (1980): Steenrod-Sitnikov homology theories on the category of compact spaces. Dokl. Akad. Nauk SSSR **254**, No. 6, 1289–1292 (Russian)

Berikashvili, N.A. (1984): On the axiomatics of Steenrod-Sitnikov homology theory on the category of compact Hausdorff spaces. Trudy. Mat. Inst. Steklov **154**, 24–37 (Russian)

Boardman, J.M. (1970): Stable homotopy theory. Mimeographed notes, John Hopkins Univ., Baltimore

Boardman, J.M., Vogt, R.M. (1973): Homotopy invariant algebraic structures on topological spaces. Lecture Notes in Math. **347**, Springer, Berlin Heidelberg New York

Borceux, F. (1994): Handbook of categorical algebra, Vol. 1-3. Cambridge Univ. Press

Borel, A., Moore, J.C. (1960): Homology theory for locally compact spaces. Michigan Math. J. **7**, 137–159

Borsuk, K. (1968): Concerning homotopy properties of compacta. Fund. Math. **62**, 223–254

Borsuk, K. (1975): Theory of shape. Polish Scientific Publishers, Warszawa

Bousfield, A.K., Kan, D.M. (1972): Homotopy limits, completions and localizations. Lecture Notes in Math. **304**, Springer, Berlin Heidelberg New York

Bredon, G.E. (1967): Sheaf theory. McGraw-Hill, New York

Brown, L.G., Douglas, R.G., Fillmore, P.A. (1973a): Unitary equivalence modulo the compact operators and extensions of C^*-algebras, Proc. Conf. Operator theory. Lecture Notes in Math. **345**, Springer, Berlin Heidelberg New York, pp. 58–128

Brown, L.G., Douglas, R.G., Fillmore, P.A. (1973b): Extensions of C^*-algebras, operators with compact self-commutators, and K-homology. Bull. Amer. Math. Soc. **79**, 973–978

Brown, L.G., Douglas, R.G., Fillmore, P.A. (1977): Extensions of C^*-algebras and K-homology. Ann. Math. **105**, 265–324

Bucur, I., Deleanu, A. (1968): Introduction to the theory of categories. Wiley-Interscience Publ., London

Burdick, R.O., Conner, P.E., Floyd, E.E. (1968): Chain theories and their derived homology. Proc. Amer. Math. Soc. **19**, 1115–1118

Butler, M.C.R., Horrocks, G. (1961): Classes of extensions and resolutions. Philos. Trans. Roy. Soc. London Ser. **A 254** (1039), 155–222

Calder, A., Hastings, H.M. (1981): Realizing strong shape equivalences. J. Pure and Appl. Algebra **20**, 129–156

Cartan, H., Eilenberg, S. (1956): Homological algebra. Princeton University Press, Princeton, N.J.

Cathey, F.W. (1981): Strong shape theory, in Shape theory and geometric topology. Proc. Dubrovnik 1981, Lecture Notes in Math. **870**, Springer, Berlin Heidelberg New York, pp. 215–238

Cathey, F.W. (1982): Shape fibrations and strong shape theory. Topology Appl. **14**, 13–30

Cathey, F.W., Segal, J. (1983): Strong shape theory and resolutions. Topology Appl. **15**, 119–130

Cathey, F.W., Segal, J. (1985): Homotopical approach to strong shape and completion theory. Topology Appl. **21**, 167–192

Cauty, R. (1973): Convexité topologique et prolongement des fonctions continues. Compositio Math. **27**, 233–271

Čech, E. (1932): Théorie générale de l'homologie dans un espace quelconque. Fund. Math. **19**, 149–183

Čerin, Z. (1993): Shape theory intrinsically. Publicacions Mathématiques, Barcelona **37**, 317–334

Chapman, T.A. (1972): On some applications of infinite-dimensional manifolds to the theory of shape. Fund. Math. **76**, 181–193

Chapman, T.A. (1976): Lectures on Hilbert cube manifolds. CBMS **28**, Amer. Math. Soc., Providence, R.I

Chapman, T.A., Siebenmann, L.C. (1976): Finding a boundary for a Hilbert cube manifold. Acta Math. **137**, 171–208

Charalambous, M.G. (1980): An example concerning inverse limit sequences of normal spaces. Proc. Amer. Math. Soc. **78**, 605–608

Charalambous, M.G. (1991): Approximate inverse systems of uniform spaces and an application of inverse systems. Comment. Math. Univ. Carolinae **32**, 551–565

Chigogidze, A. (1996): Inverse spectra. North-Holland, Amsterdam

Chogoshvili, G.S. (1940): On the homology theory of topological spaces. Soobshch. Akad. Nauk Gruzin. SSR **1**, 337–340

Christie, D.E. (1944): Net homotopy for compacta. Trans. Amer. Math. Soc. **56**, 275–308

Cohn, P.M. (1959): On free product of associative rings. Math. Z. **71**, 380–398

Cordier, J.M. (1982): Sur la notion de diagramme homotopiquement cohérent. Cahiers Topol. Géom. Diff. **23**, 93–112

Cordier, J.M. (1987): Homologie de Steenrod-Sitnikov et limite homotopique algébrique. Manuscripta Math. **59**, 35–52

Cordier, J.M. (1989): Comparaison de deux cathégories d'homotopie de morphismes cohérents. Cahiers Topol. Géom. Diff. **30**, 257–275

Cordier, J.M., Porter, T. (1986): Vogt's theorem on categories of homotopy coherent diagrams. Math. Proc. Cambridge Phil. Soc. **100**, 65–90

Cordier, J.M., Porter, T. (1989): Shape theory - Categorical methods of approximation. Ellis Horwood, Chichester

Dadarlat, M. (1993): Shape theory and asymptotic morphisms for C^*−algebras. Duke Math. J. **73**, 687–711

Deheuvels, R. (1960): Homologie à coefficients dans un antifaisceau. C. R. Acad. Sci. Paris **250**, 2492–2494

Deheuvels, R. (1962): Homologie des ensembles ordonnées et des espaces topologiques. Bull. Soc. Math. France **90**, 261–321

Dold, A. (1972): Lectures on algebraic topology. Springer, Berlin Heidelberg New York

Dold, A., Puppe, D. (1980): Duality, trace and transfer, in Proc. Intern. Conf. on Geom. Top. (Warszawa, 1978). Polish. Sci Publ., Warszawa, pp. 81–102

Dow, A., Simon, P., Vaughan, J.E. (1989): Strong homology and the proper forcing axiom. Proc. Amer. Math. Soc. **106**, 821–828

Dowker, C.H. (1947): Mapping theorems for non-compact spaces. Amer. J. Math. **69**, 200-242

Dowker, C.H. (1952): Homology groups of relations. Ann. Math. **56**, 84-95

Dugundji, J. (1966): Topology. Allyn and Bacon, Boston

Dydak, J. (1986): Steenrod homology and local connectedness. Proc. AMS, 98,153–157

Dydak, J., Nowak, S. (1991): Strong shape for topological spaces. Trans. Amer. Math. Soc. **323**, 765–796

Dydak, J., Segal, J. (1978a): Strong shape theory: A geometrical approach. Topology Proc. **3**, 59–72

Dydak, J., Segal, J. (1978b): Shape theory: An introduction. Lecture Notes in Math. **688**, Springer, Berlin Heidelberg New York

Dydak, J., Segal, J. (1981): Strong shape theory. Dissertationes Math. **192**, 1–42

Dydak, J., Segal, J. (1990): A list of open problems in shape theory, in Ed. Mill, J. van, Reed, G.M.: Open problems in topology. North-Holland, Amsterdam, pp. 457–467

Edwards, D.A., Hastings, H.M. (1976a): Čech and Steenrod homotopy theories with applications to Geometric Topology. Lecture Notes in Math. **542**, Springer, Berlin Heidelberg New York

Edwards, D.A., Hastings, H.M. (1976b): Every weak proper homotopy equivalence is weakly properly homotopic to a proper homotopy equivalence. Trans. Amer. Math. Soc. **221**, 239–248

Edwards, D.A., Hastings, H.M. (1980): Čech theory: its past, present and future. Rocky Mountain J. Math. **10**, 429–468

Eilenberg, S., Steenrod, N.E. (1952): Foundations of algebraic topology. Princeton Univ. Press, Princeton

Engelking, R. (1977): General topology. Monografie Matematyczne **60**, Polish Scientific Publishers, Warszawa

Ferry, S. (1980): A stable converse to the Vietoris-Smale theorem with applications to shape theory. Trans. Amer. Math. Soc. **261**, 369–386

Fox, R.H. (1972): On shape. Fund. Math. **74**, 47–71

Freudenthal, H. (1937): Entwicklungen von Räumen und ihren Gruppen. Compositio Math. **4**, 145–234

Fritsch, R. (to appear): A functorial description of coherent mappings between inverse systems. Glasnik Mat.

Fritsch, R., Piccinini, R.A. (1990): Cellular structures in topology. Cambridge Univ. Press, Cambridge

Fuchs, L. (1970): Infinite abelian groups, Vol. I. Academic Press, New York

Gabriel, P., Zisman, M. (1967): Calculus of fractions and homotopy theory. Springer, Berlin Heidelberg New York

Geoghegan, R., Krasinkiewicz, J. (1991): Empty components in strong shape theory. Topol. Appl. **41**, 213–233

Giraldo, A., Sanjurjo, J.M.R. (1995): Strong multihomotopy and Steenrod loop spaces. J. Math. Soc. of Japan **47**, 475–489

Goblot, R. (1970): Sur les dérivés de certaines limites projectives. Applications aux modules. Bull. Sc. Math. 2. Sér. **94**, 251–255

Godement, R. (1958): Topologie algébrique et théorie des faisceaux. Hermann, Paris

Govorov, V.E. (1965): On flat modules. Sibirski Mat. Ž. **6**, No. 2, 300-304 (Russian)

Grothendieck, A. (1959-60): Technique de descente et théorèmes d'existence en géométrie algébrique II. Séminaire Bourbaki, 12-ème année, exposé 190–195

Gruson, L., Jensen, C.U. (1981): Dimension cohomologiques reliées aux foncteurs $\lim^{(i)}$, in Séminaire d'algèbre Paul Dubreil et Marie-Paule Malliavin.

Proc. Paris 1980, Lecture Notes in Math. **867**, Springer, Berlin Heidelberg New York, pp. 234-294

Günther, B. (1989): Starker Shape für beliebige topologische Räume, Dissertation. J.W. Goethe-Universität, Frankfurt a.M.

Günther, B. (1990): Semigroup structures on derived limits. J. Pure Appl. Algebra **69**, 51–65

Günther, B. (1991a): Comparison of the coherent pro-homotopy theories of Edwards - Hastings, Lisica - Mardešić and Günther. Glasnik Mat. **26**, 141–176

Günther, B. (1991b): Strong shape of compact Hausdorff spaces. Topology App. **42**, 165–174

Günther, B. (1991c): Properties of normal embeddings concerning strong shape theory, I. Tsukuba J. Math. **15**, 261–274

Günther, B. (1992a): Properties of normal embeddings concerning strong shape theory, II. Tsukuba J. Math. **16**, 429–438

Günther, B. (1992b): The use of semisimplicial complexes in strong shape theory. Glasnik Mat. **27**, 101–144

Günther, B. (1992c): The Vietoris system in strong shape and strong homology. Fund. Math. **141**, 147–168

Günther, B. (1993): A tom Dieck theorem for strong shape theory. Trans. Amer. Math. Soc. **338**, 857–870

Günther, B. (1994a): A compactum that cannot be an attractor of a selfmap on a manifold. Proc. Amer. Math. Soc. **120**, 653–655

Günther, B. (1994b): Approximate resolutions in strong shape theory. Glasnik Mat. **29**, 109–122

Günther, B. (1995): Construction of differentiable flows with prescribed attractor. Topology Appl. **62**, 87–91

Günther, B., Segal, J. (1993): Every attractor of a flow on a manifold has the shape of a finite polyhedron. Proc. Amer. Math. Soc. **119**, 321–329

Hardie, K.A, Kamps, K.H. (1989): Track homotopy over a fixed space. Glasnik Mat. **24**, 161–179

Hastings, H.M., (1977): Steenrod homotopy theory, homotopy idempotents, and homotopy limits, Topology Proc. **2**. 461–477

Haxhibeqiri, Q., Nowak, S. (1987): Duality between stable strong shape morphisms, preprint

Haxhibeqiri, Q., Nowak, S. (1989): Stable strong shape theory. Glasnik Mat. **24**, 149–160

Henn, H.-W. (1981): Duality in stable shape theory. Archiv Math. **36**, 327–341

Heller, A. (1966): Extraordinary homology and chain complexes. Proc. Conf. on Categorical algebra, La Jolla, Springer, Berlin Heidelberg New York, pp. 355–365

Hilton, P.J., Stammbach, U. (1971): A course in homological algebra. Springer, Berlin Heidelberg New York

Hu, S.T. (1965): Theory of retracts. Wayne State Univ. Press, Detroit

Huber, M., Meier, W. (1978): Cohomology theories and infinite CW-complexes. Comment. Math. Helv. **53**, 239–257

Huber, P. (1961): Homotopy theories in general categories. Math. Ann. **444**, 361–385

Inasaridze, H.N. (1972): On exact homology. Trudy Mat. Inst. Akad. Nauk Gruz. SSR **41**, 128–142 (Russian)

Inasaridze, H.N., Mdzinarishvili, L.D. (1980): On the connection between continuity and exactness in homology theory. Soobshch. Akad. Nauk Gruzin. SSR **99**, 317–320 (Russian)

Inassaridze, H.N. (1991): On the Steenrod homology theory for compact spaces. Michigan J. Math. **38**, 323–338

Ivanšić, I., Uglešić, N. (1988): Weak fibrant compacta. Glasnik Mat. **23**, 397–415

Jensen, C.U. (1972): Les foncteurs dérivés de $\varprojlim$ et leurs applications en théorie des modules. Lecture Notes in Math. **254**, Springer, Berlin Heidelberg New York

Jensen, C.U. (1977): On the global dimension for the functor category (mod R, Ab). J. Pure Appl. Algebra **11**, 45–51

Kahn, D.S., Kaminker, J., Schochet, C. (1977): Generalized homology theories on compact metric spaces. Michigan Math. J. **24**, 203–224

Kaminker, J., Schochet, C. (1975): Steenrod homology and operator algebras. Bull. Amer. Math. Soc. **81**, 431–434

Kaminker, J., Schochet, C. (1977): K-theory and Steenrod homology: Applications to the Brown-Douglas-Fillmore theory of operator algebras, Trans. Amer. Math. Soc. **227**, 63–107

Kaplansky, I. (1954): Infinite abelian groups, Univ. Michigan Press, Ann Arbor

Kaup, L., Keane, M.S. (1969): Induktive Limiten endlich erzeugter freier Moduln. Manuscripta Math. **1**, 9–21

Kernchen, M. (1982): Bemerkungen zur Borel-Moore Homologie. Manuscripta Math. **39**, 111–118

Kleisli, H. (1965): Every standard construction is induced by a pair of adjoint functors. Proc. Amer. Math. Soc. **16**, 544–546

Kodama, Y., Koyama, A. (1979): Hurewicz isomorphism theorem for Steenrod homology. Proc. Amer. Math. Soc. **74**, 363–367

Kodama, Y., Ono, J. (1979): On fine shape theory. Fund. Math. **105**, 29–39

Kodama, Y., Ono, J. (1980): On fine shape theory, II. Fund. Math. **108**, 89–98

Kolmogoroff, A. (1936): Les groupes de Betti des espaces localement bicompacts. C. R. Acad. Sci Paris **202**, 1144–1147

Koyama, A. (1983): A Whitehead-type theorem in fine shape theory. Glasnik Mat. **18** (1983), 359–370

Koyama, A. (1984a): Coherent singular complexes in strong shape theory. Tsukuba J. Math. **8**, 261–295

Koyama, A. (1984b): An example of coherent homology groups. Proc. Japan Acad. **60**, Ser. A, 319–322

Koyama, A., Mardešić, S., Watanabe, T. (1988): Spaces which admit AR-resolutions. Proc. Amer. Math. Soc. **102**, 749 – 752

Krasinkiewicz, J. (1976): On a method of constructing ANR-sets. An application of inverse limits. Fund. Math. **92**, 95–112

Kuz'minov, V. (1967): On derived functors of the projective limit functor. Sibirski Mat. Ž. **8**, No.2, 333–345 (Russian)

Kuz'minov, V. (1971): Derived functors of inverse limits and extension classes. Sibirski Mat. Ž. **12**, No.2, 384–396 (Russian)

Kuz'minov, V., Shvedov, I.A. (1974): Covering spectra in the theory of cohomology and homology of topological spaces. Sibirski Mat. Ž. **15**, 1083–1102 (Russian)

Kuz'minov, V., Shvedov, I.A. (1975): Hyperhomology of limits of direct spectra of complexes and homology groups of topological spaces. Sibirski Mat. Ž. **16**, 49–59 (Russian)

Lazard, D. (1969): Autour de la platitude. Bull. Soc. Math. France **97**, 81–128

Lefschetz, S. (1931): On compact spaces. Ann. of Math. **32**, 521–538

Lefschetz, S. (1942): Algebraic topology. Amer. Math. Soc. Colloquium Publ. **27**, New York

Lima, E.L. (1959): The Spanier-Whitehead duality in new homotopy categories. Summa Brasil. Math. **4**, No 3, 91–148

Lisica, Ju.T. (1983): Strong shape theory and multivalued maps. Glasnik Mat. **18**, 371–382

Lisica, Ju.T. (to appear): Second structure theorem for strong homology. Topology Appl.

Lisica, Ju.T., Mardešić, S. (1983): Steenrod-Sitnikov homology for arbitrary spaces. Bull. Amer. Math. Soc. **9**, 207–210

Lisica, Ju.T., Mardešić, S. (1984a): Coherent prohomotopy and strong shape category of topological spaces. Proc. Internat. Topology Conference (Leningrad 1982). Lecture Notes in Math. **1060**, Springer, Berlin Heidelberg New York, pp. 164–173

Lisica, Ju.T., Mardešić, S. (1984b): Coherent prohomotopy and strong shape theory. Glasnik Mat. **19**, 335–399

Lisica, Ju.T., Mardešić, S. (1985a): Coherent prohomotopy and strong shape of metric compacta. Glasnik Mat. **20**, 159–186

Lisica, Ju.T., Mardešić, S. (1985b): Pasting strong shape morphisms. Glasnik Mat. **20**, 187–201

Lisica, Ju.T., Mardešić, S. (1985c): Coherent prohomotopy and strong shape for pairs. Glasnik Mat. **20**, 419–434

Lisica, Ju.T., Mardešić, S. (1985d): Strong homology of inverse systems of spaces, I. Topology Appl. **19**, 29–43

Lisica, Ju.T., Mardešić, S. (1985e): Strong homology of inverse systems of spaces, II. Topology Appl. **19**, 45-64

Lisica, Ju.T., Mardešić, S. (1985f): Strong homology of inverse systems, III. Topology Appl. **20**, 29–37

Lisica, Ju.T., Mardešić, S. (1986): Steenrod homology, in Geometric and Algebraic Topology. Banach Center Publ., Warsaw, **18**, pp. 329-343

Lisitsa, Yu.T. (1977): On the exactness of the spectral homotopy group sequence in shape theory. Dokl. Akad. Nauk SSSR **236**, 23–26 (Russian) (Soviet Math. Dokl. **18**, 1186–1190)

Lisitsa, Yu.T. (1982a): Duality theorems and dual shape and coshape categories. Doklady Akad. Nauk SSSR **263**, No. 3, 532–536 (Russian) (Soviet Math. Dokl. **25**, No. 2, 373–378)

Lisitsa, Yu.T. (1982b): Cotelescopes and the Kuratowski-Dugundji theorem in shape theory. Dokl. Akad. Nauk SSSR **265**, No. 5, 1064–1068 (Russian) (Soviet Math. Dokl. **26**, No. 1, 205–210)

Lisitsa, Yu.T. (1983a): Strong shape theory and the Steenrod-Sitnikov homology. Sibirski Mat. Ž. **24**, 81–99 (Russian)

Lisitsa, Yu.T. (1985): The theorems of Hurewicz and Whitehead in strong shape theory. Dokl. Akad. Nauk SSSR **283**, No. 1, 38–43 (Russian) (Soviet Math. Dokl. **32**, No. 1, 31–35)

Lončar, I. (1987): Some results on resolutions of spaces. Rad Jugoslav. Akad. Znan. Umjetn. Matem. Znan. **428** (6), 37–49

Lundell, A.T., Weingram, S. (1969): The topology of CW-complexes. Van Nostrand Reinhold, New York

Mac Lane, S. (1963): Homology. Springer, Berlin Göttingen Heidelberg

Mac Lane, S. (1971): Categories. For the working mathematician. Springer, Berlin Heidelberg New York

Mardešić, S. (1960): On covering dimension and inverse limits for compact spaces. Illinois J. Math. **4**, 278–291

Mardešić, S. (1973): Shapes of topological spaces. General Topology Appl. **3**, 265–282

Mardešić, S. (1981a): Inverse limits and resolutions. In Shape theory and geometric topology, Proc. Conference, Dubrovnik 1981. Lecture Notes in Math. **870**, Springer, Berlin Heidelberg New York, pp. 240–253

Mardešić, S. (1981b): Approximate polyhedra, resolutions of maps and shape fibrations. Fund. Math. **114**, 53–78

Mardešić, S. (1984a): On resolutions for pairs of spaces. Tsukuba J. Math. **8**, 81–93

Mardešić, S. (1984b): ANR-resolutions of triads. Tsukuba J. Math. **8**, 353–365

Mardešić, S. (1987a): A note on strong homology of inverse systems. Tsukuba J. Math. **11**, 177–197

Mardešić, S. (1987b): Partial continuity of strong homology groups. Rad Jugoslav. Akad. Znan. Umjetn. Matem. Znan. **428** (6), 51–58

Mardešić, S. (1991a): Strong expansions and strong shape theory. Topology Appl. **38**, 275–291

Mardešić, S. (1991b): Resolutions of spaces are strong expansions. Publ. Inst. Math. Beograd **49** (63), 179–188

Mardešić, S. (1991c): Strong expansions and strong shape for pairs of spaces. Rad Hrvat. Akad. Znan. Umjetn. Matem. Znan. **456** (10), 159–172

Mardešić, S. (1992): Strong shape of the Stone - Čech compactification. Commentationes Math. Univ. Carolinae **33**, No. 3, 533–539

Mardešić, S. (1993a): On approximate inverse systems and resolutions. Fund. Math. **142**, 241–255

Mardešić, S. (1993b): Recent advances in inverse systems of spaces. Rendiconti Ist. Mat. Univ. Trieste **25**, 317–335

Mardešić, S. (1996a): Nonvanishing derived limits in shape theory. Topology **35**, 521–532

Mardešić, S. (1996b): Strong homology does not have compact supports. Topology Appl. **68**, 195–203

Mardešić, S. (1998): Coherent and strong expansions of spaces coincide. Fund. Math. **158** (1998), 69–80

Mardešić, S. (1999a): Coherent homotopy and localization. Topology Appl. 94, 253-274

Mardešić, S. (1999b): Absolute neighborhood retracts and shape theory. Chapter 9 of History of topology, Ed. James, I.M., North Holland, Amsterdam

Mardešić, S., Miminoshvili, Z. (1990): The relative homeomorphism and wedge axioms for strong homology. Glasnik Mat. **25**, 387–416

Mardešić, S., Prasolov, A.V. (1988): Strong homology is not additive. Trans. Amer. Math. Soc. **307**, 725–744

Mardešić, S., Prasolov, A.V. (1998): On strong homology of compact spaces. Topology Appl. **82**, 327–354

Mardešić, S., Rubin, L.R. (1989): Approximate inverse systems of compacta and covering dimension. Pacific J. Math. **138**, 129–144

Mardešić, S., Segal, J. (1970): Movable compacta and ANR - systems. Bull. Acad. Polon. Sci. Sér. Sci. Math. Astron. Phys. **18**, 649–654

Mardešić, S., Segal, J. (1971): Shapes of compacta and ANR - systems. Fund. Math. **72**, 41–59

Mardešić, S., Segal, J. (1982): Shape theory. The inverse system approach. North - Holland, Amsterdam

Mardešić, S., Segal, J. (to appear): History of shape theory and its application to general topology; in Handbook of the history of general topology, Vol. 3, Ed. Aull, C.A., Lowen, R. Kluwer Acad. Publ., Dordrecht

Mardešić, S., Šekutkovski, N. (1989): Coherent inverse systems and strong shape theory, Rad Jugoslav. Akad. Znan. Umjetn. Mat. Znan. **444** (8), 63–73

Mardešić, S., Watanabe, T. (1988): Strong homology and dimension. Topology Appl. **29**, 185–205

Mardešić, S., Watanabe, T. (1989): Approximate resolutions of spaces and mappings. Glasnik Mat. **24**, 587–637

Massey, W.S. (1978): Homology and cohomology theory. M. Dekker, New York

Mdzinarishvili, L.D. (1965): On various space homology groups based on infinite coverings. Soobšč. Akad. Nauk Gruzin. SSR **38**, 23–30 (Russian)

Mdzinarishvili, L.D. (1972): On the relation between the homology theories of Kolmogorov and Steenrod. Doklady Akad. Nauk SSR **203**, No. 3, 528–531 (Russian)

Mdzinarishvili, L.D. (1978): On the relations between the homology theories of Kolmogorov and Steenrod. Trudy Tbilissk. Mat. Inst. Akad Nauk Gruzin. SSR **59**, 98–118 (Russian)

Mdzinarishvili, L.D. (1980): On Kunneth's relation and the functor lim. Bull. Akad. Nauk Gruzin. SSR **99**, No. 3, 561–564 (Russian)

Mdzinarishvili, L.D. (1981): Applications of the shape theory in the characterization of exact homology theories and the strong shape homotopic theory. Proc. Shape Theory and Geometric Topology (Dubrovnik 1981), Lecture Notes in Math. **870**, Springer, Berlin Heidelberg New York, pp. 253–262

Mdzinarishvili, L.D. (1984): Universellen Koeffizientenfolgen für den $\lim\limits_{\leftarrow}$-Funktor und Anwendungen. Manuscripta Math. **48**, 255-273

Mdzinarishvili, L.D. (1986a): On homology extensions. Glasnik Mat. **21**, 455–482

Mdzinarishvili, L.D. (1986b): On total homology, in Geometric and Algebraic Topology. Banach Center Publ., Warsaw, **18**, pp. 346–361

Mill, J. van (1989): Infinite-dimensional topology. North-Holland, Amsterdam

Milnor, J. (1960): On the Steenrod homology theory. Mimeographed Notes, Berkeley

Milnor, J. (1962): On axiomatic homology theory. Pacific J. Math. **12**, 337–341

Miminoshvili, Z. (1980): On the strong homotopy in the category of topological spaces and its applications to the theory of shape. Soobšč. Akad. Nauk Gruzin. SSR **98**, 301–304 (Russian)

Miminoshvili, Z. (1981): On the connection between shape homotopical groups and homologies for compact spaces. Soobšč. Akad. Nauk Gruzin. SSR **101**, 305–308 (Russian)

Miminoshvili, Z. (1982): On a strong spectral shape theory. Trudy Mat. Inst. Akad Nauk Gruzin. SSR **68**, 79–102 (Russian)

Miminoshvili, Z. (1984): On the sequences of exact and half-exact homologies of arbitrary spaces. Soobšč. Akad. Nauk Gruzin. SSR **113**, No. 1, 41–44 (Russian)

Miminoshvili, Z. (1991): On axiomatic strong homology theory. Rad Hrvat. Akad. Znan. Umjetn. Matem. Znan. **456** (10), 65–79

Mishchenko, E.F. (1953): On some questions in the combinatorial topology of non-closed sets. Mat. Sbornik **32** (74), 219–224 (Russian)

Mitchell, B. (1965): Theory of categories. Academic Press, New York

Mitchell, B. (1972): Rings with several objects. Advances in Math. **8**, 1–161

Mitchell, B. (1973): The cohomological dimension of a directed set. Canad. J. Math. **27**, 233–238

Mohorianu, C. (1997): Comparison between strong shape groups and shape groups for a pointed topological space. Anal. Ştiin. Univ. Al. Cuza, Iaşi, Matematică **43**, 173-190

Morita, K. (1975a): On shapes of topological spaces. Fund. Math. **86**, 251–259

Morita, K. (1975b): Čech cohomology and covering dimension for topological spaces. Fund. Math. **87**, 31–52

Morita, K. (1975c): On expansions of Tychonoff spaces into inverse systems of polyhedra. Sci. Rep. Tokyo Kyoiku Daigaku, Sect. A **13**, 66–74

Morita, K. (1984): Resolutions of spaces and proper inverse systems in shape theory. Fund. Math. **124**, 263–270

Morón M.A., Ruiz del Portal, F.R. (1994): Multivalued maps and shape for paracompacta. Math. Japonica **39**, 489–500

Mrozik, P. (1984): Chapman's complement theorem in shape theory: A version for the infinite poroduct of lines. Arch. Math. **42**, 564–567

Mrozik, P. (1985): Chapman's category isomorphism for arbitrary AR's. Fund. Math. **125**, 195–208

Mrozik, P. (1986): Hereditary shape equivalences and complement theorems. Topology Appl. **22**, 131–137

Mrozik, P. (1990): Mapping cylinders of approaching maps and strong shape. J. London Math. Soc. (2) **41**, 159–174

Mrozik, P. (1991): Finite-dimensional categorical complement theorems in strong shape theory and a principle of reversing maps. Compositio Math. **77**, 179–197

Munkres, J.R. (1984): Elements of algebraic topology. Addison-Wesley, Menlo Park

Nöbeling, G. (1961): Über die derivierten des inversen und des direkten Limes einer Modul-Familie. Topology **1**, 47–61

Nöbeling, G. (1968): Verallgemeinerung eines Satzes von Herrn E. Specker. Inventiones Math. **6**, 41–55

Nowak, S. (1987): On the relationship between shape properties of subcompacta of S^n and homotopy properties of their complements. Fund. Math. **128**, 47–60

Osofsky, B.L. (1968a): Homological dimension and the continuum hypopthesis. Trans. Amer. Math. Soc. **132**, 217–230

Osofsky, B.L. (1968b): Upper bounds on homological dimensions. Nagoya Math. J. **32**, 315–322

Osofsky, B.L. (1971): Homological dimensions of modules. American Math. Soc., CBMS 12, Ann Arbor

Osofsky, B.L. (1974): The subscript of $\aleph_n$, projective dimension, and the vanishing of $\lim^{(n)}$. Bull. Amer. Math. Soc. **80**, 8–26

Pasynkov, B.A. (1958): On polyhedral spectra and dimension of bicompacta and of bicompact groups. Dokl. Akad. Nauk SSSR **121**, No. 1, 45–48 (Russian)

Petkova, S.V. (1973): On the axioms of homology theory. Mat. Sbornik **90**, 607–624 (Russian)

Porter, T. (1973) Čech homotopy I. J. London Math. Soc. **6**, 429–436

Porter, T. (1974): Stability results for topological spaces. Math. Z. **140**, 1–21

Porter, T. (1977): Coherent prohomotopical algebra. Cahiers Topol. Géom. Diff. **18**, 139–179

Porter, T. (1978): Coherent prohomotopy theory. Cahiers Topol. Géom. Diff. **19**, 3–46

Porter, T. (1988): On the two definitions of Ho(pro-C). Topology Appl. **28**, 283–293

Porter, T., Cordier, J.-M., (1984): Homotopy limits and homotopy coherence. Lecture Notes, University of Perugia, Perugia

Prasolov, A.V. (1989): A spectral sequence for strong homology. Glasnik Mat. **24**, 17–24

Prasolov, A.V. (1998a): Strong homology is not additive, Part 2: Cardinality estimates. Math. Report **32**, Univ. Tromsø, 1–15

Prasolov, A.V. (1998b): Strong homology is not additive, Part 3: Set-theoretic cohomology. Math. Report **33**, Univ. Tromsø, 1–15

Prasolov, A.V. (to appear): Extraordinary strong homology. Topology Appl.

Przymusiński, T. (1979): On the dimension of product spaces and an example of M. Wage. Proc. Amer. Math. Soc. **76**, 315–321

Quigley, J.B. (1973): An exact sequence from the n th to $(n-1)$ st fundamental group. Fund. Math. **77**, 195–210

Quigley, J.B. (1976): Equivalence of fundamental and approaching groups of movable pointed compacta. Fund. Math. **91**, 73–83

Quillen, D.G. (1967): Homotopical algebra. Lecture Notes in Math. **43**, Springer, Berlin Heidelberg New York

Robbin, J.W., Salamon, D. (1988): Dynamical systems, shape theory and the Conley index. Ergod. Th. & Dynamic. Sys. **8**, 375–393

Roos, J.- E. (1961): Sur les foncteurs dérivés de lim. Applications. C. R. Acad. Sci. Paris **252**, 3702–3704

Saneblidze, S.A. (1983a): On the uniqueness theorem for the Steenrod-Sitnikov homology theory on the category of compact Hausdorff spaces. Soobšč. Gruzin. Akad. Nauk SSR **105**, 33–36 (Russian)

Saneblidze, S.A. (1983b): On the homology theory of paracompacta. Soobšč. Gruzin. Akad. Nauk SSR **109**, 477–480 (Russian)

Saneblidze, S.A. (1992): On the uniqueness theorem for homology theory of paracompact spaces. Trudy Mat. Inst. Gruzin. Akad. Nauk SSR **97**, 53–64 (Russian)

Sanjurjo, J.M.R. (1992): An intrinsic description of shape. Trans. Amer. Math. Soc. **329**, 625–636

Schubert, H. (1970): Kategorien I, II. Springer, Berlin Heidelberg New York

Segal, G.B. (1968): Classifying spaces and spectral sequences. Publ. Math. IHES **34**, 105–112

Segal, G.B. (1974): Categories and cohomology theories. Topology **13**, 293–312

Segal, J., Spież, S., Günther, B. (1993): Strong shape of uniform spaces. Topology Appl. **49**, 237–249

Šekutkovski, N. (1988): Category of coherent inverse systems, Glasnik Mat. **23**, 373–396

Šekutkovski, N. (1997): Equivalence of coherent theories. Topology Appl. **75**, 113–123.

Sher, R.B. (1981): Complement theorems in shape theory, in Shape theory and geometric topology. Lecture Notes in Math. **870**, Springer, Berlin Heidelberg New York, pp. 150–168

Sher, R.B. (1987): Complement theorems in shape theory, II, in Geometric topology and shape theory. Lecture Notes in Math. **1283**, Springer, Berlin Heidelberg New York, pp. 212–220

Sitnikov, K.A. (1951): Duality law for non-closed sets. Doklady Akad. Nauk SSSR **81**, 359–362 (Russian)

Sitnikov, K.A. (1954): Combinatorial topology of non-closed sets, I. The first duality law; the spectral duality. Mat. Sbornik **34**, 3–54 (Russian)

Sklyarenko, E.G. (1969): Homology theory and the exactness axiom. Uspehi Mat. Nauk **24**, No.5, 87–140 (Russian)

Sklyarenko, E.G. (1971): Uniqueness theorems in homology theory. Mat. Sbornik **85**, 201–223 (Russian)

Sklyarenko, E.G. (1979): On homology theory associated with the Aleksandrov-Čech cohomology. Uspehi Mat. Nauk **34**, No.6, 90–118 (Russian)

Sklyarenko, E.G. (1989a): General homology and cohomology theories. Present state and typical applications. Itogi nauki i tehniki, Series - Algebra. Topology. Geometry, **27**, Acad. Nauk SSSR, Moscow, pp. 125–228 (Russian)

Sklyarenko, E.G. (1989b): Homology and cohomology of general spaces. Itogi nauki i tehniki, Series - Contemporary problems of Mathematics, Fundamental directions, **50**, Akad. Nauk SSSR, Moscow, pp. 125–228 (Russian)

Sklyarenko, E.G. (1995): Hyper (co) homology left exact covariant functors and homology theory of topological spaces. Uspehi Mat. Nauk **50**, No. 3, 109–146 (Russian)

Spanier, E., Whitehead, J.H.C. (1955): Duality in homotopy theory. Mathematika **2**, 56–80

Steenrod, N.E. (1940): Regular cycles of compact metric spaces. Ann. Math. **41**, 833–851

Stramaccia, L. (1997): On the definition of the strong shape category. Glasnik Mat. **32**, 141–151

Strøm, A. (1966): Note on cofibrations I. Math. Scand. **19**, 11–14

Strøm, A. (1968): Note on cofibrations II. Math. Scand. **22**, 130–142

Thiemann, H. (1995): Strong shape and fibrations. Glasnik Mat. **30**, 135–174

Thomason, R.W. (1985): Algebraic K-theory and etale cohomology. Ann. Sci. Ecole Norm. Sup. (4. Ser.) **13**, pp. 437–552

Tsuda, K. (1982): Some examples concerning the dimension of product spaces. Math. Japonica **27**, 177–195

Tsuda, K. (1985): Dimension theory of general spaces. Thesis, University of Tsukuba

Verdier, J.-L. (1965): Équivalence essentielle des systèmes projectifs. C.R. Acad. Sci. Paris **261**, 4950-4953

Vietoris, L. (1927): Über den höheren Zusammenhang kompakter Räume und eine Klasse von zusammenhangstreuen Abbildungen. Math. Ann. **97**, 454–472

Vogt, R.M. (1972): A note on homotopy equivalences. Proc. Amer. Math. Soc. **32**, 627–629

Vogt, R.M. (1973): Homotopy limits and colimits. Math. Z. **134**, 11–52

Wallace, A.D. (1952): The map excision theorem. Duke Math. J. **19**, 177–182

Watanabe, T. (1977): On a problem of Y. Kodama. Bull. Acad. Polon. Sci. Sér. Sci. Math. Astronom. Phys. **25**, 981–985

Watanabe, T. (1987a): Approximative shape I. Tsukuba J. Math. **11**, 17–59

Watanabe, T. (1987b): Čech homology, Steenrod homology and strong homology. I. Glasnik Mat. **22**, 187–238

Watanabe, T. (1991a): Approximate resolutions and covering dimension. Topology Appl. **38**, 147–154

Watanabe, T. (1991b): An elementary proof of the invariance of $\lim^n$ on pro-abelian groups. Glasnik Mat. **26**, 177–208

Whitehead, G.W. (1962): Generalized homology theories. Trans. Amer. Math. Soc. **102** (1962), 227–283

Whitehead, G.W. (1978): Elements of homotopy theory. Springer, New York Heidelberg Berlin

Yeh, Z.Z. (1959): Higher inverse limits and homology theories. Thesis. Princeton University, Princeton

Yosimura, Z. (1972/73a): On cohomology theories of infinite CW-complexes, I. Publ. Research Inst. Math. Sci. Kyoto Univ. **8**, 295–310

Yosimura, Z. (1972/73b): On cohomology theories of infinite CW-complexes, II. Publ. Research Inst. Math. Sci. Kyoto Univ. **8**, 483–508

Zdravkovska, S. (1981): An example in shape theory. Proc. Amer. Math. Soc. **83**, 594-596

List of Special Symbols

$\boldsymbol{X} = (X_\lambda, p_{\lambda\lambda'}, \Lambda)$	inverse system, 9
id	identity morphism, 10
$\mathbf{1}_{\boldsymbol{X}}$	identity morphism, 10
$\mathcal{C}^\Lambda$	level category of systems in $\mathcal{C}$, 10
inv-$\mathcal{C}$	category of inverse systems in $\mathcal{C}$, 10
pro-$\mathcal{C}$	the pro-category of $\mathcal{C}$, 11
$\equiv$	congruence of mappings, 11
$[\boldsymbol{f}]$	congruence class of $\boldsymbol{f}$, 12
Top	category of topological spaces, 12
Top^Λ	level category of systems of spaces, 12
inv-Top	category of inverse systems of spaces, 12
pro-Top	pro-category of inverse systems of spaces, 12
Set	category of sets, 13
Grp	category of groups, 13
Ab	category of abelian groups, 13
$\mathrm{H}(\mathsf{Top})$	homotopy category, 13
$[f]$	homotopy class of f, 13
pro-$\mathrm{H}(\mathsf{Top})$	pro-category of the homotopy category, 13
H	homotopy functor, 13
M_n	set of multiindices in M of length n, 14
Λ_n	set of multiindices in Λ of length n, 14
d^j	face operator, 14
s^j	degeneracy operator, 14
Δ^n	standard n-simplex, 14
d_j	face operator, 14
s_j	degeneracy operator, 14
$\boldsymbol{f} = (f, f_\mu)$	coherent mapping, 14
$[\mathbb{N}]$	category of finite ordinals, 18
$\equiv$	congruence of coherent mappings, 22
C	coherence operator, coherence functor, 26
E	forgetful operator, forgetful functor, 27
$[\boldsymbol{X}]$	homotopy system induced by $\boldsymbol{X}$, 27
$\simeq$	homotopy, coherent homotopy, 29
$[\boldsymbol{f}]$	homotopy class of $\boldsymbol{f}$, 30

Author Index

Subject Index